The Evolution of Dasyurid Marsupials

For Nancy and Walt Krajewski. Thanks Mom and Dad.

– Carey

For Jane and my children, who shared my deep love for marsupials.

– M.W.

This family history is dedicated to the many others who have contributed to it through their research, including Basil Marlow, then Curator of Mammals at the Australian Museum, whose 1961 conference presentation of observations on reproduction in Antechinus sparked my interest to study the reproductive biology of dasyurid marsupials.

– P.A.W.

The Evolution of Dasyurid Marsupials

Systematics and Family History

Carey Krajewski, Michael Westerman and Patricia A. Woolley

CSIRO PUBLISHING

CRC Press
Taylor & Francis Group
Boca Raton London New York

CRC Press is an imprint of the
Taylor & Francis Group, an **informa** business

A catalogue record for this book is available from the National Library of Australia.

ISBN: 9781486315949 (hbk)
ISBN: 9781486315956 (epdf)
ISBN: 9781486315963 (epub)

Published in print in Australia and New Zealand, and in all other formats throughout the world, by CSIRO Publishing.

CSIRO Publishing
Private Bag 10
Clayton South VIC 3169
Australia

Telephone: +61 3 9545 8400
Email: publishing.sales@csiro.au
Website: www.publish.csiro.au
Sign up to our email alerts: publish.csiro.au/earlyalert

Published in print only, throughout the world (except in Australia and New Zealand), by CRC Press, with ISBN 9781032584546.

CRC Press
6000 Broken Sound Parkway NW, Suite 300, Boca Raton, FL 33487-2742
and
4 Park Square, Milton Park, Abingdon, Oxon, OX14 4RN
Website: www.routledge.com

CRC Press is an imprint of Taylor & Francis Group, LLC

Front cover: (left) brush-tailed phascogale (photo by Rohan Clarke); (right, top to bottom) southern ningaui (photo by Rohan Clarke), New Guinean quoll, little red kaluta (photos by Patricia A. Woolley)

Cover design by Cath Pirret
Typeset by Envisage Information Technology
Printed in China by 1010 Printing International Ltd

CSIRO acknowledges the Traditional Owners of the lands that we live and work on and pays its respect to Elders past and present. CSIRO recognises that Aboriginal and Torres Strait Islander peoples in Australia and other Indigenous peoples around the world have made and will continue to make extraordinary contributions to all aspects of life including culture, economy and science. CSIRO is committed to reconciliation and demonstrating respect for Indigenous knowledge and science. The use of Western science in this publication should not be interpreted as diminishing the knowledge of plants, animals and environment from Indigenous ecological knowledge systems.

The paper this book is printed on is in accordance with the standards of the Forest Stewardship Council® and other controlled material. The FSC® promotes environmentally responsible, socially beneficial and economically viable management of the world's forests.

FSC www.fsc.org MIX Paper | Supporting responsible forestry FSC® C016973

Aug23_01

Contents

Preface

Anyone who is willing to take the necessary trouble can learn a great deal about all the species that exist.

– Aristotle *(The Parts of Animals, Book I, Part 5)*[1]

Thus, encouraged by the Father of Biology, we have taken the necessary trouble to assemble and synthesise research on the evolution of dasyurid marsupials. Our goal is to create a reference work that will facilitate the efforts of current students and researchers to learn what has gone before in the study of these animals, with sensitivity to how past and current investigations are embedded in the scientific traditions of their times. For those just embarking on understanding dasyurid evolution, it may come as a surprise that even the seemingly simple question, 'How many species of dasyurids currently exist?', is fraught with historical, methodological and philosophical qualifications too easily glossed over in the rush to a simple answer. Indeed, there are few simple questions and virtually no simple answers in this discipline, but its problems have brought much intellectual fulfillment to those who engaged with them.

It is important to be explicit at the outset regarding our target audience. This book is for readers who have a strong conceptual foundation in biology. Because our framework is systematic biology, we have gone to some lengths to explain the tenets of that discipline, how they developed over time and how they affected dasyurid research. But we do not provide similar introductions to anatomy, animal behaviour, biogeography, ecology, evolutionary biology, mammalogy, physiology or elementary statistics. Doing so is not feasible in a single brief volume, but these subjects play a vital role in the story we have to tell, and readers should have a basic understanding of them.

One of the messages that we hope comes across clearly is that the study of evolution is an integrative science. We have experienced this first-hand in our own collaborations over the years. One of us (P.A.W.) developed expertise in the collection of dasyurids from the wild, and their maintenance and breeding in captivity, followed by studies on their reproductive biology and comparative anatomy. In the course of this work opportunities for collaboration with researchers in other disciplines emerged, including M.W., who began professional life as a cytogeneticist. C.K. started as a molecular systematist several steps removed from actual organisms. Working together for the past three decades has greatly fostered our professional growth, expanded the range of questions we have been able to ask about dasyurids and enhanced the significance of the answers we have found. It is gratifying that such collaborations have become the norm, but we encourage casting the net even wider to encompass joint efforts by systematists and ecologists, bioengineers, conservation biologists, genomicists, geographers, molecular biologists and wildlife managers. We have resisted the urge to itemise what we think are the most 'outstanding questions' about dasyurid evolution at the end of each chapter because we think such insights are the province of every active researcher. However, we do not hesitate to predict that progress on most of these questions will require integrative analyses.

Ours is a tale told by trees – phylogenetic trees – each one a hypothesis of evolutionary relationships among species, usually based on the analysis of some data. This book is replete with them and, as the saying goes, there is a danger in not seeing the forest for the trees. In our opinion, 'the forest' is the evolving interpretive framework that connects evidence to knowledge and, in turn, qualifies the reliability of that knowledge. We do not present a 'final' phylogeny of dasyurids because there probably is no such thing: only a pattern of consilience during the past few decades suggesting that some old problems have been settled even as new ones emerge. To the extent that evolutionary questions have lately been asked and answered using 'robust' phylogenies, it is worth keeping in mind that the latter have misled us before.

1 Translation quoted in: Farrington, D. (1949). *Greek Science*. Pelican Books, Baltimore.

Nevertheless, the power of this interpretive framework is truly remarkable and much remains to be resolved.

Our book has been designed for reading in both digital and print formats. The print edition contains mainly black and white figures within the chapters; however, figures for which colour is important are reproduced in colour in the colour Appendix. Digital formats are produced in colour throughout.

There is pathos in being a dasyurid biologist at a time when these animals are being pushed towards extinction by forces that science alone cannot control. Dasyurids merely exemplify what is true for many other groups of organisms. If nothing else, this book is a testament to what we stand to lose – not just a 'branch of evolutionary history', but a *family* history in both the taxonomic and the colloquial sense. We hope our narration of that history brings out the intrinsic value of its subjects and serves as a brief for efforts to maintain their existence.

Carey Krajewski
Mike Westerman
Pat Woolley
4 August 2022

Acknowledgements

We are grateful to colleagues who read and critiqued early drafts of chapters: Frank E. Anderson, Jennifer Christie, Steven J.B. Cooper, Christopher Dickman, Mark D.B. Eldridge, George A. Feldhamer, Kamal M. Ibrahim, Vicente García-Navas, Benjamin P. Kear, Mark S. Springer, Richard H. Thomas, John C.Z. Woinarski and Steve Wroe. Their efforts greatly enhanced the quality of our book, but of course we are responsible for inaccuracies that remain. Our historical research would have been impossible without literature access provided by the Biodiversity Heritage Library (www.biodiversitylibrary.org). We greatly appreciate the efforts and guidance of Briana Melideo, Mark Hamilton, Tracey Kudis and others at CSIRO Publishing, as well as Joy Window at Living Language Editing and Proofreading. C.K. also thanks Roberta Christie for her support and encouragement throughout this project. M.W. is grateful to Jane and his children for their support and understanding of his love for marsupials, and for their acceptance of the sudden appearance of small bodies in the family freezer.

1 Dasyurids and 'tree thinking'

DASYURID EVOLUTION AND SYSTEMATIC BIOLOGY

Dasyurids comprise 75 living and several extinct species of marsupials found in Australasia. They are small (up to 14 kg as adults of the largest species) insectivore-carnivores. In formal classification, they constitute the taxon Dasyuridae, one of three groups within the larger taxon Dasyuromorphia. (A 'taxon' is an explicitly recognised – i.e. named – group in biological classification.) The other two dasyuromorphian taxa are also Australasian – Myrmecobiidae (the single species of numbat) and Thylacinidae (the recently extinct Tasmanian thylacine and its fossil relatives). Dasyuromorphia is nested within taxon Marsupialia, which also includes six other major living groups. Marsupialia, Monotremata and Placentalia constitute the living Mammalia. This classification is concisely expressed in the following form, where indentation shows the nesting of groups within groups:

Mammalia
- Monotremata (platypus and echidnas)
- Placentalia (placental mammals)
- Marsupialia (marsupials)
  - Didelphimorphia (opossums)
  - Paucituberculata (caenolestids)
  - Australidelphia (microbiotheres and Australian marsupials)
    - Microbiotheria (*Dromiciops*)
    - Agreodontia
      - **Dasyuromorphia**
        - **Thylacinidae**(thylacine,recentlyextinct)
        - **Myrmecobiidae** (numbat)
        - **Dasyuridae** (dasyurids)
      - Peramelemorphia (bandicoots)
      - Notoryctemorphia (marsupial moles)
    - Diprotodontia (wombats, koalas, kangaroos, possums)

Embedded within this list of taxon names is a great deal of biological knowledge generated over thousands of years. That knowledge has been brought to bear on two foundational questions: (1) What are the species of organisms that live, or have lived, on Earth? and (2) How are those species related to one another? Behind these two questions lie two others, even more profound: (1′) What is a species? and (2′) What does it mean to be related? In the following chapters we explore how all four questions have been addressed during the history of research on dasyurids.

CLASSIFICATION, TAXONOMY AND PHYLOGENY

Biologists study all aspects of organisms and their research on different organismal features led to the development of interrelated disciplines. For animals, these features now include anatomy (the physical structure of animal bodies), physiology (how bodies function), development (how an adult body forms from a zygote), genetics (how organisms store and express hereditary information), ecology (how organisms interact with the environment) and evolution (how organisms have changed over generations since the origin of life). Our primary interest is in the last of these disciplines, but evolutionary questions draw in all other disciplines because all features of organisms have evolved. More specifically, our goal is to view the evolution of dasyurids through the lens of systematic biology. To do so, some basic tenets of modern systematics must be clearly understood.

In *Principles of Animal Taxonomy*, George Gaylord Simpson (a mammal palaeontologist and an architect of 20th century evolutionary biology) offered the following definitions of terms, delimiting the scopes of three related research programs:

> ***Systematics*** *is the scientific study of the kinds and diversity of organisms and of any and all relationships among them … Zoological* ***classification*** *is the ordering of animals into groups (or sets) on the basis of their relationships …* ***Taxonomy*** *is the theoretical study of classification, including its bases, principles, procedures, and rules. [Simpson 1961: 7–11; bold added]*

The formal practice of taxonomy and classification in Western civilisation traces to the ancient Greeks. The 18th century Swedish botanist Carol von Linné (Linnaeus) produced, among other works, *Systema Naturae*, the 10th edition of which (Linnaeus 1758) is recognised as the starting point for modern animal classification. Although Linnaeus and other early taxonomists applied various principles to produce their classifications, the basic presumption was that organisms with *similar* traits belong in the same group.[1] Similarity was the key to answering all four of the systematics questions posed above. Linnaeus applied this yardstick hierarchically such that his classification consisted of groups within groups, which he recognised at different ranks: maximally similar organisms constitute a species, a group of similar species form a genus, and so on. Linnaean classification is still used today.[2] However, taxonomists disagreed about which traits mattered most for assessing similarity and so produced conflicting classifications, as well as an ongoing and largely fruitless debate about 'principles'. Lack of consensus on how to measure similarity created a subjective component in taxonomy from which it has never entirely escaped.

Darwin's (1859) theory of evolution changed the conceptual framework of taxonomy and created the need for a more integrated discipline, systematics, to grapple with the concept of relatedness. In practice, Darwin's immediate contribution to the four questions of systematics was limited. Although it became understood that evolutionary relatedness is based on propinquity of descent (a genealogical criterion), the distinction between the evolution of traits and the organisms that bear them was very slow to be appreciated. The only illustration in *On the Origin of Species* (Fig. 1.1) embodied the concept of genealogical relatedness among species, a graphical device (a 'tree') for visualising it and Darwin's 'principle of divergence' – the notion that descendant species would steadily accumulate trait differences from their ancestor and each other. Not only do close relatives share a more recent common ancestor, but they are also expected to be more similar, compared to a distant relative. The first criterion (evolutionary relatedness = recency of common ancestry) is the cornerstone of modern systematics; the second (relatedness revealed by trait similarity) is now known to be an invalid methodological principle: closest relatives are not necessarily most similar to one another in any set of features. Nevertheless, the psychological appeal of trait similarity as a guide to ancestry is such that it colours the outlook of biologists to this day.

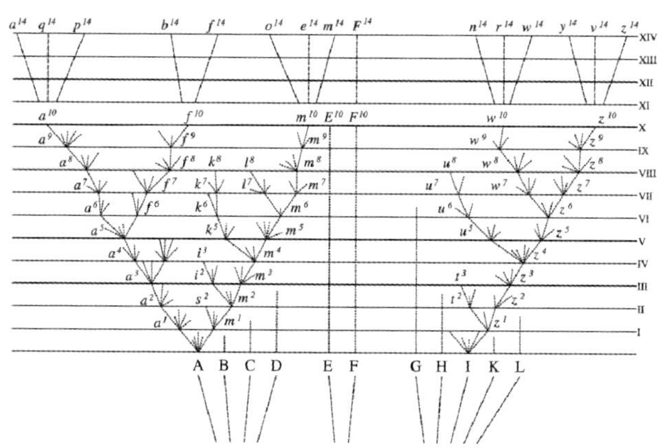

Fig. 1.1. The only illustration in *On the Origin of Species* (Darwin 1859) depicts Darwin's concept of descent with modification and presages both the concept of phylogeny and the tree-like diagrams used to depict it. Letters refer to species or 'varieties', Roman numerals to progressively more recent time horizons and the horizontal axis to trait similarity.

The term *phylogeny* (Gr., 'tribe birth') was coined by Ernst Haeckel in *Generelle Morphologie der Organismen* (Haeckel 1866), but its meaning has undergone many revisions and is still used inconsistently by systematists. In this book, phylogeny refers to the speciation history of a group of species, entailing an unambiguous statement of their evolutionary relationships. We do *not* use the term to denote histories of populations, genes or other entities that are not species, nor do we equate phylogeny with all evolutionary history. Haeckel also developed and popularised the tree metaphor to depict phylogeny and, again, tree diagrams have undergone considerable change in both construction and interpretation. However, we emphasise that a tree diagram is a *graphical device* to illustrate phylogenetic relationships. Phylogeny need not always be tree-like and, when it is not, the concept of a 'phylogenetic tree' is problematic. Nevertheless, the phylogenetic tree has become more than a mere graphic in modern evolutionary biology – it is both the conceptual and analytical framework for studying almost every aspect of evolution.

TREE INTERPRETATION

A phylogenetic tree is a directed, acyclic graph: directed in that it has a time axis from past to present; acyclic in that branches may divide, but never join in the past → present direction (Fig. 1.2). The tree originates at a *root* and ends at *terminal branches* or *tips*, the latter representing living or extinct species. Between the root and tips are *internal branches* representing ancestral species. For example, the branch labelled AB in Fig. 1.2 is the common ancestor of species A and B. The root branch is the common ancestor

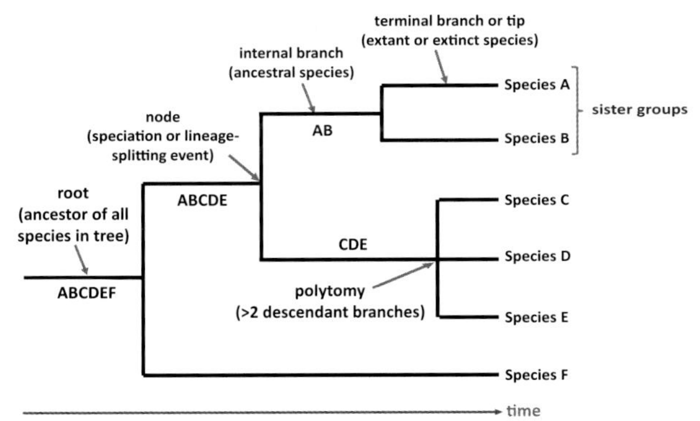

Fig. 1.2. Interpretation of a phylogenetic tree. In this tree, the time axis is unscaled, so the branch lengths have no meaning. Labels on internal branches designate ancestral species.

of all species at the tips. Internal branches ramify at *nodes*, which represent past speciation (or lineage-splitting) events.[3] For reasons not entirely biological, systematists think that most speciation events produce two daughter species, such that nodes on a phylogenetic tree are expected to *bifurcate*. It is possible, however, for a single speciation event to produce more than two daughters, in which case the node shows a *hard polytomy*. Polytomies may also indicate uncertainty about the order of speciation, and hence relationships, within a group – these are called *soft polytomies*. Because time flows from the root to the tips, the degree of relationship between any two species is measured by *recency of common ancestry*. Species A and B are more closely related to each other than either is to C precisely because their most recent common ancestor (AB) existed more recently than the common ancestor of either one with C (i.e. ABCDE). The closest relative of any species is its *sister group*. A set of species that includes all the descendants of a single ancestor, and nothing else, is a *monophyletic* ('one tribe') *group* or *clade*. Such groups are *exclusive* – any member of the group is more closely related to any other member of the group than it is to any species outside the group. In Fig. 1.2, {A, B, C, D, E} is monophyletic, but {A, C, D, E} is not; the most recent common ancestor of the latter is ABCDE, but the group does not include B, a descendant of that ancestor. Importantly, the time axis in Fig. 1.2 is unscaled, so branch lengths on the tree have no meaning (such a tree is usually called a *cladogram*); only the root-to-tip ordering of nodes along a single path has temporal significance. So, for example, the tree does not imply that the speciation event giving rise to AB occurred before that producing CDE, even though the latter node is drawn further from the root than the former. It is possible to add time or some other branch-length scale to trees, as we discuss in 'Phylogeny as an evolutionary framework' below.

This basic format of a phylogenetic tree can be extended to accommodate two common contingencies. First, tips may represent monophyletic groups: Species A and B in Fig. 1.2 could be replaced with 'Clade A' and 'Clade B'. Because monophyletic groups, by definition, include all and only the descendants of a single ancestor, a tree with clades at the tips represents a speciation history even though the speciation events *within* terminal clades are not depicted. Thus, Clade A might be Marsupialia and Clade B Placentalia (the two tips representing ca. 380 and 6100 species, respectively) while Clades C–E could represent any three reptile taxa. Second, phylogenetic trees need not be taxonomically exhaustive to be valid; they need only make clear statements of relationships among the taxa at their tips. In the marsupial-placental-reptile example just given, monotremes do not appear in the tree. Among living mammals, Monotremata is the sister to Marsupialia + Placentalia. That monotremes are omitted does not invalidate the phylogeny, which says only that marsupials and placentals are more closely related to each other than either is to the three reptile taxa included.

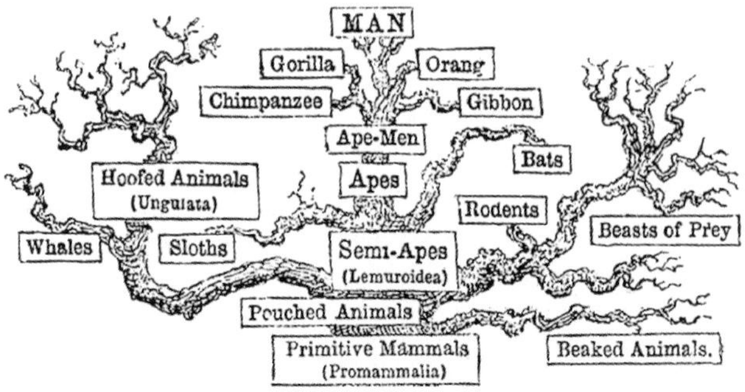

Fig. 1.3. The mammalian portion of Haeckel's (1874) 'pedigree of man'. In Haeckel's view, phylogeny meant the series of morphological types that occurred in the line leading to a particular species, in this case humans. These morphologies are represented by living or extinct groups arrayed down the trunk of the tree, whereas groups not directly involved in the ancestry of the apex species appear on side branches. This tree illustrates the abandoned hypothesis that marsupials ('Pouched Animals') are the ancestors of placentals.

PHYLOGENETIC MISCONCEPTIONS

These basic rules of tree interpretation are often unconsciously supplemented with several inappropriate presumptions. Two are especially significant in the marsupial literature. First, with reference to Fig. 1.2, a lineage like F that separates from the root node, but does not subsequently diversify as much as its sister clade, is commonly referred to as a 'basal' lineage. The implication is that such a lineage somehow resembles the root ancestor, but the tree entails no such information. Lineage F has evolved for the same length of time as any other root-to-tip pathway and its current traits were determined by historically contingent factors. Unless trait evolution occurs only during speciation events at a uniform rate (and this is not predicted by evolutionary theory), F may show anything from least to most resemblance to the root ancestor. A second misinterpretation of a modern phylogeny is deeply embedded in Western philosophy and was, until the early 20th century, implicit in the minds of many biologists. It is the notion of progressive improvement over time.

Aristotle conceived that kinds of organisms could be ordered in a lower-to-higher scale by the combination of souls (nutritive, sensitive, rational) they possessed, such that plants were 'below' animals and animals were below humans. Superimposed on these three tiers was a subjective scale of increasing anatomical complexity. This *scala naturae* persisted as a Western worldview for nearly two millennia, giving rise to the distinction between 'lower' and 'higher' life forms. Indeed, the earliest coherent evolutionary theory articulated by Lamarck (1809) embraced the *scala* and posited that evolution was the result of a 'striving for perfection' in the animate world. Humans were more perfect than non-human primates, which were more perfect than non-primate mammals, and so on down to the lowly Infusoria (Protista). Even Darwin used such language, though he was silent on its metaphysical justification. One of its champions was Haeckel, whose trees were drawn with what he considered the most perfect species at the apex, the supposed lineal ancestors of that species arrayed down the trunk and variously related groups shown as side branches (Dayrat 2003) (Fig. 1.3).

Progressivist thinking had a major impact on early discussions of mammalian phylogeny, particularly in the form of Huxley's (1880) Metatheria hypothesis for the origin of placentals. Huxley argued that the three major groups of mammals formed a single ancestor-descendant sequence: Prototheria (monotremes) → Metatheria (marsupials) → Eutheria (placentals). As summarised by Bensley (1901: 123): 'Huxley … implied the following plan of succession: the Prototheria giving rise to the Metatheria, and becoming specialised as Monotremata; the Metatheria, in turn, giving rise to the Eutheria, and becoming specialised as Marsupialia; the Eutheria becoming specialised as Placentalia.' Thus, early, unspecialised and extinct forms of prototherians and metatherians serve as ancestors both to the next major group as well as specialised modern representatives of their own. Unspecialised placentals still persist in the form of insectivorans, which were widely viewed as (at least resembling, if not actually being) the ancestors of more specialised placental groups. Although 'more specialised' replaces 'more perfect' in this language, the implication of progress remains.

The progressivist outlook is connected with what we now consider two misconceptions about phylogeny: (1)

within any lineage, 'primitive', 'unspecialised', 'general' or 'rudimentary' traits should be found in deep ancestral branches, whereas 'advanced', 'specialised' or 'more perfect' traits should occur at the tips; and (2) 'lower' groups featuring many 'primitive' traits may be directly ancestral to more 'advanced, higher' groups. Such linkage between trait composition and ancestry is implicit in the value-laden Greek roots Huxley chose for his taxon names: Prototheria = 'first beast', Metatheria = 'between beast', Eutheria = 'good beast'.[4] The idea that marsupials are second-rate mammals took deep root in 20th century biology. Nearly 60 years after Huxley, and at the dawn of modern phylogenetics, the Australian mammalogist Troughton (1959: 73) proclaimed:

> *... the Monotremata may therefore be considered as the basis of mammalian evolution. The original group spread throughout the world, and some members became* ***extremely specialized*** *to produce the platypus and spiny anteaters of today, while retaining the* ***primitive*** *egg-laying habit. The main stem, however,* ***progressed*** *as the advantages of viviparous reproduction were imposed on other survivors of the original group. The functional change entailed adaptations that characterize marsupials and which, despite their* ***intermediate limitations****, proved eminently satisfactory in evolving the unique radiation of marsupial life within the prehistoric isolation of Australian and New Guinea. Some of the* ***primitive*** *polyprotodonts [marsupials], avoiding any* ***too exclusive specialization****, evolved the mechanism for full intrauterine maturation; and with the acquisition of a corpus collosum they emerged as the ancestors of Eutheria. [Bold added.]*

Huxley could not have said it better. The supposed inferiority of marsupials to placentals was widely invoked to explain marsupial extinction whenever placentals invaded the isolated continents (South America and Australia) where marsupials had diversified. Marsupial apologists still cite with glee the possible existence of a placental fossil (*Tingamarra*) in the Oligocene of Australia, suggesting that – maybe just this once – marsupials won the class war.

Modern evolutionary biology rejects the notion of phylogenetic progress, dismissing it as an artefact of a bygone worldview. Organisms and taxa change over the course of time, but they do not inevitably get better by any universal standard. On a phylogeny, ancestral taxa have ancestral traits, not 'primitive' ones (although the latter term is still used, it is now understood as synonymous with the former). Modern systematists have abandoned the search for ancestors among living (or even extinct) taxa, such that an internal branch on a modern phylogenetic tree is largely a hypothetical construct, not classifiable at any rank below the entire clade to which it gave rise. One supraspecific taxon cannot be ancestral to another; rather, taxa share common ancestral species. Thus, Metatheria is not the ancestor of Eutheria; the two groups diverged from a common ancestor that can only be characterised as a member of Theria. Nor are marsupials intrinsically inferior to ('lower than') placentals; the outcomes of interactions between them were (and are) contingent upon ecological circumstances.

WHAT PHYLOGENY IS

In his monumental classification of mammals, Simpson (1945: 5) asserted, 'The stream of heredity makes phylogeny; in a sense, it is phylogeny.' This insight reveals the biology behind a tree diagram, namely the genealogy of individual organisms projected back in time through generations to their common ancestors with other species. Baum and Smith (2013) illustrated this connection with a series of diagrams that 'zoom out' from a single generation of parents and their offspring to a phylogenetic tree. Figure 1.4 is an adaptation of their diagrams to dasyurids. Part (a) shows five Tasmanian devil (*Sarcophilus harrisii*) parents from a single population, their sexual pairings and the offspring of each pairing. This is a standard, one-generation pedigree of biparental organisms, the arrows indicating not just parentage but transmission of genetic information (Simpson's stream of heredity). Part (b) shows a pedigree of these five devils and 11 others that form a single population, tracing lines of parent-offspring descent through five generations. In both diagrams, the devil images represent individuals but are unnecessary for tracking the pattern of genetic transmission; part (c) shows the same pedigree without the images. If we could follow descent in a larger population over more generations, the pedigree would be more complex, resembling the illustration in part (d). Zooming out farther to include all populations of the species connected by gene flow, as in part (e), shows a temporal segment of the devil lineage. Expanding the view to include the entire devil lineage

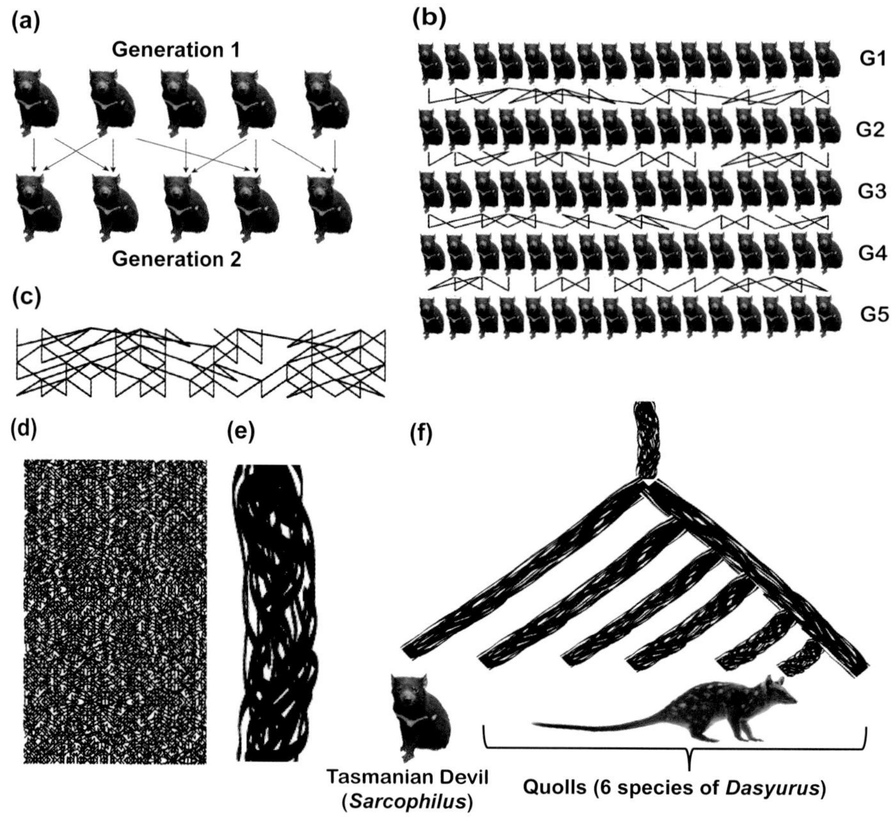

Fig. 1.4. Genealogy and phylogeny. (a) Two generations of five Tasmanian devils (*Sarcophilus harrisii*) from a single population. (b) Five generations of devils in a population of 16 individuals. (c) The same genealogy as in (b) but showing only lines of descent. (d) Many generations of Tasmanian devils in a population of many individuals. (e) Many generations of devils from several populations connected by gene flow. (f) The Tasmanian devil lineage connected to its common ancestor with quolls (*Dasyurus*). [Adapted from Baum and Smith (2013); devil cutout from a photo by Mathias Appel licensed under CC-BY-NC 2.0 via Wikimedia Commons; western quoll cutout from a photo by Jon Augier (Museums Victoria) licensed under CC-BY-SA 4.0 via Wikimedia Commons.]

and tracing it back to its common ancestor with quolls (*Dasyurus*) gives the phylogeny in part (f). These diagrams simplify reality but illustrate that our concept of phylogeny today is fundamentally the same as Darwin's – it is the genealogy of life.

SPECIES AND LINEAGES

If *On the Origin of Species* answered the question, 'What does it mean to be (evolutionarily) related?', it ironically failed to say much about its titular subject. Yes, species originate from pre-existing species (and that is important), but what exactly is a species? Again the burden of 20 centuries of Aristotelian philosophy weighed heavily in naturalists' views on the units of biodiversity: species were considered natural kinds defined by their essences or, more pragmatically, recognised by their defining traits. This is typological thinking as Mayr (1975) styled it. Broadly speaking, to determine whether a collection of organisms represents a single species, it was necessary to identify at least one trait which they all possessed and that was absent in other organisms. This is what a modern taxonomist would call 'diagnosis' and it illuminates the twofold nature of the 'species problem': (1) What is the ontological status of species? and (2) How do we identify species in nature? The first is a vital matter of philosophy, the second a pressing practical problem, and both are foundational for systematic biology.

The essentialist idea that species are defined by their traits resounded up the taxonomic hierarchy from Linnaeus almost to the present. For example, most 20th century mammalogy students were taught that Mammalia is defined by (among other things) a dentary-squamosal jaw joint, Marsupialia by a duplex uterus, and so on. Indeed, the sprawling literature of biological classification during this period is largely a record of expert opinions on which traits define which groups, both at the species and higher levels. This attempt to employ the criterion of trait similarity (described in 'Classification, taxonomy and phylogeny') was largely frustrated by the ubiquity of mosaic evolution – the reality that different traits evolve in different ways and delimit different groups, and there is no *a priori* basis for choosing among them. Hennig (1966) supplied such a basis, arguing that *shared derived* traits, or *synapomorphies*, delimit monophyletic groups. Hennig's insight gave birth to modern phylogenetics (see Chapter 3), but in one sense did little more than narrow the search for taxon-defining similarities. Modern mammalogists are reassured that the dentary-squamosal jaw seems to be a derived trait of mammals, but the exquisite fossil record of transitional cynodonts (e.g. *Diarthrognathus*) bearing both the ancestral (quadrate-articular) and derived jaw articulations in a single animal rendered even this diagnosis a matter of line-drawing. Such cases appear to be the norm and using traits to define species or higher taxa does not capture biological reality.

Defining species by their traits came to be known as the *morphological species concept* and has long been discarded as an ontological principle. Species have been otherwise defined as reproductively isolated populations (the biological species concept), groups of organisms with a unique shared niche (the ecological species concept), groups with the ability to identify other individuals as potential mates (the recognition species concept), and so on. The biological concept continues to have many devoted adherents, despite its operational limitation to sympatric sexual populations and the ubiquity of natural hybridisation. Simpson (1961) proposed the *evolutionary species concept* in which species are evolutionarily independent genealogical lineages. This captured much of what evolutionary biologists meant when they talked about species but was widely panned as a non-operational guide to the delimitation of species in nature.

The breakthrough insight about species ontology was Ghiselin's (1966) argument that species are not natural kinds (with defining properties) but historical entities, or individuals (see Hull 1978). This inspired a healthy philosophical debate, but the upshot for systematics was a focus on the centrality of *lineages* in evolution, engendering a resurgence of Simpson's evolutionary concept. As eventually argued by de Queiroz (1998), species are indeed evolutionarily independent lineages; that is one thing, but the operational task of identifying species in nature is something else. Indeed, controversy about previous species 'concepts' were mostly arguments for the priority of one type of diagnostic trait (morphology, reproduction, etc.) over others. In fact, all heritable features of organisms can provide *evidence* for evolutionary independence and species status, but no such feature serves to *define* individual species or species in general. A parallel argument applies to phylogeny: clades *are* lineages whose existence is supported by the *evidence* of trait data. What is Mammalia? One acceptable definition is as follows: 'Mammalia consists of all species descended from the most recent common ancestor of platypus and human.' How do we know Mammalia is a monophyletic group, or that *Dasyurus maculatus* is a species? At any given moment, these propositions have the status of hypotheses supported by the evidence of trait distributions. They may be wrong, but the current balance of evidence supports them.

PHYLOGENY AS AN EVOLUTIONARY FRAMEWORK

If phylogeny is tree-like and the tree is known (or at least supported by evidence), it can be used to articulate and answer many questions about evolutionary history. This is because the tree structure narrows the focus of such questions to events that happened along particular branches (i.e. during the genealogical lifetimes of ancestral or terminal species with known relations to other such species) or at particular nodes (i.e. speciation events). Collectively, these applications constitute the *tree thinking* perspective on evolution described by O'Hara (1997) as a conceptual advance equivalent to Mayr's (1959) *population thinking*. Here we introduce those applications that have been most employed in the study of dasyurids but defer discussion of their associated methodology to later chapters.

Divergence times

A branch on a phylogenetic tree represents a lineage with a temporal duration, beginning at speciation and ending at extinction, speciation, or the present. Evidence external to the phylogeny can be used to estimate these durations, thereby anchoring the tree in real time. To illustrate this with the overly simple example of a strict molecular clock, suppose some long DNA sequence is known to accumulate base substitutions at a fixed rate of 1% of sites every 10^6 years (million years = Ma). If the sequences of species A and B began diverging 2 Ma, they should be 4% different now. Reversing this calculation, if A and B are 4% different now, we estimate that they began diverging 2 Ma.

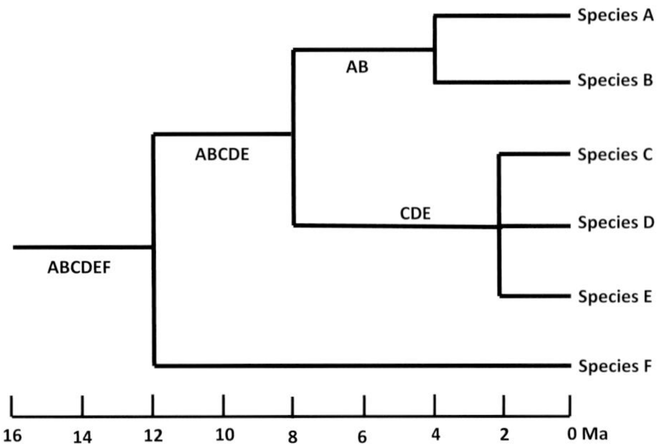

Fig. 1.5. A chronogram, or time tree, for six extant species. The scale at bottom is derived from evidence external to the phylogeny and provides dates for nodes and durations for branches. For example, the speciation event that gave rise to species A and B took place 4 million years ago (Ma) and the duration of lineage CDE was 6 Ma. In a chronogram, branch lengths represent elapsed time.

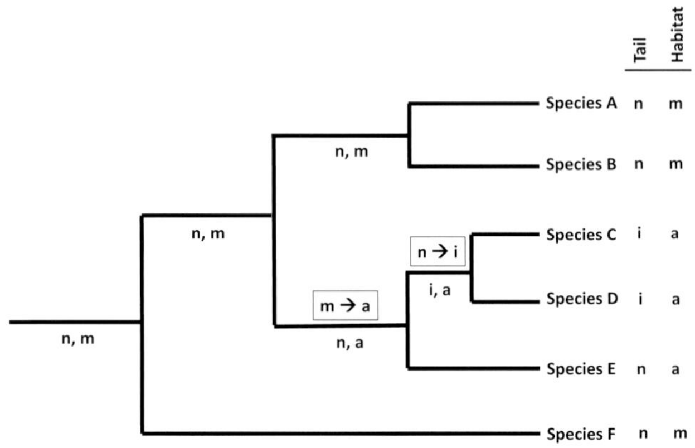

Fig. 1.6. Mapping two traits (tail structure and habitat) on a phylogeny. The character 'Tail' has states i = incrassated and n = not incrassated; the character 'Habitat' has states m = mesic and a = arid. Letters below internal branches show estimated states of ancestors; boxes above branches show the location of transitions in each character.

Applying the calculation to all species pairs dates each node and places a timescale on the tree; branch lengths now represent temporal durations (Fig. 1.5). Trees scaled in time are called *chronograms* or *time trees.* If the timescale is even approximately accurate, it connects the phylogeny to events in Earth history that were coincident with episodes of speciation, extinction, or character evolution on the tree. Because much of evolution is a response to environmental change, a dated phylogeny provides access to potential explanations of important events in the history of a clade.

Trait evolution

Interest in organismal evolution is often focused on understanding the significance of individual traits such as phenotypic characters, genomic properties or geographic distributions. If the phylogeny of a clade is known, and especially if it includes divergence times, the trait distribution among terminal species can be used to estimate the traits of ancestral species and reconstruct the trajectory of trait evolution (Fig. 1.6). This also is a powerful approach for evaluating historical correlations between traits, allowing us to test hypotheses of adaptation, coevolution, biogeography and ecology. Suppose one hypothesised that a small mammal's ability to store fat in the tail is an adaptation to living in arid environments where food abundance is unpredictable. The tree in Fig. 1.6 shows estimated ancestral states for a clade in which both tail fat storage (incrassation) and habitat vary among six species in a single clade. The hypothesis predicts that the selection pressure (arid habitat) should arise no later than the supposed adaptation to it (incrassated tail). The reconstruction suggests that the ancestor of clade CDE invaded an arid zone and that the ancestor of C and D subsequently evolved incrassated tails, consistent with the hypothesis.

Rates of evolution

Branch lengths on a phylogeny can be used to show amounts of change in some trait or set of traits, particularly quantitative ones (i.e. numerical measurements). This information allows us to calculate relative rates of change along different branches of the tree (or absolute rates if we have a chronogram). Rates of evolution, in turn, may reveal the signatures of different evolutionary

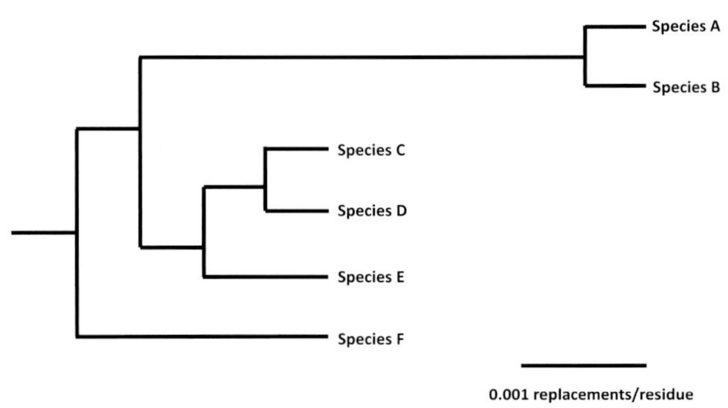

Fig. 1.7. A phylogenetic tree with branch lengths scaled to amounts of character change (phylogram). In this example, branch lengths show the amount of sequence evolution (measured as amino acid replacements/residue) in a particular protein, as indicated by the scale bar. The long branch to clade (A, B) shows accelerated evolution, possibly due to directional selection for modified protein function.

forces such as directional selection, genetic drift, or mutation. Branch lengths in Fig. 1.7 are scaled to the number of amino acid changes (replacements) per residue in a particular protein found in species A–F. The long branch of the AB ancestor shows an accelerated rate of replacement possibly due to directional selection for modified protein function. Trees drawn with branch lengths scaled to amounts of character change are called *phylograms*.

Intraspecific phylogeography

Any non-recombining stretch of DNA sequence (i.e. a locus) in the genome of an organism has a replication history that connects it with the same (homologous) locus in other members of its species and, ultimately, with all other species in which the locus occurs. Errors of DNA replication in germ cells of parents are mutations that produce new alleles in offspring which, if they continue to segregate through future generations, appear as genetic polymorphisms within species. The mutation history of alleles at a locus can be depicted as a network in which branches represent alleles and nodes represent mutations connecting related alleles. This network is not initially tree-like because ancestral alleles do not immediately go extinct after producing mutant descendants. However, over the course of generations, evolution 'prunes' enough alleles (including ancestral ones) from a population that the mutation history of a locus resembles a tree and, indeed, is commonly referred to as a *gene tree*. The same effect may be produced by drawing the relationships of only a small sample of alleles (haplotypes) from a larger population. Because gene trees can be estimated for alleles within a species, the pattern of relationships they reveal can be used to study intraspecific evolutionary history. A common application of gene trees has been to analyse the geographic structuring of genetic variation within species, a discipline called intraspecific phylogeography (Avise *et al.* 1987). When populations of a species become sufficiently isolated that gene flow among them is rare, drift and mutation eventually produce clades of alleles that are population specific (Fig. 1.8). Moreover, relationships among these clades may reveal the history of subdivision within the species or suggest the existence of previously unrecognised species boundaries (i.e. cryptic species). Similar spatial patterns of genetic subdivision in multiple co-distributed species suggest a common historical cause, such as the development of a physical boundary to migration between areas.

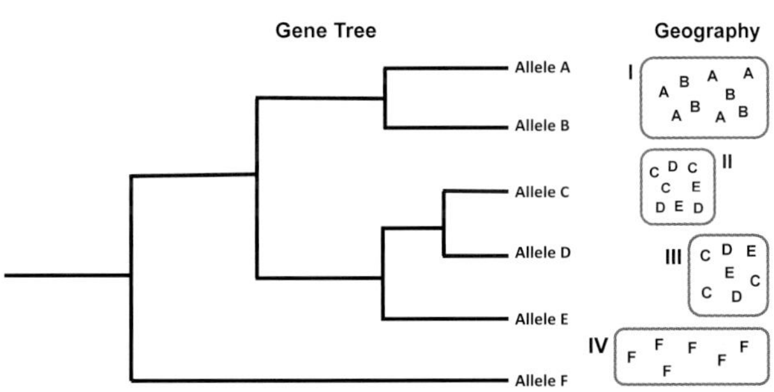

Fig. 1.8. Intraspecific phylogeography. Gene tree shows the mutation history of six alleles at a locus within a species. Geography panel shows the occurrence of alleles in individuals sampled from areas I to IV. The pattern of allele relationships suggests that populations in areas I, II + III and IV are genetically isolated and that isolation occurred in the temporal order IV – (I + II + III), then I – (II + III).

WHEN PHYLOGENY IS NOT A TREE

The tree metaphor is now so deeply rooted in systematic thought that many find it hard to conceive of phylogeny in any other form. Although no process can reverse the orientation of time's arrow in phylogeny, several well-known biological phenomena can cause forward-directed branches to join. These forms of *reticulation* are driven by movement of DNA between lineages. Interspecific hybridisation in sexual organisms produces individuals whose genomes are a mixture of loci from parental species. In some cases, such hybrids may constitute new species, a phenomenon known as *lineage fusion*. In other cases, hybrid backcrossing serves as a conduit for gene flow between parental species (introgression) such that different loci give different estimates of species phylogeny. Vector-mediated movement of DNA from one species to another (lateral gene transfer) produces the same result. Reticulation is one source of disagreement between gene trees and *species trees* (phylogenies). So long as reticulation involves only a few loci, the species tree will emerge as the majority (or dominant) gene tree in a large sample of loci from the genome, but this is not always the case. Bacterial evolution, in particular, is characterised by massive amounts of lateral gene transfer (Galtier 2007).

A second and more insidious phenomenon is *incomplete lineage sorting* (ILS). In the absence of selection, the fate of new alleles created by mutation is largely a matter of genetic drift, which randomly drives some alleles to fixation (100% frequency) and others to extinction (0% frequency) in a population. The rate of allele-frequency change due to drift is inversely proportional to effective population size (i.e. more rapid in smaller populations).

Relationships among the alleles at a locus in a focal ancestral species at the moment of its origin reflect the mutation history of those alleles in a previous ancestor (Fig. 1.9). If the focal ancestor persists long enough relative to its effective size, genetic drift will cull all but one of its original alleles (this is called *complete lineage sorting*). If the surviving allele produces mutant descendants of its own, those descendants will be transmitted through subsequent speciation events and occur in modern species. Their gene tree will show the monophyly of all modern species derived from the focal ancestor. However, if the focal ancestor does not persist long enough for complete lineage sorting, more than one of its alleles may be transmitted to its daughter species (*incomplete lineage sorting*). In this case, there is no guarantee that the gene tree of alleles will match the species tree. Lineage sorting operates independently on unlinked loci, so different loci may have different gene trees and different implications about species relationships.

Gene tree conflict due to ILS is common, but as for reticulation the presumption is that most genes in the genome will track the species tree. Based on this assumption, analytical methods have been devised to estimate the species tree even when some loci are subject to ILS. But investigations of this phenomenon at the genomic scale have only just begun. A recent analysis of hominid genomes (Mao *et al.* 2021) suggested that 36% of all loci are subject to ILS. Will this fraction remain below 50% when genomes of more species are compared? Is it safe to assume that a simple majority of gene trees will reveal the species tree? If phylogeny is the stream of heredity and the stream shows both divergence and reticulation, is there a 'species tree'? At the fundamental level of DNA sequences tracking organismal genealogy, phylogeny may be more of a tangled bush than we have long supposed.

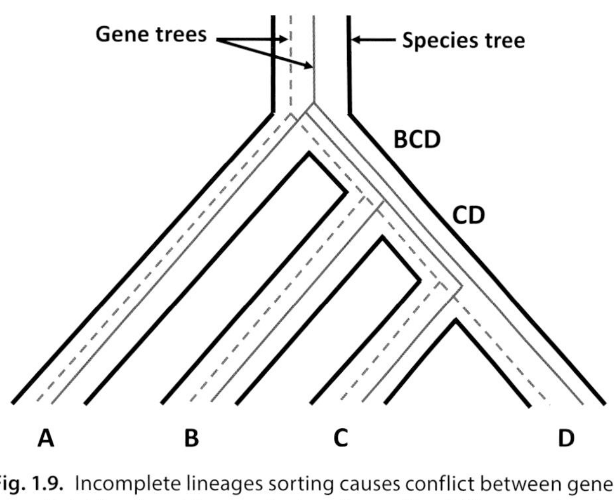

Fig. 1.9. Incomplete lineages sorting causes conflict between gene trees (light lines) and the species tree (heavy lines). The gene tree for the red locus (light dashed line) matches the species tree, but that for the blue locus (light solid line) does not. Note that lineage sorting at the blue locus in the ABCD ancestor is incomplete and two blue alleles are transmitted to ancestral species BCD and CD.

THE CHRONICLE OF DASYURID EVOLUTION

Fitted now with the lenses through which systematists see the world, we can begin to characterise the questions those intrepid scientists have asked about dasyurid evolution and how they have gone about finding answers. Along the way there is much to learn. With the modern advent of next-generation DNA sequencing and the availability of genome-scale data for many taxonomic groups, the structure of the Tree of Life is now emerging. For well-studied groups, that structure represents the integration of evidence accumulated over the course of centuries. Dasyuridae is one such group, and its history of systematic study extends from the 18th century exploration of Australasia by colonial naturalists to the modern genomics era. Indeed, the narrative of research on dasyurid diversity is a microcosm for the development of phylogenetic biology from its roots in Enlightenment taxonomy, through its transformation by the Darwinian and Hennigian revolutions, to its maturation as statistical phylogenetics and phylogenomics. This book tells the story of dasyurid evolution as it unfolded in the context of changing world views on biodiversity, biotic history and scientific methodology.

In Chapter 2 ('The diversity of dasyurids') we provide a modern phylogenetic classification of Dasyuridae and short introductions for each of its 17 genera, along with a tabular summary and distribution maps of all 75 species. The species list includes information on diet, habitat association and current conservation status. These are the actors whose historical drama we will try to reconstruct.

Chapter 3 ('Dasyurid phylogeny: morphological taxonomy from Linnaeus to Archer') begins with an overview of the earliest research on marsupials, then traces the history of dasyurid discovery and classification from the 17th century to 1984. In the context of this narrative, we explain the development of systematic thought from its Aristotelian-Linnaean beginnings to the conceptual revolutions of Darwinism, Hennigian cladistics and phenetics, culminating in early phylogenetic analyses of dasyurid anatomy.

In Chapter 4 ('Dasyurid phylogeny: molecular systematics, 1977–2003') the story of phylogenetic research continues with the birth of molecular systematics and its application to dasyurids, including comparative serology, allozyme electrophoresis, albumin microcomplement fixation, DNA hybridisation and DNA sequencing. We integrate discussion of these sources of information with descriptions of evolving systematic methodology, particularly statistical phylogenetics, to document the emerging consensus of a four-clade structure within Dasyuridae. The chapter closes with a summary of the discordant but instructive results of a 2002 morphocladistic analysis of the New Guinean genus *Murexia*.

Chapter 5 ('Dasyurid phylogeny: integrative studies') opens with an introduction to next-generation sequencing, the total evidence argument and Bayesian phylogenetic inference, then explains how the most recent molecular studies of dasyurid relationships have converged on the four-clade structure introduced in Chapter 4. We close with a review of the molecules-versus-morphology debate in modern systematics and how it informs the conflict over *Murexia*.

Chapter 6 ('Dasyurid species-level relationships') reviews research on species-level phylogeny within the clades Dasyurini, Phascogalini, Sminthopsini and Planigalini. We show how integrative phylogenetic analyses informed the modern composition of genera and resolved discordant aspects of previous results from single data types.

In Chapter 7 ('Dasyurid species recognition') we review how the evolving science of species delimitation is reflected in formal descriptions of dasyurid species and provide a critical assessment of current classifications within genera.

Chapter 8 ('Dasyuromorphian families and fossils') begins with information on the closest relatives of dasyurids – Myrmecobiidae (the numbat) and Thylacinidae (thylacines). We describe their basic biology and fossil history, followed by a summary of research on their phylogenetic positions and relationship to dasyurids. A summary of the sparse array of fossil dasyurids provides important context for the historical hypotheses of diversification and biogeography described in subsequent chapters.

In Chapter 9 ('The timing of dasyurid evolution') we recount the history of research on dating dasyurid divergence times, specifically episodes of diversification associated with the origins of major clades. Initially informed only by limited palaeontological data, the dasyurid time tree has been refined and solidified by the application of molecular and integrative dating techniques.

Chapter 10 ('Dasyurid macroevolution') addresses the question, 'What caused the rapid cladogenesis of dasyurids in the Miocene?' We discuss the evidence that this burst of speciation qualifies as an adaptive radiation, along with possible factors that explain it, including climate-induced habitat change, faunal turnover within the carnivore-insectivore niche and key innovation. We also review the origins of insular dasyurids in New Guinea and Tasmania, along with phylogeographic studies on areas of endemism, dispersal barriers and habitat refugia within the mainland.

Chapter 11 ('Evolution of the dasyurid phenotype') shows how systematists have used dasyurid phylogeny in comparative analyses to understand the evolution of key phenotypic traits described by previous generations of dasyurid biologists. We explain how phylogeny illuminates the search for craniodental synapomorphies, the evolution of penis and chromosome morphology, and the remarkable diversity of reproductive strategies (including the unusual male die-off pattern in several dasyurines).

Chapter 12 ('Dasyurids past, present and future') begins with a concise retelling of when, where and how dasyurids evolved based on the systematics research described in previous chapters. We then discuss how humans have affected dasyurid diversity since the Late Pleistocene, especially by means of introduced predators, habitat destruction, altered fire regimes and disease. We close with a discussion of how the imminent prospect of extinction for many dasyurid species may soon devastate the family's phylogenetic diversity.

In all these investigations, dasyurid systematists have applied the developing methods of their discipline to clarify not only species relationships but also some key evolutionary problems that can only be addressed with a robust phylogeny. These latter studies are at a relatively early stage but already show the power of integrative phylogenetic comparative analyses to answer questions that, 60 years ago, were matters for speculation. Dasyurid systematics has thus progressed from documenting the pattern and timing of evolution within the family to understanding the factors that caused that evolution.

ENDNOTES

1 Early taxonomists were aware that reproductive continuity ('like produces like') is a fundamental property of species, but this property did not easily lend itself to species diagnosis. Their focus on visible, physical structure reflected the Enlightenment emphasis on mechanistic explanations of natural phenomena, as well as limited knowledge of other organismal systems. Moreover, the widely held view that *natura non facit saltum* ('nature does not make a jump') led many taxonomists to suspect that no truly diagnostic traits exist; species and higher taxa, though convenient for communicating about biodiversity, are not real.

2 Virtually all modern taxonomists embrace hierarchical classification, but many have abandoned the use of Linnaean ranks (order, family, etc.). They argue that standard ranks and infra-ranks are insufficient to capture all levels of nesting in the Tree of Life, that ranks cannot be made comparable among major groups and that rank adds no additional information to a classification based on phylogeny. In this book we acknowledge, but do not emphasise or endorse, ranks associated with dasyuromorphian taxonomy.

3 It is also common for modern systematists to interpret nodes as ancestral species, and even more common for them to use phrases such as 'ancestral node'. We find this interpretation confusing and avoid it.

4 It is even more pronounced in the common, but not quite etymologically accurate, translations of *proto* and *eu* as 'early' and 'true', respectively.

2 The diversity of dasyurids

Here we introduce the species of living dasyurids in preparation for subsequent chapters that tell how researchers have come to understand their evolution. Describing biodiversity is more difficult than it seems, however, because (as the term implies) there is a great variety of relevant biological information, and we cannot summarise it all. Instead, we focus on a few key aspects of extant dasyurid species: their classification and names; their size and appearance; their general diet; their habitats and habits; and their geographic distributions. Our approach to these data requires some introduction.

CLASSIFICATION AND SPECIES NAMES

A working classification of species provides the framework for all discussions of biodiversity and contains two parts – a list of all currently recognised species and their scientific names, as well as a nested set of groups to which those species belong based on their phylogeny. Both aspects are scientific hypotheses contingent upon evidence and open to revision based on new research. While not every dasyurid biologist will agree with our classification, there is now more agreement than dissent among specialists. We describe how that consensus was achieved, as well as the points of contention, in later chapters. Our supra-specific classification of extant genera (referred to later as the 'four-clade structure') is simple to state and reflects current knowledge about phylogenetic relationships:

Family Dasyuridae
- Subfamily Dasyurinae
  - Tribe Dasyurini
    - *Dasyurus, Sarcophilus, Neophascogale, Phascolosorex, Pseudantechinus, Myoictis, Parantechinus, Dasykaluta, Dasyuroides, Dasycercus*
  - Tribe Phascogalini
    - *Phascogale, Antechinus, Murexia*
- Subfamily Sminthopsinae
  - Tribe Sminthopsini
    - *Antechinomys, Ningaui, Sminthopsis*
  - Tribe Planigalini
    - *Planigale*

Each taxon in this scheme is, we argue, a monophyletic group – i.e. its constituent species are more closely related to each other than to any member of another group at the same level. For example, Phascogalini contains all – and only – the descendants of the most recent common ancestor of *Phascogale, Antechinus* and *Murexia*. Because they are familiar to most readers, we assign each suprageneric taxon a rank (tribe, subfamily, family), though we recognise that such ranks are dispensable (de Quieroz and Gauthier 1994).

Establishing an inventory of recognised species is challenging because the International Code of Zoological Nomenclature (www.iczn.org) vests authority to describe species entirely with individual taxonomists, subject to rules about name formation, publication and type specimens. Dasyurid species descriptions form a cumulative body of literature dating to the late 1700s that has been growing vigorously to the present day. Putative species have been recognised by some authors and rejected by others, sometimes with very little evidence. Building on de Queiroz's (2007) pluralistic General Lineage Species Concept, the modern tradition of integrative taxonomy (e.g. Padial *et al.* 2010) embraces a thorough examination of phenotypic, genetic, geographic and ecological data in the delineation of species, but information from all these categories is seldom available. Thus, what we presently think about dasyurid species is only a snapshot of a dynamic discipline as field collectors, museum curators, anatomists, evolutionary geneticists and ecologists pore over the sometimes rich, but often scanty, record of specimens upon which taxonomy depends. As dasyurid systematists, we have critically reviewed this literature and here recognise those species we feel are adequately supported by published evidence. In Chapter 7 we describe current issues and ongoing research about species delimitation in several genera.

One of Linnaeus' enduring contributions to systematics is binomial nomenclature, the system of assigning two-word scientific names to formally described species. This was a huge advance over the *ad hoc* application of common names, providing much needed clarity in communication about the units of biodiversity. Nevertheless, common names of mammals are in wide use and will certainly remain so; they are, for most non-specialists, easier to hold in memory than scientific names. We therefore provide a single common name for each dasyurid species and try to use it consistently throughout this book. The names are taken from widely cited references such as *Mammals of Australia* (Van Dyck and Strahan 2008) or *Mammals of New Guinea* (Flannery 1995). These names may be different from those by which a species was most widely known for some or most of its taxonomic history, but all are in current use by mammalogists. We deliberately avoid common names that reduce dasyurids to marsupial analogues of placentals of similar appearance or habits (e.g. 'marsupial mice', 'native cats').

NATURAL HISTORY INFORMATION

Embedded in the four-clade classification of dasyurids above, we provide a short narrative description of each genus, accompanied by a table listing all species within the genus, a composite illustration including a map of their current distribution (not including areas where locally extinct species have been reintroduced), photographs of one or more species of the genus and, in some cases, the habitat in which it occurs or some aspect of the habits of the species. Our text descriptions are brief summaries of external appearance, natural habitat use, diet and distribution. The tables include scientific and common names for each species, the taxonomic authority for the scientific name (some historical information behind these is provided in Chapters 3–7), body measurements indicative of overall size and sexual dimorphism, habitat associations and current conservation status.

Tabular data on body size include B = head + body (i.e. snout–vent) length, T = tail length and W = weight. Separate measurement ranges are given for males and females. In some cases, only pooled data for the sexes are available and are given as a single range; in others, only data from one sex are available, in which case the other sex is designated 'ND' (no data). Measurements are taken from Van Dyck and Strahan (2008) for Australian species and, for New Guinean species, from records compiled by one of us (Woolley) following examination of mature adult specimens in the collections of museums. It should be kept in mind that most measurements given do not take age differences or reproductive status into account. Habitat data from the IUCN Red List (www.iucnredlist.org) are binned into eight categories as follows:

Rainforests. Includes (1) tropical rainforest with annual precipitation >168 cm, no significant seasonality and stratified vegetation, and (2) warm- and cool-temperate rainforests with mostly broadleaf evergreen trees.

Eucalypt forests. Dominated by species of tall (>30 m) *Eucalyptus, Corymbia* and *Angophora* trees with >50% crown cover, including closed (denser) and open (less dense) stands.

Wetlands. Swamps, sedgelands and areas with damp understorey.

Woodlands. Habitat dominated by shrubs (<30 m tall) separated by grasses, herbs and sparse trees (20–50% crown cover).

Shrublands. Areas with shrubs, low mallee, herbs and grasses with very sparse trees (<20% crown cover).

Savannas. Open areas dominated by unbroken stands of grass and an open canopy of widely spaced trees.

Grasslands. Treeless areas in (1) mesic regions dominated by perennial grasses or forbs that grow in clumps separated by areas that contain sparse herbs, or (2) arid regions dominated by xerophytic hummock grass (spinifex).

Deserts. Arid regions lacking significant emergent vegetation; includes gibber plains with pavements of pebble fragments and ephemeral herbs emerging after infrequent rains.

Rocky areas. Consisting of exposed rock surfaces – either solid faces, scree slopes, or loose stones; these can occur in many of the habitats listed above, especially drier areas.

The spatial distribution of habitat types (ecoregions) across Australia and New Guinea is illustrated in Fig. 2.1. Natural habitats are of course much more diverse than this, and that fine-grained variation matters very much for understanding dasyurid distributions, but ecoregions are the basis for much of the biogeographic research we describe in later chapters. Our range maps are similarly generalised, showing only broad areas of Australia and New Guinea where each species presently occurs. These are based on standard references such as Van Dyck and Strahan (2008), IUCN (www.iucnredlist.org) and Atlas of Living Australia (www.ala.org.au) for species that live in Australia, and on Flannery (1995) and Woolley (2005a, 2019) for New Guinean species. This neglects microgeographic variation in habitat and historical changes in species distributions, both of which are relevant to discussions of biogeography and conservation. IUCN conservation status abbreviations are in standard use: DD = Data Deficient, LC = Least Concern, NT = Near Threatened, VU = Vulnerable, EN = Endangered and CR = Critically Endangered. We take up biogeography and conservation in Chapters 10 and 12, respectively.

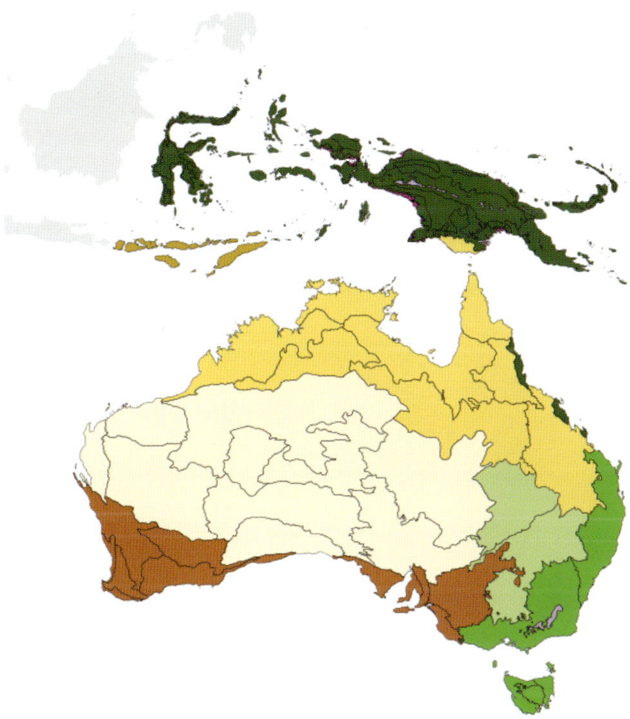

Fig. 2.1. Ecoregions of Australia and New Guinea, colour-coded by biome. Dark green = tropical and subtropical moist broadleaf forests; light brown = tropical and subtropical dry broadleaf forests; yellow = tropical and subtropical grasslands, savannas and shrublands; green = temperate broadleaf and mixed forests; light green = temperate grasslands, savannas and shrublands; light purple = montane grasslands and shrublands; brown= Mediterranean forests, woodlands and scrub; beige = deserts and xeric shrublands; magenta = mangroves. [Image by Terpsichores – own work, licensed under CC BY-SA 3.0 via Wikimedia Commons.]

DASYURINAE

Dasyurini

Dasyurus (Quolls)

The first dasyurid discovered by Europeans was likely a northern quoll (*Dasyurus hallucatus*), collected by members of Captain Cook's expedition in 1770 (see Chapter 3). Described as being like placental martens (Mustelidae), Geoffroy (1796) recognised members of this group as marsupials distinct from Linneaus's *Didelphis*, a genus that served as an 18th century catch-all for many newly described metatherians. Geoffroy introduced the genus name *Dasyurus* (Greek for 'hairy tail', a trait that distinguished quolls from then-known opossums). 'Quoll' is derived from Aboriginal words and preferred to placental analogies such as 'native cat'. Quolls are the nominal genus of dasyurids (i.e. the family name Dasyuridae is based on *Dasyurus*) and, in some ways, manifest the group's most well-known features. Dasyurids are quintessential 'carnivorous marsupials', eating other animals (vertebrate and invertebrate), but quolls and the closely related Tasmanian devil (*Sarcophilus*) consume the highest proportion of vertebrate prey among all dasyurids. This is related to their relatively large body sizes: they are heavier than all other dasyurids except the devil. The New Guinean quoll is known to consume large rodents in the wild (Menzies 2011). They are gracile animals (Fig. 2.2) with white-spotted, usually brown but occasionally black dorsal pelage, pale undersides and tails about the same length as their bodies. Males are generally larger than females of comparable age, as is the case for all dasyurid species. Four of the six species of quolls are found in Australia and two in

Dasyurus Geoffroy 1796

Species	Authority	Size range (M/F)	Habitat and habits	Status
D. albopunctatus New Guinean quoll	Schlegel 1880	B: 228–350/214–287 mm T: 212–290/204–323 mm W: 391–950/399–530 g	Rainforest; terrestrial, scansorial	NT
D. geoffroii Western quoll	Gould 1841	B: 310–400/260–360 mm T: 250–350/210–310 mm W: 710–2185/615–1130 g	Woodland, semi-arid grassland; terrestrial, scansorial	NT
D. hallucatus Northern quoll	Gould 1842	B: 270–370/249–310 mm T: 222–345/202–300 mm W: 340–1120/240–690 g	Shrubland, rainforest; terrestrial, scansorial	EN
D. maculatus Spotted-tailed quoll	Kerr 1792		Woodland, shrubland, rainforest, eucalypt forest; terrestrial, scansorial	NT
D. m. maculatus		B: 380–759/350–450 mm T: 370–550/340–420 mm W: 1500–5000/900–2500 g		
D. m. gracilis		B: 330–510/ 310–440 mm T: 310–440/285–410 mm W: 850–2540/800–1600 g		
D. spartacus Bronze quoll	Van Dyck 1988	B: 345–380/305 mm T: 250–298/230–295 mm* W: 735–1118/ 543–945 g*	Savanna, woodlands; terrestrial, scansorial	NT
D. viverrinus Eastern quoll	Shaw 1800	B: 320–450/280–400 mm T: 200–280/170–240 mm W: 900–1900/700–1100 g	Shrubland, grassland, eucalypt forest; terrestrial, scansorial	EN

* SG Hamilton (personal communication)

Fig. 2.2. *Dasyurus.* (a) Current distribution of each of the six species. (b) *D. viverrinus* colour phases. (c) *D. albopunctatus.* (d) Sleeping *D. maculatus* with erect penis and penis appendage visible (see Chapter 11 for discussion). [Photo credits: (b) T.P. Fletcher; (c) P.A. Woolley and D. Walsh; (d) J. Seebeck.]

New Guinea. Of the latter group, *D. albopunctatus*, occurs from near sea level to 3560 m, while *D. spartacus* has only been found below 60 m. While their distributions are not known to overlap, they can be readily distinguished by appearance of the tail: closely haired in *D. albopunctatus* and bushy in *D. spartacus*.

Sarcophilus (Tasmanian devil)

The largest living carnivorous marsupials, Tasmanian devils (Fig. 2.3) are nocturnal predators and scavengers, the niche from which their genus name ('flesh lover') is derived. Food items include wallabies, wombats, possums and birds, as well as introduced species such as sheep and rabbits, for which their shearing, crushing teeth and powerful jaw muscles are adapted for dismembering. Devils' black pelage (usually interrupted by a white stripe across the breast and sometimes white patches elsewhere) is unique among dasyurids. They occur throughout Tasmania but are most abundant in eucalypt forests and woodlands. Intraspecific aggressive interactions during communal feeding or mating involve much biting, which is thought to facilitate spread of devil facial tumour disease, an infectious cancer that has severely reduced devil numbers since the late 1990s.

Neophascogale (Speckled dasyure)

Very little is known about the biology of *Neophascogale lorentzii*, even though this species is well represented in museum collections (close to 200 specimens). Speckled dasyures occur in relatively high elevation (1500–3900 m) forests and montane grasslands where they forage by day, presumably on invertebrates and small vertebrates such as skinks. Their distinctive pelage features very long hairs (Woolley and Valente 1992), typically silver-tipped on the dorsum, from which the common name is derived, but melanistic specimens are known. Their claws are unusually long (Fig. 2.4). They have white hairs on the back of the ear pinnae and on the tail; either a tuft at the tip or extending for up to 70 mm of the length of the tail. Tail length is roughly equal to

Sarcophilus F.G. Cuvier 1837

Species	Authority	Size range (M/F)	Habitat and habits	Status
S. harrisii Tasmanian devil	Boitard 1841	B: 652/570 mm T: 258/244 mm W: 8–14/5–9 kg	Eucalypt forests, woodlands, shrublands; terrestrial but young animals can climb	EN

Fig. 2.3. *Sarcophilus.* (a) *S. harrisii*, the Tasmanian devil, is the only extant member of this genus and is restricted to Tasmania (inset map). (b) Woodland area where devils have been sighted. [Photo credits: (a) C. Krajewski; (b) Amy Edwards.]

Neophascogale Stein 1933

Species	Authority	Size range (M/F)	Habitat and habits	Status
N. lorentzii Speckled dasyure	Jentink 1911	B: 162–230/126–220 mm T: 161–223/141–225 mm W: 212, 237*/145*	Rainforest, alpine grassland; terrestrial, scansorial	LC

* P.A. Woolley (unpublished data).

Fig. 2.4. *Neophascogale*. (a) *N. lorentzii* (inset, underside of hind foot showing the long claws). (b) Distribution (inset map) and the village garden near Porgera, Enga Province where the animal was found. [Photo credits: (a) P.A. Woolley and D. Walsh; (b) P.A. Woolley.]

body length but may be either a little shorter or a little longer.

***Phascolosorex* (no common name)**

Members of this genus lack a common name. Because *Phascolosorex* means 'leather bag (pouched) shrew', the species are called 'marsupial shrews', but this is another placental analogy that we reject. Even the individual species have enigmatic monikers. *P. doriae* is the red-bellied phascogale: it was originally assigned to *Phascogale*, but this was abandoned over 100 years ago; 'red-bellied' describes its auburn ventral pelage contrasting with a drab dorsum (Fig. 2.5) that is more pronounced in specimens from the eastern part of its range. The ventral pelage of *P. dorsalis* is less reddish, sometimes brown. Some individuals of each species have white tipped tails. *P. dorsalis* is the narrow-striped dasyure: 'dasyure' is a generic term for dasyurids, but 'narrow-striped' identifies a distinctive longitudinal band of black hairs along the dorsal midline. Both species are endemic to the montane rainforests of New Guinea, with the larger *P. doriae* occurring at lower elevations (900–2000 m) in the west and the smaller *P. dorsalis* at higher elevations (1500–3600 m) throughout the cordillera. Virtually nothing is known of the natural history of *P. doriae*: specimens have been collected by day, in village gardens and remains of one have been found in a raptor pellet. A study of circadian activity patterns established that *P. dorsalis* is diurnal (Woolley *et al.* 1991). A diverse range of invertebrates is eaten by *P. dorsalis*, with beetles and spiders forming the bulk of the diet followed by grasshoppers and crickets. Frog bone fragments and bird eggshell have also been found in their scats (Grossek *et al.* 2010). Newly captured *P. dorsalis* ate small skinks and geckos and formed nests in long dry grass provided, pulling it over themselves when they retired for the night. Menzies (1991) reported finding one in a nest in a clump of orchids some height above the ground. Examination of specimens in museum collections indicates that both species breed throughout the year (Woolley 1994).

Phascolosorex Matschie 1916

Species	Authority	Size range (M/F)	Habitat and habits	Status
P. doriae Red-bellied phascogale	Thomas 1886	B: 193–245/168–212 mm T: 168–212/161–184 mm W: 132–145/ND g	Rainforest; probably terrestrial and scansorial	LC
P. dorsalis Narrow-striped dasyure	Peters & Doria 1876	B: 125–173/112–151 mm T: 99–143/88–123 mm W: 45–85/36–54 g	Rainforest; terrestrial, scansorial	LC

Fig. 2.5. *Phascolosorex*. (a) *P. dorsalis*. (b) North summit of Mt Kaindi, Morobe Province where this species is close to eastern limit of distribution. (c) Distribution of each of the two species. (d) Site of collection of an individual on Mt Kaindi arrowed. [Photo credits: (a) P.A. Woolley and D. Walsh; (b, d) P.A. Woolley.]

***Pseudantechinus* (Pseudantechinuses)**

First considered to be part of *Phascogale*, and later *Antechinus*, some of these 'false antechinuses' were recognised as a distinct genus – *Pseudantechinus* – by Tate (1947). More recent molecular data (Westerman *et al.* 2006) supported the inclusion of *P. bilarni* in this genus rather than in *Parantechinus*, where it had long been assigned. These small animals feed primarily on insects and all except *P. bilarni* store fat at the base of their tails (Fig. 2.6). All five species can be found in rocky areas within their ranges, and *P. macdonnellensis* is known to live in termite mounds in some parts of its range (Woolley 1991b; Umbrello *et al.* 2022).

Pseudantechinus Tate 1947

Species	Authority	Size range (M/F)	Habitat and habits	Status
P. bilarni Sandstone pseudantechinus	Johnson 1954	B: 90–115/90–115 mm T: 90–125/90–120 mm W: 20–40/15–35 g	Rocky areas of eucalypt forests; terrestrial	LC
P. macdonnellensis Fat-tailed pseudantechinus	Spencer 1895	B: 95–105/95–105 mm T: 75–80/75–85 mm W: 25–45/20–40 g	Rocky shrublands, sandplains with termite mounds; terrestrial	LC
P. mimulus Carpentarian pseudantechinus	Thomas 1906	B: 80–90/63–91 mm T: 66–70/56–76 mm W: 14–18/14–25 g	Rocky areas of shrublands, savannas, grasslands; terrestrial	NT
P. ningbing Ningbing pseudantechinus	Kitchener 1988	B: 90–100/80–90 mm T: 80–90/75–90 mm W: 20–25/15–20 g	Rocky areas of shrublands, eucalypt forests, savannas; terrestrial	LC
P. woolleyae Woolley's pseudantechinus	Kitchener & Caputi 1988	B: 107–108/100–108 mm T: 75/75–85 mm W: 35–50/30–45 g	Rocky areas of shrublands, deserts; terrestrial	LC

Fig. 2.6. *Pseudantechinus*. (a) Distribution of each of the five species. (b) *P. macdonnellensis*. This species lives in both rock outcrops and termite mounds in different parts of its range such as (c) the Sedimentaries (foreground) and at the base of Uluru (on horizon) in central Australia and (d) termite mounds in the Tanami desert. (e) *P. bilarni*. (f) Rocky habitat of *P. bilarni* in Arnhem Land. [Photo credits: (b, e) P.A. Woolley and D. Walsh; (c, d, f) P.A. Woolley.]

Myoictis (Three-striped dasyures)

Our current understanding of *Myoictis* diversity is based on Woolley's (2005a) morphological revision, later confirmed by molecular data (Westerman *et al.* 2006). Prior to the revision only two species were recognised: *M. melas*, which included all known specimens other than *M. wallacei*. *Myoictis melas* was the first dasyurid marsupial to be collected in New Guinea, by the Dutch in 1828. It was given its specific name in reference to its uniform black colour which was later found not characteristic of the species. Members of this genus are endemic to low-elevation rainforests of New Guinea and the Aru Islands. They are diurnal terrestrial predators whose diet includes small rodents and skinks. They have three black, dorsal stripes against variously grey–black, brown, or auburn pelage (Fig. 2.7). The four species are distinguishable by dorsal

Myoictis Gray 1858

Species	Authority	Size range (M/F)	Habitat & Habits	Status
M. melas Müller's three-striped dasyure	Müller 1840	B: 189–240/175–201 mm T: 158–191/145–179 mm W: 172, 223/200 g	Rainforest; terrestrial	LC
M. leucura Woolley's three-striped dasyure	Woolley 2005	B: 200–230/192, 197 mm T: 165–190/162, 165 mm W: 200–230/200 g	Rainforest; terrestrial	DD
M. wallacei Wallace's three-striped dasyure	Gray 1858	B: 185–222/176, 208 mm T: 152–183/144, 180 mm W: 206–245/192 g	Rainforest; terrestrial	LC
M. wavicus Tate's three-striped dasyure	Tate 1947	B: 165–175/140, 170 mm T: 140–146/125, 130 mm W: 110, 134/61 g	Rainforest; terrestrial	DD

Fig. 2.7. *Myoictis*. (a) *M. wavicus*. (b) *M. wallacei*. (c) Mt Missim (2962 m), as seen looking across the Wau valley from Mt Kaindi. *M. wavicus* has been found on the southern slopes at 1450 m. (d) Kobroor, one of the low-lying Aru Islands where *M. wallacei* occurs. (e) Distribution of each of the four species. (f) Site of collection of *M. wallacei*. [Photo credits: (a, b) P.A. Woolley and D. Walsh; (c, d, f) P.A. Woolley.]

coat colour, and by the length and colour of hairs on the tail, which is white tipped in *M. leucura*. Prior to the revision what little was known of their biology was reported as pertaining to *M. melas* – e.g. Woolley (1994) included observations on specimens now known as *wavicus* and *leucura* (as well as *melas*) and Woolley (2003) included observations on specimens now known as *wavicus*.

Parantechinus (Dibbler)

Presumed extinct until it was rediscovered in 1967, the dibbler is an extremely rare animal now confined to small populations (including offshore islands) in south-western Australia. The single species was placed in *Phascogale* and later *Antechinus* prior to Tate's (1947) reclassification; it is one of several 'false antechinus' genera (along with *Pseudantechinus* and *Dasykaluta*). Mostly inhabiting dense shrublands, it is a generalist insectivore that also consumes some plants including *Rhagodia* berries and nectar from *Banksia* flowers. Mainland dibblers are about 40% larger than those on islands (Mills *et al.* 2004, data for captive animals). The specific name (*apicalis*) of the dibbler refers to the tapering tail tipped with black hairs. White tipped hairs on the body give it a freckled appearance and the animals have a characteristic white eye ring (Fig. 2.8). Hair structure is distinctive (Valente and Woolley 1982).

Dasykaluta (Kaluta)

Another of the 'false antechinuses' to be placed in its own genus (by Archer 1982c), this small inhabitant of spinifex grasslands throughout its range in Western Australia has rich, rufous-brown fur over its entire body (Fig. 2.9). Kalutas are mainly nocturnal, but they do come out to bask in the sun, like many other dasyurids. They prey on arthropods and small vertebrates such as lizards. Like *Phascogale* and *Antechinus*, *Dasykaluta* has a reproductive strategy in which all males die after their first mating season due to physiological stress.

Parantechinus Tate 1947

Species	Authority	Size range (M/F)	Habitat and habits	Status
P. apicalis Dibbler	Gray 1842	B: 145/140 mm T: 105–115/95 mm W: 60–125/40–73 g	Shrubland; terrestrial, scansorial	EN

Fig. 2.8. *Parantechinus.* (a) *P. apicalis.* (b) Distribution (inset map) and habitat (low *Banksia* shrubland in foreground) at site of rediscovery (Morcombe 1967) of the species at Cheyne Beach, near Albany, WA. [Photo credits: (a) P.A. Woolley and D. Walsh; (b) P.A. Woolley.]

Dasykaluta Archer 1982

Species	Authority	Size range (M/F)	Habitat and habits	Status
D. rosamondae Kaluta	Ride 1964	B: 95–110/90–100 mm T: 55–70/60–65 mm W: 25–40/20–30 g	Grassland; terrestrial	LC

Fig. 2.9. *Dasykaluta.* (a) *D. rosamondae.* (b) Habitat and distribution (inset map). [Photo credits: (a) P.A. Woolley and D. Walsh; (b) P.A. Woolley.]

Dasyuroides (Kowari)

The kowari, once abundant on the dry gibber plains in parts of central Australia, has a striking black brush of hairs on the distal half of its tail (Fig. 2.10), hence reference to them as 'bushy tails' by local people in south-western Queensland. They are bold and inquisitive animals. By night they prey on arthropods and small mammals, and they have also been seen feeding on a cow carcass. By day they retreat to burrows which they either dig themselves or occupy those of other animals such as bilbies. Their numbers seem to have diminished, possibly a result of overstocking with cattle in parts of their range, as well as predation by introduced predators such as cats. There is anecdotal evidence for their decline on some properties following 1080 baiting of predators.

Dasyuroides Spencer 1896

Species	Authority	Size range (M/F)	Habitat and habits	Status
D. byrnei Kowari	Spencer 1896	B: 140–180/135–160 mm T: 110–160/110–160 mm W: 85–175/70–140 g	Desert, shrubland; terrestrial	VU

Fig. 2.10. *Dasyuroides.* (a) *D. byrnei* and inset map of distribution. (b) Burrow at base of shrub in sandy patch on gibber plain. White pegs mark holes associated with the burrow system. [Photo credits: (a, b) P.A. Woolley.]

***Dasycercus* (Mulgaras)**

Mulgaras are inhabitants of Australia's arid regions – the brush-tailed mulgara strongly associated with spinifex grasslands and the crest-tailed mulgara with sand dunes. Both show adaptations to these dry habitats, including the use of tunnel systems in which they shelter during the day and the ability to store fat in their tails. They prey on arthropods and small birds, lizards and mammals. The pelage is sandy or tan dorsally and pale ventrally; their moderate-length tails are distinctly tipped with black hairs (Fig. 2.11). Both forms had long been considered *Dasycercus cristicauda* prior to Woolley's (2005b) revision. The two species are readily distinguished by differences in tail morphology, including the black hairs forming a brush in *D. blythi* and a crest in *D. cristicauda*.

Dasycercus Peters 1875

Species	Authority	Size range (M/F)	Habitat and habits	Status
D. blythi Brush-tailed mulgara	Waite 1904	B: 135–165/120–140 mm T: 120–140/60–100 mm W: 75–110/60–90 g	Desert, shrubland, grassland; terrestrial, burrowing	LC
D. cristicauda Crest-tailed mulgara	Krefft 1867	B: 130–230/120–170 mm T: 80–125/80–110 mm W: 110–185/65–120 g	Shrubland, grassland, desert; terrestrial, burrowing	NT

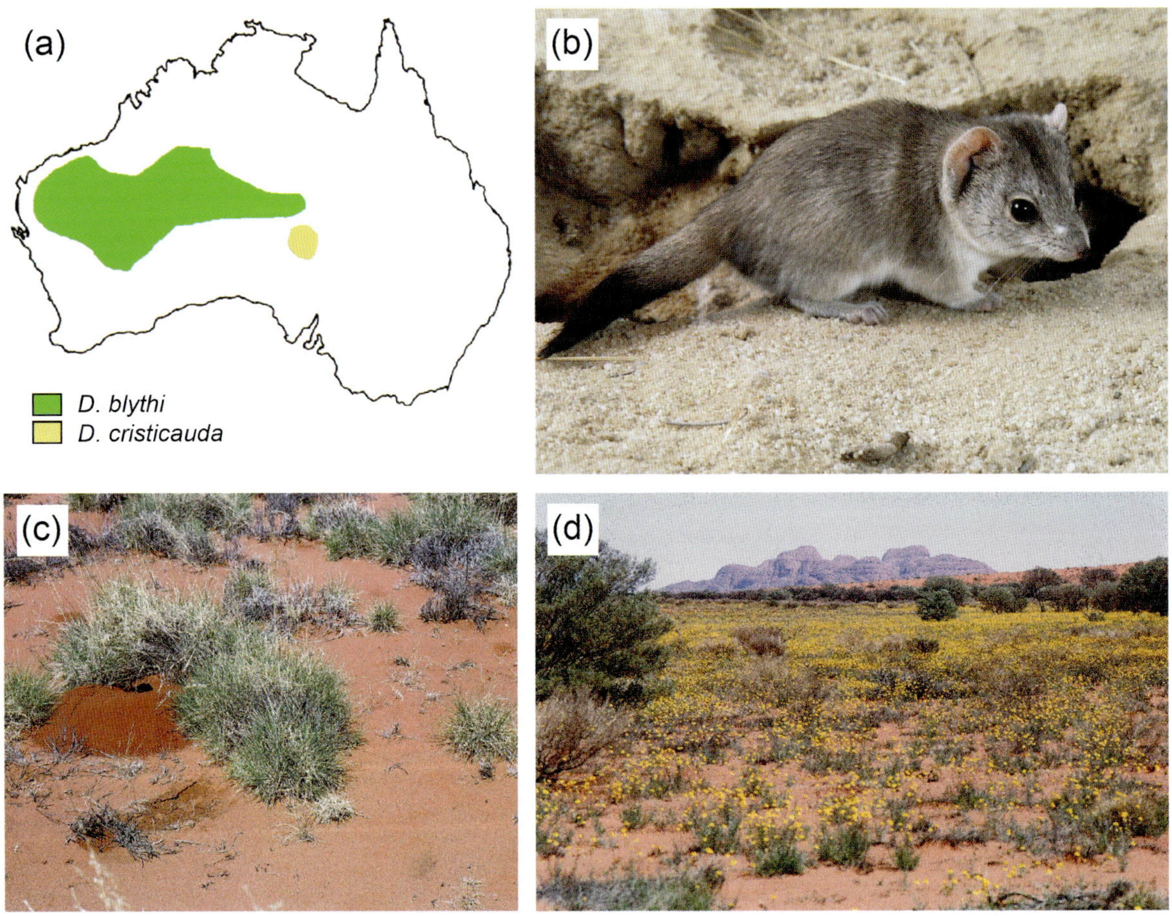

Fig. 2.11. *Dasycercus.* (a) Distribution of each of the two species. (b) *D. blythi.* (c) Entrance to burrow of a *D. blythi* female with newborn young, found on an interdune swale (d) near Uluru (Ayers Rock) in central Australia. Kata Tjuta (the Olgas) on horizon. [Photo credits: (b) P.A. Woolley and D. Walsh; (c, d) P.A. Woolley.]

Phascogalini

Phascogale (Phascogales)

Phascogales are nocturnal, primarily arboreal, predators in forests around the margins of Australia. One of the three currently recognised species, *Phascogale pirata*, is also found on Melville Island. They have long tails. Black hairs cover the terminal half of the tail in *P. calura*, the remainder being covered in reddish hairs (hence the common name), while much of the tail of both *P. pirata* and *P. tapoatafa* is covered with black hairs. These black hairs can be erected, giving the tail a bottle brush appearance (Fig. 2.12). The diet of *P. tapoatafa* consists largely of insects and spiders, but they will take small birds and mammals and are known to consume nectar. Phascogales depend largely on tree hollows for nesting sites. Like members of *Antechinus* and *Dasykaluta*, they display the 'male die-off' reproductive strategy.

Phascogale Temminck 1824

Species	Authority	Size range (M/F)	Habitat and habits	Status
P. calura Red-tailed phascogale	Gould 1844	B: 105–122/93–105 mm T: 134–145/119–144 mm W: 39–68/38–48 g	Eucalypt and sheoak forest; arboreal	NT
P. pirata Northern brush-tailed phascogale	Thomas 1904	B: 209–210/150–188 mm T: 200–208/180–192 mm W: 234/108–150 g	Eucalypt forest; arboreal	VU
P. tapoatafa Brush-tailed phascogale	Meyer 1793	B: 160–261/148–223 mm T: 175–234/160–226 mm W: 175–311/106–212 g	Eucalypt forest; arboreal	NT

Fig. 2.12. *Phascogale.* (a) *P. calura.* (b) Distribution of each of the three species. (c) *P. tapoatafa.* [Photo credits: (a, c) Rohan Clarke.]

Antechinus (Antechinuses)

Members of this speciose genus are found around the margins of Australia as well as in Tasmania, the Bass Strait islands and Melville Island. Some have exceedingly small distributions and so are especially vulnerable to habitat degradation. They are nocturnal or crepuscular and prey on insects and small vertebrates. Aspects of pelage colouration are reflected in many of the common names given to species (Fig. 2.13). The reproductive pattern, in which all males die at the end of their first short breeding season (also found in *Phascogale* and *Dasykaluta*), was first described in *Antechinus* (*A. agilis*, as *A. stuartii*) by Woolley (1966) and is now known to occur in all species for which appropriate detailed observations have been made. It is generally assumed that all 15 species exhibit the male 'die-off'. Little is known of the habits of some of the more recently described species.

Antechinus Macleay 1841

Species	Authority	Size range (M/F)	Habitat and habits	Status
A. adustus Rusty antechinus	Thomas 1923	B: 98–125/89–109 mm T: 91–103/72–101 mm W: 28–42/18–35 g	Rainforest; terrestrial, scansorial	LC
A. agilis Agile antechinus	Dickman *et al.* 1998	B: 65–124/62–128 mm T: 63–116/60–103 mm W: 20–40/16–25 g	Eucalypt forest, shrubland; terrestrial, scansorial	LC
A. argentus[1] Silver-headed antechinus	Baker *et al.* 2013	B: 94–107/81–87 mm T: 87–108/78–84 mm W: 29–38/18–23 g	Eucalypt forest; ND	
Antechinus arktos[2] Black-tailed antechinus	Baker *et al.* 2014	B: 108–145/106–111 mm T: 100–131/106–112 mm W: 60–120/46 g	Rainforest	
A. bellus Fawn antechinus	Thomas 1904	B: 121–148/110–130 mm T: 105–126/93–110 mm W: 42–66/26–41 g	Eucalypt forest, savanna; terrestrial, scansorial	VU
A. flavipes Yellow-footed antechinus	Waterhouse 1838	B: 93–165/86–127 mm T: 70–151/65–107 mm W: 26–79/21–52 g	Rainforest, eucalypt forest, wetland, savanna, shrubland; terrestrial, scansorial	LC
A. godmani Atherton antechinus	Thomas 1923	B: 120–160/90–135 mm T: 105–145/85–115 mm W: 85–125/53–73 g	Rainforest; terrestrial, scansorial	LC
A. leo Cinnamon antechinus	Van Dyck 1980	B: 142–160/109–136 mm T: 127–140/81–125 mm W: 67–124/32–74 g	Rainforest; terrestrial, scansorial	LC
A. mimetes[3] Mainland dusky antechinus	Thomas 1924	B: 89–150/85–125 mm T: 82–114/76–98 mm W: 42–112/30–48 g	Eucalypt forest, shrubland; terrestrial	
A. minimus Swamp antechinus	Geoffroy 1803	B: 103–140/98–117 mm T: 65–100/67–85 mm W: 50–129/36–68 g	Wetland, eucalypt forest, grassland, shrubland; terrestrial	LC
A. mysticus[4] Buff-footed antechinus	Baker *et al.* 2012	B: 93–117/81–114 mm T: 87–117/79–96 mm W: 32–46/18–34 g	Eucalypt forest	
A. stuartii Brown antechinus	Macleay 1841	B: 81–141/73–116 mm T: 70–124/65–116 mm W: 29–71/17–40 g	Rainforest, eucalypt forest; terrestrial, scansorial	LC
A. subtropicus Subtropical antechinus	Van Dyck & Crowther 2000	B: 95–130/95–110 mm T: 89–105/81–95 mm W: 18–61/81–95 g	Rainforest, eucalypt forest; terrestrial, scansorial	LC

Species	Authority	Size range (M/F)	Habitat and habits	Status
A. swainsonii Tasmanian Dusky antechinus	Waterhouse 1840	B: 103–188/89–140 mm T: 80–121/75–100 mm W: 43–178/37–100 g	Eucalypt forest, shrubland; terrestrial	LC
A. vandycki[3] Tasman Peninsula dusky antechinus	Baker *et al.* 2015	B: 105–133/ND mm T: 92–118/ND mm W: 46–93/ND g	Rainforest; terrestrial	

[1] Data from Baker *et al.* (2013). IUCN has not assessed *A. argentus*.
[2] Data from Baker *et al.* (2014). IUCN has not assessed *A. arktos*.
[3] Data from Baker *et al.* (2015). IUCN has not assessed *A. mimetes* or *A. vandycki*.
[4] Data from Baker *et al.* (2012). IUCN has not assessed *A. mysticus*.

Fig. 2.13. *Antechinus*. (a) *A. leo*. (b) Habitat of *A. leo*, far north Queensland. (c) *A. mimetes*. (d) Habitat of *A. mimetes*, Victoria Valley, Southern Grampians, Vic. (e) Distribution of each of the 15 species of *Antechinus*. [Photo credits: (a) P.A. Woolley and D. Walsh; (b) P.A. Woolley; (c, d) Rohan Clarke.]

***Murexia* ('Dasyures')**

Perhaps the most enigmatic genus of dasyurids, the New Guinean *Murexia* has been the subject of several attempts at taxonomic revision, the most recent by Woolley *et al.* (2020). Only *M. longicaudata* and *M. rothschildi* were included in the genus by Tate (1947). Another three species were designated *Antechinus* but moved to *Murexia* after Woolley (1984b) showed that their penis morphologies do not match those of Australian antechinuses. Molecular analyses by Armstrong *et al.* (1998) showed that the five species form a single clade, but Van Dyck (2002) rejected this conclusion and placed each species in a separate genus. We provide a detailed discussion of this research in later chapters, but it is likely that more than five species exist (Woolley *et al.* 2020). *M. habbema* is the smallest species and it has been found to nest underground (Woolley 1989, as *Antechinus habbema*). In the same study *M. naso* (as *A. naso*) was tracked to an underground nest, and nests of *M. longicaudata* were found in trees. Within *M. longicaudata* there is a very wide range in body size of the various forms recognised by Woolley *et al.* (2020). Biologically, members of *Murexia* are rainforest carnivore-insectivores with largely nocturnal (e.g. *M. habbema*, Woolley *et al.* 1991) activity patterns. *M. rothschildi* has been caught on two occasions in mist nets set for birds, suggesting that it may be crepuscular. Pelage colour of each of the five currently recognised species is distinctive (Fig. 2.14). The type specimen of *M. longicaudata*, one of the five currently recognised species, is perhaps an abnormal individual (Woolley 2020) that was collected on Kobroor, one of the Aru Islands, in 1865. No other specimens are known from this locality despite attempts to collect them.

Murexia Tate & Archbold 1937

Species	Authority	Size range (M/F)	Habitat and habits	Status
M. habbema Habbema dasyure	Tate & Archbold 1941	B: 95–131/94–117 mm T: 114–157/111–148 mm W: 28–45/23–31 g	Rainforest; terrestrial, scansorial	LC
M. longicaudata Short-furred dasyure	Schlegel 1866	B: 140–284/136–186 mm T: 147–283/150–215 mm W: 114–434/57–88 g	Rainforest; terrestrial, scansorial	LC
murex form		B: 150–165/124–150 mm T: 163–182/139–158 mm W: 88–113/57–60 g		
aspera form		B: 154–255/140–180 mm T: 166–240/150–180 mm W: – / – g		
maxima form		B: 221–284/184–230 mm T: 215–283/180–215 mm W: 255/ – g		
M. melanurus Black-tailed dasyure	Thomas 1899	B: 95–136/105–123 mm T: 116–161/128–155 mm W: 26–70/ND g	Rainforest; terrestrial, scansorial	LC
M. naso Long-nosed dasyure	Jentinck 1911	B: 114–145/106–134 mm T: 105–145/114–155 mm W: 44–62/38–48 g	Rainforest; terrestrial, scansorial	LC
naso form		B: 120–145/112–132 mm T: 145–190/126–165 mm W: 40–65/30–35 g		
tafa form		B: 110–155/100–134 mm T: 140–175/126–168 mm W: 31–76/29–49 g		
M. rothschildi Broad-striped dasyure	Tate 1938	B: 132–180/124–140 mm T: 163–195/145–170 mm W: 50/64 g	Rainforest; terrestrial, scansorial	NT

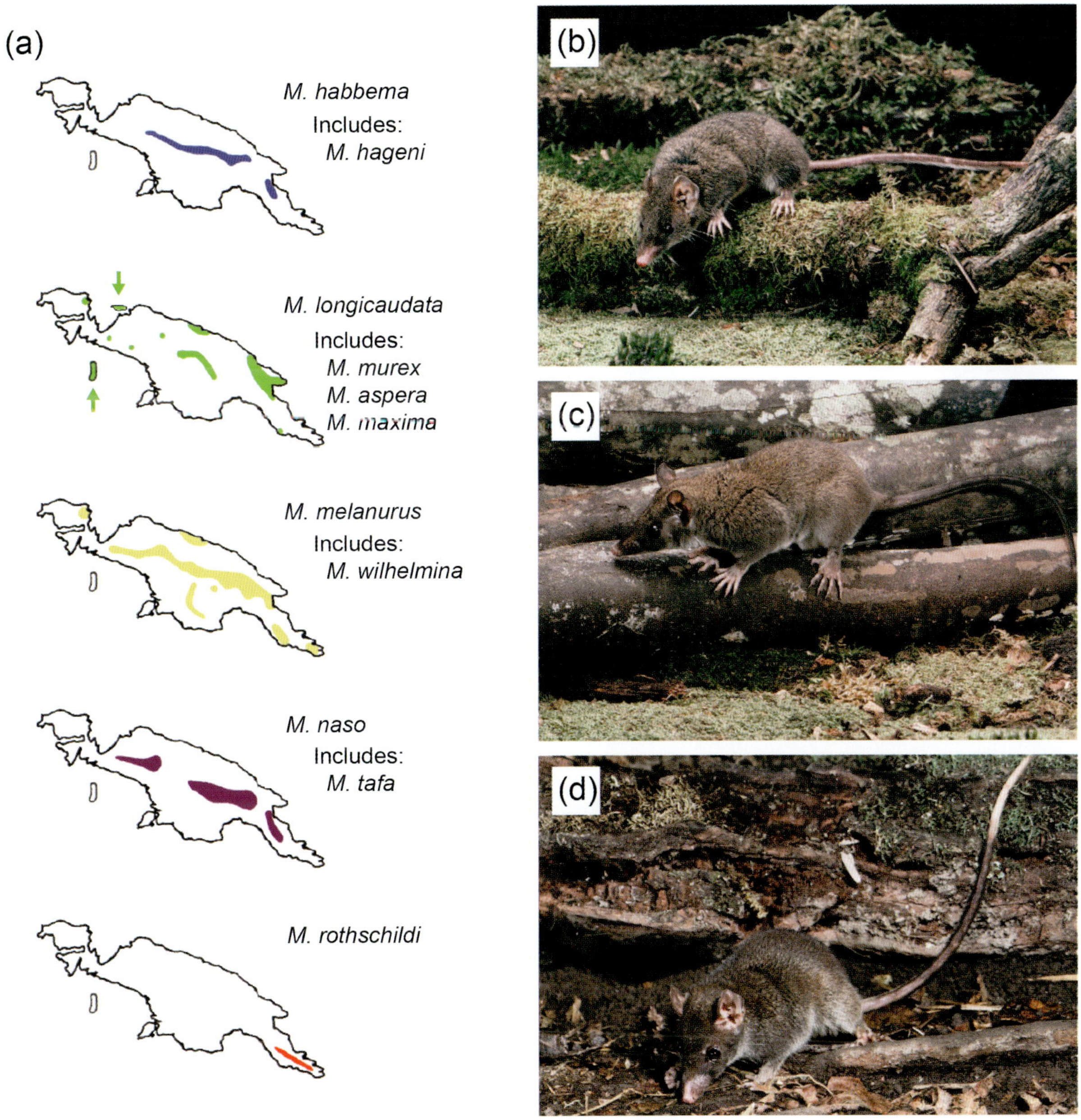

Fig. 2.14. *Murexia.* (a) Distribution of each of the five species of *Murexia* (including additional named forms) based on Woolley *et al.* (2020). (b) *M. habbema.* (c) *M. longicaudata* (*murex* form). (d) *M. naso* (*tafa* form). [Photo credits: (b, c, d) P.A. Woolley and D. Walsh.]

SMINTHOPSINAE

Sminthopsini

Sminthopsis (Dunnarts)

With 19 species, this is the most diverse genus of dasyurids. Dunnarts are small-bodied, ground-dwelling insectivores. Many species live primarily in xeric habitats in Australia but some, like *Sminthopsis archeri* and *S. virginiae* that have populations in New Guinea, are found in wetter areas. Some (such as *S. crassicaudata*) have broad distributions, while others (such as *S. douglasi*) are restricted. Some species store fat in their tails, and some have long thin tails (Fig. 2.15). Molecular data suggest that *Sminthopsis* is not monophyletic apart from *Antechinomys* and *Ningaui* (see Chapters 6 and 7).[1]

Sminthopsis Thomas 1887

Species	Authority	Size range (M/F)	Habitat and habits	Status
S. aitkeni Kangaroo Island dunnart	Kitchener *et al.* 1984	B: 84–93/72–82 mm T: 90–106/86–94 mm W: 20–25/10–15 g	Shrubland; terrestrial	ND[1]
S. archeri Chestnut dunnart	Van Dyck 1986	B: 98–107/83–85 mm T: 92–105/82–87 mm W: ND/12 g	Eucalypt forest, savanna; terrestrial	DD
S. bindi Kakadu dunnart	Van Dyck *et al.* 1994	B: 58–84/52–80 mm T: 62–105/61–97 mm W: 12–20 g	Eucalypt woodlands on stony hills; terrestrial	NT
S. butleri Butler's dunnart	Archer 1979	B: 88 mm[2] 99–116/94–107 mm[3] T: 90 mm[2] 74–94/65–85 mm[3] W: 23–33/20–30 g[3]	Shrubland, eucalypt forest; terrestrial	VU
S. crassicaudata Fat-tailed dunnart	Gould 1844	B: 60–90 mm T: 40–70 mm W: 10–20 g	Shrubland, grassland, savanna, desert; terrestrial	LC
S. dolichura Little long-tailed dunnart	Kitchener *et al.* 1984	B: 63–99/63–92 mm T: 88–109/84–97 mm W: 11–20/10–21 g	Shrubland, grassland, eucalypt forest, savanna; terrestrial	LC
S. douglasi Julia Creek dunnart	Archer 1979	B: 130–135/110–125 mm T: 120–130/110–120 mm W: 50–70/40–60 g	Savanna, shrubland, grassland; terrestrial	NT
S. gilberti Gilbert's dunnart	Kitchener *et al.* 1984	B: 81–92 mm T: 75–92 mm W: 14–25 g	Shrubland, eucalypt forest; terrestrial	LC
S. granulipes White-tailed dunnart	Troughton 1932	B: 69–88 mm T: 56–66 mm W: 18–37 g	Shrubland; terrestrial	LC
S. griseoventer Grey-bellied dunnart	Kitchener *et al.* 1984	B: 68–94/65–88 mm T: 66–98/65–95 mm W: 15–24/14–20 g	Shrubland, wetland, eucalypt forest; terrestrial	LC
S. hirtipes Hairy-footed dunnart	Thomas 1898	B: 72–83 mm T: 72–101 mm W: 13.0–19.5 g	Shrubland, eucalypt forest, savanna, grassland; terrestrial	LC
S. leucopus White-footed dunnart	Grey 1842	B: 82–112/67–117 mm T: 73–100/67–96 mm W: 16–32/16–25 g	Eucalypt forest, rainforest, savanna, shrubland; terrestrial	LC
S. longicaudata Long-tailed dunnart	Spencer 1909	B: 80–100/80–90 mm T: 200–220/180–200 mm W: 15–25 g	Grassland, shrubland; terrestrial	LC
S. macroura Stripe-faced dunnart	Gould 1845	B: 70–100 mm T: 80–110 mm W: 15–25 g	Shrubland, desert, savanna, grassland; terrestrial	LC
S. murina Common dunnart	Waterhouse 1838	B: 76–104/64–92 mm T: 70–99/68–92 mm W: 16–28/10–22 g	Eucalypt forest, savanna, shrubland; terrestrial	LC
S. ooldea Ooldea dunnart	Troughton 1965	B: 60–85/55–85 mm T: 65–95/60–90 mm W: 9–17/8–15 g	Desert, savanna, shrubland, grassland; terrestrial	LC

Species	Authority	Size range (M/F)	Habitat and habits	Status
S. psammophila Sandhill dunnart	Spencer 1895	B: 85–114 mm T: 106–118 mm W: 26–55/25–42 g	Shrubland, grassland; terrestrial	VU
S. virginiae Red-cheeked dunnart	Tarragon 1847	B: 96–135/90–133 mm T: 100–135/90–122 mm W: 31–58/18–34 g	Grassland, savanna, wetlands; terrestrial	LC
S. youngsoni Lesser hairy-footed dunnart	McKenzie & Archer 1982	B: 66–71 mm T: 62–68 mm W: 8.5–12.0 g	Shrubland, grassland, desert; terrestrial	LC

[1] *Sminthopsis aitkeni* was not assessed by the IUCN.
[2] Data for adult female holotype collected on mainland Australia.
[3] Data for animals collected on Bathurst Island (L. Potter, personal communication).

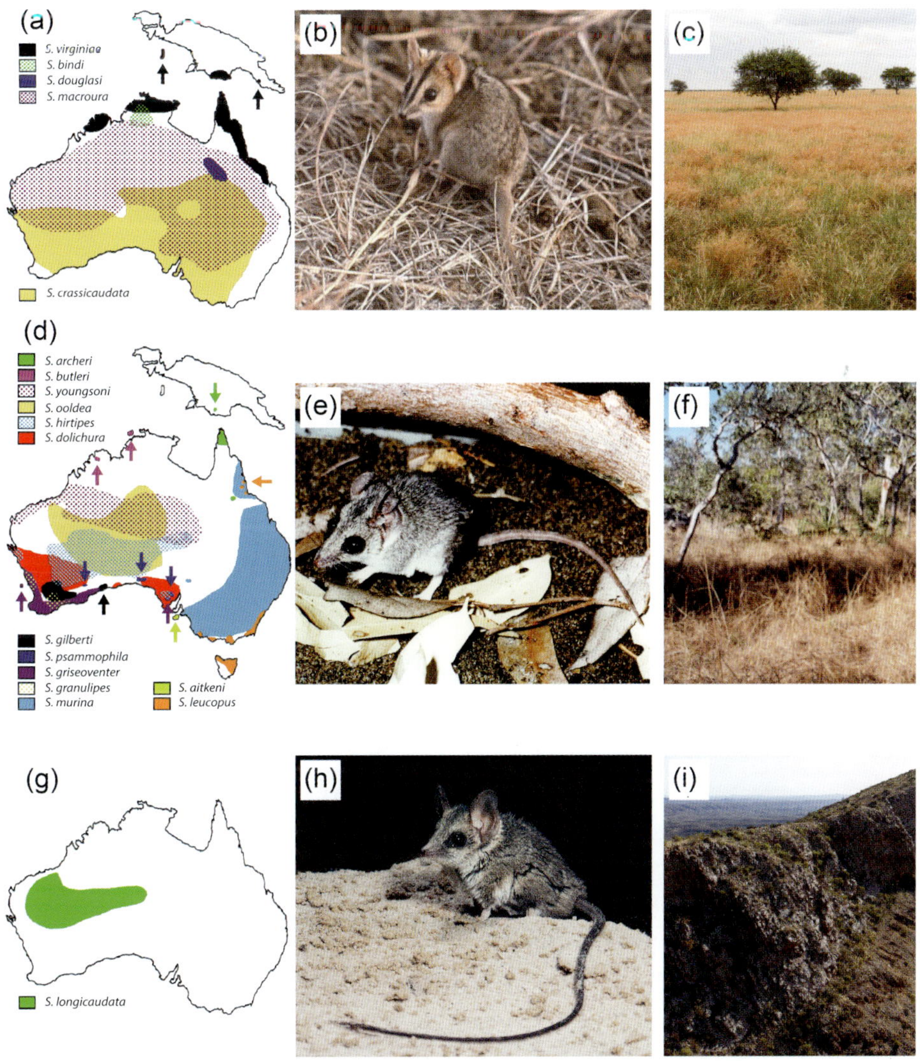

Fig. 2.15. *Sminthopsis*. (a) Distribution of each of the five species of the '*S. macroura* group'. (b) *S. douglasi*. (c) Habitat of *S. douglasi*, Mitchell grass downs with *Acacia nilotica* trees introduced for stock food. (d) Distribution of each of the 13 species of the '*S. murina* group'. (e) *S. butleri*, type specimen photographed live. (f) Habitat of *S. butleri* in the locality of collection of the type specimen at Kalumburu, WA. (g) Distribution of *S. longicaudata*. (h) *S. longicaudata*. (i) Habitat of *S. longicaudata* in the Macdonnell Ranges, central Australia. (See Chapter 6 for description of *Sminthopsis* species groups.) [Photo credits: (b, c, f) P.A. Woolley; (e) J.A.W. Kirsch; (h) P.A. Woolley and D. Walsh; (i) Rohan Clarke.]

***Antechinomys*[1] (Kultarr)**

Equipped with long hindlegs and tail, the kultarr is the only dasyurid known to employ quadrupedal hopping as its mode of locomotion (Ride 1965), and it is capable of rapid movement and abrupt changes in direction (Fig. 2.16). Kultarrs are primarily insectivorous and strictly terrestrial. Individuals have been found sheltering in spider burrows, under bushes and in soil cracks, but are not known to dig their own burrows. Kultarrs are found in a variety of habitats over their wide range, and it has long been suggested that there is more than one form (Lidicker and Marlow 1970).

Antechinomys Krefft 1867

Species	Authority	Size range (M/F)	Habitat and habits	Status
A. laniger Kultarr	Gould 1856	B: 80–100/70–95 mm T: 100–150/100–140 mm W: 21–34/13–21 g*	Savanna, shrubland, grassland, desert; terrestrial	LC

*Valente (1984).

Fig. 2.16. *Antechinomys.* (a) *A. laniger.* Pair found mating at 0200 hours beside vehicle track in central Australia. (b) Distribution. (c) Habitat, far west of range. [Photo credits: (a) Rohan Clarke; (c) P.A. Woolley.]

Ningaui (Ningauis)

Ningauis are tiny, nocturnal insectivores from arid–semi-arid parts of Australia. This genus was one of the last of the Australian marsupial genera to be erected, by Archer (1975b). The name 'ningaui' is derived from an Aboriginal story about tiny mythological beings that are hairy, have short feet and only come out at night to hunt for food, which is all eaten raw. *Ningaui timealeyi* is one of the smallest dasyurids, famous for its ability, if only in captivity, to take invertebrate prey larger than itself. In contrast, *N. ridei* is said to prefer small invertebrates less than 10 mm long. Ningauis have long thin tails (Fig. 2.17).

Ningaui Archer 1975

Species	Authority	Size range (M/F)	Habitat and habits	Status
N. ridei Wongai ningaui	Archer 1975	B: 57–75 mm T: 59–71 mm W: 6.5–10.5 g	Savanna, grassland, shrubland; terrestrial	LC
N. timealeyi Pilbara ningaui	Archer 1975	B: 56–76/56–65 mm T: 63–95/57–68 mm W: 3.6–9.5/3.5–7.5 g	Grassland, desert; terrestrial	LC
N. yvonneae Southern ningaui	Kitchener *et al.* 1983	B: 48–81 mm T: 53–71 mm W: 6–14 g	Shrubland, grassland; terrestrial	LC

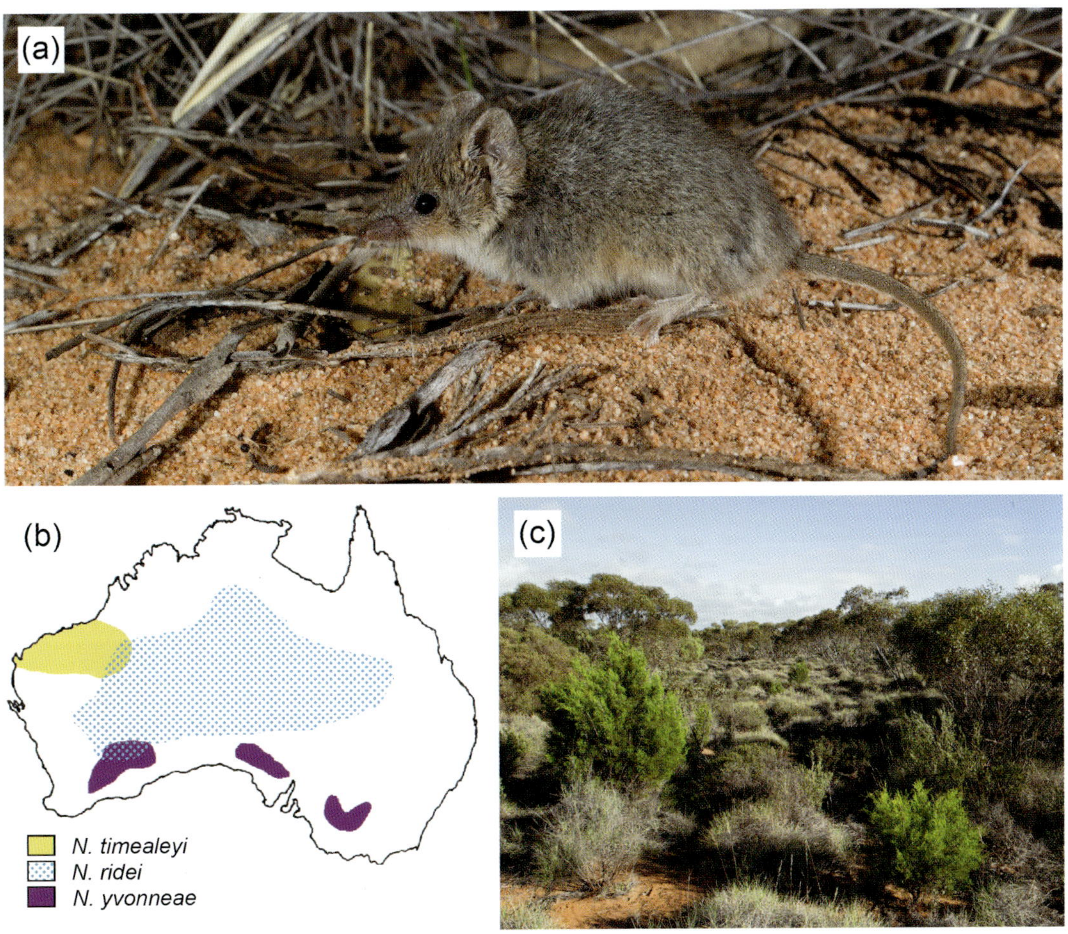

Fig. 2.17. *Ningaui.* (a) *Ningaui yvonneae.* (b) Distribution of each of the three species. (c) Mallee habitat of *N. yvonneae* on the Eyre Peninsula, SA. [Photo credits: (a, c) Rohan Clarke.]

Planigalini

Planigale (Planigales)

Known for their flat heads (found in some, but not all, species), planigales are tiny insectivores (Fig. 2.18). Indeed, at less than 7 g, *P. ingrami* is the smallest marsupial. Flat-headed forms (*Planigale ingrami, P. gilesi, P. tenuirostris*) live in areas with cracking clay soils, where the animals take shelter from external temperature extremes. An

Planigale Troughton 1928

Species	Authority	Size range (M/F)	Habitat and habits	Status
P. gilesi Giles' planigale	Aitken 1972	B: 60–80/60–70 mm T: 55–70/55–65 mm W: 9.5–16.0/5.0–9.0 g	Grassland, shrubland; terrestrial	LC
P. ingrami Long-tailed planigale	Thomas 1906	B: 50–71/49–64 mm T: 53–68/51–65 mm W: 2.8–6.6/2.6–5.8 g	Grassland, savanna; terrestrial	LC
P. maculata Common planigale	Gould 1851	B: 51.7–84/59.9–83 mm T: 48–64/46–65 mm W: 4–16.3/3.5–17.2 g	Rainforest, eucalypt forest, grassland, wetland; terrestrial	LC
P. novaeguineae Papuan planigale	Archbold 1941	B: 79/ND mm T: 80/ND mm W: 14/ND g	Grassland, savanna; terrestrial	LC
P. tenuirostris Narrow-nosed planigale	Troughton 1928	B: 50–75/50–70 mm T: 52–65/50–60 mm W: 4.5–9.0/4.0–7.0 g	Shrubland, grassland; terrestrial	LC

Fig. 2.18. *Planigale*. (a) Distribution of each of the five species. (b) Scree slope where *P. ingrami*, the long-tailed planigale shown in (c), was collected in Arnhem Land, NT, and (d) *P. ingrami* from the cracking clay soils of the Mitchell Grass Downs in Queensland (see Fig. 2.15c). [Photo credits: (b, d) P.A. Woolley; (c) P.A. Woolley and D. Walsh.]

exception to this is one specimen, identified as *P. ingrami* by the late J.H. Calaby and now held in the Australian National Wildlife Collection, trapped by P.A. Woolley on a scree slope in Arnhem Land in the Northern Territory. *P. maculata*, in contrast, occupies a wider range of mesic habitats. *P. novaeguineae* is endemic to grasslands and savanna of southern New Guinea. Although only four Australian species are currently recognised, there is strong molecular evidence for additional ones, mostly within what are now *P. maculata* and *P. ingrami* (Westerman, Blacket *et al.* 2016). The pouch of *P. 'subtilissima'* (now considered *P. ingrami* from the Kimberley Plateau) differs from that of all other dasyurids, including *P. ingrami* from other parts of its range (Woolley 1974).

ENDNOTE

1 The long-tailed dunnart was reassigned from *Sminthopsis* to *Antechinomys* by Westerman *et al.* (2023) *Australian Journal of Zoology* **70**, 95–103. doi:10.107/ZO22041.

3 Dasyurid phylogeny: morphological taxonomy from Linnaeus to Archer

PRE-DARWINIAN CLASSIFICATION AND THE ORDER OF NATURE

Discovery and early classification of marsupials

Vincente Yáñez Pinzón, explorer and captain of the *Niña* on Columbus' initial voyage to the New World, 'discovered' the first marsupial (an opossum) known to European science on an expedition to the Brazilian coast in 1500. In *De Orbe Novo* (1530), historian Peter Martyr d'Anghiera described the Spaniards' encounter:

> *Amonge these trees is fownde that monstrous beaste with a snowte lyke a foxe, a tayle lyke a marmasette, eares lyke a batte, hands lyke a man, and feete lyke an ape, bearying her whelps about with her in an outward bellye much lyke unto a greate bagge or purse. [Parrish 1997]*[1]

Thus began centuries of puzzlement about how marsupials fit into a mammalian *Weltanschauung* defined by Old World placentals. They were variously considered monstrous (Van Duzer 2013), chimeric (Eastman 1915), abnormal, incomplete, or 'second class' mammals – a legacy that lingers in popular culture to this day (Storr 1958).

World exploration during the 17th and 18th centuries challenged the capacity of taxonomists, both conceptually and practically. Knowledge of biodiversity had progressed but modestly in the two millennia since Aristotle, and most classifications were little more than idiosyncratic grouping schemes of familiar plants and animals. The notion of relationship among different kinds of organisms generally referred to their positions in a metaphysical *scala naturae* or Great Chain of Being. Existing taxonomies strained to accommodate the plethora of discoveries from newly explored continents. How to classify this 'monstrous beaste' and others like it?

The opossum's 'greate bagge' for carrying young – the pouch or marsupium – was apparent in the first specimen collected (even to Spain's Queen Isabella; Archer 1982b), but no reliable anatomical description was published until Tyson (1698)[2] dissected a female specimen (Fig. 3.1) from Virginia (Barbour 1977; Parrish 1997). Although John Smith had adapted the Algonquian word *apasum* ('white animal') as a common name (Krause and Krause 2006), Tyson (p. 107) concluded, 'I think, a *Denomination*

Fig. 3.1. The first anatomical illustration of a marsupial, the Virginia opossum (*Didelphis virginiana*) from Tyson (1698). At left is a ventral view of the sacrum and pelvic bones, including the paired epipubics ('marsupial bones') labelled 'c'.

might be best given to it, from that Particular, wherein 'tis most distinguished from all others; which is that remarkable *Pouch* or *Marsupium* it has in the *Belly* … (it being so distinguishing a Character of this Animal from all others, that as yet we know of) … nor can I think of, at present, a better, than to call it, *Marsupiale Americanum*'. Progressing through a detailed summary of dissection observations, Tyson gave the first definitive description of the marsupial duplex uterus, noting, 'I have not met with a *Parallel*, or the like in any *Animal* besides; at least of the *Quadrupede* [mammal] Kind' (p. 142). Two centuries later, it was still uncertain how (or even if) neonates moved from the birth canal to the pouch, but clearly they arrived there at a very early (altricial) stage of development. In published correspondence of 1806 and 1813, American naturalist Benjamin Smith Barton recorded the birth of an opossum litter through the urogenital sinus (Troughton 1947; Parrish 1997) and Owen (1834) published the first description of a marsupial (kangaroo) embryo *in utero*. Although several informal descriptions of marsupial neonates crawling to the pouch were recorded during the 19th century, Hartman (1920) was apparently the first eyewitness, peer-reviewed documentation of the phenomenon (observed in a Virginia opossum; see also Hartman 1952).

In the 160 years between Pinzón's discovery of 'that monstrous beaste' and the benchmark 10th edition of Linnaeus's (1758) *Systema Naturae*, several species of marsupials were proposed, most of them New World opossums. Linnaeus recognised four species, all members of *Didelphis,* a name derived from the Greek *di* (two) + *delphús* (womb) emphasising the duplex uterus documented by Tyson. These species (*D. marsupialis*, *D. philander* = *Caluromys philander*, *D. opossum* = *Philander opossum* and *D. murina* = *Marmosa murina*) were included in Linnaeus's mammalian order Bestiae along with pigs, armadillos, moles and shrews. Bestiae was defined as follows: 'front teeth of varying number in upper and lower jaws; laniariform canine teeth always more than one pair [in reference to the piercing character of the most anterior cheek teeth as well as of the canine]; snout elongate, rooting; sustenance upon succulent roots or worms' (Gregory 1910: 32). By nesting marsupials within Bestiae, Linnaeus subordinated reproductive anatomy to dental and other feeding characteristics. Indeed, Gregory (1910) considered it a typical pre-Darwinian tradition to base ordinal groups on 'adaptive' traits – i.e. those that are evolutionarily labile, variable at lower taxonomic levels and thus more prone to convergence. While that is a reasonable assessment of why Linnaean and other early classifications of mammals are not phylogenetic, it misses a deeper methodological point.

Linnaeus was an essentialist, a philosophical tradition in which the units of biodiversity have *defining* characteristics established by the Creator. The job of taxonomists was to identify these traits amid the abundant and often confusing variation of real organisms. Importantly, Linnaeus focused his essentialism on *genera*, which he thought were separated by real discontinuities in the plan of creation. Mayr (1982b) argued that Linnaeus was more nominalistic regarding higher categories, allowing (if only implicitly) that these groups may be artificial and not separated by gaps. Moreover, Linnaeus famously emphasised reproductive structures as defining characteristics in plants. Given that the known marsupials uniquely possessed a pouch and duplex uterus, it is not surprising that Linnaeus deployed this evidence at the level where it mattered most to him – the definition of *Didelphis*.

Linnaeus's contemporary, French naturalist Georges-Louis Leclerc, Comte de Buffon, rejected the essentialist view that taxa at any level are real, suggesting instead that they represent arbitrary groupings of individual organisms imposed by taxonomists – especially Linnaeus, of whom Buffon was a persistent critic (Mayr 1982b; Gregory 2008). Buffon embraced the Platonic principle of plentitude, according to which all possible forms of organisms exist with no meaningful gaps between them. His magisterial, 44-volume *Histoire Naturelle* (1749–1804), summarised the state of knowledge on all known mammals, birds, minerals, reptiles and fish – with no other classificatory structure than these broad headings and common names of species. The mammal accounts were strongly influenced by Buffon's Jardin du Roi colleague, Louis-Jean-Marie Daubenton. Their section 'The Opossum' provides a detailed argument that the specimens to which Linnaeus assigned the epithets *marsupialis* and *philander* are the same species. In this they were mistaken, but their rejection of '*philander orientalis*' (based on a specimen from Ambon Island west of New Guinea) as an opossum was correct; in fact, it was a cuscus (*Phalanger* sp.). The rest of the section, and subsequent ones on *The Marmose* (Linnaeus' '*D. murina*') and *The Cayopolin* ('*D. opossum*'), provide information on anatomy (mostly summarising Tyson), behaviour and life history.

After 1758, some taxonomists followed Linnaeus in placing the known marsupials in one or more orders with placentals (e.g. Blumenbach 1791; Vicq d'Azyr 1792), whereas others isolated them in exclusive suprageneric groups (e.g. Brisson 1762; Storr 1780; Pennant 1781; Lamarck 1809). Among the former was J.K.W. Illiger, first director of the Berlin zoological museum in 1810–1813. Illiger's (1811) *Prodromus systematis mammalium et avium* is the authoritative source for the taxon name Marsupialia, members of which he included along with primates in his Order Pollicata, though he placed kangaroos in a separate order, Salientia.

While early 19th century *nomenclateurs* (Buffon's term of derision) tinkered with the Linnaean system and Buffon himself celebrated a plentitude of organisms, other natural philosophers conceived of a metaphysical unity in the anatomical structures of living things. These included *naturphilosophen* such as Lorenz Oken, W.S. Macleay and others who struggled to articulate a distinction between anatomical resemblance due to 'affinity' (which revealed true relationship) and that due to 'analogy' (which obscured it). This distinction was the forerunner to modern concepts of homology and analogy. Among those endorsing organisational unity was French naturalist H.M.D de Blaineville, who saw biodiversity structured around the distinct body plans of major groups, upon which was superimposed variation due to adaptation. De Blaineville emphasised reproductive, brain and skull traits as revealing the deepest affinities and considered feeding and locomotor traits lower-level adaptations. In his *Prodrome d'une nouvelle distribution systematique du regne animal* (De Blaineville 1816), he boldly assigned marsupials and monotremes to a subclass Didelphes equal in rank to all placentals (Monodelphes), marsupials comprising 'normal' and monotremes 'abnormal' types. Marsupials were subdivided among *carnassiers* (carnivores) and *rongeurs* (literally 'rodents' but corresponding to known diprotodontians). Bonaparte (1837) adopted a similar scheme. In 1834, de Blaineville recognised marsupials and monotremes as distinct subclasses, differentiating marsupials into didactyl and syndactyl groups (Gregory 1910). Syndactyl species, like bandicoots and diprotodontians, have the second and third digits of their hind feet enclosed in a single sheath of skin. Many later authors accepted the didactyl/syndactyl distinction as a grouping criterion (homology) for marsupial orders – mistakenly, as it turned out, syndactyly having evolved convergently in bandicoots and diprotodontians.

Baron Cuvier's (1817) *Le Règne Animal* banished much of the classificatory shuffling of preceding decades into obscurity for non-taxonomists. The tremendous anatomical detail and comprehensive scope of Cuvier's work, his 'correlation of parts' and 'subordination of characters' principles, and his social-professional stature established *Le Règne's* authority and influence for decades to come. Cuvier rejected deep affinities; he classified animals into four unrelated *embranchements* based on the structure of their central nervous systems: vertebrates, molluscs, arthropods + annelids and 'radiata' (e.g. cnidarians and echinoderms). The nine genera of then-known marsupials constituted Marsupiata, one of several families in Order Carnassiers, distinguished by reproductive and digital characters. However, Cuvier acknowledged that marsupials might well be considered an ordinal-level group unto themselves, as indeed they were in the last (1829) edition of *Le Règne*.[3]

Despite the popularity of Cuvier's classification, many subsequent mammal taxonomists (e.g. Owen 1868; Gill 1872; Huxley 1872) followed de Blaineville in recognising marsupials as a supraordinal taxon equivalent to some or all placentals. After Huxley (1880) introduced the concept of Metatheria, treatment of marsupials as coordinate with placentals settled into stability (e.g. Simpson 1945; Aplin and Archer 1987). Unfortunately, Huxley's name Metatheria – a 'change' or 'turning point' in mammalian evolution – codified the progressivist misconception that marsupials 'constitute an intermediate type between that of the Prototheria [monotremes] and that of the higher mammals [placentals]' (Huxley 1880: 464). Monstrous beasts they remain, in popular imagination, compared to their more perfect and pouchless sisters.

Early classification of dasyurids

Captain Cook's *HMS Endeavour* expedition arrived on the east coast of Australia in 1770. Members of the party captured what was probably a northern quoll (*Dasyurus hallucatus*) in Queensland and recorded informal descriptions of the animal in their accounts of the voyage (Mahoney and Ride 1984). Presumably from this specimen (now lost), the German geographer and zoologist E.A.W. von Zimmerman (1783) constructed a formal description of '*Mustela quoll*' suggesting an affinity with placental carnivorans in Linnaeus's (1758) Order Ferae. Similarly, post-settlement collections in south-eastern Australia led to descriptions of tiger and eastern quolls as '*Viverra maculata* Kerr 1792' (later '*Mustela novaehollandiae* Meyer 1793') and '*Didelphis maculata* Anonymous 1791' (later '*Didelphis viverrina* Shaw 1800'), respectively. *Viverra*, like *Mustela*, is a Linnean genus within Ferae, and *Didelphis* included all previously described marsupials. These initial decisions to place new species from Australia into existing Linnaean genera testifies to the influence of the 'Father of Taxonomy' on practical classification well into the 19th century.

Dasyurids acquired their identity as marsupials distinct from didelphids in Étienne Geoffroy Saint-Hilaire's *Dissertation sur les animaux à bourse* (Geoffroy 1796). Geoffroy, a naturalist at the Paris Museum of Natural History, was an early collaborator and later antagonist of Cuvier. Like de Blainville, he espoused a 'unity of composition' among animals that informed his classifications and interpretations of anatomy, drawing opposition from Cuvier. Geoffroy (1796) erected the genus name *Dasyurus* ('hairy tail' in contrast to the naked tail of opossums) for specimens that had been referred to as '*Didelphis maculata*' (the Eastern Quoll, now *Dasyurus viverrinus*; Mahoney and Ride 1988). Geoffroy (1803) included '*Viverra maculata*' as *D.* '*macrourus*' (= *D. maculatus*), *D.* '*maugei*' (now a synonym of *D. viverrinus*), '*Dasyurus minimus*' (= *Antechinus minimus*) and '*Dasyurus tafa*' (= *Phascogale tapoatafa*) in the new genus (Table 3.1). In other publications prior to 1810, Geoffroy proposed generic names for bandicoots (*Perameles*), wombats (*Vombatus*), kangaroos ('*Kangurus*') and cuscuses ('*Phalangista*').

Geoffroy's marsupials, including *Dasyurus*, appear in Lamarck's (1809) *Zoological Philosophy* as members of 'Pedimana' in 'Order 4 Unguiculate Mammals' (those with claws or nails) along with edentates, rodents, insectivores, carnivorans, bats and primates (excluding humans). Lamarck's (1809) mechanism of evolution is well known as an incorrect predecessor to Darwin's, but his ranking of animals along a scale of increasing perfection from worms to humans has this to recommend it to marsupial apologists: unguiculates are second only to Bimana (man) in the ranking of orders! Marsupials were spared their usual indignity by virtue of their terrestriality, Lamarck hypothesising that mammals arose from an aquatic broodstock represented by the 'amphibian' pinnipeds and sirenians.

As noted above, Illiger (1811) included *Dasyurus* in his Marsupialia along with opossums, bandicoots, wombats and cuscuses (but not kangaroos). In *Le Règne Animal*, Cuvier (1817) listed eight members of *Dasyurus* with the epithets *cynocephalus* (= *Thylacinus*), *ursinus* (= *Sarcophilus*), *macrourus* (= *maculatus*), *maugei*, *viverrinus*, *tafa*, *penicillata* (= *Phascogale tapoatafa*) and *nain* (Table 3.1). By this time, 'Marsupiata' also included opossums (*Didelphis*), bandicoots (*Perameles*), phalangerids ('*Phalangista*'), gliders ('*Petaurista*'), potoroos (*Potorous*), kangaroos ('*Kangurus*'), koalas (*Phascolarctos*) and wombats ('*Phascolomys*').

German naturalist Georg August Goldfuss is usually credited with authority for the name Dasyuridae. In fact, Goldfuss (1820) recognised 'Familie Dasyurini' within his Order Marsupialia, but that family included bandicoots (*Perameles* or '*Thylacis*') along with *Dasyurus*. C.J. Temminck (1778–1858), zoologist and first director of the Leyden natural history museum, created *Phascogale* for what had been '*Dasyurus penicillata*' (or '*tafa*') and '*D. minimus*', based on dental differences (e.g. *Phascogale* has three premolars and unequal incisor lengths, *Dasyurus* has two premolars and equal incisor lengths; Temminck 1824; Table 3.1). Waterhouse (1838) described two new species of *Phascogale*, now known as the yellow-footed

Table 3.1. Selected classification schemes of described extant dasyurid species between 1796 and 1824.
Modern genus or species equivalents are shown in brackets.

Geoffroy (1796, 1803)	Cuvier (1817): Marsupiaux	Temminck (1824): Dasyuridae
Dasyurus	*Dasyurus*	*Dasyurus*
macrourus [*maculatus*]	*macrourus* [*maculatus*]	*macrourus* [*maculatus*]
maugei [*viverrinus*]	*maugei* [*viverrinus*]	*maugei* [*viverrinus*]
minimus [cf. *Antechinus*]	*nain* [cf. *Antechinus minimus*]	*ursinus* [cf. *Sarcophilus*]
tafa [cf. *Phascogale*]	*penicillata* [cf. *Phascogale*]	*viverrinus*
	viverrina [*viverrinus*]	*Phascogale*
	ursinus [cf. *Sarcophilus*]	*minima* [cf. *Antechinus*]
	cynocephala [cf. *Thylacinus*]	*penicillata* [*tapotafa*]
		Thylacinus
		harrisii [*cynocephalus*]

antechinus ('*P.*' *flavipes*) and the common dunnart ('*P.*' *murina*). Temminck (1824) also erected *Thylacinus* for '*Didelphis cynocephala*' (the thylacine, described in 1808), which Bonaparte (1838) placed in a separate family (Thylacinidae). Owen (1839a, b) and Waterhouse (1841) ignored Bonaparte by recognising Family Dasyuridae to include *Thylacinus*, *Dasyurus* and *Phascogale*. Owen defined marsupial tribes by diet-associated traits, with dasyurids as the only family in 'Tribe I Sarcophaga' (carnivores without an intestinal caecum); indeed, they remain the archetypal 'carnivorous marsupials'.

British naturalist George Robert Waterhouse (1810–88) provided an especially clear rationale for character evaluation in constructing his classification. His entry, *Marsupialia, Or Pouched Animals*, in Jardine's Naturalist's Library series included the following passage:

> *In analyzing the Marsupiata, with a view to determine with which group to commence the order, I was anxious to select a genus, the species of which would exhibit the most perfect dentition, and at the same time present other characters which are common in the species of that order. These characters I found combined, in the most perfect manner, in the true Opossums (Didelphys) – they possess the greatest number of teeth, (if we except the extraordinary abnormal dentition of Myrmecobius), a well developed thumb to the posterior extremities, their limbs are fitted either for climbing or walking, and their tail is muscular and capable of being used as an organ of prehension ... I may, moreover, observe that, as a general rule, animals of omnivorous or frugivorous diet are higher, in their own order, in the grade of organization, than either the carnivorous, insectivorous, or herbivorous species. Having determined to place the genus Didelphys at the head of my order, my next object was to arrange the remaining groups near to, or remote from this typical genus ... In the Dasyuri, I found the dentition and the form of the skull almost identical with that of the Opossums ...' [Waterhouse 1841: 61–62]*

This extraordinary statement lays bare a *scala* principle sufficiently important to Waterhouse that he calls attention to what might otherwise appear a minor point: 'I have mentioned these, my principal reasons, for placing the genus *Didelphys* at the head of the *Marsupiata*, because I regard the situation of this *South American* group, as one of the most important disagreements between my classification and that of Professor Owen'. The two classifications recognise essentially the same groups but differ in how they are ordered from first to last. Waterhouse rejected applying Macleay's quinarian principles to mammals and, whatever his views about 'underlying unity' might have been, his interpretive framework was strongly resonant with a Linnaean worldview less than 20 years before *The Origin*.

In 1841, John Gould described the western quoll (*Dasyurus geoffroii*; Fig. 3.2), the only new species included in Waterhouse (1841). However, when the first volume of Waterhouse's magnum opus – *A Natural History of the Mammalia* – appeared in 1846, *Phascogale* had come to include some 13 species and *Dasyurus* six. Moreover, when Macleay (1841) described the brown antechinus, he called it *Antechinus stuartii*. The new genus was based on an error in counting teeth, Macleay claiming (from Stuart's field notes, the specimen having been lost) that its incisor count was 6/6; in fact, all dasyurids have eight upper incisors. Macleay corrected the error a year later (Wakefield and Warneke 1967), but Waterhouse (1846: 404) was dismissive of *Antechinus*, noting '... some trivial differences observed among these [new species] have given rise to the establishment of new genera; as sections of minor value only can we adopt them.' His classification (Table 3.2) separated '*Phascogale* proper' (*P. penicillata*, *P. calura*) from '*Phascogale* (*Antechinus*)', again emphasising the larger size of I^1 in the former relative to the equal-sized incisors of other species. Interestingly, Waterhouse's *Phascogale* (*Antechinus*)

Fig. 3.2. A painting of western quolls (*Dasyurus geoffroii*) from Gould (1863), a species he described in 1841.

Table 3.2. Selected classification schemes of described extant dasyurid species between 1846 and 1867.
Modern genus or species equivalents are shown in brackets.

Waterhouse (1846): Dasyuridae	Gould (1863)	Krefft (1867): small Dasyuridae
Dasyurus	*Dasyurus*	*Chaetocercus* [cf. *Dasycercus*]
geoffroyi [*geoffroii*]	*Geoffroyi* [*geoffroii*]	*cristicauda*
hallucatus	*hallucatus*	*Phascogale*
maculatus	*maculatus*	*calura*
maugei [*viverrinus*]	*viverrinus*	*penicillata* [*tapoatafa*]
viverrinus	*Sarcophilus*	*Antechinus*
Dasyurus (*Sarcophilus*)	*ursinus* [*harrisii*]	*affinis* [*minimus*]
ursinus [*harrisii*]	*Phascogale*	*apicalis* [cf. *Parantechinus*]
Phascogale	*calura*	*ferruginifrons* [cf. *Sminthopsis*]
calura	*penicillata* [*tapoatafa*]	*flavipes*
penicillata [*tapoatafa*]	*lanigera* [cf. *Antechinomys*]	*leucogaster* [*flavipes*]
Phascogale (*Antechinus*)	*Antechinus*	*maculatus* [cf. *Planigale*]
affinis [*minimus*]	*albipes* [cf. *Sminthopsis*]	*minutissimus* [cf. *Planigale*]
albipes [cf. *Sminthopsis*]	*apicalis* [cf. *Parantechinus*]	*swainsonii*
apicalis [cf. *Parantechinus*]	*ferruginifrons* [cf. *Sminthopsis*]	*unicolor* [*agilis*]
crassicaudata [cf. *Sminthopsis*]	*flavipes* [incl. *stuartii, minimus*]	*Podabrus* [cf. *Sminthopsis*]
flavipes [incl. *stuartii*]	*fuliginosus* [cf. *Sminthopsis*]	*albipes* [*murina*]
leucogaster [*flavipes*]	*leucogaster* [*flavipes*]	*crassicaudatus* [*crassicaudata*]
leucopus [cf. *Sminthopsis*]	*leucopus* [cf. *Sminthopsis*]	*fuliginosus*
macroura [cf. *Sminthopsis*]	*maculatus* [cf. *Planigale*]	*leucopus*
melas [cf. *Myoictis*]	*minutissimus* [cf. *Planigale*]	*macrourus* [*macroura*]
minima [*minimus*]	*murinus* [cf. *Sminthopsis*]	*mitchelli* [*leucopus*]
murina [cf. *Sminthopsis*]	*swainsonii*	*murinus* [*murina*]
swainsonii	*unicolor* [*agilis*]	*Antechinomys*
Myrmecobius	*Podabrus* [cf. *Sminthopsis*]	*lanigera* [*laniger*]
fasciatus	*crassicaudata*	
Thylacinus	*macrourus* [*macroura*]	
cynocephalus		

section included species of what are now *Parantechinus*, *Sminthopsis* and *Myoictis* in addition to antechinuses. He similarly created a separate section of *Dasyurus* for the Tasmanian devil, '*D. (Sarcophilus) ursinus*'.

Waterhouse was aware of the distinction between affinity and analogy as sources of anatomical resemblance. He rejected the latter as a reason, among others, for recognising marsupials as more than just another mammalian order – a point to which Darwin (1847), reviewing the first volume of *A Natural History of the Mammalia*, politely objected. However, like most pre-Hennigian systematists, Waterhouse could not coherently articulate a concept of affinity as a guide to relationships. In the early pages of *A Natural History*, Waterhouse (1846: 11–12) provided a diagram illustrating 'mutual relations of the different genera of marsupials'. It is a two-dimensional array of genus names separated by vertical and horizontal lines (Fig. 3.3). As such, it goes beyond a strictly linear *scala*, being more like a subjectively weighted ordination of anatomical similarities that graphically endorses continuity among taxa. Thus, *Thylacinus*, *Dasyurus*, *Phascogale* and *Myrmecobius* are quite similar – in the characters that (to Waterhouse) best demonstrated affinity – and show close ties to *Didelphis* and monotremes. 'Of the *affinities* of the various minor groups of which the order Marsupialia is composed, with the exception of one or two cases, there can be very little difference of opinion; the modifications of the extremities, combined with the characters furnished by the digestive system, very clearly indicate the natural

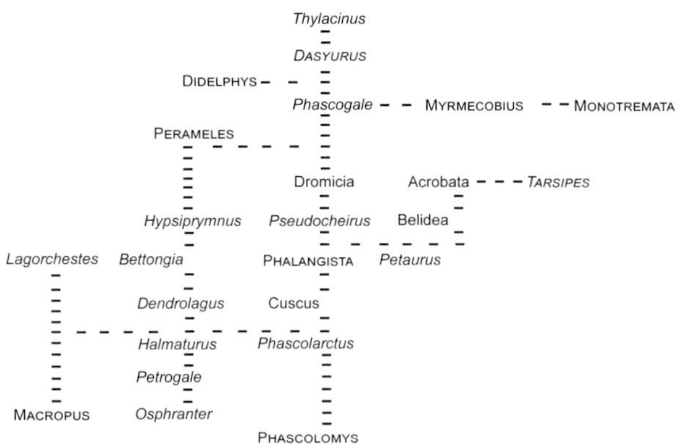

Fig. 3.3. Waterhouse's (1846) depiction of the 'mutual relations of the different genera of marsupials' (p. 11) based on 'modifications of the extremities, combined with the characters furnished by the digestive system' (p. 9). The significance of taxon names in small caps is not clear.

divisions, *and their relationship to each other*' (Waterhouse 1846: 9, italics added). Pentadactyl digits, dental formulas, well-developed canines, molar cusp structure and lack of a caecum distinguished dasyurids. Moreover, *Dasyurus* and *Phascogale* differed in body size (larger versus smaller), upper incisor structure (unequal size in oblique lateral arrays versus equal size in a semicircular array), premolar number (two versus three) and diet (carnivorous versus insectivorous). Thus, Waterhouse followed the tradition of selecting a few key characters to identify gaps among genera while at the same time using patterns of similarity to estimate generic affinities.

British naturalist John Gould, Darwin's collaborator in describing the Galápagos finches, visited Australia in 1838. From this trip would come his classic illustrated monographs on Australian birds (1840–1848) and kangaroos (1841–1842), as well as *The Mammals of Australia* (1849–1861). The latter, republished as a three-volume set (Gould 1863), had a classification of marsupials that included several new species and a genus that had been described between 1845 and 1856. Waterhouse (1846) had included the fat-tailed and stripe-faced dunnarts (now *Sminthopsis crassicaudata* and *S. macroura*, respectively) in the *Antechinus* section of his *Phascogale*, but Gould isolated them in a distinct genus, *Podabrus* – which Thomas (1886b) would rename *Sminthopsis* – based on their thick tails (Table 3.2). Moreover, he separated Waterhouse's sections into distinct genera, *Phascogale* and *Antechinus*. The former included '*Phascogale lanigera*' (the kultarr, now *Antechinomys laniger*, which Gould himself had described in 1856) along with *P.* '*penicillata*' and *P. calura*. *Antechinus* had expanded to 12 species, including representatives of the current *Antechinus* (*swainsonii*, *agilis*, *flavipes*, *minimus*), *Parantechinus* (*apicalis*), *Sminthopsis* (*leucopus*, *murina*, *fuliginosus*) and *Planigale* (*maculata*). Gould likewise separated Waterhouse's sections of *Dasyurus* into quolls (*Dasyurus*) and the devil (*Sarcophilus*), both authors commenting on the distinct pelage and body proportions of these groups. In the Introduction to *Mammals of Australia*, Gould is cautiously enigmatic on the question of affinities among taxa. After introducing *Phascogale*, he remarks, 'It might be thought that *Phascogalæ* would naturally lead to the *Antechini*, but there is no real affinity between the two groups. I find it most difficult to arrange the Australian mammals in anything like a serial order …' (Gould 1863: xviii). The *scala* framework is acknowledged, but only to say that it gives a poor fit to the observed diversity of species. This is four years after *On the Origin of Species*.

Thus, at the dawn of the Darwinian revolution, dasyurid taxonomists – like all taxonomists – struggled within the confines of a crumbling Aristotelian worldview. Are natural relationships among taxa defined by essences ordered in a progressive series (Linnaeus), by positions within genealogical streams striving for perfection (Lamarck), by functionally correlated characters (Cuvier), by similarity of body plans (Macleay and Owen), or what? Mayr (1982b: 203) felt that '[i]n the fifty years before the *Origin* most taxonomists shunned theorising altogether and … were satisfied with the simple pragmatic activity of grouping seemingly similar species and genera together'. In hindsight, this approach was not completely misguided. Careful anatomical comparisons led to the recognition of *dasyures* as a distinct group within Marsupialia, consisting of six genera (*Phascogale*, *Antechinus*, *Podabrus* = *Sminthopsis*, *Dasyurus*, *Sarcophilus*, *Thylacinus*) and 23 species. Each genus has some phylogenetic coherence by modern standards and the latter three correspond to modern groupings of species known in 1863. The conceptual chaos between Linnaeus and Darwin had brought forth an unrefined, but rapidly clarifying, shape for the dasyurid branch on the tree of life.

STRIVING TO ACHIEVE AN EVOLUTIONARY CLASSIFICATION

Darwin on marsupials

In *On the Origin of Species*, Darwin (1859) supplied the long-sought meaning of 'relationship' as it applies to species. His descent-with-modification thesis made this quite clear: 'I believe that something more is included than mere resemblance; and that propinquity of descent – the only known cause of the similarity of organic beings – is the bond, hidden as it is by various degrees of modification, which is partially revealed to us in our classifications' (Darwin 1859: 413). In modern parlance, the degree of evolutionary relationship is given by recency of common ancestry: species A and B are more closely related to each other than either is to C if, and only if, they share a more recent common ancestor with each other than either shares with C. Classification, then, should reflect this: '[W]e shall understand what is meant by the Natural System: it is genealogical in its attempted arrangement, with the grades of acquired difference marked by the terms, varieties, species, genera, families, orders and classes' (p. 497). That, however, is where Darwin's clarity ended for systematics, the threefold problem of which persists to this day: (1) How do we conceptualise the genealogy of species? (2) How do we infer that genealogy? and (3) How do we capture that genealogy in a classification?

Tracing the history of thought on these questions is beyond our scope but changing views on each profoundly affected how taxonomists practiced their craft (Mayr 1982b; Ridley 1986). Between Darwin (1859) and Hennig (1966), the dominant paradigm for species genealogy was a mixture of anagenetic transformation and cladogenic divergence in which taxa A and B could have two forms of relationship: direct ancestry (A is ancestral to B, or vice versa) or common ancestry (A and B share an ancestor that is neither A nor B). Genealogy was inferred by trying to distinguish homologous from analogous (convergent) similarity using a variety of methodological rules, none of them rigorous by current standards. Evolutionary classifications were those that grouped by homology, permitting both monophyletic and paraphyletic assemblages (i.e. groups sharing derived or primitive states, respectively, of homologous characters) but not polyphyletic ones (i.e. groups based on convergent traits). These at least were the general operating principles, articulated mostly in retrospect, of evolutionary taxonomy. The extent to which practitioners adhered to them varied considerably, as we shall see.

Darwin himself had little to say about marsupials, and his most extended published comment echoed the old familiar theme. 'It may be doubted ... whether the Australian marsupials, which are divided into groups differing but little from one another, and feebly representing, as Mr. Waterhouse and others have remarked, our carnivorous, ruminant and rodent mammals, could successfully compete with these well-pronounced orders. In the Australian mammals, we see the process of diversification in an early and incomplete stage of development' (Darwin 1859: 116). Later in the book, Darwin discussed Waterhouse's view that anatomical similarities between the South American viscacha (*Octomys mimax*) and marsupials revealed a 'relationship' of some sort between rodents and marsupials. Darwin used the example to help clarify the distinction between 'special' similarity due to common ancestry (i.e. homology) and 'general' similarity due to common function (i.e. analogy). The same example motivated Darwin, in his 1860 correspondence with Charles Lyell (Darwin 1899; Archibald 2012), to explore the question of whether marsupials are directly ancestral to placentals (Huxley's Metatheria hypothesis) or whether the two groups arose from a common ancestor that was neither one nor the other (Fig. 3.4). Although he did not choose between these hypotheses, he did declare a general

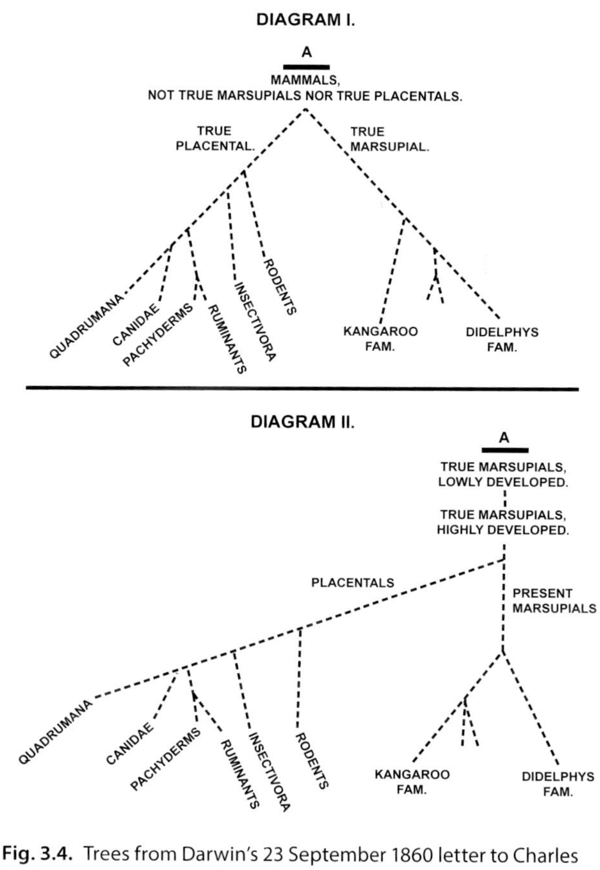

Fig. 3.4. Trees from Darwin's 23 September 1860 letter to Charles Lyell showing two hypotheses for the relationship between marsupials and placentals (Darwin 1899). Diagram I shows each group arising from a single common ancestor that was neither a marsupial nor a placental (this is the currently accepted phylogeny). Diagram II shows Huxley's Metatheria hypothesis in which the common ancestor of living marsupials and all placentals was a marsupial.

preference for single origins of taxonomic groups. In the sixth and final edition of the *The Origin*, Darwin (1872) drew attention to the convergent similarity of *Antechinus* and *Thylacinus* with mice and dogs, respectively.

Dasyurids and early Darwinians

One of Darwin's early supporters in Australia was German expatriate Gerard Krefft (1830–81), curator at the Australian Museum between 1860 and 1874. Krefft (1867: 431) undertook a revision of small-bodied dasyurids, bemoaning 'the carelessness of some authors, who will not take the trouble to examine a species properly before assigning it' to a genus. He solidified the separation of *Phascogale* (restricted to '*penicillata*' and *calura*) from *Antechinus* (still containing a mix of species from what are now multiple genera) based on the former's large first incisors (Table 3.2). Krefft diagnosed members of *Podabrus* by their slender feet and small canines, transferring into it four species from Gould's '*Antechinus*' (*fuliginosus*, *albipes*, *murinus* and *leucopus*); these, along with the original *macrourus* and *crassicaudata*, are all dunnarts (now *Sminthopsis*). Oddly, Krefft retained '*A.*' *ferruginifrons* (the mainland form of white-footed dunnart), speculating that it was a variety of *A. flavipes*. Krefft also erected two new genera. One was *Antechinomys*, for Gould's '*Phascogale lanigera*' (the kultarr), based on its four hind toes, furred upper foot surfaces, very small canines, long tail and jumping locomotion: 'an animal so different in its habits, the structure of its limbs, and the form of its teeth from *Phascogale* should have long ago been separated therefrom' (p. 434). Krefft's second new genus accommodated a species he described himself: *Chaetocercus cristicauda* (the mulgara, now *Dasycercus*) based on a single specimen in poor condition from South Australia.

Even as Darwin noted the lamentable state of marsupial development and Krefft strove for a robust classification of dasyurids, Theodore Nicholas Gill (1837–1914), curator at the Smithsonian Institution, worked to advance the conceptual clarity of taxonomic practice. Gill (1870) rejected schemes based on single adaptive characters fitting taxa into a linear series, emphasising that affinities 'are only determinable by *the sum of their agreements* in morphological characteristics' (p. 91, italics added). In his classification of mammals, Gill (1872) united marsupials and placentals as 'Eutheria' apart from monotremes (Prototheria) and grouped dasyurids and numbats as Suborder Dasyuromorphia (among other sub-orders corresponding to wombats, other diprotodonts, bandicoots and didelphimorphs). Within Dasyuridae, he recognised subfamilies Dasyurinae, Sarcophilinae and Phascogalinae, but did not extend his classification below this level.

Between 1895 and 1923, dasyurid systematics benefited from the work of two very different but equally eminent scientists. One was W. Baldwin Spencer (1860–1929), a British biologist who became professor at the University of Melbourne in 1887, had a distinguished career in Aboriginal anthropology and was a leader in the Australian scientific community (Mulvaney 1990). In 1894, the universities of Sydney, Adelaide and Melbourne sponsored an expedition by five scientists to explore the natural history of central Australia between Oodnadatta and the Macdonnell Ranges. Philanthropist W.A. Horn funded the expedition, which included Spencer as zoologist. Results of the Horn expedition included descriptions of '*Phascogale*' *macdonnellensis* (cf. *Pseudantechinus*, fat-tailed pseudantechinus) and *Sminthopsis psammophila* (sandhill dunnart) (Spencer 1895), as well as *Dasyuroides byrnei* (kowari, Spencer 1896). Over a decade later, Spencer (1909) described *Sminthopsis longicaudata*, which differs 'from all known *Sminthopsis* in the extraordinary length of its tail' (p. 451).

M.R. Oldfield Thomas (1858–1929), a contemporary of Spencer and mammalogist at the British Museum in London, was one of the most prolific mammal species describers of all time, with over 2000 species or subspecies names to his credit. He was one of the earliest mammal taxonomists to embrace C. Hart Merriam's abandonment of simple typology in favour of collecting efforts that document variation (i.e. specimen series) within and between species (Hill 1990), an approach that proved critical to the Modern Synthesis of evolutionary biology in the mid-1900s (Mayr 1982b). Thomas (1886b) replaced Gould's genus name *Podabrus* with *Sminthopsis*, due to the former's preoccupation by a genus of soldier beetles described in 1840. He also described many dasyurid species based on specimens returned from collecting expeditions to the Northern Territory and Queensland: *Sminthopsis hirtipes* (hairy-footed dunnart, Thomas 1898); '*Phascogale bella*' (cf. *Antechinus bellus*, fawn antechinus, Thomas 1904); '*Phascogale*' (cf. *Planigale*) *ingrami* (long-tailed planigale), '*Phascogale*' (cf. *Pseudantechinus*) *mimulus* (Thomas 1906b); and '*Phascogale*' (cf. *Antechinus*) *godmani* (Atherton antechinus, Thomas 1923).

So far, we have focused on the discovery and classification of Australian species. However, exploration of the New Guinea archipelago was also proceeding apace in the 19th and early 20th centuries, with concomitant description of dasyurids. By the time of Spencer and Thomas, some nine New Guinean dasyurids had been described, eight of them as members of *Phascogale*. The first formal recognition was of '*P.*' (cf. *Myoictis*) *melas* in 1840 (three-striped dasyure, see Woolley 2005a), which was included in Waterhouse (1841). Gray (1858) coined *Myoictis* and named *M. wallacei* for a specimen collected by Alfred Russell Wallace in the Aru Islands. Tarragon (1847) named '*P.*' (cf. *Sminthopsis*) *virginiae* (red-cheeked dunnart). Directors of the Leiden Museum had a significant impact on early dasyurid taxonomy, beginning with C.J. Temminck, who coined *Phascogale* (see above). Temminck's successor, ornithologist Hermann Schlegel (1804–1884), provided descriptions of '*P.*' (cf. *Murexia*) *longicaudata* (short-furred dasyure, Schlegel 1866) and *Dasyurus albopunctatus* (New Guinean quoll, Schlegel 1880). Schegel's successor, F.A. Jentink (1844–1913), described '*P.*' (cf. *Murexia*) *naso* (long-nosed dasyure) and '*P.*' (cf. *Neophascogale*) *lorentzii* (speckled dasyure) (Jentink 1911); Stein (1933) coined the subgenus *Neophascogale* for the latter. Peters and Doria (1876) named '*P.*' (cf. *Phascolosorex*) *dorsalis* (narrow-striped dasyure), for which Matschie (1916) created the subgenus *Phascolosorex*. Finally, Oldfield Thomas recognised '*P.*' (cf. *Phascolosorex*) *doriae* (red-bellied dasyure, Thomas 1886a) and '*P.*' (cf. *Murexia*) *melanurus* (black-tailed dasyure, Thomas 1899). Thus, by 1911, representatives of most modern New Guinean genera had been described, but were binned in *Phascogale*, a catch-all taxon for small dasyurids. As we have seen, Krefft, Spencer and Thomas developed genus-level classifications that captured what they saw as evolutionary relationships. This enterprise would take a giant leap forward in the work of B.A. Bensley and his successors.

Pre-Hennigian phylogenies and classifications of the 20th century

Canadian mammal anatomist Benjamin Arthur Bensley (1875–1934) was the first marsupial taxonomist to propose explicit phylogenetic hypotheses in conjunction with his classification. Bensley (1903) considered thylacines and numbats as subfamilies ('Thylacininae' and 'Myrmecobiinae', respectively) of Dasyuridae; the then-recognised genera of dasyurids (*Dasyurus*, *Sarcophilus*, *Chaetocercus*, *Dasyuroides*, *Phascogale*, *Sminthopsis*, *Antechinomys*) formed his 'Dasyurinae'. His goal was 'to construct a plan of the Australian radiation by tracing the sequence of adaptive modifications in the dentition and foot structure of the various genera' (p. 87). Dasyurid teeth in particular 'present a trend of evolution involving a progressive change from primitive insectivorous to specialised carnivorous modifications' (p. 90). Indeed, highly specialised carnivorous and myrmecophagus dentitions are what set thylacines and numbats, respectively, apart from dasyurids. In foot structure, Bensley posited 'a main line of arboreal or semi-arboreal evolution, with several subsidiary lines of terrestrial adaptation' (p. 163). Based on these two transformation series, he inferred the genus-level relationships shown in Fig. 3.5, with '*Phascogale*' (typified by '*P.*' *flavipes*) approximating the dasyurid ancestor, the main line of descent leading to arboreal *Dasyurus* (e.g. *hallucatus*) and side branches leading to specialised carnivores, anteaters and terrestrial species. This first dasyurid phylogeny exemplifies the pre-Hennigian tradition of inferring progression series in a few key characters related to adaptations and implying links of direct ancestry between living taxa.

In 1901, the Fifth International Zoological Congress adopted the *Règles Internationales de la Nomenclature*, first incarnation of the modern International Code of Zoological Nomenclature. This was an attempt to bring order to what had been, for centuries, a fundamentally haphazard system of naming taxa at and below the family level (Blackwelder 1967). Among (many) other tenets of taxonomic practice, the Code established the principles of priority for taxonomic names and typification for linking names to specimens. Iredale and Troughton

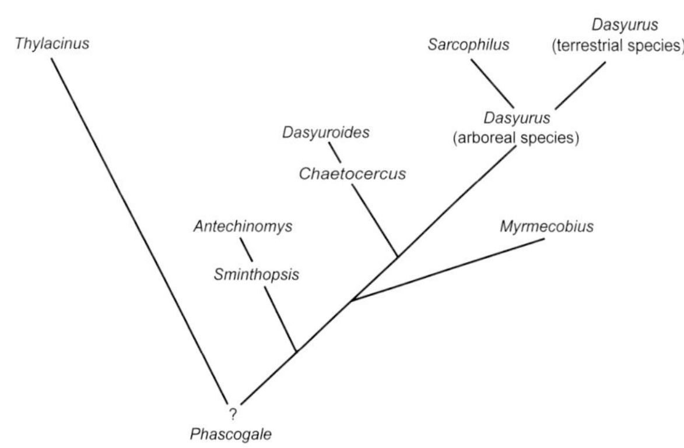

Fig. 3.5. Bensley's (1903) estimate of dasyurid phylogeny based on assumed progression series in dentition (insectivore- to carnivore-adapted) and foot structure (arboreal- to terrestrial-adapted).

(1934) provided the first comprehensive classification of Australian dasyurids in the *Règles* era (Table 3.3). Notable for its lack of any phylogenetic hypotheses, this 'checklist' revealed a more structured taxonomy than any previous attempt, as well as a thorough and consistent synonymy for each taxon. Iredale and Troughton included thylacines as a subfamily (Thylacininae) of Dasyuridae, but also used subfamilial distinction to separate the large quolls and devil (Dasyurinae) from smaller-bodied forms (Phascogalinae). In the former, each previously recognised species was allocated – without explanation – to its own genus. The latter included a new genus, *Planigale*, erected by Troughton (1928) to accommodate his new species *P. tenuirostris* (narrow-nosed planigale) along with Thomas' '*Phascogale*' *ingrami*. According to Troughton, *Planigale* is '[r]elated to *Phascogale* but differentiated by the marked flattening of the upper surface of the skull' (p. 282). The checklist also included two new species that had been described in the early 1900s, *Dasycercus blythi* (brush-tailed mulgara; Waite 1904) and *S. granulipes* (white-tailed dunnart; Troughton 1932).[4] Other novel species represent what are now regarded as subspecies, but their inclusion clearly reflects the authors' attempt at a comprehensive, up-to-date summary of taxonomic research.

George Henry Hamilton Tate (1894–1953) was a British-born mammalogist affiliated with the American Museum of Natural History (AMNH) from 1921 until his death. His extraordinary life and career included extensive professional travel in South America, Africa and Australasia (Anthony 1954). Tate's most significant contributions to marsupial biology were the result of his participation in the first two (1933–34, 1936–37) AMNH expeditions to New Guinea funded and led by zoological philanthropist Richard Archbold (1907–76). Tate and Archbold (1937) summarised the diversity of marsupials collected on the first expedition, but without specifying a particular classification beyond distinguishing subfamilies Dasyurinae and Phascogalinae. Like Waterhouse (1846), they considered *Phascogale* as a genus with multiple subgenera previously proposed – including the original *Phascogale* and *Antechinus*, as well as the New Guinean endemics *Myoictis*, *Phascolosorex* and *Neophascogale* – to which they added *Murexia* to accommodate three forms now synonymous with *Murexia longicaudata*.

Table 3.3. Classification of Dasyuridae by Iredale and Troughton (1934).
Modern genus or species equivalents are shown in brackets.

Dasyurinae	Phascogalinae	Phascogalinae (cont.)
Dasyurus	*Phascogale*	*Sminthopsis*
quoll [*viverrinus*]	*calura*	*crassicaudata*
Dasyurinus [cf. *Dasyurus*]	*tapoatafa*	*froggatti* [*macroura*]
geoffroii	*Antechinus*	*granulipes*
Satanellus [cf. *Dasyurus*]	*apicalis* [cf. *Parantechinus*]	*hirtipes*
hallucatus	*bellus*	*larapinta* [*macroura*]
Dasyurops [cf. *Dasyurus*]	*flavipes*	*leucopus*
gracilis [*maculatus*]	*godmani*	*longicaudata*
maculatus	*maculatus* [cf. *Planigale*]	*lumholtzi* [*virginiae*]
Sarcophilus	*macdonnellensis* [cf. *Pseudantechinus*]	*murina*
harrisii	*mimulus* [cf. *Pseudantechinus*]	*psammophila*
	minimus	*stalkeri* [*macroura*]
	Dasycercus	*virginiae*
	cristicauda	*Antechinomys*
	blythi	*laniger*
	Dasyuroides	*spenceri* [*laniger*]
	byrnei	
	Planigale	**Thylacininae**
	ingrami	*Thylacinus*
	tenuirostris	*cynocephalus*
	subtilissima	

Mammalian taxonomy experienced a watershed moment with publication of George Gaylord Simpson's 'The principles of classification and a classification of mammals' (Simpson 1945). The first part of this treatise formed the earliest articulation of evolutionary taxonomy as a systematic philosophy and methodology. As such, it was a prelude to foundational treatments such as *Methods and Principles of Systematic Zoology* (Mayr *et al.* 1953) and *Principles of Animal Taxonomy* (Simpson 1961). The second part of Simpson's (1945) monograph was a supra-generic classification of living and fossil mammals that endured for half a century. As significant as this work was for systematic biology, Simpson added relatively little to dasyurid taxonomy. He adopted the subfamilies Dasyurinae, Phascogalinae and Thylacininae of Iredale and Troughton (1934), but also included numbats ('Myrmecobiinae') as a subfamily of dasyurids. Simpson rejected the multiple genera of quolls, as well as recognition of *Planigale*, proposed by Iredale and Troughton (1934); nor did he separate *Antechinus* from *Phascogale* as proposed by Krefft (1867). Thus, Simpson's Dasyurinae included only *Dasyurus* and *Sarcophilus*; Phascogaline comprised *Phascogale*, *Dasyuroides*, *Dasycercus*, *Sminthopsis* and *Antechinomys*. Dasyurids were placed in a superfamily (Dasyuroidea) that also included marsupial moles (Notoryctidae); Winge (1941) had previously proposed including *Notoryctes* within Dasyuridae.

Just as Simpson revolutionised mammal taxonomy, Tate (1947) provided a landmark evolutionary classification of all living dasyurids (Table 3.4) united with a

Table 3.4. Tate's (1947) classification of Dasyuridae.
Modern genus or species equivalents are shown in brackets.

Dasyurinae	Phascogalinae	Phascogalinae (cont.)
Dasyurus	*Phascogale*	*Sminthopsis*
quoll [*viverrinus*]	*tapoatafa*	The *rufigenis* Division
Dasyurinus [cf. *Dasyurus*]	*calura*	*rufigenis* [*virginiae*]
geoffroii	*Antechinus*	*lumholtzi* [*virginiae*]
Satanellus [cf. *Dasyurus*]	*A. flavipes* Group	*longicaudata*
hallucatus	*flavipes*	*murina* [incl. *leucopus*]
Dasyurops [cf. *Dasyurus*]	*godmani*	The *crassicaudata* Division
gracilis [*maculatus*]	*melanurus* [cf. *Murexia*]	*macroura* [incl. *psammophila*]
maculatus	*mayeri* [cf. *Murexia*]	*granulipes*
Sarcophilus	*wilhelmina* [cf. *Murexia*]	*larapinta* [*macroura*]
harrisii	*A. maculatus* Group	*crassicaudata*
Neophascogale	*maculatus* [cf. *Planigale*]	*hirtipes*
lorentzii	*A. swainsonii* Group	*Antechinomys*
Phascolosorex	*minimus*	*laniger*
doriae	*swainsonii*	*spenceri* [*laniger*]
dorsalis	*A. bellus* Group	*Planigale*
Parantechinus	*bellus*	*novaeguineae*
apicalis	*Murexia*	*ingrami*
Pseudantechinus	*longicaudata*	*tenuirostris*
macdonnellensis	*rothschildi*	*Thylacinus*
mimulus		*cynocephalus*
Myoictis		
melas		
wallacei		
Dasycercus		
cristicauda		
blythi		
Dasyuroides		
byrnei		

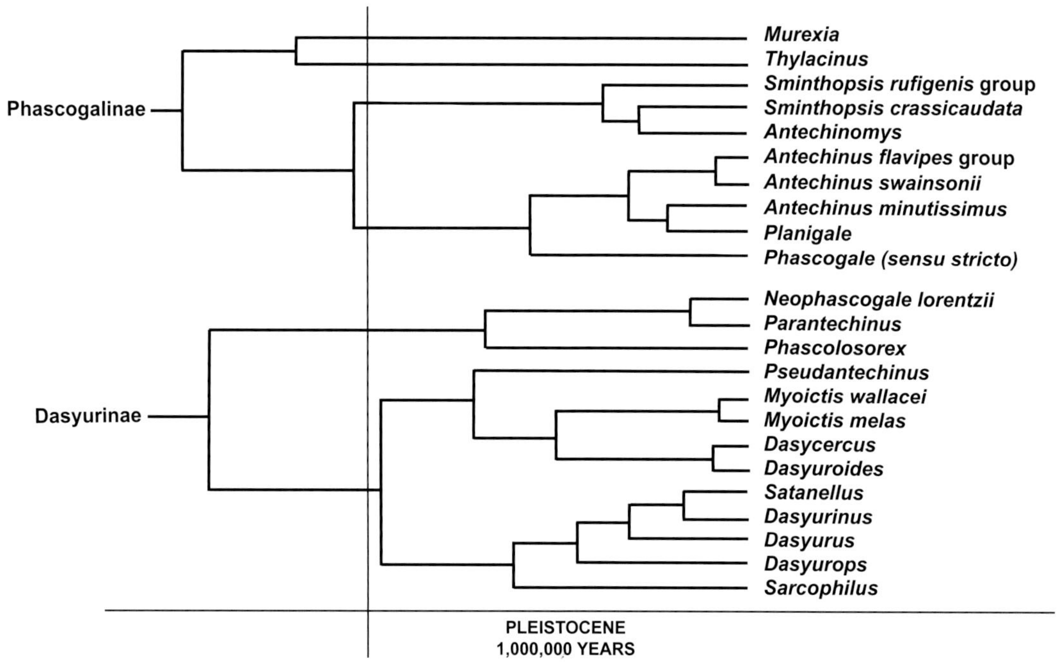

Fig. 3.6. Tate's (1947) phylogeny of dasyurids. For each pair of branches, 'the more specialised forms' are towards the bottom. [Adapted from Tate (1947).]

phylogenetic hypothesis (Fig. 3.6) based on character analysis. Tate included several new species, some of which had been described previously but not widely recognised: '*Antechinus mayeri*' (consisting of forms now variously synonymised with *Murexia naso*, *M. habbema* and *M. melanurus*), *Antechinus* (cf. *Murexia*) *wilhelmina*, *Murexia rothschildi*, *Sminthopsis rufigenis* (= *virginiae*) and *Planigale novaeguineae*. In addition, he recognised several new genera, some of them former subgenera: *Neophascogale*, *Phascolosorex*, *Parantechinus* and *Pseudantechinus*.

Tate's method of inferring relationships was based on assessing the adaptive value versus selective neutrality of similar characters, the former prone to convergence and the latter more likely homologous. Tate attached particular importance to features 'which seem to indicate ancient dichotomy' (p. 110). His summary judgment on character types is revealing in the extent to which it departs from that of previous taxonomists:

> *Because many external characters pertaining to the pelage, to locomotion, and to food getting are adaptive, morphological similarities in those cases may be the result of convergences rather than indicators of real relationships, and so can be employed in phylogeny only with discretion. Some at least of the internal characters of animals also come under the same category and must likewise be used guardedly. Others, especially those that have originated in a seemingly causeless and random way, may well point to actual phyletic history. Among those last seem to belong such characters as the size relationships of the individual teeth of a given dental group (incisors, premolars, molars) to one another, or the individual bones of the skull and skeleton to one another. But constant watch must be kept for functional significance. [Tate 1947: 111]*

Ancient dichotomies in dental characters led Tate to posit the basal division of Dasyuridae into subfamilies, the contents of which were radically different from those of earlier classifications:

> *Combinations of these supposedly phyletic characters have resulted in definitions and limitations of the two dasyurid subfamilies somewhat at variance with the standard concept. Genera that show some degree of reduction of* p^4 *to a size less than*

> *that of* p^3*, combined with* i^2 *smaller than* i^4*, are now included with the subfamily Dasyurinae. This arrangement makes the Papuan genera* Neophascogale, Phascolosorex, *and* Myoictis *primitive members of the subfamily. It also serves to include the new genera* Parantechinus *and* Pseudantechinus, *formerly treated as* Antechinus, *in the same subfamily … [A]lthough reduction of* p^4 *has commenced [in* Antechinus, Planigale, *and* Phascogale], *enlargement of* i^2 *is a new feature, incipient in* Antechinus *and culminating in* Phascogale. *To these two genera are added* Thylacinus, Murexia, Sminthopsis, *and* Antechinomys … *in which* p^4 *is as large as, or larger than,* p^3*, even though* i^{2-4} *lack the special condition of* Antechinus *and* Phascogale …' *[Tate 1947: 111]*

Tate attached so much weight to the relative sizes of incisors and premolars that he rejected Bensley's (1903) and Iredale and Troughton's (1934) separation of Thylacinidae as a lineage separate from Dasyuridae. He suggested that 'the many adaptive characters of *Thylacinus*' (p. 112) had misled Bensley and that primitive incisor and premolar sizes linked *Thylacinus* to the recently described *Murexia*. The fine structure of Tate's phylogeny was also heavily influenced by incisor-premolar sizes, though other traits are adduced as evidence for particular groupings. However, despite his unwavering sensitivity to possible convergences, Tate's selection of characters in support of his clades was ultimately subjective. In the pre-Hennigian tradition, shared primitive traits have the same evidential meaning as derived ones and species relationships are mapped onto character-progression series, with branchings to accommodate nested patterns of change. The tree, however, is modern in one respect – all named taxa appear at the tips, none on the internal branches. Tate, unlike many of his contemporaries, did not conflate degree of phylogenetic relationship with direct ancestry. His phylogeny also included a timescale, with most branching confined to the Pleistocene, though he expressed considerable doubt about such a recent radiation.

Following Tate's *tour de force*, dasyurid taxonomy was relatively quiescent until after the Hennigian revolution. Johnson (1954) described '*Antechinus*' (cf. *Pseudantechinus*) *bilarni* (sandstone false antechinus) from a single specimen collected in 1948 and Ride (1964a) described '*Antechinus*' (cf. *Dasykaluta*) *rosamondae* (little red kaluta). Wakefield and Warneke (1967) resurrected *Antechinus stuartii* from synonymy with *A. flavipes*. With these additions, the classification of Australian species presented by Ride (1970) is essentially that of Tate (1947), differing only in generic assignment of quolls (Tate, like Simpson, put them all in *Dasyurus*) and a few species-subspecies distinctions within *Sminthopsis*. Ride (1970) included *Myrmecobius*, but not *Thylacinus*, within Dasyuridae.

EVOLUTIONARY INTERPRETATIONS OF DASYURID ANATOMY AFTER HENNIG

Phylogenetic systematics

The publication of Hennig's (1966) *Phylogenetic Systematics* in English triggered the greatest conceptual revolution in biological taxonomy since Darwin. Willi Hennig (1913–76) was a German entomologist and dipteran systematist with interests in the nature of phylogeny, methods for inferring phylogeny, and the relationship between phylogeny and classification. Earlier work by botanist Walter Zimmerman (1892–1980) emphasised phylogeny as the basis for natural classification, recency of common ancestry as the measure of relatedness, the importance of character analysis to distinguish primitive and derived traits, and the limitations of overall similarity as a guide to phylogeny (Donoghue and Kadereit 1992). Hennig (1950) developed these principles into a coherent method (Box 3.1) for inferring relationships based on shared derived character states (synapomorphies), but the impact of his approach was not widespread until the 1966 translation appeared. In addition to character analysis, Hennig argued cogently that phylogeny should be the *exclusive* basis for classification – i.e. that only monophyletic groups should be recognised (named) as taxa. The early history of phylogenetic systematics (or cladistics), its confrontations with evolutionary and phenetic (Box 3.2) schools of taxonomy, and its connection to parsimony analysis and its logical underpinnings, have been amply described in the literature (e.g. Mayr 1982b; Ridley 1986; Hull 1988; Sober 1988). Although Hennigian analysis was eventually superseded by more overtly numerical and statistical methods of phylogenetic inference, the doctrine of monophyly as the sole criterion for taxon recognition had largely prevailed by the turn of the millennium.

Box 3.1. Hennigian analysis

Willi Hennig's key insight into phylogenetic inference can be summarised in the following statement: *only shared, derived characters provide evidence for monophyly.* A monophyletic group consists of an ancestral species, all of its descendants and nothing else. For example, members of the monophyletic group Mammalia have a single jaw joint between the dentary bone forming the mandible and the squamosal bone of the skull. All other vertebrates, living and fossil, have a jaw joint formed by either the quadrate and articular bones (most non-mammalian forms) or a double dentary-squamosal + quadrate-articular joint (a few derived fossil cynodonts close to the origin of mammals). Hennig's reasoning was straightforward: if a derived character evolves just once in an ancestral species, that character should be inherited by all the ancestor's descendants. Hennig called shared derived characters *synapomorphies*. In contrast, shared primitive characters (*symplesiomorphies*) are not evidence of monophyly. Monotremes and birds lay amniotic eggs, whereas therian mammals have placental gestation and live birth, but monotremes and birds are not sisters apart from therians. Amniotic oviparity arose in the common ancestor of all reptiles (including birds) and mammals, and so it is a symplesiomorphy for birds and monotremes. Placental gestation is the derived mode of reproduction found only in therians (i.e. it is a synapomorphy for therians). Thus, Hennig's view of the phylogenetic research program is a search for synapomorphies that mark out monophyletic groups in some collection of species. There are two major problems with this agenda, which effectively undermined it in favour of the statistical approaches that came later. First, Hennigian analysis requires that we know the primitive and derived states of the characters we analyse. Although Hennig thought this was possible and suggested guidelines for doing it, hypotheses of character *polarity* are almost always tenuous. Second, Hennig assumed that species sharing derived characters inherited from a common ancestor (synapomorphies) could be distinguished from those sharing derived states due to convergent evolution (*homoplasies*). This proved difficult enough for anatomical traits but is essentially impossible for molecular sequences. Although Hennig's *methodology* brought great conceptual clarity to phylogenetics (for which Hennig is justly regarded as a towering figure in systematic biology), his *method* is not generally practicable. A version of Hennigian methodology was reborn, however, in numerical parsimony analysis (see Chapter 4).

Box 3.2. Phenetics

From Linnaeus to the 1990s, there was a recurring debate in taxonomy over what constitutes a 'natural' classification and how such a classification should be constructed. Twentieth century British botanist J.S.L. Gilmour argued that natural classifications were those generally useful to all scientists, as opposed to special-purpose alternatives. This was widely accepted by systematists as an advance over the search for Aristotelian *differentia* that lingered in taxonomy even after Darwin. But how to achieve such naturalness? The debate has now been resolved in favour of phylogenetic classification (i.e. classifications should reflect best estimates of phylogeny), but during the 1960s and 1970s, advocates of phenetic classification argued for a different approach. Leaders of this movement, such as R.R. Sokal and P.H.A. Sneath, championed the position of 18th century French botanist Michel Adanson that natural affinities among organisms (and, hence, classifications that reflect them) can only be obtained by considering *all* characters, not just a few considered important *a priori*. In phenetics, this translated to the concept of *overall similarity* as a basis for grouping organisms into taxa. Volumes such as Sneath and Sokal (1973) presented a plethora of quantitative methods for doing so. Phenetics failed as a paradigm in taxonomy for a few key reasons. First, no study or set of studies could examine *all* characters of organisms. More importantly, the concept of *overall similarity* has a seemingly unlimited number of operational definitions; there are too many ways to measure similarity that give different results, and no obvious way to choose among them. In fairness to phenetics, its primary goal was to construct classifications, not infer phylogenies. However, some of its analytical methods produced tree diagrams (phenograms) based on measures of character similarity between pairs of species. These measures do not distinguish among Hennig's three kinds of similarity (synapormorphy, symplesiomorphy, homoplasy) and it soon became apparent that overall similarity would track phylogeny only when characters evolve at the same rate in all lineages under study. Empirically, such a *uniform rate* assumption rarely applies and phenetics was never a strong contender for phylogenetic supremacy. Nevertheless, many systematists of the late 20th century were drawn to the new numerical analyses and considered phenograms as one among several types of phylogeny estimators.

Early post-Hennigian work on dasyurid relationships took place between 1966 and 2000, when both phylogenetic methodology and taxonomic philosophies were in flux. Even so, these studies aimed to elucidate phylogeny in essentially its modern sense – statements of monophyly expressed as directed, acyclic graphs in the form of dendrograms ('trees') on which tips represent species or larger clades, nodes represent speciation events and internodes represent hypothetical ancestral species. Moreover, such statements were produced by explicit analyses – though not always strictly objective or Hennigian – of multiple characters. Such an approach contrasts with the subjective evidence weighting used, for example, by Tate (1947) – i.e. dasyurid phylogeny revealed by patterns of incisor-premolar size proportions. This new phase of dasyurid systematics was dominated by the work of Michael Archer of the University of New South Wales, whose seminal contributions extend across much of marsupial biology.

Archer and dasyurid relationships

Archer brought new depth and rigour to the use of comparative anatomy in studies of marsupial relationships. Like Thomas, Bensley and Tate, he saw the skull – and particularly its teeth – as a rich source of evidence for understanding evolutionary trends. Whereas earlier authors emphasised simple dental traits such as tooth formulae and relative sizes, Archer explored the implications of tooth ontogeny and cusp homology for evolutionary relationships, beginning with studies of *Antechinus flavipes* (Archer 1974, 1975a). These early investigations were quickly followed by comprehensive and detailed reviews of craniodental anatomy in all carnivorous marsupials (Archer 1976c, d). Here Archer introduced a concept that would feature strongly in his subsequent writings – that one taxon (e.g. Didelphidae) is 'structurally ancestral' to another (e.g. Dasyuridae). Structural ancestry is not genealogical ancestry – the phrase does not imply that didelphids evolved into dasyurids (though Archer independently asserted that they had; see below) – but rather that the ancestral form is 'least specialised' (i.e. generally displays primitive states) for characters that in other forms are 'specialised' (i.e. derived). The extent to which Archer often put forward strong hypotheses of character evolution while remaining tentative about species phylogeny is characteristic of the transition from evolutionary to phylogenetic systematics.

Despite asserting a criterion ('didelphids are primitive') for polarising characters, Archer (1976c, d) limited himself to one simple conclusion about dasyurid relationships: 'Dasyurids are probably derivatives of an ancestral didelphid stock' (1976c: 31). He did, however, present tree diagrams (Fig. 3.7) depicting similarities among selected genera and species in dental and basicranial traits. Both are labelled 'phenograms', implying that sister-pairs are maximally similar, but both include the following disclaimer: 'the vertical and horizontal axes have no dimensions and are not to be interpreted as structural distance' (e.g. 1976c: 30). In other words, he employed neither an objective measure of overall similarity between taxon-pairs nor an objective method of taxon clustering. These are not phenetic relationships in the sense of Sneath and Sokal (1973); instead, they exemplify the methodological subjectivity in classification and phylogenetic inference to which both pheneticists and cladists (as followers of Hennig came to be called) were opposed. If the history of animal taxonomy has one lesson, however, it is not to discount entirely the subjective assessments of those who 'know the organisms'. An influential legacy of these analyses is Archer's (1976d) contention that *Murexia*, *Neophascogale*, *Phascolosorex* and *Phascogale* are dentally primitive; notably, all but the last are New Guinean endemics.

Concomitant with his osteological research, Archer was instrumental in clarifying the alpha taxonomy of small dasyurids. Archer (1975b) described the genus *Ningaui* with two tiny species (*N. timealeyi*, *N. ridei*) from arid regions of Western Australia. His description characteristically involved detailed comparisons of crania and molar cusp patterns among *Ningaui*, *Sminthopsis* and *Planigale* (to which specimens of *Ningaui* had previously been allocated), as well as plantar pads on the hind feet. In re-diagnosing *Planigale*, Archer recognised the common planigale as *Planigale maculata*, rather than a member of *Antechinus* (cf. Tate 1947; Ride 1970), thereby eliminating 'flat-headedness' as diagnostic of *Planigale* (common planigales have domed skulls). Archer (1976b) completed the revision of *Planigale*, recognising five species in three groups: the maculata group (*Pl. maculata*, *Pl. novaeguineae*); the ingrami group (*Pl. ingrami*, *Pl. tenuirostris*); and the gilesi group (*Pl. gilesi*). The latter species had been described by Aitken (1972), who noted its unique (among planigales) possession of only two upper and lower premolars. This proved to be one of the few diagnostic traits within the genus and Archer (1976b: 341)

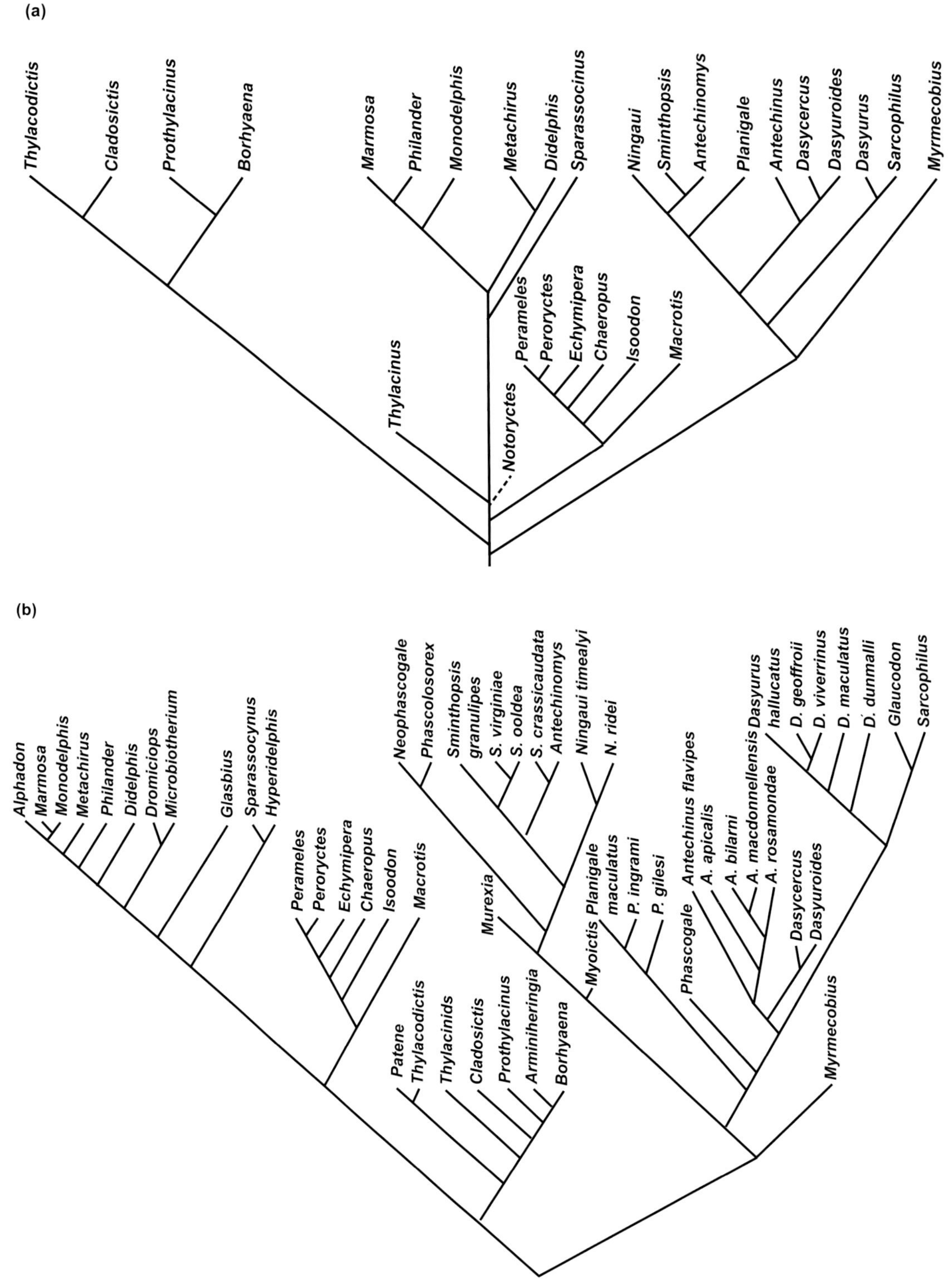

Fig. 3.7. Trees depicting cranial similarity among selected marsupials. Captions to original figures include this statement: 'In each dichotomous pair, the more derived or structurally less ancestral member … is placed to the right of the other member.' (a) Basicranial anatomy (Archer 1976c). Branch lengths are unscaled and clusters are based on subjective assessment of similarity among taxa. Dashed line shows position of Notoryctemorphia. (b) Tooth morphology (Archer 1976d). Branch lengths are unscaled and clusters are based on subjective assessment of similarity among taxa. [(a) Reprinted by permission of Oxford University Press from Archer (1976c). (b) Reprinted by permission of CSIRO Publishing from Archer (1976d). Permissions conveyed through Copyright Clearance Center, Inc.]

noted, 'Some small dasyurids have distinctive morphological dental and cranial characteristics … In other genera, such as *Planigale*, these distinctive morphological characters are less common. Only *Planigale gilesi* in lacking P4 is obviously and consistently distinct from other species. Recognition of other species of *Planigale* requires examination of size and consideration of relative cranial and dental proportions.' Indeed, Archer's revision relied heavily on morphometric ratios from skull and external body measurements. His species groups were again based on subjective assessment of overall similarity in the characters he examined. Although he interpreted similarity levels to challenge some species distinctions (e.g. *Planigale maculata* versus *P. novaeguineae*), he did not explicitly address phylogeny.

Archer (1977) examined the status and composition *Antechinomys*, specifically evaluating Tate's (1947) suggestions that kultarrs share a common ancestry with *Sminthopsis* (specifically the *crassicaudata* division) and form two species (*A. laniger* and *A. spenceri*). Archer endorsed the first hypothesis, citing as evidence, 'the fact that in both, the squamosal and frontal bones contact on the outside of the cranium, thereby excluding the more normal dasyurid alisphenoid-parietal contact. The structure of the *Antechinomys* talonid also reveals affinity to *Sminthopsis* and particularly *S. crassicaudata*, in that there is an approximation of the entoconid to the lingual end of the hypocristid' (p. 18). This is a somewhat telegraphic character analysis, but clearly implies that the rarer states (among dasyurids) shared by *Antechinomys* and *Sminthopsis* are synapomorphic, not just similar. Archer rejected the species status of *A. spenceri* without explanation.

In a larger study of *Sminthopsis*, Archer (1979) provided descriptions for *S. butleri* (Carpentarian dunnart) and *S. douglasi* (Julia Creek dunnart). Archer (1981) reclassified *Antechinomys* as a subgenus of *Sminthopsis*. This was part of a thorough revision of dunnarts based on external morphology and anatomy of the hindfoot (including plantar pads), teeth and skull (measurements as well as qualitative characters). Archer (1981: 77) noted, 'Most cranial, dental, and even external characters that differentiate *Antechinomys* [from *Sminthopsis*] do so only as a matter of degree … However, *Antechinomys* is distinguished from all *Sminthopsis* because it lacks a fifth toe on the hind foot, a condition unique among small dasyurids. This single character, although a highly derived one, is not a shared derived character and is cladistically useless in determining relationships of *Antechinomys*.' Here Archer relies on an implicit parsimony argument for assigning polarity to the fifth toe – i.e. present → absent, because a single loss entails fewer state changes than multiple gains. This argument is consistent so long as *Antechinomys* is not sister to all other dasyurids (and no previous author had suggested it was). Given its polarity, loss of the fifth toe is indeed an uninformative autapomorphy *sensu* Hennig (1966). This is the first example of cladistic reasoning in the dasyurid literature.

Archer (1981) used polarity decisions in other characters to construct the cladogram in Fig. 3.8 and made his reasoning explicit: 'Basic to many of the foregoing interpretations of character state polarity are the two assumptions that widespread aridity in central Australia is a relatively recent event and that the ancestral species of *Sminthopsis* was not arid-adapted' (p. 209). These assumptions were not gratuitous; Archer adduced three arguments in support of them: (1) the common ancestor of *Sminthopsis* likely predated Australian aridification; (2) anatomically primitive species (e.g. *S. leucopus*) now occupy non-arid habitats, whereas anatomically derived ones (e.g. *Antechinomys*) occupy arid zones; and (3) observation (2) holds for traits that are primitive among all dasyurids, not just dunnarts. The latter two arguments contain an element of circularity

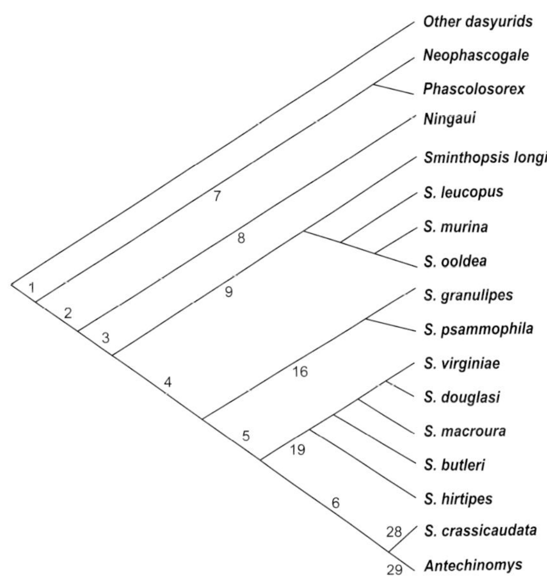

Fig. 3.8. Archer's (1981) 'Cladogram illustrating cladistic relationships of *Sminthopsis* based on shared derived (synapomorphic) characters.' The characters are anatomical features of the hindfoot, teeth, skull and external body. Internode numbers (only selected ones shown here) link to lists of synapomorphies in the original paper. [Adapted from Archer (1981).]

but are not entirely tautological in that Archer polarised some characters on grounds other than arid-adaptedness. Thus, Fig. 3.8 is a standard cladogram, to which Archer appended lists of synapomorphies for every node (e.g. 'narrow hind feet, squamosal-frontal contact, reduced metatarsal granules, transverse metacristids and hypocristids' occur at internode 3, the *Sminthopsis* ancestor). Presumably, it was constructed using Hennigian argumentation, rather than a numerical method, in that Archer makes no mention of character consistency or tree optimality. The tree suggests five species groups, understood as clades in the modern sense: (1) the *murina* group (branch 9) of *S. murina*, *S. ooldea* (which Archer elevated from a subspecies of *S. murina*), *S. leucopus* and *S. longicaudata*; (2) the *psammophila* group (branch 14) of *S. psammophila* and *S. granulipes*; (3) the *macroura* group (branch 19) of *S. macroura*, *S. virginiae*, *S. douglasi*, *S. butleri* and *S. hirtipes*; (4) *S. crassicaudata* (branch 28) and (5) *S. (Antechinomys) laniger* (branch 29). Note that, despite Archer's cladistic approach, his recognition of *S. (Antechinomys)* rendered *S. (Sminthopsis)* paraphyletic and violated a key tenet of phylogenetic classification.

A milestone in dasyurid systematics came with publication of the 1982 *Carnivorous Marsupials* volumes, especially Archer's (1982c) review of dasyurid phylogeny. In addition to the anatomy of extant species, Archer drew on information from the fossil record and early molecular studies (see Chapter 4) to produce a new phylogenetic estimate (Fig. 3.9) and corresponding suprageneric classification. The arrangement of living forms was as follows (subfamily names as revised by Archer 1984c):

Family Dasyuridae
- Subfamily Muricinae
  - *Murexia*
- Subfamily Phascolosoricinae
  - *Phascolosorex*, *Neophascogale*
- Subfamily Phascogalinae
  - *Phascogale*, *Antechinus*
- Subfamily Planigalinae
  - *Planigale*
- Subfamily Sminthopsinae
  - *Sminthopsis*, *Antechinomys*, *Ningaui*
- Subfamily Dasyurinae
  - Tribe Parantechini
    - *Parantechinus*, *Pseudantechinus*, *Dasykaluta*
  - Tribe Dasyurini
    - *Dasyurus*, *Satanellus*, *Sarcophilus*, *Myoictis*, *Dasyuroides*, *Dasycercus*

Although Archer (1982c) conscientiously explored alternative hypotheses of character evolution, his

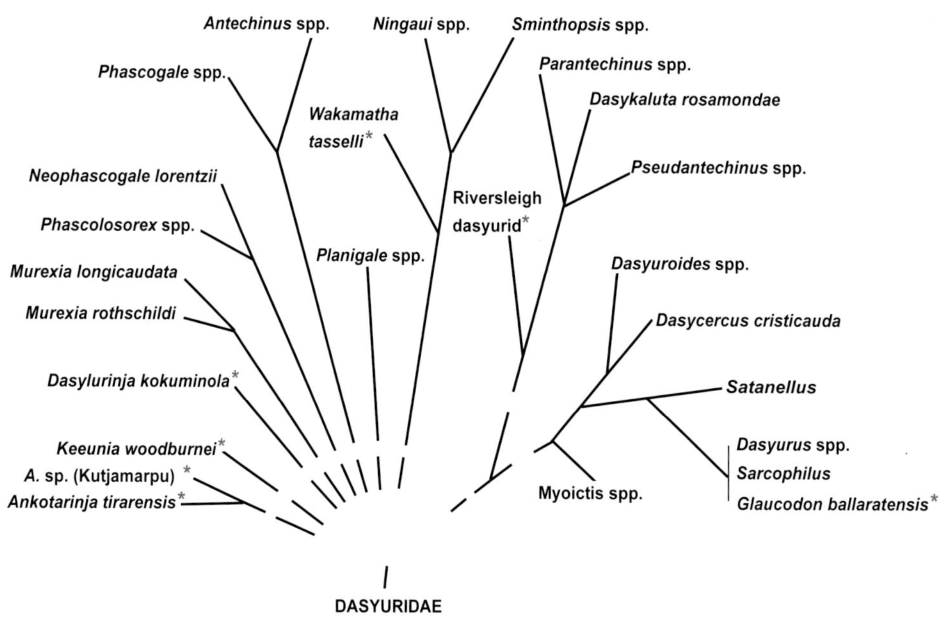

Fig. 3.9. Archer's (1982c) phylogeny of dasyurids based on cladistic interpretation of anatomy and early molecular studies. Asterisks indicate fossil taxa. [Reproduced with permission of the Royal Zoological Society of New South Wales from Archer (1982c), with minor adaptations.]

Table 3.5. Dasyurid species recognised by Mahoney and Ride (1988) and Flannery (1990) in the supraspecific groups of Archer (1984c). Asterisks indicate New Guinean endemics.

Dasyurinae	Phascogalinae	Sminthopsinae	Sminthopsinae (cont.)
Dasyurini	*Phascogale*	Sminthopsini	Planigalini
Dasyurus	*calura*	*Sminthopsis*	*Planigale*
*albopunctatus**	*tapoatafa*	*aitkeni*	*gilesi*
geoffroii	*Antechinus*	*archeri*	*ingrami*
hallucatus	*bellus*	*butleri*	*maculata*
maculatus	*flavipes*	*crassicaudata*	*tenuirostris*
*spartacus**	*godmani*	*dolichura*	*novaeguineae**
viverrinus	*leo*	*douglasi*	
Sarcophilus	*melanurus**	*fuliginosus*	**Muricinae**
harrisii	*minimus*	*gilberti*	*Murexia*
Dasycercus	*naso**	*granulipes*	*longicaudata**
cristicauda	*stuartii*	*griseoventer*	*rothschildi**
Dasyuroides	*swainsonii*	*hirtipes*	
byrnei	*wilhelmina**	*leucopus*	**Phascolosoricinae**
Myoictis		*longicaudata*	*Phascolosorex*
*melas**		*macroura*	*doriae**
Parantechini		*murina*	*dorsalis**
Parantechinus		*ooldea*	*Neophascogale*
apicalis		*psammophila*	*lorentzii**
bilarni		*virginiae*	
Pseudantechinus		*youngsoni*	
macdonnellensis		*Antechinomys*	
Dasykaluta		*laniger*	
rosamondae		*Ningaui*	
		ridei	
		timealeyi	
		yvonneae	

summary assessment of evidence for the nine major branches in Fig. 3.9 was striking: 'The boundaries of these groups are defined by the *lack* of clear synapomorphic character states' (p. 428, italics added). This is due, in part, to the paucity of characters scorable in fossil remains, but the basal polytomy in Archer's tree set the agenda for most subsequent research on dasyurid systematics. In later chapters, we discuss Archer's hypotheses of species relationships within groups, but the composition and affinities of 'Parantechini' (group 9.1) was a major departure from previous views. The genera in Archer's Dasyurinae are essentially those of Tate (1947); whereas Ride (1970) placed most parantechinin species in *Antechinus*, Archer united them with dasyurines based on molecular similarities, penis anatomy, dentition, cranial traits and lack of phascogaline synapomorphies. Having constructed a novel data matrix of 15 morphological characters and explored their compatibility with each other and penis structure, Archer settled on a basal trichotomy for Parantechini, with genus names for each clade: *Parantechinus* (*P. apicalis*, *P. bilarni*); *Pseudantechinus* (*P. macdonnellensis*, *P.* 'ningbing'); and *Dasykaluta* (*D. rosamondae*). Thus were born the 'false antechinuses', whose relationships have proven to be among the most difficult for dasyurid systematists to resolve (see Chapter 6).

Numerical analyses and a syncretic classification

Since the work of Oldfield Thomas in the 1880s, *Sminthopsis* taxonomists have struggled with complex and subtle patterns of geographic variation in morphological traits among populations – such that species boundaries were difficult to establish. McKenzie and Archer (1982) distinguished northern *Sminthopsis youngsoni* (lesser hairy-footed dunnart) from its parapatric southern sister *S. hirtipes*. We saw that Archer (1976b) utilised morphometric ratios in his revision of *Planigale*, but analysis of those data was simply by inspection. Similarly, Archer's use of overall similarity and synapomorphy in other publications was non-numerical. Darrell Kitchener of the Western Australian Museum was the first to bring numerical taxonomy methods to bear on dasyurids. Kitchener *et al.* (1984) explored possible species boundaries within what had been *Sminthopsis murina*. In addition to qualitative traits, they used a combination of univariate and multivariate statistical analyses (including the arguably cladistic 'distance Wagner' clustering method) on morphometric data from locality-grouped specimens. The results led them to recognise four new species: *Sminthopsis aitkeni* (Kangaroo Island dunnart), *S. dolichura* (little long-tailed dunnart), *S. gilberti* (Gilbert's dunnart) and *S. griseoventer* (grey-bellied dunnart).

By the late 1980s, the dominance of dasyurid systematics by morphological character analysis was ending. This is not to say that anatomical studies were no longer conducted or significant (quite the opposite), but that molecular phylogenetic analyses became more common and had greater impact after 1990. In addition to those already mentioned, several new species were described in the 1980s, including *Antechinus leo* (cinnamon antechinus; Van Dyck 1980), *Ningaui yvonneae* (southern ningaui; Kitchener *et al.* 1983), *Sminthopsis archeri*; Van Dyck 1986) and *Dasyurus spartacus* (bronze quoll; Van Dyck 1987). In his encyclopaedic review of Australian marsupial evolution, Archer (1984c) modified his syncretic 1982 dasyurid classification only by placing *Planigale* within Sminthopsinae as a distinct tribe (Planigalini, sister to Sminthopsini) based on then-unpublished serological data. Thus, intense study of morphological traits with geographically uneven sampling of specimens over some 200 years, along with scant input from serology and allozymes, yielded the species inventories of Mahoney and Ride (1988) and Flannery (1990), which we embed in Archer's (1984c) classification (Table 3.5).

ENDNOTES

1 Parrish quotes from the 1555 translation of *De Orbe Novo Decades* (Decades of the New World) by Richard Eden. The more accessible 1912 translation by F.A. MacNutt, in modern English, is less picturesque.

2 Edward Tyson (1651–1708), a London physician, was an important figure in the origin of comparative anatomy as a discipline distinct from human anatomy. In addition to his work on the opossum, Tyson provided some of the earliest detailed anatomical descriptions of a porpoise, rattlesnake, tapeworm, roundworm, peccary and chimpanzee (Cole 1944).

3 Marsupials played a part in illustrating Cuvier's 'correlation of parts' and 'subordination of characters' principles – the claims that the anatomy of an entire animal, and hence its taxonomic affinities, could be inferred from the structure of a few key parts. Moore (1993) relates the story of a 'public test' in which Cuvier predicted that a slightly exposed mammalian fossil would prove to be a marsupial rather than a placental based on his knowledge of correlated skeletal structures. Subsequent extraction of the entire fossil from its matrix revealed epipubic bones, dramatically validating his principles for the onlookers (see Cuvier 1804).

4 *Sminthopsis granulipes* Troughton 1932 was apparently first described as *Podabrus albocaudatus* by Krefft (1872), but the latter epithet has become a disused synonym (*nomen oblatum*) (Parnaby *et al.* 2015).

4 Dasyurid phylogeny: molecular systematics, 1977–2003

In the opening lines of what may be the first work on molecular phylogenetics, parasitologist G.H.F. Nuttall made a simple observation: 'In the absence of palaeontological evidence the question of interrelationship amongst animals is based upon similarities of structure in existing forms. In judging of these similarities the subjective element may largely enter …' (Nuttall 1904: 1). The centuries-old problem of distinguishing affinity from adaptation, of homology from analogy, of synapomorphy from homoplasy, had never been resolved. Consequently – and 'palaeontological evidence' notwithstanding – knowledge of species relationships was tentative at best. To improve this situation, Nuttall proposed inferring relationships from degrees of similarity in blood proteins, which he measured with an immunological technique (see below). Why should this work? '[T]he rationale of this approach and of biochemical systematics in general remains the same: that the more closely related organisms are, the more similar they ought to be chemically, as they usually are in other characters' (Kirsch 1977: 3). This phenetic argument has only weak validity by modern standards, but it was tolerated for molecular analyses much longer and more indulgently – even after Hennig – than for morphological ones. What was so special about molecules?

The decades following Nuttall's book saw the integration of Darwinian theory with Mendelian genetics to form the Modern Synthesis in evolutionary biology (Huxley 1942), the essence of which was reconciling classic Darwinian phenomena such as phenotypic variation, selection and speciation with principles of Mendelian heredity. For many, this entailed a reductionist valuation of genetic over phenotypic change as the 'real' substance of evolution. George Gaylord Simpson, a principal architect of the Synthesis, offered a lucid perspective in his paper on the classification of mammals:

> *The most direct, but unfortunately not the most useful, approach to phylogeny of recent animals is through their genetics. The stream of heredity makes phylogeny; in a sense, it is phylogeny. Complete genetic analysis would provide the most priceless data for the mapping of this stream, although it would only exceptionally provide unequivocal and conclusive criteria … [G]enotypic similarities and dissimilarities have to be interpreted in much the same way as phenotypic likenesses and differences, and the advantage of genetics lies in the fact that genes (and some other genetic factors) are the immediate physical continuants of phylogeny, while morphology is less direct, a result of these hereditary factors as modified by other influences. [Simpson 1945: 5]*

In retrospect, this statement is extraordinarily perceptive. Indeed, the criteria by which genetic characters are 'interpreted' is one of the major themes in molecular phylogenetics – a series of developments mirrored by the history of dasyurid systematics, as we will see. At the outset, however, enthusiasm for molecular data drew on the perception of its more fundamental nature and a somewhat vague notion that it allowed objective (or at least mitigated subjective) assessment of homology. The rationale for the latter argument is as follows: Proteins are sequences, often rather long ones, of 20 different amino acids. If the sequence of a 100-residue protein from species A is identical to that of a protein from species B at 99 of 100 residues, it is reasonable to assume that the proteins are homologous because ancestry is a more probable explanation of such high similarity than is convergence. As similarity declines, so does the case for homology, although no absolute lower bound can be specified. The 'sequence similarity reveals homology' assertion would be severely qualified as knowledge of molecular evolution improved later in the 20th century and Simpson's was one of but few voices tempering early enthusiasm for it.

Because 'complete genetic analysis' – however Simpson may have construed the phrase – was beyond reach in 1945, the hegemony of morphological traits in systematics would not be challenged for decades. Simpson singled out comparative serology, as pioneered by Nuttall (1904) and reviewed by Boyden (1942), as a molecular way forward: '[S]ystematic serology seems likely to become a powerful adjunct in the study of phylogeny' (Simpson 1945: 6). Over 30 years later, it would be the first chapter in the story of marsupial molecular systematics.

MOLECULAR STUDIES BEFORE DNA SEQUENCES

J.A.W. Kirsch and comparative serology

John A.W. Kirsch (1941–2007) was an American mammalogist who completed his PhD in Zoology at the University of Western Australia under W.D.L. Ride (Director of the WA Museum) in 1968, his dissertation forming the basis for 'The comparative serology of Marsupialia, and a classification of marsupials' (Kirsch 1977). At the risk of oversimplification, comparative serology involves measuring the similarity of blood plasma proteins between two species by gauging the relative strength of antigen-antibody binding (see Maxson and Maxson 1990 for a detailed explanation). In these early studies, plasma proteins were used *in toto* as antigens (i.e. the entire collection of proteins in a plasma sample constituted 'the antigen'; they were not isolated as individual chemical species). Antibodies were raised by injecting a plasma aliquot from one species into a reference vertebrate (e.g. rabbit), allowing the infected host to mount an immune response and then harvesting antibody-containing plasma (antiserum). Estimating the immunological similarity of two species A and B involved comparing the binding affinity of A antiserum for B antigen, relative to its affinity for A antigen, and vice versa. Because antigen-antibody recognition is roughly proportional to amino-acid sequence complementarity between antigen and antibody, the 'heterologous' (anti-A × B) reaction is weaker than the 'homologous' (anti-A × A) reaction. Moreover, antigen protein sequences of different species diverge in proportion to time since the species' common ancestor, so the strength of heterologous binding declines as divergence time increases. Early serologists measured binding affinity by quantifying the strength of the 'precipitin reaction' between antigens and antisera (Boyden 1942); later practitioners employed technologies such as Ouchterlony immunodiffusion, immunoelectrophoresis, nephelometery and haemagglutinin inhibition (all used by Kirsch 1977).

The results of serological comparisons were, for the most part, pairwise coefficients of similarity assembled in a species × species matrix for systematic analysis. Kirsch's (1977) analysis took the form of phenetic clustering to produce trees on which *maximally similar* species or groups of species are sisters. For phenetics to generate good estimates of phylogeny, similarity values must be *ultrametric* – i.e. the experimental measure is sufficiently accurate and precise to behave as a Euclidean distance, and the biological feature it measures changes at a uniform rate in all species. Kirsch examined these assumptions critically and found reasons why each could be questioned but not rejected. He emphasised the correspondence between his serological-phenetic results and those from other studies, especially morphology, in evaluating phylogenetic relationships and constructing an evolutionary (but not cladistic) classification.

Kirsch's (1977) primary goal was an assessment of overall marsupial relationships, so his coverage of dasyurids was limited to 19 species in eight genera. Aside from the position of *Myrmecobius*, his serological phenogram (Fig. 4.1) is fully consistent with previous trees from morphology (e.g. Tate 1947; Archer 1976b, c) and informed Archer's (1982c) syncretic phylogeny. Kirsch discounted the link of *Myrmecobius* to sminthopsines as an artifact

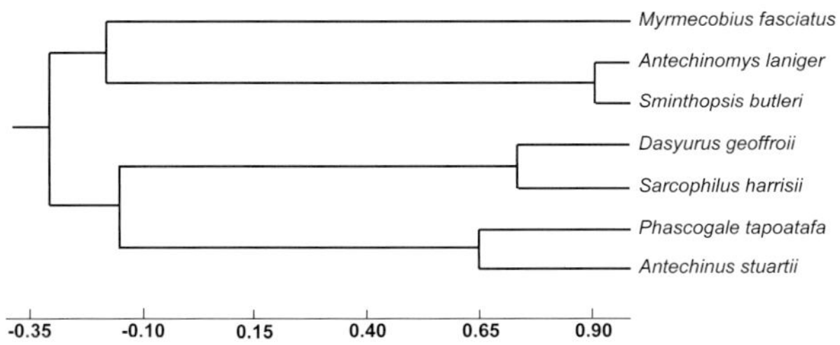

Fig. 4.1. Kirsch's (1977) phenogram of dasyurids based on immunoelectrophoretic comparisons. The tree was produced by unweighted pair-group analysis using averages (UPGMA) applied to a correlation matrix (horizontal scale) derived from strength-of-reactions scores. [Reprinted by permission of CSIRO Publishing from Kirsch (1977). Permission conveyed through Copyright Clearance Center, Inc.]

of slow evolutionary rates in numbat proteins. His summary comments on dasyurids were prophetic:

> *The genera of Dasyuridae which were studied appear to fall into three groups of closely related genera, although Antechinus at least may represent more than one genus, with species in two of the genus-groups [Kirsch's sample included 'A.' macdonnellensis and 'A.' rosamondae] … [T]he affinities of the New Guinean dasyurids [which Kirsch did not sample], when they are better known, will very likely require considerable reassessment of Dasyuridae … [Kirsch 1977: 119]*

Kirsch (1982) revisited the relative merits of molecular and morphological characters in his contribution to the *Carnivorous Marsupials* volumes. He invoked early evidence for 'molecular clocks' that would approximate the rate uniformity required for similarity to track phylogeny and concluded that '[b]iochemical characters … offer perhaps the best hope for a direct and unequivocal reconstruction of historical relationships' (p. 593). In his view, morphological traits are more prone to phylogenetically misleading convergence driven by natural selection, in contrast to what appeared at the time as a predominantly neutral mode of molecular evolution. Indeed, Kirsch and Archer (1982) could not recover any but the most obvious relationships among dasyurid taxa using numerical cladistic analyses (the 'vulgar parsimony' of Kirsch 1982) applied to craniodental characters. Interestingly, their conclusion (encapsulated in their chapter title) was that 'parsimony's not enough'. Despite Kirsch's characteristically thoughtful and clear prose, the takeaway message from these chapters is ambiguous – was it the characters, the method of analysis, or both that made early dasyurid cladograms inconsistent?

Allozymes

From the time of Darwin until the Modern Synthesis, evolutionary biologists observed phenotypic variation in natural populations and theorised about the genetic variation underlying it. That changed forever when Hubby and Throckmorton (1965) applied the technologies of protein electrophoresis and histochemical staining to *Drosophila* populations, opening the door to a plethora of studies documenting genetic polymorphism in the wild. The high levels of variation often found inspired Kimura's (1968) neutral theory of molecular evolution, which predicted molecular clocks, which in turn would partially validate genetic similarity as a guide to phylogeny (Kirsch 1982). This technique uses electrophoresis to separate enzyme proteins from a tissue sample based on their charge and size, then exploits the specific reactions they catalyse to visualise them. The method lends itself to screening many loci and organisms simultaneously, resulting in identification of the enzyme alleles (*allozymes*) at specific loci in samples of individuals from multiple species. Nei (1972) formulated a neutral model of allozyme divergence between populations, and a genetic distance based on it, that was widely used in phylogenetic studies. Other approaches treated loci or alleles as characters for cladistic analysis or used alternative distance measures (Swofford and Olsen 1990). Compared to serology, researchers knew exactly which proteins were being assayed and the contribution of each to genetic divergence.

In Australian mammalogy, P.R. Baverstock, M. Adams and B.J. Richardson were pioneers of allozyme research and used the technique to study bats, rodents and marsupials in the 1970s and early 1980s. Baverstock *et al.* (1982) scored 32 loci from 33 Australian dasyurid species, subspecies, or populations. They inferred trees with phenetic clustering, 'distance Wagner' analysis (an algorithm for finding the minimum length tree from pairwise distances) and qualitative Hennigian analysis (assuming, as a criterion for polarising characters, 'that the primary dichotomy was between *Sminthopsis, Antechinomys,* and *Ningaui* on the one hand, and all other genera on the other', p. 642). The details of each analytical approach were not clearly articulated, and the resulting trees were not identical. The authors proposed the phylogeny in Fig. 4.2 as a consensus between their Wagner and Hennigian results, noting: 'The genetic data suggest at least three major groups among

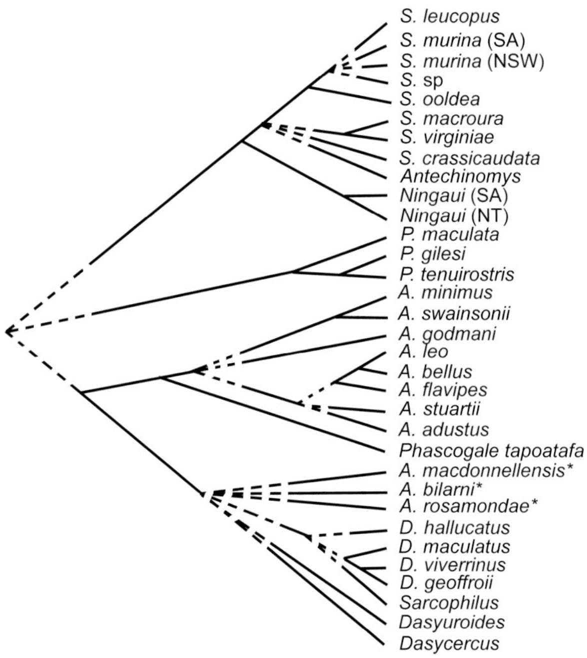

Fig. 4.2. Baverstock *et al.*'s (1982) 'proposed cladistic relationships' among selected Australian dasyurids based on allozyme data. The tree is a subjective summary of consensus between a distance Wagner tree and a cladogram derived from qualitative Hennigian analysis, each based on 32 loci. Asterisks indicate 'false antechinuses'; broken lines indicate areas of uncertain branching order. [Reprinted by permission of the Royal Zoological Society of New South Wales from Baverstock *et al.* (1982).]

Australian dasyurids tested – the *Sminthopsis* suprageneric group (*Sminthopsis, Antechinomys,* and *Ningaui*), the *Planigale* group (containing only *Planigale*), and the *Antechinus* suprageneric group (*Antechinus, Phascogale, Dasyuroides, Dasycercus, Dasyurus,* and *Sarcophilus*)' (p. 646). As indicated by the dashed branches in Fig. 4.2, many relationships within these groups were uncertain. This analysis also contributed to Archer's (1982c) syncretic classification, particularly the separation of 'parantechinins' (*Parantechinus, Pseudantechinus* and *Dasykaluta*) from *Antechinus* and the distinctness of planigales.

Microcomplement fixation

One source of concern with similarity measures from comparative serology was that multiple blood proteins contribute to antigen-antibody reactions, making the link between 'immunological distance' (*ID*) and protein sequence variation something of a black box. Less so was the immunological method of microcomplement fixation (MC′F), which compares single homologous proteins, such as albumin, from different species. Introduced into phylogenetics by Sarich and Wilson (1966), the method was subsequently put through validation procedures that documented a correlation between appropriately estimated MC′F distances and protein sequence differences, as well as a clock-like relation between distance and species divergence times estimated from the fossil record. Maxson and Maxson (1986) proposed a mechanistic model linking *ID* to amino acid differences between proteins. By 1990, the black box had largely been illuminated and an albumin immunological clock seemed plausible.

During the 1980s, Baverstock and colleagues applied albumin MC′F extensively to various marsupial groups. In a summary of those studies, Baverstock *et al.* (1990: 131) noted the widespread inconsistency of results from previous morphological analyses: 'These very discrepancies should act as a warning to treat with some suspicion all available phylogenetic hypotheses for marsupials until they have been subjected to rigorous cladistic treatment and cross-checking against independent data sets'. MC′F data certainly provided a crosscheck, not only because albumin is an 'independent data set', but also because Baverstock *et al.* used a non-phenetic optimality criterion – the method of Fitch and Margoliash (1967) – to estimate their trees. The Fitch-Margoliash method selects the tree among all possible trees on which branch lengths can be assigned to minimise the weighted, sum-of-squares deviation of path lengths between pairs of tips from measured pairwise distances (see 'Model-based phylogenetic inference' below). As such, it does not assume evolutionary rate uniformity among species and does not require that maximally similar taxa be sisters (Swofford and Olsen 1990). Moreover, Baverstock *et al.* were the first marsupial systematists to assess statistical confidence in their phylogenetic results, applying Lanyon's (1985) jackknife resampling method. Here one taxon at a time is deleted from the distance matrix, a tree inferred from those remaining and the collection of '*n* –1' trees used to form a 'jackknife strict consensus' showing only clades that occur in all resampled trees. The significance of these analytical innovations cannot be overemphasised: the first freed distance-based phylogenetic inference from the Achilles' heel of phenetics and the second freed systematists from agonising over estimated relationships not strongly supported by the data. Arguably, much of the inconsistency to which Baverstock *et al.* (1990) alluded was from precisely these sources.

Baverstock *et al.*'s (1990) MC′F phylogeny of dasyurids (Fig. 4.3) confirmed most of the emerging consensus

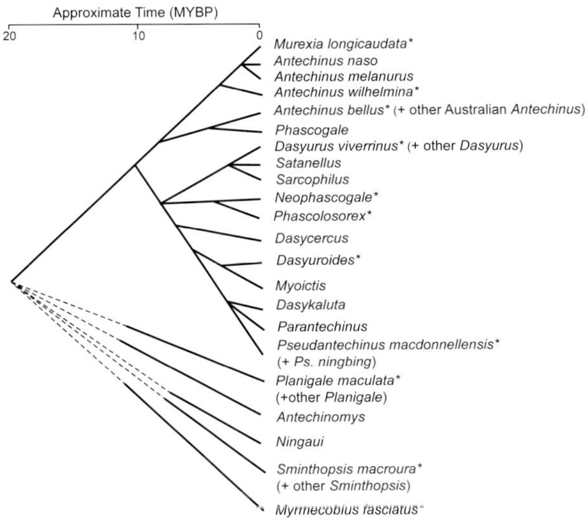

Fig. 4.3. Phylogeny of dasyurids based on albumin MC′F (Baverstock *et al.* 1990). The tree was inferred using the Fitch-Margoliash (1967) distance method. Polytomies indicate uncertainty in branching order revealed by jackknife resampling (Lanyon 1985). The timescale is based on the rate calibration of Maxson *et al.* (1975); asterisks indicate taxa to which antisera were raised; dashed lines indicate uncertain divergence times. [Reproduced with permission of CSIRO Publishing from Baverstock *et al.* (1989), with minor adaptations. Permission conveyed through Copyright Clearance Center, Inc.]

about major Australian clades, including the deep separation of sminthopsines from dasyurines, the distinctness of planigales and the association of 'parantechinins' with dasyurines. Most importantly, however, this was the first molecular study to include New Guinean species and their position in the tree could hardly have been more surprising. For example, the supposedly primitive *Murexia* formed a clade with New Guinean 'antechinuses', to which the Australian phascogalins were sister. *Phascolosorex* and *Neophascogale* were firmly embedded among dasyurins. Archer (1982c: 438) had placed *Murexia* and *Phascolosorex* + *Neophascogale* in distinct subfamilies, Muricinae and Phascolosoricinae, respectively. Moreover, he argued of Muricinae that 'its dental and basicranial morphology leave little doubt that it is distinct from other dasyurids at a fundamental level', and that Phascolosoricinae 'may be as fundamentally distinct within the Dasyuridae as is the Muricinae'. Yet the independent perspective provided by albumin MC′F strongly contradicted the hypothesis of early separation of these genera from the main body of dasyurids; rather, their constituent species were nested within Australian clades. Kirsch's (1977) prediction had come true, but the 'considerable reassessment' he anticipated would be protracted and contentious.

An important part of the MC′F tree is its timescale, which Baverstock *et al.* (1990) obtained using a marsupial-specific divergence rate proposed by Maxson *et al.* (1975), in which the divergence time (T, in millions of years or Ma) of two species scales as 60% of the albumin immunological distance (D) between them: $T = 0.6D$. The dasyurid distances suggested a much more recent diversification of dasyurine genera (ca. 10 Ma) than sminthopsine ones (ca. 20 Ma). Are dasyurines a much younger radiation than sminthopsines, or was something amiss with the albumin clock? We will return to this question in Chapter 10. In addition, MC′F suggested a deep divergence between *Antechinomys* and *Sminthopsis*, contrary to Archer's (1981) conclusion that they are congeneric. With the publication of Baverstock *et al.* (1990), molecular systematics had finally been applied to dasyurids with sufficient taxon sampling and analytical rigor to be more than vaguely confirmatory of past morphological conclusions. It was also the last major study based on protein variation; the future belonged to DNA.

DNA hybridisation

During the 1980s, ornithologists Charles Sibley and Jon Ahlquist performed an astonishing *tour de force* in systematic biology: they used the DNA–DNA hybridisation technique to construct a phylogeny of birds (virtually all of them, down to the genus level). This body of work and the comprehensive tree it produced (widely known as 'the tapestry') culminated in *Phylogeny and Classification of Birds: A Study in Molecular Evolution* (Sibley and Ahlquist 1990). Such an undertaking was necessarily accompanied by a high degree of confidence that it would produce results sufficiently robust to justify the incredible effort and expense involved. Indeed, no less a commentator than Stephen Jay Gould suggested, 'Perhaps we should mount the parapets and shout: "The problem of phylogeny has been solved"' (Gould 1985: 25). Gould's enthusiasm derived from his view that DNA hybridisation solved the problem of phylogeny by solving the problem of homology:

> *DNA differs from morphology in just those properties that capture homology. DNA is complex enough to preclude analogy as a cause of overall similarity. The items (genes) of a DNA code are also sufficiently independent to guarantee that no simple trigger can engender a large set of apparently intricate similarities. Each gene contains hundreds of*

> *nucleotides, each capable of separate change in evolution. Such a long string of independent items cannot evolve detailed similarity from initial disparity. A high percentage of item-by-item match in the DNA of two species must represent homology and shared descent. [Gould 1985: 16]*

As it turned out, this early 1980s view of molecular homology was too simple (e.g. it did not fully consider the complexities of gene duplication) and Gould was too quick to equate the apparent solution of one problem (homology) with the solution of another (phylogeny). His essay, however, gave timely expression to the optimism felt by molecular systematists that molecules in general, if not DNA hybridisation in particular, would lead to definitive phylogenetic inferences.

Britten and Kohne (1966) first used a basic DNA-DNA hybridisation technique to compare the genomes of rodent species. The method is based on the empirical observation that the temperature at which two strands of DNA dissociate from their duplex (double-helical) conformation into separate single strands is proportional to the complementarity of their nucleotide sequences: perfectly complementary sequences dissociate (melt) at a relatively high temperature and progressively less complementary sequences melt at lower temperatures. Thus, a measure of melting point depression is an estimate of sequence dissimilarity, or distance. Sibley and Ahlquist applied the technique to the entire single-copy fractions of avian genomes (i.e. excluding repetitive sequences) to obtain an 'average' distance between them. In each comparison, one species served as a reference: this perfectly complementary DNA hybridised to itself yielded a 'homologous' melting point; the same DNA hybridised to that of a different species gave a 'heterologous' (and lower) melting point. The difference in melting points (ΔT) served as a pairwise genetic distance and a matrix of such distances could be used to estimate phylogeny. Virtually every stage in this chain of inference was scrutinised and criticised during and after Sibley and Ahlquist's research. However, given several significant but plausible assumptions, the rationale of DNA hybridisation is sound (Springer and Krajewski 1989b) and many (though certainly not all) aspects of the avian 'tapestry' have withstood subsequent testing.

In the mid-1980s, John Kirsch set out to apply DNA hybridisation to mammalian phylogeny and one of the first papers to appear was an analysis of dasyurids (Kirsch

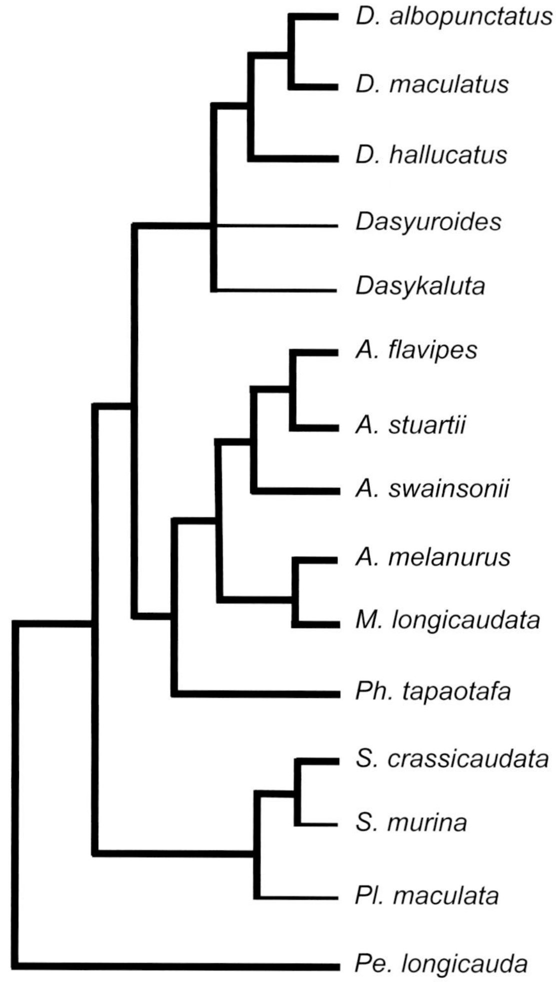

Fig. 4.4. 'Suggested cladogram' based on DNA hybridisation from Kirsch *et al.* (1990). Thick branches connect taxa that were used for both homologous and heterologous hybridisation measurements; thin lines show taxa only used in heterologous measurements. Genus abbreviations: *A.* = *Antechinus*, *D.* = *Dasyurus*, *M.* = *Murexia*, *Pe.* = *Peroryctes* (an outgroup bandicoot, now *Microperoryctes*), *Ph.* = *Phascogale*, *Pl.* = *Planigale*, *S.* = *Sminthopsis*. Branch lengths are unscaled. [Reprinted by permission of CSIRO Publishing from Kirsch *et al.* (1990). Permission conveyed through Copyright Clearance Center, Inc.]

et al. 1990). Kirsch's group made several modifications of Sibley and Ahlquist's analytical procedure, including 'corrections' to mitigate the effects of experimental error on ΔT measurements. Following Springer and Krajewski (1989b), they also used Jukes and Cantor's (1969) stochastic model of sequence evolution to transform ΔT values into distances that more accurately estimate numbers of DNA substitutions, including multiple 'hits' at the same site that are not recorded as differences in hybridised sequences (see 'Models of sequence evolution' below).

This moved the distances closer to a mathematical condition of additivity, an assumption of the Fitch-Margoliash and similar tree methods. Springer and Krajewski (1989a) had shown that a matrix of additive distances could be fit perfectly to a set of branch lengths only on the tree from which they were derived, thereby demonstrating that 'additive tree methods' are robust to rate variation. Because the data suggested experimental error in measuring ΔT was independent of distance magnitude, Kirsch *et al.* (1990) used an unweighted version of the Fitch-Margoliash method to produce the dasyurid phylogeny in Fig. 4.4. To assess confidence levels associated with each clade, they used both jackknife resampling of taxa and bootstrap resampling of replicate ΔT measures (Krajewski and Dickerman 1990). The latter approach was specifically designed to assay the effect of experimental measurement error on branch support.

As far as it goes, the DNA hybridisation phylogeny was both a validation of MC'F results regarding *Murexia* (which, with the New Guinean '*Antechinus*' *melanurus*, is sister to *Antechinus*) and aspects of Archer's (1982c) hypothesis (see Fig. 3.9). Kirsch *et al.* (1990) lacked DNA from several key genera, including *Phascolosorex*, *Neophascogale*, *Myoictis*, *Antechinomys* and *Ningaui*, so their results perhaps raised more questions than they answered. However, their tree prefigured the basic structure of all later molecular phylogenies of dasyurids: a basal separation of sminthopsines (*Sminthopsis* and *Planigale* in Fig. 4.4) and dasyurines, as well as distinct clades of (what are now) dasyurins and phascogalins within the latter.

DNA SEQUENCE ANALYSES

PCR and direct sequencing of DNA

Comparative serology, allozyme electrophoresis, MC'F and DNA hybridisation are roundabout ways of measuring sequence similarity, but the direct approach – sequencing homologous molecules from multiple species – was not feasible until the late 1980s. Although biochemical methods for sequencing proteins (Edman *et al.* 1950) and nucleic acids (Maxam and Gilbert 1977; Sanger *et al.* 1977) were available, the latter relied on prior isolation and amplification of the target locus (usually with probing and cloning technology) that was too expensive and laborious for large taxonomic surveys. All that changed after Kary Mullis developed the polymerase chain reaction (PCR, first reported in Saiki *et al.* 1985). PCR exploits prior knowledge of DNA sequences in regions flanking the locus of interest; short oligonucleotide primers will bind to these regions when duplex DNA is denatured and allowed to reanneal. A thermostable DNA polymerase then extends the primers using the target sequence as a template. Multiple rounds of denaturation, renaturation and extension lead to an exponential increase in copies of the target sequence sufficient for direct input into Sanger sequencing (Innis and Gelfand 1990). As commercial primer synthesis, amplification protocols, thermocycling machines and sequencing kits became available and affordable during the early 1990s, direct sequencing of DNA was within the grasp of molecular systematists. A proliferation of studies ensued and phylogeneticists were quick to capitalise on technological improvements that streamlined the process.

Among its other advantages, PCR can amplify DNA templates that are initially present at very low concentration and can recover short sequences that may remain after degradation of a high molecular-weight sample. Because of these properties, PCR applications in forensic and ancient DNA analysis are almost as old as the technique itself. Indeed, one of the first ancient DNA reports focused on the phylogenetic affinities of the thylacine. Thomas *et al.* (1989) extracted DNA from dry muscle tissue attached to a

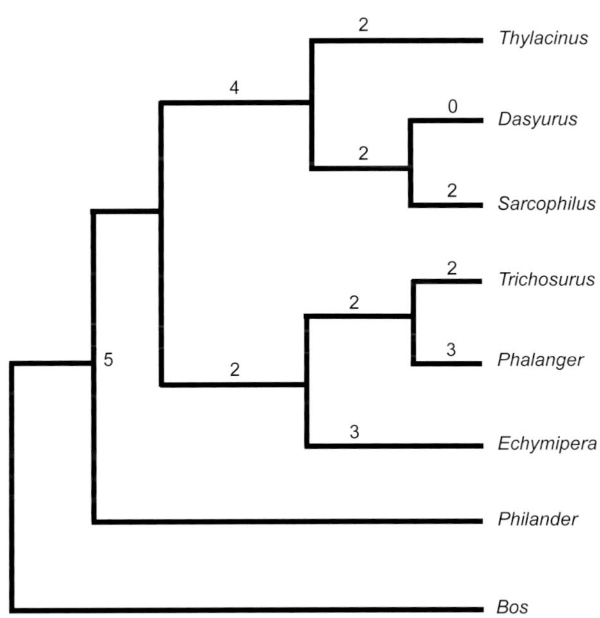

Fig. 4.5. The first phylogeny based on DNA sequences to include dasyurids showcased ancient DNA from a thylacine museum specimen (Thomas *et al.* 1989). The tree is a bootstrapped parsimony consensus based on 94 bp of the mitochondrial 12S rRNA gene. Numbers on branches are minimum counts of inferred base changes. [Redrawn with permission of Springer Nature from Thomas *et al.* (1989).]

thylacine museum specimen and used PCR to amplify short fragments of the mitochondrial cytochrome *b* (*cytb*) and 12S ribosomal RNA (*12S*) genes. They applied DNA parsimony analysis (Fitch 1971) to 94 basepairs (bp) from *12S* to test Bensley's (1903) contention that thylacines formed a clade with extinct South American borhyaenids. Their results (Fig. 4.5) showed that *Thylacinus* was sister to a dasyurid clade nested among other australidelphians and separate from the single didelphid examined (*Philander*, a proxy for borhyaenids). This was consistent with a previous MC'F analysis (Sarich *et al.* 1982); but, as recently as the *Carnivorous Marsupials* volumes, Archer (1982a) had suggested that skeletal traits alone could not rule out thylacine-borhyaenid monophyly (see Chapter 8). If such an apparently decisive result was obtained from 94 bp, what might we expect from the much longer sequences of modern DNA?

Statistical phylogenetics

Before delving into dasyurid relationships based on DNA sequences, we must pause to understand something of the revolution that took place in analytical phylogenetics during the 1980s and 1990s. Because of its reliance on firm hypotheses of character evolution, and the relative rarity of such hypotheses, Hennigian argumentation had largely been supplanted by numerical parsimony analyses around the time the earliest dasyurid molecular results appeared. Instead of purging a data matrix of homoplastic characters à la Hennig, parsimony methods sought to find the tree (among all possible trees) that required the fewest state changes for all characters scored, in a sense 'minimizing hypotheses of homoplasy'. This maximum-parsimony (or minimum-length) tree would then be the best phylogenetic explanation of the data (Box 4.1). Justification for the parsimony criterion crystalised around the arguments of Farris (1983) identifying cladistics within the hypothetico-deductive epistemology of Popper (1959), a framework popularly equated with 'the scientific method'. In contrast, Felsenstein (1983) placed phylogenetic inference in the context of statistical estimation: phylogeny is an unknown parameter value that must be estimated from available data by some rule that connects the data to the processes that generated them. Since character evolution is complex and historically contingent, it is understood as a stochastic process, features of which must be captured by an appropriate stochastic model. The usefulness of a rule for converting data into a parameter estimate is evaluated by considering its performance as a statistical estimator. Statistical phylogenetics focused on two such criteria, consistency and maximum likelihood.

Box 4.1. Parsimony

Hennig's 'auxiliary principle' (see 'Molecules and morphology: the problem of *Murexia*' below) is an *a priori* assumption that derived characters shared by species are more likely to be synapomorphies (due to common ancestry) than homoplasies (due to convergence). If the assumption holds, then synapomorphies should, on average, outnumber homoplasies as characters evolve on a phylogeny. The caveat 'on average' is important – any finite character set might, by chance or bias in trait choice, include more homoplasies than synapomorphies. The auxiliary principle describes the properties of very large (tending to infinite) samples of characters, or to a very large number of smaller character samples. The expectation that synapomorphies outnumber homoplasies suggests a numerical implementation of Hennig's logic: find the phylogenetic tree, among all possible trees for a set of species, that requires the fewest homoplastic character-state changes. An ideal Hennigian synapomorphy is one in which the derived state of a character evolves just once in an ancestor (i.e. an internal branch of the tree) and is inherited by all descendants of that ancestor; homoplasies require at least two state changes. Tree topologies constrain the minimum number of changes ('steps') required for a character with a particular state distribution. For example, if a character has distribution 011 (with 0 the ancestral state) for species A, B and C, tree A(BC) requires one evolutionary step (0 → 1 in the BC ancestor), whereas trees (AB)C and (AC)B require two (e.g. 0 → 1 in the terminal branches leading to both B and C).[1] This *minimum length* optimality criterion generalises to unpolarised characters (i.e. those for which ancestral states are unknown) mapped onto unrooted trees with four or more species. In this context, 'minimizing homoplastic changes' is the same as minimising (all kinds of) changes. A character distributed 0011 for species A–D requires only one change on tree (AB)(CD), but at least two on trees (AC)(BD) and (AD)(BC), and therefore counts as evidence favouring (AB)(CD). If we assume that each character evolves independently, then our optimality criterion becomes, 'find the tree among all possible trees that requires the fewest state changes in *all* characters'. This is cladistic parsimony. Fitch (1971) developed an algorithm for determining the minimum number of changes required for a particular state-distribution on a particular tree and

> 'Fitch parsimony' has been implemented by various computer programs that analyse multiple characters – including multistate ones such as DNA sequences – and search among all possible trees. The latter becomes computationally intractable for more than about 20 species because the number of possible trees increases to astronomical levels as more species are added (Felsenstein 1978). Thus, most parsimony (and other) tree-searches are heuristic (i.e. not guaranteed to find the shortest, or otherwise optimal, tree).

Roughly speaking, a consistent estimator is one that converges on the true parameter value as the sample size of observations increases. For example, the sample mean converges on the mean of the population sampled as the number of sampled scores increases. An inconsistent estimator is biased – it converges on something *other than* the true parameter value as sample size increases. (Imagine using the sample mean + 1 to estimate the population mean.) In phylogenetics, one can ask whether an estimation rule (like parsimony) is generally consistent, or under what conditions it is inconsistent. Does it converge on the true tree as more and more characters are sampled? The likelihood (L) of an estimated parameter value (P) is the probability of observing the data (D) if P were the true value in whatever process model (M) is assumed to have generated the data. This is often expressed with algebraic shorthand: $L = \Pr(D|P, M)$, where the vertical bar (|) denotes conditional probability (i.e. this equation can be read as 'likelihood equals the probability of the data *given* the parameter value and the process model'). A good rule of estimation should return a value of P that maximises L. Or, to say the same thing differently, choosing the value of P that maximises L is a good rule of estimation. For example, if a coin flipped 100 times comes up heads 55 times, the maximum likelihood (ML) estimate of P = probability of heads is 0.55 because, according to the binomial process model of coin-flipping, 0.55 is the value of P that maximises $[100!/(65!)(100-65!)]P^{65}(1 - P)^{100-65}$. One way of understanding the ML criterion is that it makes the best use of the available data in estimating P, whereas consistency is a desirable asymptotic property with increasing sample size. In phylogenetics, the parameter value is a tree and the data are character distributions, but what is the process model?

Models of sequence evolution

As noted above, Kirsch *et al.* (1990) used the Jukes and Cantor (1969) model of sequence evolution to 'correct' measured distances for unmeasurable divergence caused by multiple base changes at the same site. Suppose that homologous DNA sequences of 1000 bases from species A and B are identical. Does that mean that *no substitutions* have occurred in the lineages leading to A and B since their common ancestor? It might, or it might be the result of *both* sequences experiencing the *same substitutions at the same sites*. The relative probabilities of these two scenarios depend on the rules by which the sequences evolve. Of course, the rulebook of sequence evolution is complicated and few of its strictures are hard and fast. For morphological characters, it seems to be utterly idiosyncratic, to the point that there may be no rules (and precious few regularities) at all. DNA and protein sequences are another matter. From the earliest comparative sequence data sets, researchers noted recurring patterns of change and used these patterns to refine stochastic models of sequence evolution. After decades of such work, the repertoire of models is quite diverse, as are the options for deploying them in phylogenetic inference. Baum and Smith (2013) provide a readable introduction to this topic.

Sequence models are usually described as elaborations of the Jukes-Cantor (JC) scheme just mentioned (Fig. 4.6).

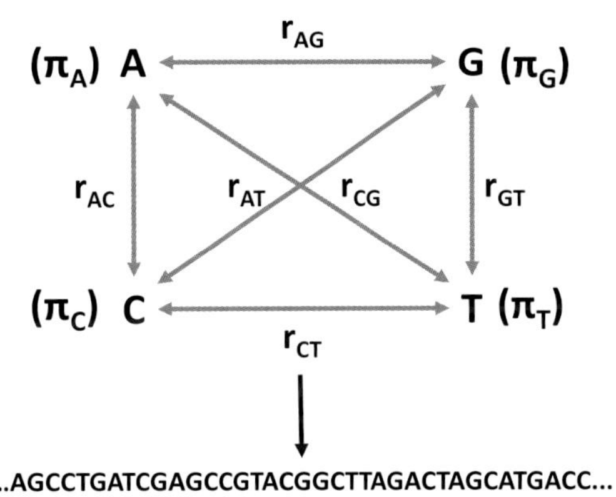

Fig. 4.6. Framework for most stochastic models of DNA sequence evolution. At top, A, G, C and T represent possible nucleotides at a single position in a sequence, and r_{ij} values are substitution rates between each pair of nucleotides. Double-headed arrows indicate that rates are symmetric (e.g. A → G and G → A occur at the same rate, r_{AG}). A portion of a sequence is shown at bottom, with a single-headed arrow showing that the rate model is being applied to one position at a time. Values of π are the equilibrium frequencies of each base in the evolving sequences.

The JC model specifies that all pairwise substitutions of one nucleotide for another at a given site in a sequence take place at the same rate and that rate is the same for all sites in the sequence, which has the consequence that all four nucleotides will be present at equal frequencies (all $\pi_i = 0.25$) in all sequences. The rates r_{ij} are stochastic, expressing the probability that one base will change to another in a unit of time. JC is a model of mathematical convenience, and few real sequences evolve so simply. One of the earliest elaborations of the model was inspired by the observation that transitional substitutions (i.e. those between purines A $\leftrightarrow$ G or pyrimidines C $\leftrightarrow$ T) occur more frequently than transversional ones (i.e. purine $\leftrightarrow$ pyrimidine). Kimura's (1980) two-parameter (K2P) model set the horizontal (transition) rates in Fig. 4.6 equal to one another (i.e. $r_{AG} = r_{CT}$) and potentially different from the vertical (transversion) rates, all of which were equal. A second observation was that most real sequences have unequal base frequencies; Felsenstein's (1981), or F81, model weights each substitution rate by the frequency of the target base (e.g. rate of A $\rightarrow$ G = $r_{AG}\pi_G$). Hasegawa *et al.*'s (1985) HKY model includes both unequal transition-transversion rates *and* unequal base frequencies. Allowing all the rates to be different and base frequencies to vary yields the 'general time-reversible' (GTR) model of Lanave *et al.* (1984). If different subsets of sites within the sequence evolve faster or slower than other sites, different sets of rates can be applied to these 'rate categories', usually by a scalar multiple of the slowest rate category (e.g. first codon positions evolve four times faster than second positions). Alternatively, if rates vary among sites but no *a priori* categories are known, the model can be adjusted to include a distribution of rates, usually a discrete gamma distribution, specifying probabilities that any site evolves at a particular rate. These are the main variations of the simple JC model that were used in early dasyurid sequence analyses, but many other versions have been developed. In all of them, the relevant parameters are estimated to maximise the probability of the data. Statistical model-testing procedures then allow a particular model to be matched to a given data set. Swofford *et al.* (1996) provide a more detailed introduction to this topic.

Model-based phylogenetic inference

In the early literature on dasyurid DNA systematics, models of sequence evolution were applied in two broad categories of phylogenetic inference – additive distance and maximum likelihood tree methods. Additive tree methods are exemplified by the FM approach (Fitch and Margoliash 1967) described above, in which the best tree is the one on which branch lengths can be assigned with minimal distortion of observed pairwise distances. This method is statistically defensible if distances are additive or nearly so (and errors in estimating individual distances are independent of one another; Felsenstein 1984); i.e. fitted branch lengths can be assigned on a particular tree topology that add up to the observed pairwise distances. Distances that fail to account for multiple substitutions will not meet this expectation, particularly when many such multiple hits have occurred (e.g. with large divergences). The role of the model is to allow calculation of distances that are additive in expectation by converting observed mismatch values (p = differences/site) into distances (d = substitutions/site). For example, the JC model entails the following distance formula: $d = -(3/4)\ln[1 - (4/3)p]$; other models have more complex distance equations. Saitou and Nei (1987) developed the neighbour-joining (NJ) method, which uses a stepwise algorithm to construct an additive tree from pairwise distances. Although NJ was not often employed for dasyurid analyses, it is the only additive tree method still widely used.

The other model-based approach employed to early dasyurid phylogenies is maximum-likelihood (ML) estimation. We noted above that the ML estimate of a parameter P given data D and process model M is the value of P that maximises $L = \Pr(D|P, M)$. In ML estimation of phylogeny, P is a tree with associated branch lengths. The mathematics of calculating tree likelihoods under the various sequence evolution models are complex (see Swofford *et al.* 1996 for an overview). However, by the end of the 20th century, the vast majority of molecular systematists were persuaded that ML estimation of phylogeny is more powerful and robust than either parsimony or additive distance approaches.

Like any statistical parameter estimate, a phylogenetic tree inferred from character data has some uncertainty that derives from the limited sample of observations (e.g. sites in a sequence) on which it is based. Quantifying this error is a critical part of the inference process. The minimum-length, best-fit, or maximum-likelihood tree is a statistical *point estimate* that is difficult to interpret without some form of *confidence interval*. Just as a sample mean of 2.0 with a standard error of ±0.5 cannot exclude 2.2 as the true population mean, an estimated tree with branching order (AB)C may not exclude A(BC) as the true phylogeny if support for AB is weak. Felsenstein (1985a) proposed the bootstrap method of resampling sites to place confidence limits on estimated trees. Bootstrap and jackknife resampling methods are widely used in situations where observations (e.g. characters) are drawn from an unknown sampling distribution. The frequency of a clade among trees inferred from pseudoreplicate samples of characters approximates its confidence interval, at least with respect to character-sampling error. Systematists generally refer to these frequencies, expressed as percentages, as *bootstrap* or *jackknife support*. The closer they are to 100, the stronger the support. Although there is no absolute lower threshold, values less than 75% probably indicate very weak resolution in the data. Bootstrap analysis has been widely applied with parsimony, distance and likelihood methods.

It perhaps goes without saying that searching for, say, an ML tree under a complex evolutionary model for long sequences from many taxa is computationally intense. Indeed, statistical phylogenetics grew in parallel with increasing computer processor speed, particularly that of desktop computers. Software to implement phylogenetic analyses tapped into this computing power and proliferated during and after the 1970s. The programs most often used in dasyurid studies were Felsenstein's (1993) PHYLIP package, Swofford's (1999) PAUP* and Kumar *et al.*'s (1993) MEGA suite.

Gene trees and species trees

Molecular phylogenetics, particularly when based on analysis of individual protein or DNA sequences, involves a deeper premise than the pattern of nested similarity described by Kirsch (1977). When a phylogenetic tree is obtained, say, from sequences of locus *L* sampled across some group of species, systematists traditionally assumed that branching order among the sequences is a good estimate of relationships among the species from which they came. However, this need not be the case. A homologous segment of non-recombining DNA in the genomes of species within a clade has a tree-like history of mutation, lineage extinction and transmission through speciation events; this history is termed a *gene tree* (see Chapter 1). Phenomena such as recombination, hybridisation and *incomplete lineage sorting* (ILS) can cause a gene tree to differ from the species tree in which it resides. ILS is particularly problematic when the time between consecutive speciation events is insufficient for fixation of a single allele lineage in descendant species and is more likely when effective population sizes are large. Retention of ancestral polymorphism may prevent alleles in sister species from showing *reciprocal monophyly* (i.e. alleles from one species are more closely related to each other than to any allele from the sister species and vice versa). In this scenario, the mutation history of the alleles may not match the speciation history of the taxa. Moreover, gene trees of unlinked loci may differ, depending on how each locus was affected by ILS and other factors. Avise (1994) provides a good introduction to these phenomena. The gene- versus species-tree distinction served to check naïve overconfidence in early dasyurid sequencing results and led investigators to emphasise gene-tree concordance when multiple, independent loci were analysed.

DNA sequence studies of dasyurid family structure: four tribes

Mitochondrial DNA

Before PCR made direct DNA sequencing affordable, animal systematists became familiar with mitochondrial DNA (mtDNA) through widespread application of restriction-site analysis. Mammal mtDNA (Fig. 4.7) is a circular molecule of approximately 16.5 kilobases (kb) that contains 13 protein-coding genes (whose products are involved in mitochondrial respiration), two rRNA subunits (*12S* and *16S*), 23 short transfer RNA (tRNA) genes and a non-coding 'control region' (*CR*, sometimes called the 'D-loop'). This molecule has both practical and theoretical advantages as a

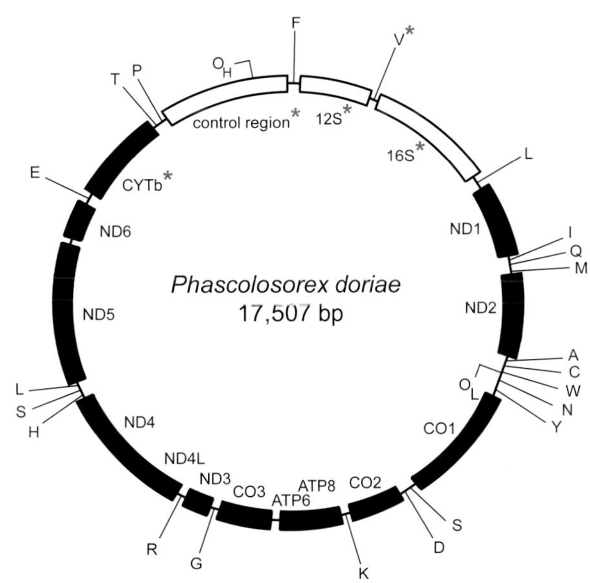

Fig. 4.7. Map of dasyurid (*Phascolosorex doriae*) mitochondrial DNA. Protein coding genes are shaded black, rRNA and control-regions are white, tRNA loci are thin lines indicated by their one-letter amino acid code. Note that ATPase6 and ATPase8 genes overlap by 46 nucleotides. Asterisks indicate the genes most often sequenced in dasyurid studies.

marker for phylogenetic inference. Prior to PCR, it was often physically isolated from the nuclear genome by ultracentrifugation. More importantly, vertebrate mtDNA has the following properties: (1) it is structurally stable (e.g. it rarely experiences recombination that affects gene content or position); (2) it shows almost no allelic variation within individuals (i.e. is homoplasmic); (3) it is maternally inherited as a clonal lineage through the egg cytoplasm; and (4) it evolves rapidly due to lack of a DNA repair mechanism (Moritz *et al.* 1987). Thus, the concept of a single gene tree for the entire mtDNA molecule is generally valid and the mtDNA tree should be especially useful for inferring relatively recent species divergences. Moreover, because mtDNA is haploid and uniparentally inherited, its effective population size is one-quarter that of nuclear loci; this accelerates its rate of genetic drift and make it less likely than a nuclear locus to experience ILS (Moore 1995).

The advantages of mtDNA for phylogenetic inference are partially offset by some disadvantages. One is the tendency for mitochondrial sequence fragments of various sizes to be copied into the nuclear genome where they evolve as pseudogenes, non-functional paralogs of the original mtDNA. First documented in the 1960s, 'nuclear mitochondrial' (numt) pseudogenes are now known to be common in animal genomes (Bensasson *et al.* 2001). In fact, they are more prevalent in the two complete dasyurid genomes published to date (*Sarcophilus harrisii* and *Antechinus flavipes*) than in 68 other vertebrate species examined by Hazkani-Covo (2022). The main problem numts pose for phylogenetics is that caused by any mistaking of paralogs for orthologs in a single analysis – paralogs track a duplication history that is decoupled from the speciation history tracked by orthologs.[2] Numts can sometimes, but not always, be identified by features such as altered base compositions or (for protein-coding sequences) the presence of stop codons or indels. A second limitation of mtDNA, at least in mammals, is that it is more prone to substitutional saturation – and thus loss of phylogenetic resolving power – at deep divergences than is nuclear DNA (Springer *et al.* 2001). For this reason, it is most effective at estimating relatively recent (e.g. intrafamilial) relationships.

Kocher *et al.* (1989) developed a set of 'universal' PCR primers that targeted conserved regions of vertebrate (and a few invertebrate) mtDNAs. Using these and other published primer sequences, Krajewski *et al.* (1992) performed a follow-up study to Thomas *et al.* (1989) on the phylogenetic position of thylacines relative to dasyurids. This study recovered 574 bp of *cytb* from *Thylacinus cynocephalus*, 13 dasyurid species and an outgroup bandicoot; trees were estimated with distance (Fig. 4.8) and

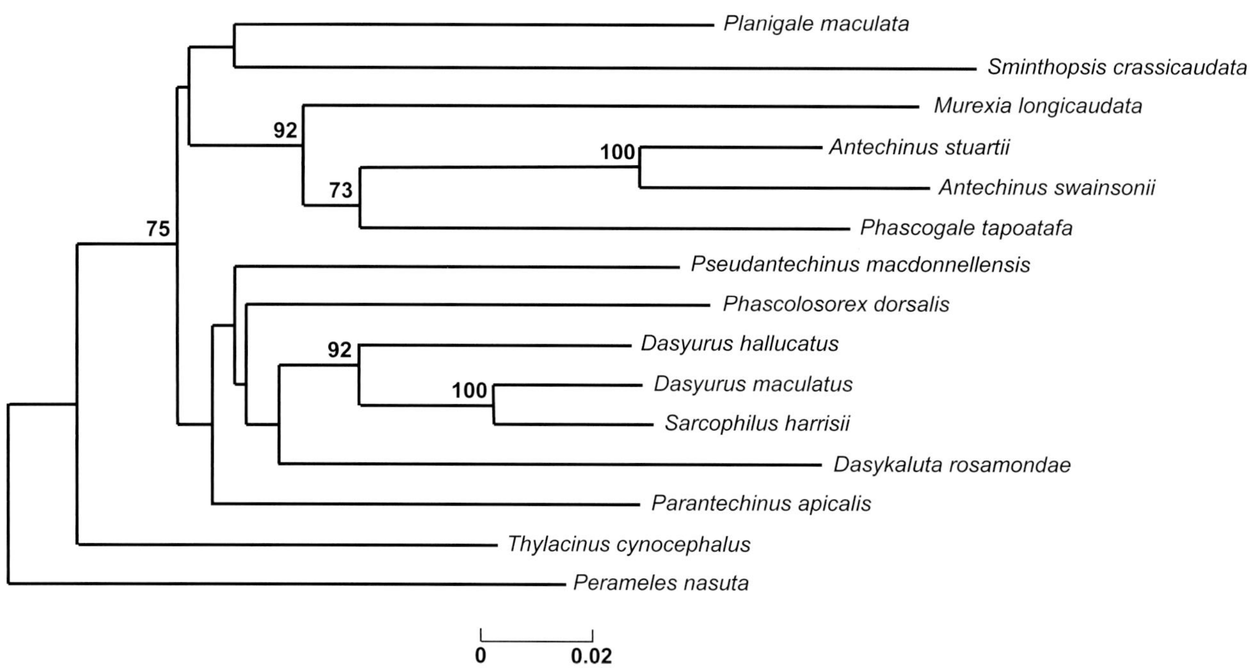

Fig. 4.8. Cytochrome *b* (*cytb*) phylogeny from Krajewski *et al.* (1992). The tree was estimated with the Fitch-Margoliash method operating on Kimura (1980) two-parameter distances derived from 574 bp of *cytb* sequence. Numbers at nodes are bootstrap support values >50%; branch length scale is substitutions per site. *Perameles* is an outgroup bandicoot. [Adapted from Krajewski *et al.* (1992).]

parsimony methods, both using bootstrap assessment of branch support. The trees were poorly resolved, with only a few bootstrap values >90%. Like previous molecular trees, the *cytb* fragment placed *Murexia* in a clade with *Antechinus* and *Phascogale*. *Parantechinus apicalis*, *Pseudantechinus macdonnellensis*, *Dasykaluta* and *Phascolosorex dorsalis* clustered with *Dasyurus* and *Sarcophilus*. *Planigale maculata* was sister to *Sminthopsis crassicaudata*. Dasyuridae was monophyletic apart from *Thylacinus*, supporting recognition of Thylacinidae, but with only moderate (75%) resolution. Anomalously, *Dasyurus* appeared paraphyletic relative to *Sarcophilus* with strong support.

Krajewski *et al.* (1993) attempted to clarify the relationships of New Guinean dasyurids using a portion of *cytb* and distance analysis identical to that of Krajewski *et al.* (1992). Data from 29 species yielded the tree in Fig. 4.9. Reciprocal monophyly of Dasyurinae and Sminthopsinae is moderately to weakly supported (77% and 67% bootstraps, respectively). Within Dasyurinae, clades Dasyurini and Phascogalini are moderately to strongly supported (88% and 98%, respectively); within Sminthopsinae, support is strong for Planigalini (90%) and weak for Sminthopsini (62%). New Guinean endemics are part of larger Australian clades: *Murexia* and 'antechinus' species are part of Phascogalini; *Neophascogale*, *Phascolosorex dorsalis*, *Dasyurus albopunctatus*, *D. spartacus* and *Myoictis melas* are part of Dasyurini. The *cytb* fragment does not resolve Archer's (1982c) parantechinin, phascolosoricine, or muricine clades, nor does it suggest that species of the latter two diverged early from other dasyurids.

Using a sample of 31 species and a larger fragment of *cytb* sequence, Krajewski *et al.* (1994) again attempted to resolve the major clades of Dasyuridae. The results (Fig. 4.10) were like those of the previous study, showing strong support for some groups (e.g. Phascogalini including *Murexia* and New Guinean 'antechinuses' = 99%, Sminthopsinae = 95%, Planigalini = 96%) and weak support for others (e.g. Dasyurini = 69%). Again, however,

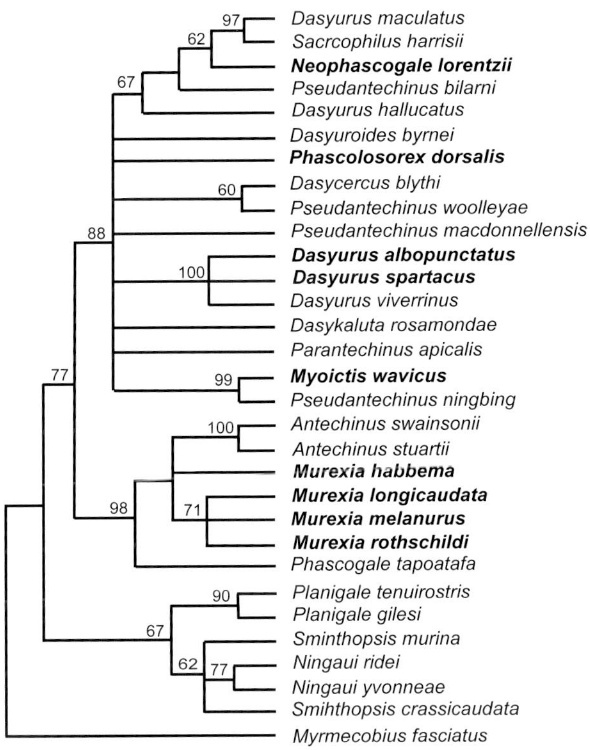

Fig. 4.9. Cytochrome *b* (*cytb*) phylogeny from Krajewski *et al.* (1993). The tree was estimated with the Fitch-Margoliash method using Kimura (1980) two-parameter distances derived from 574 bp of *cytb* sequence. Numbers at nodes are bootstrap support values; branches with bootstrap values <50% were collapsed to polytomies. *Myrmecobius* is an outgroup. Branch lengths are not to scale. New Guinean endemic species are in boldface. Genus and species assignments of specimens used have been updated to match current classification. [Adapted from Krajewski *et al.* (1993).]

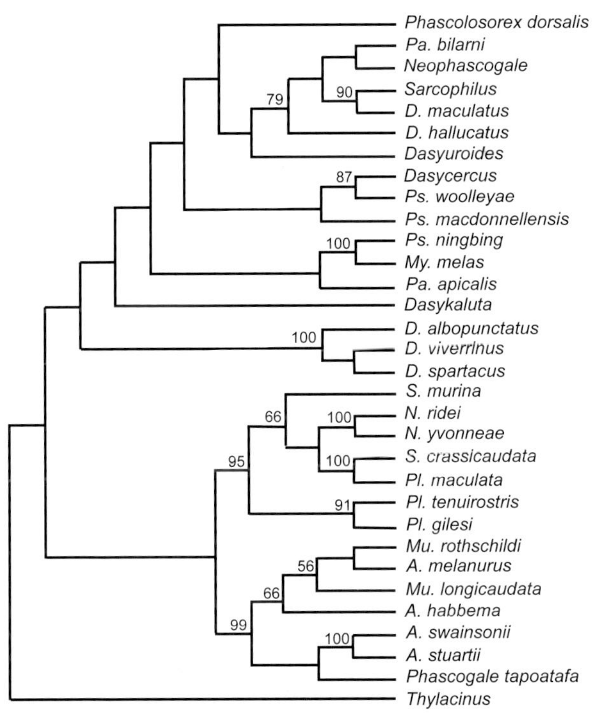

Fig. 4.10. *Cytb* phylogeny from Krajewski *et al.* (1994). The tree was estimated with the Fitch-Margoliash method operating on Kimura (1980) two-parameter distances derived from 657 bp of *cytb* sequence. Numbers at nodes are bootstrap support values >50%. Genus abbreviations as in Fig. 4.4 *Thylacinus* is the outgroup. Branch lengths are not to scale. [Reprinted by permission of Springer Nature from Krajewski *et al.* (1994).]

phascolosoricines and 'false antechinuses' (parantechinins) associated with dasyurins, though the data provided essentially no resolution of branching order within this clade.

Nuclear gene information: protamine P1

While mtDNA analyses of relationships within major clades proceeded apace in the late 1990s, dasyurid systematists also began to explore the nuclear genome. Retief *et al.* (1995b) described evolutionary variation in the protamine P1 gene (*P1*) of marsupials. Protamines are short proteins that replace histones during condensation of sperm nuclei during spermatogenesis. The locus includes 5′ and 3′ flanking regions and two exons separated by an intron (Fig. 4.11). Protamine exon sequences are rich in arginine codons and show modest length variation due to insertions and deletions ('indels') among marsupials. In contrast to eutherians, none of the marsupials examined by Retief *et al.* (1995b) contained cysteine residues. These cysteines, which occur at conserved positions in placentals, form intra- and intermolecular disulfide bridges that stabilise sperm chromatin. Indeed, Retief *et al.* (1995c) cited evidence that sperm nuclei are more easily decondensed in marsupials than in placentals. Marsupial *P1* introns are extremely variable in both sequence and length, ranging from 118 bp in a bandicoot to 317 bp in *Notoryctes*. Dasyurid introns were 223–231 bp, including a 40-base duplication in the two species studied (*D. hallucatus* and *S. crassicaudata*).

P1 introns posed an analytical challenge for Retief *et al.* (1995b). On the one hand, this portion of the locus included most of the phylogenetic information, especially for recent divergences. On the other, intron sequences are so prone to substitutional and indel variation that *aligning* them is difficult, especially for older divergences. Alignment is the determination of site-by-site homology among sequences of the same locus in different organisms. When species have similar sequences of the same length, alignment is easy. But as sequences diverge and accumulate numerous indels, alignment becomes an exercise in computational bioinformatics. Graur (2016) gives an introductory overview of modern alignment options. Beyond alignment, however, Retief *et al.* (1995b) faced the problem of recovering phylogenetic information from indel-rich *P1* introns. Most available models of sequence evolution do not include indel events (Loewenthal *et al.* 2021), so use of model-based methods required a choice among bad options: (1) all alignment sites with gaps could be discarded; or (2) gap sites could be retained and coded as 'missing data'. The first approach discards all phylogenetic information present in the gap sites and the second makes ambiguous use of that information (Machado *et al.* 2021). Faced with this dilemma, Retief *et al.* (1995b) employed parsimony analysis in conjunction with a novel method called 'homologous gap coding'.

> *First, a set of informative sites is identified using only nucleotide bases (these are base-informative sites). We then proceed through the alignment to identify homologous gaps, defined as those which are collinear and possess identical 5′ and 3′ boundaries. Homologous gaps are treated as single characters, equivalent to a nucleotide position. When two or more taxa share a homologous gap at a base-informative site, the gap is coded as present at that site in those taxa by a dash in the data matrix. Thus a base-informative site of this type will include a fifth state. When several base-informative sites occur within the same homologous gap, only one site is given the fifth state; others are coded as missing data by a question mark. In some cases, homologous gaps provide information not present in the set of base-informative sites. Such gap-informative sites are added to the data matrix as single characters. When non-homologous gaps overlap, informative sites must be coded carefully to avoid confusion. This is accomplished by assigning a fifth state to one homologous gap and a question mark to the other at one informative site, then assigning the opposite code at a second site. [Retief et al. 1995b: 9]*

We describe homologous gap coding in detail because it was applied extensively in subsequent analyses of dasyurid *P1* data sets, as well as of matrices that combined *P1* with mtDNA sequences. These papers, therefore, shifted emphasis from model-based, additive-distance methods to DNA parsimony. Without this analytical innovation, however, the phylogenetic information in *P1* sequences would have been difficult to

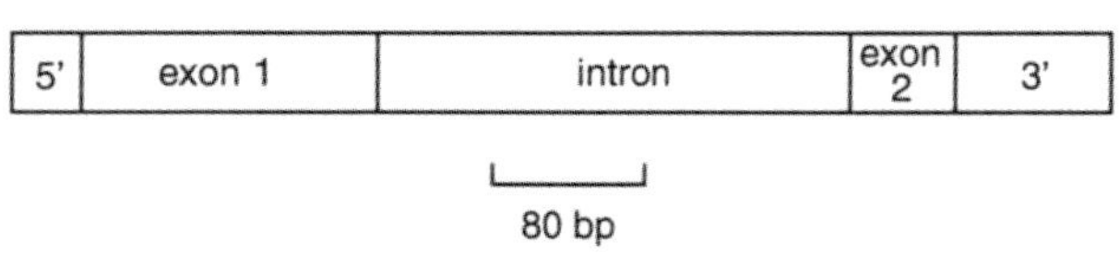

Fig. 4.11. Structure of the marsupial protamine P1 locus. [Reprinted by permission of CSIRO Publishing from Krajewski and Westerman (2003). Permission conveyed through Copyright Clearance Center, Inc.]

assess. Even with homologous gap coding, Retief *et al.*'s (1995a) analysis of *P1* sequences from 13 dasyurid species found that most-parsimonious trees were weakly resolved. They assessed potential gene-tree conflict between *cytb* and *P1* by asking whether any clades with bootstrap support >70% on both gene trees were incompatible. In fact, this '70% combinable component' rule served as the criterion for gene-tree concordance in the remaining studies described in this chapter. Because the *P1* and *cytb* trees did not conflict according to this rule, Retief *et al.* (1995a) concatenated sequences from the two loci to estimate the phylogeny in Fig. 4.12. Again, no deep branches were strongly supported, but the familiar four-clade structure of the family was recovered with weak-moderate resolution.

Dasyurid *P1* data also revealed a striking example of molecular convergence: *Planigale ingrami* has five cysteine residues, unlike any other marsupial known at the time but resembling placental mammals (Retief *et al.* 1995a). Because cysteines are absent in other non-eutherian vertebrates, their occurrence in planigales is likely a derived trait. Planigale species contain variable numbers of cysteines, which occur at locations within the protein similar to those of eutherians. Retief *et al.* (1995c) showed that each was derived by a single point mutation from its ancestral codon. Krajewski, Blacket *et al.* (1997) found that the number of cysteines varies from zero to six among all planigale species but shows no phylogenetic pattern. A follow-up study by Rooney *et al.* (2000) documented the role of stabilising selection in protamine evolution.

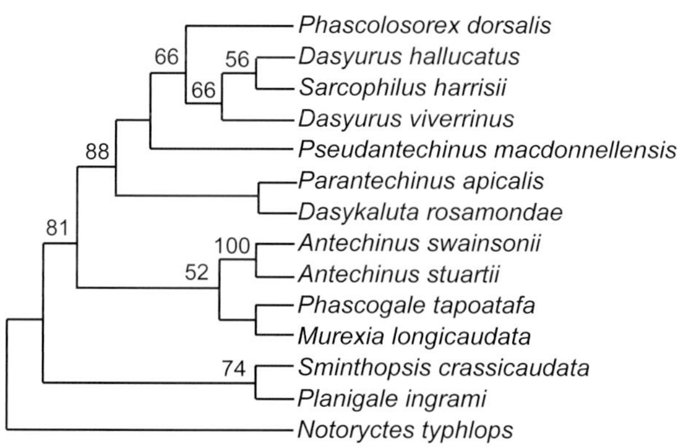

Fig. 4.12. Maximum-parsimony tree for combined *cytb* and *P1* DNA sequences (1408 bp) from Retief *et al.* (1995a). Informative sites from *P1* were extracted with the homologous gap coding method (Retief *et al.* 1995b). Numbers at nodes are bootstrap support values >50%. *Notoryctes typhlops* is the outgroup. [Redrawn with permission of Springer Nature from Retief *et al.* (1995a).]

Multilocus mitochondrial and nuclear analyses

The partial *cytb* sequences used in the foregoing studies did a relatively poor job of resolving the mtDNA gene tree. As research progressed, investigators began increasing the length of sequences analysed from each species. The first such attempt was by Krajewski, Buckley *et al.* (1997), who set out to resolve the position of *Thylacinus* among marsupial orders using complete *cytb* gene sequences and partial *12S*, along with *P1*. Their parsimony tree (Fig. 4.13) showed Thylacinidae as a member of Dasyuromorphia with high support and improved resolution on several key nodes within Dasyuridae. Among the latter were monophyly of Phascogalini (including *Murexia*) and monophyly of Dasyurinae (including false antechinuses).

All DNA sequence trees presented so far have some nodes with bootstraps <50%. Additive-distance and parsimony algorithms search for – and find – optimal trees that are fully bifurcating, even if the evidence for some splits is trivial. For example, the tree in Fig. 4.13 shows *Myrmecobius* as sister to *Sminthopsis crassicaudata*, a result any dasyurid biologist would consider bizarre. However, the low bootstrap (44%) tells us that this pairing has insignificant support from the sequences and should not be taken seriously. Indeed, if we collapse nodes with <50% bootstraps, the tree would show *Sminthopsis*, *Myrmecobius*, *Planigale* and dasyurines forming a soft polytomy at the base of Dasyuridae. Bizarre, but unsupported, branches on optimal trees do not imply that the data or the analyses are unreliable, only that they cannot resolve branching order with confidence. What *is* misleading, at least in modern systematics, is to present an optimal tree with no assessment of evidential support.

Throughout the late 1990s, studies of individual dasyurid clades (discussed in Chapters 6 and 7) generated complete sequences of mitochondrial *cytb* and *12S* genes, as well as the nuclear *P1* locus. Krajewski, Wroe and Westerman (2000) analysed this entire collection of data with a view to establishing both the phylogenetic structure and diversification chronology of Dasyuridae. We summarise their findings on the first topic here and the second in Chapter 9. The accumulation of sequence data from the protein-coding genes of many different species and groups revealed that codon positions typically evolve at different rates, with third positions changing most rapidly and second positions most slowly. This rate disparity reflects selective constraints at the protein level coupled with the genetic code's degeneracy. In most codons, third positions are two- or fourfold degenerate (e.g. TTY = phe-

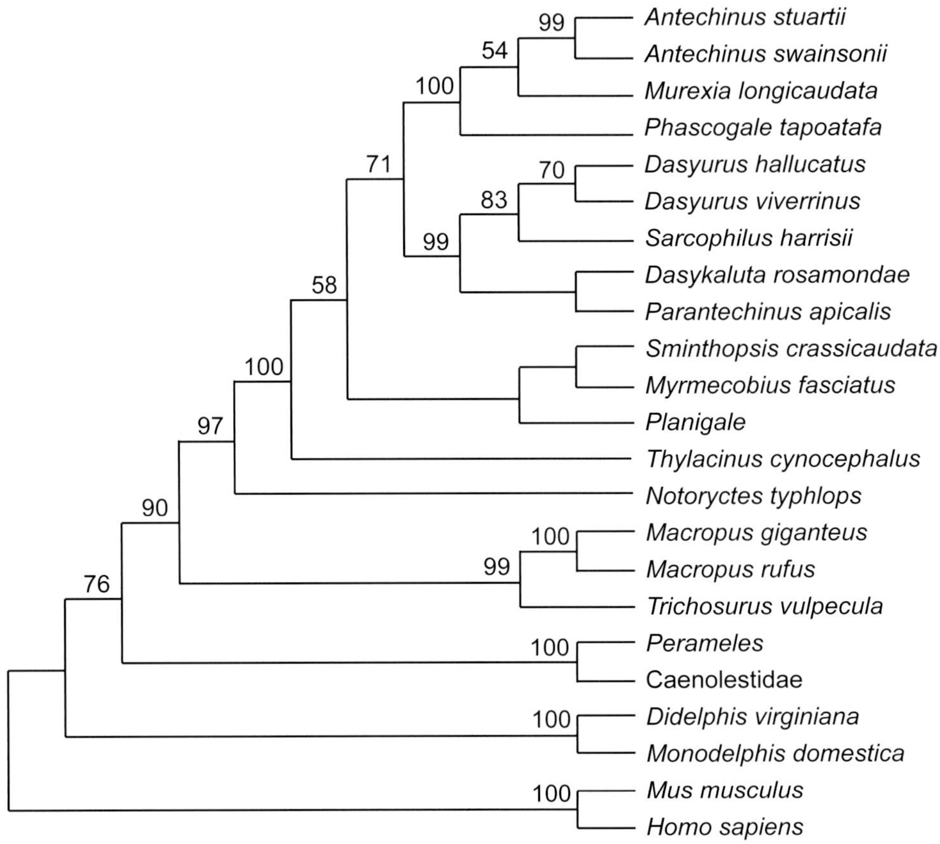

Fig. 4.13. Most parsimonious tree for combined *cytb* (1149 aligned sites), *12S* (397 aligned sites) and *P1* (814 aligned sites) sequences (Krajewski, Buckley and Westerman. 1997). Informative sites were identified with homologous gap coding. Numbers at nodes are bootstrap support values >50%. *Planigale* is a composite of *cytb* + *12S* from the 'Pilbara planigale' and *P1* from *Pl. ingrami*. Caenolestidae is a composite of *cytb* from *Lestoros inca* and *12S* + *P1* from *Caenolestes fuliginosa*. [Adapted from Krajewski, Buckley and Westerman (1997).]

nylalanine, where Y = C or T; TCN = serine, where N = A, G, C, or T). Thus, many third-position substitutions are synonymous or silent and shielded from strong selection at the protein level. In contrast, only a few first positions are twofold degenerate (e.g. TTR = CTN = leucine) and all second positions are nondegenerate. Moreover, most twofold degenerate codons differ by transition differences, such that the rate of third-position transitions can be quite high. As divergence between homologous sequences increases, many third position sites accumulate a high proportion of multiple transitions, which compromises (and eventually eliminates) their phylogenetic information. Ongoing investigation of dasyurid *cytb* sequences revealed that third positions for all but the most closely related taxa were thus 'saturated'. Therefore, Krajewski, Wroe and Westerman (2000) eliminated third-position transitions from parsimony analyses and considered only transversions at third positions to be informative ('3TV' weighting), which considerably improved resolution in the estimated phylogeny (Fig. 4.14).

Krajewski, Wroe and Westerman's (2000) results were the first time that the deep phylogenetic structure of Dasyuridae was *strongly* supported by molecular data. Dasyurinae (100%) comprises clades Dasyurini (100%) and Phascogalini (100%); Sminthopsinae (94%) comprises Sminthopsini (98%) and Planigalini (100%). Krajewski and Westerman (2003) used this phylogeny as the basis for the subfamily and tribe designations we recognise in this book. *Murexia* species are phascogalins and form a clade with New Guinean 'antechinuses', so all the latter were transferred to *Murexia*. New Guinean endemics *Phascolosorex* and *Neophascogale* are sisters within Dasyurini. Thus, 'Muricinae' and 'Phascolosoricinae' were rejected as taxonomic designations. After redefinition of *Murexia*, all polytypic dasyurid genera – except one – evaluated by Krajewski, Wroe and Westerman (2000) received strong

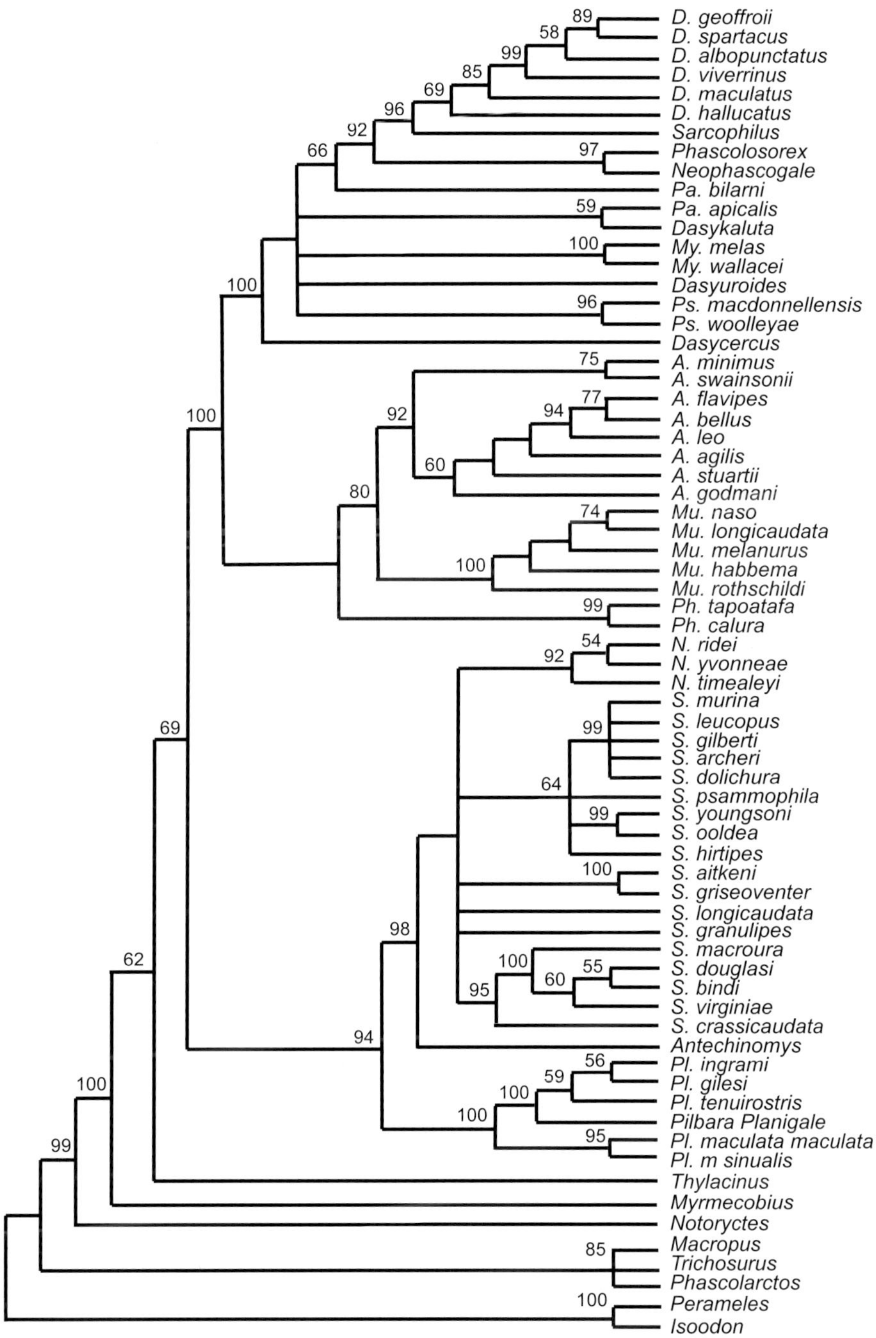

Fig. 4.14. Strict consensus of eight most parsimonious trees from combined *cytb*, *12S* and *P1* genes (complete sequences, 1015 informative sites) from Krajewski, Wroe and Westerman (2000). Transitions at *cytb* third-codon positions were given zero weight ('3TV' weighting). Informative sites in *P1* were identified with homologous gap coding. Numbers at nodes are bootstrap support values >50%. Genus abbreviations as in Fig. 4.4. *Perameles* and *Isoodon* are outgroups. [Reprinted by permission of Oxford University Press from Krajewski, Wroe and Westerman (2000).]

support as monophyletic groups. The exception was *Sminthopsis*, the most speciose dasyurid genus, which could not be resolved apart from *Antechinomys* and *Ningaui*. The tree in Fig. 4.14 set the agenda for dasyurid phylogenetics research in the new millennium. No subsequent molecular study has contradicted its four-clade structure, but much progress has been made on genus- and species-level relationships, as we discuss in Chapter 6.

MOLECULES AND MORPHOLOGY: THE PROBLEM OF *MUREXIA*

The series of results from morphological and molecular analyses carried out in the 1970s and early 1980s to DNA phylogenies put forward in the 1990s and early 2000s seems to show progress towards a consensus. Anatomical and allozyme data suggested seven groups of closely related extant species and genera of dasyurids but could not resolve relationships among them (see Fig. 3.9). Three results from immunological and DNA hybridisation data hinted that these seven groups represent four clades: (1) *Planigale* appeared as sister to the sminthopsin genera (DNA hybridisation); (2) 'phascolosoricines' clustered with dasyurins; and (3) *Murexia* and New Guinean 'antechinuses' were monophyletic with phascogalins. Subsequent DNA sequence results were confirmatory, beginning with short fragments of a single mitochondrial gene from a few species, but eventually including multilocus data from nuclear and mitochondrial genomes of nearly all dasyurids. This consilience of evidence would continue through much larger and more complex data sets analysed after 2010 (see Chapter 5).

Amid all this emerging clarity, a strong note of discord was struck by Van Dyck's (2002) morphological revision of *Murexia* and *Antechinus*. Van Dyck carefully reviewed character analyses by previous authors (e.g. Archer 1976d, 1981, 1982a, 1982c; Kirsch and Archer 1982; Reig *et al.* 1987) and made extensive examinations of museum specimens. Most of his results are a detailed evaluation of species-level taxonomy in the two genera, which (among other things) confirmed that New Guinean 'antechinuses' are not members of *Antechinus*. However, his monograph concludes with a phylogenetic analysis in which the New Guinean species (here, *Murexia*) appear as successive sister groups to the rest of Dasyuridae (Fig. 4.15). Reviewing the character evidence and phylogenetic results, Van Dyck (2002: 324) concluded: 'I consider these reasons sufficient to warrant revision at the generic levels.' He placed each species in a monotypic genus, four of them new: *Murexia* (for *Mu. longicaudata*), *Paramurexia* (for *Mu. rothschildi*), *Murexechinus* (for 'A.' *melanurus*), *Phascomurexia* (for 'A.' *naso*) and *Micromurexia* (for 'A.' *habbema*). Van Dyck's phylogeny and its implications for dasyurid classification could hardly be more different from the emerging molecular consensus. What happened, and what does this episode tell us about dasyurid evolution and phylogenetics?

Considering the studies that came before and after, one might naively conclude that so many molecular trees cannot all be wrong. Nevertheless, Van Dyck (2002) suggested two reasons – inappropriate outgroups and base composition biases – why early DNA analyses might have been misled. Both were dismissed by Krajewski, Wroe and Westerman (2000) and later studies, but one could expand the list to include gene-tree discordance hidden by analysis of concatenated loci (Kubato and Degnan 2007). Although mtDNA and *P1* both support monophyly of *Murexia* (*sensu* Krajewski and Westerman 2003) and its placement within Phascogalini, bootstrap support levels vary between the loci: 93% and 99%, respectively, for mtDNA; 80% and 79%, respectively, for *P1* (Krajewski, Wroe and Westerman 2000). Readers should wonder, therefore, whether other loci give the same result. (They do; see Chapter 5.) At what point does the weight of evidence favour one phylogenetic hypothesis over another? Systematists have no standard answer to this question, except perhaps insofar as one tree becomes the basis of an accepted classification. Van Dyck's five genera were accepted by Groves (2005) in the influential *Mammal Species of the World* but rejected in favour of a monophyletic *Murexia* by authors of 2016 species accounts in the IUCN Red List. Given the importance of classifications as a framework for studies of biodiversity, this is not merely a nomenclatural issue. We return to it in Chapter 5.

Van Dyck (2002: 324) expressed concern that 'genetic phylogenies would suggest New Guinea "murexias" have undergone massive and successive reversals and losses, and that *Phascogale* and *Antechinus* have accrued states in parallel'. This seems to be the crux of the problem, because in Van Dyck's view, '[t]he putative relationships between *Phascogale* with *Murexia* as revealed by genetic studies … will continue to constitute one of the most bitter pills for dasyurid morphologists to swallow' (p. 324). Whence the bitterness? Van Dyck is not specific, but perhaps it derives from the possibility that (at least these 73) morphological characters violate the fundamental assumption of cladistics – that similarity in derived traits is, in general, more likely due to homology than to homoplasy. This is Hennig's *auxiliary principle*:

> *… that the presence of apomorphous characters in different species 'is always a reason for suspecting kinship [i.e., that species belong to a monophyletic group], and that their origin by convergence should not be assumed a priori' … This was based*

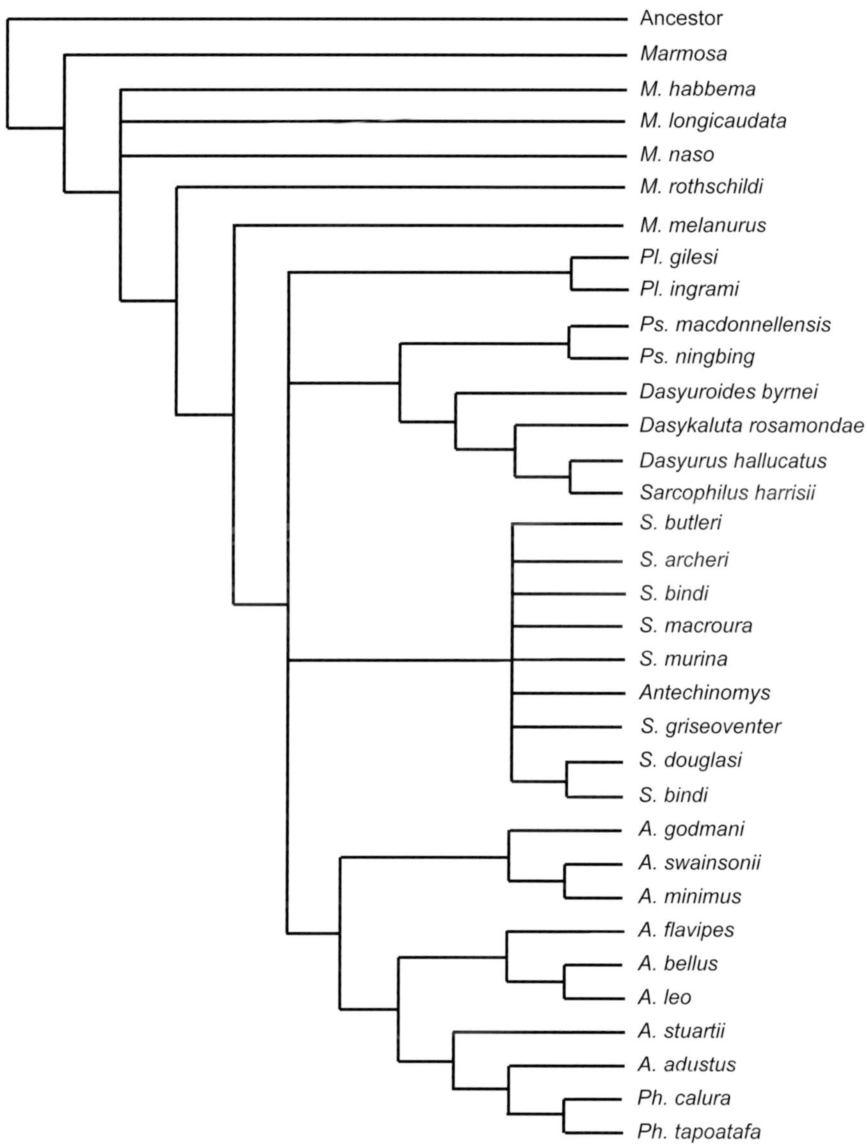

Fig. 4.15. Van Dyck's (2002) phylogeny based on 73 dental, cranial and external traits. Characters were polarised prior to analysis and the tree rooted with a hypothetical 'Ancestor' having all primitive states. Tree shown is a strict consensus of 15 minimum-length trees. Genus abbreviations: *A.* = *Antechinus*; *M.* = *Murexia*; *Ph.* = *Phascogale*; *Pl.* = *Planigale*; *Ps.* = *Pseudantechinus*; *S.*= *Sminthopsis*. [Redrawn with permission of the Queensland Museum from Van Dyck (2002).]

on the conviction that 'phylogenetic systematics would lose all the ground on which it stands' if the presence of apomorphous characters in different species were considered first of all as convergences (or parallelisms), with proof to the contrary required in each case. Rather the burden of proof must be placed on the contention that 'in individual cases the possession of common apomorphous characters may be based only on convergence (or parallelism)'. [Hennig 1966: 121–122]

If the evolution of dasyurid craniodental and external traits can be as homoplastic as Van Dyck feared, there is a dual methodological threat. First, parsimony (and arguably other) analyses of these traits will be an unreliable guide to phylogeny. Second, similarity will be an unreliable guide to homology. Yet few alternatives are available in the toolkit of morphological phylogenetics. How far problems with Van Dyck's traits can be generalised to all of morphology is unclear but lingering in the aftertaste of this 'bitter pill' is a hint of much stronger poison.

In a fascinating parallel to the problem of *Murexia*, Archer (1982a) reversed his earlier (Archer 1976d) assessment that thylacinids were sister to the mid-Tertiary borhyaenoids of South America, opting instead for affinities with dasyuroids. (The latter opinion, of course, reflects a solid modern consensus, but this is hindsight and beside the point. See Chapter 8 for details.) Archer's initial opinion was based on dental and cranial characters, including some of the same ones used by Van Dyck (2002). Between 1976 and 1982, two new pieces of evidence had been adduced showing that thylacines are dasyuromorphians – Szalay's (1982) analysis of tarsal morphology supporting the clade Australidelphia (including thylacines) and Sarich *et al.*'s (1982) MC'F finding that thylacine albumin is most similar to that of dasyurids. Archer (1982a) undertook an extensive review of all the evidence and analytical options then available and came to the following subjective but discerning conclusions:

> *If we ignore the most parsimonious conclusion based on the analysis of morphology, and conclude instead that the Thylacinidae is the sister-group of Dasyuroidea, we must simultaneously assume a remarkable amount of convergent (or parallel) evolution in order to account for the otherwise striking resemblance of thylacinids to borhyaenids … And yet, comparable examples of remarkable convergence are known such as the similarity of marsupial moles (notoryctids) and placental moles (e.g. chrysochlorids). For this reason, the thylacinid-dasyuroid sister-group hypothesis should not be abandoned on the basis of parsimony alone. [p. 470]*
>
> *Dental morphology provides one of the main data sets supporting the thylacinid-borhyaenid hypothesis … Despite this seemingly strong support … I have no illusions (persistent ones anyway) about the immunity of dental characters to convergence. [p. 471]*
>
> *If just one conclusion is drawn from this study, it should be that hypotheses about phylogenetic relationships of a group of extant organisms, particularly if the group has a poor fossil record, will almost certainly be doomed to early obsolescence unless the data from many tissue systems have been taken into account … Some tissue systems are undoubtedly more reliable indicators of phylogeny than others, but all appear to be subject to either convergence, glitches in evolutionary rates of divergence … or limitations in applicability … For the moment, based on all available information, it would seem that the most probable interpretation about thylacinid relationships is that they are the plesiomorphic sister-group of the Dasyuroidea and have convergently developed a borhyaenid-like dentition. Because this syncretic hypothesis is based on the data from five rather than the three previous tissue systems, it might last a bit longer than the last! [p. 474]*

And indeed, it has: the position of Thylacinidae within Dasyuromorphia was confirmed by multiple lines of evidence in subsequent studies (see Chapter 8). With Archer's insights, we close this chapter on early molecular studies of dasyurid phylogeny and move to more recent and integrative approaches. We return in Chapter 5, however, to the question of how to weigh the evidence from molecules and morphology, as well as the question of *Murexia* relationships.

ENDNOTES

1 Throughout this book we use a simplified version of 'Newick format' to express phylogenetic relationships in a line of text with nested parentheses. For example, '(A (B, C))' means that species B and C are sisters apart from A; (D (E (F, G))) represents a tree on which D splits from the ancestor EFG above the root, EFG then splits into E and FG, and finally the sister-species F and G diverge.

2 Molecular evolutionists distinguish two kinds of homology when discussing gene sequences. *Paralogous* loci ('paralogs') are those in the genome of a single species that are homologous by virtue of gene duplication in an ancestor. For example, α-globin and β-globin genes in *Antechinus swainsonii* are paralogous because they were derived by duplication of a single globin gene in a vertebrate ancestor. *Orthologous* loci ('orthologs') are those in the genomes of two or more species that are homologous by virtue of their presence in the most recent common ancestor of those species. Thus, the α-globin genes of *A. swainsonii* and *A. minimus* are orthologs because the most recent common ancestor of these species had an α-globin gene. Orthologs in sister-lineages begin diverging at the time of the speciation event that gave rise to those lineages and therefore track phylogeny. Paralogs begin diverging at the time of gene duplication and do not track phylogeny. Estimating *Antechinus* relationships from the α-globin genes of all species in the genus would be a reasonable endeavour, but comparison of α-globin genes from some species with β-globin genes of others would not.

5 Dasyurid phylogeny: integrative studies

In the decade after Krajewski and Westerman's (2003) review, most dasyurid molecular systematists were working to clarify relationships of species within the four tribes (see Chapters 6 and 7), while research on morphology emphasised phylogenetic analyses of dasyuromorphian fossils (see Chapter 8). The latter efforts did little to clarify dasyurid systematics. For example, Wroe *et al.* (2000), Wroe and Musser (2001) and Murray and Megirian (2006) published parsimony trees based on skeletal data that included representative species from each dasyurid tribe. However, these trees had virtually no resolution of intergeneric relationships as indicated by bootstrap support or consensus analysis. During this interval, however, molecular phylogenetics at large was expanding into new areas of data gathering and analysis. Next-generation sequencing (NGS) technologies, in which millions of short DNA fragments are sequenced and computationally assembled into genome-scale alignments, superseded Sanger sequencing (Schuster 2008). Bayesian inference of phylogeny allowed a fuller exploration of the parameter space associated with sequence-evolution models and multilocus alignments (Felsenstein 2004). These methodological advances, along with a growing collection of mitochondrial and nuclear data sets available for marsupials, powered studies that integrated large alignments with geographic, morphological and temporal information. In this chapter, we focus on the implications of these papers for the four-clade structure of Dasyuridae, then return to the striking discordance between molecules and morphology on the affinities of *Murexia* species. Eldridge *et al.* (2019) summarised progress in marsupial systematics, including the four-clade structure described here, as well as other aspects of dasyuromorphian phylogeny (see Chapters 6 and 9), species diversity (see Chapter 7) and palaeontology (see Chapter 8).

METHODOLOGY

Next-generation sequencing (NGS)

Dasyurid DNA sequences published in the 1990s were derived from Sanger sequencing of PCR-amplified targets ≤1000 bp. The assays involved radioactively labelled fragments separated by size on polyacrylamide slab gels, which were then visualised on autoradiograms and read by eye. By the early 2000s, the sequencing process had evolved into size-separation of fluorescently labelled fragments in a capillary tube, with the sequence recorded digitally as different colours of terminal bases were excited by a laser and detected with a fluorimeter. Although much faster than its predecessor, automated Sanger sequencing still operated on individual PCR products, which placed a practical limit of just a few kilobases per taxon on any reasonably large systematic study.

Next-generation (or 'high-throughput') technologies do not proceed one fragment at a time, but instead generate sequences in parallel from many template fragments derived from many loci across an organism's genome. The fragments are very short (usually <1000 bp) and the accuracy of each sequence determination ('read') is high; millions of reads are generated in a single round of sequencing that usually requires less than a day at relatively little cost. Individual reads are recorded digitally and computational algorithms assemble the sequences of overlapping fragments into longer (ultimately, chromosome-length) 'scaffolds'. By comparing assembled sequences to the reference genome of a closely related taxon, gene boundaries and other structural features can be identified. Reference sequences, of course, must be obtained *de novo* and usually require library construction, assembly and annotation phases; however, for mammals, many such genomes have been completed. Phylogenetic researchers may employ some form of genomic screening prior to NGS, such as using probes for ultra-conserved elements (UCEs) to recover DNA regions with high levels of variation among species (Faircloth *et al.* 2012). The UCE approach also provides an efficient means of obtaining complete mitochondrial genomes (Raposo do Amaral *et al.* 2015).

'Total evidence' analysis

As early molecular systematic studies revealed a high frequency of incongruence with morphological results for the same taxa, researchers began to contemplate the relative merits of these data types. Are molecular characters better than morphological ones? If so, is it because they are genetic, as asserted in qualified terms by Simpson (1945; see Chapter 4)? Are they less prone to misleading convergence, as claimed by Gould (1985; see Chapter 4)? Are they more useful because they lend themselves to model-based statistical analyses in the inferential framework of Felsenstein (1983)? Many systematists have answered all four questions in the affirmative, with and without caveats. Others are reluctant to devalue a category of evidence with such a long pedigree in systematics, especially given that morphology is sometimes the only source of information available on relationships (e.g. fossil taxa). Moreover, Gatesy *et al.* (1999) demonstrated that inclusion of morphological characters in a data matrix could provide 'hidden support' for relationships not resolved by sequence data alone. The molecules-versus-morphology tension has sometimes been felt acutely in dasyurid phylogenetics, as exemplified by Van Dyck's (2002) conclusions about *Murexia* species (see Chapter 4 and 'Morphology, molecules and *Murexia*' below).

Kluge (1989) reminded systematists that inductive inferences (e.g. of phylogeny) should consider all available evidence. On this basis, he argued against the use of separate analyses for morphological and molecular data from the same taxa (partitioning) followed by emphasis on consensus between the results, as well as the practice of 'conditional combination' of the two data types (i.e. when their results are congruent). Instead, Kluge advocated what came to be known as 'total evidence' analysis – unconditional combination of all characters (whatever their type) into a single data matrix for analysis. The validity of this argument and its extensions have been debated (e.g. Fitzhugh 2006), but many systematists embrace total evidence as rationale for integrating morphological and molecular information. At the time, Kluge's (1989) requirement would have restricted phylogenetic estimation to parsimony analysis, but Lewis's (2001) formulation of the Mk model for morphological characters seemingly opened the door to overtly statistical treatment of combined data sets.

Bayesian inference of phylogeny

Stochastic models of DNA evolution allowed systematists to use observed regularities of the substitution process (e.g. base compositions, pairwise substitution rates, among-site rate variation) to inform and strengthen their inferences of gene trees, at first using additive-distance and maximum-likelihood (ML) frameworks. An alternative inferential framework in statistics is founded on Bayes's theorem relating the probability of a hypothesis given data to the likelihood of that hypothesis and the unconditional ('prior') probabilities of the data and hypothesis. Rannala and Yang (1996) pioneered the application of Bayesian inference to phylogenetics, adapting Bayes's theorem as follows:

$$\mathrm{Pr}(\mathrm{Tree}\,|\,\mathrm{Data}) = \left[\mathrm{Pr}(\mathrm{Data}\,|\,\mathrm{Tree}) \times \mathrm{Pr}(\mathrm{Tree})\right] / \mathrm{Pr}(\mathrm{Data})$$

The quantity on the left is the *posterior probability* of a given tree; those on the right are (in order) the likelihood of the tree, the *prior probability* of the tree and the prior probability of the data. Although not included explicitly in the equation above, some model of evolution is assumed. The best tree for a given data set is the one with maximum posterior probability; the problem is how to evaluate that probability by calculating the quantities on the right.

In maximum-likelihood estimation, parameters of the process model are assigned their ML values (given the data) and the ML tree is conditional on those values. Bayesian inference, however, evaluates the tree likelihood by integrating over all parameter values. The tree prior may be constrained if there is good reason to believe that some trees are more probable than others, but without such expectation one may assign all possible trees the same probability (i.e. a uniform or 'flat' prior). Calculating the data prior would involve integrating over all possible trees, but this is computationally infeasible for even a moderate number of sequences. Larget and Simon (1999) introduced the 'Markov chain Monte Carlo' (MCMC) method for exploring the space of parameter values (including tree topology and branch lengths) that determine the three right-side probabilities in Bayes's formula above. From an initial set of parameter values and corresponding posterior, the MCMC algorithm randomly alters some values, updates the posterior and applies a decision rule to accept or reject the new values based on the difference in posteriors. This constitutes a single MCMC 'generation' and the process is repeated for millions of generations (a 'chain') with trees sampled at regular intervals throughout. As more and more parameter combinations are evaluated, the chain converges on those that maximise the posterior probability relative to other combinations. The frequency of sampled parameter combinations traces out their posterior distribution and the frequency of specific clades in this distribution estimates posterior probabilities of those clades, an obvious measure of evidential support. Implementation of Bayesian phylogenetics (and other evolutionary inferences) was made feasible by high-speed computers and software development, including packages such as MrBayes (Huelsenbeck and Ronquist 2001) and BEAST (Suchard *et al.* 2018).

Species tree methods

In the foregoing discussion, we have assumed that a single tree is estimated from concatenated sequences of individual loci and that this tree is the species phylogeny – in effect, that the genome evolves like a single linkage group. But this is not the case for nuclear loci; extensive recombination ensures that segments of chromosomal DNA can have independent histories as evolutionary forces act on them between speciation events. Indeed, the mutation histories ('gene trees') of different linkage groups can differ from one another (gene tree discordance) and from the phylogeny or 'species tree' (Maddison 1997). Such discordance is caused by incomplete lineage sorting or some form of reticulation (e.g. introgression). That gene tree conflict is common quickly became apparent when genome-scale data sets began to be used for phylogenetic inference. An initial, and still important, methodological response was to assume that most gene trees would match the species tree. Theorists soon connected the gene-species tree problem to coalescent theory in population genetics (Maddison and Knowles 2006) and began to develop analytical methods that would estimate a phylogeny from gene trees for multiple loci (e.g. Edwards *et al.* 2007). Because co-estimation of all gene trees and the species tree is computationally intractable, many such species tree methods employ the simplifying assumption that estimated gene trees are correct, then use the topologies and branch lengths of those trees to estimate a phylogeny with a likelihood, Bayesian, or other criterion ('summary coalescent methods'). Strong criticism of this assumption (e.g. Gatesy and Springer 2014) initiated a 'coalescence versus concatenation' debate that led to more circumspect application of both methods, an appreciation that concatenation may be more robust than was first suspected and acceptance that properly applied multispecies coalescent methods have great potential in phylogenomics (Jiang *et al.* 2020).

RESOLVING THE FOUR-CLADE STRUCTURE OF DASYURIDAE

Cardillo *et al.* (2004) performed supertree analysis on 158 estimated marsupial phylogenies published between 1980 and 2002, including both morphological and molecular studies. Supertrees are assembled algorithmically from source trees that include overlapping but nonidentical sets of species, treating clades present in the source trees as data, but not utilising the characters on which those trees were based (see Bininda-Emonds *et al.* 2002 for a review). The Dasyuromorphia supertree (Fig. 5.1) provides an integrative representation of the research described in Chapters 3 and 4 and sets the stage for studies that followed. The tree shows ((Dasyurini, Phascogalini), (Sminthopsini, Planigalini)), reflecting most source trees, with *Murexia* monophyletic within Phascogalini, 'phascolosoricines' and 'parantechinins' within Dasyurini and *Sminthopsis* monophyletic (although the authors followed Archer (1981) in recognising the kultarr as '*Sminthopsis*' *laniger*, rather than as *Antechinomys*).

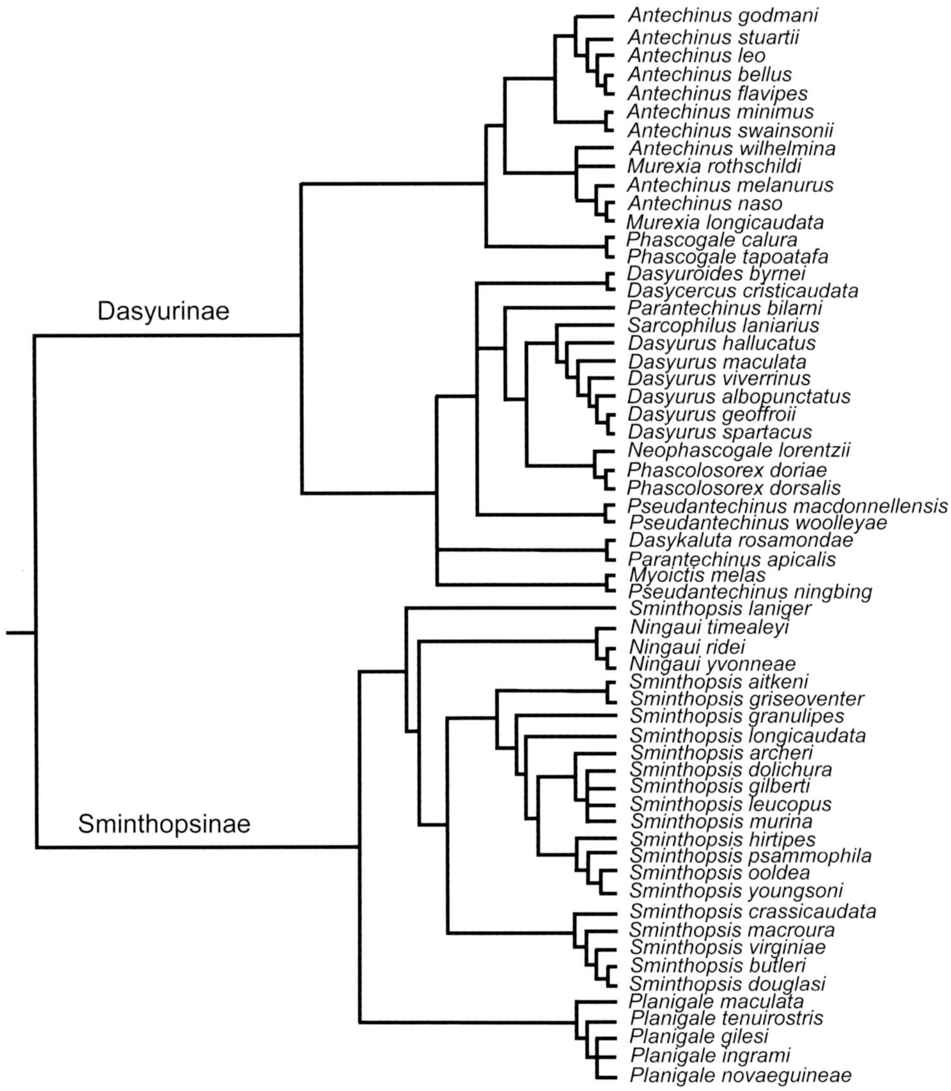

Fig. 5.1. Dasyurid supertree based on larger analysis of 150 source trees published between 1980 and 2002 (Cardillo *et al.* 2004). Non-dasyurid taxa in the original analysis have been removed for clarity. [Redrawn by permission of John Wiley and Sons from Cardillo *et al.* (2004).]

Large nuclear and mitochondrial datasets

Using NGS methods, Mitchell *et al.* (2014) added 69 mitochondrial genomes to an existing set of 32 mitogenome and 26 nuclear-locus sequences (43 616 bp) from 193 marsupials. The alignment was 39% complete (i.e. 11 of the 27 loci were available for all 193 species); the other 61% of loci were missing from 12 to 88% of the species. Sites were placed into 72 bins that included codon position within a locus (third positions of mtDNA protein-coding genes were deleted), introns and stems and loops of RNA genes. These bins were grouped into partitions with distinct evolutionary models using likelihood-based model selection. Trees were estimated with parsimony, maximum-likelihood and Bayesian methods; clade support was assayed with bootstrapping and posterior probability. All trees had similar structure but varying levels of support for individual clades (Fig. 5.2). However, the four-clade structure of Dasyuridae was strongly supported by all analyses. Although Mitchell *et al.* (2014) used Van Dyck's (2002) genus names for *Murexia* species, those species clearly form a single clade that is part of Phascogalini. False antechinuses ('parantechinins') and 'phascolosoricines' are part of Dasyurini and *Sminthopsis* is not monophyletic apart from *Ningaui* and *Antechinomys*.

May-Collado *et al.* (2015) assembled previously published sequences of six nuclear (*ApoB*, *vWF*, *IRBP*, *Rag1*,

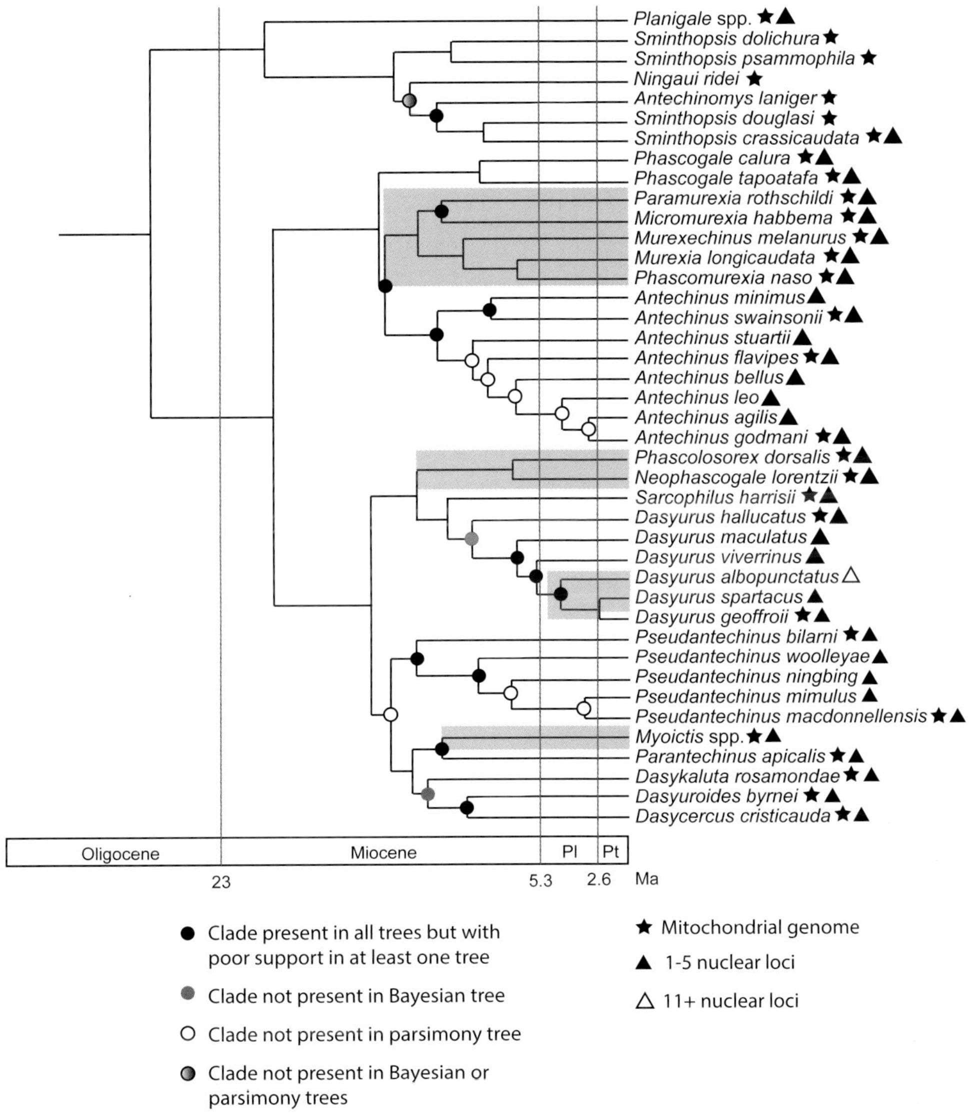

Fig. 5.2. Dasyuromorphian portion of the Mitchell *et al.* (2014) phylogeny from mitogenomes and 26 nuclear loci. All nodes are supported by ≥90% bootstrap (parsimony, maximum likelihood) and ≥0.95 posterior probability (Bayesian analysis). Shading indicates inferred ancestral areas: unshaded = Australia; grey = New Guinea and Wallacea; white–grey = uncertain. Divergence times derived from Bayesian analysis with fossil calibrations (see Chapter 9). Pl = Pliocene; Pt = Pleistocene. [Adapted from Mitchell *et al.* (2014).]

BRCA1, P1) and five mitochondrial genes (*cytb, 12S, 16S, CO1, ND2*), along with presence/absence data on retroposons, from some 276 marsupial species (80% of extant diversity). Not surprisingly, the authors noted alignment difficulties with *P1* sequences. Retroposon insertions are rare genomic events, such that the shared presence of an insertion at a site in different species is likely to be homologous. Moreover, retroposons may insert into each other to produce a nested pattern that tracks phylogeny (Nilsson *et al.* 2010). Different subsets of the data were analysed independently and compared; the model-selection procedure for each gene and codon position was like that in Mitchell *et al.* (2014). Trees were estimated with Bayesian and ML methods, as well as parsimony for the retroposon data. Analyses recovered all but one of 22 'benchmark clades' (i.e. well-established monophyletic groups) in Marsupialia, including Dasyuromorphia and Dasyuridae; the exception was Phalangeriformes. Within Dasyuridae (Fig. 5.3), several key branching patterns are strongly supported: the four tribal clades; a monophyletic *Murexia* (*sensu* Krajewski and Westerman 2003) within Phascogalini; paraphyletic genera *Sminthopsis*, *Parantechinus* and *Pseudantechinus*.

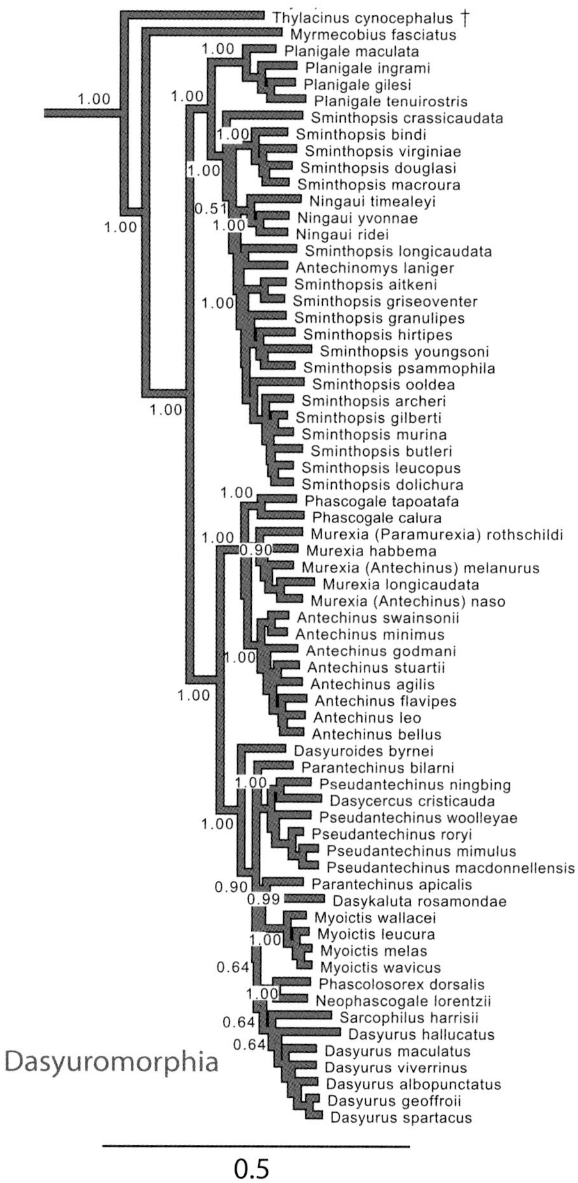

Fig. 5.3. Bayesian tree of dasyuromorphian species based on five mitochondrial and five nuclear gene sequences, as well as presence/absence of retroposons (May-Collado *et al.* 2015). Nodal values are posterior probabilities >0.50. Non-dasyuromorphian taxa in the original analysis have been removed for clarity. [Reprinted from May-Collado *et al.* (2015) under CC BY license.]

Westerman, Krajewski *et al.* (2016) augmented the data set of May-Collado *et al.* (2015) with sequences from additional species and included two new nuclear (ε-globin = *εglob*, beta-fibrinogen intron 7 = *bfib7*) and two new mitochondrial (*CO2*, *ND1*) loci. Concatenated sequences were analysed with ML and Bayesian methods, assuming the GTR + Γ + I evolutionary model (ML) or independent models for each locus (Bayesian). The ML estimate with bootstrap values (Fig. 5.4) strongly resolved the four-clade structure of Dasyuridae with expected placements of *Murexia*, 'phascolosoricines' and false antechinuses. *Sminthopsis* appeared paraphyletic, but *Pseudantechinus* monophyly received high support. Westerman *et al.* (2008) had pre-empted concern over generic boundaries among false antechinuses by showing that '*Parantechinus*' *bilarni* is part of the *Pseudantechinus* clade apart from *Parantechinus apicalis* (see Chapter 6). Although these authors suggested the appropriate reclassification, it was not immediately adopted.

In the only phylogenomic study of marsupial relationships to date, Duchêne *et al.* (2018) sampled exon sequences at 1550 loci from 45 species representing 18 of 22 families, including eight dasyurids and the numbat. They estimated ML gene trees individually, which fell into two clusters (1046 and 504 loci), differing principally in the relative positions of Phalangeroidea and Petauridoidea. Species tree estimation used the program ASTRAL (Mirarab and Warnow 2015), which finds the species tree that agrees with the maximum number of four-taxon (quartet) trees in the set of gene trees. The phylogeny (Fig. 5.5) obtained from all loci resolved the expected four-clade structure within Dasyuridae. We discuss the divergence-time aspect of this study in Chapter 9.

García-Navas *et al.* (2020) expanded the data set of Westerman, Krajewski *et al.* (2016) to include Ω-globin, *RUNX2* and *col2A1* gene sequences, and obtained an identical phylogeny (Fig. 5.6) featuring Dasyurini, Phascogalini, Sminthopsini and Planigalini. We describe the details of this tree in our discussion of species-level relationships (Chapter 6). Phylogenomic estimates of dasyurid relationships (i.e. based on thousands of loci throughout the genome) are expected soon through efforts of the Oz Mammal Genomics initiative (Eldridge *et al.* 2020).

Combined molecules and morphology

Kealy and Beck (2017) collected published sequence data for 31 extant dasyuromorphian species and combined it with morphological characters in a total evidence analysis of phylogeny. Their molecular dataset included mitochondrial (*cytb*, *CO1*, *CO2*, *ND1*, *ND2*, *12S*, *16S*) and nuclear (*ApoB*, *bfib7*, *BRCA1*, *ε-glob*, *IRBP*, *P1*, *RAG1*, *vWF*) loci. In addition to rescoring 58 craniodental characters from Wroe *et al.* (2000), Wroe and Musser (2001) and Murray and Megirian (2006), they examined 115

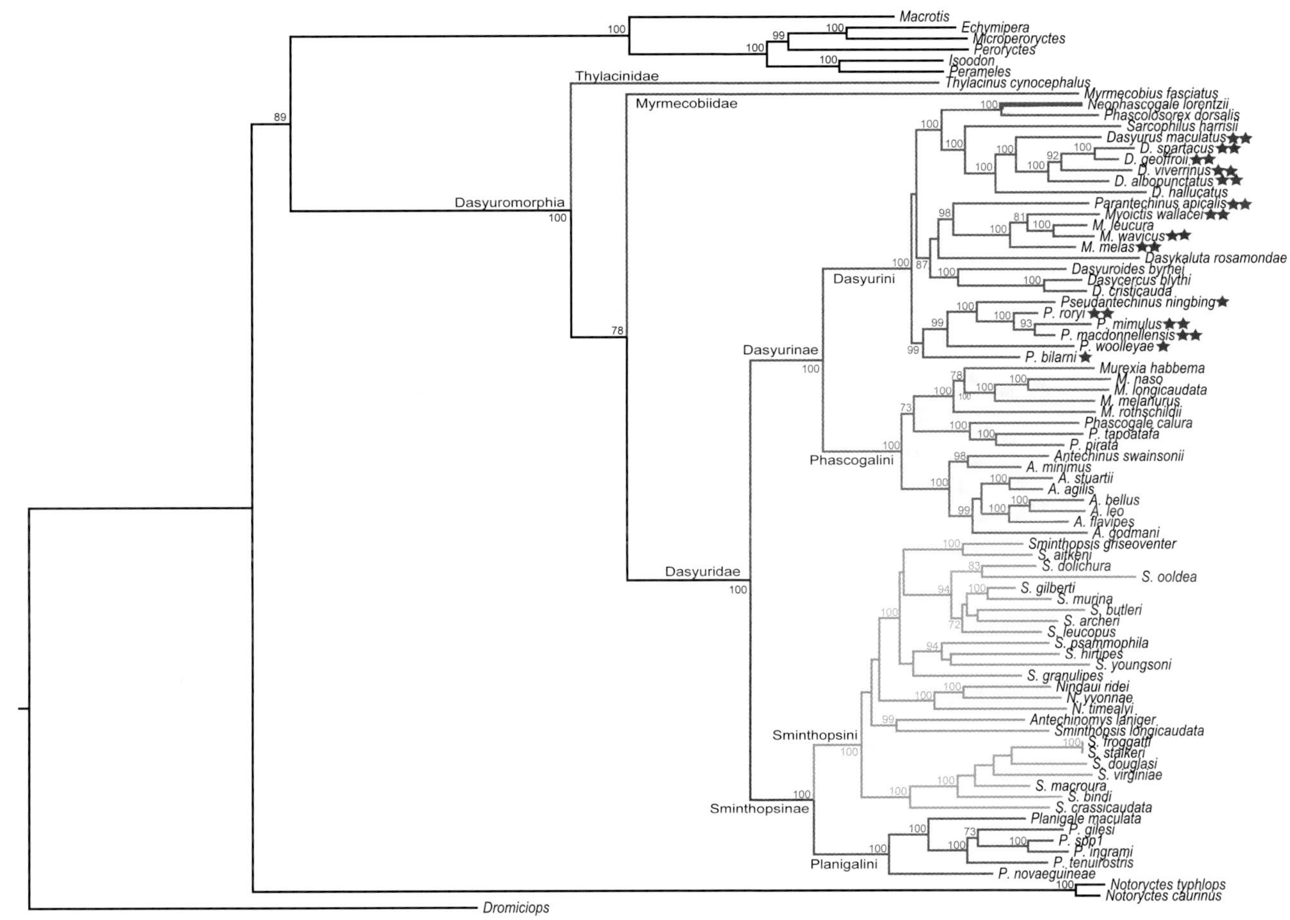

Fig. 5.4. Dasyuromorphian phylogeny from Westerman, Krajewski *et al.*'s (2016) analysis of 17 kb of sequence from eight nuclear (*BRCA1, ApoB, RAG1, vWF, ε-globin, P1, IRBP, bfib7*) and seven mitochondrial loci (*cytb, 12S, tRNA-Val +16S, ND1, ND2, CO1, CO2*). Tree was estimated with the ML method under the GTR + Γ + I evolutionary model. Numbers at nodes are bootstrap support values >70%. Branch length scale is substitutions/site. Stars refer to penis anatomy: double = penis has an appendage containing erectile tissue; single = penis has accessory erectile tissue that does not form an appendage. [Reprinted by permission of John Wiley and Sons from Westerman, Krajewski *et al.* (2016).]

postcranial traits. Bayesian analysis of morphology (assuming the Mk model with gamma-distributed evolutionary rates among characters) produced the tree in Fig. 5.7a, in which most *Murexia* species – labelled with Van Dyck's (2002) genus names – appear basal, but resolution (indexed by posterior probabilities on nodes) was very poor. In contrast, Bayesian analysis of molecular and combined data sets yielded the four-clade structure with high support (Fig. 5.7b, c). *Parantechinus* as originally defined was paraphyletic; the authors therefore recognised *Pseudantechinus bilarni* and *Parantechinus apicalis* (see Chapter 6). *Sminthopsis* could not be resolved apart from *Ningaui* and *Antechinomys*. Kealy and Beck (2017: 15) assessed the starkly contrasting results from morphological versus molecular and total-evidence data as follows:

Morphological data alone might not always be capable of fully resolving relationships within a clade, even in principle … which may explain, at least in part, the relatively low resolution and low support values for most clades. Areas of actual incongruence between our morphological phylogenies and our molecular and total evidence phylogenies, meanwhile, may be due to factors such as non-independence and/or saturation of morphological characters … rather than simply homoplasy, and this warrants further investigation … [D]ental characters have been shown to perform worse than the rest of the skeleton at recovering mammalian phylogeny, as measured by their ability to recover clades that are strongly supported by molecular data.

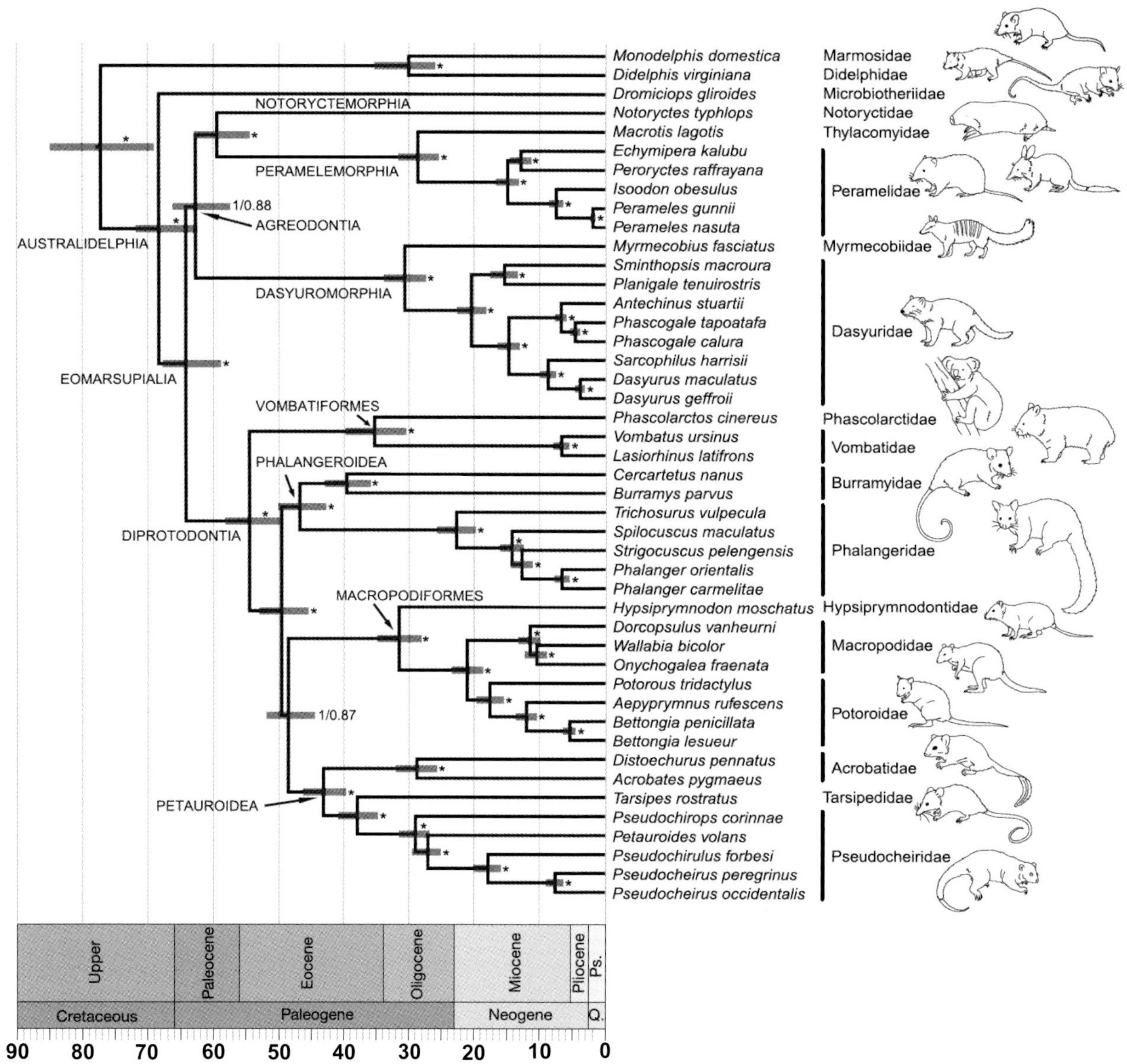

Fig. 5.5. Phylogeny and divergence times of marsupial families from Duchêne *et al.* (2018). Species tree estimated with ASTRAL-II from gene trees for 1550 loci. Divergence time estimates used the fossil calibration dates in Mitchell *et al.* (2014) and the Bayesian MCMCTREE program with a GTR + Γ substitution model and an uncorrelated lognormal relaxed-clock model. Blue bars at nodes represent 95% credibility intervals on divergence dates. [Reprinted by permission of Oxford University Press from Duchêne *et al.* (2018).]

These results were replicated by Beck *et al.* (2022) in an integrative analysis of complete mtDNA and six nuclear gene sequences from 97 extant marsupials, along with 180 craniodental characters from these and an additional 42 fossil taxa. The findings of this study regarding placement of dasyuromorphian fossils and divergence times are discussed in Chapters 8 and 9, respectively.

More than any other factor, the disagreement among trees from morphology and molecules has prevented consensus in modern dasyurid systematics. Do these alternative sources of evidence have equal reliability? If not, why not? Are the limitations of each character type context-specific? At what point should strong consensus from one type take precedence over lack of resolution in

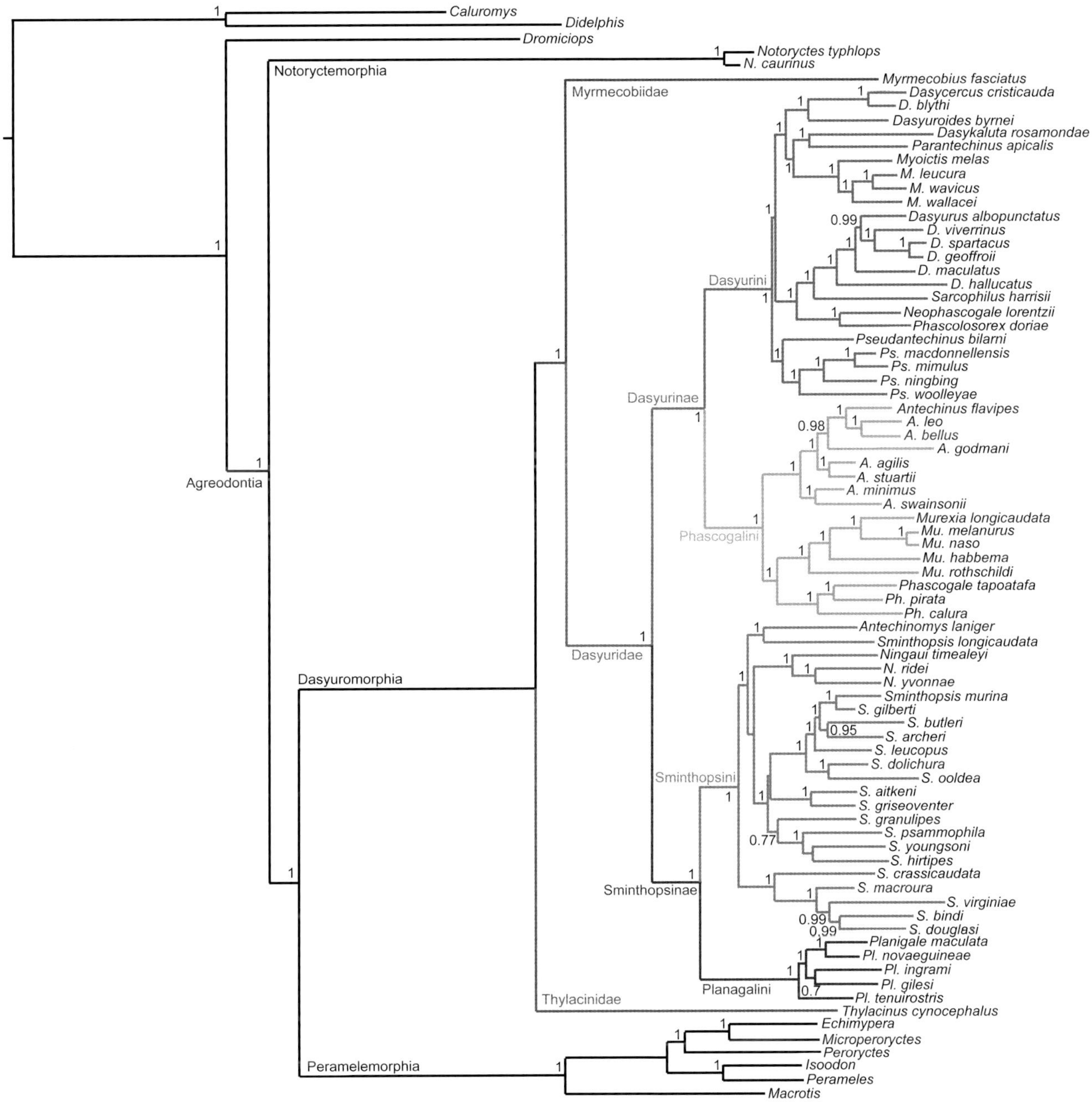

Fig. 5.6. Dasyuromorphian phylogeny from García-Navas *et al.* (2020). Tree is based on partitioned (by locus) Bayesian analysis of the mitochondrial and nuclear DNA sequence alignment of Westerman, Krajewski *et al.* (2016) plus *Ω-globin*, *RUNX2* and *col2A1* loci (17 968 bp). Nodal values are posterior probabilities. [Redrawn with permission of John Wiley and Sons from García-Navas *et al.* (2020).]

the other? How should irreconcilable phylogenetic conflict be reflected in a working classification? What, for example, is the value in recognising each species of *Murexia* as a monotypic genus if those species form a single, recently evolved clade? We explore these issues in the next section.

MORPHOLOGY AND MOLECULES IN PHYLOGENETICS

Reliability of morphological characters

One year after Van Dyck's (2002) analysis of *Murexia*, Scotland *et al.* (2003) offered a provocative evaluation of the role of morphology in phylogeny reconstruction.

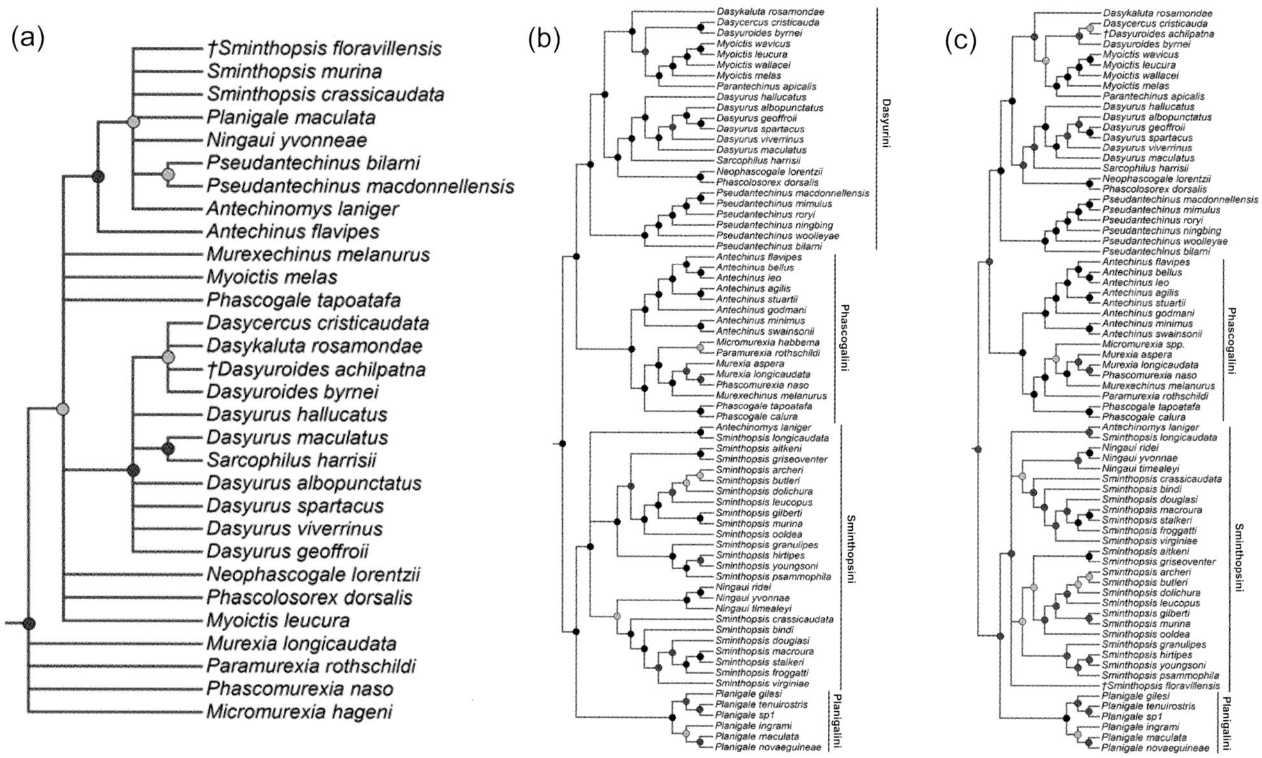

Fig. 5.7. Bayesian majority-rule consensus trees from Kealy and Beck (2017). (a) Tree from 173 morphological characters analysed with the Mk + G model of evolution. (b) Tree from 16.4 kb of partitioned DNA sequence data. (c) Tree from combined morphological and molecular data. Dots at nodes indicate ranges of posterior probabilities: black = ≥0.95; dark grey = 0.75–0.94; light grey = 0.5–0.74. Branches are unscaled. Non-dasyurid taxa in the original analyses have been removed for clarity. [Reprinted from Kealy and Beck (2017) under CC BY 4.0 license.]

They argued that relatively few anatomical traits could be found in which homology assessment and character coding were unproblematic. Presuming that the accuracy and precision of phylogenetic estimates increase asymptotically with increasing numbers of characters, assembling morphological data matrices large enough to obtain robust phylogenetic estimates is undermined by a growing fraction of ambiguous characters. In contrast, much larger numbers of unambiguous characters can be obtained from molecular sequences, although molecular data also present homology issues (see below). The authors concluded that phylogenies of the future should be based on large sequence data sets, with morphological analysis refocused to search for synapomorphies of well-supported molecular clades.

Responses to Scotland *et al.* (2003) were swift and firm. Jenner (2004) rejected the claim that morphological characters have lower phylogenetic accuracy than molecular ones, pointing out that known trees against which to test this claim are unavailable, that gene trees often conflict and that even congruent gene trees are often based on overly simplistic DNA evolution models. Jenner also took issue with Scotland *et al.*'s assertion that morphological data yield lower support values than molecular ones, citing the 'hidden support' phenomenon of Gatesy *et al.* (1999). Moreover, molecular sequence alignment may be just as challenging as morphological homology assessment. In fact, 'there is no evidence to suggest that morphology generally performs more poorly than molecules in estimating phylogeny' (Jenner 2004: 337).

Wiens (2004) emphasised that accurate phylogenetic placement of fossil taxa is essential for assembling the tree of life and – since fossils usually possess only morphological characters – Scotland *et al.*'s (2003) limited role for morphology in systematics is unsupportable. Deemphasising morphology in phylogenetics may also lead to its neglect in other critical endeavours such as species description. Like Jenner (2004), Wiens found Scotland *et al.*'s (2003) pessimism about the efficacy of morphological characters to have little supporting evidence. Smith and Turner (2005) went further, asserting that there is 'no fixed relationship between character sampling and accuracy' and that Scotland *et al.*'s (2003) claim of a small proportion of 'unproblematic' morphological characters is baseless. Like Wiens (2004), they attached great significance to the importance of morphology for placing fossil taxa.

Documenting morphological failures

Following the exchange of views engendered by Scotland *et al.* (2003), the time was ripe for a more careful assessment of morphology's role in phylogeny reconstruction. Springer *et al.* (2007) asked whether ordinal and superordinal clades of placental mammals that were well supported by numerous molecular data sets could be recovered with confidence by the large morphological data matrix of Asher *et al.* (2003). Springer *et al.*'s 'pseudo-extinction analysis' involved assembling a total evidence matrix of 196 anatomical characters, as well as ca. 14 kb of DNA sequence data from 20 nuclear loci, for 44 species representing all placental orders. They sequentially removed the molecular data for each individual order and for each of three superorders (Afrotheria, Euarchontoglires, Laurasiatheria), then estimated trees with parsimony and Bayesian (Mk model) criteria. This analysis mimicked the situation in which fossil species represented only by anatomical traits are placed among living taxa for which molecular data are available. Pseudoextinction analysis failed to recover any superordinal clades and four of the 18 orders, often misplacing key families or genera with superficially high confidence. Springer *et al.* (2007: 682–683) concluded that 'morphological studies of eutherian interordinal relationships have failed to separate homology and homoplasy and have consistently been misled by the latter', and 'combined data only mask the problem for extinct taxa where molecular data are still missing'.

Asher *et al.* (2008) complained that Springer *et al.* (2007) had ignored uncertainty about monophyly of the three superorders, over-generalised results from analysis of a single data set and failed to acknowledge adequately the importance of morphology for fossil taxa. Springer *et al.* (2008: 501–502) rebutted these criticisms, noting with respect to the latter that 'the primacy of morphological data for reconstructing relationships of extinct forms does not excuse morphology from critical assessments of its performance' and '[t]he problem of teasing homoplastic morphological markers from homologous morphological markers will not simply disappear if it is ignored'. Whatever the general import of Springer *et al.*'s (2007) study, it put the issue of morphological adequacy in phylogenetics on an empirical footing. What is more, Springer *et al.* (2007, 2008) suggested an explanation for its inadequacy, namely that excess homoplasy in morphological data matrices – the 'problematic characters' of Scotland *et al.* (2003) – too often supports incorrect trees with high confidence. Subsequent research investigated these ideas from several perspectives.

Perhaps one of the most novel follow-up studies was by Cascini *et al.* (2019), who compared the phylogenetic performance of morphological traits with ancient DNA (aDNA) from two species of Pleistocene kangaroos, *Simosthenurus occidentalis* and *Protemnodon anak*. They used Bayesian analysis to estimate relationships inferred from 119 skeletal traits scored for these plus nine other fossil and 21 extant macropods (all with >70% character-sampling completeness); they compared the results to those from nearly complete mtDNA genomes and a total evidence matrix. Molecular and combined data strongly supported sister-group relationships of *Simosthenurus* with *Lagostrophus* (hare wallabies) and *Protemnodon* with *Macropus* (terrestrial kangaroos, wallaroos and wallabies), whereas morphology alone placed both genera in unresolved positions within a clade sister to *Lagostrophus*. Morphological data were inconsistent in recovering several benchmark clades of kangaroos established by previous multigene analyses. The authors concluded that these morphological characters were positively misleading in the inference of macropod phylogeny and that '[c]ombining morphological and molecular data has democratic appeal, but we should be wary of the possibility of masking rather than resolving discordance' (Cascini *et al.* 2019: 531). Consistent with the suggestion of Springer *et al.* (2008), they speculated that dietary convergence was an important driver of morphological homoplasy.

Saturation, convergence and hemiplasy of morphological traits

Somewhat prior to the debate engendered by Scotland *et al.* (2003), palaeontologist Peter Wagner investigated the possibility that morphological characters, like their sequence counterparts, might experience evolutionary saturation – i.e. multiple changes (including reversals) between states that effectively erase phylogenetic information as clades diverge (Wagner 2000). Such 'character state exhaustion' should be more common when characters have a relatively small number of states. Wagner developed a method to evaluate saturation (curve fitting to plots of the number of states observed as a function of the number of steps on a maximum parsimony tree) and found evidence for it in many fossil data matrices. He concluded that 'the basic premise of cladistic analysis (i.e. that maximum congruence reflects homology rather than homoplasy) becomes increasingly less sound as clades age' (Wagner 2000: 365). Davalos *et al.* (2012, 2014) used Wagner's method to demonstrate state saturation in phyllostomid bat characters.

Kelley and Motani (2015) tabulated dietary information from 69 species of marine mammals (cetaceans, pinnipeds, sirenians), chelonians and crocodylians, along with 17 morphometric craniodental traits. They showed that species clustering by morphometric similarity recovered groups with similar diets (plants, benthic invertebrates, zooplankton, fish, squid, or tetrapods) at high levels of confidence and that discriminant analysis of skull morphometrics classified 87% of species into correct diet categories. Parsimony optimisation of trophic guilds on a published phylogeny of these species suggested that diets (and, by extension, correlated skull dimensions) evolved with a high degree of parallelism. Although this was not a test of the phylogenetic accuracy of morphology, it supported the assertion that adaptive evolution in diet can result in homoplastic similarity of craniodental traits.

Inspired partly by the debate over placental interordinal relationships, Zou and Zhang (2016) investigated the nature of morphological homoplasy in 4541 morphological characters scored for 40 fossil and 46 extant mammal species. They used parsimony to optimise character evolution on trees from morphological, molecular (5722 amino acid residues) and total evidence data. Regardless of which tree or homoplasy measure was used, morphological traits showed significantly higher levels of convergence than molecular ones. Moreover, morphology had higher ratios of convergent to consistent characters, as well as higher ratios of convergent to divergent changes. In view of the latter, the authors concluded that 'the higher convergence of morphological characters is caused by having fewer states rather than intrinsically higher susceptibilities to adaptive convergent evolution, because morphological characters are no more prone to convergence than molecular characters once the number of states is controlled for' (Zou and Zhang 2016: 4). They claimed that the same conclusion holds for DNA sequence data despite the much lower number of possible states (four) relative to proteins (20).

In one of the most provocative considerations of potentially misleading morphological traits, Feng *et al.* (2022) partitioned the genomes of six representatives of Didelphimorphia, Microbiotheria, Dasyuromorphia and Diprotodontia into loci whose gene trees matched their estimated species tree (i.e. *Dromiciops* sister to a clade of dasyuromorphians + diprotodontians) and those that did not (i.e. paired *Dromiciops* with either of the latter two groups). The authors demonstrated that the discordant gene trees were due to incomplete lineage sorting (ILS). Among these loci they identified two (*WFIKKN1* and *PAPSS2*) that could plausibly affect two phenotypic traits in which *Dromiciops* resembles one or the other Australian order (relative heights of spinous processes on first and second thoracic vertebrae and curvature of the humerus, respectively). When alternative versions of these sequences were expressed in mice, they indeed produced the phenotypic differences observed between marsupial groups. The authors argue that misleading phenotypic similarity due to ILS among genotypic determinants represents *hemiplasy* (Avise and Robinson 2008), a pitfall to phylogenetic inference originally associated only with molecular data.

The studies described thus far identify several potential problems in morphological data that are not easily mitigated during phylogenetic analysis: homologous character definition, character saturation by state exhaustion, adaptive convergence and hemiplasy. Recall that the fundamental premise of parsimony is that consistent sets of shared, derived characters are more likely to be homologous than non-homologous. When non-homology predominates, parsimony will be positively misleading about relationships. In contrast, the ability of Bayesian analysis to improve phylogenetic estimates from morphology is contingent upon the adequacy of the process model employed. To date, however, only the simple Mk model (see 'Total evidence analysis' above) has been developed for morphology and its adequacy has been questioned (see below).

Developmental correlation of dental characters

A second assumption of parsimony – and all other criteria for phylogenetic inference – is that characters in a data matrix evolve independently of one another. If two shared derived characters support joining species A and B, but a third supports joining A with C, parsimony supports clade AB. However, if the first two characters are both the result of a single evolutionary event, they should really be counted as one. If that single evolutionary event occurred independently in A and B, and is responsible for the correlated evolution of *several* characters, those characters could provide apparently strong support for the incorrect relationship AB. It is difficult to detect such correlation unless the phylogeny is already known, but one cause of correlated change in morphology is pleiotropic mutation in a developmental pathway – i.e. a single genetic change early in ontogeny causes multiple anatomical changes later.

Kangas *et al.* (2004) demonstrated that dental characters in lab mice (*Mus musculus*) are developmentally correlated. Building on the understanding of developmental signalling pathways in tooth morphogenesis described by Thesleff (2003) and Mustonen *et al.* (2003), Kangas *et al.* examined the effect of knockout and overexpression mutations of the *ectodysplasin A1* gene and reached the following conclusion:

> *We found that with increasing expression levels of this one gene, the number of cusps increases, cusp shapes and positions change, longitudinal crests form, and number of teeth increases. The consistent modification of characters related to lateral placement of cusps can be traced to a small difference in the formation of an early signaling centre at the onset of tooth crown formation. Our results suggest that most aspects of tooth shape have the developmental potential for correlated changes during evolution which may, if not taken into account, obscure phylogenetic history. [Kangas et al. 2004: 211]*

As knowledge of the regulatory networks in tooth development increased (Bei 2009), Jernvall and Thesleff (2012) reviewed how genetic modification of those networks could explain major transitions in tooth morphology during the evolution of vertebrates. Harjunmaa *et al.* (2014) showed that minor manipulations of *ectodysplasin A* expression during development could produce adult tooth morphologies very like those observed among rodent species (including ancient fossils). Indeed, this group of authors has sounded multiple warnings about evolutionary-developmental correlation among dental traits. To what extent have systematists taken heed?

Mounce *et al.* (2016) asked whether the predominance of craniodental over postcranial characters in vertebrate data sets had any consequences for estimated trees. Using a test based on tree-to-tree distances, they found that these two partitions imply different relationships more often than expected by chance, possibly due to distinct selection pressures. Although true trees were unknown, the result provides another caution against over-reliance on traits from a single body region (e.g. teeth). Sansom *et al.* (2017) compared measured levels of homoplasy between dental and non-dental skeletal characters, as well as the ability of each partition to recover molecular benchmark clades, in mammalian groups for which all three types of data were available. They found that dental and non-dental traits implied significantly different relationships in 21 of 40 data sets. Moreover, dental data had significantly higher homoplasy levels (lower retention indices) and recovered significantly fewer benchmark clades than non-dental data. This contradicted an earlier investigation of the same question by Sánchez-Villagra and Williams (1998). Assuming the molecular benchmark clades are valid, Sansom *et al.* (2017: 7) concluded 'that dental data are generally less-reliable indicators of phylogenetic history than osteological data. Not only do dental characters exhibit elevated levels of homoplasy, but they are also unlikely to be sufficient for accurate phylogeny reconstruction on their own'. As a follow-up to this work, Brocklehurst and Benevento (2020) fit models of character evolution to dental and other morphological traits on time-calibrated molecular phylogenies of mammals. For mammals in general, they found that dental characters have higher rates of change – but are no less independent – than other classes of anatomical traits. In marsupials, dental features evolved less rapidly but were more correlated with each other than were other characters, possible due to earlier ontogenetic development of the marsupial jaw and face.

In a study of dental variation among invasive populations of sibling voles (*Microtus rossiaemeridionalis*) in northern Asia, Markova *et al.* (2019) documented evolutionarily rapid changes in molar morphology possibly related to expanded dietary ranges. They suggested that these populations might be experiencing 'despecialisation' away from herbivory, with recently evolved molar traits approximating those found in ancestral arvicolines. Although the impact of developmental correlation on phylogenetic inference from other osteological characters has not been as extensively investigated, the therian skull is thought to be composed of six 'modules' plus the mandible within which morphometric traits (at least) tend to covary (Goswami 2006).

Molecules versus morphology

Early in this debate, some provocateurs made the seemingly outrageous suggestion to abandon morphology as a source of evidence in phylogenetic inference. After all, the most exciting advances in resolving the tree of life have come from recent molecular studies, most of which eventually converged on the same relationships. Putting aside the problem of fossil taxa for the moment, is it naïve to think that molecular data alone can carry phylogenetics forward? Virtually all the authors cited above as critical of morphology also recognised corresponding issues with molecular sequences. Is molecular homology ever problematic? Yes, when orthologous loci must be identified within diverse gene families and when divergent orthologs must be aligned. Can molecular sequences evolve convergently? Yes, there are many examples, including in the phyllostomid bats studied by Davalos *et al.* (2012, 2014). Do molecular characters ever evolve in correlated fashion? Sometimes, as when compensating substitutions occur in RNA stems to maintain stable base pairing. And, of course, sequences can diverge to the point of saturation and lose phylogenetic information. Perhaps the most serious reality check is that gene trees often conflict within the same species tree, considerably complicating inference of the latter from the former.

In one respect, however, morphological and molecular systematics have always differed starkly. For a variety of historical, philosophical and practical reasons, morphologists have clung to cladistic parsimony as their primary analytical tool. Indeed, morphological systematics became virtually synonymous with 'morphocladistics'. In contrast, the analytical tools of molecular systematics have steadily diversified over the past 50 years to include robust inferential frameworks (e.g. maximum likelihood, Bayesian probability), more realistic models of character evolution and sensitive statistical tests for all manner of methodological assumptions. Much of this progress is predicated on stochastic regularities in how DNA evolves – regularities that, so far, have been easy to discern and account for in phylogenetic analysis. Moreover, the genome is big and diverse; large samples of sites can be obtained efficiently from loci with properties conducive to a specific phylogenetic problem. Even the challenge of conflicting gene trees can be addressed with species-tree methods grounded in coalescent theory (Nute *et al.* 2018).

The limitation of morphological analysis to parsimony seemed to have been removed when Lewis's (2001) Mk model allowed likelihood and Bayesian inference of trees from anatomical data. However, this model treats trait evolution as a simple Markov process with symmetric rates of change among states, analogous to the most elementary model of sequence evolution (i.e. Jukes and Cantor 1969). Although some early assessments (e.g. Wright and Hillis 2014) found that analysis with the Mk model outperformed parsimony, few elaborations of the model to capture regularities of trait evolution have been suggested. Indeed, few regularities of anatomical evolution have been identified and therein lies the problem for morphological phylogenetics. Moreover, Goloboff *et al.* (2019: 494) found that key simplifying assumptions make 'the Mk model too unrealistic and inadequate for the analysis of most morphological data sets'. If so, the shadow of parsimony may endure a while longer, despite growing awareness that progress will require methodological advances in character coding and tree estimation beyond the morphocladistic status quo.

Most recent commentaries on the 'molecules versus morphology' issue de-emphasise the role of anatomy in estimating relationships among extant taxa. Lee and Palci (2015) stressed the importance of gathering morphological information from living and fossil taxa to facilitate the 'tip dating' method (Stadler and Yang 2013) of estimating divergence times (see Chapter 9). Pyron (2015) endorsed total evidence analysis for placement of fossil taxa among extant clades but acknowledged the methodological challenges described above. Wanninger (2015: 6) doubted that morphology-based trees would have much future impact in phylogenetics but called for integrated anatomical and developmental research: 'I see the prime contribution of morphology to phylogeny in making sense of trees rather than building them … [O]nce all organisms have been sequenced and once we have agreed on the "true" phylogenetic tree, we will still need morphology to understand how phenotypic diversity evolved.' In this spirit, we now return to the vexing problem of *Murexia* relationships. If molecular results are correct in resolving *Murexia* as a monophyletic group within Phascogalini, what does that imply about the evolution of the craniodental characters in which these species appear primitive?

MORPHOLOGY, MOLECULES AND *MUREXIA*

Morphological character encoding

The basis of Hennig's (1966) method of phylogenetic inference is that clades are identified by shared derived characters (synapomorphies); but before synapomorphies can be adduced, characters and their states must be coded, scored and polarised. The first two steps (but not the last) are equally important for parsimony analysis. As noted in the previous section, all three can be problematic and indeed have been in morphological analyses of *Murexia* (Van Dyck 2002).

State concatenation, in which one or more states of a character are coded to correspond to multiple biological traits, can be a pitfall for cladistic analysis. For example, Van Dyck (2002) defined his character 48 ('C48'), relating to the lower third premolar, as 'P_3 single-rooted or absent' with states '0 = neither' and '1 = yes, single-rooted or absent'. Presumably the 0 state corresponds to species that have a double-rooted P_3, but state 1 encompasses two very different traits – a single-rooted P_3 *and* absence of P_3. Coding the latter into a single, derived state creates an *a priori* synapomorphy uniting species with either trait. C64 (tail length) also includes a concatenation of derived states: 0 = slightly shorter than head + body, 1 = very short or very long. State concatenation is even more insidious when combined with *logical correlation*, in which state variation in one character is necessarily linked to state variation in others. In addition to C48, Van Dyck used five other traits associated with P_3 (characters 40, 41, 43, 45 and 46). For example, C46 is 'P_3 cingulated' with states '0 = not cingulated or P_3 absent' and '1 = cingulated'. State 0 entails a trait concatenation (species without P_3 are treated as identical to those with an uncingulated P_3) but is also correlated with C48 in a complex way. Parsimony computes tree lengths by algorithmically constructing hypothetical ancestors independently for each character (Fitch 1971). An allowed state set for C46 and C48 is 00, but this could be either of two combinations, including one that is impossible (P_3 absent + P_3 present with >1 root). A similar problem applies to the 11 state-combination. Impossible ancestral state combinations undermine tree-length computations in parsimony analyses (Maddison 1993). Logical correlation also affects four other pairs of Van Dyck's characters (4–6, 11–14, 24–25 and 41–42).

For some correlated traits, Van Dyck coded states of the dependent characters (C40 and C43 for P_3, C24 for P^3) as missing. For example, since P_3 is absent in *Planigale gilesi* and seven other species, Van Dyck assigned the missing state '–' to these species for C43 (P_3 orientation). Parsimony handles missing states by optimising through them, using whatever assignment of present states to internal nodes requires the fewest steps on candidate trees. This too may result in impossible ancestral state-combinations – e.g. a node that has P_3 in a longitudinal orientation even though P_3 is absent! Maddison (1993) suggested ways to deal with logical correlations, but Van Dyck (2002) did not explore them.

State definition and delimitation are particularly challenging when scoring traits for which variation among species is continuous rather than discrete (Wiens 2001). Distinguishing between continuous and discrete variation is difficult when the magnitude of discontinuities among observed traits in different species is small. Nevertheless, some morphocladistic analyses attempt to capture continuous variation with discrete states, often without explaining the scoring criteria. In such cases, investigators risk overly subjective scoring that obscures or biases phylogenetic information in the characters. Van Dyck (2002) is a case in point: 42 of 62 craniodental traits divide a range of variation into discrete state values, but no operational definitions of the states are provided. For example, C31 captures size variation of the M^4 paracone with the following states: 0 = large, 1 = slightly reduced, 2 = reduction greater than 1, 3 = reduction greater than 2, 4 = reduction greater than 3, 5 = reduction greater than 4, 6 = reduction greater than 5, 7 = reduction greater than 6. Scoring of such vaguely defined states is unlikely to be replicated by an independent investigator. Indeed, Kealy and Beck (2017) examined 12 traits similar or identical to those reported by Van Dyck (2002), but defined states differently in all but one of them. State-matching in Hennigian analysis and change-counting in parsimony may be severely compromised by subjectively discretised states. For example, M^4 paracones are considered identical in *Antechinus bellus* and *Sminthopsis virginiae* ('more reduced than 3') but differ from those in *S. douglasi* ('more reduced than 2') by one step and from those in *Planigale gilesi* ('more reduced than 4') by two steps, if states are ordered. In an unordered analysis, transitions between all pairs of states – including the most extreme ones – count as a single step! Although Van Dyck (2002) reported no significant difference in parsimony results for ordered versus unordered analyses, bootstrap values are generally higher with the former than with the latter

(Fig. 5.8), including ≥90% at key nodes 53 and 51 of Fig. 4.13. Wiens (2001) recommended a gap-weighting strategy for parsimony analyses of quantitative traits, but this would require measurements of trait values rather than subjective relative assessments.

An easy-to-overlook aspect of character encoding is that states must correctly capture the biological variation present within and among species. Two external features defined by Van Dyck (2002) are problematic in this regard. In his data matrix, *Murexia rothschildi* and *Dasyuroides byrnei* uniquely share the derived state of C67 (body with longitudinal stripe, 0 = absent, 1 = present). However, no such stripe is present in *Dasyuroides*. C72 (males die soon after mating) has just two states, 0 = no and 1 = yes. However, reproductive patterns in dasyurids are considerably more complex than this, with six strategies typically recognised (Lee *et al.* 1982). Not only does Van Dyck's coding scheme treat species showing strategies III-VI as identically primitive, it bins some strategy II species (e.g. *Sarcophilus*, *Pseudantechinus*) with strategy III–VI forms in state 0. Oddly, *Dasykaluta* is also coded as state 0, despite previous documentation that it shows the male die-off pattern (Woolley 1991a).

Evaluation of putative synapomorphies

Although Hennigian argumentation is no longer used as a standard approach to phylogenetic inference, Hennig's legacy of associating inferred clades with specific synapomorphies remains an important component of phylogenetic systematics. In the case of Van Dyck's (2002) tree, the relevant synapomorphies are those marking a clade of all dasyurids apart from *Murexia* (which, for brevity, we call 'non-murexias'). Although Van Dyck did not identify them, there are four traits in his matrix for which all non-murexias possess a derived state and all *Murexia* a more primitive one (these are 'unique, unreversed synapomorphies'; Table 5.1). Three of these are features of the upper first molar (M^1) and one involves the lower fourth molar (M_4). In polarising these characters, Van Dyck cited decisions by Archer (1976d) and Reig *et al.* (1987).

C27 (M1 anterior margin) is not described in Reig *et al.* (1987), which Van Dyck (2002: 317) cites as the basis for his polarity decision: 'Reig *et al.* (1987) added that a bulbous, non-compressed protocone on the upper molars (Chs 27 and 31) was representative of the primitive condition …' It is not clear how this statement bears on the straight-versus-indented anterior margin of M^1; since no other authors discuss this character, the basis for Van Dyck's polarisation is uncertain. Archer (1976d) and most subsequent authors agree that the presence of stylar cusp B on M^1 (C30) is a marsupial symplesiomorphy, but Wroe (1999b) noted that it was independently lost in microbiotheriids and myrmecobiids, as well as some

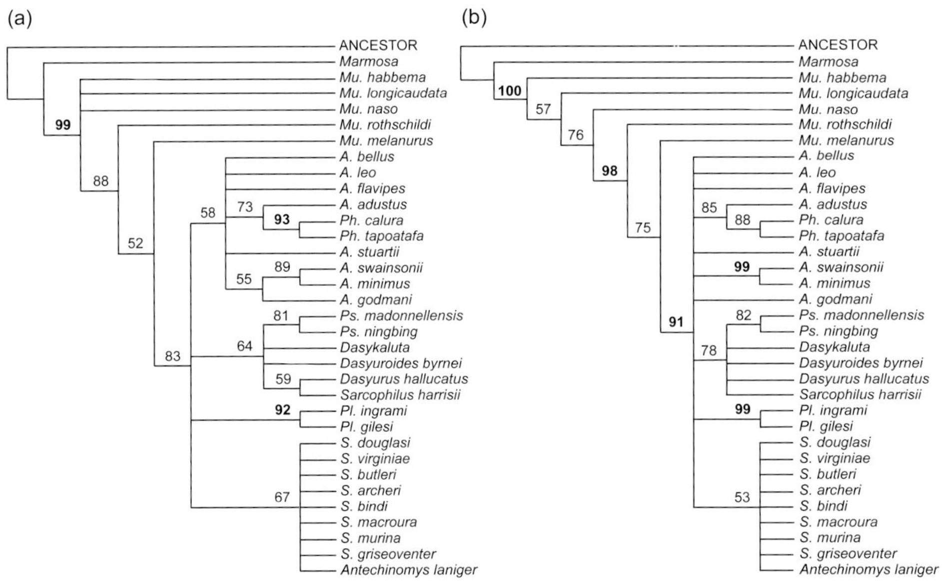

Fig. 5.8. Bootstrapped parsimony analysis of Van Dyck's (2002) morphological data matrix. (a) Unordered states. (b) Ordered states. Analysis performed by PAUP* (Swofford 1999) using heuristic searches with 500 bootstrap replicates, 10 random addition sequences per replicate and TBR branch-swapping. Numbers at nodes are bootstrap support values >50%; values ≥90% are in bold.

Table 5.1. Putative synapomorphies for non-murexia dasyurids from Van Dyck (2002).
For each character, 0 is the primitive state and derived states are ordered 1, 2, 3, etc.

Character number and description	*Murexia* species	Non-murexias
27. M^1 anterior margin	0 = convex or straight	1 = indented or concave
30. M^1 stylar cusp B	0 = no reduction, almost coplanar with stylar cusp D	1 = reduced 2 = greatly reduced
37. M^1 paracone & stylar cusp B	0 = widely separated	1 = approximated 2 = more approximated than 1 3 = fused
51. M_4 talonid cusps	0 = 3 cusps, well developed, or 1 = 3 cusps, poorly developed	2 = 2 cusps 3 = 1 cusp 4 = loss of talonid

thylacinids and dasyurids. Thus, although Van Dyck's polarity for C30 is conventional, the character is known to evolve with considerable homoplasy. Van Dyck considered a widely separated paracone and stylar cusp B on M^1 (C37) to be primitive relative to approximation or fusion of these cusps, based on Kirsch and Archer (1982). Again, however, this character is not discussed in the cited reference. Indeed, Van Dyck (2002: 244) mentions it only once and this with respect to *Antechinus*: 'it [the paracone] is never fused with stylar cusp B'. Polarity for C51 (M_4 talonid cusps) was established by Archer (1976d: 14): 'In forms interpreted to be ancestral marsupials (e.g. *Holoclemensia*) the talonid of m4 has three cusps …' The phylogenetic position of *Holoclemensia* is now uncertain and it may be a stem therian rather than a metatherian (Averianov 2015). However, the Cretaceous marsupialiaformes *Alphadon* and *Pediomys*, as well as extant didelphids and *Dromiciops*, have three talonid cusps, supporting this as the primitive state in dasyurids. Interestingly, Archer (1976d) scored *Murexia longicaudata* as having just two cusps, whereas Van Dyck (2002) shows all *Murexia* species with '3 cusps, well-developed'.

The purpose of this critique is not to impugn Van Dyck's (2002) anatomical acumen, but to view the strongest evidence for 'primitive' murexias in a cladistic framework. Hennigian argumentation would make much of the consistency among these four putative synapomorphies of non-murexias, subject to the caveat that antecedent character analysis had established their homology and polarity. Scotland *et al.* (2003) worried that few anatomical traits would allow for such certainty and even the brief review presented here casts doubt on one or both aspects of these putative synapomorphies. Moreover, the kind of developmental correlation in dental traits found by Kangas *et al.* (2004) could apply to virtually all tooth characters scored by Van Dyck (2002). In view of such considerations, we suggest that C27, C30, C37 and C51 provide extremely weak evidence for monophyly of non-murexias. Of course, Hennigian analysis was abandoned precisely because homoplasy is a real feature of character evolution. If we cannot confidently identify unique and unreversed synapomorphies for non-murexias, perhaps the results of parsimony analysis offer stronger support for this conclusion.

Evaluation of morphological parsimony

Parsimony analysis requires that characters in a data set contain phylogenetic information. Archie's (1989) randomisation, or permutation tail probability (PTP), test was developed to make this assessment. It is now widely appreciated that PTP lacks statistical power, but certainly a data set that fails the test should not be used for phylogenetic inference with parsimony (Slowinski and Crother 1998). Van Dyck's (2002) original data pass the test, as do reduced matrices in which the hypothetical outgroup and one member of congeneric sister-pairs are deleted (our unpublished results).

Springer *et al.* (2007) asked whether morphological data matrices could recover benchmark clades. For dasyurids, reasonable benchmarks not involving *Murexia* might be tribes Dasyurini, Sminthopsini and Planigalini, subfamily Sminthopsinae and a clade of *Phascogale* and *Antechinus* (i.e. Phascogalini minus *Murexia*). Although his taxon sampling was incomplete, Van Dyck's (2002) published consensus tree (see Fig. 4.15) does recover clades of the dasyurin, sminthopsin and planigalin species he included, as well as *Phascogale* + *Antechinus* (though *Antechinus* is not monophyletic). It does not resolve sminthopsins and planigalins in a single clade of sminthopsines.

Van Dyck (2002) did not provide node-support values for his parsimony consensus tree, but the results of a bootstrapped parsimony analyses are shown in Fig. 5.8. With unordered states (the analysis reported by Van Dyck), only three clades receive strong (≥90%) bootstrap support: all dasyurids (99%), *Phascogale* (93%) and *Planigale* (92%). With ordered states, more clades are well-resolved: dasyurids (100%); *M. rothschildi*, *M. melanurus* and non-murexias (98%); non-murexias (91%); *Planigale* (99%); *Antechinus swainsonii* + *A. minimus* (99%). Benchmark clades on the parsimony consensus tree receive variable support: Dasyurini (64% unordered, 78% ordered); *Phascogale* + *Antechinus* (58%, <50%); Sminthopsini (67%, 53%); Planigalini (92%, 99%), Sminthopsinae (0%, 0%). Given that so many characters are continuous with discretised states (described in the preceding section), analysis of ordered states is appropriate. Despite Van Dyck's (2002: 231) comment that unordered and ordered 'runs produced basically the same result', the ordered bootstrap results make a stronger case for his tree.

Strongly supported nodes in Fig. 5.8b that separate murexias from non-murexias are isolated among nodes with weaker (<90%) support. One may ask, then, whether rearrangements involving mostly the latter would result in significantly worse trees for this data set. Tree-comparison tests were developed to make this assessment. Parsimony-based tests include the Wilcoxon signed ranks test (Templeton 1983), winning sites test (Prager and Wilson 1988) and *t*-test of length differences (Kishino and Hasegawa 1989), each designed to evaluate whether the minimum length of one tree is significantly less than the minimum length of another for a single data matrix. We applied these tests to the Van Dyck (2002) data, comparing the set of minimum-length trees from unconstrained analyses of ordered and unordered states with two sets of minimum-length constrained trees: those requiring monophyly of all *Murexia* species and those requiring monophyly of Phascogalini. The results are shown in Table 5.2.

In some cases (shown with asterisks in Table 5.2), *P*-values of the test statistic straddle a nominal 0.05 significance threshold, marginally rejecting some or all of the constrained trees. In other cases, the null hypothesis of no significant difference is not rejected. Interpretation of these simple tests is complicated by multiple comparisons (among equally parsimonious trees) and selection bias (the unconstrained trees were not chosen *a priori*, but after analysis of the data on which the tests were performed), but both problems should increase Type I error (rejection of a true null). In Table 5.2, however, we see marginally significant or insignificant comparisons. Thus, tree-comparison tests provide a check to the confidence suggested by bootstrap values on Van Dyck's (2002) tree.

Murexia: what's in a name?

Romeo's question has perpetual relevance for biological classification, including what to call the clade we refer to as *Murexia*. Virtually all molecular analyses of dasyurid phylogeny place *Murexia* species as a monophyletic group within the larger clade Phascogalini (Eldridge *et al.* 2019). Van Dyck's (2002) morphological parsimony analysis suggested that they form five separate lineages that diverged prior to the common ancestor of all other dasyurids. The two hypotheses could not be more different, but which one is correct? And, if one is feebler than the other, what causes it to be so? In the foregoing review, we suggested the following reasons why Van Dyck's (2002) results are weak: (1) they suffer from several flaws of character encoding (state concatenation, logical and developmental correlation); (2) they are influenced by subjective discretisation of continuous traits; and (3) their support for non-murexia monophyly is at best only marginally better than for the alternative hypotheses of *Murexia* and Phascogalini monophyly.

Kealy and Beck's (2017) total-evidence analysis of dasyurid relationships recovered a monophyletic group of murexias within Phascogalini. Nevertheless, they recognised the murexias by Van Dyck's (2002) novel genus names and this nomenclature has also been adopted in several prominent marsupial classifications: Wilson and Reeder's (2005) *Mammal Species of the World*, the *Encyclopedia of Life* (https://eol.org) hosted by the Smithsonian Institution, the U.S. government's Integrated Taxonomic Information System (https://itis.gov), the University of Michigan's Animal Diversity Web (animaldiversity.org), the Discover Life website (www.discoverlife.org), the Universal Protein Resource's taxonomy database (www.uniprot.org/taxonomy), the MorphoBank database (https://morphobank.org) and many others. Only a few equally prominent classifications have retained the single genus *Murexia*, including the IUCN Red List (www.iucnredlist.org), the National Center for Biotechnology Information's Taxonomy Browser (www.ncbi.nlm.nih.gov/taxonomy) and the American Society of Mammalogists Mammal Diversity Database (www.mammaldiversity.org) based on Burgin *et al.* (2018). Which classification should be adopted and why does it matter?

Table 5.2. Results of tree-comparison tests for Van Dyck's (2002) morphological data.

Analyses were performed with PAUP* (Swofford 1999) using heuristic searches with random stepwise addition and TBR branch-swapping. An asterisk (*) indicates significance at the 0.05 level. MP = maximum parsimony (i.e. minimum length), KH = Kishino-Hasegawa.

Unordered states. Unconstrained analysis = 3 MP trees (513 steps)

		Tree-comparison test *P*-values		
Constraint	**MP trees (length)**	**Templeton**	**Winning Sites**	**KH**
Murexia monophyly	21 (519)	0.0093*–0.0581	0.0106*–0.0614	0.0113*–0.3338
Phascogalini monophyly	48 (526)	0.1797– 0.3231	0.2632– 0.5572	0.1815–0.3338

Ordered states. Unconstrained analysis = 24 MP trees (612 steps)

		Tree-comparison test *P*-values		
Constraint	**MP trees (length)**	**Templeton**	**Winning Sites**	**KH**
Murexia monophyly	15 (622)	0.2753–0.3205	0.2002–0.4404	0.2724–0.3804
Phascogalini monophyly	6 (629)	0.0610–0.0904	0.0177–0.0449*	0.1385–0.1903

The central tenet of phylogenetic classification is that only monophyletic groups should be named. This was Van Dyck's (2002) justification for erecting monotypic genera for each species of murexia – if his tree were correct, this would be the only defensible scheme. At the time, the only other comprehensive estimates of murexia relationships were the mitochondrial and limited nuclear DNA analyses summarised by Krajewski, Wroe and Westerman (2000). Van Dyck suggested several reasons why these molecular results might have been incorrect and, on these grounds, dismissed them in favour of his own. However, as summarised in Chapter 4 and above, subsequent molecular studies with multiple, independent loci and broader taxon sampling have uniformly supported the monophyly of *Murexia*. Assuming the latter tree is correct, what should we name the species in this clade? Provided their boundaries are correctly delineated, there is no absolute prohibition against placing each species in a monotypic genus. What are the costs and benefits associated with alternative classifications?

One perspective has to do with taxonomic information retrieval. If we recognise a genus for every species of murexias, why not do the same for any (or every) other genus? There are currently some 19 species of *Sminthopsis*, each of which could be recognised in a monotypic genus. While 'legal', this would result in complete classificatory redundancy at the genus level, merely duplicating the species inventory without adding any information about relationships. Recognising a single genus for the clade eliminates such redundancy and provides an unambiguous indication of relationship.

A second basis on which to evaluate alternative classifications is taxonomic consistency. Systematists have long debated whether groups recognised at specific ranks can be made comparable in any objective feature, including species richness, morphological diversity, genetic diversity, adaptive trait-complexes and age. Hennig (1966) devoted 20 pages of *Phylogenetic Systematics* to this topic, but most modern workers consider it a futile endeavour. Nevertheless, classificatory allowances are regularly made for individual species or small clades that show marked divergence from their sisters. There is only one living species of Tasmanian devil (*Sarcophilus harrisii*), for example, and it is clearly the sister group of quolls (*Dasyurus*), so why not recognise it as part of the latter genus? This has not been done because the devil is phenotypically distinct from quolls in several traits adaptive for a distinct ecological niche (stout-bodied predator-scavenger). Are the species of *Murexia* as adaptively distinct from one another as are, say, quolls and the devil? Murexias are relatively little-studied, but no published report has ever suggested adaptive divergence comparable to that between *Sarcophilus* and *Dasyurus*: they are all medium-sized, terrestrial-scansorial insectivore-carnivores with aseasonal breeding. Genetic distances among *Murexia* species are comparable to those within other dasyurid genera, as are the divergence times among species (see Kealy and Beck 2017 and references therein).

Finally, there is the widespread perception that elevating a taxon to a higher rank enhances popular and political recognition of its conservation value (Isaac *et al.* 2004). Morrison *et al.* (2009) found no consistent effect of taxonomic revision on conservation status but did note that all cases they found of improved conservation effort were associated with splitting. Indeed, discussion of 'taxonomic inflation' (Isaac *et al.* 2004; Zachos 2015) has focused mostly on revision of single polytypic species into multiple phylogenetic ones, but the same rationale (such as it is) applies to the elevation of one species to a monotypic genus. Some authors (e.g. Padial and De La Riva 2006) argue that ever-changing species lists are an inevitable consequence of vigorous taxonomic research, while others (e.g. Garnett and Christidis 2017) bemoan the 'taxonomic anarchy' resulting from unregulated classificatory revisions. Either way, there is a major difference between accurate species or clade delimitation and arbitrary taxonomic elevation. The latter constitutes manipulation of biological nomenclature for political purposes, undermining the integrity of taxonomic research. Will recognising five monotypic genera of murexias, rather than five species of *Murexia*, improve the conservation outlook for these animals? This question has never been addressed, but it is at least debatable whether 'playing games with names' is justified by even such a laudable goal. We stress that this strategy played no part in Van Dyck's (2002) original revision and argue that it should be similarly irrelevant to evaluating his classification.

The conflict over *Murexia* exemplifies many similar situations in systematic biology and has been instructive for dasyurid taxonomists. We conclude that the balance of evidence strongly rejects Van Dyck's (2002) tree in favour of a monophyletic *Murexia* and that classification of murexias into five monotypic genera is unjustified. It cannot be rescued by appeal to information retrieval, evolutionary divergence, or conservation efficacy and should be abandoned.

6 Dasyurid species-level relationships

The history of research leading to the current consensus that Dasyuridae consists of four major clades was traced in Chapters 3–5. The four clades are represented by two subfamilies, each containing two tribes, as shown in the classification presented in Chapter 2. Here we consider studies of the phylogeny of the species within each tribe, the results of which form the basis for current definitions of genera as well as inter- and intrageneric relationships. Frequent reference will be made to the large-alignment and integrative phylogenies described in Chapter 5 and for brevity we abbreviate these and link them to their corresponding figures as follows: M14 = Mitchell *et al.* (2014; Fig. 5.2); MC15 = May-Collado *et al.* (2015; Fig. 5.3); W16 = Westerman, Krajewski *et al.* (2016; Fig. 5.4); KB17 = Kealy and Beck (2017; Fig. 5.7); García-Navas *et al.* (2020) = GN20 (Fig. 5.6).

DASYURINAE

When Archer (1984c) summarised the state of dasyurid systematics just before the contributions of molecular data began to be felt, he recognised five major extant clades: *Planigale* + *Ningaui* + *Sminthopsis* including *Antechinomys* (Sminthopsinae); *Murexia* ('Muricinae'); *Phascolosorex* + *Neophascogale* ('Phascolosoricinae'); *Phascogale* + *Antechinus* (Phascogalinae); and the remaining seven genera ('Dasyurinae') (Fig. 6.1, branches 4–8). Relationships among them, however, were unclear. In the previous three chapters we showed how molecular studies found that the latter four lineages coalesce into a more inclusive Dasyurinae. The molecular consensus codified by the classification in Chapter 2 and achieved by the research described in Chapters 4 and 5 shows that Dasyurinae also consists of two clades: Dasyurini includes Archer's dasyurine genera plus the phascolosoricines; and Phascogalini includes Archer's phascogalines plus *Murexia*. In this section we examine species-level relationships within these four tribes.

Dasyurini

The first indication of dasyurin monophyly came from Baverstock *et al.*'s (1990) albumen MC′F study (see Fig. 4.3). Archer (1984c) had identified quolls and the Tasmanian devil (*Dasyurus*, '*Satanellus*', *Sarcophilus*), false antechinuses (*Pseudantechinus*, *Parantechinus*, *Dasykaluta*) and phascolosoricines as separate lineages. MC′F also recovered these lineages, but united them with *Dasycercus*, *Dasyuroides* and *Myoictis* in what is now Dasyurini. Although early DNA sequence studies (Krajewski *et al.* 1992, 1993, 1994; Retief *et al.* 1995a) supported the association of phascolosoricines with other dasyurins, it was not until the multigene analysis of Krajewski, Wroe and Westerman (2000) that this result received strong statistical support (compare Figs 4.8–4.10 and 4.12 with Fig. 4.14). All subsequent molecular studies confirmed the monophyly of Dasyurini, as well as the sister-pairing of *Phascolosorex* and

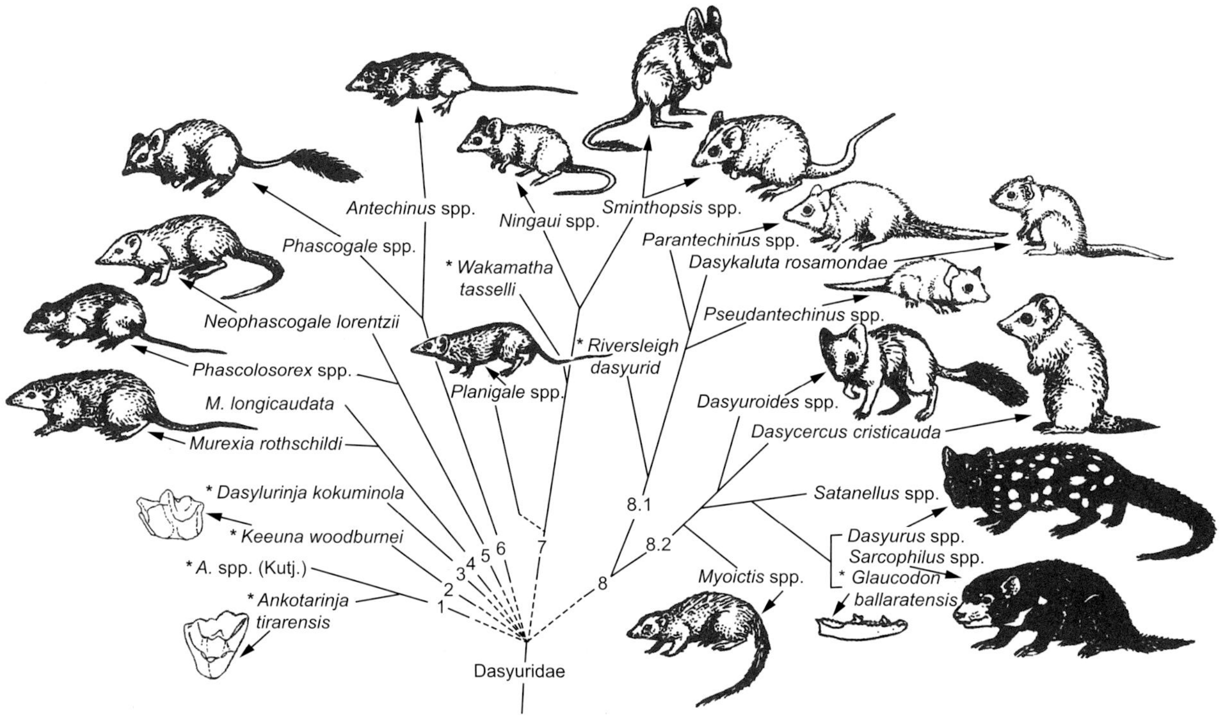

Fig. 6.1. Archer's (1984c) view of dasyurid phylogeny based on previous morphology and allozyme studies. Extinct taxa are indicated with asterisks (*) and discussed in Chapter 8. [Reproduced with permission of the Royal Zoological Society of New South Wales from Archer (1982c), with minor adaptations.]

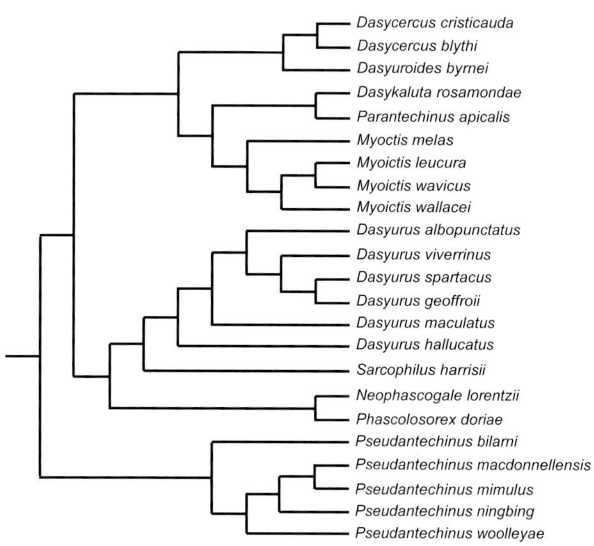

Fig. 6.2. Phylogeny of Dasyurini from García-Navas *et al.* (2020). This tree is part of a larger phylogenetic estimate for dasyuromorphian species. The branching order is derived from Bayesian analysis of ca. 18 kb of mitochondrial and nuclear loci; all clades have posterior probabilities ≥0.99. [Adapted from García-Navas *et al.* (2020).]

Neophascogale (see Chapter 5). For reference to the discussions in subsequent sections, we show the dasyurin portion of the GN20 phylogeny in Fig. 6.2.

Phascolosorex and *Neophascogale*

Baverstock *et al.* (1990) placed the New Guinean endemic phascolosoricines as one of three unresolved branches at the base of Dasyurini, the other two being quolls + devil and the remaining six genera (see Fig. 4.3). Krajewski, Young *et al.* (1997) and Krajewski, Wroe and Westerman (2000) resolved this trichotomy in favour of a sister relationship between phascolosoricines and *Dasyurus* + *Sarcophilus* (Fig. 6.3; see also Fig. 4.14). This was surprising in that Archer (1982c: 429) had characterised phascolosoricines as follows: 'Dental morphology suggests that these species are either very autapomorphically specialised dasyurids, or that they represent another major phylogenetic branch, hierarchically equivalent taxonomically to the rest of the dasyurids.' Archer (1984c: 636) noted, 'Some of the highland New Guinean living forms such as species of *Murexia* and *Neophascogale* resemble the fossil forms from the middle Miocene … This

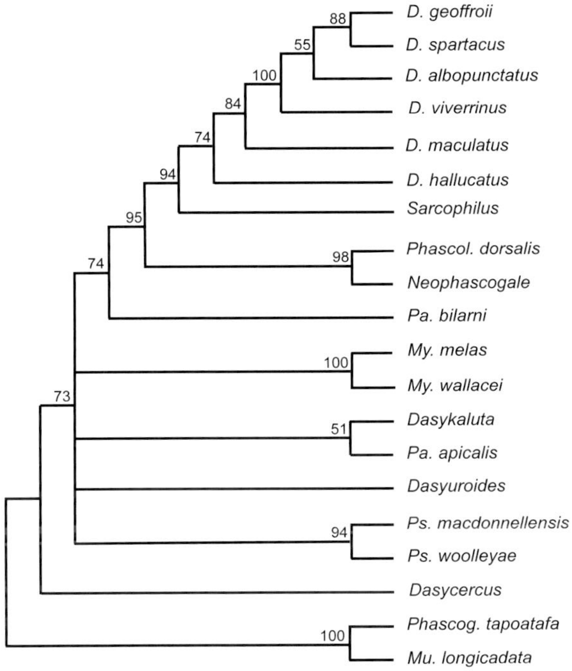

Fig. 6.3. Maximum parsimony tree for *cytb*, *12S* and *P1* DNA sequences from dasyurines (Krajewski, Young *et al.* 1997). Numbers at nodes are bootstrap support values. [Reprinted by permission of Springer Nature from Krajewski, Young *et al.* (1997).]

again suggests that these living genera are in a way relics …' Molecular juxtaposition of such apparently primitive species with specialised carnivores like *Sarcophilus* and *Dasyurus* was anatomically jarring.

Tate (1947) included phascolosoricines in his 'Dasyurinae' (= Dasyurini), a group he defined in part on reduction of the third premolars to accommodate a shorter rostrum in animals with more carnivorous diets (see Fig. 3.6). Among dasyurins, *Phascolosorex* and *Neophascogale* show the least reduction, whereas *Dasyurus* and *Sarcophilus* show the most (in fact, they lack third premolars). In contrast, Archer's (1976d) interpretive 'phenogram of dental similarities' among dasyurids (see Fig. 3.7) positioned phascolosoricines as sister to Sminthopsini, a result apparently driven by the bilobed I_3 that the former species share with *Sminthopsis granulipes*. In his revision of *Sminthopsis*, Archer (1981) included a Hennigian analysis of diverse morphological characters and came to the same conclusion (see Fig. 3.8) but based on a different putative synapomorphy: I^4 generally longer crowned than I^2 (caption to his fig. 74). Oddly, Tate (1947) considered $I^2 < I^4$ primitive for dasyurids, as did Archer (1976d), who equated plesiomorphic states with those found in didelphids. Archer (1981) did not comment on this inconsistency, but also claimed that *Neophascogale* and *Phascolosorex* 'share many derived character states with *Sminthopsis*' (p. 211). Other than the bilobed I_3 and squamosal-frontal contact, however, he did not identify these characters.

In an attempt to determine relationships of the dasyurid fossil *Archerium chinchillensis* (see Chapter 9), Wroe and Mackness (2000: 323) adduced six 'derived dasyuromorphian features that are common to all dasyurines … except "phascolosoricines"': reduction of P^3, reduction of stylar cusp B on M^1; reduction of metaconids; approximation of stylar cusp B and metacone on M^1; alignment of M^1 postmetacrista to the long axis of the toothrow; oblique orientation of hypocristids to the long axis of the dentary. The authors did not explain how these characters were polarised, nor did they estimate a phylogenetic tree, but simply noted, '… the interpretation of phylogeny presented here for *A. chinchillensis* is based on subjective analysis alone, although the fundamental principles of phylogenetic analysis are adhered to' (Wroe and Mackness 2000: 323). Despite finding apparent support for a clade of non-phascolosoricine dasyurins, they concluded, 'much homoplasy must be accepted in any interpretation of the subfamily's phylogeny' (p. 324).

Wroe *et al.* (2000) used numerical parsimony analysis of 77 polarised craniodental traits to estimate relationships among 24 dasyuromorphian species (including *Neophascogale lorentzii* as an exemplar of phascolosoricines), with living and fossil outgroups from several marsupial orders. Parsimony trees placed *Neophascogale* in a clade with phascogalins and sminthopsines, supported by two homoplastic state-changes, but with bootstrap support <50%. Monophyly of two quolls and the devil (the only dasyurins included) also had <50% support. Wroe and Musser (2001) reanalysed the same data from 29 taxa and found that *Neophascogale* was sister to *Murexia*, again based on two homoplastic traits (different from those in Wroe *et al.* 2000) and without bootstrap support.

Krajewski *et al.* (2004) attempted to resolve this situation by constructing a molecular phylogeny for 19 of the then-recognised 22 species of dasyurins with ca. 7 kb of DNA sequence data from four mitochondrial and three nuclear loci using parsimony, additive distance, maximum likelihood and Bayesian methods. This study also included parsimony re-analyses of morphological data from Archer (1976d) and Wroe and Mackness (2000), the latter supplemented with states for phascolosoricines. DNA parsimony analyses used homologous gap-coding (*P1*, *bfib7*) and 3TV weighting (*cytb*). Additive distances were calculated with the weighted-average method of Krajewski *et al.* (1999) to account for rate-variation among codon positions, RNA stems and loops, and introns. Maximum likelihood searches employed discrete rate categories for these partitions. Bayesian analysis used the best-fit model for a concatenated, unpartitioned alignment of sequences from all species. The parsimony tree in Fig. 6.4 is compatible with distance, likelihood and Bayesian results for all well-supported nodes. Consistent with the findings of Krajewski, Wroe and Westerman (2000), phascolosoricines are sister to *Dasyurus* + *Sarcophilus*. The large alignment analyses of Cardillo *et al.* (2004), M14, MC15, W16 and GN20 (Fig. 6.2) – as well as the integrative KB17 analyses – obtained the same result.

Parsimony reanalysis of Archer's (1976d) 45-character morphology matrix with unordered states failed to recover a well-supported Dasyuridae or any benchmark clades within it other than *Phascogale* (Krajewski *et al.*

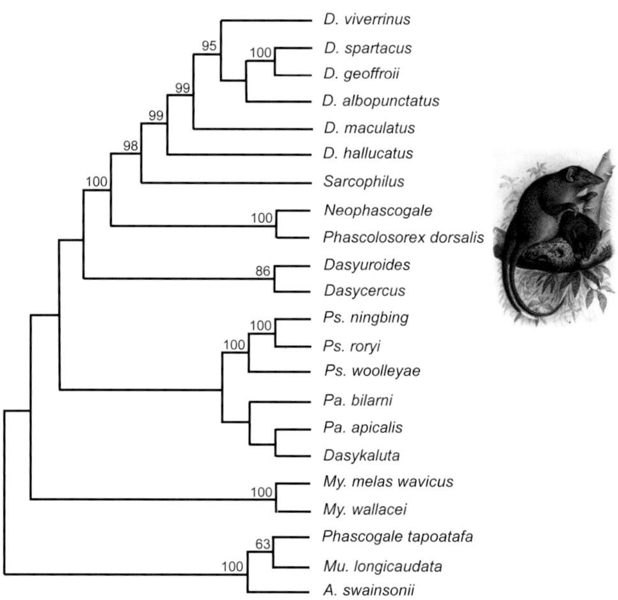

Fig. 6.4. Phylogeny of Dasyurini from Krajewski *et al.* (2004). This tree is based on parsimony analysis of 1043 informative sites from *cytb, 12S, Val, 16S, P1, IRBP and bfib7* loci with 3TV coding of *cytb* and homologous-gap coding of *P1* and *bfib7*. Numbers at nodes are bootstrap support values >50%. [Reprinted by permission of CSIRO Publishing from Krajewski *et al.* (2004). Permission conveyed through Copyright Clearance Center, Inc.]

2004). Analysis of only the 36 informative characters for dasyurins, using polarised and ordered states, produced a tree on which *Neophascogale*, *Phascolosorex dorsalis* and *Myoictis* are excluded from a highly supported clade of other dasyurins. A single Hennigian synapomorphy (M_1 metaconid reduction) maps to the latter branch. A bootstrapped parsimony tree from the ordered but unpolarised characters of Wroe and Mackness (2000) recovered non-phascolosoricine dasyurins in a moderately resolved (79% bootstrap) clade supported by the six synapomorphies cited in the original paper (see above). These results highlight a striking conflict between molecular and morphological results comparable to that for *Murexia* discussed in Chapters 4 and 5. We suggested that the *Murexia* conflict was heavily influenced by problems of morphological character definition, coding, polarisation and correlation, as well as overconfidence in parsimony results. Do the same factors influence our understanding of phascolosoricine relationships?

Krajewski *et al.* (2004) argued that molecular placement of phascolosoricines was robust to several possible biases: limited taxon sampling, contamination of phascolosoricine DNA samples, dominance of a single gene, nonstationary base composition, rooting artifacts and model misspecification. What about the morphological results? Krajewski *et al.* (2004: 409) gave the following assessment.

> *Are the results affected by limited taxon sampling? Clearly they are in one important sense, because phylogenetic use of these dental features diminishes greatly when many taxa other than dasyurins are included. Is the result an artefact of specimens examined? Unlikely, because most authors examined series of specimens and coded polymorphic characters as such in their data matrices. Is the result biased by correlated characters? All the relevant evidence presented to date has been from teeth, and five of the six simple synapomorphies for 'Dasyurinae' in [Wroe and Mackness (2000)] are from molars. Archer (1976[d]: p. 21) made the following assessment of the dental features he studied: 'Almost every character tabulated or discussed above is phenetically related to one or more of the other characters and very few vary independently' ... Such covarying characters will inflate the bootstrap support for internodes on which they transform. Is the morphological result an analytical*

artefact? We have not considered the strength of the character analyses, including polarisations, made by Archer (1976[d]) and subsequent authors. In general, however, these are among the most compelling aspects of the original papers, usually involving explicit arguments based on outgroup and fossil criteria. Nor is it clear that changing polarities of the dental characters would mitigate the conflict with DNA. Finally, we have emphasized numerical parsimony and bootstrap methods for obtaining a phylogenetic estimate from the dental data. Those methods reveal that, with few exceptions, the characters evolve in a highly homoplastic fashion and that much of their resolving power comes from parallel or reverse state changes. That these dental traits are prone to homoplasy is uncontroversial and was noted explicitly by Tate (1947), Archer (1976[d]), and Wroe and Mackness (2000).

In other words, yes – the conflict between morphological and molecular results for phascolosoricines and *Murexia* have several common causes, the most significant being the apparently extreme evolutionary lability and interdependence of dental characteristics in dasyurids. In the case of *Phascolosorex* and *Neophascogale*, however, no aberrant reclassification obfuscates the modern consensus that these genera form a single clade that is sister to *Dasyurus* + *Sarcophilus*.

Dasyurus and *Sarcophilus*

Archer's (1984c) dasyurid phylogeny (Fig. 6.1) reflected the long-standing consensus that species of *Dasyurus* and *Sarcophilus* form a single clade within Dasyurini. However, the tree reveals uncertainty about the monophyly of quolls, with Archer following Tate (1947) in assigning *D. hallucatus* and *D. albopunctatus* to Pocock's (1926) *Satanellus* and placing them as sister to a branch bearing *D. maculatus*, *D. viverrinus*, *D. geoffroii* and *Sarcophilus*. Tate's union of northern and New Guinean quolls was based on what he considered primitive dentition (e.g. $I^4 > I^2$). Archer (1976d) expanded this view and placed '*Satanellus*' species at the primitive end of a 'structural trend' in dental traits that culminated in *Sarcophilus*. Inclusion of the devil with dentally derived quolls was based on Archer's (1982c) contention that dentition of the Pliocene *Dasyurus dunmalli* (see Chapter 9) implied independent loss of P3 in '*Satanellus*' and *Dasyurus-Sarcophilus*. Moreover, allozyme data (Baverstock *et al.* 1982) showed that *D. maculatus-viverrinus-geoffroii* formed a clade apart from *D. hallucatus* (*D. albopunctatus* was not included; see Fig. 4.2).

Van Dyck (1987) described *Dasyurus spartacus* from specimens collected in savanna habitat of the Trans-Fly Plains in southern New Guinea. Waithman (1979) had allocated this material to *D. geoffroii* which 'resulted from not only its strong superficial resemblance to *D. geoffroii* but also from a reasonable expectation that *D. geoffroii* with its extensive pre-European distribution throughout Australia … could occur in New Guinea' (Van Dyck 1987: 152). Van Dyck diagnosed the Bronze Quoll by its exceptionally narrow rostrum, greatly reduced hallux and short pinnae. More significantly, he undertook a Hennigian analysis of quolls based on 19 ordered and polarised anatomical traits (14 of them dental). Although he explored two scenarios for interpreting character evolution, with and without the influence of *D. dunmalli*, Van Dyck (1987) presented only one branching order of species (Fig. 6.5a). In contrast to Archer (1982c, 1984c), this tree placed *D. albopunctatus* as sister to *D. spartacus* in a clade with *D. maculatus* but retained *D. hallucatus* as sister to all other species. In their redescription of *D. dunmalli*, Wroe and Mackness (1998) performed a numerical parsimony analysis with 22 similar characters and obtained a similar tree (Fig. 6.5b). Two things about these analyses are noteworthy: (1) the basal separation of northern quoll from other species is supported by 9–10 putative synapomorphies, whereas other internal branches are marked by only 1–4 changes; and (2) most character state definitions are nonoperational (e.g. character 5 = 'premolars narrow, small').

The albumin MC′F study of Baverstock *et al.* (1990) showed *Dasyurus*, '*Satanellus*' and *Sarcophilus* as a trichotomy (see Fig. 4.3) which early DNA analyses were unable to resolve. Based on dasyurin *cytb*, *12S* and *P1* sequences, Krajewski, Young *et al.* (1997) recovered a monophyletic *Dasyurus* (Fig. 6.3), as did Firestone (2000) using partial sequences of the mitochondrial control region (*CR*, Fig. 6.6). Both studies, however, lacked strong support for the genus, but both resolved *D. hallucatus* as sister to all other species. Krajewski *et al.* (2004) added three additional loci (*Val-16S*, *IRBP*, *bfib7*) and obtained a strongly supported ladder-like tree for *Dasyurus* species (Fig. 6.3), with *D. hallucatus*, *D. maculatus* and *D. viverrinus* + *D. albopunctatus* + *D. geoffroii-spartacus* forming successive splits from the common ancestor.

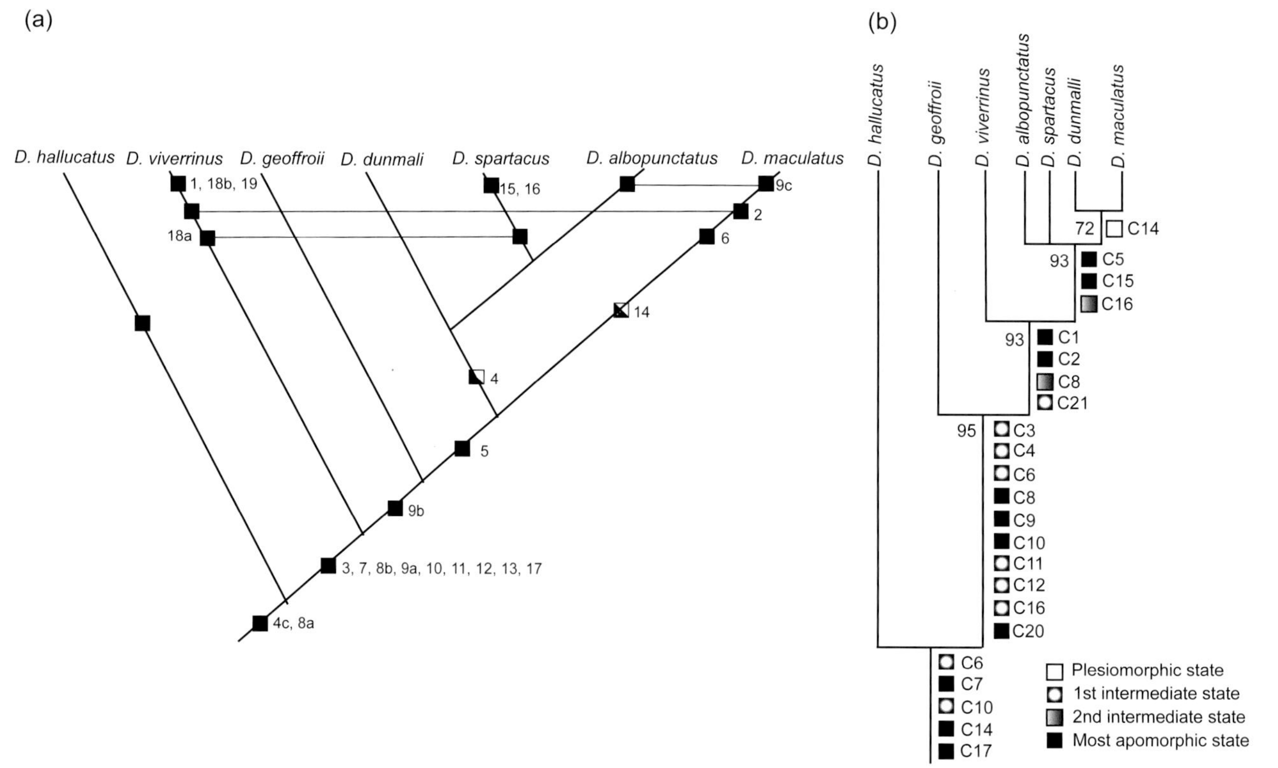

Fig. 6.5. (a) Van Dyck's (1987) cladogram for *Dasyurus* species based on Hennigian analysis of 19 morphological characters. Numbered boxes on branches refer to the origin of derived states for the numbered characters. Horizontal lines show parallel state gains. Putative synapomorphies supporting monophyly of all species except *D. hallucatus* are as follows: 3 = C^1 not proportionate; 7 = incisor row U-shaped; 8b = paracones not greatly reduced; 9a = metaconids reduced; 10 = metacone on M^2 not perpendicular to stylar cusp D; 11 = metacrista of M^3 shorter than that of M^4; 12 = M^5 reduced; 13 = M^5 lacking metacone; 17 = penis not simple. (b) Wroe and Mackness's (1998) most parsimonious tree for *Dasyurus* species based on analysis of 22 morphological characters. States of specific characters ('C1', etc.) reconstructed with DELTRAN optimisation are shown for each internal branch. Numbers at nodes are bootstrap support percentages. [(a) Redrawn with permission of Steve Van Dyck from Van Dyck (1987). (b) Redrawn with permission from the Queensland Museum from Wroe and Mackness (1998).]

Woolley *et al.* (2015) combined the alignments of Firestone (2000) and Krajewski *et al.* (2004) with sequences of breast cancer 1 (*BRCA1*), apolipoprotein B (*apoB*), embryonic globin (*e-globin*), recombination activating gene 1 (*RAG1*) and von Willebrand factor (*vWF*) for a total data set of ca. 17 kb from multiple individuals of most quoll species. The tree estimated from this alignment (Fig. 6.7) is consistent with previous DNA trees and resolves *D. viverrinus* and *D. albopunctatus* as successive sisters to *D. geoffroii-spartacus.* The single bronze quoll nests within a clade of western quoll haplotypes, casting doubt on the validity of the former as a distinct species. Indeed, the 'strong superficial resemblance' of these animals may be more phylogenetically informative than Van Dyck's (1987) craniodental data. Woolley *et al.* (2015) also provided a formal description of the penis in *D. hallucatus*, documenting its lack of an appendage (see Chapter 12) and arguing that the gain of this structure is a synapomorphy for all other quolls (as anticipated by Van Dyck 1987). The same branching order of *Dasyurus* and *Sarcophilus* species was recovered from the M14, MC15, W16, KB17 and GN20 (Fig. 6.2) multigene and integrative analyses.

Dasycercus and *Dasyuroides*

Tate (1947) considered mulgaras and kowaris to be sisters, in part based on their expanded auditory bullae, their petrosals and mastoids. Archer (1976c, d) found these genera most similar in dental and basicranial traits, with Archer (1982c) noting that their teeth represent the primitive end of a structural trend towards adaptation to carnivory that runs through *Dasyurus* and culminates in *Sarcophilus*. Archer's (1984c) phylogeny (Fig. 6.1) reflects this interpretation of morphology. The pairing was not confirmed by allozyme (Baverstock *et al.* 1982; see Fig. 4.2), albumin MC′F (Baverstock *et al.* 1990;

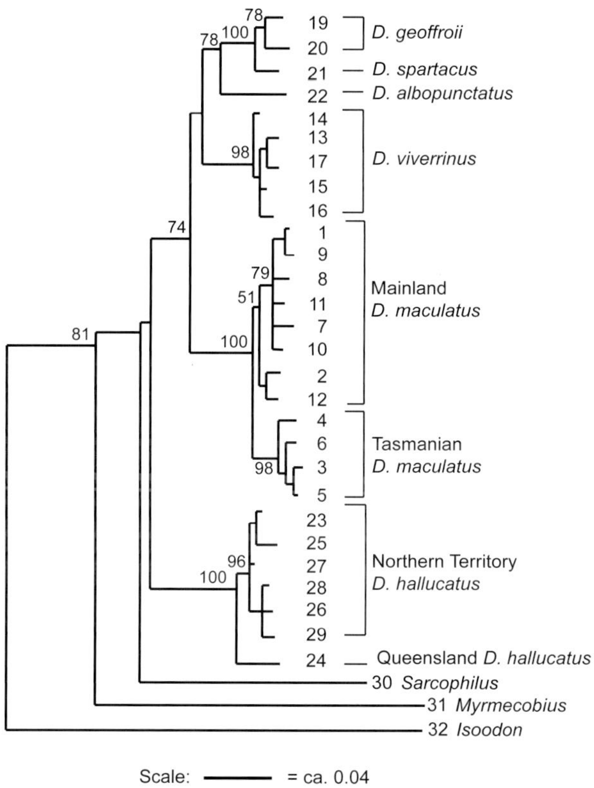

Fig. 6.6. Firestone's (2000) phylogeny of quolls based on Domain I of the mitochondrial control region (*CR*). Neighbour-joining tree based on Tamura distances. Numbers at tips identify specific haplotypes; numbers at nodes are bootstrap support values >50%; branch-length scale is substitutions/site. [Redrawn with permission of Springer Nature from Firestone (2000).]

see Fig. 4.3), DNA hybridisation (Kirsch *et al.* 1990; see Fig. 4.4), or early DNA sequence (Krajewski *et al.* 1993, 1994; Krajewski, Wroe and Westerman 2000) studies (see Figs 4.9, 4.10 and 4.14). For the most part, these papers simply failed to resolve confidently the position of either genus within Dasyurini. Indeed, it was not until analyses of large molecular data sets by M14, W16 and KB17 that strong support for a *Dasycercus* + *Dasyuroides* clade emerged (Eldridge *et al.* 2019). This reflects the need for long sequences to resolve short branches deep within the dasyurin radiation. Not only did these studies unite mulgaras and kowaris, they also located this clade in a larger group with *Myoictis*, *Parantechinus* and *Dasykaluta*; the most recent analysis (GN20) shows *Dasycercus-Dasyuroides* as sister to these three genera (Fig. 6.2) with high support.

Although only one species of kowari (*Dasyuroides byrnei*) has ever been recognised, the taxonomy of mulgaras (*Dasycercus*) is more unsettled. Tate (1947) and previous authors accepted two or more species, but Ride's (1970) influential book synonymised all into *Dasycercus cristicauda*, a decision formalised by Mahoney and Ride (1988) and almost universally adopted during the 1990s. However, Adams *et al.* (2000) obtained partial *cytb* and *CR* sequences from mulgara specimens collected across much of west-central Australia. They discovered two moderately supported haplotype clusters in the gene tree representing specimens that could be distinguished by a combination of external and cranial traits. Adams *et al.* (2000) recognised these as *D. cristicauda* and *D. hillieri*. Woolley (2005b), after reviewing available museum specimens and published descriptions, concluded that the appropriate names for Adams *et al.*'s (2000) groups are *D. blythi* (brush-tailed mulgara) and *D. cristicauda* (crest-tailed mulgara), respectively. They can be diagnosed by the form of the tail, number of upper premolars and number of nipples. Woolley (2006) clarified the species identity of mulgara samples that had been used in previous anatomical, physiological and molecular systematic studies.

Myoictis

Archer (1984c: 435) succinctly expressed the morphological basis for inclusion of three-striped dasyures (*Myoictis*) among dasyurins (Fig. 6.1): '*Myoictis*... is placed here because the penis has accessory corpora cavernosa [see Chapter 12], and there is a reduction or loss of P_3. Tate (1947) groups the species of *Myoictis* specifically with those of *Dasycercus* and *Dasyuroides*. In terms of dental morphology this is plausible. However, a detailed gross morphological analysis is needed, as are of course serological and enzyme data.' Allozyme data were not obtained for many years, but albumin MC′F (Baverstock *et al.* 1990) did link *Myoictis* with *Dasyuroides* (see Fig. 4.3). Partial *cytb* DNA sequences (Krajewski *et al.* 1993, 1994) showed strong support for *M. melas* + *Pseudantechinus ningbing* (see Figs 4.9 and 4.10), but this anomalous pairing disappeared in analyses of complete *cytb*, *12S* and *P1* genes (Krajewski, Wroe and Westerman 2000; see Fig. 4.14). As noted above, large multilocus and integrative studies (M14, MC15, W16, KB17) locate *Myoictis* in a clade with *Dasycercus*, *Dasyuroides*, *Dasykaluta* and *Parantechinus*. Within this group, GN20 (Fig. 6.2) found strong support for *Myoictis* as sister to *Dasykaluta* + *Parantechinus*.

Tate (1947), Archer (1982c) and many authors in between recognised two species of three-striped

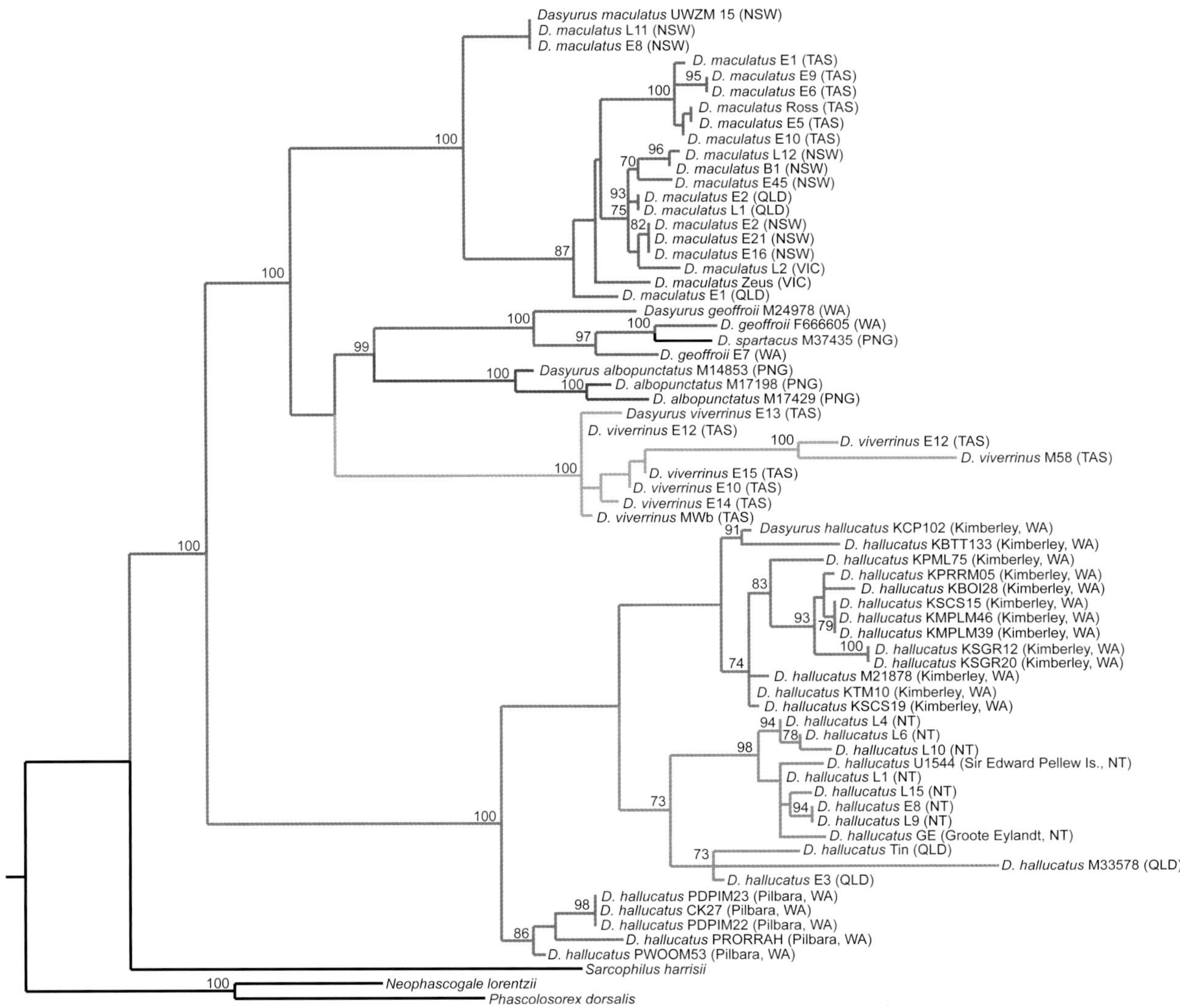

Fig. 6.7. Phylogeny of quolls from Woolley *et al.* (2015) based on ca. 17 kb of mitochondrial (*cytb, 12S, Val-16S, CR*) and nuclear (*P1, IRBP, bfib7, BRCA1, RAG1, apoB, vWF, e-globin*). Tree was estimated with maximum likelihood analysis of a partitioned alignment. Branches are scaled to substitutions/site; numbers at nodes are bootstrap support values >50%. [Redrawn by permission of Oxford University Press from Woolley *et al.* (2015).]

dasyures, *Myoictis melas* and *M. wallacei*. However, other reputable authorities (Ziegler 1982; Menzies 1991; Flannery 1990, 1995) listed only *M. melas*, perhaps with *wallacei* as a subspecies. Woolley (2005a) undertook a revision of the genus and recognised four species – *M. melas, M. wallacei, M. wavicus* and *M. leucura* – with nonoverlapping ranges and diagnosable by the form of their tails. Westerman *et al.* (2006) used *cytb, 12S, 16S* and *e-globin* DNA sequences, along with allozyme data, to assess the validity and relationships of these groups. Their results showed that *cytb-12S* haplotypes from each of Woolley's (2005a) taxa form discrete groups on the gene tree and that fixed allozyme differences exist between all species. Moreover, species relationships were fully resolved by DNA sequences, with the widespread and genetically diverse *M. melas* as sister to a clade of the more localised species (Fig. 6.8a). The same branching order was recovered by W16, KB17 and GN20 (Fig. 6.2) and the follow-up study of Phillips *et al.* (2021) (Fig. 6.8b).

Parantechinus, Pseudantechinus and *Dasykaluta*

Tate (1947) coined the genus names *Parantechinus* and *Pseudantechinus*, here abbreviated '*Pa.*' and '*Ps.*', respectively. He distinguished the genera principally by loss of P_3

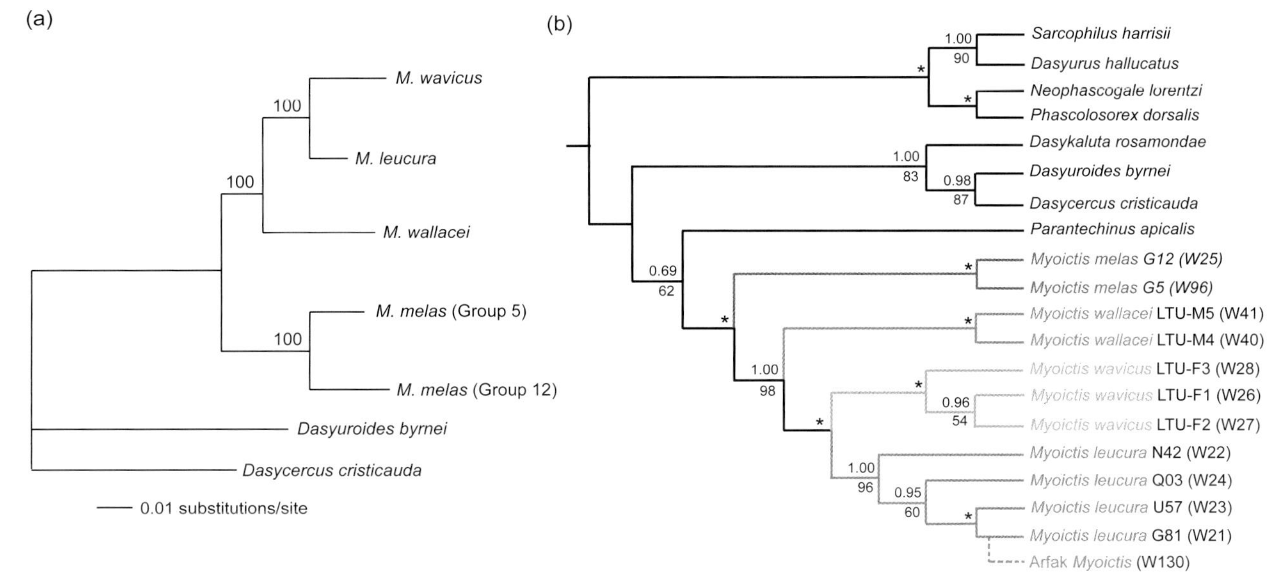

Fig. 6.8. Molecular phylogenies of *Myoictis* species. (a) Westerman *et al.* (2006) tree from *cytb, 12S, 16S* and *e-globin* sequences obtained with Bayesian analysis partitioned by gene and assuming a GTR model with gamma-distributed rates and a fraction of invariable sites (GTR+Γ+I). Nodal values are posterior probabilities × 100. (b) Phillips *et al.* (2021) tree from *cytb, tPHE-12S-tVAL-16S, IRBP, vWF, ε-globin and bfib7* obtained with ML and Bayesian analyses partitioned by stem-loop and codon positions with models for each partition. Numbers on internodes are posterior probabilities (above) and bootstrap percentages (below); asterisks indicated full support from both indices. Each *Myoictis* tip label includes a specimen number. [(a) Redrawn with permission of Springer Nature from Westerman *et al.* (2006). (b) Redrawn with permission of Elsevier from Phillips *et al.* (2021).]

and inflation of the alisphenoid + mastoid bullae in *Pseudantechinus*. In describing '*Antechinus*' *rosamondae*, Ride (1964a) lumped *Parantechinus* and *Pseudantechinus* into *Antechinus*, partly to stimulate more taxonomic research. Woolley (1982) showed that the 'false antechinus' genera have a penis morphology quite distinct from 'true' antechinuses (see Chapter 11). Archer (1982c) created *Dasykaluta* for Ride's new species, resurrected *Parantechinus* and *Pseudantechinus* and collected all three genera in his clade 'Parantechini' (Fig. 6.9) based on one putative synapomorphy (M_4 entoconid reduction) and weak evidence from allozymes (Baverstock *et al.* 1982; see Fig. 4.2). His morphocladistic analysis united *Parantechinus apicalis* and '*Pa.*' *bilarni* because they share the derived condition of a trifid corpora cavernosa in the penis (the homology of this trait was challenged by Cooper *et al.* 2000a). *Ps. macdonnellensis* and *Ps. ningbing* share the derived states of slight nasal-frontal contact and a bifid corpora cavernosum. Archer (1982c) agonised over which two parantechinin genera were sisters, ultimately opting for a soft trichotomy because all pairwise resolutions implied awkward convergences in craniodental or penile characters. As with many of Archer's early interpretations of dasyurid phylogeny, this was prescient in that the parantechinin polytomy proved difficult for later systematists to resolve.

Following the allozyme analysis of Cooper and Woolley (1983), Kitchener (1988) provided a formal description of *Ps. ningbing*. Kitchener and Caputi (1988) described *Ps. woolleyae* from specimens previously assigned to *Ps. macdonnellensis*. These authors also performed a parsimony analysis of 36 craniodental, external and penile traits but could not resolve relationships among members of *Pseudantechinus*. They did, however, argue that *Ps. bilarni* is more closely related to members of *Pseudantechinus* than to *Pa. apicalis* and Groves (2005) adopted this now-accepted classification (Eldridge *et al.* 2019). Kitchener (1991) redescribed *Ps. mimulus* after the species was rediscovered on North Island in the Gulf of Carpentaria. Cooper *et al.* (2000a) described *Ps.* '*roryi*', another putatively overlooked species, the specimens of which had been assigned to *Ps. macdonnellensis*. These authors discoursed at length on the relationships and taxonomy of false antechinuses but provided no new estimates of the former or revisions to the latter beyond their species description. Meanwhile, Mahoney and Ride (1988) had placed all parantechinins in *Parantechinus*!

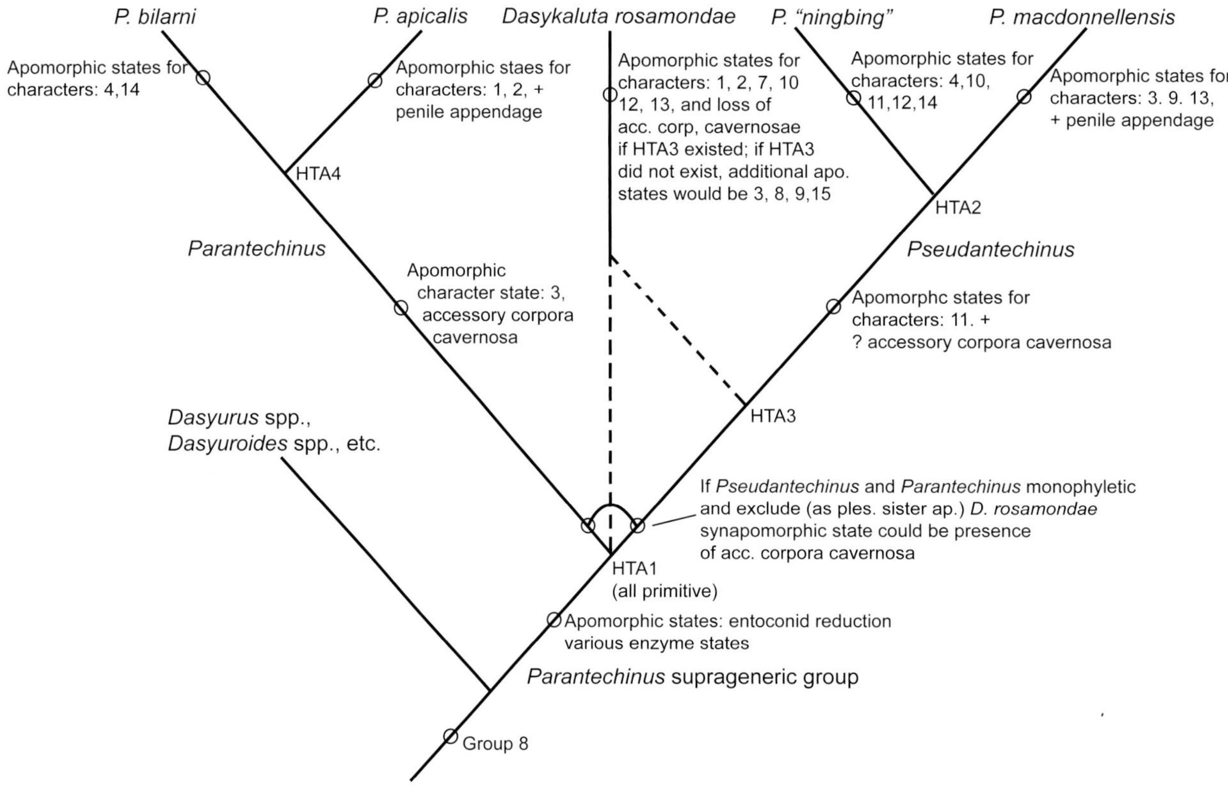

Fig. 6.9. Archer's (1982c) phylogeny of parantechinins (false antechinuses = '*Parantechinus* suprageneric group') based on Hennigian analysis of 15 morphological traits. Character numbers refer to traits described in the original text. HTA = hypothetical ancestor. [Redrawn with permission of the Royal Zoological Society of New South Wales from Archer (1982c).]

Albumin MC′F (Baverstock *et al.* 1990; see Fig. 4.3) recovered false antechinuses in a single clade (with *Pseudantechinus* represented only by *Ps. macdonnellensis* and *Ps. ningbing*). However, other than the previously noted anomalous pairing of *Ps. ningbing* with *Myoictis* from partial *cytb* (Krajewski *et al.* 1993, 1994; see Figs 4.9 and 4.10), early DNA sequence studies (Retief *et al.* 1995a; Krajewski, Young *et al.* 1997; Krajewski, Wroe and Westerman 2000) failed to achieve any firm resolution of relationships among parantechinins (see Figs 4.12–4.14). Westerman *et al.* (2008) reported *cytb*, *12S*, *16S*, *IRBP*, *bfib7* and *e-globin* DNA sequences from a wide geographic sample of parantechinin specimens (16 from *Pseudantechinus*) along with other representative dasyurins. Their phylogenetic analysis (Fig. 6.10) demonstrated that the three genera do not form a single clade. Rather, *Dasykaluta* and *Parantechinus* were resolved in a group with *Myoictis* and *Dasycercus-Dasyuroides*; *Pseudantechinus* was monophyletic and appeared as sister to this group. The sequence of branching within *Pseudantechinus* was resolved as (*Ps. bilarni* (*Ps. woolleyae* (*Ps. ningbing* (*Ps. mimulus* (*Ps. macdonnellensis*-'*roryi*'))))). Haplotypes from *Ps.* '*roryi*' specimens did not form a single branch but were interspersed among those from *Ps. macdonnellensis*. Thus, Westerman *et al.* (2008) endorsed the generic assignment of *Ps. bilarni* but rejected the validity of *Ps.* '*roryi*' as a distinct species. Umbrello *et al.* (2017) added sequences from an additional 24 individuals to the *Pseudantechinus* data set of Westerman *et al.* (2008) and reached the same conclusion regarding *Ps.* '*roryi*' (Fig. 6.11). Large alignment and integrative analyses have refined the phylogeny of Westerman *et al.* (2008) somewhat. Although M14 found limited resolution on intergeneric relationships of false antechinuses, W16, KB17 and GN20 (Fig. 6.2) resolved *Dasykaluta* and *Parantechinus* as sisters, while *Pseudantechinus* is sister to all other dasyurins.

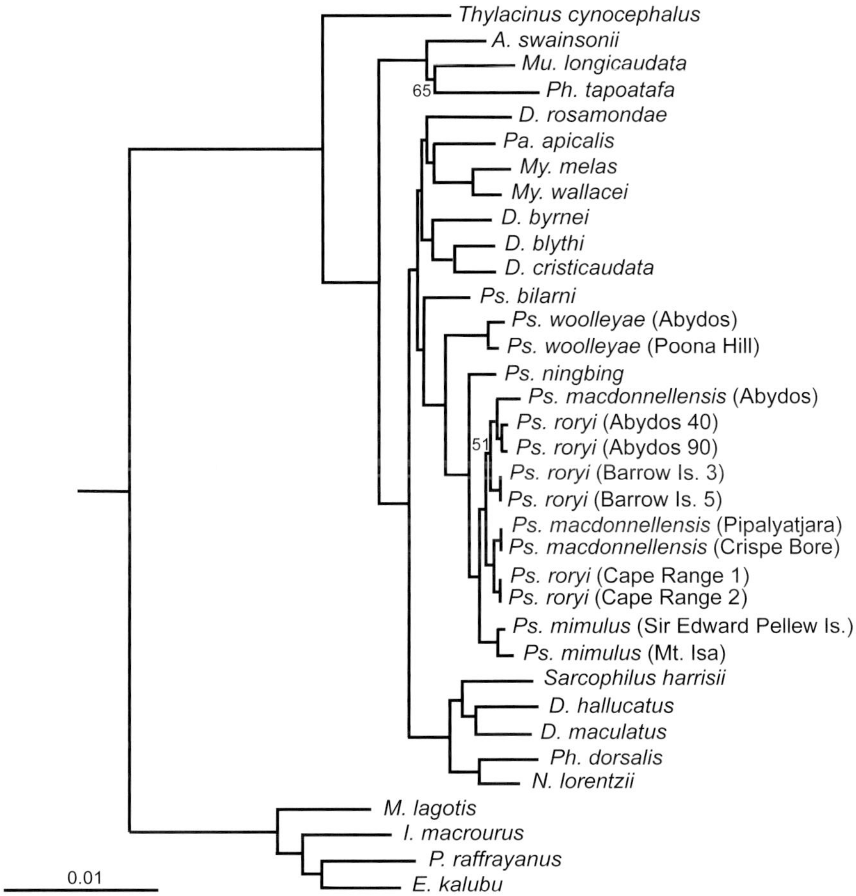

Fig. 6.10. Phylogeny of false antechinuses from Westerman *et al.* (2008). Tree estimated by Bayesian analysis of concatenated *cytb, 12S, 16S, IRBP, bfib7* and *e-globin* DNA sequences; each locus was treated as a partition with distinct parameter values under the GTR + Γ + I model. Branch length scale is substitutions/site; all clades have posterior probabilities >0.98 except as indicated. Two outgroup macropodids have been omitted. [Redrawn with permission of CSIRO Publishing from Westerman *et al.* (2008).]

Phascogalini

Archer's (1984c) review of dasyurid systematics reflected the morphological consensus, bolstered by allozyme studies (Baverstock *et al.* 1982; Fig. 6.2), that *Phascogale* and *Antechinus* are sisters – a clade that Archer named 'Phascogalinae'. As described in Chapter 4, albumin MC′F (Baverstock *et al.* 1990), DNA hybridisation (Kirsch *et al.* 1990), *cytb* (Krajewski *et al.* 1992, 1993, 1994, 1996), protamine (Retief *et al.* 1995a) and multilocus (Krajewski, Young *et al.* 1997; Krajewski, Wroe and Westerman 2000) DNA sequences all supported the inclusion of *Murexia* in this clade (see Figs 4.3–4.4, 4.8–4.10, and 4.12–4.14), despite the conflicting morphological assessment of Van Dyck (2002; see Fig. 4.15). Large alignment analyses (M14, MC15, W16 and KB17) resolved this conflict convincingly in favour of a *Phascogale-Antechinus-Murexia* clade, which Krajewski and Westerman (2003) called Phascogalini. The integrative analysis of KB15 and the multilocus study of GN20 (Fig. 6.12) found strong support for sister-pairing of *Phascogale* and *Murexia* apart from *Antechinus* within Phascogalini.

Phascogale

As described in Chapter 3, 20th century taxonomists distinguished two species of phascogales: *Phascogale tapoatafa*, with isolated populations in coastal regions of eastern, northcentral (Arnhem Land), north-eastern (Kimberley Plateau) and south-western Australia; and *P. calura* in more inland areas of the south-west corner. Within *P. tapoatafa*, Rhind *et al.* (2001) documented skull morphometric

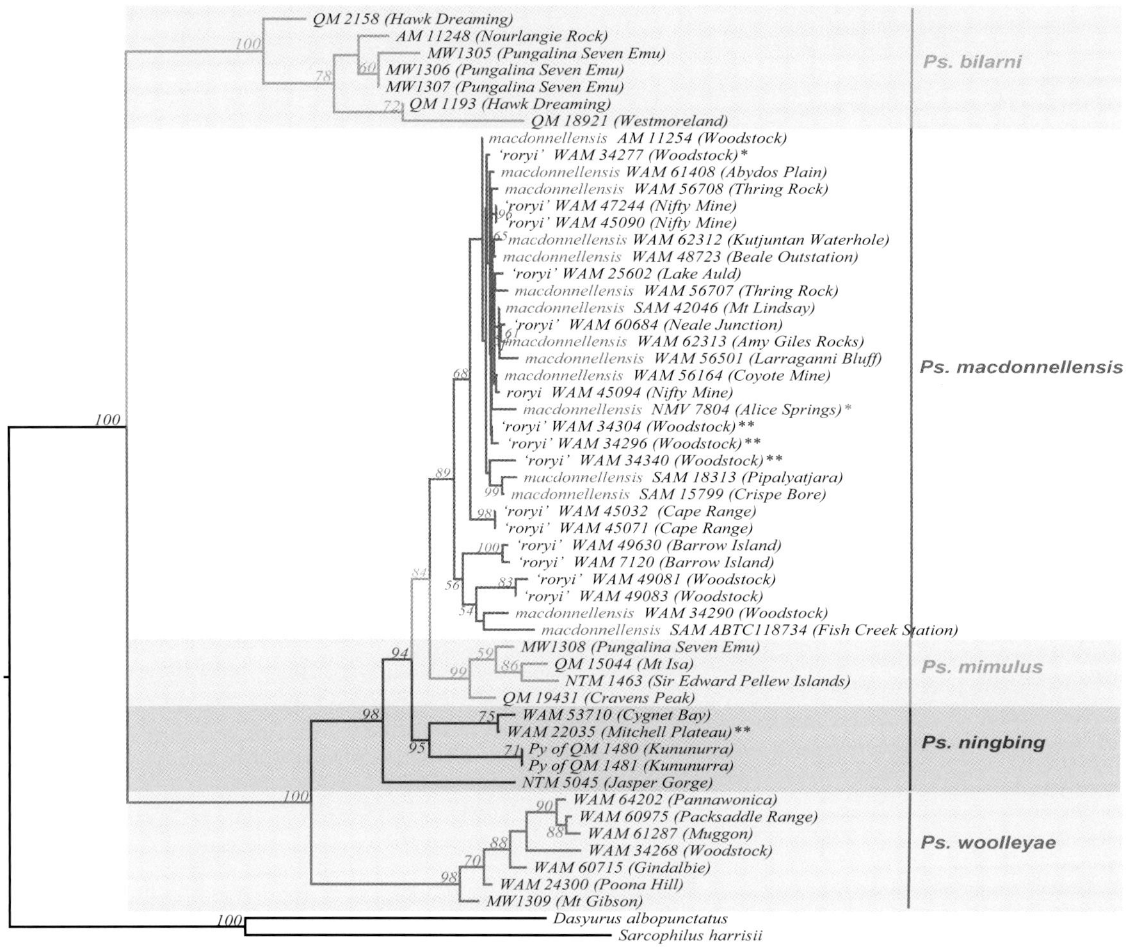

Fig. 6.11. Phylogeny of *Pseudantechinus* from Umbrello *et al.* (2017). Tree estimated by Bayesian analysis of concatenated *cytb*, *12S*, *16S*, *IRBP*, *bfib7* and *e-globin* DNA sequences; each locus was treated as a partition with distinct parameter values under the GTR + Γ + I model. Tip labels are specimen identification numbers with localities in parentheses. Asterisk (*) indicates sequence from a holotype specimen; double asterisks (**) indicate paratypes. Branch lengths are scaled to substitutions/site; nodal values are posterior probabilities × 100. [Reprinted by permission of CSIRO Publishing from Umbrello *et al.* (2017). Permission conveyed through Copyright Clearance Center, Inc.]

differences among northern, south-eastern and south-western forms, which Spencer *et al.* (2001) confirmed as showing reciprocal monophyly of partial *cytb* haplotypes. Aplin *et al.* (2015) expanded the data sets of both foregoing studies with additional individuals and provided a formal revision of *P. tapoatafa* that raised the Arnhem Land population to species status as *P. pirata*. Both subsequent analyses of dasyurid phylogeny that included *P. pirata* (W16, GN20) found it to be sister to *P. tapoatafa* (Fig. 6.12).

Antechinus

Based on craniodental morphology (Van Dyck 1982b) and allozymes (Baverstock *et al.* 1982; see Fig. 4.2), Archer (1982c) posited a tentative hypothesis of relationships among *Antechinus* species (Fig. 6.13). Chief among these were: (1) a clade of *A. bellus*, *A. leo* and *A. flavipes*, to which *A. stuartii* was sister; and (2) sister-pairing of *A. swainsonii* and *A. minimus*. The same evidence led Archer to anticipate later taxonomic recognition of *A. stuartii adustus* as a distinct species (see below). Archer's (1982c: 429–430) explanation for the uncertain position of *A. godmani* exemplifies the incomplete transition from phenetic to phylogenetic interpretations of character similarities in the early 1980s:

> *Dental morphology at least would suggest that it* [A. godmani] *is more similar to* A. flavipes *than to the* A. swainsonii-A. minimus *cluster, but only because* A. swainsonii *and* A. minimus *have as a*

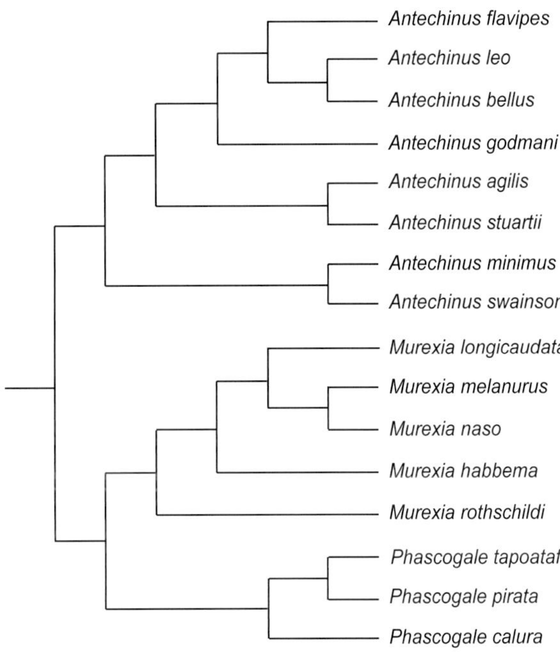

Fig. 6.12. Phylogeny of Phascogalini from García-Navas *et al.* (2020). This tree is part of a larger phylogenetic estimate for dasyuromorphian species. The branching order is derived from Bayesian analysis of ca. 18 kb of mitochondrial and nuclear loci; all clades have posterior probabilities ≥0.98. [Adapted from García-Navas *et al.* (2020).]

group, autapomorphic features that make them dissimilar to other antechinuses. On the other hand, A. godmani, A. swainsonii *and* A. minimus *are externally similar, with small eyes and long pointed heads. They are also similar ecologically, being essentially ground-living crepuscular dasyurids. The enzyme data do not help to resolve these alternatives although there is a very slight suggestion of closer affinity between* A. godmani *and the* A. swainsonii-A. minimus *group. Very tentatively, I have accepted the ecological and external morphological data over the dental morphological data as a guide to the relationships of* A. godmani.

Baverstock *et al.*'s (1990) albumin MC′F study did not distinguish individual *Antechinus* species and early DNA studies (Kirsch *et al.* 1990; Krajewski *et al.* 1992, 1993, 1994; Retief *et al.* 1995a; Krajewski, Young *et al.* 1997) sampled too few of them to test Archer's (1982c) hypothesis. Armstrong *et al.* (1998), however, had more complete sampling and resolved both the *A. bellus-leo-flavipes* and *A. swainsonii-minimus* clades with complete sequences of *cytb*, *12S* and *P1* loci (Fig. 6.14). Meanwhile, Dickman *et al.* (1998) described *A. agilis* from what had been

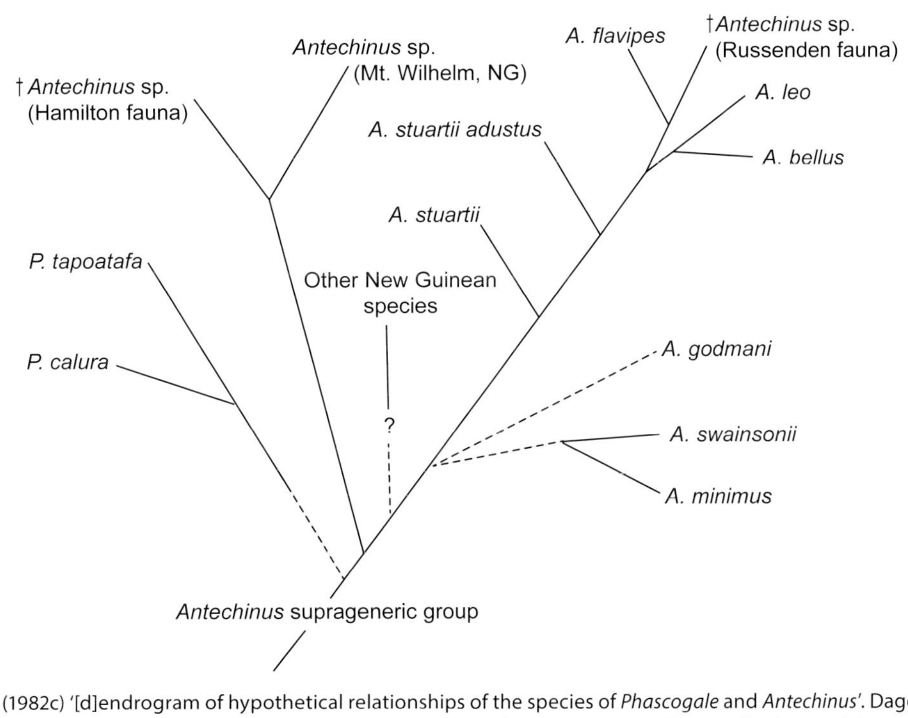

Fig. 6.13. Archer's (1982c) '[d]endrogram of hypothetical relationships of the species of *Phascogale* and *Antechinus*'. Daggers (†) indicate fossil taxa; '*Antechinus*' sp. (Mt Wilhelm, NG) and 'Other New Guinean species' refer to taxa now in *Murexia*. [Redrawn with permission of the Royal Zoological Society of New South Wales from Archer (1982c).]

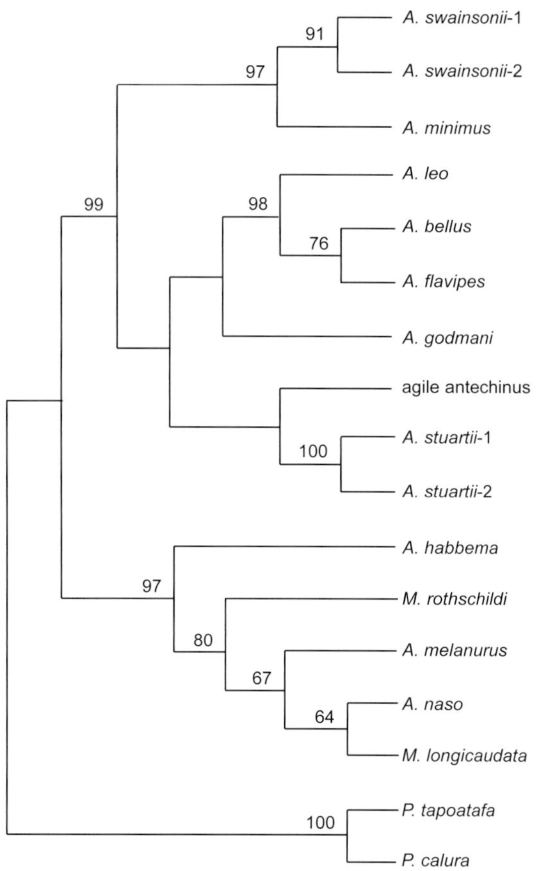

Fig. 6.14. Phylogeny of *Antechinus* and *Murexia* species from Armstrong *et al.* (1998). Maximum parsimony tree based on complete DNA sequences of *cytb, 12S and P1* loci. Numbers at nodes are bootstrap support values >50%. 'A. habbema', 'A. melanurus' and 'A. naso' are now recognised as members of *Murexia*; 'agile antechinus' refers to *A. agilis*. [Redrawn with permission of Oxford University Press from Armstrong *et al.* (1998).]

south-eastern populations of *A. stuartii*. Sumner and Dickman (1998) showed that these species are distinguished by fixed differences at two allozyme loci, but Crowther (2002b) found extensive overlap between them in bivariate and multivariate plots of cranial and external measurements. The *cytb-12S-P1* dasyurid phylogeny of Krajewski, Wroe and Westerman (2000; see Fig. 4.14) included *A. agilis* but essentially replicated the *Antechinus* relationships of Armstrong *et al.* (1998).

While phylogenetic studies of *Antechinus* experienced a hiatus in the early 2000s, work on species boundaries within the genus proceeded with considerable vigour. Van Dyck and Crowther (2000), considering only external and craniodental traits, recognised two more new species among populations previously considered *A. stuartii* – *A. subtropicus* and *A. adustus* – from rainforest areas of south-eastern and north-eastern Queensland, respectively. Following initial work by Sumner and Dickman (1998), Crowther (2002a) used bioclimatic modelling to show that predicted habitats and distributions differed significantly among *A. stuartii*, *A. flavipes*, *A. agilis* and *A. subtropicus*, reinforcing the ecological distinctness of the new species. Crowther *et al.* (2003) further examined the separation of *A. subtropicus* from adjacent *A. stuartii* populations with craniometric and molecular (partial *CR*) data. Specimens from each species formed nonoverlapping clusters in a principal components analysis of cranial and dental measurements but lacked fixed allozyme differences; although *CR* sequences formed reciprocally monophyletic groups on an estimated gene tree (Fig. 6.15a), only the *subtropicus* cluster had strong bootstrap support. The authors interpreted these

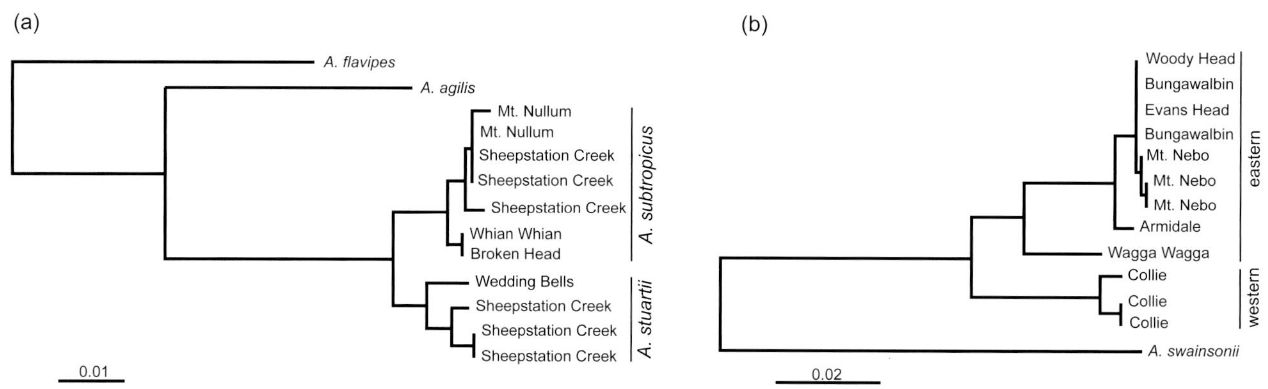

Fig. 6.15. (a) Gene tree for partial *CR* (470 bp) haplotypes of *Antechinus stuartii* and *A. subtropicus* (Crowther *et al.* 2003). Tree was estimated with neighbour-joining analysis of Tamura-Nei distances. (b) Gene tree for partial *cytb* (455 bp) haplotypes of *Antechinus flavipes* specimens from eastern and western Australia (Crowther *et al.* 2002). Tree estimated with neighbour-joining analysis of Kimura two-parameter distances. In both trees, nodal values are bootstrap percentages and branch lengths scales are substitutions/site. [Redrawn with permission of CSIRO Publishing from (a) Crowther *et al.* (2003) and (b) Crowther *et al.* (2002). Permissions conveyed through Copyright Clearance Center, Inc.]

findings as evidence for a recent speciation event. How *et al.* (2002) investigated whether northern and southern populations of *A. flavipes leucogaster* in south-western Australia, which differ in nipple number, could be distinguished by cranial morphometric data. Their results were equivocal, but Crowther *et al.* (2002) documented the distinction of *A. f. leucogaster* from eastern populations of *A. flavipes* in craniometric, molecular (partial *cytb* DNA sequences), reproductive timing and habitat (prediction by bioclimatic mapping) traits. Although western and eastern forms showed moderately supported (82% bootstrap) reciprocal monophyly (Fig. 6.15b) and moderate sequence divergence (ca. 6%), Crowther *et al.* (2002) favoured their continued recognition as subspecies.

In a series of papers beginning in 2012, Andrew Baker and collaborators undertook an extensive revision of *Antechinus* using craniodental and external morphometric data along with partial *cytb* and *IRBP* gene sequences (see Eldridge *et al.* 2019). Baker *et al.* (2012; see also Baker 2012) distinguished specimens of a new eastern Queensland species, *A. mysticus*, from *A. flavipes*. Bivariate and multivariate (principal components and canonical variates) analyses showed moderate to no overlap between these forms, or between them and *A. stuartii*. Baker *et al.* (2013) described *Antechinus argentus*, also distinct from *A. flavipes* and apparently restricted to the vicinity of Kroombit Tops National Park in south-eastern Queensland. Baker *et al.* (2014) recognised *A. arktos* as distinct from *A. swainsonii* in the wet montane habitat of Tweed Volcano caldera near the easternmost border of Queensland and New South Wales. Although an early allozyme study (Smith 1983) had endorsed the conspecific status of *A. swainsonii* populations from the mainland and Tasmania, Baker *et al.* (2015) elevated former subspecies in these areas to full species – *A. swainsonii* in Tasmania and *A. mimetes* on the mainland. In addition, they recognised a new species, *A. vandycki*, restricted to the Tasman Peninsula. Like Smith (1983), Baker and Van Dyck (2015) found no evidence for unrecognised species boundaries within *A. minimus*. Thus, the revisionary work of Baker and colleagues increased the diversity of *Antechinus* from 10 to 15 species.

In parallel with the studies described above, Mutton *et al.* (2019) estimated the phylogeny of *Antechinus* using DNA sequences of *cytb*, *16S*, *IRBP*, *vWF*, *apoB* and *BRCA1* loci (5.5 kb total). The concatenated alignment was partitioned and analysed with maximum likelihood and Bayesian methods, as well as a coalescent-based species tree approach (Mirarab and Warnow 2015) developed to accommodate incomplete lineage sorting. Results of the concatenated analyses (Fig. 6.16) showed a monophyletic *Antechinus* consisting of the familiar '*stuartii* complex' and '*swainsonii* complex' groups. Within the former, the authors designated three clades: (C2) *A. godmani* is sister to all other species; (C3) *A. stuartii*, *A. agilis* and *A. subtropicus*; and (C4) *A. bellus*, *A. leo*, *A. flavipes*, *A. adustus*, *A. mysticus* and *A. argentus*. The species tree had a similar structure but could not separate clades C3 and C4. Except for the position of *A. godmani*, this phylogeny is congruent with that of GN20.

Murexia

In Chapters 4 and 5 we recounted the history of research leading to recognition of *Murexia* as currently conceived – i.e. a single genus including the original *M. longicaudata* and *M. rothschildi*, as well as three other New Guinean species that had been placed in *Antechinus* (*M. melanurus*, *M. naso*, *M. habbema*). These studies provided little clarity, however, on relationships of species within the genus. Indeed, the *cytb-12S-P1* phylogeny of dasyurids reported by Krajewski, Wroe and Westerman (2000) could not provide strong support for a single node within *Murexia* (see Fig. 4.14). Krajewski *et al.* (2007) tackled this problem directly with *cytb-12S-16S-P1-bfib7-IRBP* sequences from representatives of each recognised species and multiple outgroups. The result (Fig. 6.17) was the first well-resolved branching order for *Murexia*, a ladder tree on which *M. rothschildi* is sister to all other species, with *M. longicaudata* and *M. naso* as each other's closest relatives. The M14, MC15, W16 and KB17 large alignment analyses recovered trees on which strongly supported nodes were consistent with this pattern.

This seemed like the successful end of a challenging quest for phylogenetic clarity, but *Murexia* once again proved enigmatic to systematists. Notwithstanding Van Dyck's (2002) conclusion that the genus comprises just those five species named above, Woolley *et al.* (2020) reassessed species boundaries with morphological, morphometric and molecular data. The obstacles to such a study were substantial. The relatively small number of museum specimens, some collected over a century ago, are distributed among institutions around the world and data from them required several decades to assemble. Original identifications of those specimens were often unreliable and many of their localities were described

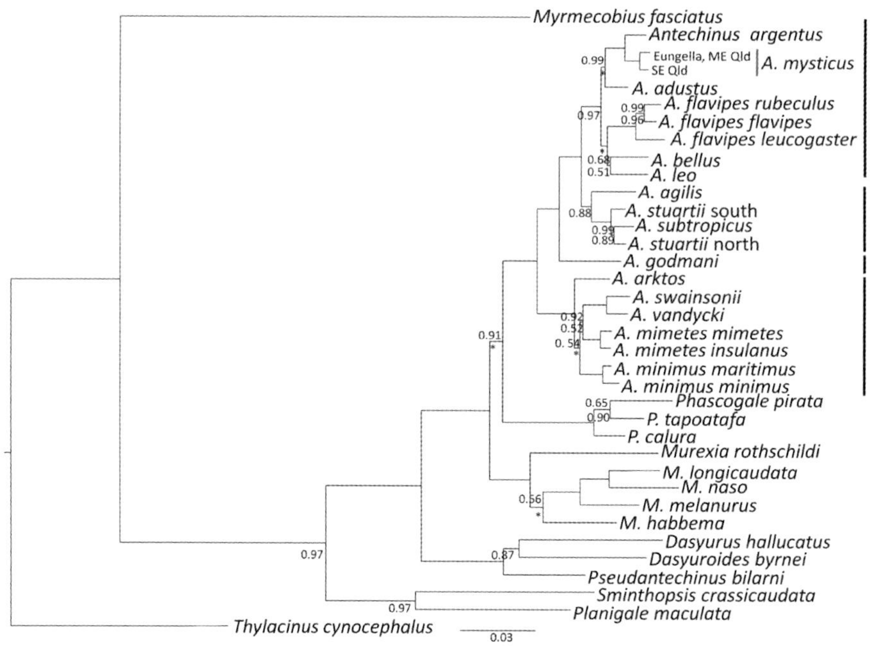

Fig. 6.16. Phylogeny of *Antechinus* based on *cytb, 16S, IRBP, vWF, apoB and BRCA1* gene sequences (5508 bp) from Mutton *et al.* (2019). Tree was estimated with Bayesian analysis of partitioned sequences with separate models for each partition. Nodal values are posterior probabilities; nodes without values have probability = 1.0; asterisks (*) indicate internal branches not recovered by ML analysis; branch length scale is substitutions/site. [Reprinted by permission of Oxford University Press from Mutton *et al.* (2019).]

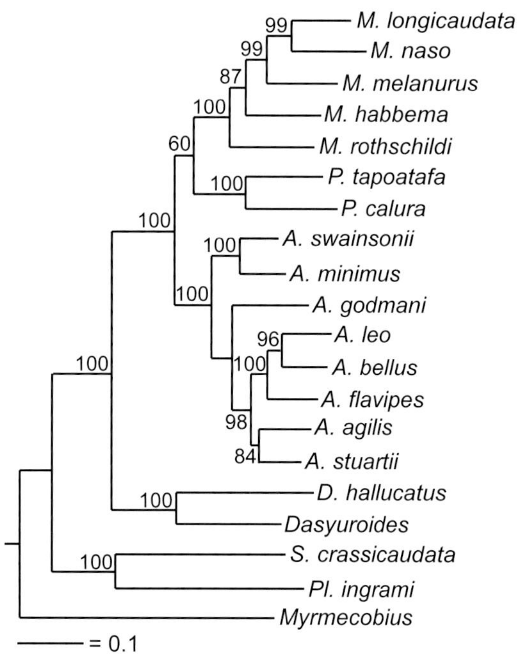

Fig. 6.17. Phylogeny of *Murexia* based on DNA sequences of *cytb, 12S, 16S, P1, bfib7 and* IRBP loci (Krajewski *et al.* 2007). Tree estimated from partitioned sequences with maximum likelihood under the F84 model of DNA evolution with discrete rate categories for each partition. Numbers at nodes are bootstrap support values; branch length scale is substitutions/site. [Redrawn by permission of Oxford University Press from Krajewski *et al.* (2007).]

vaguely. More recent collecting efforts in New Guinea, including any that would allow comprehensive geographic tissue sampling, have been extremely limited, such that molecular data could only be obtained opportunistically from older or museum tissues with variable preservation quality. Moreover, Woolley (2020) showed that the holotype skull and dentaries of *M. longicaudata* from the Aru Islands used by Schlegel (1866) to describe the species are abnormal or mismatched, rendering the application of this name to other specimens uncertain. Despite these limitations, the findings of Woolley *et al.* (2020) provided an eye-opening exposé of how little we know about *Murexia* diversity.

A key goal of Woolley *et al.* (2020) was to evaluate the status of previously named forms. For example, '*Antechinus wilhelmina*' was described by Tate (1947) but synonymised with *Murexia melanurus* by Van Dyck (2002) and others. Can specimens of *wilhelmina* be distinguished from other *M. melanurus* by morphological, molecular, or distributional information? Woolley *et al.* analysed discrete morphological traits, craniodental and external measurements and sequence data from *CR, 12S, 16S, IRBP, bfib7* and *e-globin* to make these evaluations. A phylogeographic tree estimated for composite haplotypes from individual specimens (Fig. 6.18) provided a glimpse

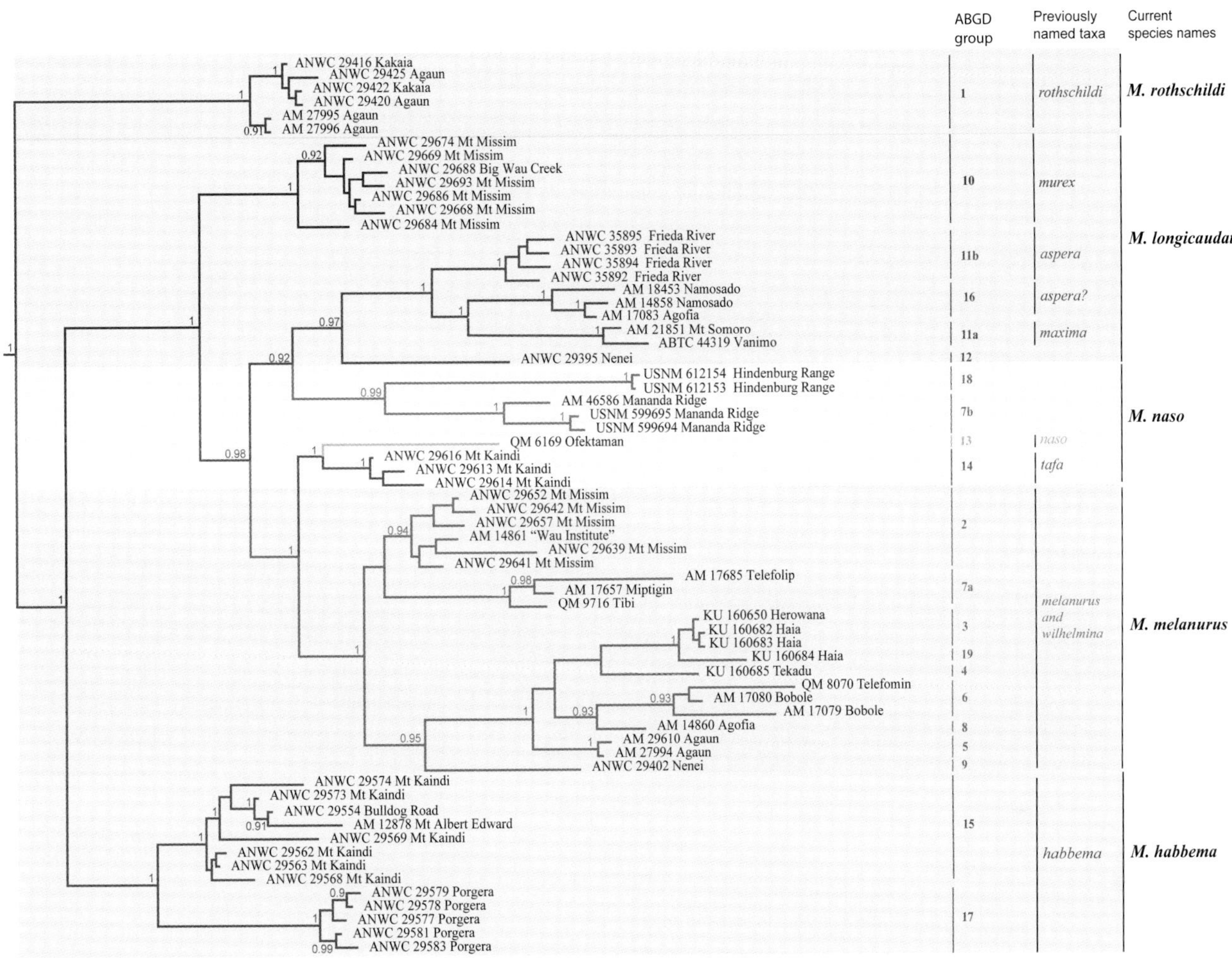

Fig. 6.18. Phylogeographic tree for *Murexia* specimens (Woolley *et al.* 2020). Tree estimated with Bayesian analysis of concatenated, partitioned *cytb, 12S, 16S, IRBP, bfib7 and e-globin* DNA sequences (ca. 12 kb). Tip labels are specimen numbers followed by collecting localities; nodal values are posterior probabilities; branch length scale is substitutions per site. 'ABGD group' shows haplotype clusters identified as potential species by the ABGD method. 'Previously named taxa' identifies tips that correspond to forms described in earlier taxonomic literature. 'Current species names' shows the assignments of specimens to the five species currently recognised. [Reprinted by permission of CSIRO Publishing from Woolley *et al.* (2020). Permission conveyed through Copyright Clearance Center, Inc.]

of genetic structure within and among species. The results were complex. No structured variation was apparent in *M. habbema* or *M. rothschildi*. *M. wilhelmina* could be diagnosed apart from *M. melanurus* based on molar size and showed limited overlap in morphometric space but did not correspond to any phylogeographic group. *M. tafa* could be distinguished from *M. naso* by its white-tipped tail, and specimens from each formed distinct haplogroups, but the two showed broad morphometric overlap. More interesting was a haplogroup of specimens identified as *M. naso* that was more closely related to *M. longicaudata* than to putative conspecifics. Within *M. longicaudata*, all three analyses suggested that *M. murex*, *M. aspera* and *M. maxima* may be distinct, along with an unnamed formed from western New Guinea. Finally, a heuristic method of molecular species delimitation – Automatic Barcode Gap Discovery (ABGD; Puillandre *et al.* 2012) applied to *CR* sequences – identified 19 haplotype clusters as potential species! The result of this study was not a revised taxonomy of *Murexia* or an improved phylogenetic estimate; rather, it was a sobering reminder that species trees and all the conclusions we draw from them are utterly dependent on a robust species-level taxonomy. We discuss the problems of species recognition in Chapter 7 but, clearly, they cannot be solved for *Murexia* without considerably more effort.

SMINTHOPSINAE

In terms of their natural history, sminthopsines are rather different from dasyurines. Whereas the latter include three lineages that represent evolutionary radiations endemic to New Guinean rainforests, sminthopsines are almost exclusively Australian (only one New Guinean endemic). Much of, though by no means all, dasyurine diversity occurs in moist habitats, but most sminthopsines are adapted to drier conditions. With some 19 species, *Sminthopsis* is the most diverse dasyurid genus, but recent studies strongly suggest that it is not monophyletic and that sminthopsin genus boundaries must be redrawn to reflect phylogeny. Moreover, *Planigale* almost certainly harbors cryptic species that render its current taxonomy inadequate.

Sminthopsins

As a reference for discussion of sminthopsin relationships, we provide a portion of the GN20 molecular phylogeny in Fig. 6.19.

Ningaui

When Archer (1975b) erected *Ningaui* it contained only two species, *N. ridei* and *N. timealeyi*, both described based on qualitative morphology. Morphometric analyses by Kitchener *et al.* (1983) led to discovery of a third species, *N. yvonneae*. Although these authors constructed phenograms and distance Wagner trees from their measurements, data from male and female specimens suggested different patterns of relationships between species. The *cytb-12S-P1* analysis of Krajewski, Blacket *et al.* (1997) was the first molecular study to include all three species but could not resolve relationships among them (Fig. 6.20); this was reflected in the larger studies of Blacket *et al.* (1999; Fig. 6.21) and Krajewski, Wroe and Westerman (2000; see Fig. 4.14) involving the same genes. Krajewski *et al.* (2012), analysing *cytb-12S-16S-CR-P1-bfib7-IRBP* sequences from all sminthopsins, resolved the trichotomy in favour of *N. ridei* and *N. yvonneae* as sisters apart from *N. timealeyi* (Fig. 6.22). This relationship was also recovered in the large-alignment and integrative analyses of MC15, W16, KB17 and GN20 (Fig. 6.19).

Antechinomys and *Sminthopsis*

Archer (1981) performed a Hennigian analysis of >30 external and craniodental characters to produce the

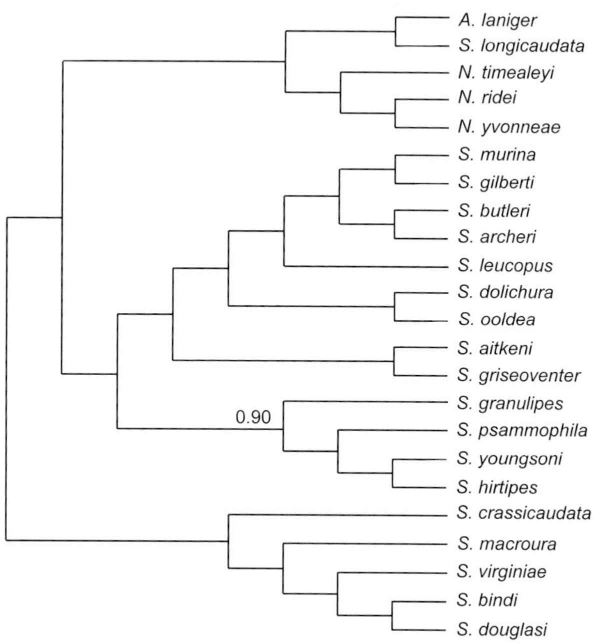

Fig. 6.19. Phylogeny of Sminthopsini from García-Navas *et al.* (2020). This tree is part of a larger phylogenetic estimate for dasyuromorphian species. The branching order is derived from Bayesian analysis of ca. 18 kb of mitochondrial and nuclear loci; all clades have posterior probabilities ≥0.95 except as noted. [Adapted from García-Navas *et al.* (2020).]

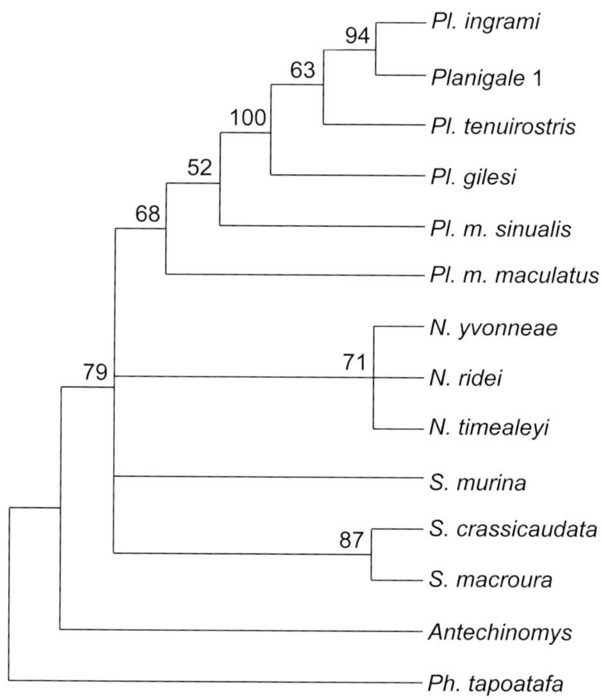

Fig. 6.20. Sminthopsine phylogeny from Krajewski, Blacket *et al.* (1997). Tree estimated by parsimony analysis of *cytb*, *12S* and *P1* gene sequences with homologous gap coding. Nodal values are bootstrap percentages. [Redrawn with permission of Elsevier from Krajewski, Blacket *et al.* (1997).]

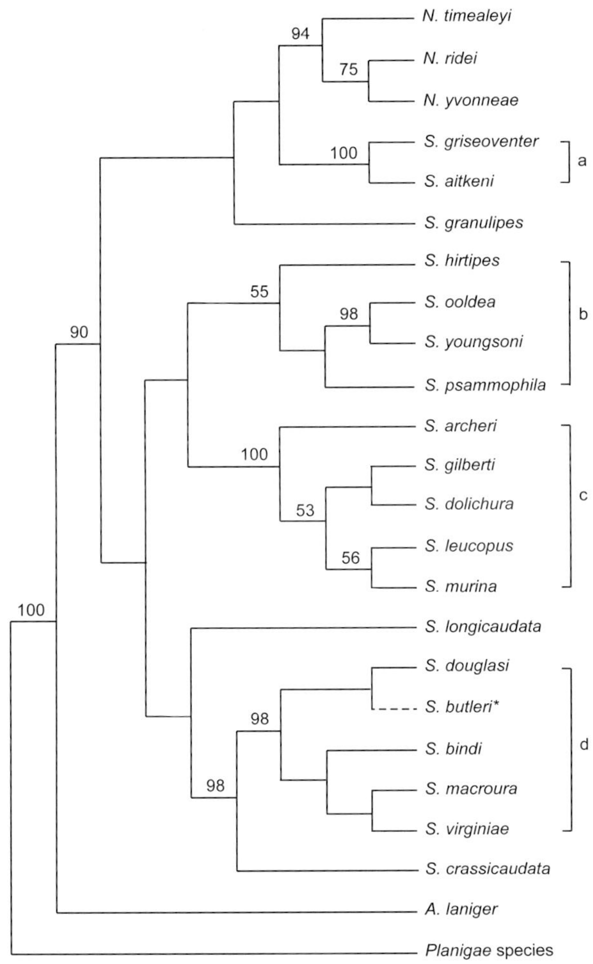

Fig. 6.21. Sminthopsin phylogeny of Blacket *et al.* (1999). Tree estimated with parsimony analysis of *cytb*, *12S*, partial *CR* and *P1* DNA sequences with homologous gap coding. Numbers at nodes are bootstrap support values >50%; groups labelled 'a'–'d' are discussed in the text. An asterisk (*) indicates that the position of *S. butleri* was inferred from partial *12S* sequence only. [Redrawn with permission of Elsevier from Blacket *et al.* (1999).]

sminthopsin cladogram in Fig. 3.8. He polarised characters based on the assumption that states associated with arid-adapted forms are derived, as well as other criteria explained in Archer (1976a, b, c). As noted above regarding phascolosoricines, Archer's interpretation of these traits is sometimes difficult to reconstruct, involving occasional use of primitive states as evidence for monophyly, vague state definitions, or missing justifications of polarity. For example, three putative synapomorphies – loss of posterior cingula on upper molars, narrow hind feet, narrow nasals – supported monophyly of Sminthopsini. Archer (1976d) provided an analysis of the first trait, but none of the cited references address the other two. Similarly, a clade of *Sminthopsis* + *Antechinomys* species shares five derived states: narrow hind feet, squamosal-frontal contact, reduced metatarsal granules, transverse metacristids and hypocristids. Only metacristid orientation was explicitly polarised (in Archer 1976d), but transverse orientation was there identified as 'ancestral'. Issues of character analysis aside, Archer summarised his phylogenetic hypothesis by recognising five major clades: the *Sminthopsis murina* group, the *S. psammophila* group, the *S. macroura* group, *S. crassicaudata* and *Antechinomys* (branches 9, 16, 19, 28 and 29 on Fig. 3.8). Sister-pairing of the latter two (branch 6) renders *Sminthopsis* paraphyletic and is supported by seven putative synapomorphies, all explicitly polarised in Archer (1981); see Chapter 3, 'Archer and dasyurid relationships' for discussion. Archer (1982c) updated his phylogeny with the allozyme results of Baverstock *et al.* (1982; see Fig. 4.2), which resulted in some rearrangements within the *S. murina* group and the dissociation of *S. crassicaudata* and *Antechinomys* (Fig. 6.23).

Early molecular studies did not sample many *Sminthopsis* species, but Van Dyck *et al.*'s (1994) description of *S. bindi* included a numerical parsimony analysis of 79 polarised, external and craniodental traits of all dunnarts, including the new species described by McKenzie and Archer (1982), Kitchener *et al.* (1984) and Van Dyck (1986) (see Chapter 3, 'Numerical analyses and a syncretic classification'). Their cladogram (Fig. 6.24) bears little resemblance to that of Archer (1981), preserving only his *S. psammophila* group and sister-pairing of *S. virginiae* with *S. douglasi*. Like Archer (1981), Van Dyck *et al.* (1994) placed *Antechinomys* within *Sminthopsis*, but as sister to *S. longicaudata* rather than to *S. crassicaudata*. This clade was supported by six 'non-homoplasious forward changes' (unique, unreversed synapomorphies), four associated with the upper canines and two with the alisphenoid bone, as well as several traits that showed multiple changes on the tree. It was the first indication of a sister-pairing now strongly supported by molecular data (see Chapter 5 and below).

The limited sampling of *Sminthopsis* species in Krajewski, Blacket *et al.*'s (1997) *cytb-12S-P1* study resolved nothing about dunnart relationships (Fig. 6.20), but Blacket *et al.* (1999) found support for several clades (Fig. 6.21) – one of which was *Sminthopsis* and *Ningaui* apart from *Antechinomys* (but not *Sminthopsis* alone). Within the former, Blacket *et al.* (1999) drew attention to four groups, designated a–d: (a) *S. griseoventer* +

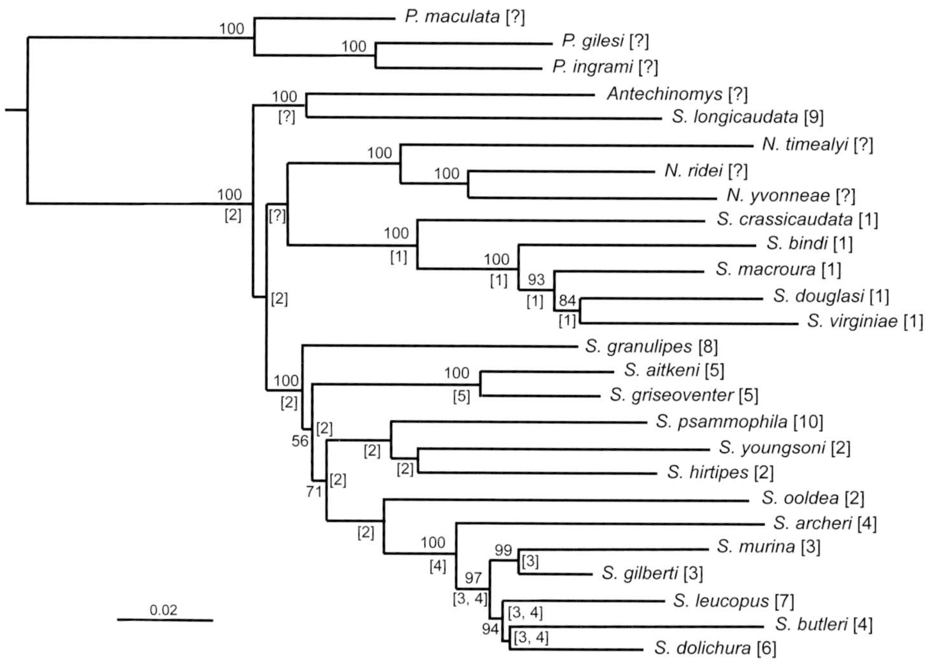

Fig. 6.22. Sminthopsin phylogeny from Krajewski *et al.* (2012). Tree is based on maximum likelihood analysis of concatenated, partitioned *cytb*, *12S*, *16S*, partial *CR*, *P1*, *IRBP* and *bfib7* DNA sequences under the GTR + I + Γ model. Numbers at nodes are bootstrap support values >50%; branch length scale is substitutions/site. Numbers in brackets are penis morphotypes; those at nodes are ancestral states estimated with parsimony and maximum likelihood. See Chapter 11 for discussion of penis evolution. [Redrawn with permission of Springer Nature from Krajewski *et al.* (2012).]

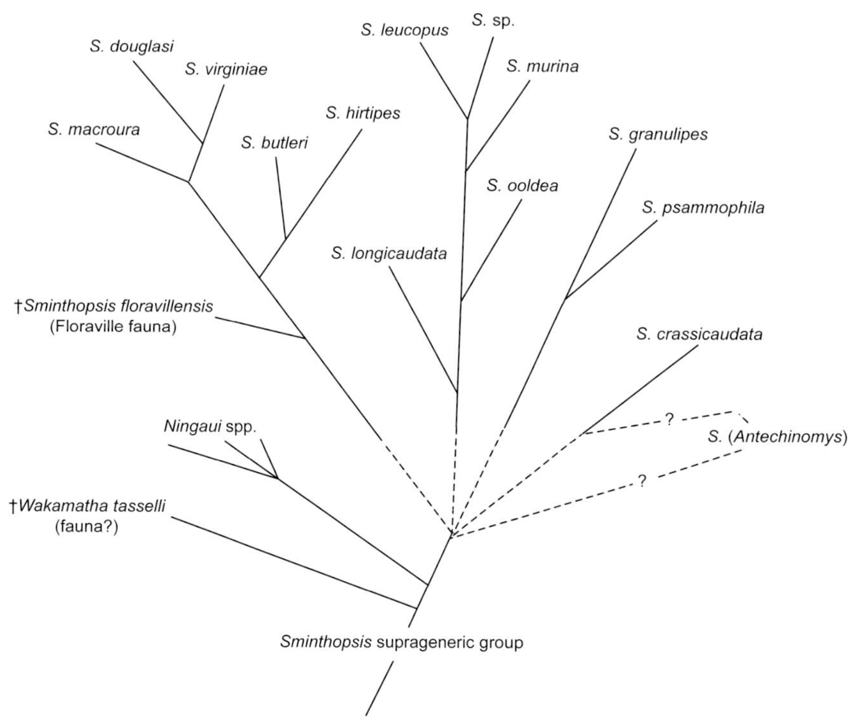

Fig. 6.23. Phylogeny of sminthopsins from Archer (1982c), combining results of morphology (Archer 1981) and allozymes (Baverstock *et al.* 1982). Daggers (†) indicate fossil taxa; dashed branches indicate uncertain relationships. [Redrawn with permission of the Royal Zoological Society of New South Wales from Archer (1982c).]

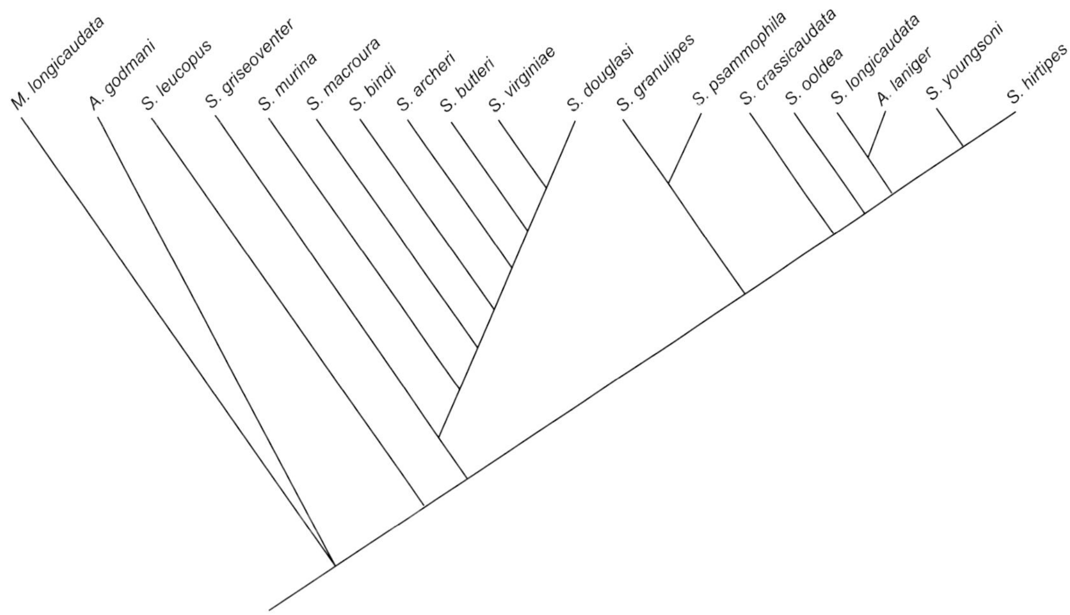

Fig. 6.24. Phylogeny of *Antechinomys* and *Sminthopsis* from Van Dyck *et al.* (1994). Tree is based on a numerical parsimony analysis of 79 polarised, external and craniodental characters. *M. longicaudata* and *A. godmani* are outgroups. [Redrawn with permission of the Queensland Museum from Van Dyck *et al.* (1994).]

S. aitkeni; (b) *S. ooldea* + *S. youngsoni*, loosely associated with *S. hirtipes* and *S. psammophila*; (c) a modified *S. murina* group including *S. archeri*, *S. gilberti*, *S. dolichura* and *S. leucopus*; and (d) a modified *S. macroura* group including *S. douglasi*, *S. butleri*, *S. bindi* and *S. virginiae*, to which *S. crassicaudata* was resolved as sister. *S. granulipes* and *S. longicaudata* were not clearly related to any of these groups. The conclusions of Blacket *et al.* (1999) were confirmed by Krajewski, Wroe and Westerman (2000; see Fig. 4.14). Blacket *et al.* (2001) used *12S* and partial *CR* sequences to improve resolution within the Macroura group, showing *S. bindi* as sister to all other species and five haplotype clusters within *S. macroura* (Fig. 6.25). Blacket *et al.* (2006) undertook a similar analysis for the Murina group using *12S*, partial *CR* and partial *ω-globin* gene sequences. They recovered a well-resolved tree with *S. archeri* and *S. dolichura* as successive sisters to the remaining species, among which *S. gilberti* and *S. murina* formed a clade (Fig. 6.26). Woolley *et al.* (2007) performed a comparative analysis of penis morphology in *Sminthopsis* (not including *Antechinomys*) and concluded that groups of species with similar penile anatomy were more congruent with clades recovered from molecular phylogenies by Blacket *et al.* (1999, 2006) than with morphological trees from Tate (1947), Archer (1981), or Van Dyck *et al.* (1994). Studies of dasyurid penis morphology are described further in Chapter 11.

Krajewski *et al.* (2012), using a *cytb-12S-16S-P1-IRBP-bfib7* alignment, produced a much better supported set of relationships (Fig. 6.22): *Antechinomys* + *S. longicaudata*; Blacket *et al.*'s (1999) Macroura group (minus *S. butleri*), to which *S. crassicaudata* is the sister; *S. griseoventer* + *S. aitkeni*; all remaining species of *Sminthopsis* (the 'deep clade' of Krajewski *et al.* 2012); and Blacket *et al.*'s (1999) Murina group (plus *S. butleri*). Within the 'deep clade', the positions of *S. granulipes*, *S. psammophila*, *S. youngsoni*, *S. hirtipes* and *S. ooldea* were unresolved. These relationships are consistent with the large-alignment results of W16, KB17 and GN20 (Fig. 6.19). Moreover, the large-alignment trees recover *S. granulipes-psammophila-youngsoni-hirtipes* as a clade and associate *S. ooldea* with *S. dolichura* in the Murina group.

The current consensus on species-level relationships among sminthopsins reveals a problem in classification – namely, that *Sminthopsis* as presently constituted is not monophyletic (Eldridge *et al.* 2019). A minimal revision to correct this would involve transferring *S. longicaudata* to *Antechinomys*[1] and erecting a new genus for the 'deep clade' of Krajewski *et al.* (2012). Because *S. crassicaudata* is the type species of *Sminthopsis* (Archer 1981), clade C of Fig. 6.22 would retain that genus name.

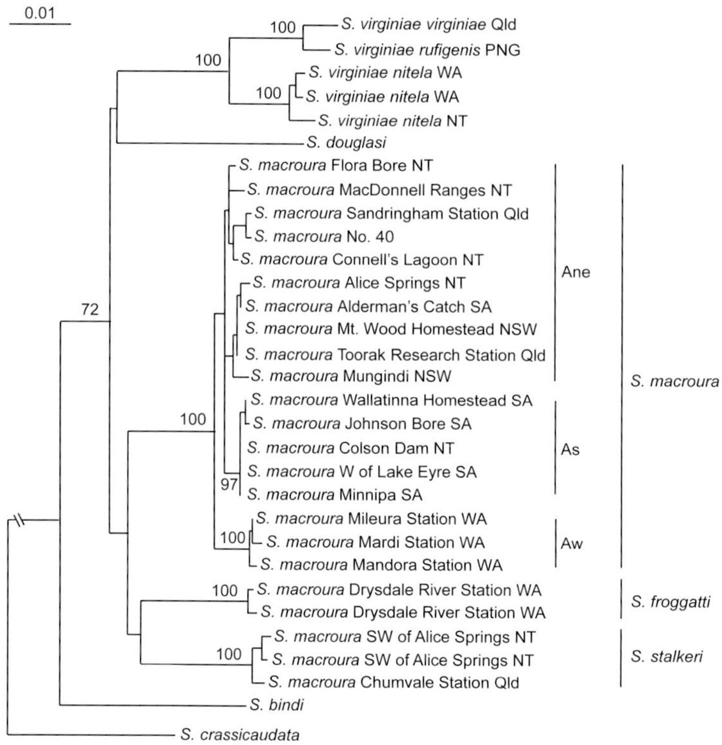

Fig. 6.25. Phylogeny of the *S. macroura* group from Blacket *et al.* (2001). Tree is based on additive best-fit analysis of weighted-average distances for *12S* and *CR* loci assuming the F84 model of DNA evolution. Numbers on nodes are bootstrap support values >50%. Branch length scale is substitutions/site. 'Ane', 'As' and 'Aw' delimit haplotype clusters within *S. macroura*. Proposed species designations are shown at right. [Redrawn with permission of Springer Nature from Blacket *et al.* (2001).]

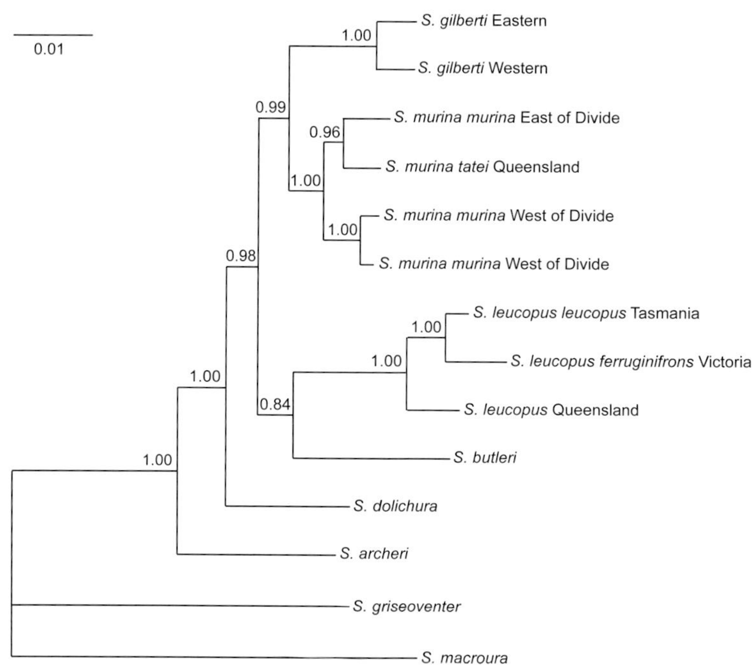

Fig. 6.26. Phylogeny of the Murina group from Blacket *et al.* (2006). Tree is based on Bayesian analysis of complete *12S*, partial *CR* and partial *ω-globin* DNA sequences assuming the GTR + I + Γ model of evolution. *S. macroura* and *S. griseoventer* are outgroups. Nodal values are posterior probabilities; branch length scale is substitutions/site. [Redrawn with permission of Springer Nature from Blacket *et al.* (2006).]

Planigalins

Although Archer's (1976b) revision of *Planigale* did not include a phylogeny, he did recognise the following species groups based on anatomical similarities: *Planigale maculata* group (*P. maculata*, *P. novaeguineae*); *Planigale ingrami* group (*P. ingrami*, *P. tenuirostris*); *Planigale gilesi*. Archer (1982c: 430) noted that '*P. maculata* is the most plesiomorphic species', which he thought was corroborated by Baverstock *et al.*'s (1982) finding that *P. gilesi* and *P. tenuirostris* form a clade apart from *P. maculata* based on allozymes. This reasoning about implications of the allozyme tree is flawed, however. The fact that a lineage with one extant species is sister to a lineage with more than one species in no way implies that the singleton lineage is primitive, since characters change over time in all lineages.

Krajewski *et al.* (1994) found that partial *cytb* sequence from a specimen of *S. crassicaudata* united it with *P. maculata* rather than other dunnarts (see Fig. 4.10). Painter *et al.* (1995) replicated this result (Fig. 6.27) and found that a very short *cytb* sequence from *P. novaeguineae* resolved in a haplogroup with *P. ingrami* and *P. gilesi*. These early anomalies were likely due to the relatively small amount of sequence data analysed, as well as the sparse ingroup sampling. The most interesting findings in Painter *et al.* (1995) were *cytb* sequences from two specimens collected in the Pilbara region of Western Australia and identified as *P. maculata* that did not associate with any other vouchered haplotypes (Fig. 6.27). Painter *et al.* (1995) named these 'Planigale 1' and 'Planigale 2'. The latter was subsequently shown to be a misidentified specimen of *Ningaui*, but Planigale 1 (the 'Pilbara Planigale') appears to be a new species that is yet to be formally described. Krajewski, Blacket *et al.*'s (1997) analysis of sminthopsine *cytb-12S-P1* sequences (Fig. 6.20) resolved *P. maculata* as sister to other planigales (*P. novaeguineae* was not included), with Planigale 1 as sister to *P. ingrami*. The former result, but not the latter, was replicated in the *cytb-12S-P1* analysis of all dasyurids by Krajewski, Wroe and Westerman (2000).

Blacket *et al.* (2000) examined genetic variation within *P. maculata* using allozymes and *12S* sequences but included data from all other planigale species. Their *12S* phylogeny (Fig. 6.28) showed little resolution of interspecific relationships but did provide moderate to strong support for placing *P. novaeguineae* in a clade with *P. maculata*, possibly rendering the latter paraphyletic. This study also found a novel *Planigale* haplotype from the Mt. Tom Price area of Western Australia. Westerman, Blacket *et al.* (2016) expanded the geographic and gene sampling

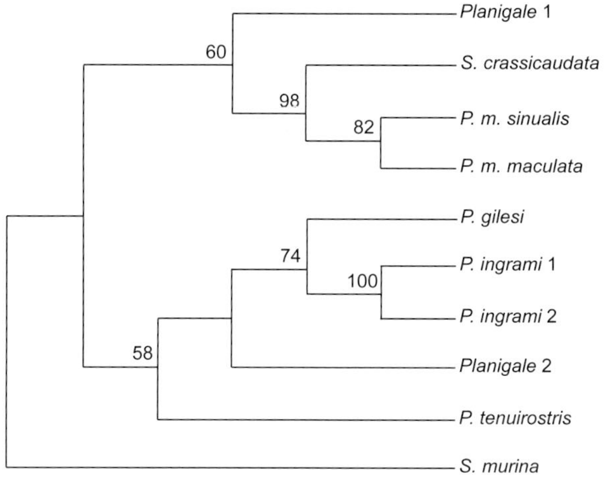

Fig. 6.27. Phylogeny of *Planigale* from Painter *et al.* (1995). Tree is based on parsimony analysis of partial *cytb* sequences. Numbers at nodes are bootstrap support values >50%. *S. murina* is the outgroup. [Redrawn with permission of Oxford University Press from Painter *et al.* (1995).]

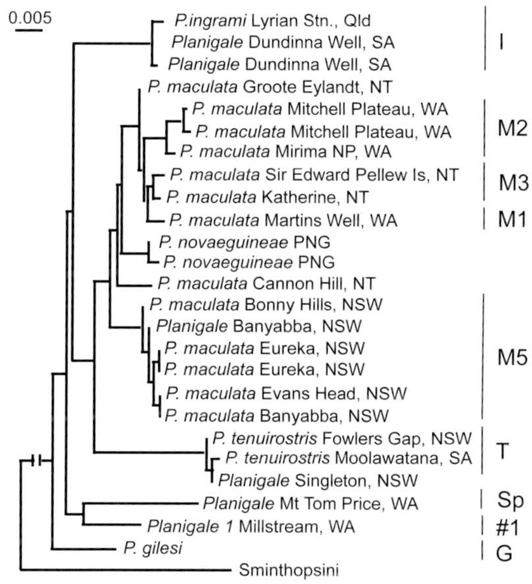

Fig. 6.28. Phylogeny of *Planigale* from Blacket *et al.* (2000). Tree was estimated with additive best-fit analysis of weighted-average distances from *12S* stem and loop partitions, assuming the F84 model of DNA evolution. Annotations at right designate taxonomically relevant haplogroups discussed in the original paper. Only monophyly of the *P. maculata* ('M') haplogroups had bootstrap support >90%. Branch length scale is substitutions/site. [Redrawn with permission of CSIRO Publishing from Blacket *et al.* (2000). Permission conveyed through Copyright Clearance Center, Inc.]

of Blacket *et al.* (2000) to include *cytb*, *12S*, *16S*, *CR*, *P1*, *IRBP*, *bfib7* and *e-globin* DNA sequences from 16 additional planigale specimens. Their results (Fig. 6.29) provided the best-supported phylogeny of the genus to date. Among the traditional five species, relationships were

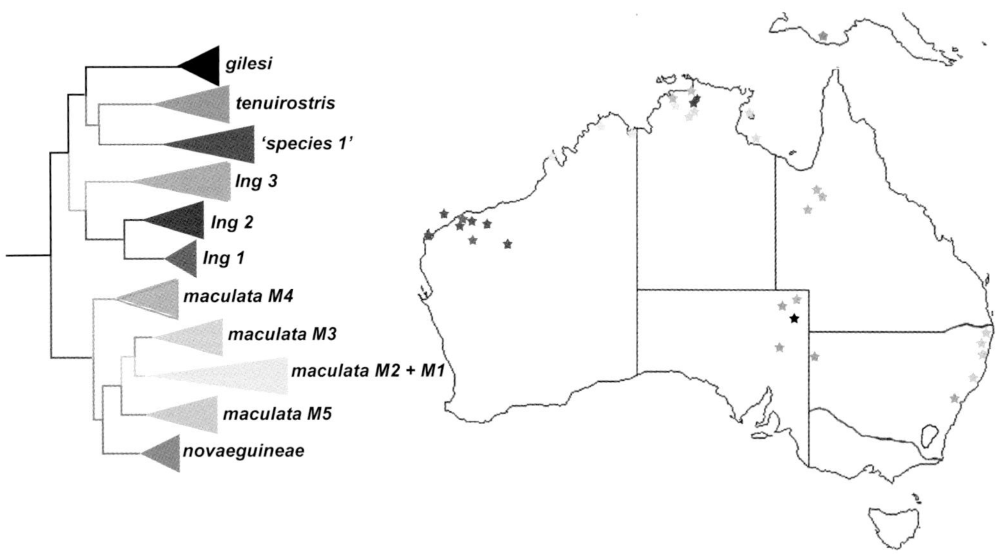

Fig. 6.29. Phylogeny of *Planigale* from Westerman, Blacket *et al.* (2016). The tree summarises Bayesian analyses of partitioned *cytb, 12S, 16S, CR, P1, IRBP, bfib7* and *e-globin* DNA sequences from multiple specimens representing each species. All clades on this tree have posterior probabilities ≥0.98 except that linking 'species 1' to *P. tenuirostris* (0.94). Each terminal triangle represents a resolved haplogroup. Stars on map correspond to collecting localities of specimens associated with each haplogroup. 'Ing' = *Planigale ingrami*. [Redrawn with permission of CSIRO Publishing from Westerman, Blacket *et al.* (2016). Permission conveyed through Copyright Clearance Center, Inc.]

(*P. maculata-novaeguineae*, (*P. ingrami* (*P. gilesi* (*P. tenuirostris*)))). 'Planigale 1' was part of the *P. ingrami-gilesi-tenuirostris* clade and the Mt. Tom Price haplotypes cluster within *P. ingrami* (clade 'Ing 1' in Fig. 6.29). Like Blacket *et al.* (2000), Westerman, Blacket *et al.* (2016) found evidence of geographically structured genetic subdivision within *P. maculata* (including exemplars of *P. novaeguineae*) that calls for further taxonomic investigation, and a similar situation in *P. ingrami* (see Chapter 7). Within the limits of geographic sampling and branch support, the large-alignment analyses of MC15, W16, KB17 and GN20 (Fig. 6.30) are consistent with the *Planigale* phylogeny of Westerman, Blacket *et al.* (2016). As is the case for *Murexia*, however, our current understanding of *Planigale* phylogeny is obscured by an inadequate species-level taxonomy.

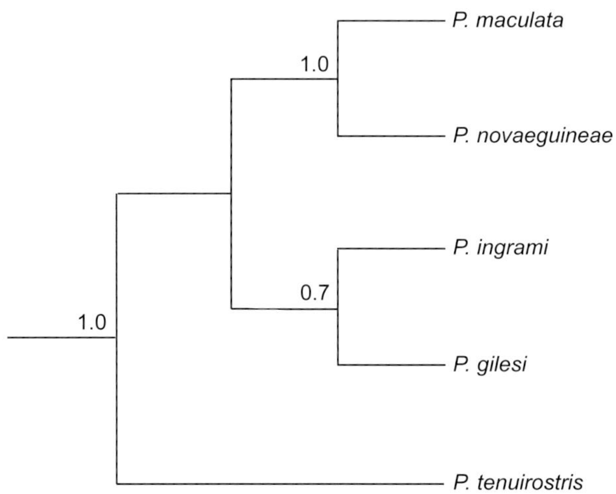

Fig. 6.30. Phylogeny of traditional *Planigale* species from García-Navas *et al.* (2020). This tree is part of a larger phylogenetic estimate for dasyuromorphian species. The branching order is derived from Bayesian analysis of ca. 18 kb of mitochondrial and nuclear loci. Nodal values are posterior probabilities >0.7. [Adapted from García-Navas *et al.* (2020).]

RETROSPECTIVE

The long struggle to resolve phylogenetic relationships among dasyurid species shows some regularities consistent with those revealed in earlier chapters. First, analyses based on dental, cranial and external characters have not converged on a single, robust estimate of relationships within any genus. Just the opposite is true for analyses of molecular data. Although early molecular work included some anomalous results (e.g. *Planigale maculata* + *S. crassicaudata*), these were rectified by the increased gene- and taxon-sampling of later studies. Enthusiasm for the 'total evidence' approach of combining morphological and molecular data in a single analysis must be tempered by the observation that morphology has little positive impact on the resulting tree estimate, at least for extant dasyurids, as exemplified by the results of KB17. On the other hand, dasyurid morphological systematists have put relatively little effort into character systems other than craniodental and external features. The higher congruence of penile morphology with molecular phylogenies suggests that soft anatomy may be a fruitful source of characters for future research at the species level (but see Callender-Crowe and Sansom 2022). Second, there can be no clear understanding of species-level relationships until species themselves are adequately delimited, as evidenced by the nomenclatural confusion within both *Murexia* and *Planigale*. Here the prospects for morphological-molecular synergy are much greater, as we describe in the next chapter.

ENDNOTE

1 The long-tailed dunnart was reassigned from *Sminthopsis to Antechinomys* by Westerman *et al.* (2023) *Australian Journal of Zoology* **70**, 95–103. doi:10.107/ZO22041.

7 Dasyurid species recognition

THE SCIENCE OF SPECIES RECOGNITION

A legacy of morphospecies

The idea of species as units of biodiversity has been part of Western civilisation's worldview at least since Aristotle. Loosely speaking, species are the most basic groups into which organisms can be classified. From a Judeo-Christian perspective, they have been understood as the forms of life created by God, the individuals of which were to propagate 'after their kinds'. Aristotle's kinds were defined by their essences, though it was left to natural historians to identify features that delimit them. The practicability of this research program was challenged as knowledge of Earth's tremendous biodiversity expanded up to and beyond the Enlightenment. Linnaeus's nomenclatural system was a manifestation of the strain that an explosion of species discovery put on biological classification, but equally important were the struggles of Linnaeus and his contemporaries to *recognise* the natural kinds into which, as they believed, the Creator had breathed life. Nature is complicated; patterns of trait variation often failed to sort individuals into tidy groups and taxonomists had to make judgments about which traits to use or ignore in recognising species. Disagreements about trait significance arose among specialists, leading to conflict about species recognition and competing classifications. Which classifications were natural and which artificial? This situation characterised the early taxonomic history of marsupials described in Chapter 3.

Although some prominent naturalists (e.g. Buffon) rejected the idea that species are discrete entities separated by gaps in trait values, Darwin inherited the mainstream tradition of Aristotle and Linnaeus. By this time, the practice of parsing trait variation among individuals into intraspecific ('varieties') versus interspecific (species) was common but no less arbitrary than in previous eras. Natural classifications came to be understood as those based on all traits rather than a chosen few, but this did not eliminate the subjective element in species recognition. Early in *On the Origin of Species*, Darwin (1859: 48) voiced one of his key motivating observations: 'I was much struck by how entirely vague and arbitrary is the distinction between species and varieties.' This rings true more than 150 years later. In Darwin's wake, species became the entities that form the Tree of Life, that originate by speciation and vanish by extinction. After Mendelian principles were brought to light in 1900, theorists focused on populations as fundamental units of evolution – the gene pools in which natural selection and other forces produce descent with modification. For population geneticists, species were collections of populations connected by gene flow, and speciation was the permanent interruption of that flow. All these intellectual advances, however, left the science of species recognition relatively unchanged. What are species and how can we tell them apart? Is this one question or two?

Among the most influential architects of the Modern Synthesis were Ernst Mayr and George Gaylord Simpson. Both were taxonomists engaged in the practical business of species recognition and both made major contributions to understanding the ontological status of species. Mayr (1942) crystalised the long-affirmed centrality of reproductive compatibility in his biological species concept (BSC): a species is a population or group of populations whose members are reproductively isolated from other populations. This concept connected naturalists' understanding of species with population genetics and provided an operational criterion for species recognition. If two populations are different species, their members should not be able to produce fertile offspring. Reproductive isolation means no effective gene flow, which in turn entails genetic and phenotypic divergence. Vertebrate zoologists embraced the BSC as fundamental despite its apparent shortcomings (e.g. restriction to sexual organisms, inapplicability to allopatric populations, indeterminate meaning of stable hybridisation) and it remains so in the minds of many vertebrate taxonomists. Indeed, complete reproductive isolation in the wild could be the only *sufficient* criterion for species recognition, though it is not generally considered necessary (because many valid species hybridise). Although widespread breeding experiments involving wild organisms have never been practical, differences among populations in heritable phenotypic traits is taken as evidence against gene flow, thus legitimising the traditional practice of recognising morphospecies.

Simpson's (1951b: 289) evolutionary species concept (ESC) was more expansive: a species is 'a phyletic lineage (an ancestral-descendant sequence of interbreeding populations) evolving independently of others, with its own separate evolutionary role and tendencies'. This seems to capture in words what evolutionary biologists *mean* when they talk about species, but by itself provides no operational guide for practicing taxonomists. As for the BSC, however, patterns of trait variation in natural populations have implications for evolutionary independence such that trait discontinuities provide evidence for species boundaries. When cladistics rose to prominence after 1966, systematists put forward several species concepts grounded in Hennigian principles. The most durable has been the phylogenetic species concept (PSC) articulated by Cracraft (1983: 170): 'A species is the smallest diagnosable cluster of individual organisms within which there is a parental pattern of ancestry and descent.' The PSC builds on the ESC by incorporating Hennig's notion of monophyly and operationalises it by emphasising *diagnosis*. The bridge to morphospecies practice is obvious and the PSC has been broadly endorsed.

Although the BSC, ESC and PSC were the most popular species concepts of the 20th century, the 1980s and 1990s saw a diversification of alternatives – Mayden (1997) counted 22 of them – that emphasised various ontological or operational aspects. Stepping back from this cacophony, O'Hara (1993) suggested that defining species may be an example of what philosopher John Dewey would consider a problem we get over rather than solve! By this time, however, most systematists had at least abandoned the essentialist idea that species are natural kinds, embracing instead Ghiselin's (1966) proposal that they are individuals – spatiotemporally limited historical entities (see Hull 1976). This conceptual shift aligned well with the ESC-PSC emphasis on the genealogical nature of species and set the stage for De Queiroz's (1998, 2007) general lineage concept (GLC). De Queiroz argued that previous attempts to resolve the species problem confused *concepts* (what we mean by species) with *criteria* (how we recognise them) and that, in fact, modern systematists agree that species are evolutionary lineages (ancestor–descendant populations that exist between speciation and/or extinction events). The disagreement has been over what recognition criteria are most fundamental. De Queiroz's solution was simply to admit that phenotypic traits, reproductive compatibility, ecological distinctness and other organismal data are sources of evidence in deciding whether a population or group of populations is a distinct species. The GLC has been widely, but not universally, adopted as a working solution to the species problem.

Molecular markers and species boundaries

When allozyme methods were introduced to the study of wild populations in the 1960s, it became possible to look for direct evidence of genetic isolation. The most straightforward marker of a barrier to gene flow was an allelic difference between two populations. Suppose, for example, allele 1 of locus A (A1) occurs at a reasonable frequency in population I but A2 is absent, whereas population II has A2 but lacks A1. That suggests I and II are not exchanging genes and may be distinct species. If A1 and A2 are fixed in populations I and II, respectively, they also serve as diagnostic traits and provide even stronger evidence for a species boundary. This interpretation is predicated, of course, on adequate sampling of both populations. Alternatively, genetic distances based on estimated allele frequencies – such as Nei's (1972) *D* – are expected to increase with time after populations are isolated. Finding congruent evidence for multiple unlinked loci strengthens the case for, or against, genetic isolation (the genealogical concordance principle of Avise and Ball 1990). Thus, the allozyme studies of species boundaries that proliferated in the 1970s and 1980s looked for fixed allelic differences and large genetic distances to complement morphological diagnoses. Importantly, however, these criteria were not *necessary* for claims of species distinction.

Much the same logic applies to mtDNA haplotypes, microsatellites, or nuclear linkage groups (as in the genetic species concept advocated by Baker and Bradley 2006), but gene-tree analyses proved more powerful. Here the simplest situation was finding that haplotypes from two well-sampled populations form reciprocally monophyletic groups on the gene tree (evidence for species status) or that they are intermixed (evidence against species). Evolutionary geneticists realised early on that reciprocal monophyly requires time to become established after the cessation of gene flow between populations. How much time depends on the rate of drift and nature of selection. The process of attaining reciprocal monophyly for a molecular marker became known as *lineage sorting*. The time lag between population isolation and complete lineage sorting meant that reciprocal monophyly could not be a necessary condition for species recognition. Genetically isolated species may indeed show *incomplete lineage sorting* (ILS). Conversely, the inevitable limitations of sampling wild populations preclude both fixed differences and reciprocal monophyly from being sufficient conditions for documenting species.

Gene tree analyses of species boundaries are more complicated when multiple markers give different results. It is common, for example, to find that two populations show reciprocal monophyly for mtDNA but not for nuclear loci. Usually this is due to mtDNA's higher rate of mutation and drift, and consequently shorter time to complete lineage sorting (clonally and maternally inherited mtDNA has only one-quarter the effective population size of a nuclear locus), but the same conflict can occur for unlinked nuclear markers. The expansion of coalescent theory following Kingman's (1982) model of gene genealogies (see Wakeley 2006) led to the development of quantitative tools for inferring species boundaries from multiple gene trees. Starting with the pioneering work of Pons *et al.* (2006), these methods use the expectations of a multispecies coalescent model (with assumptions about selection, ancestral populations sizes and other unknown parameters) to find an assignment of species boundaries that maximises the probability of observed ILS patterns among gene trees. The utility of molecular species delimitation methods is currently bounded by their assumptions ('priors' in the case of Bayesian approaches), but they are an active area of research, and their use is likely to increase rapidly.

Integrative taxonomy

Recognising that modern research on species boundaries involves diverse categories of characters, and in response to tension between advocates of specific categories, Padial *et al.* (2010) appropriated the phrase *integrative taxonomy* to describe a reasoned, pluralistic approach to species description. It has the following goals: 'On the one hand, to provide empirical rigor to species hypotheses and stability to their names, which requires a careful and often painstaking and time-consuming labour of species delimitation. On the other hand, an acceleration in the pace of species description, with the peril of erroneous species hypotheses and thus of unstable names' (Padial *et al.* 2010: 1). Typical categories of taxonomic evidence in mammalogy are qualitative morphology, morphometrics, reproduction, distribution, habitat and genetics. According to Padial *et al.* (2010), integrating them usually takes one of two forms: (1) congruence, in which acceptance of a new species requires supporting evidence from multiple categories; and (2) cumulation, which 'defends the view that because all taxonomic characters are contingent in existence, order of appearance and magnitude of divergence during speciation, the only way for true integration is allowing any source of evidence – even a single one – to form the basis for species discovery' (p. 7). Congruence is the more conservative position but risks overlooking recent or otherwise cryptic speciation events; cumulation is more progressive but prone to over-splitting. Dasyurid taxonomists have taken both approaches, but recent workers have at least looked for congruence among multiple evidence categories; the extent to which they have found it is variable.

Both traditional and modern taxonomy consider diagnosability as fundamental to species recognition. If an individual organism is examined, can we unequivocally assign it to the correct species, at least in theory? ('In theory' because an actual specimen may not have all the necessary information required for diagnosis.) To establish a character as diagnostic requires a sufficiently large and geographically dispersed sample of specimens to demonstrate that the ranges of trait variation in candidate species do not overlap. Sample sizes are often quite limited in studies of dasyurids, because the animals are usually hard to catch, such that even museum holdings of skin, skeleton and spirit specimens collected over many decades may amount to no more than a handful of exemplars from a few localities. In the following paragraphs, we describe the taxonomic characters most often used in dasyurid taxonomy and provide examples of diagnostic traits.

Qualitative morphology. Discrete anatomical traits of dasyurids have mostly been scored from just four sources: external structures (pelage markings, foot pads, pinna morphology, vibrissae count, nipple number, pouch structure); dentition (e.g. presence/absence of a cusp, number of premolars); cranium (e.g. position of a foramen); penis morphology (e.g. presence/absence of accessory corpora cavernosa). For example, Woolley (2005b: 213) examined 65 specimens from 20 localities and diagnosed the two species of *Dasycercus* as follows (Fig. 7.1): '*Dasycercus blythi* has a non-crested tail, two upper premolars (and a diastema between P^2 and M^1) and six nipples; *D. cristicauda* has a crested tail, three upper premolars (the third very small and sometimes present on one side only) and eight nipples.' As another simple example, *Planigale gilesi* differs from all other planigale species in having two rather than three premolars per quadrant (Archer 1976b). The issue of operational character definitions discussed in Chapters 4–6 with respect to phylogenetic analysis applies with equal force to species recognition. Traits such as 'body size large' may be descriptive in narrow comparative contexts but render diagnoses subjective.

Morphometrics. Measurements – usually taken with calipers but more often recently from digitised images – between anatomical landmarks reflect the size and shape of body parts. Species diagnoses can be made with one (univariate), two (bivariate), or ≥3 (multivariate) measurements when these data show individuals occupying nonoverlapping regions of morphospace. Multivariate scores are usually subjected to some form of ordination, such as principal components analysis (PCA) or discriminant function analysis (DFA), that reduce their dimensionality by finding linear combinations of scores that account for the variance between individuals (PCA) or pre-designated groups (DFA). Truly diagnostic measurements show no overlap among similar species, but slightly overlapping trait values may be taken as evidence of species status. As for qualitative traits, dasyurid morphometric studies have largely focused on external and craniodental measures. Among closely related species, univariate diagnoses can be difficult to establish. However, Wakefield and Warneke (1963) measured 155 and 140 crania of *Antechinus swainsonii* and *A. minimus*, respectively, and found no overlap in lengths of their anterior palatine foramina (4.60–6.65 mm versus 3.30–4.40 mm). Bivariate measures are often used in the form of ratios. For example, Archer (1975b) found that the ratio of tail-vent/nose-vent length diagnosed *Ningaui timealeyi* (TV/NV = 1.13–1.36) from *N. ridei* (TV/NV = 0.96–0.97), though sample sizes were small (five and two, respectively). Figure 7.2a shows measurements taken by Van Dyck and Crowther (2000) for descriptions of *Antechinus subtropicus* and *A. adustus*; these are typical for many dasyurid multivariate analyses. When Kitchener *et al.* (1983) discovered *Ningaui yvonneae*, they applied DFA to 19 cranial and five external measurements (including some of those in Fig. 7.2a); cranial, but not external, discriminant functions showed complete separation of *Ningaui* species (Fig. 7.2b).

Reproduction. To our knowledge, no description of a dasyurid species has included direct evidence of reproductive isolation between wild populations formerly considered conspecific. However, in the 1960s one of us (P.A.W.) observed mating attempts in captivity between

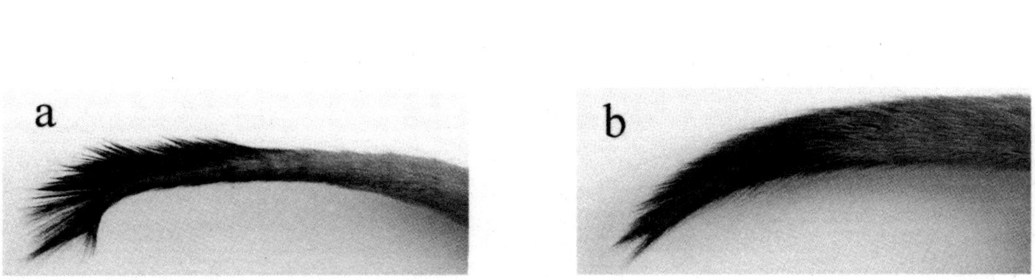

Fig. 7.1. Example of a qualitative morphological trait that diagnoses species. Crested and non-crested tails distinguish (a) *Dasycercus cristicauda* from (b) *D. blythi*. [Reprinted from Woolley (2005b).]

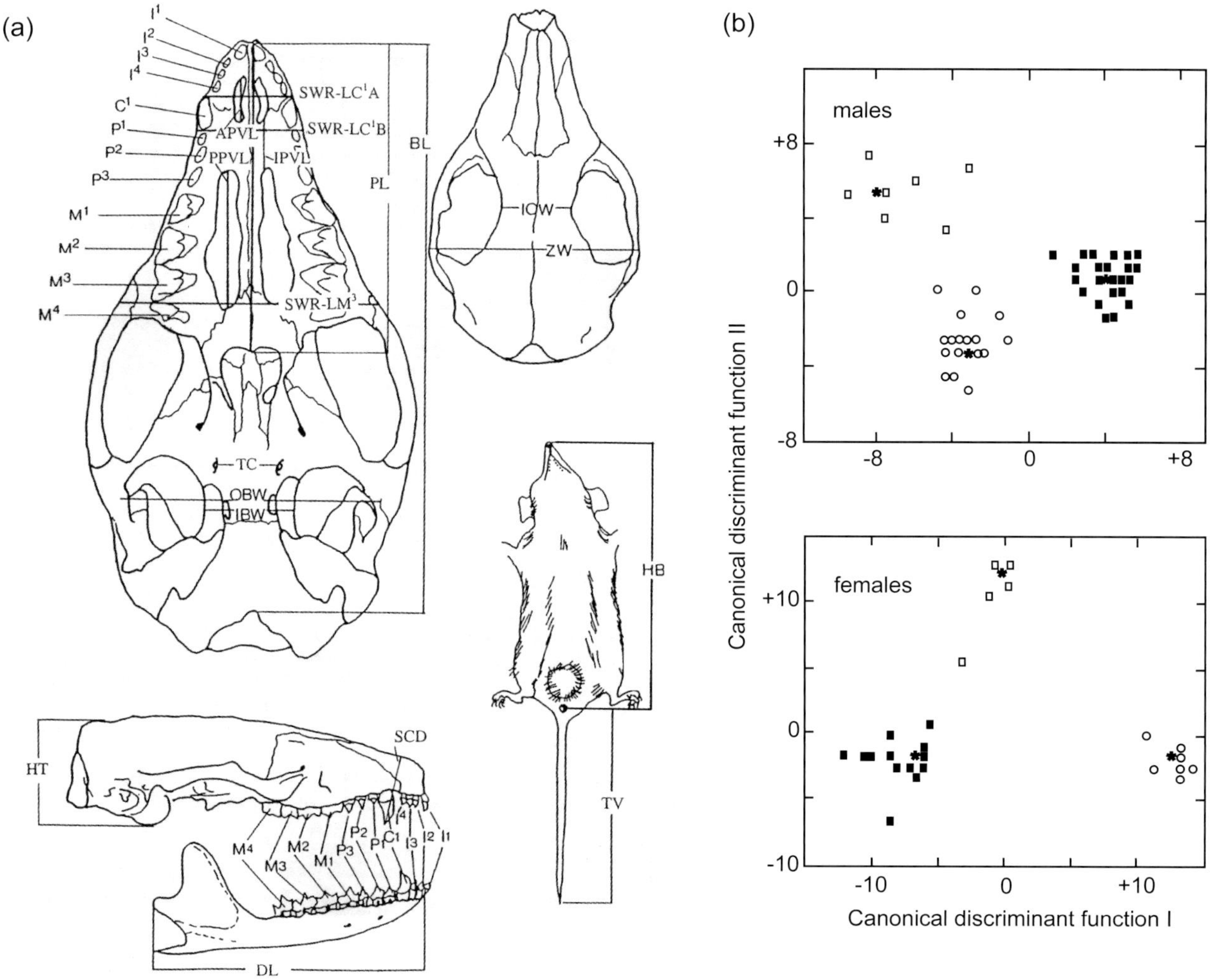

Fig. 7.2. (a) Cranial and external measurements used by Van Dyck and Crowther (2000) in recognition of *Antechinus subtropicus* and *A. adustus*. These measurements are typical for dasyurid morphometric studies. (b) Discriminant function analysis showing separation of *Ningaui yvonneae* (circles), *N. ridei* (black squares) and *N. timealeyi* (white squares) based on 19 cranial measurements (Kitchener *et al.* 1983). Asterisks (*) are group means. [(a) Reprinted with permission of the Queensland Museum from Van Dyck and Crowther (2000). (b) Redrawn with permission of CSIRO Publishing from Kitchener *et al.* (1983). Permission conveyed through Copyright Clearance Center, Inc.]

male *Antechinus flavipes* and female *A. agilis* (then part of *A. stuartii*). Although males forced females into copulation, only undivided eggs were subsequently recovered from the uteri of females, consistent with post-mating isolation. Crowther *et al.* (2002) argued that the timing of ovulation as a function of latitude was different in *Antechinus flavipes leucogaster* and *A. f. flavipes* (Fig. 7.3) but did not consider this decisive evidence for a species boundary.

Distribution and habitat. Allopatric populations are, by definition, geographically isolated and likely to display little or no gene flow. But have they been isolated long enough to achieve evolutionary independence, and will they remain so in the future? Addressing this question involves documenting divergence in phenotypic or genetic traits, though a definitive answer is unlikely to be forthcoming. One feature indicative of divergent adaptation relates to the habitats occupied by allopatric populations. Different habitat affinities suggest distinct ecological niches consistent with speciation. In their descriptions of *Antechinus subtropicus* and *A. adustus*, Van Dyck and Crowther (2000) showed that these allopatric taxa (both formerly *A. stuartii*) have strikingly different habitat preferences – sclerophyll woodland in southern Queensland for *A. subtropicus* versus subtropical rainforest in northern Queensland for *A. adustus*.

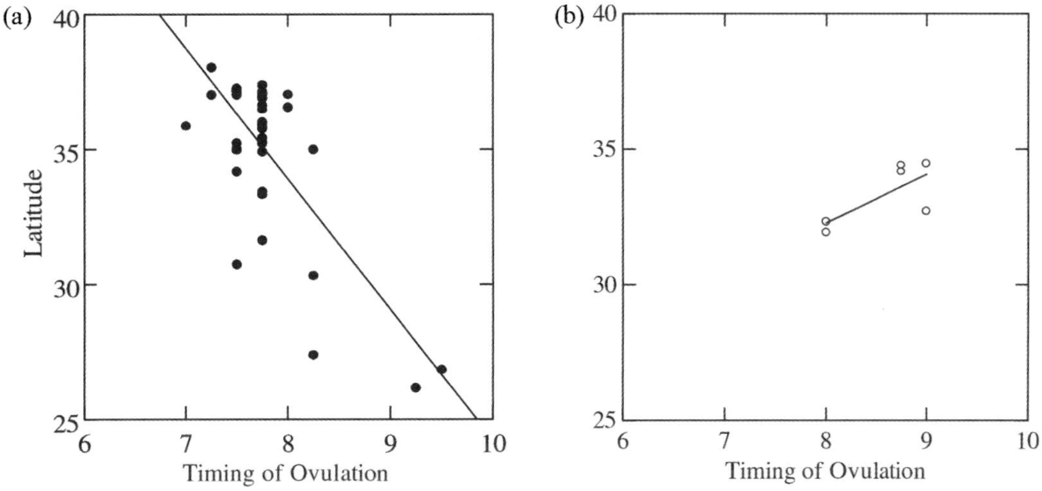

Fig. 7.3. The relationship between timing of ovulation and latitude in (a) *Antechinus flavipes flavipes* and (b) *A. f. leucogaster*. Horizontal axis scaled in calendar months. [Reprinted with permission of CSIRO Publishing from Crowther *et al.* (2002). Permission conveyed through Copyright Clearance Center, Inc.]

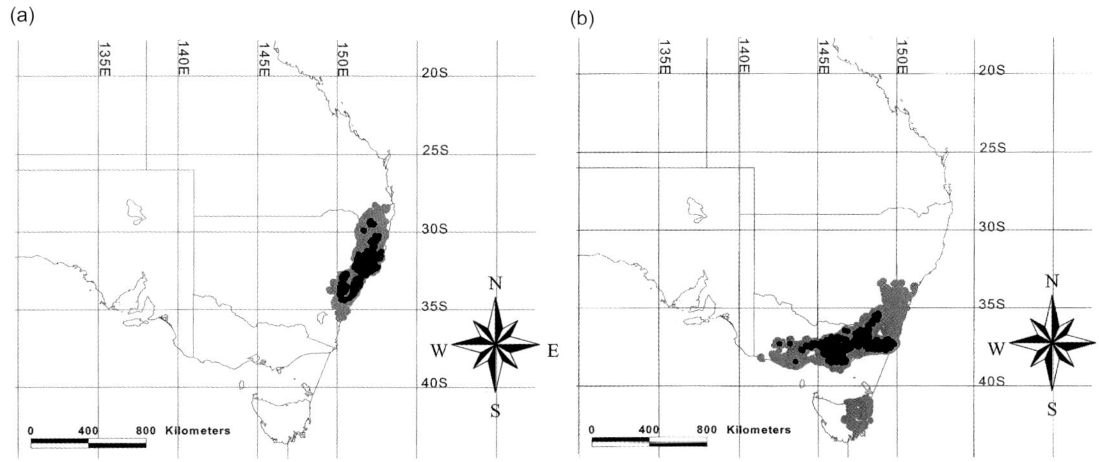

Fig. 7.4. Distributions of (a) *Antechinus stuartii* and (b) *A. agilis* predicted by bioclimatic modelling from habitat data (Crowther 2002a). Grey shading indicates predicted distributions; black indicates predicted core areas. Limited overlap of predicted ranges supports the species status of these populations. [Reprinted by permission of CSIRO Publishing from Crowther (2002a). Permission conveyed through Copyright Clearance Center, Inc.]

Evidence of different habitat usage in sympatric or parapatric populations is even more telling in that it suggests niche partitioning, as may be the case for *A. flavipes* and *A. stuartii* in southern Queensland (Van Dyck 1982b), as well as *A. flavipes* and *A. agilis* at Stoney Creek in New South Wales (Dickman 1986). (With regard to the latter case, however, P.A.W. has records of capturing individuals of both species at this locality in the same trap on consecutive nights.) A relatively new approach to investigating distributions and habitat preferences of candidate species is bioclimatic modelling, in which a statistical model is 'trained' with habitat information from known localities of species and then used to predict all areas within a region where these species could occur. Sumner and Dickman (1998) used this method to predict only slight overlap in the distributions of *A. stuartii* and *A. agilis*. Crowther (2002a) updated these results to show more, but still only partial, range overlap (Fig. 7.4).

Genetics. As noted above, genetic data have been brought to bear on species delimitation in four ways: fixed allelic differences, magnitude of genetic divergence, reciprocal monophyly of gene trees and molecular species delimitation under a coalescent model. In their confirmation of species distinction for *Pseudantechinus ningbing* apart from *P. macdonnellensis*, Cooper and Woolley (1983) found fixed differences of haemoglobin and albumin alleles. Moreover, the genetic identity (proportion of all loci with identical alleles) was 50%, comparable to or lower than that between accepted species (e.g. 67% for *Dasycercus blythi* and *P. bilarni*). Westerman *et al.* (2006) showed that exemplars from Woolley's (2005a) four *Myoictis* species exhibit reciprocal monophyly in a mtDNA gene tree estimated from *cytb* and *12S* (Fig. 7.5). To date, the only attempt to use molecular species delimitation in dasyurids is Woolley *et al.*'s (2020) analysis of *Murexia*, which we discuss later in this chapter.

Subspecies and cryptic species

During most of the 20th century, and especially while the BSC dominated systematists' view of species, geographic and morphological distinctness of populations were subordinate to reproductive isolation as criteria for species boundaries. This provided a conceptual justification for recognising polytypic species: those composed of multiple, geographically localised and phenotypically distinct – but interfertile – populations. These populations were often designated as *subspecies*, a taxonomic rank just below species. Ideally, members of different subspecies are diagnosable and have allopatric breeding ranges (Mayr and Ashlock 1991). Subspecies might, or might not, be 'incipient species' (Mayr 1942); only time will tell. A practical flaw in this conceptualisation is that direct, or strong indirect, evidence of reproductive compatibility was seldom obtained and biological species boundaries were inferred with the same criteria (phenotypic and geographic distinctness) as subspecies boundaries. Amadon (1949) suggested a '75% rule' for subspecies (75% of individuals from one subspecies must be diagnosable from 99% of individuals from another and vice versa for some set of characters), but this operational approach was more often discussed than implemented (Patten 2015). Indeed, taxonomy at both species and subspecies levels continued to be dominated by the morphospecies approach under the BSC paradigm. Not surprisingly, a substantial literature developed around the question of whether subspecies could be more than 'categories of convenience' (e.g. Mayr 1982a; Reydon and Kunz 2021).[1]

In the 1990s, the rise of the PSC with its emphasis on diagnosability led to the elevation of many subspecies to species status, including in dasyurids. This trend was also driven by the proliferation of phylogeographic studies, many of which found that allopatric populations harboured distinct variants of genetic markers. If two subspecies are phenotypically distinct and reciprocally monophyletic for mtDNA haplotypes, why not call them species? Why indeed. This was the rationale for recognising *Dasycercus cristicauda* and *D. blythi* as species rather than subspecies (Adams *et al.* 2000). Of course, not all subspecies meet these criteria, but are those that do confirmed as evolutionarily independent? Integrative taxonomists recognise that the peculiar inheritance pattern of mtDNA makes it susceptible to overinterpretation if used in isolation. For example, if gene flow between populations involves only males, maternally inherited mtDNA might well show reciprocal monophyly even though the populations are actively exchanging nuclear genes. Thus, modern practice emphasises the use of multiple independent genetic markers (Avise and Ball 1990). Despite numerous phylogeographic analyses of dasyurids in the

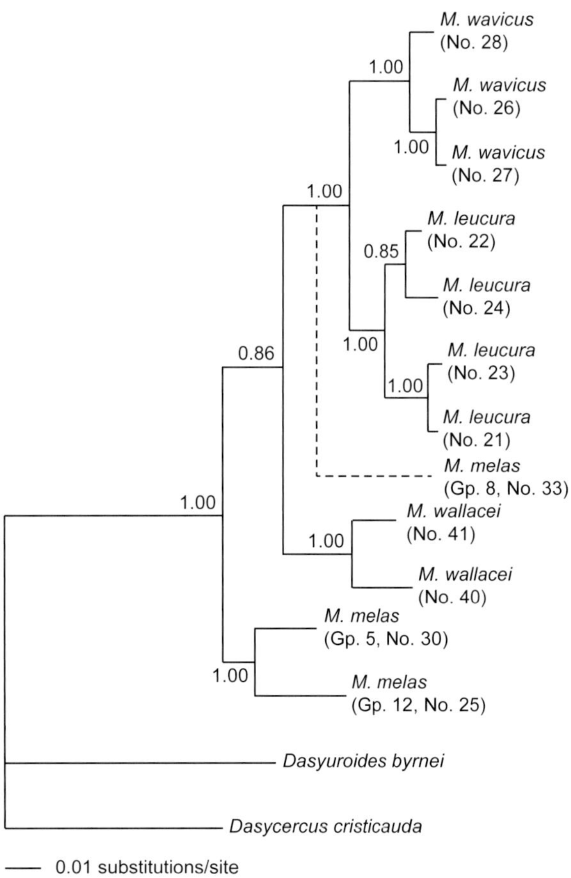

Fig. 7.5. Mitochondrial gene tree showing reciprocal monophyly of *Myoictis* species (Westerman *et al.* 2006). Tree was estimated with Bayesian analysis of *cytb* and *12S* loci under the GTR + I + Γ model. Nodal values are posterior probabilities. Branch length scale is substitutions/site. [Redrawn with permission of Springer Nature from Westerman *et al.* (2006).]

past several decades, many polytypic species are still recognised. What is the status and significance of their subspecies? We explore this question in 'Current issues in dasyurid species taxonomy' below.

Given the complex causes of phenotypic character divergence, it is possible for isolated populations to achieve genetic isolation while still being phenotypically quite similar (or perhaps identical). The result is *cryptic species*, an ensemble of which is often referred to as a *species complex* (Bickford *et al.* 2007). Genetic data have been critical for discovering cryptic species, again subject to the caution against drawing strong conclusions from single markers (e.g. mtDNA). Typically, a phylogeographic survey reveals evidence of genetic structure within what was considered a widespread species, then subsequent investigations with additional data types address the question of species status for individual populations. Chaplin *et al.* (2020) described an integrative approach to investigate cryptic species.

Strong evidence for a species boundary

The principles of integrative taxonomy suggest that congruent evidence from multiple data types and analytical approaches makes the most robust case for species-level distinction. Underlying this straightforward assertion are several less obvious and often problematic considerations to which we have alluded in previous sections. Unmack *et al.* (2021) provided a more in-depth discussion of these and other considerations.

- Adequate numerical sampling of specimens or individuals. To claim that a trait is diagnostic, or that populations are reciprocally monophyletic, or that distinct habitat preferences exist when data from only a few individuals have been gathered is unsupportable. This is a chronic issue for dasyurid studies, in which the animals are rare or hard to catch. Nevertheless, taxonomy must be based on whatever relevant information is available and the option of 'lumping' distinct populations into one polytypic species has no logical priority over 'splitting' them into two monotypic ones. Taxonomic stability is, however, a legitimate concern, a point to which we return below.
- Adequate geographic sampling. The geographic range of a species is one of its fundamental characteristics. The fact that specimens from just two localities within a much broader range can be diagnosed is weak support for their status as different species. The considerations just discussed for sample size apply here as well and taxonomists are often caught between the rock of analytical rigor and the hard place of collecting from wild populations.
- Operationally defined diagnostic traits. Operational definitions can be a slippery slope, but certainly extremes of vagueness must be avoided in identifying characters that distinguish species. In particular, the practice of expressing quantitative characters (e.g. tail–vent length 2–3 cm) in qualitative terms (e.g. 'tail short') is unhelpful.
- Diagnostic traits should be diagnostic. Univariate traits for which population means are significantly different, or ordinations in which individuals from two populations overlap slightly, or large frequency differences among shared alleles may be *evidence* of species distinction, but they are not diagnostic. Recognising populations displaying only these kinds of differentiation may be the most defensible use of subspecies rank.
- Gene trees are not enough, and they must be interpreted cautiously. Most taxonomists would agree that haplotypes within a nominal species that cluster by geography on a gene tree are not by themselves definitive evidence for species boundaries. This assumes that haplogroups are strongly supported (e.g. by high bootstrap or posterior probability values). When the same clusters appear in multiple gene trees (genealogical concordance), the inference is stronger, but in this case complementary analyses of the data with molecular species delimitation approaches provides valuable 'methodological' concordance.

The formalities of species description

The naming of animal species is governed by the International Code of Zoological Nomenclature (ICZN 1999). The Code provides rules for valid grammatical construction of names, adherence to the principle of priority, the practice of typification and other considerations. Priority dictates that the valid name of a species is the oldest available name that applies to that species. Typification fixes a species name to one or several 'type' specimens usually designated by the describer. These principles restrict the freedom of taxonomists to apply names in revisionary work. Woolley (2005b) used both to determine the correct names of the two *Dasycercus* species as *D. cristicauda* and *D. blythi*, the former of which Adams *et al.* (2000) had designated *D. hillieri*. The Code does not, however, establish criteria for species recognition and hence does not restrict taxonomists' freedom to propose new, or synonymise old, species. The fate of such proposals is governed initially by the peer review process and later by the extent to which new classifications are accepted in the zoological community.

Although its authority is confined to nomenclatural issues, the ICZN recognises the general importance of taxonomic stability 'to insure that scientists and other users do not get confused' (www.iczn.org/about-the-iczn/why-is-the-iczn-important/). At the level of species, the virtue of stability must be balanced against that of accuracy. Thus, *Sminthopsis murina* had come to include a wide range of populations since Waterhouse (1838) described it. Kitchener *et al.* (1984) destabilised *Sminthopsis* classification for a while by fracturing *S. murina* into five species, but subsequent research showed that most of the new species boundaries are robust. Poorly supported species descriptions, in contrast, are less likely to be validated and introduce needless instability into a classification. In short, recognition of a new species is a scientific hypothesis that has significant consequences for users of biological classifications. Modern standards for accepting such a hypothesis, even provisionally, are understandably high.

Taxonomy and conservation

May's (1990) argument for 'taxonomy as destiny' raised biologists' awareness that successful conservation depends on robust classification: 'Without taxonomy to give shape to the bricks, and systematics to tell us how to put them together, the house of biological science is a meaningless jumble' (p. 130). Not only does taxonomy identify the units of biodiversity, but it also informs policy and management options. For example, when taxonomists conclude that allopatric populations are conspecific, they imply the potential for fertile interbreeding, which then becomes a possible rescue strategy. That strategy is off the table if the populations are different species. Even for conspecifics, however, allopatry may lead to genetic divergence and local adaptation that would make mixing populations imprudent. The integration of systematics with phylogeography and population genetics should lead to the recognition of meaningful units of taxonomy (species, subspecies) and conservation (evolutionarily significant units = ESUs, management units) (Moritz 1994). Thus, the current biodiversity crisis has raised the stakes for species-level and intraspecific taxonomy, putting a premium on the kind of thorough and objective revisionary work described above. Firestone (2003) discussed the application of systematics and conservation genetics to the management of dasyurid populations. Willi *et al.* (2022) provide an up-to-date review of conservation genetics that highlights the importance of taxonomy and genetic analyses in defining conservation units, assessing the potential for outbreeding depression and managing population translocations. The species-level research summarised in Chapter 6 and below is directly relevant to this endeavour, though we do not explore its management implications.

CURRENT ISSUES IN DASYURID SPECIES TAXONOMY

In this section we describe research showing that recent or current dasyurid species classifications within selected genera may require revision. We include examples in which formal revisions have been proposed (e.g. *Antechinus*) as well as cases in which current observations significantly challenge existing species boundaries (e.g. *Murexia*, *Planigale*). Our treatment is not exhaustive but highlights areas we feel exemplify sound taxonomic practice or are most significant for future work.

Dasyurini

Dasyurus spartacus. Early DNA studies (Krajewski, Young *et al.* 1997; Firestone 2000) found that this species was most closely related to *D. geoffroii* (see Figs 6.3 and 6.6), the two showing the smallest genetic divergence observed between any dasyurid sister-species known at the time. Indeed, the earliest specimens of the bronze quoll were assigned to *D. geoffroii* (Waithman 1979). When Woolley *et al.*'s (2015) concatenation of mitochondrial and nuclear DNA sequences resolved *D. spartacus* within a clade of three *D. geoffroii* exemplars (see Fig. 6.7), they suggested that the species status of the bronze quoll be reinvestigated. No such studies have been published to date, but preliminary mtDNA data gathered by one of us (MW) from a larger sample of *D. geoffroii* and *D. spartacus* individuals suggest that they are not reciprocally monophyletic.

Dasyurus geoffroii. The western quoll was long considered polytypic, with the extant *D. g. fortis* confined to south-western Australia and the recently extinct *D. g. geoffroii* occurring across most of the Australian mainland. Thomas (1906a) distinguished these forms based on skull size and colour, but their recognition was largely abandoned following Ride (1970). As pointed out by Woolley *et al.* (2015), however, no careful morphological or molecular study of *D. g. geoffroii* has ever been conducted, despite the availability of ca. 30 museum specimens. Although a moot point from the perspective of modern biodiversity, the phylogeographic structure of *D. g. geoffroii* would have implications for our understanding of *Dasyurus* evolutionary history.

Dasyurus maculatus. Ramsay (1888) described *D. 'gracilis'* from northern Queensland, but Tate (1952) reduced it to a subspecies of *D. maculatus*, the nominate form of which occurs in the south-eastern mainland and Tasmania. Firestone *et al.* (1999) obtained the partial *CR* sequences of 20 spotted-tailed quolls from across the species range and found that Tasmanian and mainland haplotypes were reciprocally monophyletic, though bootstrap support for the mainland haplogroup was weak (Fig. 7.6a). Woolley *et al.* (2015) also recovered a Tasmanian lineage with combined mtDNA and nuclear sequences, but not a single mainland clade; again, however, bootstrap resolution on the latter pattern was only moderate (see Fig. 6.7). Firestone *et al.*'s (1999) analysis of six microsatellite loci also showed divergence among populations separated by >50 km and indeed Woolley *et al.*'s (2015) tree shows some structuring of mainland exemplars. Grealy *et al.* (2019) again replicated the mtDNA results and showed that a new haplotype from Kangaroo Island subfossils clustered with moderate support as sister to the Tasmanian lineage (Fig. 7.6b). Although Firestone *et al.* (1999) and later authors called for expanded studies of *D. maculatus* phylogeography, no such study has yet been published.

Dasyurus hallucatus. Northern quolls currently exist in isolated populations in Pilbara, Kimberley, Arnhem

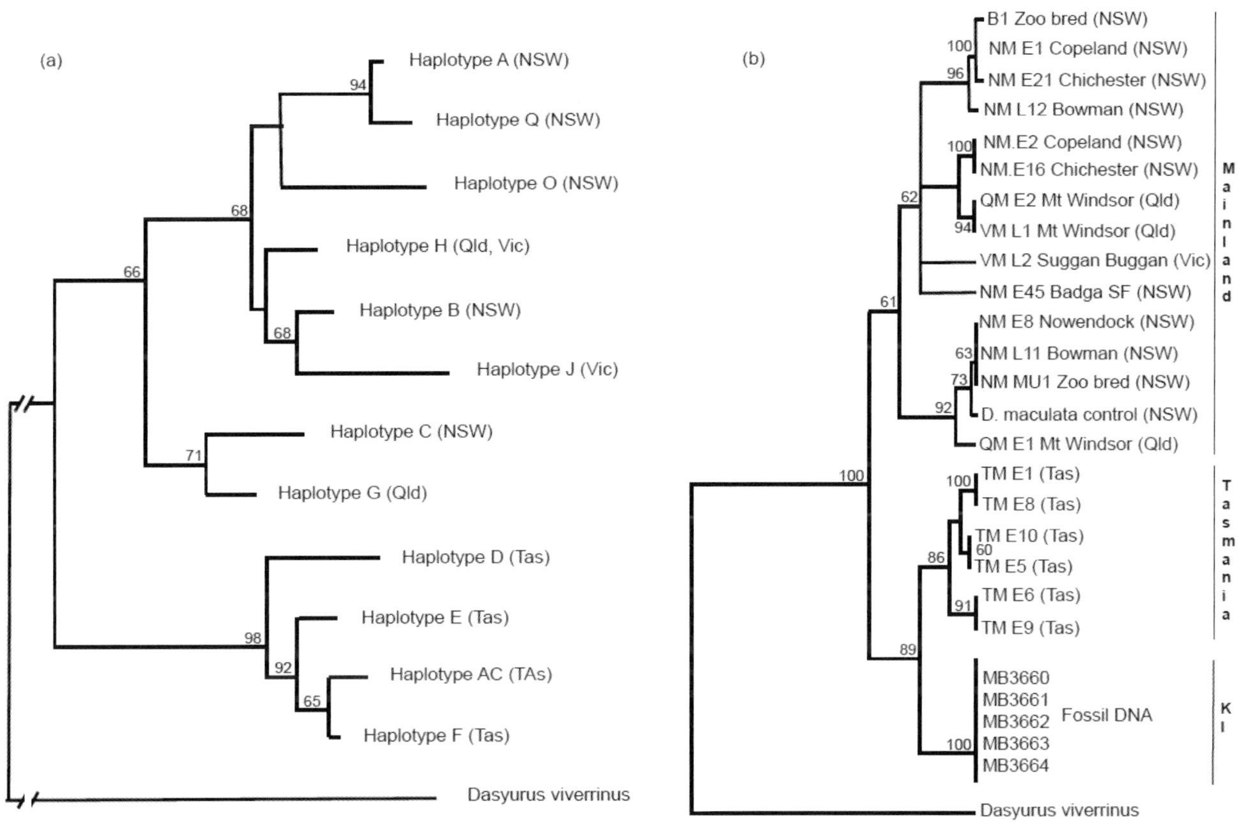

Fig. 7.6. MtDNA phylogeography of *Dasyurus maculatus* based on partial *CR* sequences. (a) Firestone *et al.*'s (1999) neighbour-joining tree from Tamura (1992) sequence distances. (b) Grealy *et al.*'s (2019) UPGMA tree from HKY (Hasegawa *et al.* 1985) sequence distances. Tip labels are haplotype designations and localities, or (for fossils) specimen numbers. In both trees, numbers at nodes are bootstrap support values >50%; branch length scale is substitutions per site. KI = Kangaroo Island. [(a) Redrawn with permission of John Wiley and Sons from Firestone *et al.* (1999). (b) Adapted from Grealy *et al.* (2019).]

Land and eastern Queensland regions (see Fig. 2.2), though some of this range may have been continuous in the past. Early taxonomists (see Thomas 1926) recognised some of these as subspecies: *D. h. exilis* in Western Australia, *D. h. hallucatus* in the Northern Territory, *D. h. nesaeus* on Groote Eylandt and *D. h. predator* in Cape York. These subspecies were later abandoned, but Firestone (2000) and Firestone *et al.* (2000) found mtDNA and microsatellite evidence, respectively, for isolation between the *hallucatus* and *predator* forms (the only areas sampled). How *et al.* (2009) similarly showed genetic separation of Pilbara and Kimberley populations. Woolley *et al.* (2015) sampled all populations and found strong phylogeographic structure corresponding to the Pilbara, Kimberley, Northern Territory (including Groote Eylandt) and Queensland regions separated by known dispersal barriers (see Fig. 6.7). Umbrello (2018) undertook a comprehensive morphological and morphometric analysis of northern quolls from across their range and concluded that the data fit better to the lineages identified by Woolley *et al.* (2015) than to previous subspecies groupings. Together, these results call into question whether, and for how long, *D. hallucatus* populations have been isolated, which in turn has implications for their taxonomic status.

Dasyurus viverrinus. The eastern quoll is now mostly restricted to Tasmania, but historical populations were found in the south-east mainland. No taxonomic distinction has ever been made between them. Woolley *et al.* (2015) analysed partial *12S* and *cytb* sequences from *D. viverrinus* subfossils collected on Kangaroo Island, SA, and in Victoria (Haouchar *et al.* 2014), which showed little divergence from their Tasmanian counterparts. In contrast, Frankham *et al.* (2017) analysed short *cytb* sequences and found that four mainland haplotypes resolved apart from a single Tasmanian sequence (Fig. 7.7); mainland (NSW) haplotypes were much more similar to one another (0.9% mismatch) than to the Tasmanian one (8.4% mismatch).

Sarcophilus. The Tasmanian devil, like the eastern quoll, had a trans-Bassian distribution until recent times, with mainland populations known from across southern Australia. Brüniche-Olsen *et al.* (2018) gathered mtDNA sequence data from teeth and bones of 17 subfossil mainland devils and a larger sample (172 individuals) of Tasmanian populations. Phylogeographic analysis revealed that haplotypes from these regions formed reciprocally monophyletic groups (Fig. 7.8). The data also revealed a genetic separation between eastern and north-western Tasmanian populations that had previously been reported by Jones *et al.* (2004). Thus, *D. maculatus*, *D. viverrinus* and *S. harrisii* all suggest the possibility of isolation across Bass Strait that could have taxonomic implications.

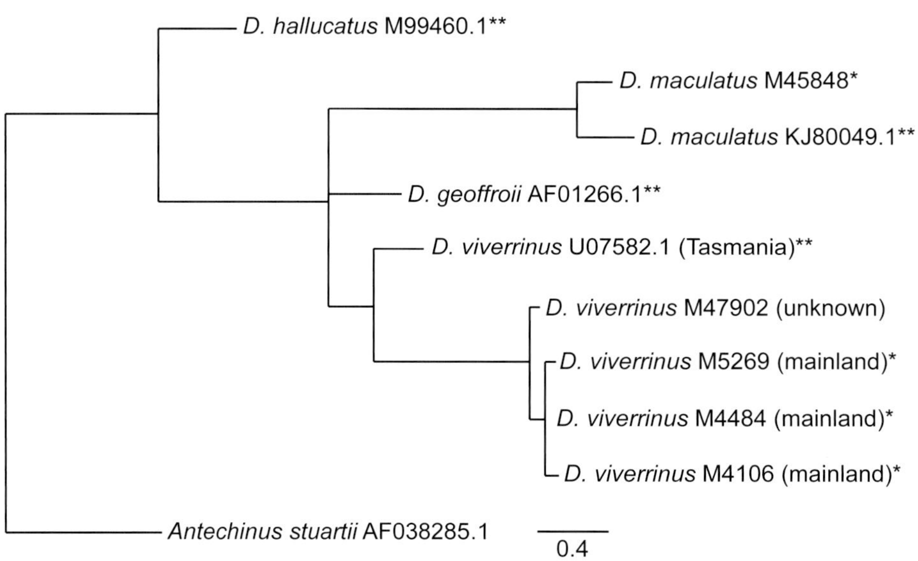

Fig. 7.7. Phylogeographic analysis of partial *cytb* haplotypes from mainland and Tasmanian exemplars of *Dasyurus viverrinus* (Frankham *et al.* 2017). Tree was estimated with Bayesian analysis; nodal values are posterior probabilities; branch length scale is substitutions/site. Tip labels include GenBank accession numbers. The 'unknown' haplotype is from a specimen collected at Barrington Tops, NSW. [Redrawn by permission of CSIRO Publishing from Frankham *et al.* (2017).]

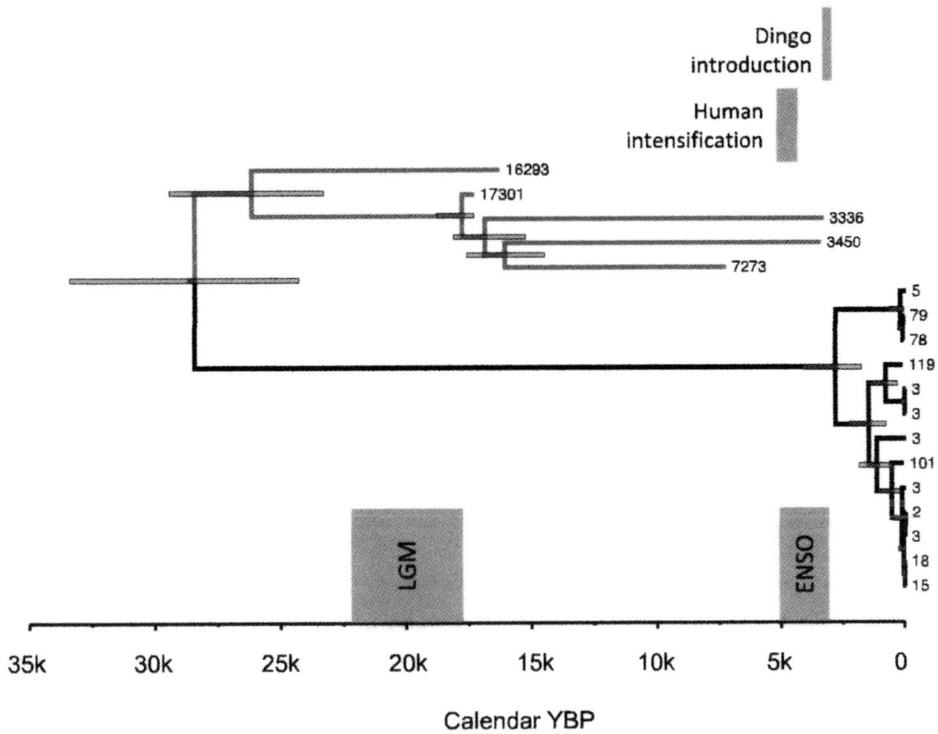

Fig. 7.8. Phylogeography of *Sarcophilus harrisii* (Brüniche-Olsen *et al.* 2018). Tree was estimated by Bayesian and maximum likelihood analysis of complete mtDNA sequences. Branch lengths are scaled to divergence time in years before present (YBP); nodal rectangles are 95% highest probability densities on estimated divergence times. Grey branches are mainland, and black branches Tasmanian, haplotypes; tip labels are arbitrary haplotype identifiers. Boxes show the timing of historical events; LGM = last glacial maximum, ENSO = El Niño-Southern Oscillation. [Reprinted by permission of John Wiley and Sons from Brüniche-Olsen *et al.* (2018).]

***Phascolosorex* and *Neophascogale*.** The three recognised phascolosoricine species have not been the subject of recent species-level research. However, Woolley *et al.* (2016) made the following comment: '*Phascolosorex dorsalis* represents a species complex that will likely be split into *P. dorsalis*, endemic to the Arfak Mountains, and *P. brevicauda* from the remaining range (with *P. whartoni* as a distinctive subspecies) (K. Helgen, pers. comm.).' The proposed species distinction corresponds to disjunct populations from the Vogelkop (*dorsalis*) and central New Guinea (*brevicauda*); *P. d. whartoni* specimens were collected near Mt Tafa and Murray Pass on the Papuan Peninsula (Tate and Archbold 1937). No follow-up research has been published on this question, nor on the widely distributed *P. doriae* and *N. lorentzii*. Eldridge *et al.* (2019) noted the likely existence of cryptic *Phascolosorex* species.

***Myoictis*.** Woolley's (2005a) revision of New Guinean three-striped dasyures and its molecular validation by Westerman *et al.* (2006) were described in Chapter 6. Prior to delimiting species boundaries, Woolley (2005a) assigned 131 specimens and their collecting localities to 13 groups based on tail and pelage characters (Fig. 7.9). Only 12 specimens from eight groups were represented in the molecular data set, but Westerman *et al.* (2006) nevertheless discerned hints of phylogeographic structure within species. Thus, Woolley's (2005a) 'geo-morphological' groups provide a starting point for more detailed genetic analyses of subdivision in all four species (see Eldridge *et al.* 2019).

***Pseudantechinus macdonnellensis*.** As noted in Chapter 6, Cooper *et al.* (2000a) erected *P.* '*roryi*' for Pilbara populations of *P. macdonnellensis* based on a single putatively fixed allozyme difference and morphometric distinctness. The allozyme analysis involved four '*roryi*' and seven *macdonnellensis* individuals (all from South Australia); only 11 of 53 loci were variable. Discriminant analysis of 19 cranial measurements showed no overlap of 20 '*roryi*' and 51 *macdonnellensis* individuals in a space defined by the first two canonical variates axes (Fig. 7.10a).

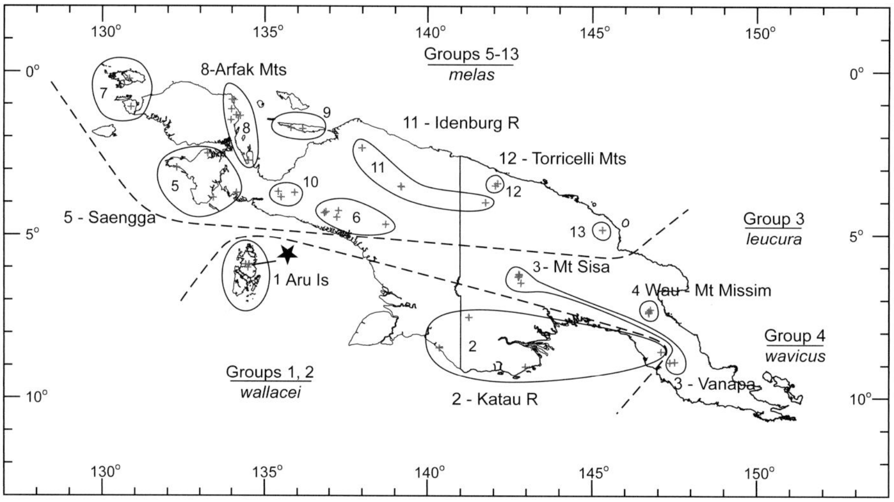

Fig. 7.9. Distribution of *Myoictis* species and their constituent groups (1–13) of specimens and localities distinguished by tail and pelage traits (Woolley 2005a). Stars show localities of specimens used for molecular analyses by Westerman *et al.* (2006). [Reprinted by permission of Springer Nature from Westerman *et al.* (2006).]

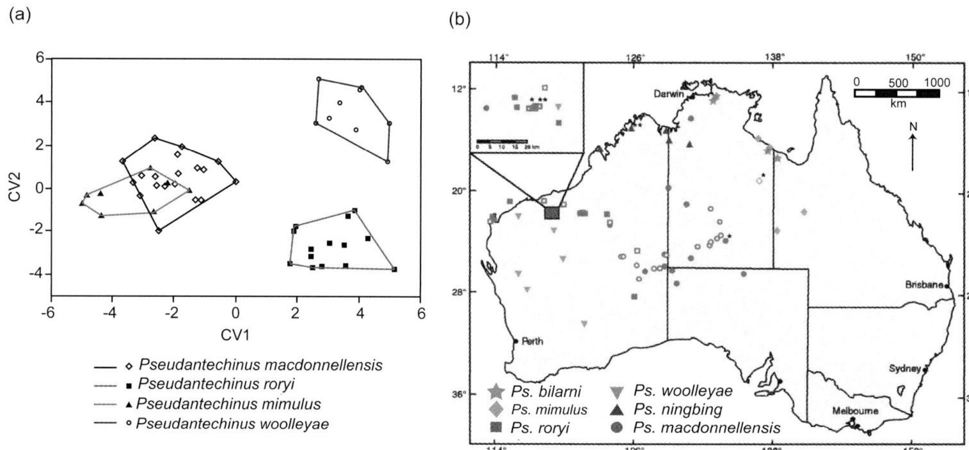

Fig. 7.10. (a) Discriminant analysis of *Pseudantechinus* exemplars based on five cranial characters statistically selected for best group separation from 19 total measurements (Cooper *et al.* 2000a). (b) *Pseudantechinus* collecting localities from Umbrello *et al.* (2017) that outline the distributions of five currently recognised species and *P. 'roryi'*. Open symbols are localities from Cooper *et al.* (2000a). [(a) Redrawn with permission of the Western Australian Museum from Cooper *et al.* (2000a). (b) Reproduced with permission of CSIRO Publishing from Umbrello *et al.* (2017), with minor adaptations. Permission conveyed through Copyright Clearance Center, Inc.]

Umbrello *et al.* (2017), however, failed to find any phylogeographic structure in nuclear and mitochondrial DNA sequences from a sample of 30 individuals collected across the range of *P. macdonnellensis* (see Fig. 6.11) and rejected the validity of *P. 'roryi'*. These authors noted that, although several dasyurid genera have Pilbara-endemic species, the pattern of genetic differentiation within *P. macdonnellensis* is more consistent with isolation-by-distance of the Pilbara from other populations. Because of its wide distribution and the still-limited geographic sampling for genetic analyses, future work might reveal phylogeographic structure in the fat-tailed pseudantechinus (Eldridge *et al.* 2019). At present, *Pseudantechinus* taxonomy appears to be accurately reflected in the five species whose ranges are outlined in Fig. 7.10b (see also Fig. 2.6a), though some authorities (e.g. IUCN Red List) continue to recognise *P. 'roryi'*.

Phascogalini

Phascogale tapoatafa. Spencer *et al.*'s (2001) phylogeographic study of *Phascogale* revealed three strongly supported mitochondrial (*cytb*) haplogroups within what was then *P. tapoatafa* (Fig. 7.11); the 'north' haplogroup was elevated to species status (*P. pirata*) by Aplin *et al.* (2015). The latter study sampled all major disjunct populations of *P. tapoatafa* and found substantial genetic and some morphometric divergence among them. After reviewing species and subspecies concepts, the authors argued that – following recognition of *P. pirata* – the remaining populations were most appropriately recognised as subspecies or 'forms' of *P. tapoatafa*. Their justification included the following reflection:

> *The question of how much divergence is required to warrant subspecies recognition is one that begs further theoretical consideration. At present, however, it seems that any decision would of necessity be an arbitrary one … [A]ll we can offer is that good application of the subspecies rank should follow the same principles as good species level taxonomy – with a strong emphasis on congruence of signal among multiple lines of evidence, on clear diagnoses, and where possible, on analytical methods able to estimate historical and current levels of gene flow, as well as the role of adaptation in shaping patterns of morphological variation. However, we can also note that species and subspecies are, for nomenclatural purposes, one and the same thing (both are 'species-level' names), with the rank assigned to any particular taxon being a matter of individual taxonomic opinion. In this regard, the somewhat arbitrary decision of whether to recognize the allopatric populations of* P. tapoatafa *as species or subspecies is relatively unimportant compared with the conclusion that each of these populations has experienced long and divergent evolutionary histories in the context of very different biogeographic regions and even biomes. [Aplin et al. 2015: 55]*

This statement expresses the modern version of Darwin's 'species problem' as clearly as anything in the dasyurid literature. While we do not necessarily agree that species and subspecies are, in any sense, 'one and the same thing' or that the distinction is 'relatively unimportant', the case of *P. tapoatafa* illustrates the challenge of objectively using the subspecies rank when data suggest that populations are evolutionarily independent. In any case, *Phascogale* systematics will benefit from further investigation with nuclear markers.

Antechinus. In many ways the *Antechinus* revision initiated by Andrew Baker and colleagues in 2012 (see Chapter 6) exemplifies the spirit of integrative taxonomy.

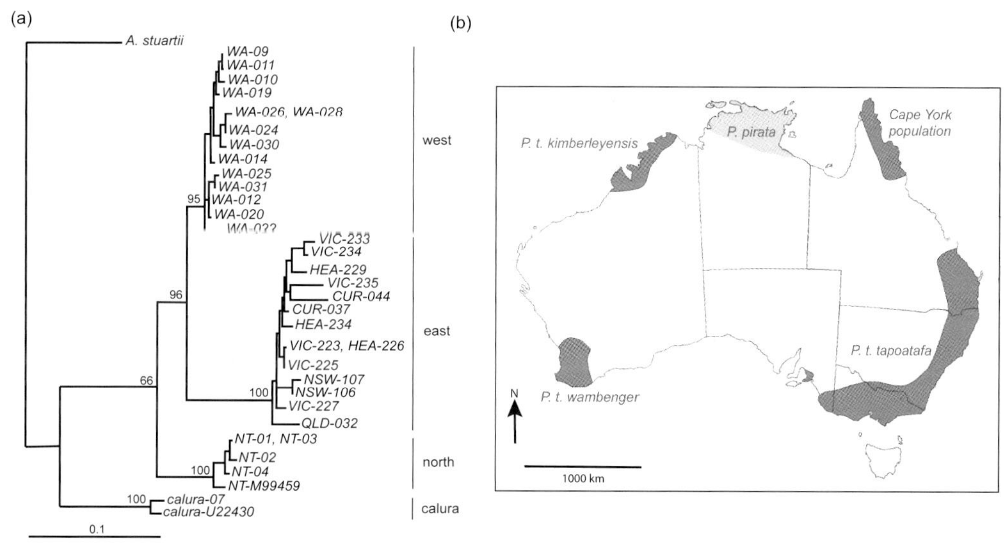

Fig. 7.11. (a) Mitochondrial gene tree for *Phascogale* (Spencer *et al.* 2001). Tree estimated by neighbour-joining analysis of Kimura (1980) distances from partial *cytb* DNA sequences. Tip labels are specimen numbers; groups 'west', 'east' and 'north' refer to allopatric populations of *P. tapoatafa*. Nodal values are bootstrap percentages; branch length scale is substitutions/site. (b) Distributions of *P. pirata* and the subspecies of *P. tapoatafa*. *Cytb* mismatch percentages are 7.3–16.3% between *P. pirata* and *P. tapoatafa*, and 3.0–7.8% among *P. tapoatafa* subspecies. [(a) Redrawn with permission of CSIRO Publishing from Spencer *et al.* (2001). Permission conveyed through Copyright Clearance Center, Inc. (b) Adapted from Aplin *et al.* (2015).]

Descriptions of *A. mysticus* (Baker *et al.* 2012), *A. argentus* (Baker *et al.* 2013), *A. arktos* (Baker *et al.* 2014), *A. mimetes* and *A. vandycki* (Baker *et al.* 2015), as well as revalidation of *A. bellus* (Baker and Van Dyck 2013) incorporate qualitative morphological, morphometric, genetic and ecological data, along with thorough quantitative analyses. One shortcoming in this work, often unavoidable in dasyurid alpha-taxonomy, is the relatively small number of specimens available for some species: eight for *A. argentus*, nine for *A. arktos*, seven for *A. vandycki*. Usually even fewer individuals were available for molecular analyses. Even for species with larger samples of specimens examined, evidence for morphological divergence is sometimes marginal. For example, Baker *et al.* (2012) based their description of *A. mysticus* on 34 specimens but found non-overlapping variation for only a few uni- or bivariate measures. Although DFA classified specimens to species with ≥94% accuracy, *A. mysticus* showed overlap with *A. stuartii* in the morphospace defined by canonical variate axes. Nevertheless, synthesis of these studies within the phylogenetic framework of Mutton *et al.* (2019; see Fig. 6.16) provides a much sharper picture of *Antechinus* diversity (Fig. 7.12) and solid framework for future studies of the genus. Several lingering questions persist. One involves the status of *A. flavipes leucogaster*, the south-western form of yellow-footed antechinus. As described in Chapter 6 and in Baker and Van Dyck (2015), there is evidence for phenotypic and phylogeographic separation between this population and eastern forms (see Fig. 6.15b). Another relates to the

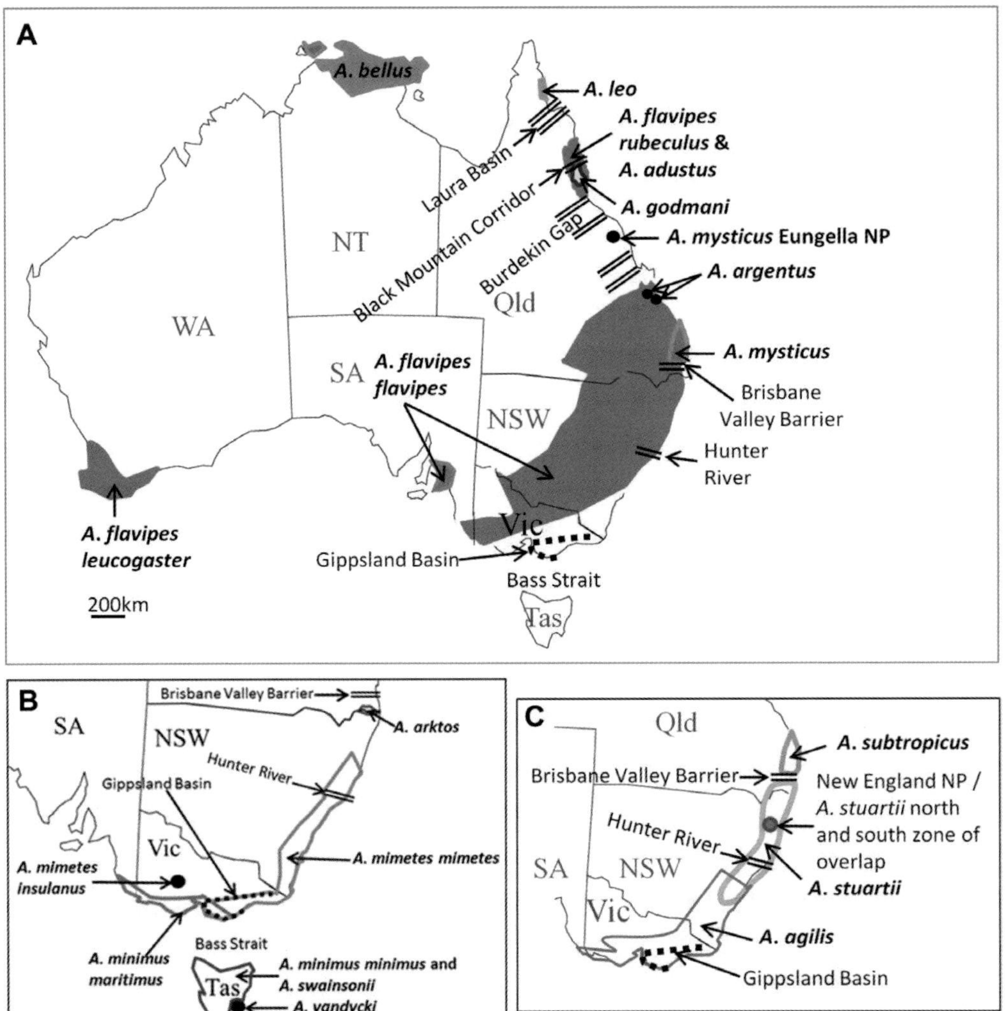

Fig. 7.12. Geographic distributions of currently recognised *Antechinus* species (Mutton *et al.* 2019). [Reprinted by permission of Oxford University Press from Mutton *et al.* (2019).]

finding of Mutton *et al.* (2019) that *A. stuartii sensu stricto* is paraphyletic, with northern populations forming the sister-clade to *A. subtropicus*. Viacava *et al.* (2021) used geometric morphometrics to show that skulls from the two forms of *A. stuartii* have significantly different shapes.

Murexia. The review of *Murexia* taxonomy by Woolley *et al.* (2020) described in Chapter 6 involved examination of over 700 museum specimens, for which the authors tabulated qualitative data on footpads, dentition and pelage, along with nine craniodental and external measurements. These were supplemented by DNA sequences of three mitochondrial (*CR, 12S, 16S*) and four nuclear (*IRBP, bfib7, ε-globin, P1*) loci from 6 to 66 individuals. The molecular tree was shown in Fig. 6.18. Here we describe the results for each nominal species; specimen distributions are shown in Fig. 7.13 (see also Fig. 2.14).

- Although specimens of *M. habbema* appeared morphologically homogenous, the DNA tree resolved two divergent and strongly supported haplogroups – one from the Porgera locality, the other from Mt. Kaindi + Mt. Albert-Edward – both in the highlands of Papua New Guinea (PNG). The same two groups were recovered by the Automatic Barcode Gap Discovery (ABGD) method of molecular species delimitation. Neither haplogroup is associated with the once recognised *hageni* form.

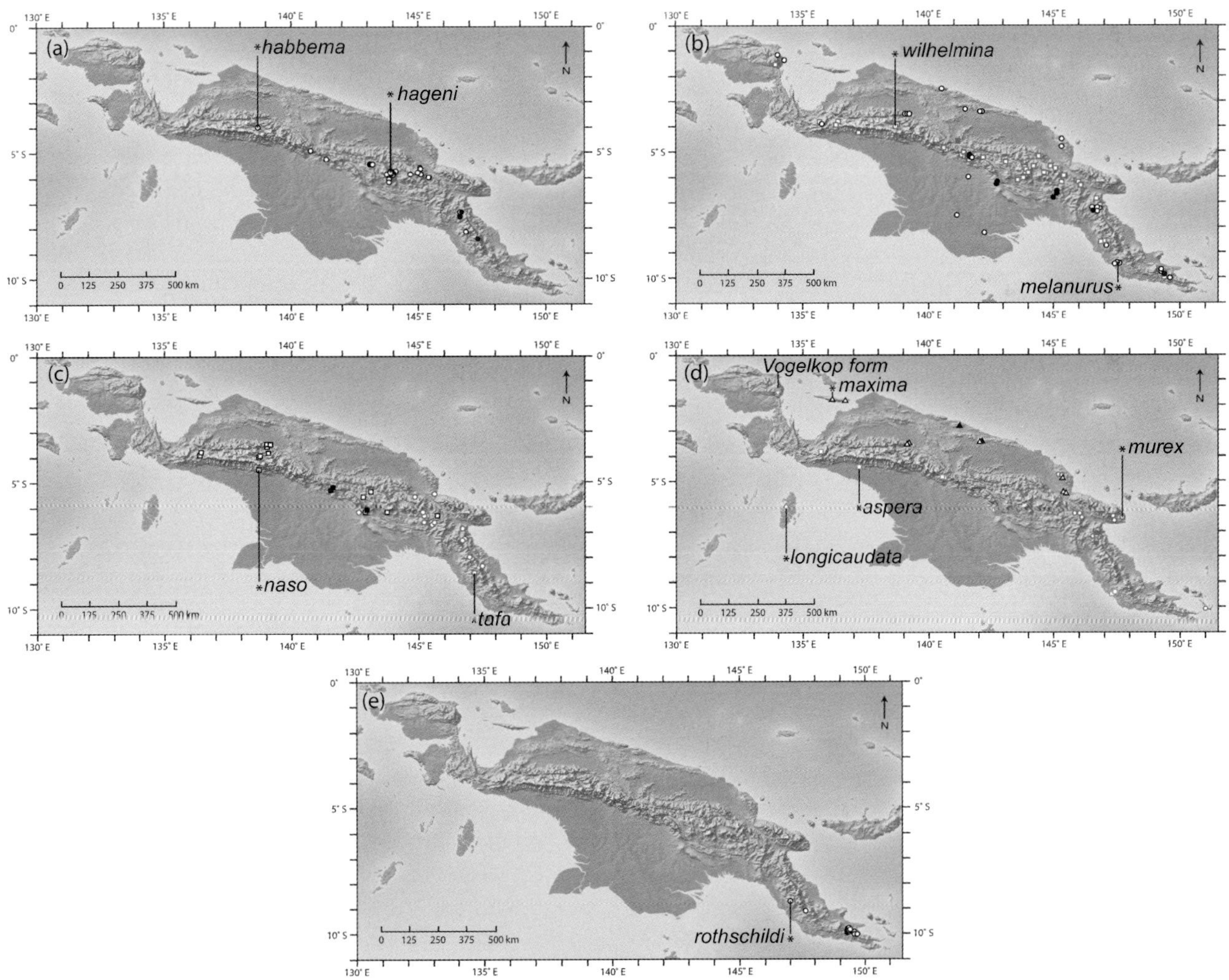

Fig. 7.13. Distribution of *Murexia* species and forms (Woolley *et al.* 2020): (a) *M. habbema*; (b) *M. melanurus*; (c) *M. naso*; (d) *M. longicaudata*; (e) *M. rothschildi*. Type localities are indicated by asterisks (*) followed by form name. Symbol shapes correspond to named forms. Solid symbols show localities of specimens included in molecular analyses. [Reprinted by permission of CSIRO Publishing from Woolley *et al.* (2020). Permission conveyed through Copyright Clearance Center, Inc.]

- Molar tooth measurements and PCA separate specimens of *M. melanurus* into two groups, a large-toothed *melanurus* form and a small-toothed *wilhelmina* form. The DNA tree, however, resolves two major groups, each with internal structure, that do not correspond to *melanurus* or *wilhelmina*.
- Tail pelage distinguished two forms within *M. naso*, a *naso* form with long reddish hairs at the tip of the tail and a *tafa* form with white hairs (see Fig. 2.14d). These forms were not separated in morphometric analyses. The DNA tree resolved one individual of *naso* from two of *tafa*, but also found a third strongly supported group from the Hindenberg Ranges + Mananda Ridge localities of central PNG that corresponded to neither form. This third group formed a clade with *M. longicaudata* sequences and its identity remains in doubt.
- Molars and morphometrics distinguished three forms of *M. longicaudata*: an eastern *murex* form, a southern *aspera* form and a northern *maxima* form. A single specimen from the Vogelkop of West Papua is distinct from all other representatives of this species, as is the holotype specimen from the Aru Islands. The DNA phylogeny and ABGD analyses also recover these groups, but the *murex* clade is sister to a larger group including all other *M. longicaudata* and *M. naso* sequences.
- *Murexia rothschildi* showed no obvious internal structure with morphological or molecular data.

Given the complexity of these results, Woolley *et al.* (2020) did not propose a revision of *Murexia*. Even so, the study shows that current taxonomy of the genus is inadequate in at least two ways. First, the existing boundary between *M. naso* and *M. longicaudata* is probably unreliable. Second, all nominal species except *M. rothschildi* include phylogeographic subdivisions if not cryptic species (Eldridge *et al.* 2019).

Sminthopsini

Antechinomys. Eastern and western populations of *A. laniger* have been considered subspecies (*A. l. laniger* and *A. l. spenceri*, respectively), distinct species (Liddicker and Marlow 1970), or parts of a cline within a monotypic species (Archer 1977). Recent analyses of four mitochondrial and five nuclear loci (Westerman *et al.* 2023) revealed distinct phylogeographic clusters in Western Australia, South Australia and New South Wales + Queensland, none of them corresponding to the *laniger-spenceri* distinction.

Ningaui. No phylogeographic studies of ningauis have yet been carried out, even though *N. ridei* has a broad distribution across interior Australia and *N. yvonneae* exists in fragmented populations across the south (see Fig. 2.17). Baverstock *et al.*'s (1983) allozyme and chromosome survey of the genus did not reveal any clear genetic subdivisions within species, but this issue should be addressed with DNA markers.

Sminthopsis virginiae. Archer (1981) recognised three subspecies of red-cheeked dunnart corresponding to isolated populations in Queensland (*S. v. virginiae*), the Kimberley and Northern Territory (*S. v. nitela*) and New Guinea (*S. v. rufigenis*) (Fig. 7.14). He suggested that the *nitela* form is morphologically diagnosable from the other two by the absence of a continuous anterior cingulum on the upper molars. Woolley's (1995b) attempts to breed a male *nitela* with a female *rufigenis* in the laboratory were unsuccessful, whereas the other two forms produced offspring. Moreover, wild populations of *nitela* appear to have different reproductive strategies than those of *virginiae* and *rufigenis* (Taplin 1980; Woolley 1984b; Morton *et al.* 1987). Blacket *et al.* (1999) noted that mtDNA sequence divergence between *nitela* and the other two forms was large and comparable to those between some recognised *Sminthopsis* species. Blacket *et al.* (2001) observed the same pattern in *12S* and *CR* distances and found that *nitela* haplotypes and *virginiae* + *rufigenis* haplotypes resolved as reciprocally monophyletic on the mtDNA tree (see Fig. 6.25); these two groups

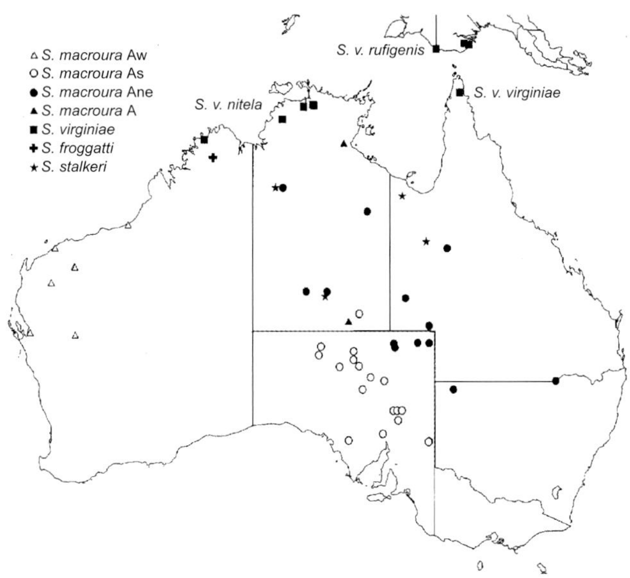

Fig. 7.14. Distribution and taxonomic affiliation of *S. macroura* and *S. virginiae* specimens in Blacket *et al.* (2001). [Reprinted by permission of Springer Nature from Blacket *et al.* (2001).]

also showed a single fixed allozyme difference and relatively large Nei's *D*. The authors recommended recognising *S. nitela* as a distinct species and called for additional research on variation within it (e.g. Kimberley versus Northern Territory), as well as investigation of *S. v. rufigenis* isolated on the Aru Islands. Neither the species description nor the follow-up studies have been published to date.

Sminthopsis macroura. The taxonomic history of stripe-faced dunnarts is complex, but Archer (1981) settled on two subspecies – *S. m. macroura* at the eastern edge of the species range and *S. m. froggatti* throughout central and western Australia. The latter form included *S. 'stalkeri'* of Thomas (1906b). Blacket *et al.* (2001) obtained *12S* and *CR* sequences, and assayed allozyme variation, from specimens collected across the range of *S. macroura* (Fig. 7.14). They recovered three lineages in their mtDNA tree, which they designated 'A', 'B' and 'C' (see Fig. 6.25). These groups showed fixed differences at 4–11% of the allozyme loci scored, large allozyme distances and no overlap in a principal coordinates analysis (PCoA, Fig. 7.15). Group A showed geographic substructure in both the mtDNA tree and allozyme PCoA, and Blacket *et al.* designated these groups as Aw (western), As (southern) and Ane (northern + eastern). Based on unpublished morphometric analyses, Blacket *et al.* (2001) argued that Groups B and C represent distinct species *S. froggatti* and *S. stalkeri*, respectively. They also cited the relatively small genetic divergences and data on successful interbreeding (Woolley 1990a, b) as evidence that the Aw, As and Ane groups within *S. macroura* should be recognised as subspecies, though they did not suggest names. Westerman *et al.* (unpublished) find that *S. m. macroura*, *S. m. froggatti* and *S. m. stalkeri* form distinct haplogroups on a tree of combined mitochondrial and nuclear loci. Some authors now recognise these as distinct species.

Sminthopsis leucopus. Archer (1981) distinguished *S. l. leucopus* from Tasmania and *S. l. ferruginifrons* from the mainland, the latter with widely allopatric populations in coastal areas of south-eastern Australia and rainforests of northern Queensland (Fig. 7.16). Smith (1983) interpreted allozyme variation as supporting the subspecies distinction. Baverstock *et al.* (1984) found no fixed differences between them, but there were two fixed differences between northern and southern *ferruginifrons* populations. Van Dyck (1985) documented several craniodental differences between northern and southern specimens of *ferruginifrons*. DNA sequences from *12S*, *CR* and *ω-globin* analysed by Blacket *et al.* (2006) showed that single haplotypes from Tasmania (*leucopus*) and Victoria (*ferruginifrons*) were sisters apart from a Queensland individual (see Fig. 6.26). They suggested that the Queensland population warrants recognition as a subspecies but also called for a more comprehensive study to

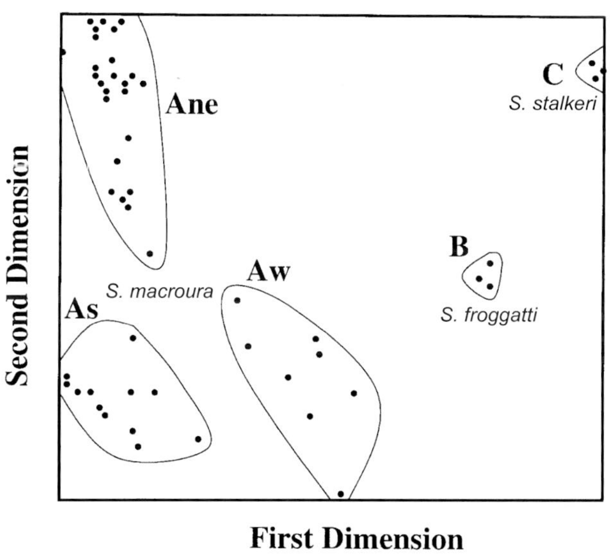

Fig. 7.15. Principal coordinates analysis of allozyme frequencies at 37 loci from 51 individuals of *S. macroura* (Blacket *et al.* 2001). The two axes shown explain 46% of the frequency variation. Group abbreviations correspond to those in Fig. 6.25. [Reprinted by permission of Springer Nature from Blacket *et al.* (2001).]

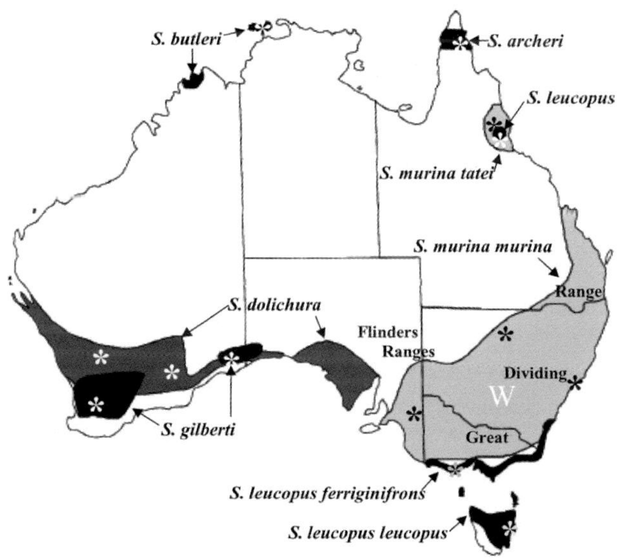

Fig. 7.16. Distribution of species and subspecies in the *Sminthopsis murina* species complex. Asterisks (*) show collecting localities for specimens studied by Blacket *et al.* (2006). 'W' marks the location of the Wellington Caves fossil site. [Reprinted by permission of Springer Nature from Blacket *et al.* (2006).]

clarify species and subspecies boundaries. Lavery *et al.* (2022) indeed found that Queensland white-footed dunnarts are genetically and morphologically distinct from other forms, recognising them as a subspecies (*S. leucopus janetzkiae*).

Sminthopsis murina. Like *S. leucopus*, the common dunnart has disjunct populations in south-eastern Australia and northern Queensland (Fig. 7.16) which have been recognised since Archer (1981) as *S. m. murina* and *S. m. tatei*, respectively. Read *et al.* (1983) found these forms to be interfertile and Baverstock *et al.* (1984) saw little allozyme divergence between them. Kitchener *et al.*'s (1984) morphometric analyses separated *S. m. murina* into groups on east and west sides of the Great Dividing Range, the latter including *tatei*, that could be distinguished by tail length and interpalatal vacuity distance. Using *12S*, *CR* and *ω-globin* DNA sequences from four individuals representing these three populations, Blacket *et al.* (2006) found two strongly supported clades – one consisting of two individuals from west of the Dividing Range and the other of one individual each from east of the Dividing Range plus *tatei*. Such sparse sampling precludes any taxonomic revision (other than abandoning the current subspecies), but both genetic and morphometric data suggest significant subdivision within *S. murina* that awaits more thorough investigation.

S. griseoventer* and *S. aitkeni. Kitchener *et al.* (1984) distinguished these species morphometrically, but at the time grey-bellied dunnarts were known only from Western Australia (WA). Kemper *et al.* (2011) included individuals from more recently discovered populations in South Australia (SA) in an integrative study involving morphometric, allozyme and mtDNA sequence data. They found that *S. aitkeni* and SA *S. griseoventer CR* haplotypes formed a well-resolved clade apart from WA *S. griseoventer* (Fig. 7.17) and that morphometric differentiation between the former two populations was minimal. They suggested that Kangaroo Island dunnarts might better be regarded as a subspecies of *S. griseoventer*, possibly showing size and pelage differentiation consistent with island effects. This classification has been adopted by the Australian Department of Climate Change, Energy, the Environment and Water (www.dcceew.gov.au) and the Atlas of Living Australia (www.ala.org.au). Alternatively, there may be a species boundary between WA *griseoventer* and SA *griseoventer* + *aitkeni*, but no further analyses of this situation have been published. Krajewski *et al.* (2012) found that the multilocus genetic divergence between *griseoventer* and *aitkeni* haplotypes was comparable to that of other *Sminthopsis* sister-species. Crowther *et al.* (1999) recognised *S. g. boullangerensis* from Boullanger Island as a separate subspecies from mainland *S. g. griseoventer* based on allozymes and morphology. However, a reassessment by Start *et al.* (2006) emphasised the genetic homogeneity of these populations found by Labrinidis *et al.* (1998) and rejected their taxonomic distinction.

Other members of the *S. murina* complex. Blacket *et al.* (2006) sequenced a single specimen each from the disjunct populations of *S. gilberti* in south-western Australia and Roe Plain on the Nullarbor; they found sequence divergence between these animals comparable to that observed between *S. murina* populations. Kemper *et al.* (2011) reported no strong phylogeographic separation between WA and SA populations of *S. dolichura* but sequenced few specimens from WA. No genetic studies have yet been conducted on allopatric populations of *S. butleri* (Bathurst and Melville Islands versus Kimberley) or *S. archeri* (Cape York versus PNG).

Sminthopsis crassicaudata. Thomas (1902) recognised two subspecies of fat-tailed dunnart – *S. c. crassicaudata* from mesic areas at the periphery of the species range and *S. c. centralis* from the arid interior – with no clear geographic boundary between them. Archer (1981) rejected the distinction, and Morton and Alexander (1982) documented a north-south cline in ear and tail length, two of the traits used to diagnose subspecies. Hope *et al.* (1986) found large allozyme frequency differences at two loci between populations north and south of the Murray River. Using mtDNA sequences from a moderate number of specimens, Cooper *et al.* (2000b) established a phylogeographic break between populations in south-eastern South Australia and Victoria versus all other localities (Fig. 7.18a). These authors then assayed a much larger sample of individuals for a *CR* restriction site that diagnoses south-eastern and north-western haplotypes to estimate the geographic distribution of these groups (Fig. 7.18b). Given some evidence of interbreeding between animals from the two regions (Bennett *et al.* 1990), Cooper *et al.* (2000b) retained both in a single species but recognised them as distinct ESUs. Subsequently, Thomas's (1902) subspecies were abandoned and no further phylogeographic studies of *S. crassicaudata* have been performed.

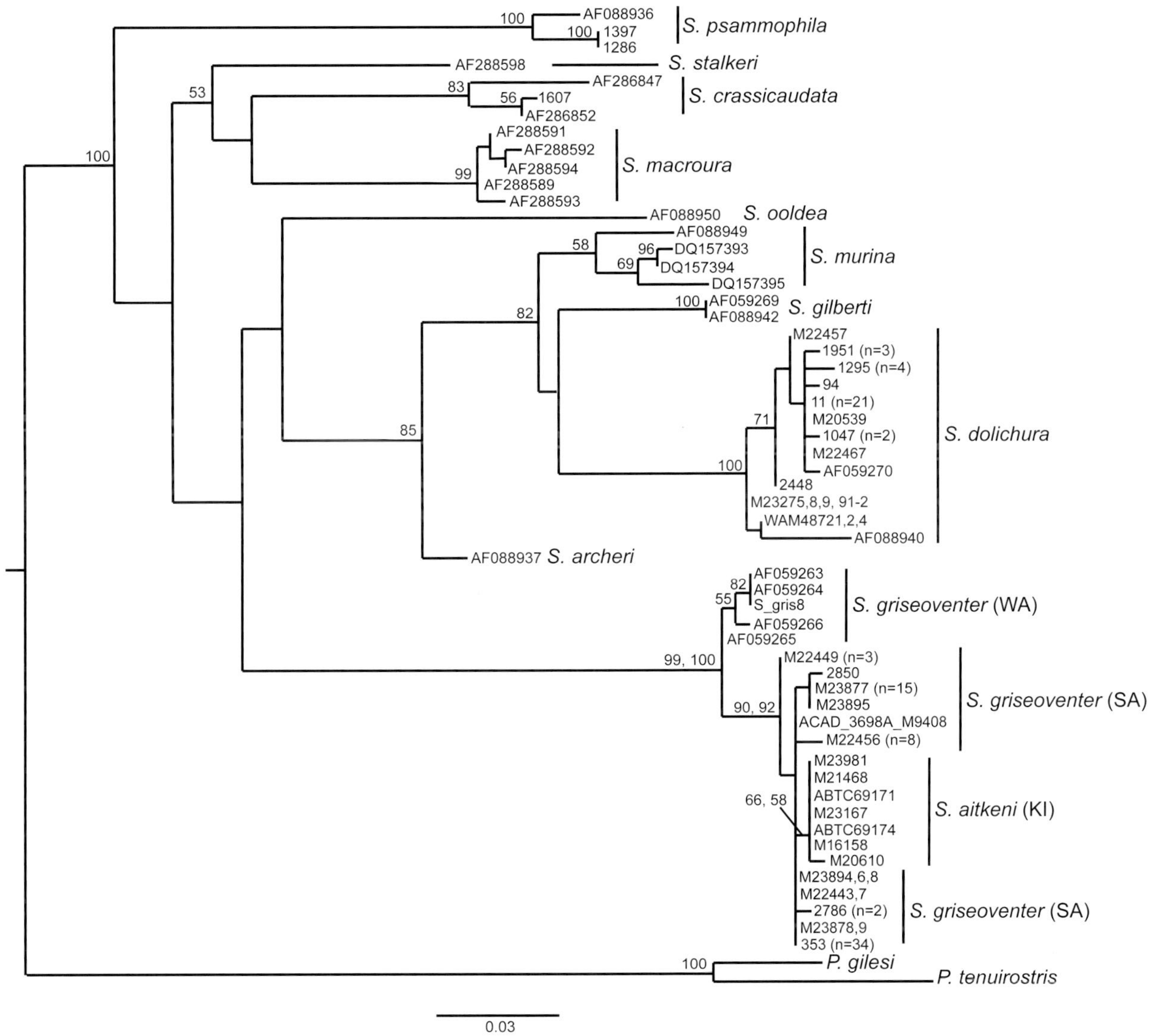

Fig. 7.17. Maximum likelihood (ML) tree for *CR* haplotypes of *Sminthopsis* species from Kemper *et al.* (2011). Values on branches are bootstrap percentages >50%; when two numbers appear, the first is from ML analysis and the second is from maximum parsimony analysis. Tip labels are specimen numbers with species identities to the right (KI = Kangaroo Island, SA = South Australia, WA = Western Australia). Branch length scale is substitutions/site. [Redrawn by permission of CSIRO Publishing from Kemper *et al.* (2011). Permission conveyed through Copyright Clearance Center, Inc.]

Other *Sminthopsis* species. Two other dunnarts have distributions that include disjunct populations: *S. hirtipes* (coastal WA, inland and Eyre Peninsula) and *S. psammophila* (three or four small patches in SA and WA) (see Fig. 2.15). Three more (*S. longicaudata*, *S. ooldea*, *S. youngsoni*) have relatively wide and continuous distributions. McLean *et al.* (2019) found no phylogeographic structure, but significant genetic differentiation, among populations of *S. psammophila*. None of the other species have yet been the subject of phylogeographic analysis.

Planigalini

Planigale maculata. Although Archer (1976b) designated the Groote Eylandt population of common planigales a subspecies (*P. m. sinualis*) distinct from the widespread mainland form (*P. m. maculata*), Blacket *et al.*'s (2000) *12S* DNA sequence + allozyme study showed a much more complicated phylogeographic structure within this species (see Fig. 6.28). These authors identified five genetic subdivisions designated M1-M5 but could not resolve their monophyly apart from *P. novaeguineae* on the mtDNA

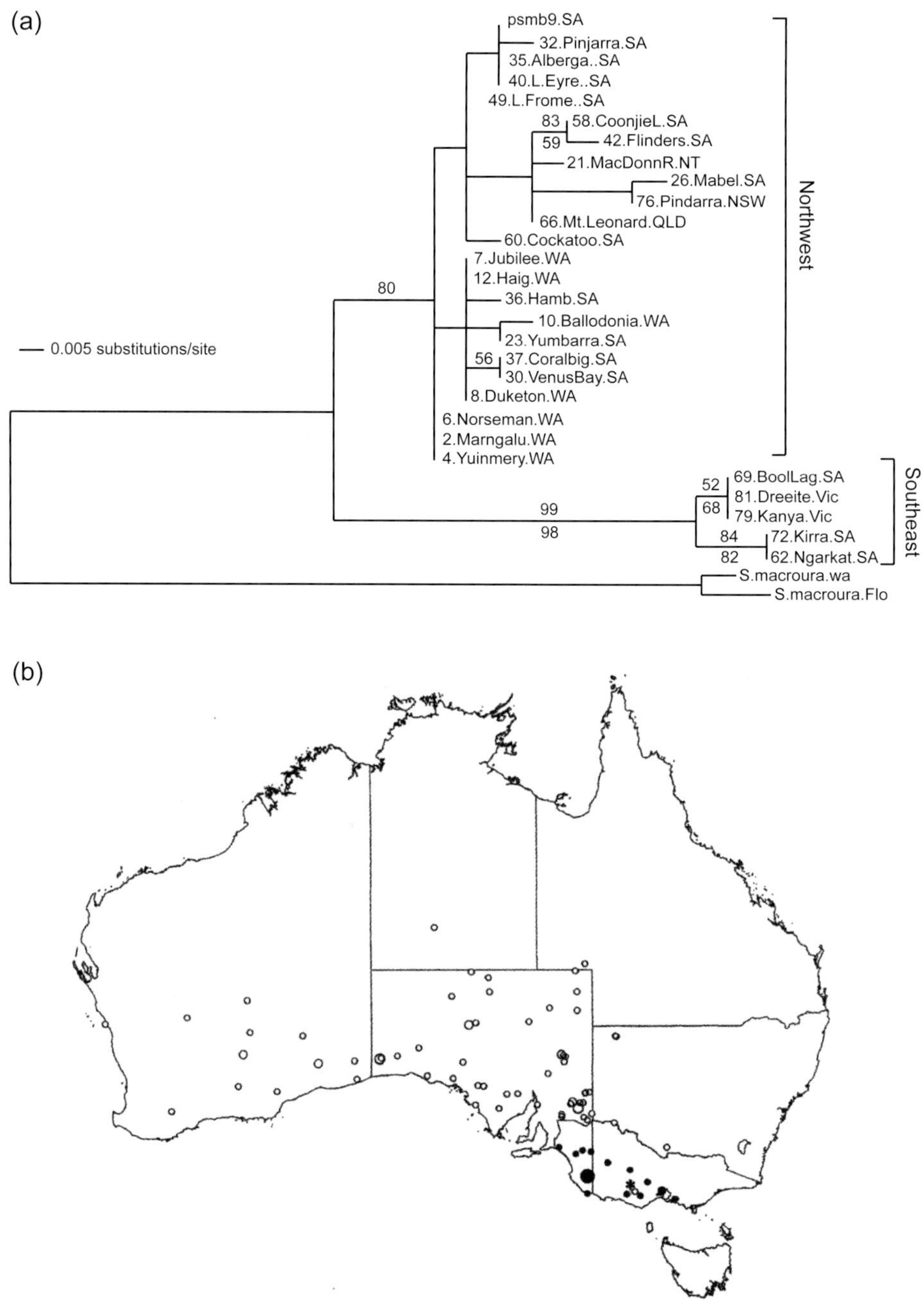

Fig. 7.18. Phylogeography of *Sminthopsis crassicaudata* from Cooper *et al.* (2000b). (a) Maximum likelihood tree of *CR* haplotypes estimated with the HKY model. Values above and below branches are bootstrap percentages from maximum parsimony and maximum likelihood analyses, respectively. Tip labels are haplotype localities. (b) Distribution of north-west (open circles) and south-east (filled circles) haplotypes based on restriction site analysis. [(a) redrawn and (b) reprinted with permission of CSIRO Publishing from Cooper *et al.* (2000b). Permission conveyed through Copyright Clearance Center, Inc.]

gene tree. With larger samples of specimens, localities and loci for DNA sequence analysis, Westerman, Blacket *et al.* (2016) recovered clades corresponding to M1 + M2, M3, M4 and M5, and showed that *P. maculata* is indeed paraphyletic with respect to *P. novaeguineae* (Fig. 7.19). Sampling localities corresponding to these lineages (Fig. 7.20) are geographically coherent: M1 + M2 occurs across the Top End region of NT and WA; M3 is found on islands in the Gulf of Carpentaria and adjacent mainland sites, and probably corresponds to the *sinualis* form; M4 is

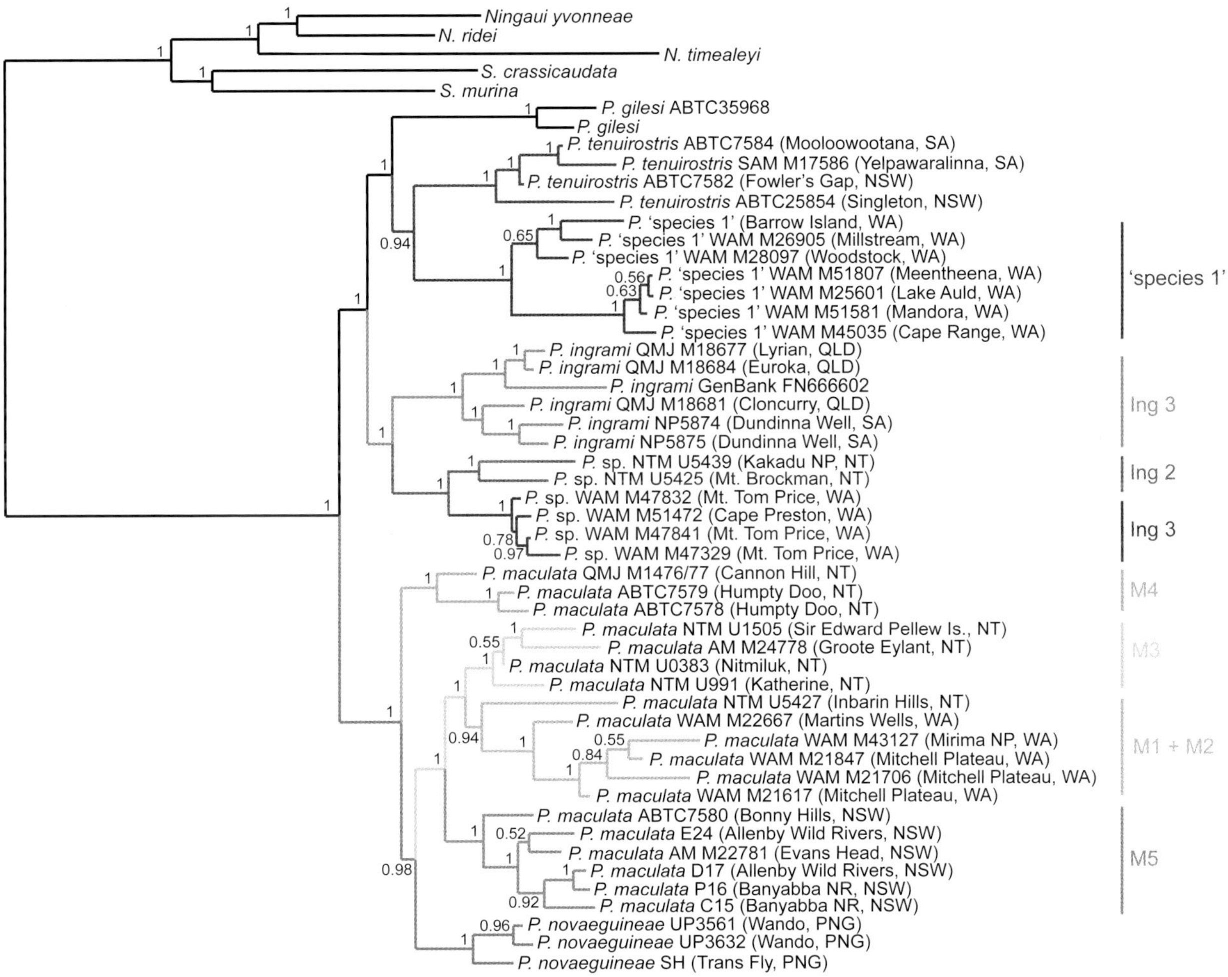

Fig. 7.19. Phylogeography of *Planigale* from Westerman, Blacket *et al.* (2016). DNA sequence phylogeny estimated with Bayesian analysis of partitioned *cytb, 12S, 16S, CR, P1, IRBP, bfib7* and *e-globin* loci. Tip labels are specimen numbers followed by collecting localities. Nodal values are posterior probabilities; clade names given at right. [Redrawn by permission of CSIRO Publishing from Westerman, Blacket *et al.* (2016). Permission conveyed through Copyright Clearance Center, Inc.]

another NT form from localities west of Arnhem Land; M5 represents eastern populations in Queensland and New South Wales. Westerman, Blacket *et al.* (2016) argued that these groups are cryptic species and warrant formal description as such, but there has been no morphological follow-up study to evaluate this claim.

Planigale ingrami. Westerman, Blacket *et al.*'s (2016) DNA sequence analysis recovered three clades (Ing1, Ing2, Ing3) from specimens assigned to the long-tailed planigale (Fig. 7.19). One of these includes the 'Mt. Tom Price' (WA) animal noted by Blacket *et al.* (2000). Ing1 is sister to Ing2, the latter representing animals from the Top End and perhaps Archer's (1976b) *P. i. subtilissima*. Ing3, from localities in western Queensland and north-eastern SA represents Archer's *P. i. brunnea*. Westerman, Blacket *et al.* (2016) lacked samples from the nominate subspecies *P. i. ingrami* but suggested that *P. ingrami* as currently conceived includes three or four cryptic species.

Pilbara planigale. Originally identified by a partial *cytb* sequence from one individual collected at Millstream Station, WA (Krajewski *et al.* 1994), '*Planigale* 1' has appeared in numerous molecular phylogenetic studies of dasyurids and planigales (e.g. Painter *et al.* 1995; Blacket *et al.* 2000). Westerman, Blacket *et al.* (2016) included DNA sequences from a wider set of sampling localities in the Pilbara and found that they formed a single clade ('species 1' in Figs 7.19 and 7.20). Some of these specimens were originally assigned to *P. maculata* but are clearly a distinct and geographically isolated species. In the nearly three decades since '*Planigale* 1' was recognised, the population

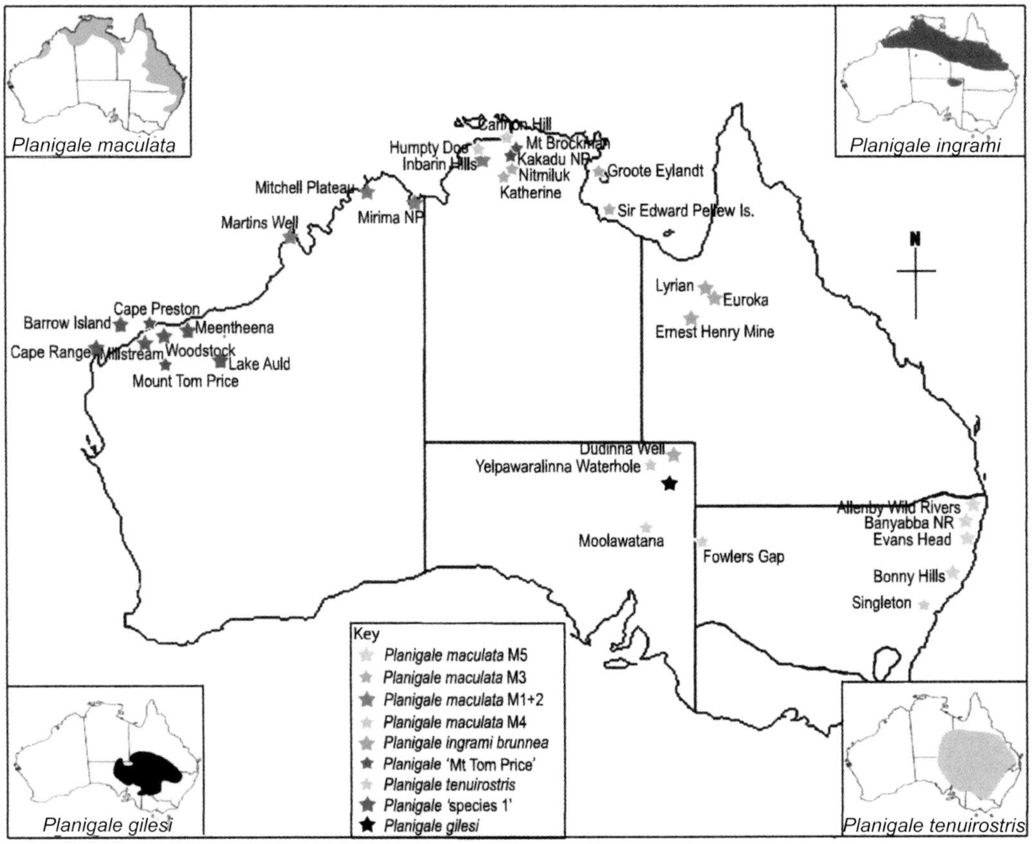

Fig. 7.20. Collecting localities that map the geographic distribution of *Planigale* clades shown in Fig. 7.19 from Westerman, Blacket *et al.* (2016). Inset maps show distributions of currently recognised species. [Reprinted by permission of CSIRO Publishing from Westerman, Blacket *et al.* (2016). Permission conveyed through Copyright Clearance Center, Inc.]

has become known colloquially as the Pilbara planigale but no formal species description has been published.

DOCUMENTING DASYURID BIODIVERSITY

Dasyurid systematics has come a long way since Kerr described the first quoll in 1792. In this and the previous four chapters, we have seen how knowledge of dasyurid diversity and relationships grew because of efforts in field collecting, morphological studies of museum specimens, application of molecular methods and use of increasingly sophisticated quantitative analyses. It is remarkable, however, that the great inferential power of modern systematics is impotent without the availability of specimens obtained from the wild with adequate locality data, professionally curated in natural history collections and studied by taxon specialists. As we have seen in this chapter, most vividly for *Murexia* and *Planigale*, the notion of understanding phylogenetic relationships within genera is meaningless when species boundaries are poorly established. The science of species recognition begins with specimen-based research informed by knowledge of natural history. Its significance extends beyond phylogenetics to all human endeavours for which knowledge of biodiversity is required, most urgently conservation biology and environmental policy. Arguably the greatest concern for the future of systematic biology – and perhaps biology in general – is the declining interest in taxon-specific expertise and the growing shortage of taxonomists trained to perform specimen-based research (Drew 2011), not to mention the scarcity of funding, facilities and professional positions to support such research. At the time of this writing, dasyurid species have been known to science for 228 years and many of them now face extinction. The first step towards saving them – or understanding their loss – is to know their identities and positions on the tree of life.

ENDNOTE

1 There is an important distinction between recognising that species show geographic variation among populations and capturing that variation in formal classification. The latter practice – naming of subspecies – has been perennially controversial in animal systematics. Use of trinomial nomenclature for this purpose was introduced in the mid-1800s, championed by American ornithologists in the late 1800s and formalised in 20th century iterations of the International Code of Zoological Nomenclature. Proponents of the practice emphasise its convenience for communicating about geographically distinct forms of the same species but insist that subspecies are not units of evolution. Opponents argue that multiple traits often vary independently with geography such that subspecies based on one or a few traits are not even meaningful units of taxonomy. This debate has been complicated by the popularity of the phylogenetic species concept (in which 'taxa' below the species level have no ontological status) and the use of evolutionarily significant units in conservation (in which local populations have exactly the evolutionary role that was denied to subspecies). No resolution of the issue appears imminent.

8 Dasyuromorphian families and fossils

Dasyuridae is part of a larger clade – Dasyuromorphia – that includes two other Recent marsupial families, Thylacinidae (thylacines) and Mymecobiidae (numbat). Confirmation that these families form a monophyletic group was achieved only in the late 20th century and relationships among them were not known with confidence until the 21st, both due in large part to DNA sequence data. Inclusion of thylacines and numbats in systematic studies of dasyurids has been crucial to developing a timeline of cladogenic events and macroevolutionary hypotheses, topics we discuss in Chapters 9–11. Here we introduce the numbat and thylacines, summarise the fossil record of Dasyuromorphia and describe research on the phylogeny of families and extinct taxa.

The fossil record of Australian marsupials begins with the Late Paleocene *Djarthia murgonensis* from south-eastern Queensland, represented by isolated teeth and jaw fragments (Godthelp *et al.* 1999). The anatomical features of these remains are plesiomorphic, resembling anatomically primitive didelphimorphians, but sharing one otherwise unique dental feature with two dasyuromorphians (*Ankotarinja* and *Keeuna*; see 'Dasyuromorphia *incertae sedis*' below). On balance, *Djarthia* cannot be assigned to any specific marsupial order. It is often characterised as a possible 'structural ancestor' of Australidelphia (e.g. Long *et al.* 2002). Dasyuromorphian fossils date from the Oligocene, but their sparse and fragmentary nature has made phylogenetic and functional interpretations difficult.

MYRMECOBIIDAE

The single species of numbat (*Myrmecobius fasciatus*) is a termite-eating specialist currently confined to south-western Australia, though its historical range extended over much of southern Western Australia (WA) and South Australia (SA) (Friend 2008). Numbats are medium-sized (average head–body length 234–251 mm, tail length 175–176 mm, weight 478–597 g) marsupials with distinctive transverse white stripes on reddish-brown dorsal pelage, pale belly and a bushy tail (Fig. 8.1a). They live in eucalypt forest areas with termite mounds, use hollow logs for shelter and have a diurnal-crepuscular activity cycle. Like other myrmecophagus species, numbats have long sticky tongues used for foraging in termite galleries (Cooper 2011). Their dramatic range contraction in the 20th century seems to have been caused by fox predation and they are now listed as Endangered with <1000 individuals remaining (Woinarski and Burbidge 2016).

Europeans discovered numbats near the Avon River, WA, in 1831 (Friend 1989) and *Myrmecobius fasciatus* was formally described by Waterhouse (1836). Although it was moved in and out of Dasyuridae for over a century, its placement within the monotypic Myrmecobiidae has been stable since Archer and Kirsch (1977). The affinity of the numbat with dasyurids was initially inferred from its polyprotodont incisor count and pentadactyl hindfeet, but its dentition and basicranium also have unique features. Unlike any other marsupial, the numbat's dental formula is I 4/3, C 1/1, P 3/3,

(a)

(b)

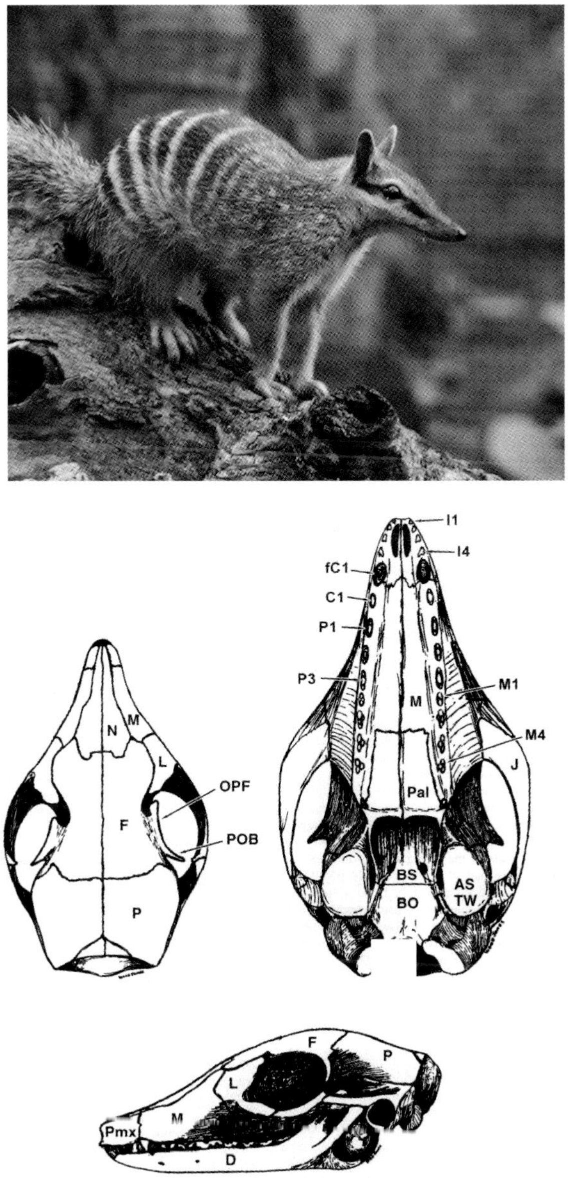

Fig. 8.1. (a) The numbat *Myrmecobius fasciatus.* (b) Numbat skull in dorsal (top left), palatal (top right) and lateral (bottom) views. Abbreviations: I1–4 = upper incisors; fC1 = fossa for tip of lower canine; C1 = upper canine; P1–3 = upper premolars; M1–4 = upper molars; M = maxilla; N = nasal; Pal = palatine; J = jugal; ASTW = alisphenoid tympanic wing; BS = basisphenoid; BO = basioccipital; Pmx = premaxilla; L = lachrymal; F = frontal; P = parietal; D = dentary; OPF = orbital process of frontal; POB = postorbital bar. [(a) Photo by M. Westerman. (b) Reprinted by permission of Michael Archer from Archer and Hand (2006).]

M 4–5/5–6 (Fig. 8.1b); the postcanine teeth are simple and their numbers vary among, and sometimes within, individuals (Berkovitz and Shellis 2018). There is no fossil record of myrmecobiids prior to the Pleistocene (Cooper 2011).

THYLACINIDAE

The Tasmanian thylacine

The last known living thylacine (*Thylacinus cynocephalus*) died in captivity in 1936 and the species is now considered extinct. Prior to this, and since its discovery by Europeans near the turn of the 19th century, *T. cynocephalus* occurred throughout the eucalypt forests and woodlands of Tasmania (Paddle 2000; Mooney and Rounsevell 2008). However, Aboriginal records, subfossil remains and trace fossils indicate that thylacines inhabited mainland Australia and New Guinea prior to European settlement. White, Saltré *et al.*'s (2018) quantitative analysis of the fossil record suggested that mainland populations of *Thylacinus* (and *Sarcophilus*) went extinct ca. 3200 years ago. The Tasmanian thylacine was the largest living carnivorous marsupial (15–35 kg) at the time of its extinction. It was a nocturnal predator with acute olfactory sensitivity and craniodental adaptations for hypercarnivory, preying mostly on macropodids and wombats. The thylacine's distinctive external morphology included its large size (head–body length = 100–130 cm, tail length = 50–65 cm), pronounced sexual dimorphism, transverse stripes on the posterior part of the back, a posteriorly opening pouch and wide-gaping jaws (Fig. 8.2). Reputed to be a threat to sheep introduced by European settlers, thylacines were hunted for bounty from 1830 until the early 1900s. This may have been the last straw for a species that was never abundant, declining due to disease in the early 20th century and had relatively low genetic variability (White, Mitchell and Austin 2018; Feigin *et al.* 2022).

Fig. 8.2. The Tasmanian thylacine, *Thylacinus cynocephalus.* [Colourised version of image dated 1933 by an unknown photographer, from Wikimedia Commons.]

Harris (1808) described the thylacine as *Didelphis cynocephala*, utilising Linnaeus's catch-all genus for marsupials. When Geoffroy Saint-Hilaire (1796) erected *Dasyurus* for Australian marsupials, he included the thylacine as *D. cynocephalus*. Temminck (1824), recognising the anatomical distinctions between thylacines and dasyurids, placed the former in its own genus *Thylacinus*. Like the numbat, thylacines were often considered members of Dasyuridae (e.g. Simpson 1945); universal adoption of Bonaparte's (1838) Thylacinidae for *Thylacinus* and its fossil relatives seems to have taken place after Aplin and Archer's (1987) syncretic classification.

Fossil thylacinids

The earliest fossil remains attributed to *Thylacinus cynocephalus* come from the Pliocene Chinchilla Local Fauna (LF) in south-eastern Queensland (Mackness *et al.* 2002; Louys and Price 2015), though Rovinsky *et al.* (2019) questioned this attribution. In any case, the Tasmanian thylacine made a relatively late appearance in the fossil record of thylacinids, which currently comprises nine genera and 14 species (Table 8.1). Thylacinids first appeared in the latest Oligocene some 24 Ma with *Badjcinus*, diversified considerably in the Miocene (ca. 5–23 Ma) with 11 species of the other eight genera, then declined in the Plio-Pleistocene (Fig. 8.3). In the following paragraphs

Table 8.1. The fossil record of Thylacinidae.

Genus	Species	Age[1]	Type locality local fauna	Authority
Badjcinus[2]	*B. turnbulli*	Late Oligocene-Early Miocene 16–26 Ma	White Hunter LF; Riversleigh, Qld	Muirhead and Wroe 1998
Maximucinus	*M. muirheadae*	Middle Miocene 13–14 Ma	Ringtail LF; Riversleigh, Qld	Wroe 2001
Muribacinus	*M. gadiyuli*	Middle Miocene 13–15 Ma	Dwarnamor LF; Riversleigh, Qld	Wroe 1996
Mutpuracinus[3]	*M. archibaldi*	Middle Miocene 12–17 Ma	Bullock Creek LF; NT	Murray and Megirian 2000
Ngamalacinus	*N. timmulvaneyi*	Early Miocene 16–18 Ma	Inabeyance LF; Riversleigh, Qld	Muirhead 1997
Nimbacinus	*N. dicksoni*	Late Oligocene 23–26 Ma[4]	Henk's Hollow LF; Riversleigh, Qld	Muirhead and Archer 1990
	N. richi[5]	Middle Miocene 12–17 Ma	Bullock Creek LF, NT	Murray and Megirian 2000
Thylacinus	*T. cynocephalus*	Pliocene ca. 4 Ma[6]	Chinchilla LF, Qld	Harris 1808
	T. macknessi	Early Miocene 16–18 Ma	Neville's Garden LF; Riversleigh, Qld	Muirhead 1992
	T. megiriani	Late Miocene 4–8 Ma	Ongeva LF, NT	Murray 1997
	T. potens	Late Miocene 5–8 Ma	Alcoota LF, NT	Woodburne 1967
	T. yorkellus	Pliocene 3–5 Ma	Curramulka LF, SA	Yates 2015
Tyarrpecinus	*T. rothi*	Late Miocene 5–8 Ma	Alcoota LF, NT	Murray and Megirian 2000
Wabulacinus	*W. ridei*	Early Miocene 17–18 Ma	Camel Sputum LF; Riversleigh, Qld	Muirhead 1997

[1] Source: Rovinsky *et al.* (2019).
[2] Beck *et al.* (2022) considered *Badjcinus* to be Dasyuromorphia *incertae sedis*.
[3] Rovinsky *et al.* (2019) considered *Mutpuracinus* to be Dasyuromorphia *incertae sedis*.
[4] Oldest remains from D Site in Riversleigh's Faunal Zone A. The type locality, Henk's Hollow LF, is mid-Miocene.
[5] Wroe and Musser (2001) and Rovinsky *et al.* (2019) considered *N. richi* synonymous with *N. dicksoni*.
[6] Rovinsky *et al.* (2019) questioned the attribution of Chinchilla LF fossils to *T. cynochephalus*; excluding this material, the oldest remains of *T. cynocephalus* are Pleistocene.

we briefly describe the currently accepted thylacinid genera and in the next section summarise understanding of their evolution (see also Eldridge *et al.* 2019).

Badjcinus. Known from skull fragments including upper and lower jaws with teeth (Fig. 8.3b), as well as a complete basicranium and parts of the dorsal cranium, *B. turnbulli* was the earliest and one of the smallest thylacininds (Muirhead and Wroe 1998). Its resemblance to quoll-sized dasyurids suggests that it may have been a carnivore-insectivore in the moist forest habitats of the Oligo-Miocene at Riversleigh in north-western Queensland. Phylogenetic analyses by Kealy and Beck (2017) and Beck *et al.* (2022) found only equivocal support for *Badjcinus* as a thylacinid.

Maximucinus. Known from a single M^2, *M. muirheadae* was the largest thylacinid prior to the Late Miocene (Fig. 8.4a). Wroe (2001b) estimated its body weight at 18 kg, near the low end of the range reported for *T. cynocephalus*. Long *et al.* (2002) suggest that its dentition was less specialised for carnivory than was that of other large thylacinids.

Muribacinus. *M. gadiyuli* was a small thyalcinid; at ca. 1.5 kg, its weight was comparable to that of *Dasyurus maculatus* (Wroe 1996, 2001b). Its size and dentition, known from fragments of upper and lower jaws (Fig. 8.3a), suggest a diet of small vertebrates (Wroe 1996). Long *et al.* (2002) characterised it as the most primitive thylacinid.

Mutpuracinus. The smallest of the thylacinid fossils with an estimated weight of just over 1 kg (comparable to *Dasyurus viverrinus*; Wroe 2001b), *M. archibaldi* was described from two fragments of upper jaw in the Bullock Creek LF (Murray and Megirian 2000), but Murray and Megirian (2006) found additional material in the Jaw Junction LF at Riversleigh sufficient to reconstruct most of the skull (Fig. 8.4b). *Mutpuracinus* was likely sympatric and contemporaneous with *Nimbacinus richi*, the other, larger thylacinid found at Bullock Creek (Long *et al.* 2002). Murray and Megirian (2006) argued that *Mutpuracinus* is anatomically most similar – and perhaps phylogenetically closer – to the Miocene dasyurid *Barinya* (see 'Barinyainae' below) and Kealy and Beck (2017) failed to recover it as a thylacinid. Rovinsky *et al.* (2019) and Beck *et al.* (2022) concurred that there is insufficient evidence for inclusion of *Mutpuracinus* in Thylacinidae and considered it Dasyuromorphia *incertae sedis.*

Ngamalacinus. This small dog-sized (ca. 5.7 kg) thylacinid (Fig. 8.3d) is known from partial upper and lower jaws and an isolated M^2 (Muirhead 1997). It was a contemporary of the comparably sized *Wabulacinus* in the early Miocene at Riversleigh, but less specialised for carnivory.

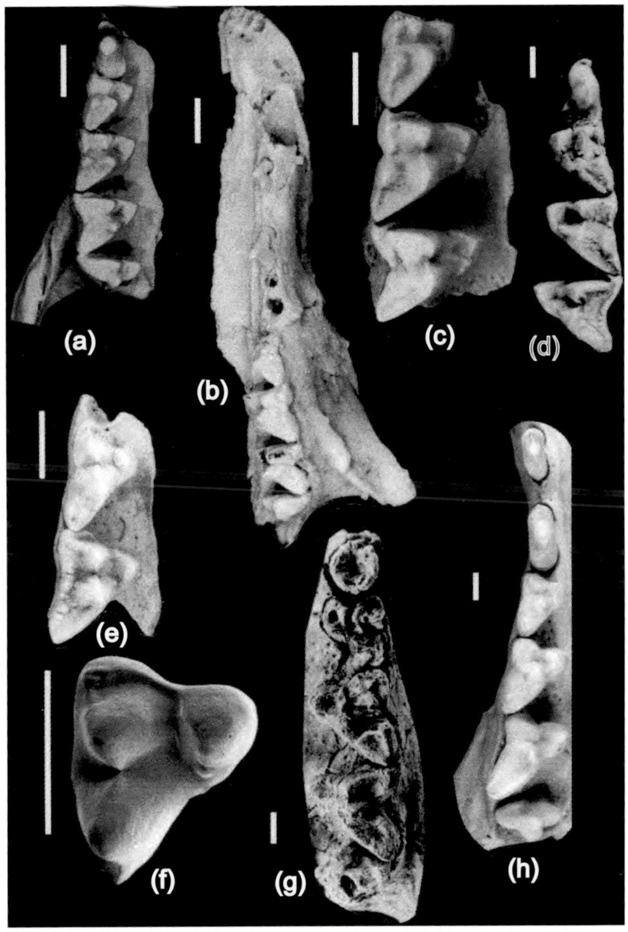

Fig. 8.3. Maxillary bones and teeth of fossil thylacinids in occlusal view. (a) *Muribacinus gadiyuli* (QMF 30386) P^2, M^{1-4}. (b) *Badjcinus turnbulli* (QMF 30488) P^{1-2}, M^{1-4}. (c) *Nimbacinus dicksoni* (QMF 16803) M^{1-3}. (d) *Ngamalacinus timmulvaneyi* (QMF 30300) P^3, M^{1-3}. (e) *Wabulacinus ridei* (QMF 16851) M^3. (f) *Thylacinus macknessi* (QMF 16850) M^1. (g) *Thylacinus potens* (CPC 6746) P^3, M^{1-4}. (h) *Thylacinus cynocephalus* (AM 217) P^{2-3}, M^{1-4}. Scale bar = 5 mm. [Reprinted by permission of CSIRO Publishing from Wroe (2003). Permission conveyed through Copyright Clearance Center, Inc.]

Nimbacinus. Originally described from jaw fragments (Muirhead and Archer 1990), *N. dicksoni* is now known from complete skull and skeleton remains (Figs 8.3c and 8.5). It was another quoll-sized thylacinid (Wroe 2001b) with relatively primitive dentition (Wroe and Musser 2001). Attard *et al.* (2014) performed a biomechanical analysis of the skull and concluded that the feeding niche of *N. dicksoni* was like that of *Dasyurus maculatus*, including the ability to take prey larger than itself. Wroe

(a)

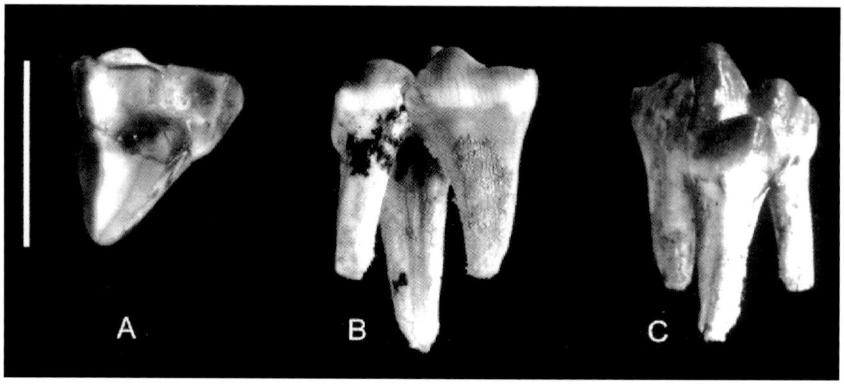

(b)

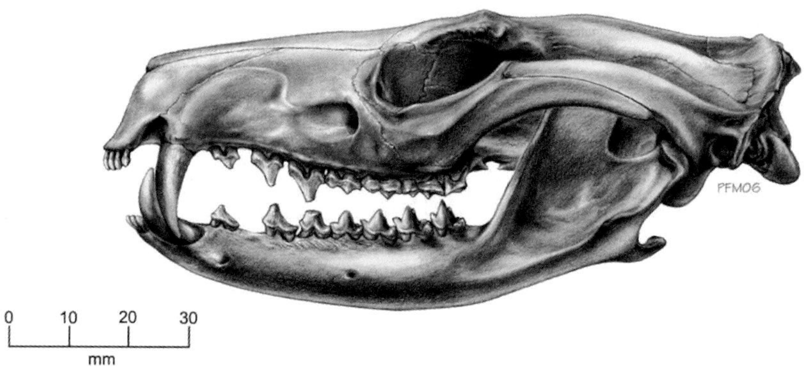

Fig. 8.4. (a) *Maximucinus muirheadae* holotype, right M^2 (QM F30331) in occlusal (A), lateral (B) and lingual (C) views. The tooth is approximately 9 × 11 mm. (b) Reconstruction of *Mutpuracinus archibaldi* cranium and dentary. [(a) Reprinted by permission of CSIRO Publishing from Wroe (2001b). Permission conveyed through Copyright Clearance Center, Inc. (b) Reprinted by permission of Taylor and Francis from Murray and Megirian (2006).]

and Musser (2001) questioned the distinctness of *N. richi*, which Murray and Megirian (2000) described from a partial left dentary.

Thylacinus. Comprising five described species, *Thylacinus* is the most diverse and long-lived thylacinid genus, its temporal range extending from the Early Miocene to Holocene. The oldest species, *T. macknessi* and *T. potens*, were medium- and large-sized carnivores, respectively (Fig. 8.3f, g). The Late Miocene *T. megiriani* was larger than the modern species and showed more dental indications of hypercarnivory. With body sizes estimated at 40–50 kg, *T. potens* and *T. megiriani* were the largest thylacinids (Rovinsky *et al.* 2019). The recently described *T. yorkellus* (Fig. 8.6a) of the Pliocene was somewhat smaller (16–18 kg) (Yates 2015). Dawson *et al.* (1999) assigned a dentary fragment with partial M_3 and M_4 from the Pliocene Big Sink LF to *Thylacinus* but could not identify its species. As noted above, *T. cynocephalus* may have been present in the late Pliocene but is widespread in Pleistocene and later deposits (Johnson and Wroe 2003).

Tyarrpecinus. *T. rothi* is known only from a partial left maxilla that Murray and Megirian (2000) suggested was enclosed in a crocodile coprolite. Wroe (2001b) estimated the animal's size at ca. 5.45 kg and Long *et al.* (2002) speculated that it may have been sympatric with *T. potens* in Early Miocene habitats of the Northern Territory.

Wabulacinus. Comparable in size to *Tyarrpecinus*, *W. ridei* (Fig. 8.6b) was a generalised thylacinid of the Early Miocene forests at Riversleigh (Muirhead 1997). It is known only from tooth-bearing fragments of a maxilla and dentary.

Relationships and evolution of thylacinids

As discoveries of new fossil thylacinids accumulated during the 1990s and 2000s, so did morphocladistic estimates of their relationships (e.g. Muirhead 1997; Muirhead and Wroe 1998; Murray and Megirian 2000; Wroe and Musser 2001; Wroe 2003; Yates 2014, 2015; Archer, Hand *et al.* 2016; Kealy and Beck 2017). Rovinsky *et al.* (2019) provided an integrative analysis of anatomical characters

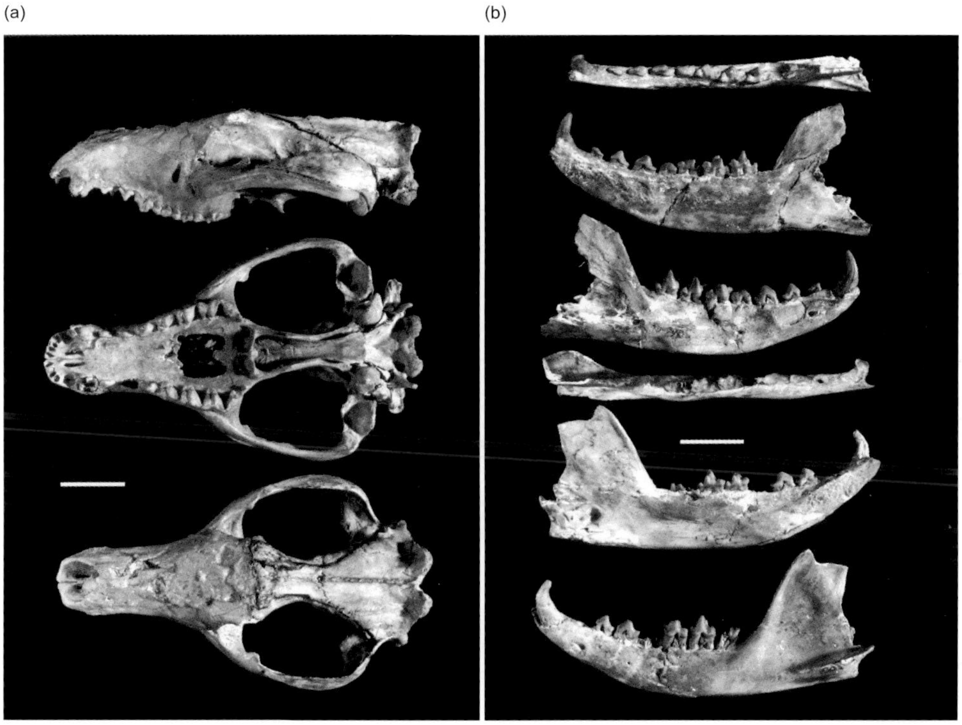

Fig. 8.5. Skull of *Nimbacinus dicksoni* (QM F36357). (a) Cranium in (top to bottom) lateral, ventral and dorsal views. (b) Right (top three) and left (bottom three) mandibles in (top to bottom) occlusal, medial and lateral views. Scale bar = 2 cm. [Reprinted by permission of CSIRO Publishing from Wroe and Musser (2001). Permission conveyed through Copyright Clearance Center, Inc.]

and *12S* sequences (from Recent taxa) to estimate thylacinid phylogeny and divergence times. Their results (Fig. 8.7) place *Mutpuracinus* as a stem dasyurid but all other thylacinids in a single clade (with moderate or no statistical support, respectively). The Oligocene *Badjcinus* appears as sister to the remaining six genera, which form two clades: *Muribacinus* + *Ngamalacinus* + *Nimbacinus* and *Thylacinus* + *Tyarrpecinus* + *Wabulacinus*. *Thylacinus* is not resolved as monophyletic but the sister-pairing of *T. cynocephalus* and *T. yorkellus* is well-supported in some analyses (as also found by Yates 2015). Exclusion of *Mutpuracinus* from Thylacinidae, early separation of *Badjcinus* and association of *Thylacinus* with *Tyarrpecinus* and *Wabulacinus* are consistent with several previous estimates of thylacinid relationships.

Rovinsky *et al.*'s (2019) tip dating analysis (see Chapter 9) suggests that the most recent common ancestor of Thylacinidae appeared in the Eocene some 36–42 Ma, consistent with earlier studies (Mitchell *et al.* 2014; Westerman, Krajewski *et al.* 2016; Kealy and Beck 2017). The split involving *Badjcinus* occurred at 32–35 Ma in the Oligocene, with all subsequent cladogenesis in the later Oligocene or Miocene. Yates (2015) argued that the Late Miocene diversification of *Thylacinus* contradicts the scenario that thylacinids suffered a steady decline after their Early-Middle Miocene radiation (Wroe 2003), perhaps with the most recent *Thylacinus* species representing a single lineage culminating in *T. cynocephalus*.

The tree in Fig. 8.7 includes a mapping of body size and molar traits associated with hypercarnivory, both of which show distinct trends. Rovinsky *et al.* (2019) characterised early thylacinids (*Badjcinus*, *Ngamalacinus*, *Muribacinus*) as 'unspecialized faunivores' feeding on invertebrates and small vertebrates, with the more robust *Nimbacinus* capable of taking larger prey (Attard *et al.* 2014). The transition to hypercarnivory begins with *Wabulacinus* and increases through the diversification of *Thylacinus*, reaching a maximum in *T. cynocephalus*. Body size increases dramatically in the wolf-like *Thylacinus potens*, *T. megiriani* and *T. cynocephalus*. The timing of this transition may not be coincidental:

> *The dramatic shift in mass within* Thylacinus *coincides with the ~15–14 Ma middle Miocene climatic transition (MMCT), a period of dramatic global cooling and aridification, and the*

(a)
(b)

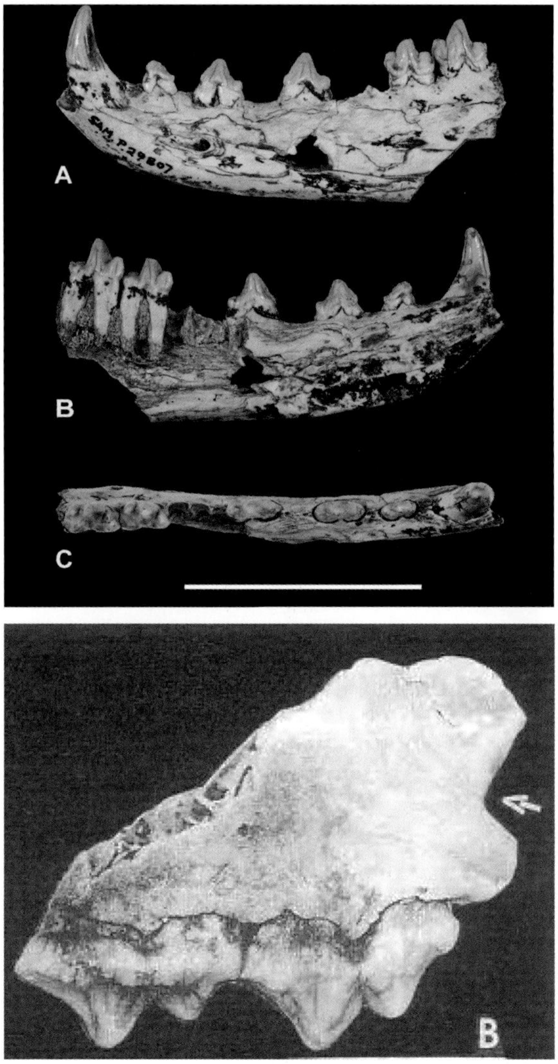

Fig. 8.6. (a) The holotype of *Thylacinus yorkellus* (SAM P29807) in lateral (A), medial (B) and dorsal (C) views. Scale bar = 50 mm. (b) The holotype of *Wabulacinus ridei* (QMF16851), a right dentary in buccal view; arrow points to the infraorbital foramen. [(a) Reprinted from Yates (2015) licensed under CC BY 4.0. (b) Reprinted by permission of the Queensland Museum from Muirhead (1997).]

> *subsequent radiation of the Dasyuridae ... The MMCT therefore marks a point of both significant expansion in thylacinid body mass as well as a possible reduction in generic diversity, during a time of drastic shifts toward the aridification and cooling of the Australian environment and potentially rapid expansion of competitors. [Rovinsky et al. 2019: 26]*

In Chapter 10 we discuss Wroe's (1997a, b) hypothesis that the decline of thylacinids may have been due to competitive displacement by dasyurids after the MMCT, possibly facilitated by the latter's middle ear morphology being better adapted to open habitats. Although many early thylacinids occupied moist forest environments at Riversleigh during the Oligocene and Early Miocene, it is less clear that all members of the family were restricted to such habitats. Both before and after the MMCT, several species are found in areas thought to have been more open forests with seasonal rainfall (Rovinsky *et al.* 2019).

MALLEODECTIDAE

Arena *et al.* (2011) described *Malleodectes mirabilis* from a partial left maxilla with P^2 and P^3 found in Middle Miocene deposits at Riversleigh. The authors interpreted its tooth structure, including the exceptionally large and conical P^3, as evidence for a specialised diet of snails convergently like that of certain blue-tongue skinks (Fig. 8.8). A second species (*M. moenia*) was named based on an isolated P^3. Archer, Hand *et al.* (2016) assigned subsequently discovered cranial fragments – including a partial maxilla bearing C^1, dP^3, M^1, M^2 and an unerupted P^3 – to *M. mirabilis*. They estimated the body mass of this individual at 896 g and argued that its dentition reflects a diet somewhat broader than snails alone.

Archer, Hand *et al.* (2016) evaluated specific morphological features and performed a numerical parsimony analysis of *Malleodectes* to determine its affinities:

> *We consider that it is probably a dasyuromorphian rather than a representative of some other order of marsupicarnivores for the following reasons. First, StD [stylar cusp D] is larger than StB on both M1 and M2, which is the common condition in dasyuromorphians, but is rare amongst other marsupials. Second, there is no 'central cusp' lingual to the ectoloph on M1 or M2 ... Third, except for its huge size, dP3 of Malleodectes broadly resembles those of some dasyurids (such as* Sminthopsis *spp.) in that the protocone is very small and closely adpressed to the base of the metacone, a distinct, functional postmetacrista is present, but a well-developed stylar shelf and stylar cusps are absent ... [Archer, Hand et al. 2016: 6–7]*

This conclusion was supported by parsimony analysis (Fig. 8.9), which showed *Malleodectes* as part of a basal polytomy with other dasyuromorphian families. Given

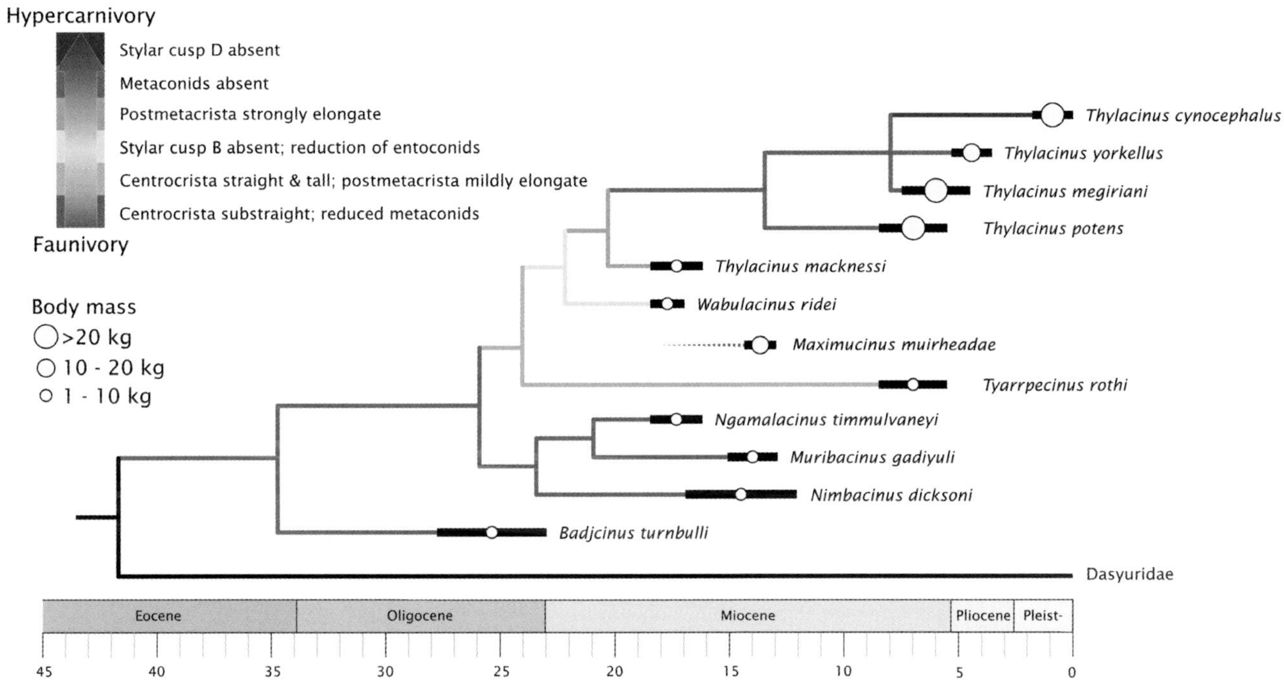

Fig. 8.7. Time tree of thylacinids estimated with Bayesian analysis of 80 morphological characters and *12S* DNA sequences (for *T. cynocephalus* and selected extant outgroups, not shown). Tip dating employed a gamma-rates clock model and fossilised birth-death process model. [Reprinted from Rovinsky *et al.* (2019), licensed under CC BY 4.0.]

these results, the authors erected the new family Malleodectidae: 'The combination of unusual dental specialisations found in *Malleodectes* – namely its hypertrophied P^3, the enormous dP^3, narrow canine, unusual morphology of P^2 and robust nature of M1 – is not found in representatives of any of the three dasyuromorphian families described to-date (Dasyuridae, Myrmecobiidae and Thylacinidae) and it lacks obvious synapomorphies that would specifically link it to one of these three families' (Archer, Hand *et al.* 2016: 8). They note, however, the possible phylogenetic relationship of *Malleodectes* to extinct dasyuromorphian genera *Ganbulyani* (Wroe 1998), *Whollydooleya* (Archer, Christmas *et al.* 2016), or *Barinya* (Wroe 1999b). We discuss these genera in 'Dasyuridae' below.[1]

DASYURIDAE

Barinyainae

Archer's (1982c) review of the dasyurid fossil record recognised several taxa (*Ankotarinja*, *Keeuna*, *Dasylurinja* and what is now *Apoktesis*) known primarily from dental remains in what are currently considered Late Oligocene deposits. Wroe (1997a) argued, however, that the only valid synapomorphies for Dasyuridae identified at the time were basicranial traits (see Chapter 11). Lack of dental synapomorphies resulted in the Oligocene genera being reassigned as Dasyuromorphia *incertae sedis* (see below). Murray and Megirian (2006) disputed these conclusions, but most subsequent treatments (e.g. Black *et al.* 2012) followed Wroe's (1997a) recommendations. Thus, the dasyurid fossil record (Table 8.2) begins with *Barinya wangala* (Wroe 1999b) of the Early Miocene Neville's Garden Site at Riversleigh, Qld, known from a nearly complete skull (Fig. 8.10a) and several skull fragments. Because the specimens display dasyurid basicranial synapomorphies but lack other derived traits that unite modern dasyurids, Wroe (1999b) erected a separate subfamily, Barinyainae, for *Barinya*. At 300–500 g, this rainforest carnivore was smaller than *Dasyurus hallucatus*. Binfield *et al.* (2017) described a second species, *B. kutjamarpensis*, from the Early-Middle Miocene Kutjamarpu LF in north-eastern South Australia. Though generally resembling *B. wangala*, Binfield *et al.* (2017) interpreted molar wear patterns on *B. kutjamarpensis* as indicative of a diet richer in hard substances such as bone. Phylogenetic analyses by Kealy and Beck (2017) and Beck *et al.* (2022) did not specifically associate *Barinya* with dasyurids, though the genus was resolved as a dasyuromorphian.

Fig. 8.8. The Middle Miocene *Malleodectes mirabilis* from Riversleigh. Artist's reconstruction of live animal consuming a snail and illustrating the 'hammer tooth' morphology of P^3. [Illustration by Peter Schouten. Reprinted from Archer, Hand *et al.* (2016) under CC BY 4.0 International License.]

Dasyurinae

The oldest and only known Miocene fossil of crown-group Dasyuridae is *Ganbulyani djadjinguli*, described originally from a single upper molar and hypertrophied premolar in the 10–12 Ma Encore LF, Riversleigh (Fig. 8.11). Wroe (1998) characterised this as a 'bone-cracking' dasyurid because the broad, blunt, conical cusps on both teeth resemble those of hyaenas (*Crocuta*) and Tasmanian devils (*Sarcophilus*). However, at ≤7 kg, *Ganbulyani* may be the smallest mammal with bone-cracking adaptations yet discovered.

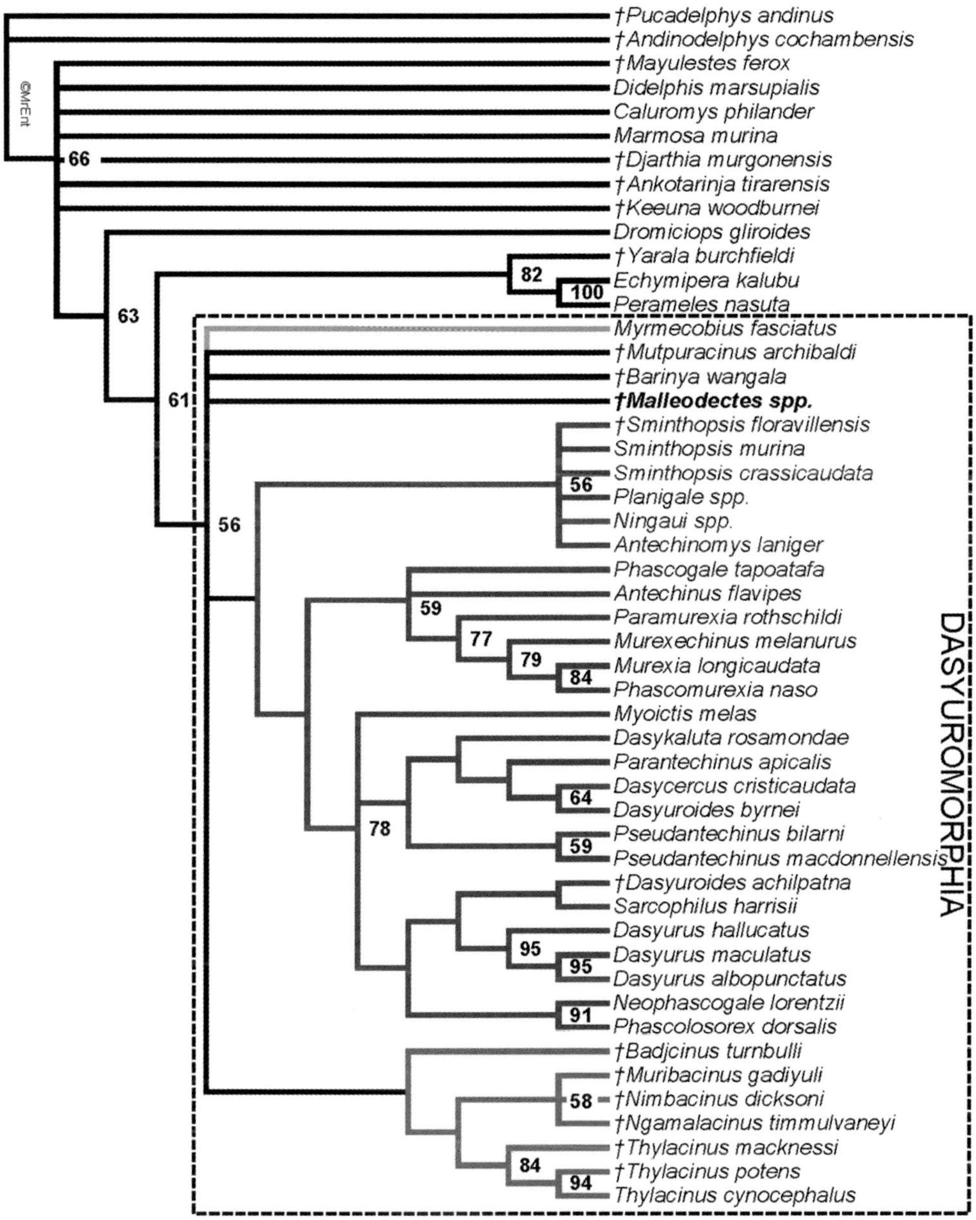

Fig. 8.9. Phylogeny of *Malleodectes*. Tree is a strict consensus of 18 619 minimum-length trees for 62 craniodental characters, with relationships among extant taxa constrained to those recovered from molecular analyses by Mitchell *et al.* (2014), Westerman, Krajewski *et al.* (2016) and Jansa *et al.* (2014). Coloured branches indicate recognised families Myrmecobiidae (orange), Dasyuridae (blue) and Thylacinidae (red). Nodal values are bootstrap percentages ≥50%. [Reprinted from Archer, Hand *et al.* (2016) under CC BY 4.0 International License.]

'*Dasycercus*' *worboysi* from the Pliocene Big Sink LF in New South Wales was described by Dawson *et al.* (1999) from a partial left dentary with the molar row (Fig. 8.12a). Derived dental features place this taxon within Dasyurinae, but the basis of its generic assignment was unusual: 'this new species has been assigned to *Dasycercus* Peters 1875, the earliest available generic name in its clade' (p. 270). Reduction of the m1 paraconid suggested an animal that took large prey and loss of the entoconid has been correlated with arid habitats. '*D.*' *worboysi* was marginally bigger than living members of *Dasycercus*. Bartholomai (1971) described *Dasyurus dunmalli* from the Pliocene Chinchilla LF in Queensland and additional material from the Bow LF in New South Wales (Archer 1982c) and Bluff Downs LF in Queensland (Wroe and Mackness 1998) has been referred to this species

Table 8.2. The fossil record of Dasyuridae.
Extant species and undescribed Pleistocene species assigned to extant genera are excluded.

Clade/genus	Species	Age[1]	Type locality local fauna	Authority
Barinyainae				
Barinya	*B. wangala*[2]	Early Miocene 16–18 Ma	Neville's Garden LF; Riversleigh, Qld	Wroe 1999
	B. kutjamarpensis	Early Miocene 18–20 Ma	Kutjamarpu LF, SA	Binfield *et al.* 2017
Dasyurinae: Dasyurini				
Archerium	*A. chinchillensis*	Pliocene 3.4 Ma[3]	Chinchilla LF, Qld	Wroe and Mackness 2000
Dasycercus	*D. worboysi*	Pliocene 3.5–4.5 Ma	Big Sink LF, NSW	Dawson *et al.* 1999
Dasyurus	*D. dunmalli*	Pliocene 3.4 Ma[4]	Chinchilla LF, Qld	Bartholomai 1971
	Dasyurus sp.	Pliocene	Floraville LF, Qld	Archer 1982
Dasyuroides	*D. achilpatna*	Pliocene 2.5–2.9 Ma[5]	Fisherman's Cliff LF, NSW	Archer 1982
Ganbulyani	*G. djadjinguli*	Late Miocene 10–12 Ma	Encore LF; Riversleigh, Qld	Wroe 1998
Glaucodon	*G. ballaratensis*	Pliocene 3.2 Ma[6]	Smeaton LF, VIC	Stirton 1957
Myoictis	*Myoictis* sp.	Pliocene 3 Ma	Awe LF, PNG	Archer 1982
Sarcophilus	*S. moornaensis*	Pliocene 2.5–2.9 Ma[5]	Fisherman's Cliff LF, NSW	Crabb 1982
	S. laniarius	Pleistocene	Wellington Caves LF, NSW	Owen 1838
Dasyurinae: Phascogalini				
Antechinus	*A. puteus*	Pleistocene	Russenden Cave LF, Qld	Van Dyck 1982
	Antechinus sp.	Pliocene[7]	Floraville LF, Qld	Archer 1982
	Antechinus sp.	Pliocene 4.5 Ma[8]	Hamilton LF, VIC	Archer 1982
	Antechinus sp.	Pliocene 3.5–4.5 Ma	Big Sink LF, NSW	Dawson *et al.* 1999
	A. yammal	Pleistocene	Speaking Tube Cave LF, Qld	Cramb and Hocknull 2010
	A. yuna	Pleistocene	Elephant Hole Cave LF, Qld	Cramb and Hocknull 2010
Sminthopsinae: Sminthopsini				
Sminthopsis	*S. floravillensis*	Pliocene	Floraville LF, Qld	Archer 1982
	Sminthopsis sp. 1	Pliocene 3.5–4.5 Ma	Big Sink LF, NSW	Dawson *et al.* 1999
	Sminthopsis sp. 2	Pliocene 3.5–4.5 Ma	Big Sink LF, NSW	Dawson *et al.* 1999
Sminthopsinae: Planigalini				
	Planigale sp.	Pliocene 3.6 Ma[9]	Bluff Downs LF, Qld	Archer 1982

[1] Woodhead *et al.* (2016) except as noted.
[2] Kealy and Beck (2017) and Beck *et al.* (2022) consider *B. wangala* as Dasyuromorphia *insertae sedis*. Fossil age estimate from Rovinsky *et al.* (2019).
[3] Wroe and Mackness (1998).
[4] Wroe and Mackness (2000).
[5] Travouillon (2016).
[6] Webb (2013).
[7] Wroe (2003).
[8] Turnbull *et al.* (2003).
[9] Mackness *et al.* (2000).

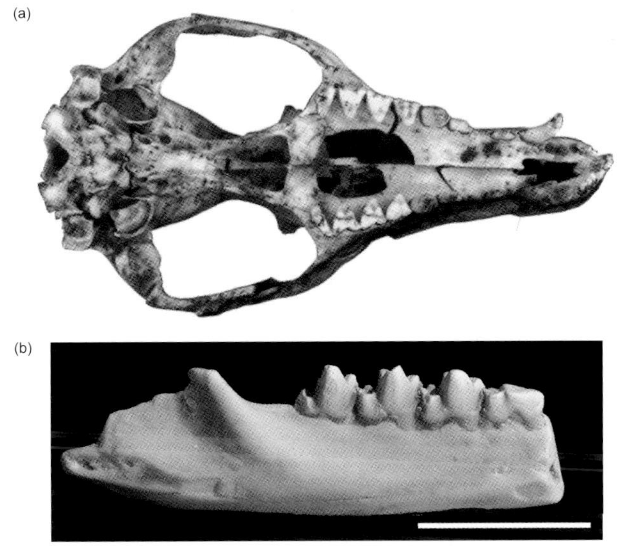

Fig. 8.10. Fossil remains of *Barinya*. (a) Skull of the Early Miocene *Barinya wangala*, ventral view (QM F31408). Approximate length 60 mm. (b) Right dentary of Middle Miocene *Barinya kutjamarpensis*, lateral view (SAM P53348). Scale bar = 5 mm. [(a) Reprinted by permission of John Wiley and Sons from Wroe (1999b). (b) Reprinted by permission of Taylor and Francis from Binfield *et al.* (2017).]

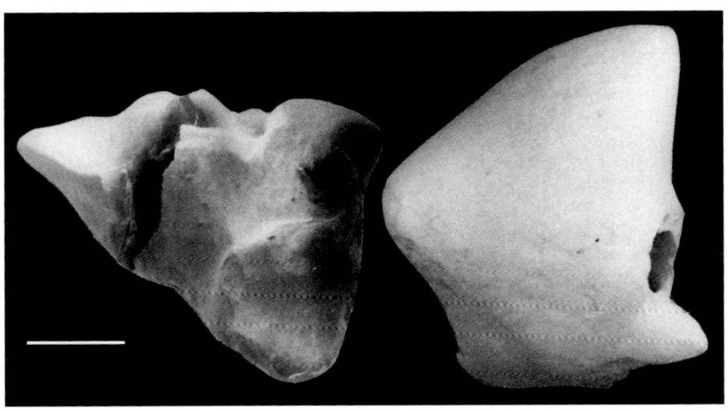

Fig. 8.11. The Miocene dasyurid *Ganbulyani djadjinguli*. Upper molar in lingual view (left, QMF 24537) and upper premolar in lateral view (right, QMF 30464). Scale bar = 2.0 mm. [Reprinted by permission of Taylor & Francis from Wroe (1998).]

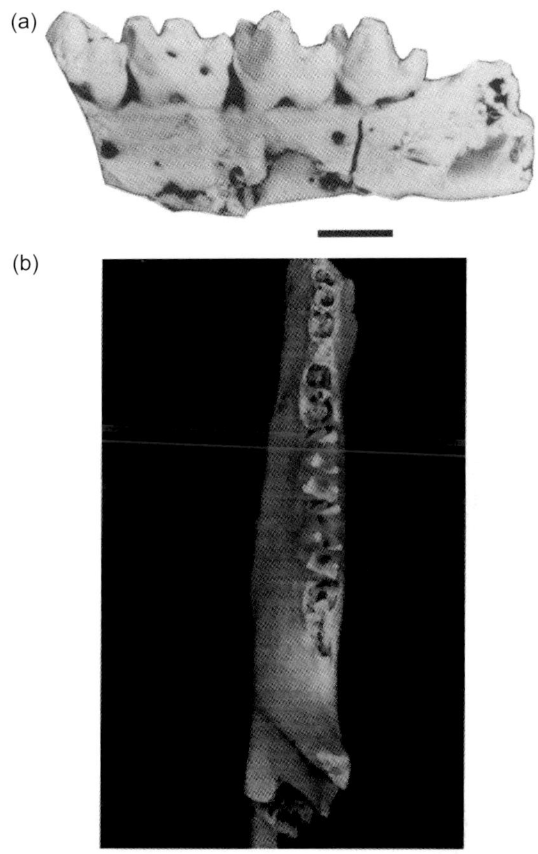

Fig. 8.12. Pliocene dasyurins. (a) *'Dasycercus' worboysi* partial left dentary with M_1–M_4 (AM F69805), lateral view. Scale bar = 2.1 mm. (b) *Dasyurus dunmalli* partial left dentary with M_{2-3} and alveoli for P_{1-3}, M_1 and M_4 (QM F12682); occlusal view. [(a) Reprinted by permission of the Western Australian Museum from Dawson *et al.* (1999). (b) Reprinted by permission of the Queensland Museum from Wroe and Mackness (1998).]

(Fig. 8.12b). Its dental morphology resembles *Dasyurus viverrinus*, except for the presence of a reduced P_3, a tooth not found in living quolls.

Two monotypic genera from the Pliocene have been placed within Dasyurini. *Archerium chinchillaensis* (Wroe and Mackness 2000) from the Chinchilla LF, Qld, is represented by a maxillary fragment and three isolated teeth (Fig. 8.13a). Long *et al.* (2002) characterised it as a kowari-sized insectivore-carnivore that lived in open woodlands with a seasonal climate. Stirton (1957) described the lower jaw of *Glaucodon ballaratensis* (Fig. 8.13b) from Smeaton, Vic, which Ride (1964b) and Archer (1976d) considered to be anatomically – if not phylogenetically – intermediate between *Dasyurus* and *Sarcophilus*. A more recently discovered right mandible from *Glaucodon* supports this interpretation (Gerdtz and Archbold 2003).

Dasyuroides achilpatna (Archer 1982c) is known from several dentary fragments and isolated molars in the Fisherman's Cliff LF of New South Wales (Fig. 8.14a). Erosion patterns on the holotype mandible suggest that the fossil was part of a quoll coprolite. Dental similarities led Archer (1982c) to place the species in *Dasyuroides* but rejected Marshall's (1973) contention that it was the common ancestor of *Dasyuroides* and/or *Dasycercus*. However, *D. achilpatna* did form a clade with *D. byrnei* and

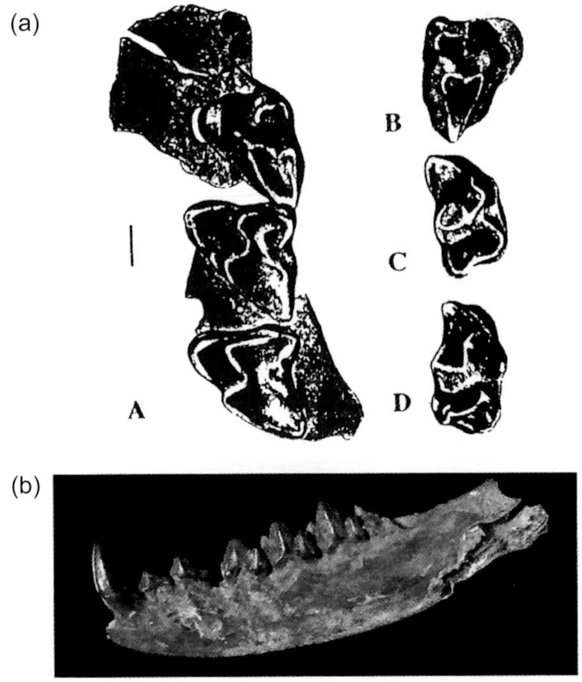

Fig. 8.13. More Pliocene dasyurins. (a) *Archerium chinchillaensis*: A, left maxilla fragment with M^{1-3} (QM F39847); B, right M^1 (QM F39848); C, right M_3 (QM F39850); D, right M_2 (QM F39849); all occlusal views. The scale bar at left is 1 mm. (b) Mandible and lower dentition of *Glaucodon ballaratensis*; lateral view. [(a) Reprinted by permission of Taylor & Francis from Wroe and Mackness (2000). (b) Reprinted by permission of Michael Archer from Long *et al.* (2002).]

Dasycercus cristicauda in the phylogenetic analyses of Kealy and Beck (2017; see Fig. 5.7c). Archer (1982c) allocated a right dentary fragment with M_2 (Fig. 8.14b) from the Awe LF in Papua New Guinea to *Myoictis*, but no species description has yet been published. Crabb (1982) erected *Sarcophilus moornaensis* for an isolated M^3 (or M^4) and partial left dentary with M_{2-5} and alveoli for anterior teeth recovered from the Fisherman's Cliff LF in New South Wales (Fig. 8.14c). Based on his assessment of molar similarities, Crabb (1982: 511) concluded that '*S. moornaensis* is probably in the direct evolutionary line leading to *S. harrisii* and possibly shares a common ancestor with *Glaucodon ballaratensis*'. Remains of a large Pleistocene devil from Wellington Caves LF, New South Wales were referred to *Sarcophilus laniarius* by Owen (1877), but the specific distinction of this taxon from the smaller *S. harrisii* has been questioned (Dawson 1982; Werdelin 1987).

Among phascogalins, two isolated teeth – left and right M^2s from Hamilton LF, Vic, and Floraville LF, Qld, respectively – were assigned to *Antechinus* by Archer (1982c). Genus-level affinities of these Pliocene teeth are problematic, however, especially considering subsequent changes in the taxonomy of living *Antechinus*, *Murexia* and parantechinin species. The Hamilton LF specimen, for example, most closely resembles *Murexia naso* (Turnbull *et al.* 2003). Dawson *et al.* (1999) assigned five isolated molars to an undetermined *Antechinus* species. Van Dyck (1982a) named *Antechinus puteus* from dentary fragments found in Pleistocene deposits at Russenden Cave in Queensland (Fig. 8.15a). The dentition resembles that of *A. flavipes* (remains of which were found in the same deposit), though *A. puteus* was larger and had several distinctive features. More recently, Cramb and Hocknull (2010) recognised *A. yammal* and *A. yuna* from multiple mandibles and maxillae preserved in cave deposits at Mt Etna, Qld (Fig. 8.15b, c). *A. yammal* resembled *A. flavipes*; *A. yuna* may have been a close relative of *A. leo*. Both appear to have occupied rainforest and gone extinct when that habitat was lost in the mid-Pleistocene.

Sminthopsinae

Sminthopsis floravillensis (Fig. 8.16) from the Pliocene Floraville LF, Qld, is based on two right dentaries and several isolated molars (Archer 1982c). Based on the size of its alveolus, P_3 was larger than in extant dunnarts, leading Archer (1982c) to speculate that *S. floravillensis* might be a primitive sister group to all other *Sminthopsis*. Kealy and Beck (2017) found that it was one branch of a polytomy at the base of Sminthopsini (see Fig. 5.7c). Dawson *et al.* (1999) assigned a dentary fragment and five isolated molars to one unnamed species of *Sminthopsis* and an isolated m1 to another. The only planigalin known from the Pliocene is a *Planigale* from the Bluff Downs LF, Qld (Archer 1982c). Represented by a right dentary fragment with m4, this appears to have been a medium-sized planigale, larger than *P. ingrami* but smaller than *P. maculata*.

Dasyuromorphia Incertae Sedis

During the 1970s and early 1980s, several fossils from Oligocene or Miocene deposits were assigned to Dasyuridae but are now considered of uncertain affiliation (*incertae sedis*) within Dasyuromorphia (Table 8.3), largely because of Wroe's (1997a) demonstration that no dental synapomorphies of dasyurids had yet been identified. *Incertae sedis* designation does not exclude these

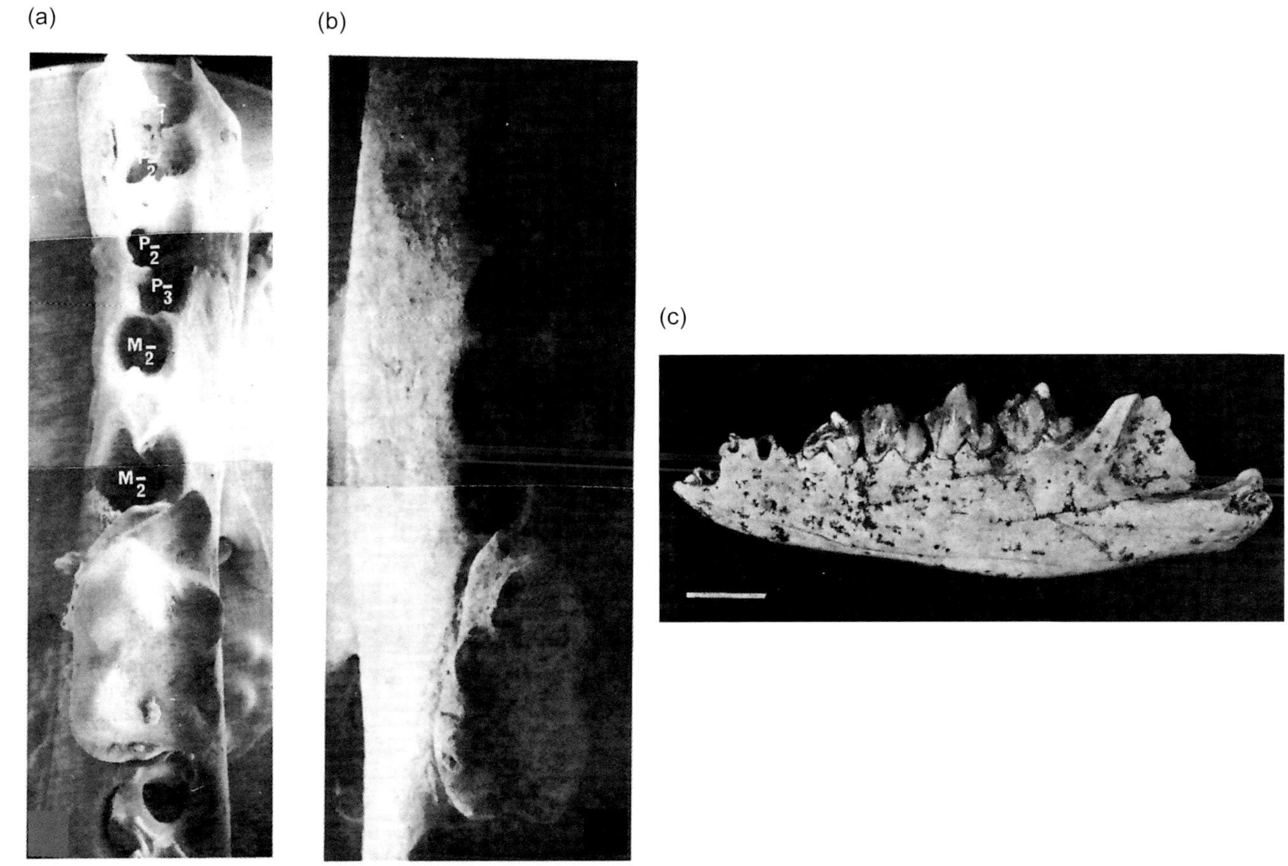

Fig. 8.14. Even more Pliocene dasyurins. (a) *Dasyuroides alchilpatna* dentary fragment with M_3 and alveoli for other teeth as labelled (NMV P38334); occlusal view. (b) *Myoictis* sp. right dentary fragment with M_2 and alveoli for anterior teeth (UCMP P63640); occlusal view. The tooth is 2.8 mm in length. (c) *Sarcophilus moornaensis* left dentary with M_{2-5} (NMV P28268); lateral view. Scale bar = 10 mm. [(a, b) Reprinted by permission of the Royal Zoological Society of New South Wales from Archer (1982c). (c) Reprinted by permission of the Royal Zoological Society of New South Wales from Crabb (1982).]

taxa from membership in Dasyuridae or imply that they could only be stem dasyurids. Their current significance is in demonstrating that small, dasyurid-like animals were present in some Australian ecosystems prior to the diversification of crown-group lineages.

Ankotarinja. *A. tirarensis* (Archer 1976a) is known from dentary fragments and isolated molars in the Ditjimanka LF of South Australia (Fig. 8.17a). Its relatively unreduced P_3 and molar morphology led Archer (1982c: 398) to consider it 'essentially a very primitive type of dasyurid' and Wroe (1996: 1039) called it 'the most plesiomorphic Australian taxon described to date'. Although some have suggested *Ankotarinja* may be a peradectid (Reig *et al.* 1987) or microbiotheriid (Marshall 1987), Wroe (1996) rejected these arguments in favour of dasyuromorphian affinities. Nevertheless, Kealy and Beck's (2017) analyses placed it outside Dasyuromorphia.

Keeuna. From the same Late Oligocene deposits that yielded *Ankotarinja*, Archer (1976a) described *K. woodburnei* (Fig. 8.17b). The *Antechinus*-sized species is known from isolated teeth and jaw fragments, the molars showing structurally primitive features resembling those found in *Ankotarinja* and *Murexia*. As with *Ankotarinja*, Kealy and Beck (2017) failed to recover *Keeuna* as a dasyuromorphian.

Apoktesis. Described by Campbell (1976) from a left dentary in the Ngapakaldi LF of South Australia, *A. cuspis* (Fig. 8.17c) is listed among dasyuromorphians by some authors (e.g. Long *et al.* 2002) but not others (e.g. Wroe 2003). Archer (1982c) dismissed its suggested relationship to *Dasyurus* as based on dental traits that are found in most dasyurids. Campbell (1976) also described two putative dasyuromorphian genera from the Kutjamarpu LF in South Australia: *Achilpa* from a partial left

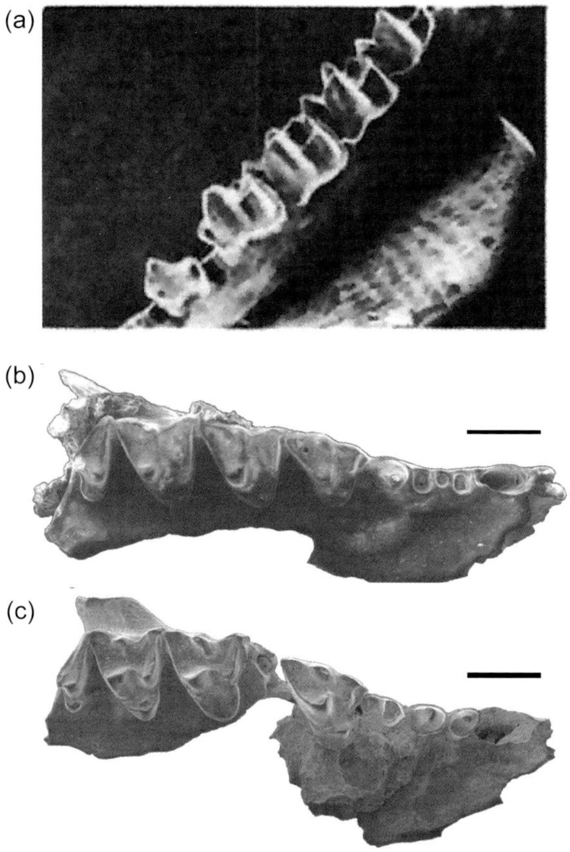

Fig. 8.15. Pleistocene fossils of *Antechinus*. (a) *A. puteus* right dentary with P_3 and molars (QM F8038); lateral view. (b) *A. yammal* right maxilla with P^3 and M^{1-4} (QM F53690); occlusal view. Scale bar = 2 mm. (c) *A. yuna* right maxilla with P1–3 and M1–4 (QM F53888); occlusal view. Scale bar = 2 mm. [(a) Reprinted by permission of Steve Van Dyck from Van Dyck (1982a). (b, c) Reprinted by permission of CSIRO Publishing from Cramb and Hocknull (2010).]

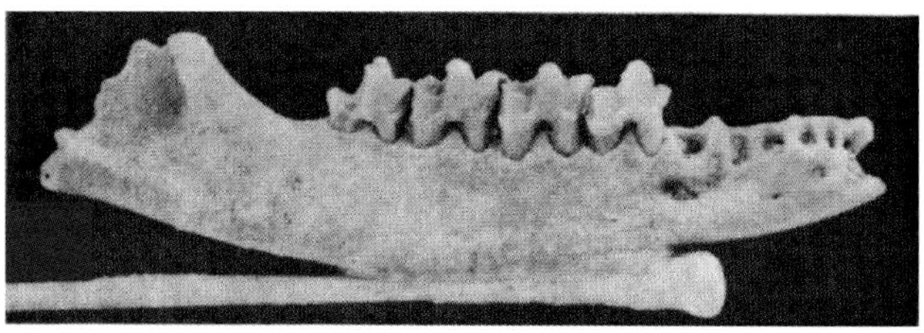

Fig. 8.16. *Sminthopsis floravillensis* right dentary with M_{1-4} (QM F10628); lateral view. The specimen is ca. 13 mm long. [Reprinted by permission of the Royal Zoological Society of New South Wales from Archer (1982c).]

dentary with P_2-M_4 and *Tikawara* from a partial left dentary with P_2 and M_{1-4}. No subsequent evaluations of these fossils have been published, however, so we do not list them in Table 8.3.

Dasylurinja. *D. kokuminola* (Archer 1982c) is represented only by a single M_3 from the late Oligocene Ngama LF of South Australia (Fig. 8.18a). The elongate postmetacrista on this tooth forms a shearing crest like that of carnivorous dasyurids (e.g. *Dasyurus*, *Sarcophilus*). Long *et al.* (2002) extrapolate from the 3.2 mm length of the tooth to an animal approximately the size of *Dasycercus cristicauda*.

Joculusium. Based on a left dentary from the mid-Miocene Dwornamor LF, Wroe (2001a: 53) characterised *J. muizoni* (Fig. 8.18b) as 'the least derived dasyuromorphian yet described from Oligocene-Miocene deposits of Riversleigh'. He argued that the oblique orientation of its metacristid is an apomorphy for dasyurids and thylacinids, but no other features were available to determine its familial affinities.

Mayigriphus. Wroe (1997b) described the small-bodied *M. orbus* (Fig. 8.18c) from two right dentary fragments in the Encore LF of Riversleigh, Qld. The fragments preserve P_{2-3} and M_{1-4}, and the molars – like the *Dasylurinja* M^3 – have developed longitudinal shearing crests. Although it shares two apparently derived features with *Planigale* (involving reduction of paraconids and metaconids), lack of cranial remains precludes its placement in Dasyuridae.

Wakamatha. The partial left dentary representing *W. tasselli* (Fig. 8.19a) was found in the bed of Billeroo Creek in eastern South Australia (Archer and Rich 1979). It is unclear whether these sediments were derived from erosion of nearby Oligocene or Pleistocene deposits, so the age of *Wakamatha* is uncertain. Archer (1982c) suggested that its dentition was convergent on *Sminthopsis crassicaudata* and that phylogenetic affinities with Sminthopsini were likely, but Wroe (1996) argued that it could plausibly be closer to bandicoots.

Whollydooleya. The most recently described fossil taxon that is potentially a dasyuromorphian is *W. tomnpatrichorum* (Archer, Christmas *et al.* 2016), known from a partial lower molar (M_2 or M_3) (Fig. 8.19b) collected in a Miocene limestone south-west of Riversleigh, Qld. The tooth is about the size of a *Sarcophilus harrisii* m3 and shows features associated with hypercarnivory. Although described as a dasyurid, we adhere to Wroe's (1997a) finding that no molar synapomorphies for the family are known and include it among the Dasyuromorphia *incertae sedis*. If it is a dasyurid, however, Archer, Christmas *et al.* (2016) pointed out that it is the oldest of the large carnivores represented by *Glaucodon* and *Sarcophilus*.

Table 8.3. The fossil record of Dasyuromorphia *incertae sedis*.

Genus	Species	Age	Type locality local fauna	Authority
Ankotarinja	*A. tirarensis*	Late Oligocene 23–26 Ma[1]	Ditjimanka LF, SA	Archer 1976a
	Ankotarinja sp.	Early Miocene 18–20 Ma	Kutjamarpu LF, SA	Archer 1982c
Apoktesis	*A. cuspis*	Late Oligocene 23–26 Ma[1]	Ngapakaldi LF, SA	Campbell 1976
Dasylurinja	*D. kokuminola*	Late Oligocene 24–26 Ma[2]	Ngama LF, SA	Archer 1982c
Joculusium	*J. muizoni*	Middle Miocene 13–15 Ma[3]	Dwornamor LF; Riversleigh, Qld	Wroe 2001a
Keeuna	*K. woodburnei*	Late Oligocene 23–26 Ma[1]	Ditjimanka LF, SA	Archer 1976a
Mayigriphus	*M. orbus*	Late Miocene 10–12 Ma[4]	Encore LF, Riversleigh, Qld	Wroe 1997b
Wakamatha	*W. tasselli*	Late Oligocene or Pleistocene	Billeroo Creek LF, SA	Archer and Rich 1979
Whollydooleya	*W. tomnpatrichorum*	Mid-Late Miocene 13–16 Ma	Wholly Dooley, Qld	Archer, Christmas *et al.* 2016

1 Travouillon *et al.* (2013).
2 Woodburne *et al.* (1994).
3 Rovinsky *et al.* (2019).
4 Woodhead *et al.* (2016).

COMPOSITION AND PHYLOGENETIC STRUCTURE OF DASYUROMORPHIA

Numbats

The extra molars of *Myrmecobius* led several early workers to posit an ancient origin for numbats. Owen (1871) imagined that the Jurassic stem mammal *Amphitherium* was the 'prototype' for *Myrmecobius*: 'If Australia possessed Marsupials as far back in time as did America and Europe, analogy would lead us to suppose that the primitive diminutive multimolar insectivorous type prevailed. It has not yet there become extinct; but it seems to have been reduced to the solitary exceptional form of *Myrmecobius*.' Bensley (1903) rejected suggestions by Cope (1888) and Poulton (1889) that the molars of *M. fasciatus* were derived from an ancestral platypus or multituberculate, arguing that they were instead a result of 'retrogressive' evolution within Dasyuridae. He considered numbats as branching from the 'main line' of dasyurid dental progression (from insectivory to carnivory) after *Antechinomys* and *Sminthopsis* (see Fig. 3.4). Gregory (1920) concurred with Bensley (1903), but Osgood (1921) again raised the possibility of a Mesozoic metatherian ancestor. Pocock's (1926) assessment of external characters led him to conclude that *Myrmecobius* is more distant from Dasyuridae than is *Thylacinus*. Regardless of its sister group, the numbat remained phylogenetically associated with dasyurids for the rest of the 20th century, both on the basis of morphology (Archer 1976c, d; Wroe *et al.* 2000; Wroe and Musser 2001) (see Fig. 3.7) and molecules (Kirsch 1977; Baverstock *et al.* 1990; Krajewski, Buckley and Westerman 1997; Krajewski, Blacket and Westerman 2000) (see Figs 4.1, 4.3, 4.13 and 4.14). In recognising numbats as dasyuromorphians, Aplin and Archer (1987) cited two derived craniodental features they share with dasyurids and thylacinids: incisors 4/3 and an enlarged squamosal epitympanic sinus. As final confirmation, Zemann *et al.* (2013) showed that numbats and dasyurids share 14 insertions of two retrotransposons that other marsupials lack and that dasyurid monophyly is supported by another 11 insertions (Fig. 8.20). Retroposon insertions are examples of 'rare genomic changes' that accumulate with minimal homoplasy in evolving lineages and, as such, are exceptionally reliable indicators of phylogeny (Rokas and Holland 2000). Gallus *et al.* (2015) repeated this analysis, adding a third retrotransposon, in a study of marsupial orders; they also recovered numbat + dasyurids.

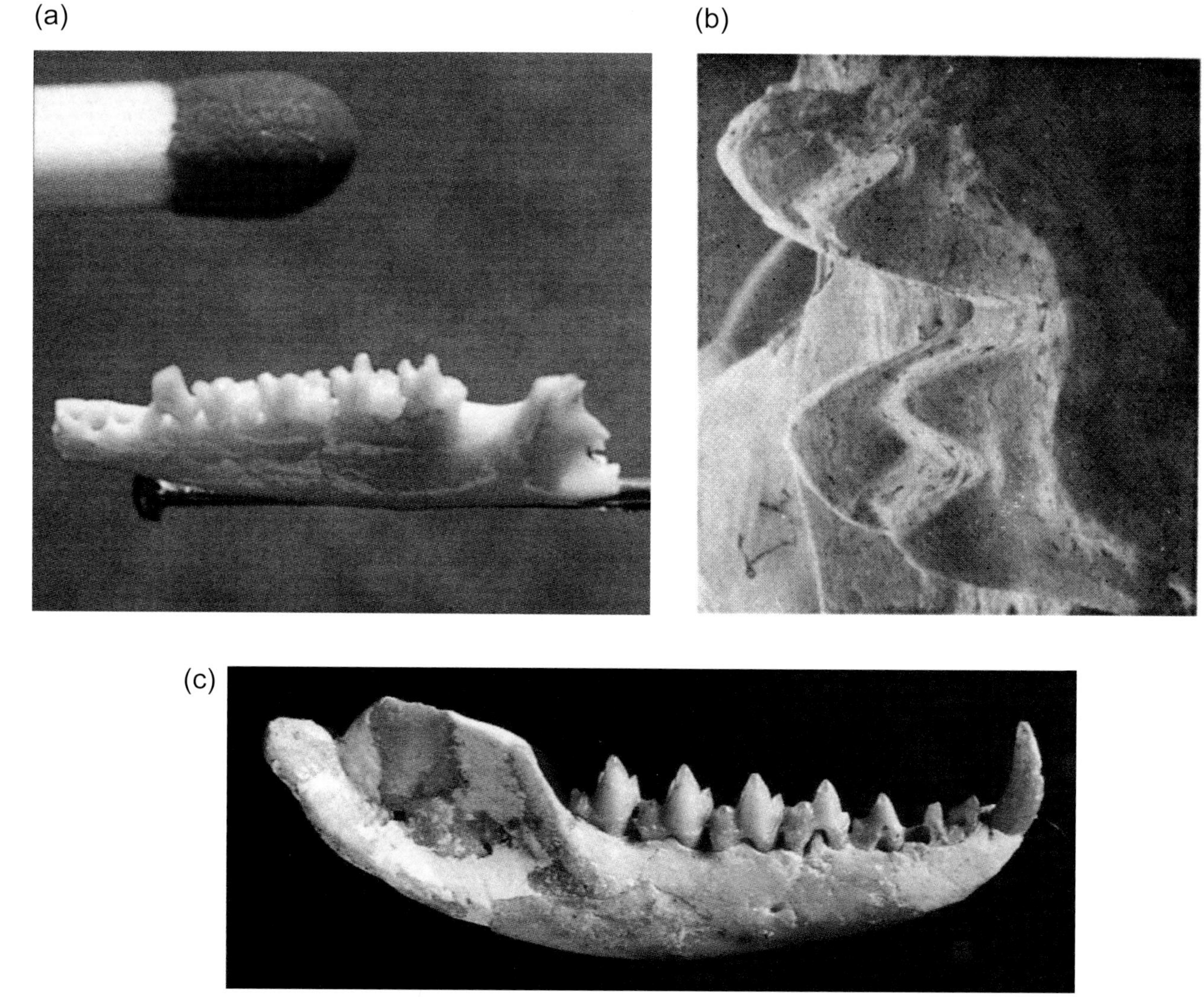

Fig. 8.17. Oligocene dasyuromorphians. (a) *Ankotarinja tirarensis* right dentary, superior medial view (SAM P18190). Photo by M. Archer. (b) *Keeuna woodburnei* left maxilla fragment showing M^{3-4} in occlusal view (QM F7333). (c) *Apoktesis cuspis* dentary. Photo by C. Campbell. [(a, c) Reprinted by permission of Michael Archer from Long *et al.* (2002). (b) Reprinted by permission of the Royal Zoological Society of New South Wales from Archer (1982c).]

Thylacines and borhyaenids

As discussed in Chapter 3, Bensley (1903) argued that the history of dasyurid evolution was revealed by two anatomical trends – a 'main line' of dental evolution from primitive insectivores (e.g. *A. flavipes*) to derived carnivores (e.g. *Sarcophilus*) and several side branches in which ancestral forms with arboreal-adapted hind feet evolved into terrestrial ones. Although Bensley followed Thomas (1888) in classifying *Thylacinus* and *Myrmecobius* as members of Dasyuridae, these genera did not fit comfortably into his evolutionary scheme and were placed in subfamilies separate from 'Dasyurinae'. Thylacines were especially problematic in that their carnivorous dental adaptations clearly evolved along a different path from those in *Sarcophilus*.

> *[T]he Dasyurinae present successive phases of an extremely homogenous dental evolution, the chief features of the transformation being: (a) reduction of the protocone in the upper molars and of the talonid in the lower; (b) approximation of the external styles of the upper molars to their respective cusps; and (c) reduction with final obliteration of the posterior premolars and of the*

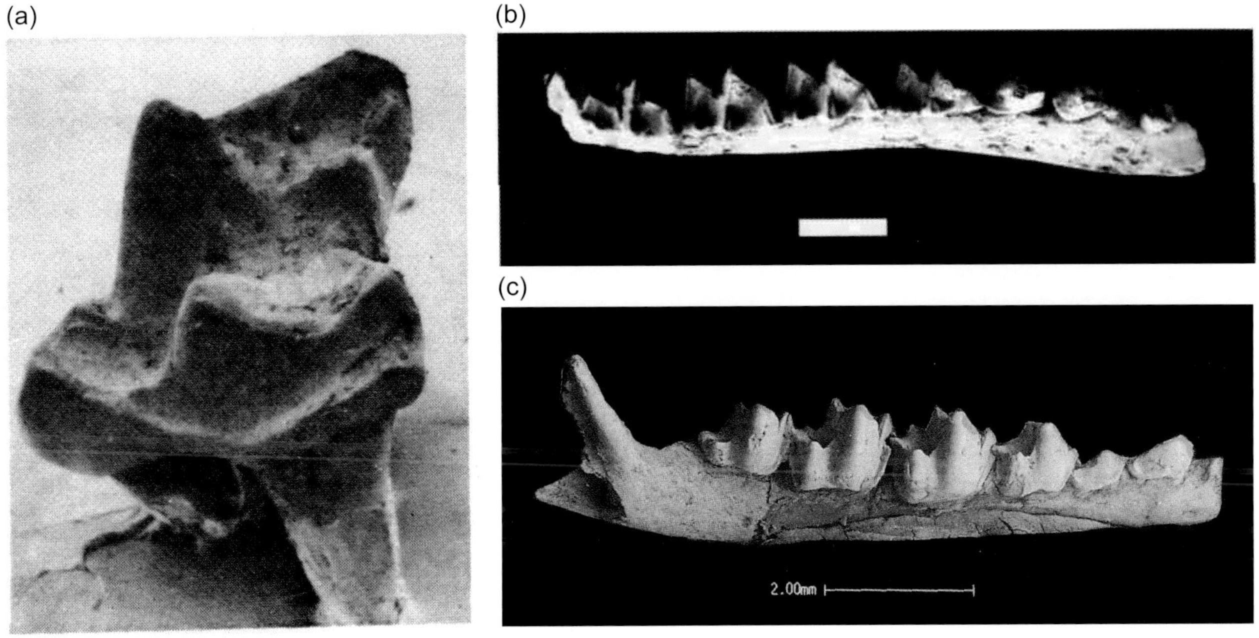

Fig. 8.18. Miocene dasyuromorphians. (a) *Dasylurinja kokuminola* right M^3 in anterior oblique view (SAM P23623). (b) *Joculusium muizoni* left dentary, occlusal view (QMF 36442). Scale bar = 2 mm. (c) *Mayigriphus orbus* right dentary in medial view (QM F23780). [(a) Reprinted by permission of the Royal Zoological Society of New South Wales from Archer (1982c). (b) Reprinted by permission of the Australasian Association of Palaeontologists from Wroe (2001a). (c) Reprinted by permission of the Queensland Museum from Wroe (1997b).]

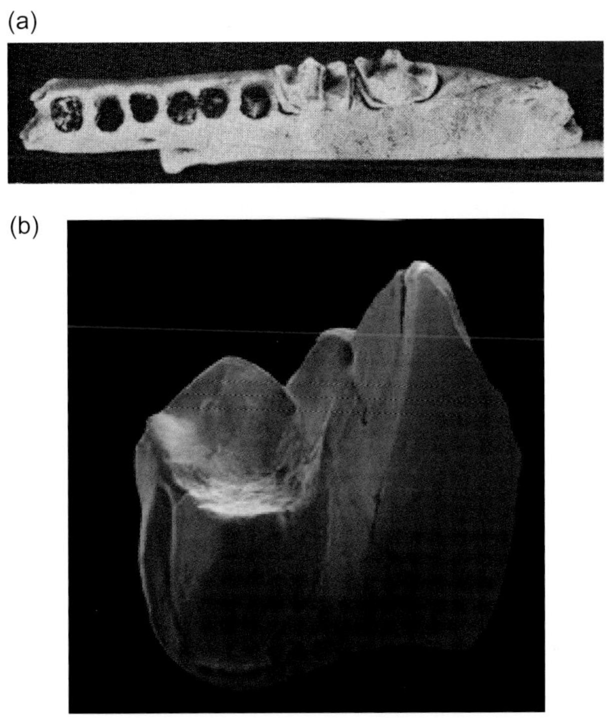

Fig. 8.19. More fossil dasyuromorphians. (a) *Wakamatha tasselli* left dentary, occlusal view (NMV33253). (b) *Whollydooleya tomnpatrichorum* lower molar in posterobuccal view (QMF 57892). [(a) Reprinted by permission of the Royal Zoological Society of New South Wales from Archer (1982c). (b) From Archer, Christmas *et al.* (2016a).]

> *paraconid in the first lower molars. [Bensley 1903: 107]*

In contrast, the protocones of *Thylacinus* were well-developed, talonids broad, stylar cusps reduced and third premolars robust (Fig. 8.21). In these and other dental traits, however, the thylacine resembled fossils of *Prothylacinus* and *Sipalocyon* (= *Amphiproviverra*) – Miocene genera belonging to the metatherian taxon Sparassodonta. His classification and phylogeny notwithstanding, Bensley (1903: 195) drew the following conclusion about *Thylacinus*:

> *The evidence of the dentition, however, favors a relation with the Sparassodonta of the South American Miocene, the most important consideration being, first that the dentition is well-advanced in the carnivorous evolution, but its characters are in many respects different from those gradually developed in the Dasyurinae and perfected in* Sarcophilus, *so that* Thylacinus *comes to occupy an isolated position in the Australian family, and, secondly, that the dentition resembles that of the carnivorous Sparassodonta in exactly those characters in which it*

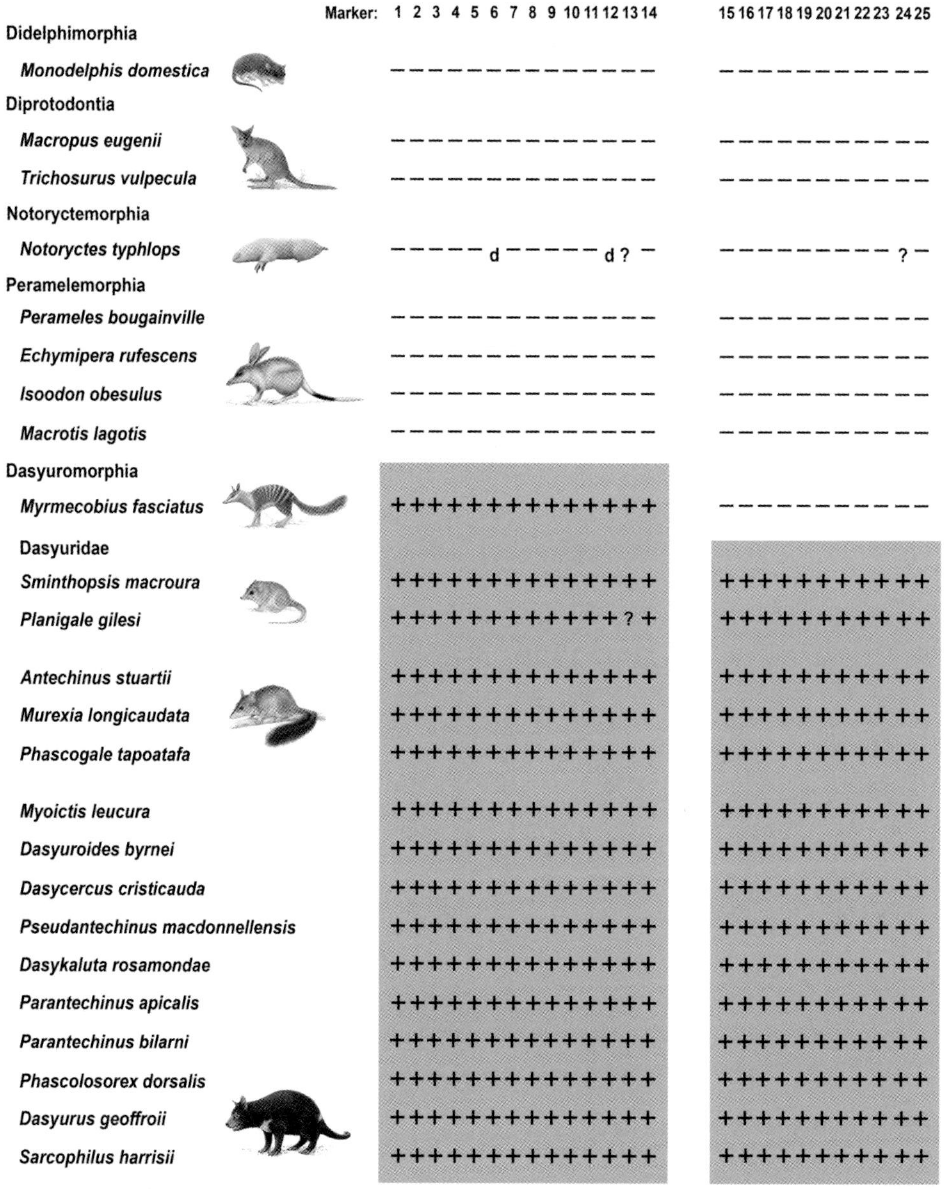

Fig. 8.20. Shared retroposon insertions among selected marsupials demonstrating that *Myrmecobius* forms a clade with a monophyletic Dasyuridae apart from other families. Markers 1–25 (top) are specific insertion loci. A dash '–' indicates that the retroposon is absent from the locus; '+' indicates that the retroposon is present. A 'd' indicates deletion of the locus; '?' indicates experimental failure to amplify the locus. Shaded areas highlight markers shared by numbats and dasyurids (left) and dasyurids only (right). [Reprinted by permission of Oxford University Press from Zemann *et al.* (2013).]

> *differs from that of* Sarcophilus. *This may not be regarded as indicating that the Dasyuridae are descendants of Sparassodonta through forms like* Thylacinus, *but that* Thylacinus *is a foreign element in the Australian series.*

The last sentence rejects Lydekker's (1899) hypothesis that prothylacinid sparassodonts were the ancestors of dasyurids. Bensley's view of *Thylacinus* epitomises the ambiguous relationship among concepts of ancestry, phylogeny and classification in early 20th century systematics. He made the situation even more perplexing by concluding:

> *The dental resemblances of* Thylacinus *to the Sparassodonta are sufficiently close to warrant a*

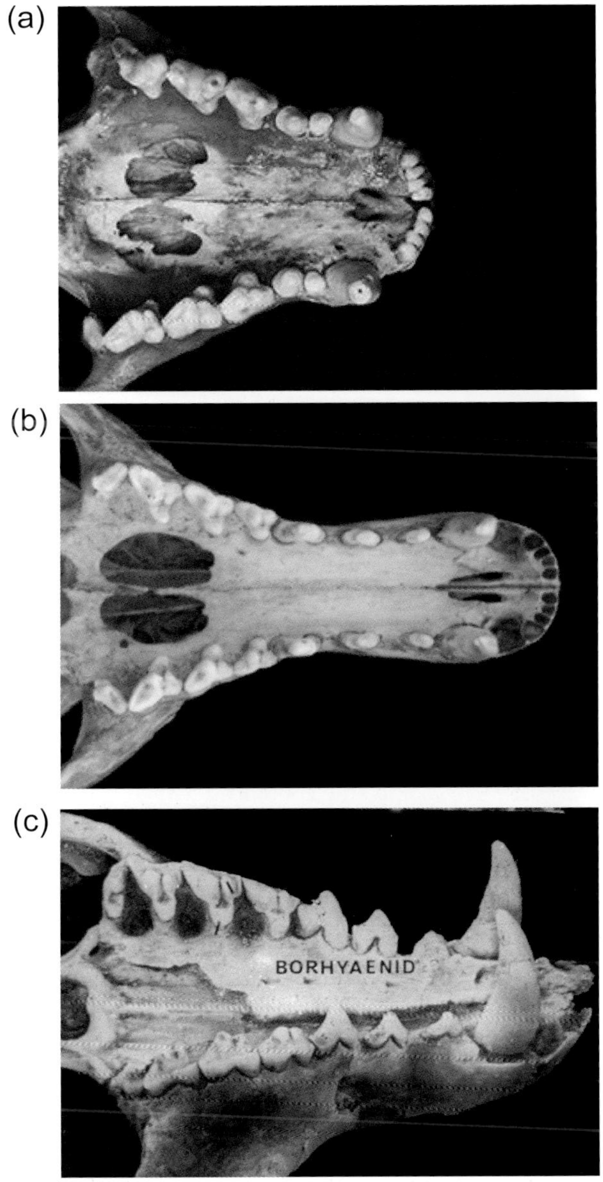

Fig. 8.21. Comparison of the upper teeth in (a) *Sarcophilus*, (b) *Thylacinus cynocephalus* and (c) an adult borhyaenid (a sparassodont of unspecified identity). Compare the reduced protocone, absent P^3 and shortened snout in *Sarcophilus* with the prominent protocone, robust P^3 and elongated snout in *Thylacinus* and the borhyaenid. [(a) Reprinted by permission of Oxford University Press from Rose *et al.* (2017). (b) Reprinted from Pask *et al.* (2008) under CC BY license. (c) Reprinted by permission of Michael Archer from Archer and Hand (2006).]

> *belief in their common origin. Even so, however, there is no evidence that the evolution of* Thylacinus *has been connected in any way with that of the Dasyuridae, the reverse being indicated by the fact that the various forms of the Dasyuridae present successive stages of a dental evolution, which, carnivorous in its character as in* Thylacinus, *is otherwise of a totally different facies. [Bensley (1903: 210)]*

These remarks laid the foundation for almost a century of debate over the affinities of thylacines. Sinclair (1905) went so far as to classify four genera of sparassodonts from the Miocene Santa Cruz Formation of Patagonia as members of Thylacinidae, a family 'sharply separated from the Dasyuridae and all other existing carnivorous marsupials by the absence of the metaconid in the lower molars and by the great reduction of the outer cingulum and styloid cusps of the upper teeth' (pp. 73–74). Sinclair's taxonomy was adopted in the prominent mammal classifications of Gregory (1910) and Osborn (1910). Taxonomy aside, the thylacines-as-sparassodonts hypothesis begged the question of how *Thylacinus*, or any other marsupial, got to Australia. Early speculation favoured dispersal from Asia (of which Alfred Russell Wallace was a proponent; Wallace 1876), but even in the early 1900s phylogenetic linkages between Australian and South American marsupials were becoming mainstream. Of course, acceptance of continental drift and the reconstruction of past continental positions were decades away and the debate over thylacines and sparassodonts became one of many flashpoints for discussions of ephemeral land bridges and other mechanisms of long-distance dispersal.

Biogeography was clearly in Matthew's (1915) mind when he argued that resemblances between sparassodonts and thylacines were parallel adaptations to carnivory (like those documented for placental carnivorans) as the two groups evolved from 'didelphid' ancestors. (Early concepts of Didelphidae included many fossil genera now excluded from the family.) This was consistent with his denial of a past land connection between South America and Australia (contra Scott 1913). Wood (1924) rejected Matthew's conclusion based on anatomical resemblance: among 49 skeletal traits he examined, 28 were shared by thylacines and sparassodonts, and only seven by didelphids and sparassodonts. This pattern persisted after Wood accounted for correlated characters; moreover, among the remaining thylacine-sparassodont matches, eight were not obviously associated with carnivory. Pocock (1926) examined external characters (vibrissae, rhinarium, ear, feet and pouch) and reached a different conclusion, expressed in the boldest pre-Hennigian terms:

> *Judging from the characters discussed in this paper, it appears to me obvious that we have complete evidence of tolerably close affinity between* Thylacinus *and the Dasyuridae. To set these resemblances aside as of no importance is to deny the fundamental axiom of taxonomy that likenesses imply affinity. [p. 1076]*

Clearly, Gill's (1870) dictum – discussed in Chapter 3 – that classification should be based on *all* available characters had not yet become established in mammalian taxonomy. In one of his earliest papers on South American taxa, the great Spanish mammologist and cataloguer Angel Cabrera considered a wider and older range of sparassodont taxa (Cabrera 1927). He concluded that their affinities were with Australian marsupicarnivores, but not specifically with *Thylacinus*. Cabrera's 'Dasyuroidea' included separate families for dasyurids, thylacinids and sparassodonts.

By 1940, sparassodont taxa resembling *Thylacinus* had been placed in the family Borhyaenidae; indeed, the debate is still often presented as one of 'borhyaenid' affinities for thylacines, despite recognition of several sparassodont families (Prevosti and Forasiepi 2018). Simpson (1941) considered three possibilities: (1) sparassodonts and thylacinids were derived independently from 'didelphoid' ancestors (vide Matthew 1915); (2) sparassodonts are thylacinids (vide Sinclair 1905; Wood 1924); and (3) sparassodonts are dasyuroids (vide Cabrera 1927). Simpson endorsed the first hypothesis, in which the similarity of sparassodonts and thylacinids evolved in parallel. His approach was to expand Wood's (1924) data set to include more recently described and geologically older fossil material, which reduced the evidence for thylacinid-sparassodont affinity to just one character – absence of epipubic bones. Simpson (1948) noted two other supporting traits – vertebral formula and a thickened posterior palate – but dismissed them as probably not homologous. With characteristic bluntness, Simpson (1941: 5) assessed the more general possibility that sparassodonts arose from dasyuroid ancestors:

> *If one postulates a dasyuroid primitive enough to be ancestral to the borhyaenid group as a whole, it is, practically speaking, indistinguishable from a didelphoid ... As for* Thylacinus, *there seems to be abundant, conclusive evidence that it is merely a specialized dasyurid.*

As noted in Chapter 3, Tate (1947) attached great significance to reduction of third premolars. Considering *Murexia* (which he had described in 1937) as exemplifying primitive dasyuroid dentition, Tate linked this genus to *Thylacinus* based on unreduced third premolars (see Fig. 3.5). His hypothesis was simply that the impressive dental specialisations of thylacinids were derived from ancestral, *Murexia*-like dasyuroid teeth independently of all other dasyurids. Putting aside Tate's tree, his explanation of dental evolution is essentially that of modern palaeontologists. In any case, thylacinids had nothing to do with sparassodonts.

Matters stood thus until Archer's (1976c, d) detailed investigations of basicrania and teeth. The intervening years had seen the birth of cladistics and phenetics, as well as a growing discomfort with the notion of direct ancestry of one clade from another. As we saw in Chapter 3, Archer's early work on dasyurid morphology embraced the cladistic distinction of primitive and derived traits, but still employed overall similarity as a guide to phylogeny and entertained the possibility of direct ancestry as an inferable relationship. Archer (1976d) rejected Simpson's (1941, 1948) interpretation of dental traits on two levels. First, that *Thylacinus* (the contents of which now included *T. potens*) is 'merely a specialized dasyurid', against which he adduced evidence that thylacinids have some features (e.g. missing metaconid and large paraconid on M_1) that all dasyurids lack. Second, that thylacinid teeth show maximal resemblance to sparassodonts (see Fig. 3.7). However, he found no such resemblance in basicrania of the two groups. 'Although ... the basicranium of thylacinids is most like that of didelphids, the possibility remains that borhyaenids and thylacinids had a common ancestor whose teeth were thylacinid-like but whose basicranium was didelphid-like' (Archer 1976d: 31).

Inevitably, cladistic analysis would be brought to bear on thylacinid affinities. Marshall's (1977b) monograph on Borhyaenidae endorsed Simpson's (1941) view that their similarities with thylacinids were parallelisms. In his response to Archer (1976c, d), Marshall (1977a) evaluated 14 skeletal (12 craniodental) traits and performed a cladistic analysis of them, each trait weighted by its degree of presumed phylogenetic value. The weighting procedure followed Hecht (1976) and Hecht and Edwards (1976) by ranking putative anatomical synapomorphies in order of increasing 'information content' as follows: 1 = state loss; 2 = reduction in complexity; 3 = change correlated with growth or physiology; 4 = change in a functional complex of characters; and 5 = origin of unique or innovative structure. For example, Marshall gave loss of an incisor (from the ancestral 5/4 in didelphids to 4/3 in other groups) a weight of 1. At the other end of the spectrum, the posterior squamosal epitympanic sinus of thylacinids and dasyurids was weighted 5. Marshall (1977a) argued that Archer's (1976d) evidence for thylacinid-borhyaenoid affinities consisted entirely of variable, primitive (0 weight) and 1- or 2-weighted derived traits, most of the latter correlated within molars evolving towards carnassialisation. His weighted analysis produced two trees (Fig. 8.22a), both showing thylacinids as sister to dasyurids supported by three synapomorphies: incisors 4/3 (weight = 1), reduced dP3 (weight = 1 or 2) and large epitympanic squamosal sinuses (weight = 5).

Archer's (1982a) review of thylacinid phylogeny included cladistic rather than phenetic interpretation of similarities; he also attempted to separate the inferentially tractable problem of identifying the thylacine's sister with the more difficult one of tracing its ancestry to a particular taxon. His critique of Marshall (1977a) anticipated much of the debate about character weighting in later years (see Goloboff *et al.* 2008). Referring to 'the problem of apriorisms in character analysis', Archer cited several logical issues (paraphrased here for brevity): (1) categories of 'phylogenetic informativeness' are not operationally defined; (2) there are many possible combinations of weights but no justifiable way to choose among them; (3) functional complexes are difficult to identify because character correlation is not all-or-nothing; and (4) cases of reduction in one trait can often be re-described as gains in another. These concerns, though valid, do not establish a logical priority for uniform character weighting; the 'principle of indifference' remains a fallacy (Sober 1988). Archer (1982a) also took issue with Marshall's (1977a) definition, scoring and polarisation of characters – all problems now recognised as significant challenges for morphocladistics (see Chapter 5). Archer's data set consisted of 11 dental, 15 cranial and three postcranial characters (including the 14 used by Marshall 1977a), 18 of which were informative. His cladistic analyses again supported thylacinids + sparassodonts (Fig. 8.22b), but his interpretation of results did not discount thylacinids + dasyuroids.

In the same volume as Archer's (1982a) review, Szalay (1982) published his analysis of tarsal morphology that established Australidelphia, a clade consisting of South American microbiotheres (*Dromiciops*) and all Australasian marsupials. Australidelphians share the derived 'continuous lower ankle joint pattern' of facets on the astragalus and calcaneum (along with three other tarsal synapomorphies), whereas Szalay's ameridelphians (including sparassodonts) have versions of the primitive 'separate lower ankle joint pattern'. Szalay (1982: 621) left no doubt about the implications of tarsal morphology for thylacine affinities: 'Thylacinids show no special, homologously derived, shared similarities with any of the Borhyaeniformes. Their most recent ties are with other australidelphians, particularly within the Dasyurida.' Szalay's findings were one reason why Archer (1982a) considered his own craniodental results inconclusive.

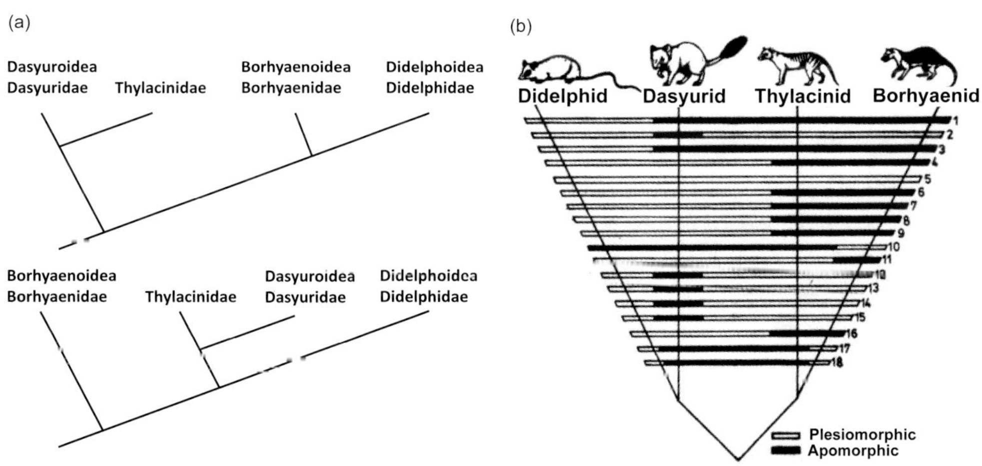

Fig. 8.22. Cladistic estimates of thylacinid phylogeny. (a) Cladograms from Marshall's (1977a) analysis of weighted anatomical traits. (b) Distribution of plesiomorphic and apomorphic states of 18 skeletal characters analysed by Archer (1982a). [(a) Adapted from Marshall (1977a). (b) Reprinted by permission of the Royal Zoological Society of New South Wales from Archer (1982a).]

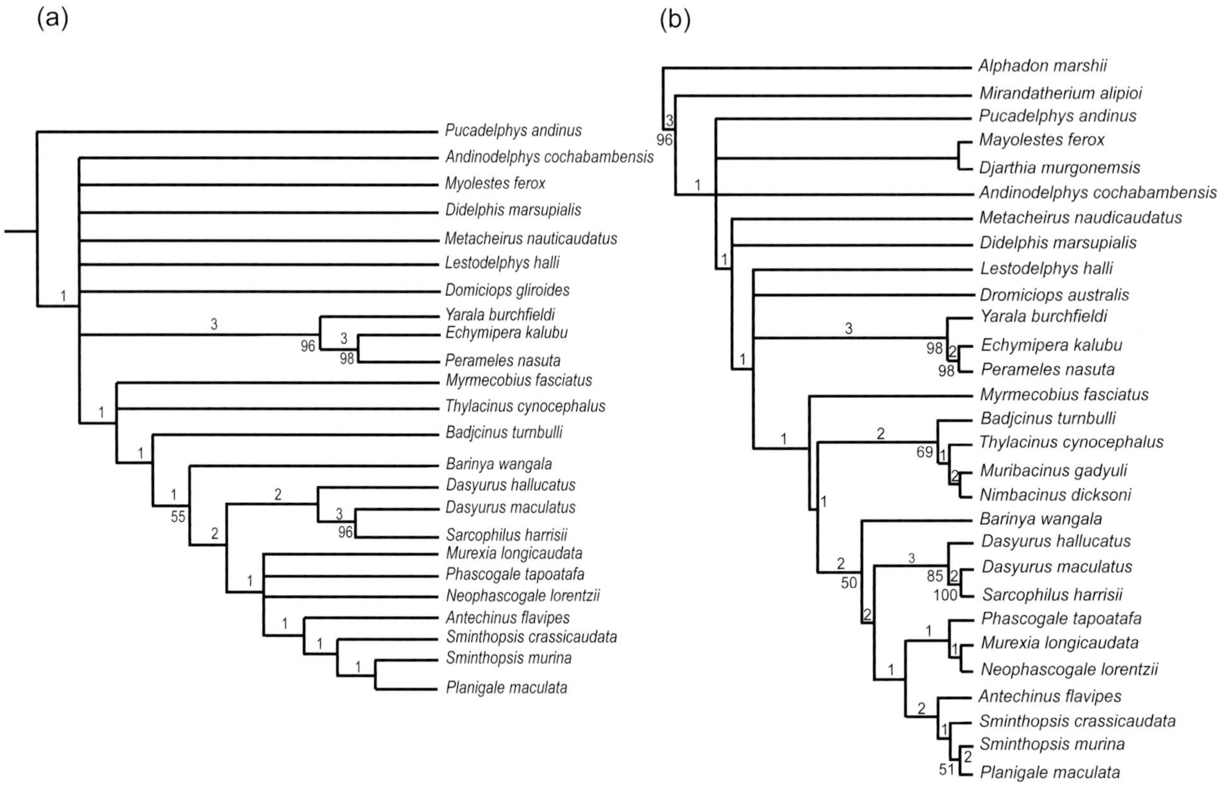

Fig. 8.23. Dasyuromorphian phylogenies estimated from 77 polarised craniodental characters. (a) Strict consensus of 64 most-parsimonious trees from Wroe *et al.* (2000). Values above branches are bootstrap percentages; those below branches are Bremer support indices. (b) Strict consensus of 10 most-parsimonious trees from Wroe and Musser (2001). Values above branches are Bremer support indices; those below branches are bootstrap percentages. [(a) Reprinted by permission of Oxford University Press from Wroe *et al.* (2000). (b) Reprinted by permission of CSIRO Publishing from Wroe and Musser (2001).]

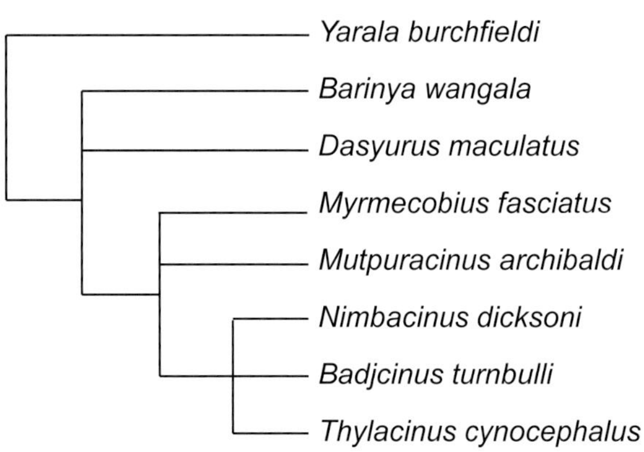

Fig. 8.24. Strict consensus of five equally parsimonious trees from Murray and Megirian's (2006) analysis of 33 polarised craniodental traits in selected dasyuromorphians. *Yarala burchfieldi* is the outgroup. [Adapted from Murray and Megirian (2006).]

The other and stronger reason was that Lowenstein *et al.* (1981) and Sarich *et al.* (1982) used radioimmunoassay (RIA) and microcomplement fixation (MC′F; see Chapter 4) to estimate albumin immunological distances (ID) of 11 between *Thylacinus* and dasyurids versus >>80 between *Thylacinus* and didelphoids. This was one of the earliest applications of molecular systematics to an extinct species, thylacine albumin having been obtained from untanned museum skins. Archer (1982a) correctly pointed out that, lacking ID measurements for sparassodonts, Sarich *et al.* (1982) could not rule out inclusion of this group in a larger clade of dasyurids and thylacinids. Nevertheless, he conceded:

> *… I am now convinced that the thylacinid-borhyaenid dental similarity, although marked, is probably convergent. So must be the common reduction of epipubic bones and loss of the hallux. This reversal of my 1976 opinion is based not on a simple recounting of potential synapomorphic character-states, but on the conclusion that immunological data are probably more reliable indicators of phylogenetic history than others involved in this particular debate. [Archer 1982a: 471]*

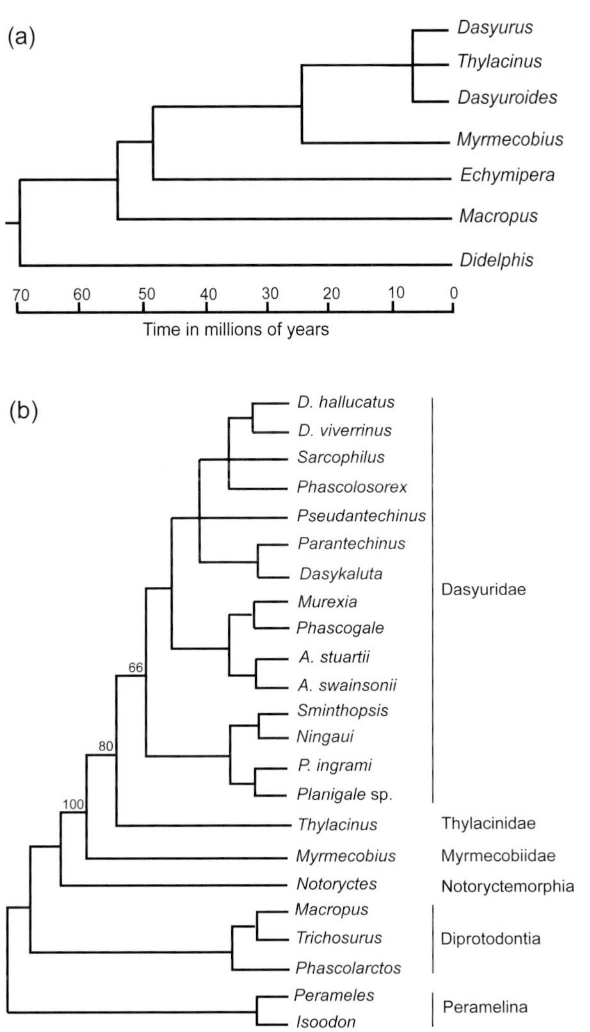

Fig. 8.25. Early molecular trees showing thylacines as sister to dasyurids. (a) Time tree from Lowenstein *et al.* (1981) based on albumin similarities estimated with radioimmunoassay and calibrated as explained in the text. (b) Maximum parsimony tree from *cytb*, *12S* and *P1* from Krajewski, Blacket and Westerman (2000). Analysis based on 3TV coding of *cytb* and homologous gap coding of *P1* sites. Nodal values are bootstrap percentages. [(a) Redrawn with permission of Springer Nature from Lowenstein *et al.* (1981). (b) Redrawn with permission of Springer Nature from Krajewski, Blacket and Westerman (2000).]

After 1982, consensus on the status of thylacinids as a dasyuromorphian family came swiftly. Thomas *et al.* (1989) obtained short sequences of *cytb* and *12S* mitochondrial genes from *T. cynocephalus* museum tissue and found that they grouped with other australidelphians apart from a didelphid (see Fig. 4.5). Longer *cytb* fragments (Krajewski *et al.* 1992) excluded *Thylacinus* from Dasyuridae and complete *cytb*, *12S* and *P1* sequences (Krajewski, Buckley and Westerman 1997; Krajewski, Blacket and Westerman 2000) resolved a monophyletic Dasyuromorphia (including *Thylacinus*) well apart from Didelphimorphia (see Figs 4.13 and 4.14). Moreover, as new fossil discoveries expanded Thylacinidae beyond *T. cynocephalus* and *T. potens*, the family was seen to include forms with less specialised dentitia than those of the original two species. *Nimbacinus*, for example, had larger paracones, stylar cusps, metaconids, entoconids, paraconids and talonid basins – as well as less elongated molar shearing crests – than the hypercarnivorous *Thylacinus* (Muirhead and Archer 1990). Meanwhile, phylogenetic analysis of South American fossils (e.g. Ladeveze and De Muizon 2010) have led to a consensus that sparassodonts are non-marsupial metatherians.

The sister group of Dasyuridae

Setting aside Malleodectidae, dasyuromorphian family relationships must take one of three forms: ((numbat, thylacines) dasyurids); (numbat (thylacines, dasyurids)); or (thylacines (numbat, dasyurids)). These possibilities have been debated for over a century. We saw above that Bensley's (1903) model of dental evolution led him to favour *Myrmecobius* as phylogenetically closer to dasyurids, whereas external morphological similarity let Pocock (1926) to place *Thylacinus* in that position. Although Tate (1947: 109) recognised Myrmecobiidae and Notoryctidae as 'relatives of the Dasyuridae', his subjective assessment of morphological evolution placed *Thylacinus* as a 'very ancient form' related to *Murexia* within his 'Phascogalinae'. Archer's (1976c, d) subjective phenograms of basicranial and dental similarity found *Thylacinus* most like didelphoids or sparassodonts, respectively, but numbats adjacent to dasyurids (see Fig. 3.7). The earliest molecular studies that included all three families were the *cytb*, *12S*, *P1* sequence analyses of Krajewski, Buckley and Westerman (1997) and Krajewski, Wroe and Westerman (2000), but these could not resolve interfamilial relationships with confidence, nor could the parsimony analyses (Fig. 8.23) of craniodental characters by Wroe *et al.* (2000) and Wroe and Musser (2001). Although Murray and Megirian (2006) did not include any measure of statistical support, their analysis of 33 polarised craniodental traits returned five minimum-length trees, the consensus of which shows numbats and thylacinids as sisters apart from dasyurids (Fig. 8.24).

Lowenstein *et al.* (1981) and Sarich *et al.* (1982) included *Mymecobius* in their albumin RIA and MC'F comparisons (see 'Thylacines and borhyaenids' above) and found that IDs from *Thylacinus* were 11 to dasyurids

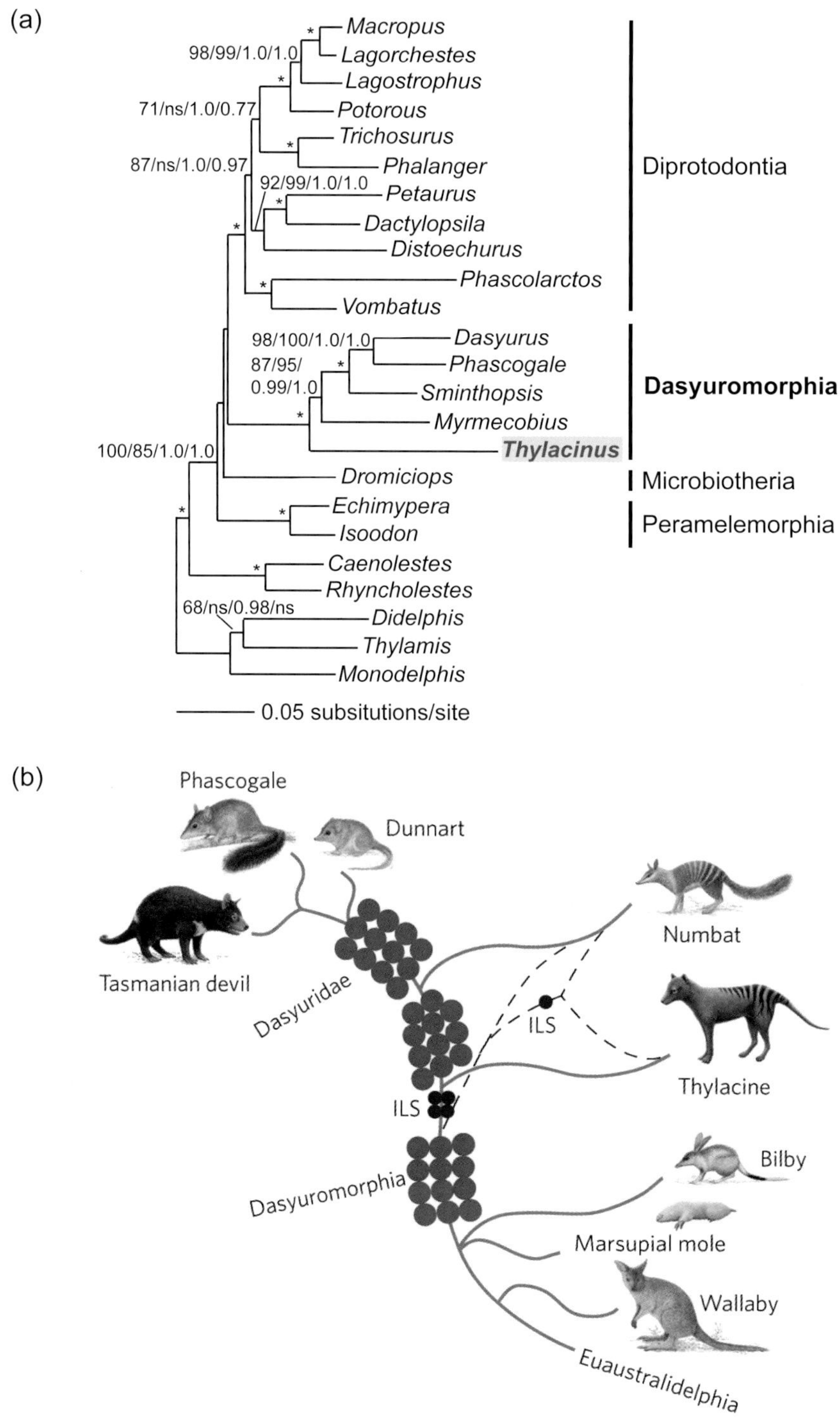

Fig. 8.26. Genomic resolution of dasyuromorphian family relationships. (a) Maximum likelihood tree based on complete mitochondrial genomes from Miller *et al.* (2009). Analysis based on GTR +I + Γ model with 'reduced bias coding' (i.e. omitting third positions of protein-coding genes, considering transversions only in RNA genes and transitions only at first positions of protein-coding genes). Nodal values are (left to right): bootstrap percentage with reduced bias coding, bootstrap percentage with raw sequences, posterior probability with reduced bias coding, and posterior probability with raw sequences. Asterisks (*) indicate branches with bootstrap = 100 and posterior probability = 1.0. (b) Retroposon phylogeny from Feigin *et al.* (2018). Circles indicate shared insertions of retroposons WSINE1/1a (red) and WALLSI1A (green). ILS indicates inferred instances of incomplete lineage sorting suggesting relationships discordant with most insertions. [(a) Redrawn from Miller *et al.* (2009) under CC BY 4.0 license. (b) Reprinted from Feigin *et al.* (2018) under CC BY 4.0 license.]

but 40 to numbat; from this they inferred that thylacines and dasyurids were sisters (Fig. 8.25a). Interestingly, however, the thylacine-dasyurid divergence time, estimated with the calibration ID 1 = 600 ka, was anomalously small (7 Ma). Archer (1982a) noted that such a recent divergence was difficult to reconcile with the fossil record and distribution of character states among the three families. Krajewski, Blacket and Westerman (2000), however, obtained the same pattern of relationships from complete *cytb*, *12S* and *P1* sequences, with moderate support for the thylacine-dasyurid clade (Fig. 8.25b). Complete mtDNA sequences (Miller *et al.* 2009), however, decisively reversed this pattern in favour of a numbat-dasyurid pairing (Fig. 8.26a). When Feigin *et al.* (2018) obtained the complete genome of an alcohol-preserved *Thylacinus* pouch young, they were able to score the same retroposon insertions identified by Zemann *et al.* (2013) in numbats (see 'Numbats' above). These markers showed with great clarity that *Myrmecobius* is indeed sister to dasyurids, with *Thylacinus* forming an earlier branch (Fig. 8.26b). Thus, although analyses of different character types and taxon samples have historically supported all three possible phylogenies of dasyuromorphian families, recent genomic data sets leave little doubt that numbats are the sister group of dasyurids.

ENDNOTE

1 Churchill *et al.* (2023) describe a third species of malleodectid – *Malleodectes*? *wentworthi* – that may be phylogenetically closer to dasyurids than to thylacines or numbats.

9 The timing of dasyurid evolution

A LITERAL READING OF THE FOSSIL RECORD

Driven by seafloor spreading, the northward-drifting continent of Australia made its final separation from Antarctica sometime during the Eocene (56–34 Ma). Prior to this, the ancestors of Australian marsupials had dispersed across Antarctica from South America, the first unequivocal record of an australidelphian being *Djarthia murgonensis* from the early Eocene (55 Ma) Tingamarra Local Fauna (LF) in south-eastern Queensland (Beck *et al.* 2008). As we saw in Chapter 8, the oldest taxa that may have been dasyuromorphians (perhaps even dasyurids) are the late Oligocene (26–23 Ma) *Apoktesis*, *Ankotarinja* and *Keeuna* from South Australian deposits (Fig. 9.1). The oldest putative dasyurids are the early Miocene *Barinya* species, the most ancient remains assigned to a crown-group tribe are of *Ganbulyani* (Dasyurini) from the Late Miocene (12–10 Ma) Encore LF at Riversleigh and the oldest records of crown-group genera are Pliocene (5.3–2.5 Ma) from several sites across Australia.

Fossils provide minimum dates of origin for the clades to which they belong. The dates are minima because the process of fossilisation is such that only a small proportion of organismal remains in any time horizon will be preserved and an unknown but probably small fraction of preserved remains have been discovered (Foote *et al.* 1999). Thus, the occurrence of *Barinya* seems to tell us only that Dasyuridae arose *at least* by 18 Ma. Unfortunately, fossil-based divergence times are associated with sources of error beyond the vagaries of taphonomic processes. Fossil ages are not known with certainty but are based on the ages of rock strata in which they occur, and stratigraphic ages are themselves estimates based on radiometric, biocorrelation or paleomagnetic analyses. Long *et al.* (2002: 12–13) explain the basics of fossil dating; Woodhead *et al.* (2016) and Megirian *et al.* (2010) have more in-depth discussions about radiometric and biocorrelation studies, respectively, of Australian local faunas. If the age of a fossil-bearing stratum cannot be dated directly, its maximum or minimum age may be estimated by dated layers below or above it. Errors associated with stratigraphic dates are often difficult to quantify, especially when those dates are inferred from indirect evidence.

Using fossils to date the origin of a clade also requires that the phylogenetic affinities of fossil taxa be correctly estimated. For example, if *Barinya* is a dasyurid genus but *Ankotarinja* is not, then the minimum fossil age of Dasyuridae is Early Miocene (Wroe 1997a, 1999b). Murray and Megirian (2006), however, argued that Wroe's (1997a) exclusion of *Ankotarinja* (and other taxa) from Dasyuridae was unwarranted. If *Ankotarinja* is a dasyurid, then the minimum date for Dasyuridae is Late Oligocene. If *Barinya* is not a dasyurid (Beck *et al.* 2022), then the minimum age of Dasyuridae is Late Miocene. An accurate phylogeny also allows fossils of one clade to help date the origin of its sister. For example, since Thylacinidae has a minimum date of Late Oligocene (*Badjcinus*), its sister group (Myremecobiidae + Dasyuridae) must be at least this old, even

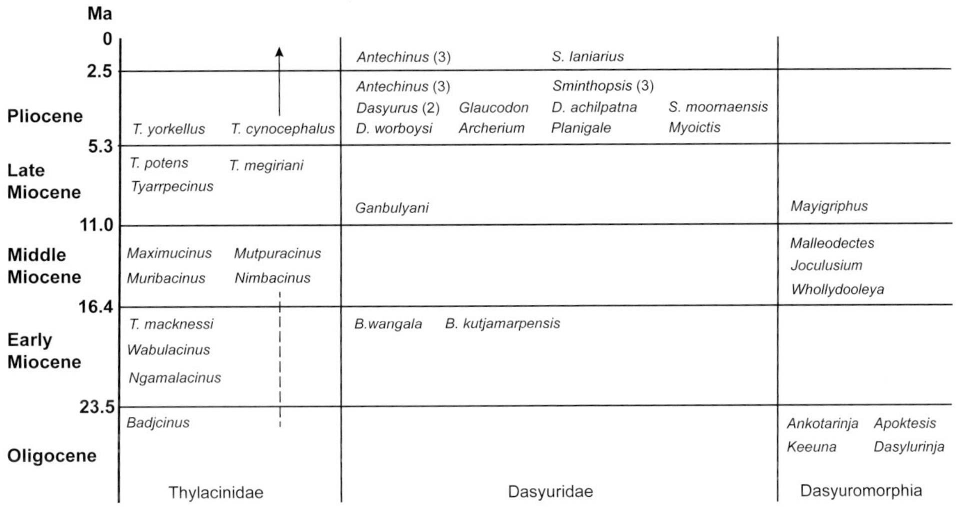

Fig. 9.1. Temporal distribution of dasyuromorphian fossil taxa, grouped by major divisions of geological time. Unnamed and extant taxa are not shown. Post-Pliocene taxa are Pleistocene. 'Dasyuromorphia' includes *incertae sedis* and Malleodectidae. The dasyuromorphian *Wakamatha* (Late Oligocene or Pleistocene) is not shown. Numbers in parentheses next to taxon names refer to the number of fossil species in genera with more than one fossil species.

though the first record of the latter clade is Early Miocene. The missing trace of the numbat-dasyurid clade in the Late Oligocene is a 'ghost lineage' (Norell 1992). As discussed in Chapter 4, phylogenetic placement of fossil taxa relies exclusively on morphology, often with very few characters due to the fragmentary remains of fossils, and this makes it extremely difficult to obtain well-supported trees. Nevertheless, using the best available descriptions, dates and clade assignments of fossil taxa have been, and remain, critical components in estimating phylogenetic divergence times and the macroevolutionary hypotheses they inform.

Given the information on dasyuromorphian palaeontology and family relationships summarised in Chapter 8 and depicted in Fig. 9.1, minimum fossil ages for the origin of major clades are listed below. These dates refer to origins of entire clades, including extinct ('stem') lineages that are more closely related to the clade containing all extant species (the 'crown group') than to any other clade.

- Dasyuromorphia = Late Oligocene (26–23 Ma) represented by *Ankotarinja*, *Apoktesis* and *Keeuna*
- Thylacinidae = Late Oligocene to Early Miocene represented by *Badjcinus* (26–16 Ma)
- Myrmecobiidae + Dasyuridae = Late Oligocene to Early Miocene (26–16 Ma) as ghost lineage to Thylacinidae
- Dasyuridae = Early Miocene (20–16 Ma) represented by *Barinya*; Late Miocene (12–10 Ma) if *Barinya* is not a dasyurid
- Myrmecobiidae = Early Miocene (20–16 Ma) as ghost lineage to Dasyuridae; Late Miocene (12–10 Ma) if *Barinya* is not a dasyurid
- Dasyurinae = Late Miocene represented by *Ganbulyani* (12–10 Ma)
- Sminthopsinae = Late Miocene (12–10 Ma) as ghost lineage to Dasyurinae.

Again, these are *minimum* ages; actual divergence times are expected to be older, but how much older is unclear. As we will see, estimated molecular dates usually exceed these fossil minima, sometimes by a considerable amount. In the few cases where they have not, doubt is cast on the accuracy of the molecular ages.

EARLY MOLECULAR DIVERGENCE TIME ESTIMATES

The strict molecular clock concept

When protein sequences from a range of species became available in the 1960s, several researchers observed an approximate linearity between the measured dissimilarity of orthologous proteins (p = proportion of mismatched amino acid residues in two aligned sequences) and the divergence times of the species from which they were derived (Zuckerkandl and Pauling 1962; Margoliash 1963; Doolittle and Blombäck 1964). Based on these observations, Zuckerkandl and Pauling (1965) articulated the 'molecular clock hypothesis' that the evolutionary rate of amino acid replacement in an individual protein (e.g. haemoglobin) is roughly the same among lineages within major clades. Molecular clocks received a theoretical foundation from Kimura's (1968) 'neutral theory of molecular evolution' which posited that most amino acid replacements were due to a balance between mutation and genetic drift without the influence of natural selection. For neutral mutations, the rate of replacement is determined by the mutation rate independent of effective population size. To the extent that mutation rate is stable, rates of amino acid replacement are *uniform* among lineages and *constant* over time – i.e. there is a molecular clock. The neutral theory and its clock corollary can also be applied at the DNA level (King and Jukes 1969). Rate variation among proteins or their corresponding genes reflect different degrees of functional constraint enforced by purifying selection (e.g. histones evolve more slowly than fibrinopeptides because a lower proportion of possible replacements in histone than in fibrinopeptide residues are neutral).

While neutral theory became an integral part of population models of allele frequencies (Kimura 1983), it also engendered a new focus on evolving sequences themselves. An early manifestation of this was the development of probabilistic models of sequence evolution, starting with that of Jukes and Cantor (1969) described in Chapter 4. These models conceived amino acid replacement (or nucleotide substitution) as a stochastic process driven by mutation and drift that played out over evolutionary time. Jukes and Cantor (1969) treated replacement as a Poisson process with a single average rate that applies to all residues in a sequence, all possible replacements at each residue and all orthologous sequences sampled. Soon thereafter, protein models began to incorporate empirically estimated replacement rates among the 20 amino acids, largely based on their biochemical properties (Dayhoff *et al.* 1978). When DNA sequences became widely available in the 1980s, theorists began to incorporate observed substitutional asymmetries (e.g. between transition and transversion rates; Kimura 1980) into evolutionary models.

Almost from the moment of its proposal the molecular clock hypothesis was controversial and, as a global generalisation, is now considered to be false (Kumar 2005). Nevertheless, compelling evidence has been found for approximately uniform rates for some sequences in some taxonomic groups. The earliest assay for such uniformity was Sarich and Wilson's (1967) 'relative rate test', based on the expectation that if *A* and *B* are orthologous sequences from sister species relative to outgroup sequence *O*, then distances d_{AO} and d_{BO} should be the same. This test was originally proposed as a subjective comparison, but various statistical versions were developed (e.g. Fitch 1976). Later rate-uniformity tests employed likelihood and are discussed below, but early applications of the molecular clock to dasyuromorphians used the original relative rate test. The test, however implemented, is more reliable when sequence distances or branch lengths are estimated with a correct evolutionary model.

The final component of estimating divergence time with a molecular clock is *calibration* – i.e. determining the actual rate of sequence divergence, usually in units of substitutions/site/unit-time (but often expressed as percentage sequence change per unit time). Suppose, for example, that three sequences are related as ((A, B) C), d_{AB} = 0.10 and $d_{AC} = d_{BC}$ = 0.20, the A-B split is known to have occurred 10 Ma and the date of the (A, B)-C split is unknown. The rate of divergence is (0.10 substitutions/site)/(10 Ma) = 0.01 substitutions/site·Ma (i.e. 1% per Ma) and the (A, B)-C date is estimated as (0.20 substitutions/site)/(0.01 substitutions/site·Ma) = 20 Ma. (Note that the *divergence* rate as defined in this example is twice the substitution rate along any single branch of the tree.) Of course, the dates of evolutionary splits for calibration are never known with certainty but must be estimated from fossils or historical events; if neither is available, rates calibrated for another group may be used (by assuming that they are the same in both groups). The A-B split date above, for example, may be based on a 10 Ma age for the oldest fossil of lineage AB, the vicariant event thought to have caused the A-B split or application of a 1% per Ma rate calibrated for orthologous sequences in a clade outside ABC. As simple as these calculations seem, they involve a concatenation of estimates (such as age of strata, age of fossils or historical events, and rate of evolution), each with associated uncertainties that propagate to create error in the final divergence times. In the early days of molecular dating, it was difficult or impossible to quantify this error, but later and more sophisticated approaches allow us to place approximate confidence intervals on estimated dates. We discuss these in later sections.

The dasyuromorphian immunological clock

Drawing on previous calibrations of clock-like albumin MC′F distances in mammals (Maxson *et al.* 1975; Wilson *et al.* 1977), Lowenstein *et al.* (1981) produced the chronogram described in Chapter 8 (see Fig. 8.25a). Sarich *et al.* (1982: 709) gave the albumin rate as 'One unit of distance is equivalent to 600 000 years since separation' (or $T = 0.6D$, with time T in millions of years and D the albumin distance). Given measured albumin distances, dasyuromorphian origins trace to 24–48 Ma (mid-Eocene to Late Oligocene) between their separation from bandicoots and divergence of the *Myrmecobius* lineage, a timeframe consistent with the fossil record. As noted by Archer (1982a), however, the late Miocene divergence date (7 Ma) between *Thylacinus* and dasyurids is anomalous, even more so in light of subsequently described fossils documenting these groups as distinct in the Late Oligocene. No later immunological work on *Thylacinus* explained this anomaly.

By the time Baverstock *et al.* (1990) applied albumin MC′F to marsupial phylogeny it had become clear that albumin rates vary among mammal groups, being notably slower in carnivorans and primates than in other placentals (Sarich 1985). Distrusting the taxonomic assignment of fossil marsupials, Baverstock *et al.* (1990) opted to apply the $T = 0.6D$ rate used in the thylacine work and calibrated by Maxson *et al.* (1975) from the separation of American and Australian marsupials by continental drift. However, they noted a rate slowdown in Sminthopsinae that made estimation of divergence dates in this clade problematic. With some caution, therefore, they put the basal dasyurid diversification at 20 Ma (see Fig. 4.3), which is coincident with the minimum fossil date based on *Barinya*. Baverstock *et al.* (1990: 137) estimated the early radiation of Dasyurinae at 10 Ma but cryptically remarked, '[d]ating of divergence times more recently than this is not possible with albumin, because it is below the limits of the technique'. This date, however, crowds the 10–12 Ma age of the putative dasyurin *Ganbulyani*. Baverstock *et al.* (1990: 144) concluded, '[W]e wish to stress that the dates of divergence we have given for the Australian marsupial groups should be treated as tentative. They are provided here as a hypothesis against which to test fossil data.' Some 30 years later and excepting *Thylacinus*, the albumin divergence times approximate minimum clade ages from the fossil record.

Aplin *et al.* (1993) used albumin MC′F to date the divergence of Australian and New Guinean marsupial sister-clades, including four dasyurid pairs identified by Baverstock *et al.* (1990): (1) *Murexia* and *Antechinus* + *Phascogale*; (2) phascolosoricines and a '*Dasyuroides* group' (*Dasyuroides*, *Dasycercus*, *Myoictis*, false antechinuses); (3) *Myoictis* and *Dasyuroides*; and (4) *Dasyurus albopunctatus* and *D. maculatus*. Based on current understanding, none of these comparisons involve actual sister groups (see Chapter 6), but the estimated dates are relevant for comparison with other studies. To deal with variable albumin evolutionary rates, Aplin *et al.* (1993) obtained correction factors for measured distances using comparison of all taxa to an outgroup didelphid (*Caluromys lanatus*). Dasyurid albumin appeared to evolve some 27% slower than the average rate. *Murexia* and phascolosoricines diverged from their putative sisters an estimated 10–12 Ma, *Myoictis* from *Dasyuroides* 2.7–4.7 Ma and *D. albopunctatus* from *D. maculatus* sometime in between. These dates are approximately consistent with Baverstock *et al.* (1990) and the fossil record, but younger than predicted by Flannery's (1989) hypothesis of a 20 Ma vicariant event separating Australian and New Guinean marsupials. We discuss this further in the context of dasyurid biogeography (Chapter 10).

DNA HYBRIDISATION AND THE UNIFORM AVERAGE RATE OF GENOME EVOLUTION

Based on several qualitative relative rate tests among bird groups, Sibley and Ahlquist (1983) argued that the '$\Delta T_{50}H$' measure of genetic distance from DNA hybridisation experiments scaled linearly with time at approximately the same slope in most avian taxa (i.e. the rate was uniform). Because, in theory, $\Delta T_{50}H$ is collectively determined by all single-copy loci in the genome, it represents a sort of 'average' distance over these loci. Hence, Sibley and Ahlquist (1983) claimed the existence of a 'uniform average rate' (UAR) of genome evolution measured by DNA hybridisation distances. They calibrated $\Delta T_{50}H$ with fossil evidence for the earliest divergences among living birds, as well as estimated dates of Africa–South America separation and the opening of the Tasman Sea isolating New Zealand. The result was a UAR of $T = 4.5D$, with T in Ma and $D = \Delta T_{50}H$. Following Sibley and Ahlquist's (1983) analysis, the effects of generation time (e.g. Wu and Li 1985) and DNA repair efficiency (Britten 1986) on the rate of DNA evolution received much attention. The former is correlated with many life-history characters that vary among species and should indeed affect genome-wide rates (Bromham *et al.* 1996; Bromham 2011). When Sibley and Ahlquist (1990) summarised their vast DNA hybridisation data set of the previous decade, they accepted a generation time effect in birds. Using age at first breeding as a proxy for generation time, they reached a reassuring conclusion about their divergence time estimates: 'Fortunately, many of our avian comparisons require no correction because they have been based on the assumption of a uniform average rate for birds with a one-year generation time. For species with delayed maturity and long generation times, a correction must be made' (Sibley and Ahlquist 1990: 178).

Springer and Kirsch (1989) showed that DNA hybridisation distances are rate-variable in marsupials, a point echoed by Kirsch *et al.* (1990) who – after some rather complex rate calculations – concluded that endemic New Guinean dasyurids were isolated <20 Ma. Westerman *et al.* (1990) interpreted their DNA hybridisation distances to suggest isolation of *Phascolosorex* at 8.5 Ma. Kirsch and Springer (1993) calibrated marsupial $\Delta T_{50}H$ against a fossil-based 20 Ma divergence of rat kangaroos and macropodids, as well as phalangerid clades Phalangerini and Trichosurini. Applied to two dasyurid splits, this rate suggested a *Murexia-Antechinus* divergence at 12.5 Ma and a *Murexia–Dasyurus* divergence at 21.2 Ma. All these dates are consistent with those from albumin MC′F and fossil minimum ages. Springer (1995) showed that DNA hybridisation distances for marsupials and monotremes are consistent with a molecular clock (0.4% divergence per Ma) when uncertainty in fossil calibration dates is accounted for. Kirsch *et al.* (1997) used the calibration dates in Kirsch and Springer (1993) and corrected $\Delta T_{50}H$ distances for experimental error in a study of deep divergences across Marsupialia. They estimated split dates from branch lengths on best-fit ultrametric trees and concluded that Dasyuromorphia diverged from a Diprotodontia + Notoryctemorphia clade at 65.7 Ma, in the Paleocene. This was the first molecular divergence time estimate that greatly exceeded the minimum fossil age of a dasyuromorphian clade (in this case, the *incertae sedis* genera from the Oligocene, 26–23 Ma).

Mitochondrial clocks: the 2% rule and *12S* transversions

As restriction-enzyme studies of whole mtDNA accumulated during the 1980s, researchers discovered a divergence rate of approximately 2% per Ma across many animal groups (Wilson *et al.* 1985). Despite growing awareness of the many exceptions to, and subtle complexities of, molecular clocks (Hillis and Moritz 1990), this '2% rule' was widely applied in early studies of mitochondrial *cytb* sequences (Lovette 2004). Its first use in dasyurids was by Krajewski *et al.* (1993), who found that Kimura (1980) two-parameter (K2P) distances from partial *cytb* sequences between six sister groups on opposite sides of the Torres Strait yielded divergence times in the range of 4.3–9.3 Ma. That these dates overlap those from MC′F (Aplin *et al.* 1993) and DNA hybridisation (Kirsch and Springer 1993) was taken as evidence that the 2% rule applies – at least approximately – to dasyurid *cytb*.

After Irwin *et al.*'s (1991) detailed investigation of *cytb* evolutionary dynamics, several studies of birds (Krajewski and King 1996; Randi 1996; Nunn and Stanley 1998) estimated rates of substitution at each codon position, from which overall rates of 0.6–1.5% per Ma were calculated. These lower rates could reflect long generation times in the avian groups examined and many studies of mammals continued to apply the 2% rule. Painter *et al.* (1995) found maximum *cytb* distances (K2P) within *Planigale* of 22–30%, implying a basal diversification 11–15 Ma, much older than the Pliocene *Planigale* fossil from Bluff Downs. Krajewski *et al.* (1996) drew attention to a similarly large *cytb* distance (12.79%) between *Phascogale tapoatafa* and *P. calura*, suggesting a Miocene divergence.

While analyses of mammalian *cytb* expanded rapidly during the 1990s, several studies began to explore the evolution of *12S* sequences (Springer *et al.* 1994, 1997; Springer and Douzery 1996). Springer *et al.* (1994: 108) assumed a molecular clock for *12S* transversions calibrated to the median of fossil divergence time estimates for placental orders (70 Ma) and inferred 'that the divergence of extant marsupial orders would have peaked 52.5 million years ago, although extant didelphids at least appear to have diverged from other marsupials somewhat earlier, at approximately 62 million years ago'. Springer *et al.* (1997) regressed *12S* transversion percentages against 21 therian divergence dates (in Ma) from fossils or vicariant events and obtained the rate equation $D = 0.085T + 0.058$. This gave a 63 Ma mean divergence for non-didelphimorphian marsupial orders. Within Dasyuridae, *12S* transversions suggested split times of 29.5 Ma between subfamilies ((*Phascogale*, *Murexia*)-*Planigale*) and 21.1 Ma within Phascogalini (*Phascogale-Murexia*).

Krajewski, Buckley and Westerman's (1997) analysis of *Thylacinus* phylogeny used a more sophisticated evolutionary model to estimate *cytb* distances, including empirical estimates of base composition, transition bias and relative rates of codon positions in Felsenstein's (1993) generalisation of his F84 model. The authors referred to this as the 'maximum likelihood distance with rate categories' or d_{MLC}. They obtained a mean d_{MLC} + standard deviation of 43.6% ± 7.9% from *Thylacinus* to numbats and dasyurids, which the 2% rule converted to 18–26 Ma, essentially equivalent to the age of *Badjcinus*. This study also included partial *12S* sequences; using the marsupial *12S* transversion calibration of Springer *et al.* (1997), these gave a thylacine divergence date of 18–19 Ma. The authors cautioned that this might be an underestimate, however, because the short fragment of *12S* used in the study evolves more slowly than the entire gene sequence calibrated by Springer *et al.* (1997).

Molecular clocks, when they operate, are most accurate if sequence distances (*d*-values) increase linearly with the number of substitutions per site. Multiple substitutions at the same site cause observed mismatch percentages (*p*-values) to underestimate true distance. As explained in Chapter 4, probabilistic models of DNA sequence evolution were developed to overcome this problem. However, models that are too simple to capture the actual dynamics of sequence evolution may underestimate true distance. The K2P model, for example, assumes all nucleotides occur at equal frequencies (25%), and all sites have the same transition and transversion rates, in all sequences being compared. By the mid-1990s it was obvious that mtDNA evolves by more complicated rules than those in the K2P model. Distinct 'partitions' exist within loci that have their own dynamics, such as codon positions in protein-coding genes (Krajewski *et al.* 1999) and stem versus loop regions in rRNA genes. Krajewski *et al.* (1996) used d_{MLC} to capture rate variation among partitions in the calculation of sequence distances, but this approach could not account for distinct base compositions and transition/transversion ratios among partitions. Indeed, at the time, no method was able to do so.

In a study of sminthopsine relationships, Krajewski, Blacket *et al.* (1997) introduced the *weighted-average distance* (d_{MLA}) to address the problem of dynamically heterogeneous partitions. Distances were calculated separately for each partition under the F84 model, each with its own base composition and transition/transversion ratio. Overall distances were taken as the weighted (by proportion of total sites) average of partition distances. Krajewski, Blacket *et al.* (1997) showed that sminthopsine d_{MLC} values regressed with a slope of 0.9 on d_{MLA}, confirming that the former underestimate the latter. A series of studies (Krajewski, Buckley and Westerman 1997; Krajewski, Young *et al.* 1997; Blacket *et al.* 1999) used the 2% rule for *cytb* and the Springer *et al.* (1997) calibration for *12S* with d_{MLA} to obtain the dasyurid divergence dates in Table 9.1. As shown graphically in Fig. 9.2, these early estimates for miscellaneous split dates did not produce a coherent timeline for dasyurid evolution.

Mitochondrial clocks for dasyuromorphians

Dasyurid divergence time estimates prior to 2000 relied on the 2% rule for *cytb* and a *12S* rate calibration derived from all mammals. Part of the motivation for such gratuitous or broad assumptions was investigators' lack of confidence in clade-specific applications of the traditional relative rate test. During the 1990s, however, software became available that implemented a likelihood ratio test (LRT) of the molecular clock (Felsenstein 1981). Consider two unrooted trees, T_0 and T_1, each with *s* tips (i.e. sequences) and the same branching order but estimated with and without the assumption of a clock, respectively. For T_0, all pathways from an internal node to descendant tips must have the same length, which means that only the $s - 1$ split dates are free to vary. For T_1, however, each of the $2s - 3$ branches may have a different length. Thus, the clock model (T_0) is a special case of (nested within) the no-clock model (T_1). In this situation, the fit of the sequence data under a specific evolutionary model to T_0 and T_1 can be evaluated with an LRT. If L_0 and L_1 are the likelihoods of T_0 and T_1, the test statistic $2\Delta L = 2(L_1 - L_0)$ has a χ^2 distribution with $(2s - 3) - (s - 1) = s - 2$ degrees of freedom. If $2\Delta L$ is large enough to fall in the 5% tail of that distribution, the clock hypothesis is rejected. This provides a more robust assessment of the clock model than the relative rate test (Yang 2014).

Applications of the clock LRT soon led to the realisation that few sequence data sets were free of rate variation. Nei and Kumar (2000: 203–206) described a method proposed by Takezaki *et al.* (1995) to identify the largest subset of rate-uniform sequences in an alignment, then use the molecular clock to estimate divergence times on this 'linearized tree'. The method involves four steps: (1) obtain a reliable estimate tree topology without the clock assumption; (2) test the clock model for this tree (e.g. using an LRT) and remove sequences that deviate most from the average rate of change; (3) re-estimate the tree and branch lengths with a clock model; and (4) calibrate sequence divergence and estimate split times on this tree.

Krajewski, Wroe and Westerman (2000) combined weighted-average distances for complete *cytb* and *12S* sequences, the clock LRT and a linearised tree approach to place divergence times on a phylogeny of dasyurids that was nearly complete at the species level. Because of software limitations, LRTs assumed the F84 model with rate categories. Linearised trees were obtained by excluding

Table 9.1. Dasyurid divergence dates estimated from weighted-average distances of *cytb* (assuming 2% per Ma) and *12S* (assuming the transversion calibration of Springer *et al.* 1997).
Data from Krajewski, Buckley and Westerman (1997); Krajewski, Young *et al.* (1997); and Blacket *et al.* (1999). 'Macroura' and 'Murina' refer to species groups within Sminthopsis (see Chapter 6).

Split	*Cytb* date estimate (Ma)	*12S* date estimate (Ma)
Phascogale-Sminthopsinae	22–28	38–50
Sminthopsinae		
Planigale-(*Sminthopsis, Ningaui*)	17–22	28–48
Planigale-Sminthopsini	19.4	32.7
(Macroura, *S. crassicaudata*)-Murina	16.5	13.3
S. crassicaudata-Macroura	15.9	13.9
Antechinomys-(*Sminthopsis, Ningaui*)	15.5	11.6
Antechinomys-Sminthopsis	15.6	10.1
Ningaui-Sminthopsis	15–18	17–24
N. timealeyi-(*N. ridei, N. yvonneae*)	13–16	6–28
Within Macroura	11.6	9.2
N. ridei-N. yvonneae	11	10
P. maculata-other planigales	10–14	5–22
Within Murina	7.5	7.7
Dasyurini		
(*Phascogale, Murexia*)-Dasyurini	14	14
Phascolosoricines-(*Dasyurus, Sarcophilus*)	12	–
Dasyurus-Sarcophilus	11	–
P. macdonnellensis-P. woolleyae	10	–
Neophascogale-Phascolosorex	8	–
M. melas-M. wallacei	6	–
D. geoffroii-D. spartacus	4	–

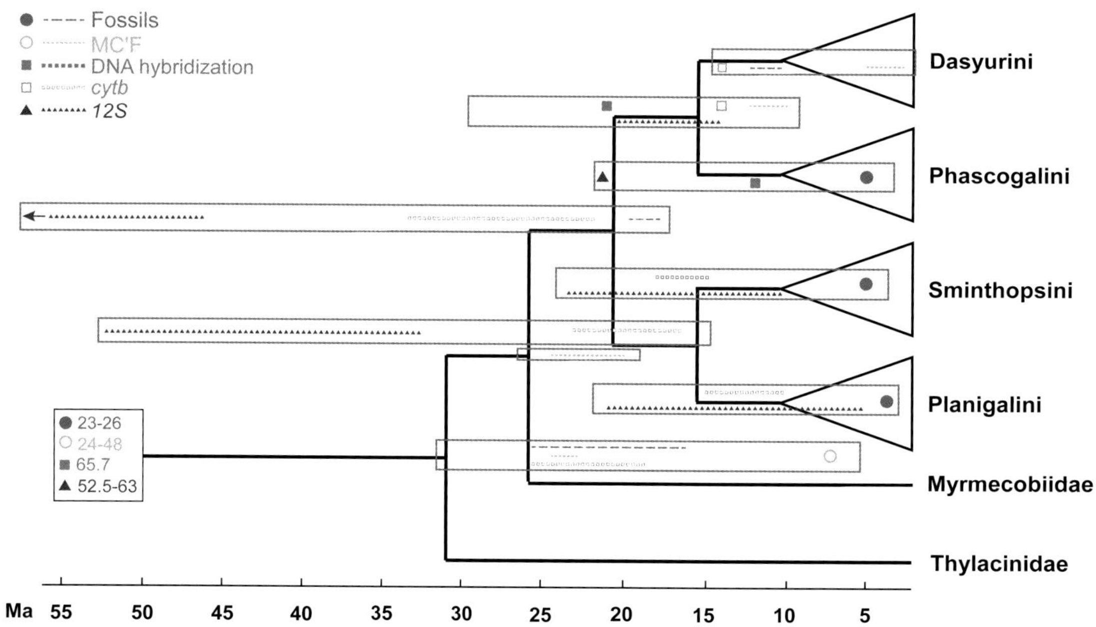

Fig. 9.2. Estimates of divergence times made prior to 2000. Each box encloses a phylogenetic split and its estimated dates (dots and lines). Fossil dates are earliest clade occurrences. Dates at the root are estimates for the origin of Dasyuromorphia (including stem lineages). Apparent dates of branch points on the tree itself have no significance. See text for description of sources.

sequences showing the largest deviations from root-to-tip pathlengths until a rate-uniform subset remained. Krajewski, Wroe and Westerman (2000) used a single calibration point – the ca. 25 Ma age of *Badjcinus* for the thylacinid-dasyurid divergence – to estimate *cytb* and *12S* rates from d_{MLA} values between *Thylacinus* and members of each dasyurid tribe. Rates were 1.6–2.1%/Ma for *cytb* and 0.56–0.59%/Ma for *12S*. The average of *cytb* and *12S* dates was used to estimate divergence times on a maximum-parsimony tree obtained from *cytb*, *12S* and *P1* sequences (see Fig. 4.14) and the result presented as a chronogram (Fig. 9.3).

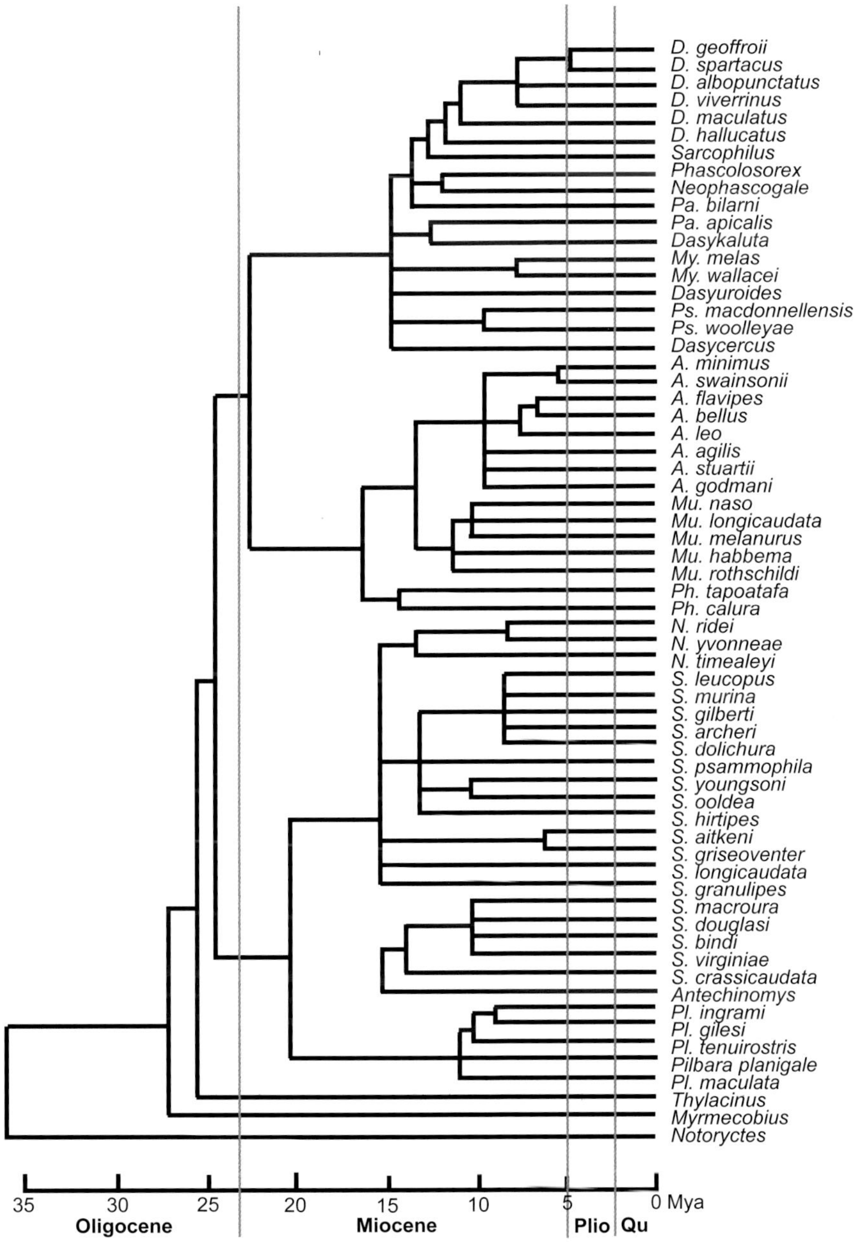

Fig. 9.3. Time tree of dasyurid evolution from Krajewski, Wroe and Westerman (2000). Tree topology estimated with maximum parsimony analysis of *cytb*, *12S* and *P1* sequences. Divergence times based on weighted-average distances (d_{MLA}) for *cytb* and *12S* calibrated to a 25 Ma divergence of dasyurids from *Thylacinus*. Divergence times rounded to the nearest 1 Ma, causing some branch points to collapse into polytomies (compare to Fig. 4.14). [Reprinted by permission of Oxford University Press from Krajewski, Wroe and Westerman (2000).]

Krajewski, Wroe and Westerman's (2000) time tree was the most comprehensive analysis of dasyurid cladogenesis that had been published at the time. Taking the marsupial mole *Notoryctes* (Notoryctemorphia) as a sister group, *cytb-12S* distances suggested a late Eocene origin and Oligocene diversification of Dasyuromorphia. Within Dasyuridae, the separation of Dasyurinae and Sminthopsinae dated to the late Oligocene some 24.3 Ma. Basal splits within these took place in the early Miocene: Dasyurini-Phascogalini at 21.6 Ma and Sminthopsini-Planigalini at 19.6 Ma. Radiations within the first three tribes began 13.8–16.0 Ma, with planigales slightly later at 10.7 Ma. The New Guinean *Murexia* was isolated from its apparent Australian sister (*Antechinus*) some 13 Ma and diversified 10–11 Ma. The only post-Miocene divergence for this set of species was *Dasyurus geoffroii-D. spartacus* at 4.2 Ma in the Pliocene.

Subsequent genus- and species-level studies produced divergence time estimates consistent with those in Fig. 9.3. Blacket *et al.* (2000) found that *12S* d_{MLA} values within *Planigale* indicated a late Miocene radiation, though some splits within the *P. maculata-novaeguineae* clade occurred in the Pleistocene. Blacket *et al.* (2001) took the same approach to infer that the most recent divergences among *Sminthopsis macroura* haplogroups were Pliocene, but those between *S. virginiae* subspecies were Pleistocene. In a parallel study, Blacket *et al.* (2006) found basal divergences within the *S. murina* species group were late Miocene to early Pliocene, with some subspecies separations (in *S. murina* and *S. leucopus*) extending into the early Pleistocene. In their study of phascolosoricine relationships Krajewski *et al.* (2004) repeated the dating analysis of Krajewski, Wroe and Westerman (2000) but used branch lengths estimated with maximum-likelihood and additive distance methods from combined *cytb* and *12S* d_{MLA} values to estimate split times. To obtain linearised trees, this study also supplemented the clock LRT with an *F*-ratio test of sum-of-squares residuals on additive distance trees with

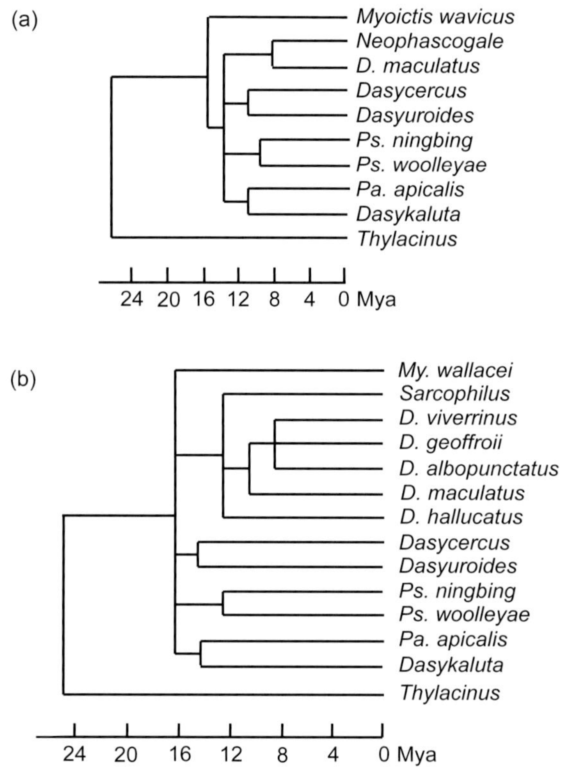

Fig. 9.4. Time trees of dasyurine species estimated from combined *cytb* and *12S* sequences (Krajewski *et al.* 2004). (a) Linearised tree obtained with maximum-likelihood analysis under the F84 model with rate categories for codon-position and stem-loop partitions. (b) Linearised tree obtained with analysis of weighted-average distances under the F84 model with codon-position and stem-loop partitions. [Redrawn by permission of CSIRO Publishing from Krajewski *et al.* (2004), with minor adaptations. Permission conveyed through Copyright Clearance Center, Inc.]

Table 9.2. *Myoictis* divergence dates estimated from weighted-average distances of *cytb* + *12S* calibrated at 1.26% per Ma (Westerman *et al.* 2006).

Comparison	Mean date estimate (range)* in Ma
M. leucura – *M. melas*	8.10 (7.24–8.59)
M. leucura – *M. wallacei*	7.68 (6.42–8.17)
M. leucura – *M. wavicus*	4.06 (3.56–4.38)
M. melas Group 5 – *M. melas* Group 12	3.35
M. melas – *M. wallacei*	8.65 (8.44–8.78)
M. melas – *M. wavicus*	7.80 (7.25–8.47)
M. wavicus – *M. wallacei*	7.46 (6.80–7.94)
Myoictis – other dasyurins	14.59 (10.87–16.38)

* Range over all pairwise comparisons of sequences for each taxon.

and without a clock assumption. The composite *cytb-12S* divergence rate (1.26% per Ma) applied to linearised trees (Fig. 9.4) suggested that the dasyurin radiation began 15–16 Ma and that phascolosoricines separated from *Dasyurus* + *Sarcophilus* 8–13 Ma. Westerman *et al.* (2006) used the 1.26% per Ma *cytb-12S* rate to infer that *Myoictis* was isolated from other dasyurins some 14.6 Ma with splits among *Myoictis* species spanning 8.6–3.3 Ma (Table 9.2).

RELAXED CLOCKS AND STATISTICAL CALIBRATIONS

Given the many factors that affect substitution rates in evolving DNA sequences, strict molecular clocks should be relatively rare, especially in large and ancient clades – and this is what empirical research revealed during the 1990s and 2000s, as exemplified by Krajewski, Wroe and Westerman (2000) and subsequent studies of dasyurids described above. Does this mean that estimates of divergence time must be qualified by the assumption of a global clock? Not necessarily. DNA sequences, analysed in a phylogenetic context, contain information about substitution rate variation among the lineages from which they are sampled. Theoreticians found ways to extract this information and use it to obtain divergence time estimates that do not rely on a strict clock. Approaches using likelihood and Bayesian inference have been developed, though the latter have been applied much more extensively to dasyuromorphians. The fundamental challenge in relaxed-clock dating is that a branch length on a DNA phylogeny is the product of substitution rate and elapsed time; this product can be estimated from the sequences, but its individual factors cannot (Yang 2014). A strict clock ensures that one rate applies to all branches and allows divergence time to be estimated as described above. If rates vary from branch to branch, more assumptions and computations are required to estimate time.

Rate-smoothing and penalised likelihood

Operating on the assumption that the rate of evolution itself evolves from ancestral to descendant branches on a phylogeny, Sanderson (1997) used likelihood to estimate rates and times from sequence data on a tree with branch lengths. If **T** and **R** are the sets of time durations and branch rates, respectively, on a tree with branch lengths (b_k) estimated from sequence data set $\mathbf{X}^1$, Sanderson found rate values for each branch (r_k) that minimise the function W:

$$W(\mathbf{T},\ \mathbf{R}) = \Sigma_k \left(r_k - r_{anc(k)}\right)^2 \quad \text{given } r_k t_k = b_k$$

The variable $r_{anc(k)}$ is the rate on the branch ancestral to k. Minimising W is equivalent to finding values of r_k and t_k that minimise the rate difference between ancestral and descendant branches, hence the characterisation of this as a 'rate-smoothing' method. Adding at least one calibration allows estimation of absolute divergence times. Sanderson (2002) modified the optimality criterion to maximise the following log-likelihood function:

$$L(\mathbf{T},\mathbf{R},\lambda \mid \mathbf{X}) = \log[\Pr(\mathbf{X} \mid \mathbf{T},\mathbf{R})] - \lambda\Sigma_k \left(r_k - r_{anc(k)}\right)^2$$

The first term on the right side is the likelihood of the estimate, the summation term is W (from the previous equation) and λ is a rate-smoothing parameter. If λ is small, rates may vary a lot; if λ is large, excessive rate variation severely 'penalises' the likelihood and is not tolerated. Sanderson (2002) subsampled the data to find a value of λ that minimised the coefficient of variation of estimated numbers of substitutions among branches. However, Yang (2014: 368) felt that '[t]here is no rigorous criterion for choosing λ'. The penalised likelihood method was implemented by Sanderson's (1997) R8S program.

Bayesian relaxed clock models

Thorne *et al.* (1998) and Kishino *et al.* (2001) also approached this problem from the perspective of a correlated rates model, based on the idea that substitution rates are likely to be more similar among closely related species – or between an ancestral lineage and its immediate descendants – than among distantly related ones. The Bayesian framework implemented by an MCMC algorithm (see Chapter 5, 'Bayesian inference of phylogeny') allows divergence times to be estimated with this assumption using prior distributions. Working through Thorne *et al.*'s (1998) relatively simple formulation provides insight into the essential aspects of Bayesian relaxed-clock dating. A modest amount of mathematical abstraction now will pay dividends by creating a conceptual framework for understanding the more complex methods described later.

Let **X** be a set of sequences related by a tree as shown in Fig. 9.5. For simplicity in allowing us to focus on estimating dates, suppose this tree is known and its branch lengths are maximum likelihood estimates (MLEs) based on **X**. The sequences evolve according to some model (e.g. F84) with its associated parameters (e.g. rate, transition

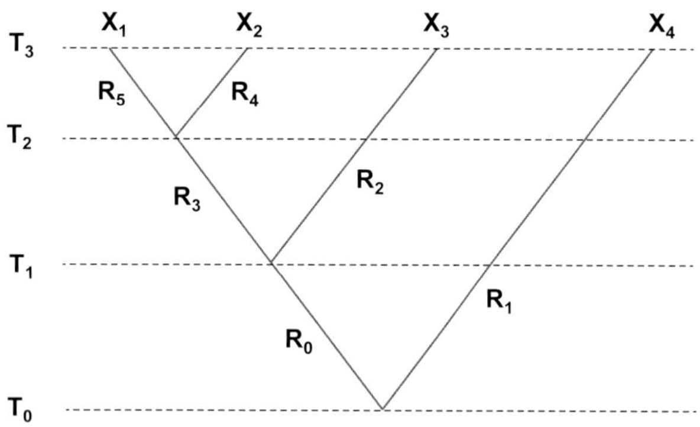

Fig. 9.5. Sample phylogeny for Bayesian relaxed-clock dating from Thorne *et al.* (1998). X_{1-4} are sequences, R_{1-5} are substitution rates and T_{0-3} are divergence times. See text for discussion. [Redrawn with permission of Oxford University Press from Thorne *et al.* (1998).]

bias, base frequencies), though again for simplicity we will not include the model parameters in what follows. **T** is the set of divergence times and **R** the set of rates associated with each branch; the values of R can vary among branches. The model includes the following assumptions in the form of prior distributions:

1. $\log(R_d) \sim N[\log(R_a), v\Delta T]$, where R_d is the rate of a descendant branch, R_a is the rate of its ancestor, v is a constant that scales inversely with the correlation between R_a and R_d and ΔT is the time difference between the midpoints of R_a and R_d. In words, we assume that the logarithm of the rate on any branch is drawn from a normal distribution whose mean is the logarithm of the rate of its ancestor and whose variance is $v\Delta T$. A small v means that there is strong correlation between R_a and R_d (evolution is clock-like) and a large v means the correlation is weak (much rate variation). The expected rate on any branch is the rate of its ancestor but its variance increases with time and v. Kishino *et al.* (2001) removed the effect of branch duration on rate correlation by assigning rates to nodes rather than branches.
2. v is drawn from an exponential or gamma distribution with user-specified parameter values.
3. R_0, the rate on one of the branches from the root, has an exponential or gamma distribution with specified parameter values and is independent of v and T. All other rates are conditional on R_0.
4. $\log(R_1) \sim N\{\log(R_0), v[(\Delta T_0 + \Delta T_1)/2]\}$. R_1 is the rate on the other branch from the root; $\log(R_1)$ is drawn from a normal distribution with mean $\log(R_0)$ and variance equal to v times the average duration of the two root branches.
5. The prior on split dates in **T** is given by a Yule 'pure birth' process. This is simply a continuous stochastic branching process that generates splits along branches at 'random' time points given an expected root depth (T_0). Thorne *et al.* (1998) used ML branch lengths to obtain an initial relative value of T_0. Kishino *et al.* (2001) assumed a Dirichlet distribution to assign prior split times recursively along each path from root to tip.

Assumptions 1–5 allow us to write the expression for the posterior probability of a particular set of split times, rates and v, given the sequence data:

$$\begin{aligned}\Pr(\mathbf{T},\mathbf{B},v|\mathbf{X}) &= \left[\Pr(\mathbf{X}|\mathbf{T},\mathbf{R})\times Pr(\mathbf{R}|\mathbf{T},v)\times\Pr(\mathbf{T})\times\Pr(v)\right]/\Pr(\mathbf{X})\\ &= \left[\Pr(\mathbf{X}|\mathbf{B})\times\Pr(\mathbf{R}|\mathbf{T},v)\times\Pr(\mathbf{T})\times\Pr(v)\right]/Pr(\mathbf{X})\end{aligned}$$

Note that $\Pr(\mathbf{X}|\mathbf{T},\mathbf{R}) = \Pr(\mathbf{X}|\mathbf{B})$ where **B** is the set of branch lengths, because branch length = rate × time. Expressed in words, we have:

$$\begin{aligned}&\text{Posterior probability }(\text{times, rates, } v \text{ given the sequences}) =\\ &\left[\text{Likelihood }(\text{sequences})\times\text{Prior }(\text{rates})\times\text{Prior }(\text{times})\times\text{Prior }(v)\right]/\text{Prior}(\text{sequences})\end{aligned}$$

The model is now specified and what remains is to find the combination of **T**, **R** and v that maximises the posterior probability, informed by constraints on the values in **T** (i.e. calibrations). This will yield estimates of the split times even though rates vary among branches.

As described in Chapter 5, calculating the prior probability of the data (the denominator in the last two expressions above) is computationally intractable, but the posterior distribution can be found using an MCMC algorithm. Analogous to the tree-search problem, the MCMC algorithm proceeds as follows:

1. Draw initial values in **R**, **T** and v from their priors as specified above; calculate the likelihood L_0.
2. Propose a new set of parameter values and calculate a new likelihood L_1. If $L_1 > L_0$, accept the new parameter values. If $L_1 < L_0$, accept the new parameter values with probability L_1/L_0 (or other reasonable *proposal ratio*). The latter step ensures that the algorithm does not 'get stuck' on a local maximum of the posterior distribution.
3. Repeat step 2 many times (e.g. millions). Each iteration is an *MCMC generation*.
4. Discard results from some number (e.g. 10%) of initial generations ('burn in') before the search has converged on the posterior distribution. At some regular interval (e.g. every 1000 generations), save the parameter values.
5. Summarise the posterior distribution by calculating the frequency of each combination of parameter values in the saved generations. The Bayesian point estimate is the combination that occurs most frequently; the posterior probability of that combination is the frequency at which it occurs among the MCMC samples; and a 'credibility interval' or 'highest probability density' (HPD) is the range of parameter values comprising some fraction (e.g. 95%) of the probability density around the point estimate.

Implementing this search involves computation strategies and approximations that we do not discuss. However, the robustness of results to the prior distributions chosen is an important question in any analysis.

Kishino *et al.* (2001) incorporated age constraints into this model by modifying the posterior distribution to reflect 'hard bounds' imposed by fossils or other evidence on divergence times, along with the requirement that descendant nodes must be younger than their ancestors. These constraints are enforced by disallowing in the MCMC routine any proposed set of divergence times that violates them. Use of hard bounds alone makes it difficult to incorporate age uncertainties (e.g. of fossils) into the dating analysis, a challenge that would be addressed in later studies (see below). Given age constraints **C**, the posterior probability takes the following form:

$$\Pr(\mathbf{T},\mathbf{B},v|\mathbf{X},\mathbf{C})=\left[\Pr(\mathbf{X}|\mathbf{B})\times\Pr(\mathbf{R}|\mathbf{T},v)\times\Pr(\mathbf{T}|\mathbf{C})\times\Pr(v)\right]/\Pr(\mathbf{X})$$

Thorne and Kishino (2002) extended this model to the case of multiple loci with the same divergence times, comparing situations in which each locus has a different level of rate correlation over the tree (independent values of v) to that in which all loci have the same correlation. The Kishino *et al.* (2001) approach was implemented in the software package MULTIDIVTIME.

Incorporating calibration uncertainty

As described in 'A literal reading of the fossil record', fossil ages are seldom known with certainty, so their use as constraints on molecular divergence times should account for this source of error. Thorne *et al.* (1998) and Kishino *et al.* (2001) used maximum and minimum age estimates for fossil ages as upper and lower bounds, respectively, on split dates. These were 'hard' in the sense that node dates outside them had 0 prior probability and dates between them had equal probability (i.e. a uniform prior). Yang and Rannala (2006) developed a strict-clock Bayesian method to incorporate 'soft' bounds – i.e. age constraints with more flexible priors to reflect uncertainties associated with fossil dates. To accomplish this, they used a 'birth–death' (BD) process to generate priors on divergence times for calibrated and uncalibrated nodes (this takes the place of the Yule and Dirichlet priors described above). The BD process includes parameters for 'birth rate' (speciation), 'death rate' (extinction) and the fraction of all clade lineages sampled in the analysis. In this model, calibration dates and uncertainties are independent of the time prior, allowing investigators to choose any reasonable probability density for calibrated node ages. Yang and Rannala (2006) used those shown in Fig. 9.6, but others are possible. These authors also noted the importance of having at least one upper and one lower bound somewhere on the tree. Rannala and Yang (2007) extended the approach to include relaxed-clock calibrations for multiple loci with correlated (as in Thorne *et al.* 1998) or uncorrelated rates, including a separate rate parameter for each locus. This method was implemented by the program MCMCTREE in Yang's PAML package. Drummond *et al.* (2006) developed a similar approach

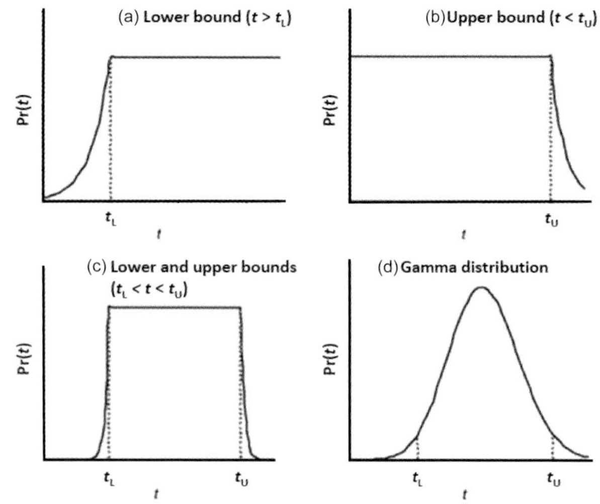

Fig. 9.6. Probability distributions for node ages to reflect uncertainty in fossil ages used by Yang and Rannala (2006). Symbols: t = time; t_L = lower bound; t_U = upper bound; Pr(t) = probability of t. In all graphs, the tail probabilities are 2.5%. (a) Left tail below t_L showing power decay and uniform density above t_L. (b) Right tail above t_U showing exponential decay and uniform density below t_U. (c) Distribution combining (a) and (b). (d) A gamma density with mode fixed to the most likely age of the fossil. [Adapted from Yang (2014).]

(implemented by BEAST software) in which phylogeny and divergence times are co-estimated from sequences and calibration priors. This study found little evidence for correlated rates in several data sets (including marsupial nuclear protein-coding genes).

Rannala and Yang (2007: 458) concluded from simulation studies 'that typical sequence data sets appear to be nearly as informative as infinitely long sequences and most of the uncertainties in the posterior time estimates are therefore due to uncertainties in the fossils'. This places a high value on including as many accurate calibration points as possible in Bayesian molecular dating. Benton and Donoghue (2007) argued that fossil ages are usually able to place hard lower bounds (minimum ages) and sometimes soft upper upper bounds (maximum ages) on clade divergences. They called for a new rigor in establishing these bounds:

> *[M]inimum age constraints should be justified on the basis of a phylogenetic hypothesis, with reference to the oldest integral member of the clade – on which the date is ultimately based, the justification for the membership of the clade, the means by which a correlation is achieved to a section in which a chronostratigraphic date may be obtained and the source of the chronostratigraphic date. Maximum age constraints should likewise be justified on the basis of a phylogenetic hypothesis, with reference to fossils belonging to outgroups, and to the next oldest fossil horizon that lacks relevant fossils. [Benton and Donoghue 2007: 29]*

Priors on soft upper bounds may be arbitrary (as for the methods described in the previous section) or based on information about patterns of cladogenesis, taphonomy, or other determinants of the probability of fossil occurrences. Benton and Donoghue (2007) used these principles to obtain 23 maximum and minimum date estimates for vertebrate divergences. For example, they placed a lower bound for the 'Ameridelphia'-Australidelphia split at 61.7 Ma based on the stratigraphic age of the presumed oldest fossil of either group (the ameridelphian *Cocatherium*) and a soft upper bound of 71.2 Ma based on the absence of fossils from either clade in Late Cretaceous strata.

RELAXED CLOCK ESTIMATES OF DASYUROMORPHIAN DIVERGENCE TIMES

Nilsson *et al.* (2004) applied Sanderson's (2002) rate-smoothing and penalised likelihood methods to date splits on a tree of 19 marsupial taxa (representing all living orders) derived from complete mtDNA sequences and rooted with placental outgroups. They used three fixed calibration points, two among placentals and a presumed 135 Ma divergence of metatherians and eutherians. Alternative values of the rate-smoothing parameter did not strongly affect estimated dates. The results (Fig. 9.7) suggested separation of Dasyuromorphia from a Notoryctemorphia sister some 39 Ma, with a split between Dasyurinae and Sminthopsinae some 20 Ma. These dates are comparable to the ones (36.1 and 24.3 Ma, respectively) obtained by Krajewski, Wroe and Westerman (2000).

Drummond *et al.* (2006) used BEAST to co-estimate the phylogeny and divergence times of marsupials from sequences of five nuclear protein-coding loci (Amrine-Madsen *et al.* 2003) in an unpartitioned GTR + Γ model with uncorrelated log-normal (UCLN) distributions of rates. Four calibration points were used: 61.5 Ma for elephants-sirenians, 64 Ma for dasyurids-diprotodontids, 17 Ma for *Phascogale-Dasyurus* (each of these with a normal prior) and 145 Ma for marsupials-placentals (translated lognormal prior). Results (Fig. 9.8) indicated

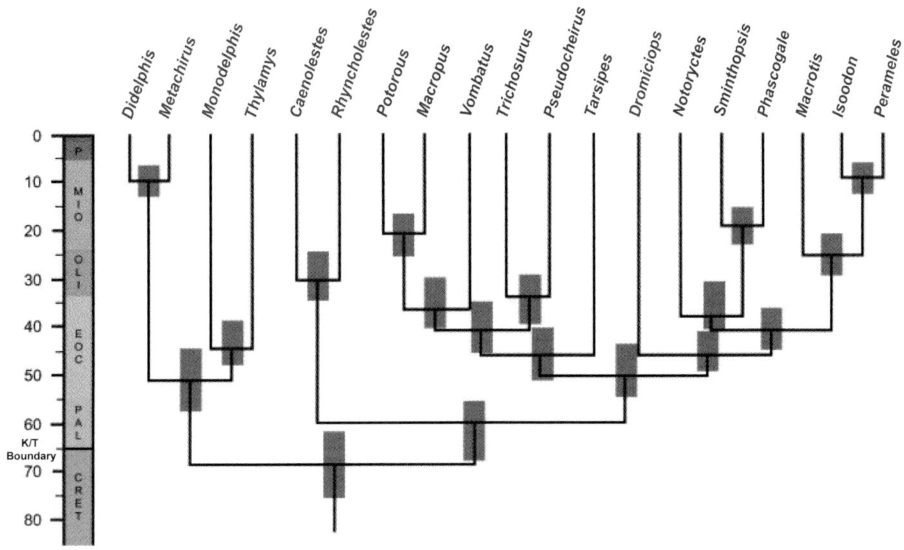

Fig. 9.7. Time tree of marsupials based on complete mtDNA sequences (Nilsson *et al.* 2004). Phylogeny was obtained with maximum likelihood analysis under a variety of models. Dates estimated with rate-smoothing and penalised-likelihood methods implemented in R8S. Analysis included a range of values for the rate-smoothing parameter. Grey shaded boxes represent confidence intervals. [Reprinted by permission of Elsevier from Nilsson *et al.* (2004).]

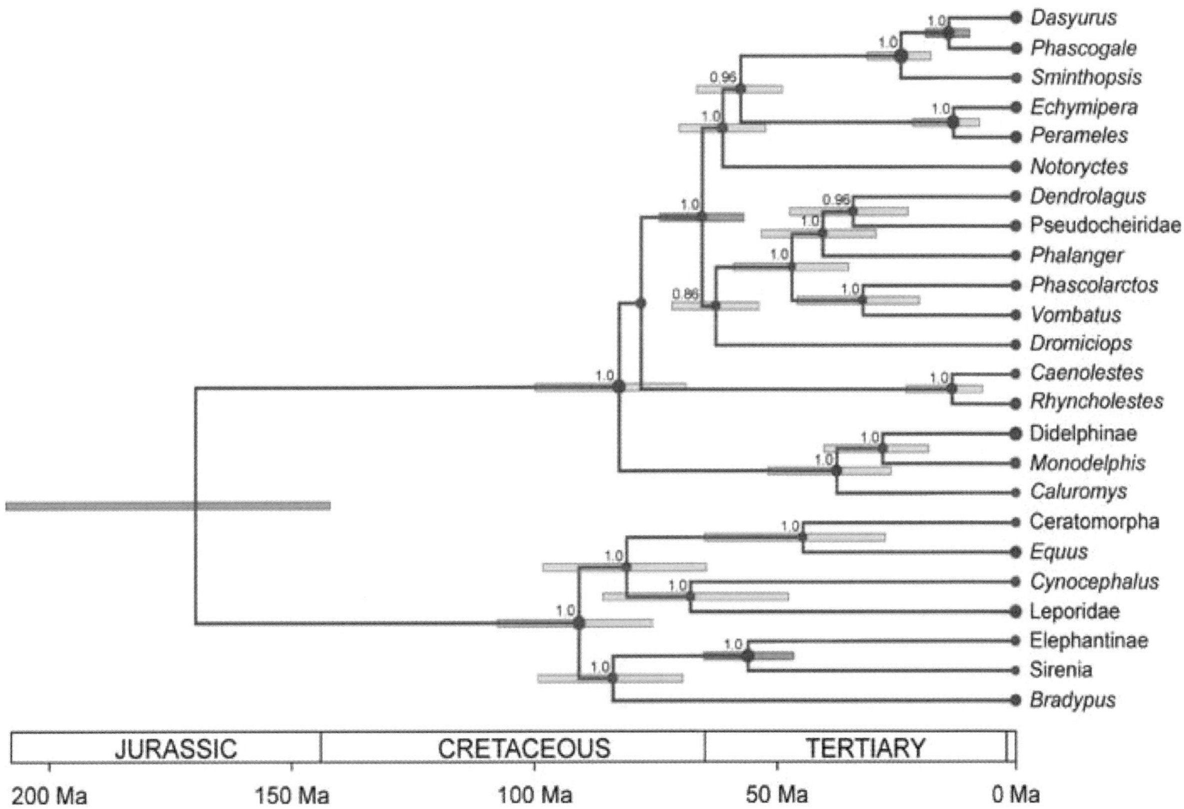

Fig. 9.8. Marsupial time tree based on DNA sequences of *APOB*, *RAG1*, *IRBP*, *vWF* and *BRCA1* (Drummond *et al.* 2006). Topology and dates estimated with BEAST 1.4 assuming an unpartitioned GTR + Γ model and uncorrelated log-normal rate distributions among branches. Nodal values are posterior probabilities >0.5. Yellow bars are 95% highest probability densities for divergence dates. Orange bars show nodes with normal calibration priors. Red branches have above-average, and blue branches below-average, rates. [Reprinted from Drummond *et al.* (2006) under CC BY license.]

separation of Dasyuromorphia from a Peramelemorphia sister at ca. 60 Ma and the Dasyurinae-Sminthopsinae split at 25 Ma.

In another analysis of the Amrine-Madsen *et al.* (2003) nuclear sequences, Meredith *et al.* (2008) used both MULTIDIVTIME and BEAST approaches to date the divergences of marsupial orders. This study was remarkable for using the principles of Benton and Donoghue (2007) to identify 26 calibration points with upper and lower bounds from the marsupial fossil record. Among these were three involving dasyurids: 4.46–33.9 Ma for Dasyurinae-Sminthopsinae (based on *Antechinus* sp. from the Hamilton LF and the base of the Oligocene); 4.46–23.03 for both Dasyurini-Phascogalini and *Phascogale-Antechinus* (based on the Hamilton *Antechinus* and the base of the Miocene). The root prior had a mean of 75 Ma and the rate prior a mean given by the median root-to-tip pathlength (substitutions/site) divided by 75 Ma. The authors also explored the effects of two calibrations (25.5–54.6 Ma and 25.5–65 Ma) on the deepest split within Diprotodontia, treating the calibration ranges as hard or soft bounds (the latter using normal priors with 2.5% tails extending beyond the ranges) and single-rate versus gene-specific rate models of sequence evolution. Rates were correlated in MULTIDIVTIME, but uncorrelated in BEAST, analyses. The results (Fig. 9.9 and Table 9.3) show Dasyuromorphia diverging from a Notoryctemorphia sister 56.2–59.9 Ma, Dasyurinae-Sminthopsinae at 14.4–22.9 Ma, Dasyurini-Phascogalini at 9.7–10.2 Ma and *Phascogale-Antechinus* at 5.5–7.7 Ma. The first date concurs with Drummond *et al.* (2006) in suggesting a Paleocene origin of the order; the latter three are younger than those in Krajewski, Wroe and Westerman (2000), but still in the Miocene.

Bininda-Emonds *et al.* (2007) constructed a supertree of mammals with divergence times estimated from

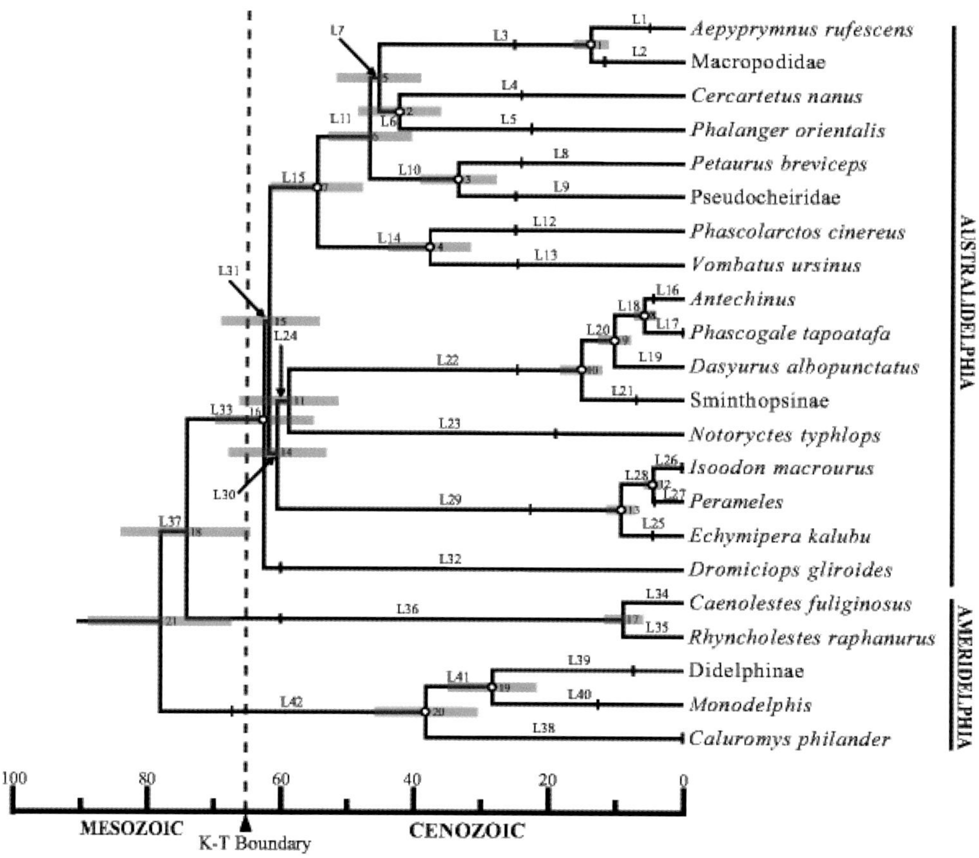

Fig. 9.9. Marsupial time tree based on DNA sequences of *APOB, RAG1, IRBP, vWF* and *BRCA1* (Meredith *et al.* 2008). Topology derived from Bayesian analysis of sequences partitioned by gene. Time estimates from MULTIDIVTIME analysis with gene-specific rates, hard bounds and a 65 Ma maximum for the origin of Diprotodontia. See text for other analytical details. Branch and node numbers are arbitrary; horizontal grey bars are confidence intervals; vertical bars indicate the oldest known fossil of each lineage; open circles indicate calibrated nodes. L = lineage; K-T = Cretaceous-Tertiary. [Reprinted by permission of Springer Nature from Meredith *et al.* (2008).]

fossils, 'local' (clade-specific) molecular clocks and interpolation that dated the origin of Dasyuromorphia to 58.5 ± 8 Ma and the separation of dasyurids from numbat at 32.8 ± 20.0 Ma. Meredith *et al.* (2011) obtained similar dates for these splits (59.6 and 30.0, respectively) in MCMCTREE analyses of mammalian families with extensive and well-documented fossil calibrations. Beck (2008) used MULTIDIVTIME to date marsupial nodes based on a Bayesian tree derived from seven nuclear and 15 mitochondrial genes (the latter with third codon positions omitted), along with 23 lower- and three upper-bounded nodes provided by fossil information. These included minima at 23.03 Ma (the age of *Badjcinus*) for the origin of Dasyuromorphia and 4.45 Ma (the Hamilton LF *Antechinus* and possible *Dasyurus*) for the Dasyurini-Phascogalini split. The results (Fig. 9.10) show a Paleocene

Table 9.3. Relaxed clock estimates of selected dasyuromorphian splits from Meredith *et al.* (2008). Dates obtained with calibration ranges treated as hard bounds.

	Divergence estimates (and confidence intervals) in Ma			
	MULTIDIVTIME		BEAST	
Diprotodontia maximum	**54.6 Ma**	**65 Ma**	**54.6 Ma**	**65 Ma**
Phascogale-Antechinus	5.5 (4.5–7.0)	5.7 (4.6–7.5)	7.8 (4.5–11.3)	7.7 (4.5–11.4)
Dasyurini-Phascogalini	9.7 (7.9–12.0)	10.2 (8.1–12.8)	10.0 (6.4–14.5)	9.9 (5.8–14.4)
Dasyurinae-Sminthopsinae	14.4 (12.0–17.2)	15.0 (12.8–18.3)	22.6 (16.6–29.3)	22.9 (16.6–29.7)
Dasyuromorphia-Notoryctemorphia	56.2 (51.2–61.3)	59.1 (51.7–66.4)	59.1 (52.0–67.1)	59.9 (52.6–67.9)

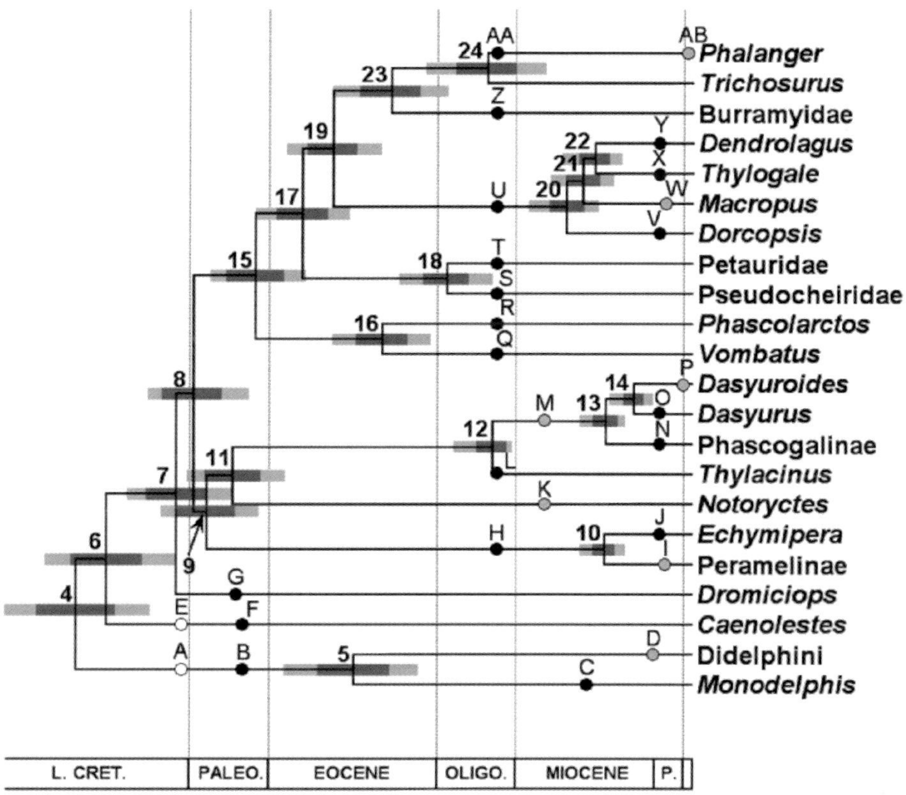

Fig. 9.10. Marsupial time tree based on 15 mitochondrial and seven nuclear genes (*ApoB, BRCA1, IRBP, PGK1, P1, RAG1* and *vWF*) (Beck 2008). Tree topology obtained via Bayesian analysis of partitioned sequences with third positions of mtDNA protein-coding genes omitted. Dates obtained with MULTIDIVTIME using 23 minimum and three maximum age constraints (indicated by black dots with letters) and mean of the root prior set to 162.5 Ma (monotreme-therian split, not shown). Grey and white circles show ages of other fossils assigned confidently or tentatively, respectively, to lineages. Black bars at nodes show ±1 SD of the estimate date; grey bars show 95% confidence intervals. [Reprinted by permission of Oxford University Press from Beck (2008).]

(60.1 Ma, 95% CI = 53.2–66.1 Ma) origin of dasyuromorphians, the thylacinid-dasyurid split in the Oligocene (26.0 Ma, 95% CI = 23.2–31.4 Ma) and the dasyurin-phascogalin split in the Miocene (11.6 Ma, 95% CI = 8.8–15.2 Ma).

Five subsequent large-alignment studies described in Chapter 5 dated many more dasyurid lineages. Mitchell *et al.* (2014) used MCMCTREE with 14 calibration points (priors with hard minima and soft maxima) and rate variation among loci. Their results (see Fig. 5.2) showed an origin of Dasyuromorphia near the Cretaceous-Paleocene boundary (ca. 66 Ma), separation of Thylacinidae in the mid-Eocene (ca. 40 Ma), the numbat-dasyurid split in the late Eocene (ca. 35 Ma), Dasyurinae-Sminthopsinae in the late Oligocene (ca. 25 Ma), tribal splits within each subfamily in the early Miocene (ca. 18 Ma) and tribal radiations beginning in the mid-Miocene (ca. 13–14 Ma). These dates are marginally older than those estimated in previous studies.

Westerman, Krajewski *et al.* (2016) applied the UCLN rates model in BEAST to date dasyuromorphian splits with eight nuclear and seven mitochondrial genes. Uniform priors on soft upper and lower bounds (with 2.5% probability tails) were set for 12 calibration points, including a minimum 24.7 Ma (based on *Badjcinus*) age of Dasyuromorphia. The results are shown in Fig. 9.11 and Table 9.4. Dasyuromorphia originated in the mid-Eocene (ca. 49 Ma) and Thylacinidae in the later Eocene (ca. 40 Ma), with the numbat-dasyurid split near the Eocene-Oligocene boundary (36.5 Ma). Subfamily and tribal separations date to the Oligocene (29.3 and 24.1–24.5 Ma, respectively). Tribal radiations began in the Miocene, the

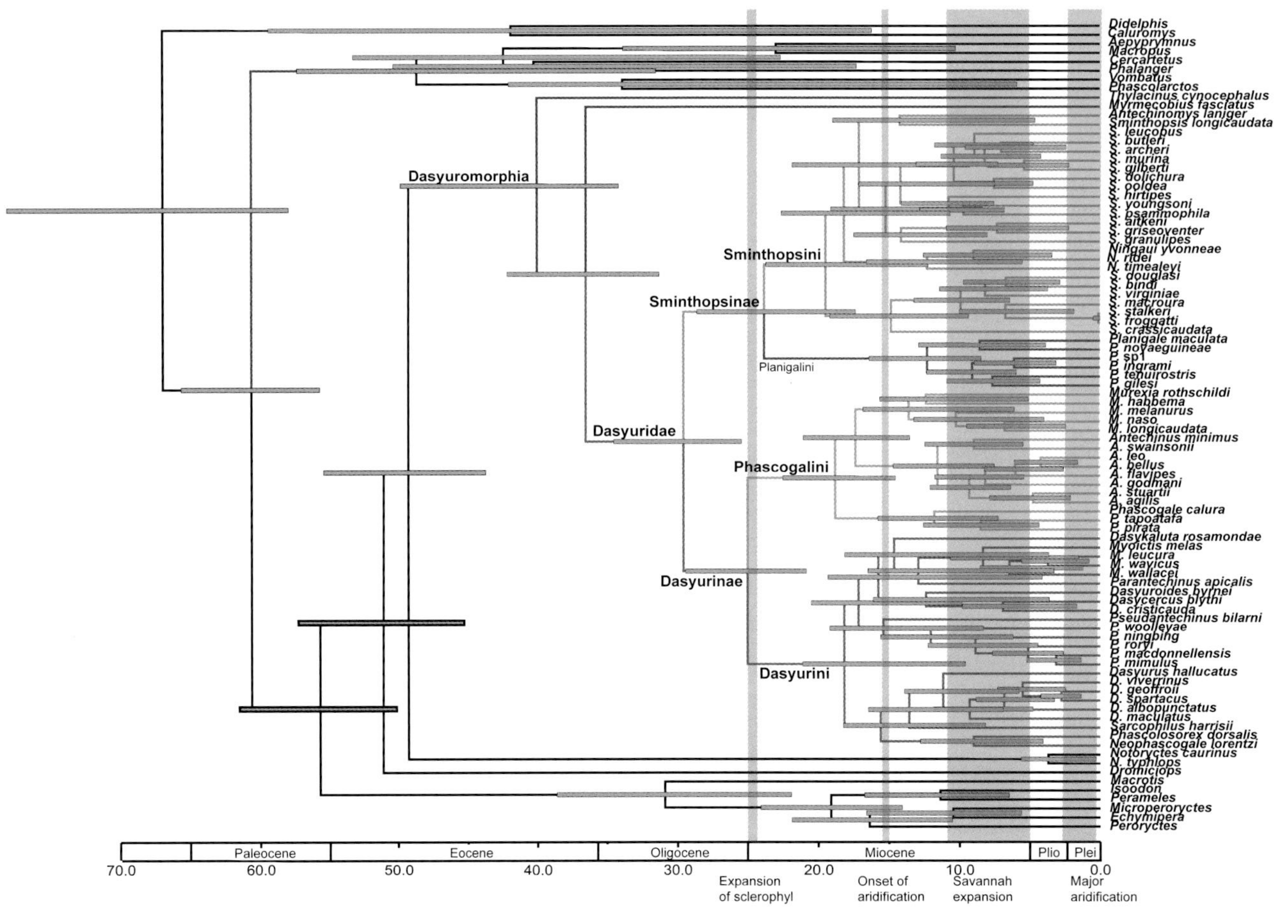

Fig. 9.11. Dasyuromorphian time tree from Westerman, Krajewski *et al.* (2016). Tree topology inferred with maximum likelihood analysis of eight nuclear (*BRCA1, ApoB, RAG1, vWF, ε-globin, P1, IRBP, bfib7*) and seven mitochondrial (*cytb, 12S, tVal-16S, ND1, ND2, CO1, CO2*) loci. Divergence dates estimated with BEAST using the UCLN rates model and 12 soft-bounded calibration points. Grey bars show 95% highest probability densities for each node. [Reprinted by permission of Oxford University Press from Westerman, Krajewski *et al.* (2016).]

Table 9.4. Estimated divergence times from Westerman, Krajewski *et al.* (2016).
Point estimates are followed by 95% highest posterior density (HPD) intervals.

Split	Date (95% HPD) in Ma
Dasyuridae – Myrmecobiidae	35.6 (32.1–41.0)
Dasyurinae – Sminthopsinae	29.3 (25.5–33.1)
Dasyurini – Phascogalini	24.5 (21.2–28.0)
Sminthopsini – Planigalini	24.1 (20.5–27.7)
(*Dasycercus, Dasyuroides*) – (*Dasykaluta, Parantechinus, Myoictis*)	15.7 (13.4–18.3)
Dasykaluta – (*Parantechinus, Myoictis*)	14.7 (12.4–17.2)
Myoictis – *Parantechinus*	13.0 (10.6–15.5)
Myoictis melas – other *Myoictis*	8.4 (6.2–10.8)
Dasyuroides – *Dasycercus*	12.3 (9.7–15.0)
Dasycercus blythi – *D. cristicauda*	7.1 (4.4–9.7)
Dasyurus – *Sarcophilus*	13.5 (11.3–15.3)
Dasyurus hallucatus – other *Dasyurus*	11.2 (0.0–13.4)
Pseudantechinus bilarni – other *Pseudantechinus*	15.5 (12.7–18.6)
Phascogale – (*Antechinus, Murexia*)	18.8 (15.7–21.9)
Murexia – *Antechinus*	17.4 (14.6–20.4)
Phascogale calura – (*P. tapoatafa, P. pirata*)	11.8 (8.2–15.2)
(*Murexia rothschildi, M. habbema*) – other *Murexia*	13.7 (11.0–16.6)
(*Antechinus swainsonii, A. minimus*) – other *Antechinus*	11.9 (9.5–14.5)
(*Planigale maculata, P. novaeguineae*) – other *Planigale*	12.3 (9.4–15.5)
Ningaui timealeyi – (*N. ridei, N. yvonneae*)	12.5 (9.1–15.9)
Sminthopsis crassicaudata species group – other Sminthopsini	19.7 (16.7–23.0)
Sminthopsis crassicaudata – Macroura group	15.5 (12.0–18.4)
Sminthopsis murina group basal divergence	15.4 (12.6–18.3)

earliest in Sminthopsini (ca. 20 Ma) and the latest in Planigalini (ca. 12 Ma). In a study focused primarily on the dasyurid adaptive radiation (see Chapter 10), García-Navas *et al.* (2018) generated a time tree for dasyurids from the data of Westerman, Krajewski *et al.* (2016), excluding seven species and using the UCLN rates model in BEAST. Not surprisingly, their results (Fig. 9.12) resemble those of the original study.

The phylogenomic study by Duchêne *et al.* (2018) used fossil calibration points provided by Mitchell *et al.* (2014) to estimate divergence times on a species tree derived from 1550 gene trees. They applied the relaxed clock method in MCMCTREE with UCLN rates and estimated a Paleocene (60–75 Ma) origin for Dasyuromorphia, an Oligocene (ca. 32 Ma) split of Dasyuridae and Myrmecobiidae, early Miocene split (ca. 20 Ma) of dasyurid subfamilies and mid-Miocene (ca. 15 Ma) divergence of tribes (see Fig. 5.5).

Mutton *et al.*'s (2019) study of *Antechinus* phylogeny included UCLN dating in BEAST for their alignment of two mitochondrial and four nuclear loci, with the fossil age constraints of Westerman, Krajewski *et al.* (2016). Their results (Fig. 9.13) place the separation of dasyurids from myrmecobiids in the late Oligocene (ca. 27 Ma), the dasyurine-sminthopsine split in the early Miocene (ca. 22 Ma) and tribal divergences in the middle Miocene (ca. 14–17 Ma). In their presentation of a chromosome-level genome assembly for *Antechinus flavipes*, Tan *et al.* (2021) estimated the Dasyurini-Phascogalini separation at 11.0–17.3 Ma using MCMCTREE analysis of 6090 protein-coding gene sequences from six dasyurids and six outgroups.

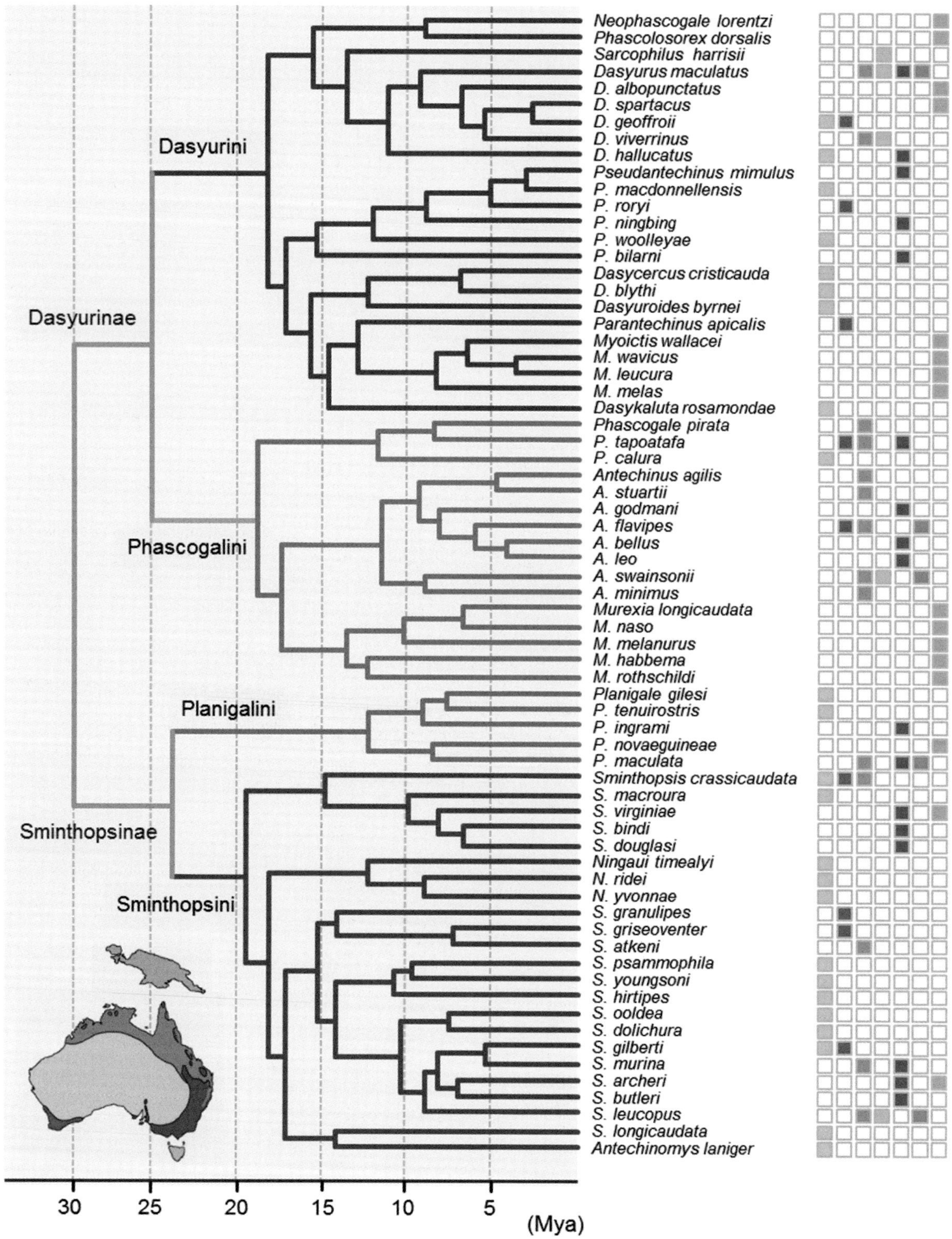

Fig. 9.12. Dasyurid time tree with mapping of species by geographic region (García-Navas *et al.* 2018). Branches are coloured by tribe. Biogeographic regions inhabited by species are shown in the grid at right: orange = Eremean; red = south-west; purple = Bassian; pink = Tasmania; blue = Torresian; dark green = Irian; light green = New Guinea. Note: print book readers refer to the colour version of this figure in the Appendix. [Reprinted by permission of Oxford University Press from García-Navas *et al.* (2018).]

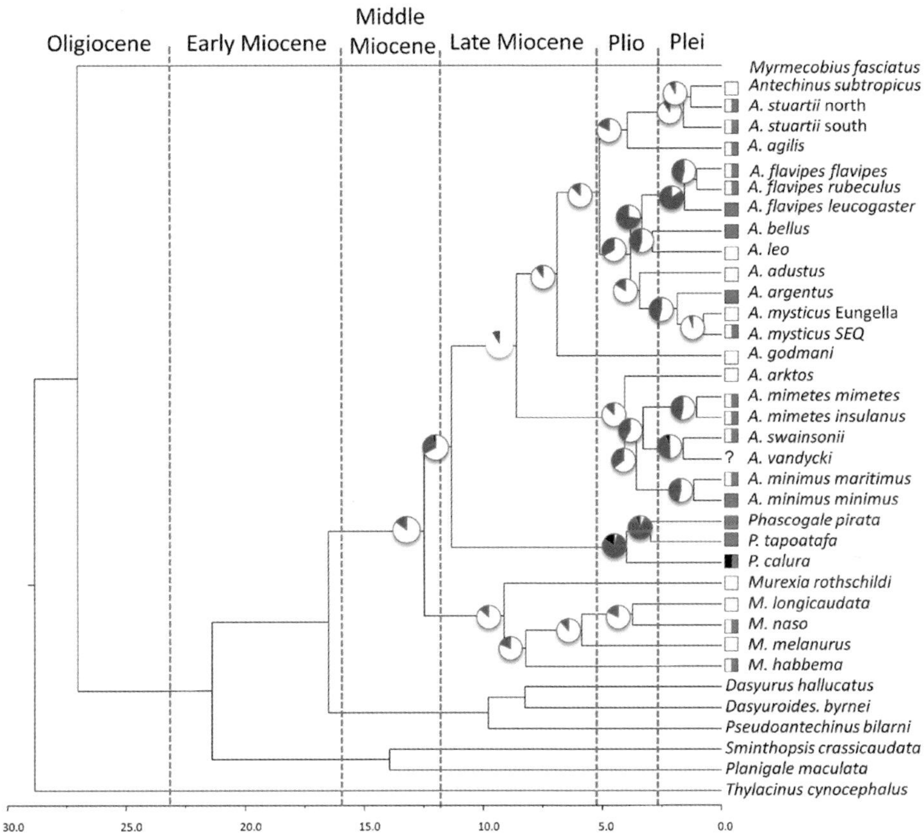

Fig. 9.13. Time tree of Mutton *et al.* (2019). Results based on analysis of two mitochondrial (*cytb, 16S*) and four nuclear (*IRBP, ApoB, vWF, BRCA1*) gene segments with a UCLN rates model in BEAST, using fossil age constraints from Westerman, Krajewski *et al.* (2016). Squares at tips are habitat codes for current species, circles on internodes are pie charts of ancestral habitat probabilities estimated with BayesTraits. White = wet-closed habitat; blue = wet-open habitat; black = arid habitat. [Reprinted by permission of Oxford University Press from Mutton *et al.* (2019).]

TOTAL EVIDENCE APPROACHES TO DIVERGENCE DATING

Relaxed clock models and the use of prior distributions for calibration points significantly extended our ability to obtain divergence dates beyond the strict clock assumption of early molecular studies. It is prudent, however, to remember the limitations of this 'node dating' approach, which were bluntly summarised by O'Reilly *et al.* (2015: 638):

> *Node calibrations have been considered unsatisfactory because they require a prior phylogenetic hypothesis and they fail to integrate uncertainty in the phylogenetic affinity of the calibrating fossils. This is problematic because the earliest fossil occurrences are often fragmentary, and therefore of uncertain affinity, and they are often ignored in favor of younger, better-known and therefore phylogenetically-secure species. However, this leads to less-certain and less-informative calibrations – and dismisses an effectively infinite amount of other rate-informative fossil evidence. Some consider maximum age constraints based on fossil evidence or, rather, its absence, as unjustifiable and establishing the nature of a probability density function spanning minimum and maximum constraints has little justification beyond gut-feeling. Unfortunately, arbitrary choices between competing parameters have an almost overwhelming impact on divergence-time estimates.*

Tip dating

An alternative to node dating is inclusion of fossil taxa in the phylogeny to be dated and allowing their ages to serve as calibration points. Theoretically, all relevant fossil taxa could be included, not just those representing first occurrences of extant clades. This approach is reasonable if the phylogenetic position of fossil taxa can be estimated accurately or if uncertainty about such placement can be incorporated into time estimates. Because phylogenetic placement of fossils generally requires analysis of morphological characters, combining it with molecular data from extant taxa entails the sort of 'total evidence' approach described in Chapter 5. Taking advantage of the Bayesian inferential framework, however, means that a probabilistic model of morphological trait evolution must be employed. To date, the only ones widely available are versions of Lewis's (2001) simple Mk model. Nevertheless, total evidence dating – or 'tip dating' – has been used extensively in recent years.

Pioneered by Rambaut (2000) and implemented substantially by Pyron (2011), tip dating was formalised by Ronquist *et al.* (2012), who incorporated their method into the MrBayes software package. Options for relaxed clock models include two with autocorrelated rates – one being that of Kishino *et al.* (2001) – and an Independent Gamma Rates (IGR) model like that of Drummond *et al.* (2006). Although Ronquist *et al.* (2012) used a uniform tree prior, rather than the Yule or birth-death processes used in node dating, later workers favoured the 'fossilized birth-death (FBD) process' (Heath *et al.* 2014) for this purpose. The appeal of the FBD process is that it allows a single prior to encompass both the tree-wide distribution of split times and the uncertainty associated with calibrated nodes. The FBD has five parameters (speciation rate, extinction rate, fossil recovery rate, proportion of total species sampled and root age) set by users, preferably based on relevant data rather than educated guesses. Zhang *et al.* (2016) modified the tip-dating approach of Ronquist *et al.* (2012) in MrBayes to include an FBD prior that considers taxon-bias introduced by investigators – namely 'diversity sampling' in which a deliberate attempt is made to include representatives of each major lineage within a clade, even though some are much less speciose than others.

Despite early optimism that tip dating might be relatively robust to errors in phylogenetic placement of fossils, O'Reilly *et al.* (2015) noted that the tendency of phylogenetic analyses to place fossil taxa in basal positions due to missing character data ('stem-ward slippage') would influence rate and time estimates. They also expressed concern about the computationally convenient assumption that morphological and molecular rates of evolution are correlated along branches. Most significantly, they suggested that tip dating and node dating used together might provide better estimates than either one used separately.

Tip and node dating

Among the problems of tip dating noted by O'Reilly *et al.* (2015) was a consistent tendency to return anomalously ancient divergence times, as well as times that were incompatible with fossil ages. O'Reilly and Donoghue (2016) attributed this to a lack of constraints on split times, a problem that could be alleviated by node dating, which they suggested should be used in conjunction with tip dating. Such a strategy allows one set of fossils to place minimum, while another set (those at tips) constrain maximum, divergence times.

> *[W]hen they are implemented alone, tip calibrations can yield divergence time estimates that violate empirical fossil evidence or place exaggerated probability on overly ancient age estimates. Combining node and tip calibrations obviates these effects with the hard minima of node calibrations constraining the uncertainty associated with tip calibrations that, in turn, serve to objectively define the maxima of node age constraints. This approach is appealing because of the positive complementary interaction between the two classes of calibration, but also because it makes the best use of palaeontological data in the construction of evolutionary timescales. [O'Reilly and Donoghue 2016: 4]*

Tip and node dates for dasyuromorphians

Kealy and Beck (2017) used node, tip and tip-and-node approaches to estimate dasyuromorphian split times (see Chapter 5). Node dating used the IGR clock model, an FBD branching prior, diversity sampling (sampling probability = 80% for extant taxa, 0% for fossil taxa) and six calibration points (two inside and four outside Dasyuromorphia). Node calibration priors were either exponential with hard minima and soft maxima (5% tail) ('NodeCalib1') or uniform with hard minima and maxima for four of the six points ('NodeCalib2'). Tip dating used similar priors, but with age ranges of fossils taken from the literature. Age constraints were placed on the root and the Didelphimorphia-Australidelphia split with the same exponential or uniform priors used for node dating ('TipCalib1' and 'TipCalib2'). Tip-and-node analyses combined both approaches ('TipNodeCalib1' and 'TipNodeCalib2') with calibrated nodes constrained to monophyly. The results of all six analyses were similar (Table 9.5); that

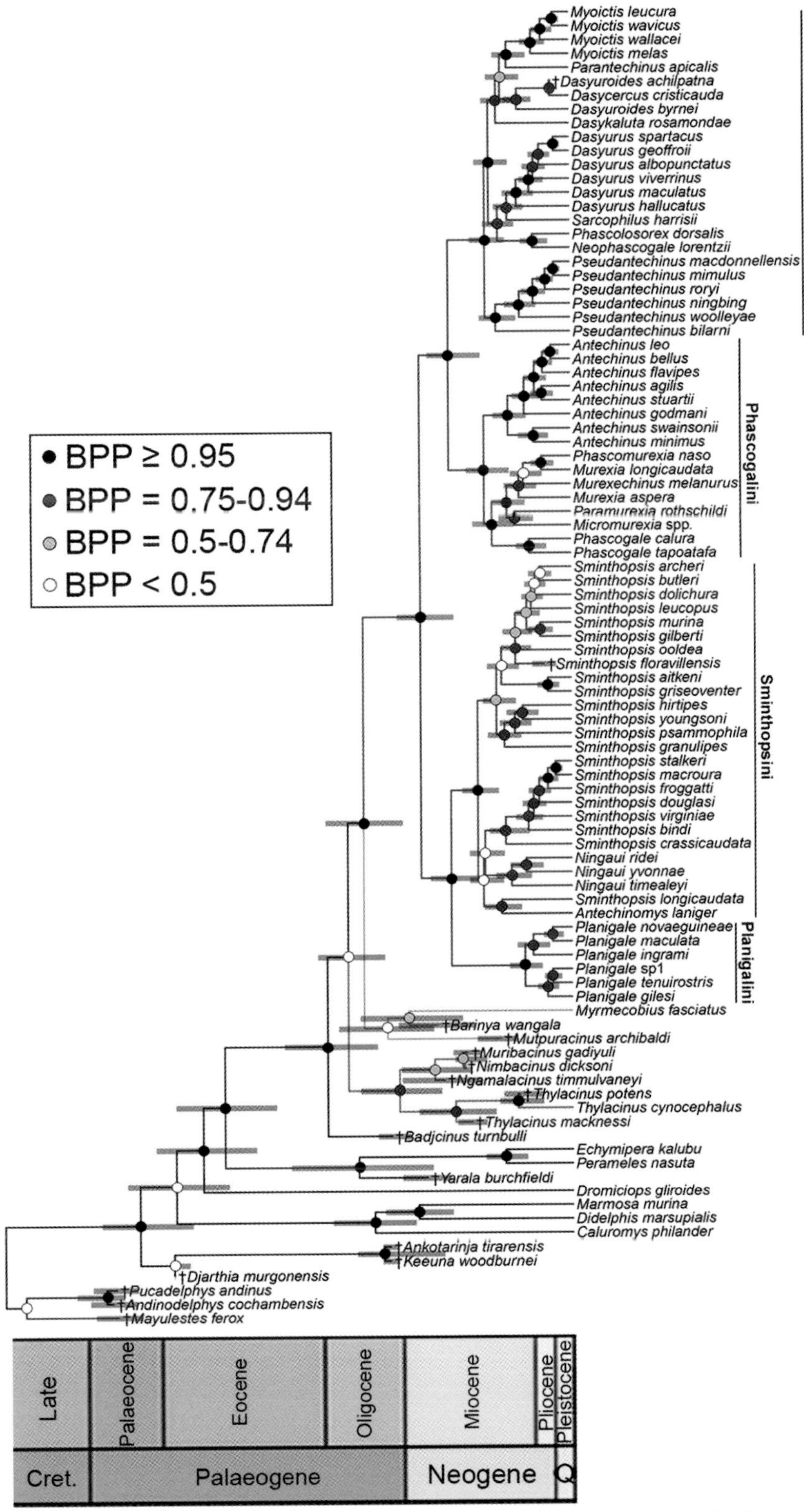

Fig. 9.14. Dasyuromorphian time tree from Kealy and Beck (2017). Tree and divergence times obtained with Bayesian inference (MrBayes analysis of 16.4 kb of DNA sequence data and 173 morphological characters) using tip and node dating with IGR clock model and 'TipNodeCalib1' calibration priors (see text for explanation). Bars at nodes are 95% HPDs. Branch colour coding: orange = Myrmecobiidae; red = Thylacinidae; blue = Dasyuridae. [Reprinted from Kealy & Beck (2017) under CC BY 4.0 license.]

for 'TipNodeCalib1' is shown in Fig. 9.14. Tip and node dates suggest that Dasyuromorphia originated in the mid-Eocene ca. 48 Ma and Dasyuridae in the early Oligocene ca. 31 Ma, with the Dasyurinae-Sminthopsinae split in the early Miocene ca. 21 Ma. Tribal separations within each subfamily were simultaneous in the mid-Miocene 16–17 Ma. Dasyurini, Phascogalini and Sminthopsini also experienced basal radiations during a narrow mid-Miocene window of 12–13 Ma, with Planigalini in the late Miocene ca. 6.5 Ma. Only a few splits among sister-species within tribes were as recent as the Pliocene (2.6–5.3 Ma) and none was Pleistocene.

In a geographic study of dasyurid speciation, García-Navas *et al.* (2020) used tip and node dating to estimate divergence times from expanded versions of Westerman, Krajewski *et al.*'s (2016) molecular, and Kealy and Beck's (2017) morphological, data sets. Divergence estimates employed five node and 12 tip calibrations, along with the 'TipNodeCalib1' priors and associated parameter settings from Kealy and Beck (2017). The results are shown in Fig. 9.15 and Table 9.6. (Because these are the most comprehensive dasyurid divergence date estimates, we provide a complete listing in the table.) In general, these split times are more recent than those of Kealy and Beck (2017). The origin of Dasyuridae is in the early Miocene (ca. 22 Ma), with mid-Miocene separation of subfamilies (ca. 15 Ma) and their constituent tribes (12–13 Ma). Dasyurin, phascogalin and sminthopsin radiations again appear roughly simultaneous in the late Miocene (ca. 9 Ma), with Planigalini delayed until the early Pliocene (ca. 5 Ma).

The phylogenomic study of Feng *et al.* (2022) estimated that the radiation of Australian marsupial orders took place 45.8 Ma based on a subset of loci that showed complete lineage sorting, roughly consistent with Kealy and Beck's (2017) tip-and-node date of 48 Ma for the origin of Dasyuromorphia. The most recent set of dasyuromorphian divergence dates come from the marsupial phylogeny study of Beck *et al.* (2022) based on 97 extant species, 42 fossils and five outgroup non-marsupial metatherians scored for 180 craniodental characters, with 24.5 kb of nuclear and mitochondrial DNA sequence from extant forms. Their dated total-evidence analysis used separate IGR models for morphological and molecular characters, normally distributed clock rate priors (mean = 0.01, SD = 0.1), uniform age priors on fossil terminals and the root, an FBD tree prior, an estimated sampling diversity of 25% and minimum-maximum bounds on 25 nodes. They estimated separation of dasyuromorphians from a bandicoot sister group in the Eocene (43.7 Ma), divergence of thylacinids from the numbat-dasyurid lineage in the Oligocene (31.2 Ma) and the numbat-dasyurid split at 24.9 Ma. Within Dasyuridae, separation of dasyurines and sminthopsines took place in the Early Miocene (ca. 20 Ma), followed by tribal splits in the mid-Miocene (ca. 13–14 Ma).

Table 9.5. Dasyuromorphian divergence times (and 95% HPDs) in Ma estimated by Kealy and Beck (2017).

Split/clade	Calibration priors					
	NodeCalib1	NodeCalib2	TipCalib1	TipCalib2	TipNodeCalib1	TipNodeCalib2
Dasyuromorphia	52.1 (43.5–64.9)	64.4 (49.9–74.4)	51.8 (42.8–61.7)	–	47.6 (41.2–55.0)	48.4 (42.6–54.0)
Dasyuridae*	34.7 (28.4–44.1)	42.9 (33.4–51.2)	36.4 (30.3–44.0)	36.5 (30.2–43.1)	30.7 (26.9–36.1)	31.6 (27.7–35.5)
Dasyurinae-Sminthopsinae	23.4 (18.8–29.4)	28.6 (22.2–34.5)	22.6 (18.7–26.7)	22.6 (18.7–25.5)	20.6 (17.9–25.6)	21.1 (18.2–23.7)
Dasyurini-Phascogalini	18.8 (15.2–23.6)	22.9 (17.6–27.7)	18.0 (14.8–21.7)	17.9 (15.7–20.9)	16.7 (14.1–21.5)	16.9 (14.7–19.4)
Dasyurini base	13.0 (10.4–16.5)	15.8 (12.1–19.3)	12.4 (10.2–15.0)	12.3 (10.8–14.6)	11.5 (9.6–14.5)	11.7 (9.9–13.3)
Dasyurus base	8.1 (6.1–10.5)	9.7 (7.0–12.3)	7.7 (5.9–9.7)	7.6 (6.1–9.3)	7.2 (5.4–9.6)	7.2 (5.6–8.7)
Phascogalini base	13.3 (10.6–17.0)	16.2 (12.2–19.8)	12.6 (10.0–15.4)	12.4 (10.7–14.7)	11.8 (9.5–15.4)	11.9 (10.2–13.6)
Antechinus base	9.5 (7.2–12.4)	11.5 (7.9–14.7)	9.0 (6.9–11.3)	8.9 (7.2–10.8)	8.5 (6.6–11.1)	8.5 (6.6–9.9)
Murexia base	9.9 (7.7–13.0)	12.0 (9.0–15.0)	9.3 (7.3–11.8)	9.1 (7.8–10.7)	8.7 (6.7–11.2)	8.8 (7.1–10.5)
Sminthopsini-Planigalini	18.9 (14.9–23.8)	22.8 (17.6–27.8)	17.9 (14.6–21.6)	17.9 (14.9–20.8)	16.5 (14.1–20.8)	16.9 (14.3–19.2)
Planigalini base	7.2 (5.2–9.7)	8.8 (5.9–11.6)	6.8 (5.1–9.0)	6.8 (5.1–8.6)	6.5 (4.4–8.9)	6.5 (4.8–8.2)
Sminthopsini base	14.7 (11.8–18.5)	17.8 (13.7–21.8)	13.8 (11.3–16.5)	13.8 (11.5–15.9)	12.8 (10.8–15.9)	13.1 (11.4–15.2)

* *Thylacinus-Myrmecobius*-Dasyuridae split.

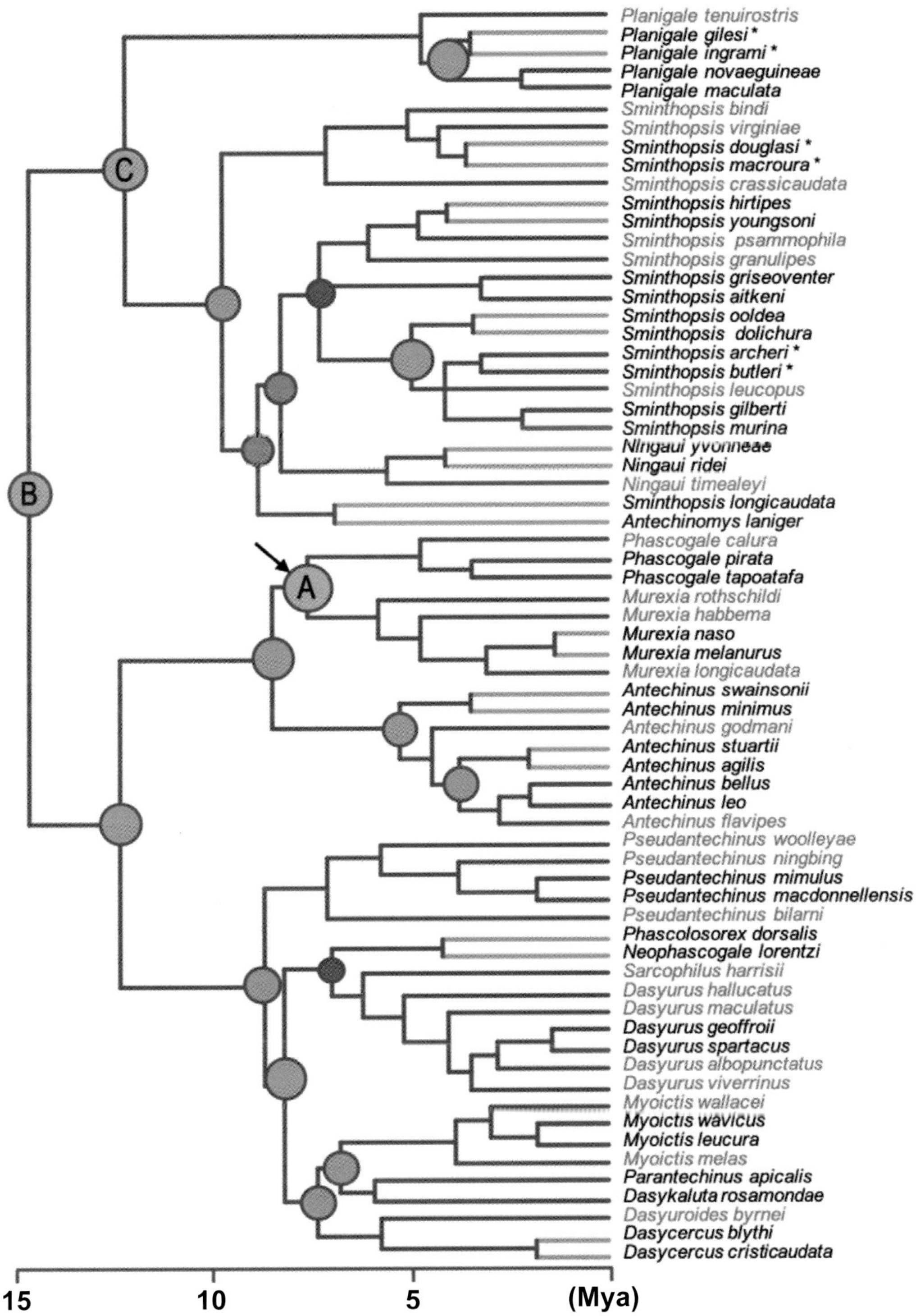

Fig. 9.15. Dasyurid time tree from García-Navas *et al.* (2020). Tree based on Bayesian analysis of the DNA sequence data set of Westerman, Krajewski *et al.* (2016) expanded to include 80 extant dasyurids and the *RUNX2* and *col2A1* loci, along with the morphological data set of Kealy and Beck (2017) expanded to include five additional fossil taxa. Divergence times estimated with tip and node dating following the 'TipNodeCalib1' scheme of Kealy and Beck (2017). Only extant taxa are shown. Arrow, dots, coloured branches and tip-label shading refer to biogeographic analyses in the original paper. [Reprinted by permission of John Wiley and Sons from García-Navas *et al.* (2020).]

Table 9.6. Estimated average divergence times within Dasyuromorphia (with 95% HPDs) in Ma from García-Navas *et al.* (2020). Clades correspond to those in Fig. 9.14.

Split	Divergence estimate
(Dasyuridae, Myrmecobiidae) – Thylacinidae	24.47 (23.98–35.42)
Dasyuridae – Myrmecobiidae	22.41 (18.79–27.86)
Dasyurinae – Sminthopsinae	14.78 (4.41–19.54)
Dasyurini – Phascogalini	12.62 (9.76–16.23)
(*Dasycercus, Dasyuroides, Dasykaluta, Parantechinus, Myoictis, Dasyurus, Sarcophilus, Neophascogale, Phascolosorex*) – *Pseudantechinus*	8.92 (6.81–11.75)
(*Dasycercus, Dasyuroides, Dasykaluta, Parantechinus, Myoictis*) – (*Dasyurus, Sarcophilus, Neophascogale, Phascolosorex*)	8.58 (6.51–11.19)
(*Dasycercus, Dasyuroides*) – (*Dasykaluta, Parantechinus, Myoictis*)	7.8 (5.07–10.15)
Dasycercus – Dasyuroides	6.01 (4.38–8.47)
Dasycercus cristicauda – Dasycercus blythi	1.91 (1.17–2.69)
(*Dasykaluta, Parantechinus*) – *Myoictis*	7.17 (5.32–9.38)
Dasykaluta – Parantechinus	6.21 (4.29–8.17)
Myoictis melas – (*M. leucura, M. wavicus, M. wallacei*)	3.95 (2.63–5.64)
(*M. leucura, M. wavicus*) – *M. wallacei*	3.01 (1.88–4.46)
Myoictis leucura – M. wavicus	1.79 (1.06–2.82)
(*Dasyurus, Sarcophilus*) – (*Neophascogale, Phascolosorex*)	7.65 (5.66–10.19)
Dasyurus – Sarcophilus	6.71 (4.86–8.79)
Dasyurus hallucatus – (*D. maculatus, D. albopunctatus, D. viverrinus, D. spartacus, D. geoffroii*)	5.42 (4.15–7.41)
Dasyurus maculatus – (*D. albopunctatus, D. viverrinus, D. spartacus, D. geoffroii*)	4.24 (3.07–5.85)
Dasyurus albopunctatus – (*D. viverrinus, D. spartacus, D. geoffroii*)	3.68 (2.57–5.12)
Dasyurus viverrinus – (*D. spartacus, D. geoffroii*)	2.95 (2.01–4.33)
Dasyurus spartacus – D. geoffroii	1.41 (0.91–2.08)
Neophascogale – Phascolosorex	4.25 (3.01–5.81)
Pseudantechinus bilarni – (*P. woolleyae, P. ningbing, P macdonnellensis, P. mimulus*)	7.31 (5.5–9.67)
Pseudantechinus woolleyae – (*P. ningbing, P. macdonnellensis, P. mimulus*)	5.98 (4.43–8.15)
Pseudantechinus ningbing – (*P. macdonnellensis, P. mimulus*)	4.08 (2.78–5.87)
Pseudantechinus macdonnellensis – P. mimulus	1.82 (1.02–2.73)
Antechinus – (*Murexia, Phascogale*)	8.96 (6.93–11.38)
(*Antechinus minimus, A. swainsonii*) – (*A. godmani, A. stuartii, A. agilis, A. flavipes, A. leo, A. bellus*)	5.7 (4.08–7.65)
Antechinus minimus – A. swainsonii	3.91 (2.61–5.47)
Antechinus godmani – (*A. stuartii, A. agilis, A. flavipes, A. leo, A. bellus*)	4.62 (3.22–6.4)
(*Antechinus stuartii, A. agilis*) – (*A. flavipes, A. leo, A. bellus*)	3.98 (2.73–5.48)
Antechinus stuartii – A. agilis	2.19 (1.24–3.43)
Antechinus flavipes – (*A. leo, A. bellus*)	2.93 (1.9–4.18)
Antechinus leo – A. bellus	2.11 (1.3–3.12)
Murexia – Phascogale	7.84 (5.88–10.09)
Murexia rothschildi – (*M. habbema, M. longicaudata, M. melanurus, M. naso*)	6.15 (4.45–8.26)
Murexia habbema – (*M. longicaudata, M. melanurus, M. naso*)	5.1 (3.43–7.17)
Murexia longicaudata – (*M. melanurus, M. naso*)	3.28 (1.88–4.65)
Murexia melanurus – M. naso	1.35 (0.82–1.92)
Phascogale calura – (*P. tapoatafa, P. pirata*)	4.82 (3.3–6.9)

Split	Divergence estimate
Phascogale tapoatafa – P. pirata	3.49 (2.14–5.11)
Sminthopsini – Planigalini	12.17 (9.84–15.9)
(Antechinomys, Ningaui, Sminthopsis longicaudata, S. ooldea, S. dolichura, S. archeri, S. butleri, S. leucopus, S. gilberti, S. murina, S. griseoventer, S. aitkeni, S. granulipes, S. psammophila, S. youngsoni, S. hirtipes) – (S. crassicaudata, S. bindi, S. douglasi, S. virginiae, S. macroura)	9.53 (7.88–12.55)
Antechinomys – Sminthopsis longicaudata	7.11 (5.53–9.64)
Ningaui – (Sminthopsis ooldea, S. dolichura, S. archeri, S. butleri, S. leucopus, S. gilberti, S. murina, S. griseoventer, S. aitkeni, S. granulipes, S. psammophila, S. youngsoni, S. hirtipes)	8.22 (6.82–10.81)
Ningaui timealeyi – (N. ridei, N. yvonneae)	5.71 (3.8–7.87)
Ningaui ridei – N. yvonneae	4.16 (2.55–6.17)
(Sminthopsis ooldea, S. dolichura, S. archeri, S. butleri, S. leucopus, S. gilberti, S. murina) – (S. griseoventer, S. aitkeni, S. granulipes, S. psammophila, S. youngsoni, S. hirtipes)	7.27 (5.78–9.38)
(Sminthopsis ooldea, S. dolichura) – (S. archeri, S. butleri, S. leucopus, S. gilberti, S. murina)	5.08 (3.8–6.59)
(Sminthopsis archeri, S. butleri, S. leucopus) – (S. gilberti, S. murina)	4.22 (3.14–5.72)
Sminthopsis gilberti – S. murina	2.24 (1.25–3.47)
Sminthopsis ooldea – S. dolichura	3.59 (2.38–5.17)
Sminthopsis griseoventer – S. aitkeni	3.45 (2.09–5.09)
Sminthopsis psammophila – (S. youngsoni, S. hirtipes)	5.07 (3.49–6.57)
Sminthopsis youngsoni – S. hirtipes	4.31 (2.85–5.92)
Sminthopsis crassicaudata – (S. bindi, S. douglasi, S. virginiae, S. macroura)	7.39 (5.52–9.52)
Sminthopsis bindi – (S. douglasi, S. virginiae, S. macroura)	5.26 (3.75–7.01)
Sminthopsis douglasi – (S. virginiae, S. macroura)	4.46 (3.06–6.91)
Planigale tenuirostris – (P. gilesi, P. ingrami, P. novaeguineae, P. maculata)	4.72 (3.45–6.73)
(Planigale gilesi, P. ingrami) – (P. novaeguineae, P. maculata)	4.08 (2.82–5.84)
Planigale novaeguineae – P. maculata	2.2 (1.18–3.44)

SUMMARY: THE TIMELINE OF DASYURID CLADOGENESIS

Table 9.7 compares date estimates for the origin of Dasyuridae, subfamilial and tribal divergences and the onset of tribal radiations based on all the methods discussed in this chapter. As 'ballpark' estimates, these numbers are remarkably consistent, despite the presence of a few outliers. Differences among them are due to the different data sets and estimation methods employed, but the most recent and sophisticated tip and node estimates are often close to the older and cruder strict clock values. It appears that Dasyuridae split from Myrmecobiidae in the Oligocene, although estimated dates of this split span the entire period. Dasyurinae and Sminthopsinae most likely diverged in the latest Oligocene or early Miocene, with tip and node dates favouring the latter. Both García-Navas *et al.* (2020) and Beck *et al.* (2022) found nearly simultaneous mid-Miocene tribal splits within each subfamily, roughly coinciding with the mid-Miocene Climatic Optimum, which has been suggested as a causal factor in dasyurid diversification (see Chapter 10). Tribal radiations that produced extant genera began in the mid to late Miocene, with Planigalini later by some 2–5 Ma than the other three. Most molecular clock studies found that cladogenesis responsible for the extant dasyurid species was completed by the Pliocene, but García-Navas *et al.* (2020) dated several intrageneric splits to <5 Ma (Table 9.6). While more precise dates may be needed to answer questions about the determinants of dasyurid macroevolution, it appears that palaeontology, morphology and molecular clock analyses have converged on a reasonably clear timeline for dasyurid cladogenesis.

ENDNOTE

1 In keeping with standard linear algebra notation, bold-face variables denote matrices or vectors.

Table 9.7. Comparison of estimated dates for major dasyurid divergences.
Dates for tribal radiations (last four columns) refer to the deepest splits within each tribe.

Data	Method	Split dates (Ma)							
		Dasyuridae	Dasyurinae-Sminthopsinae	Dasyurini-Phascogalini	Sminthopsini-Planigalini	Dasyurini	Phascogalini	Sminthopsini	Planigalini
Fossils	Stratigraphy	>25	>20	>12	>4.5	>12	>12	>4.5	>4.5
Albumin[1]	Strict clock	24	20	10	<20	≈8	≈8	<20	<20
DNA hybridisation[2]	Strict clock		35.9				25.5		
cytb[2]	Strict clock		22–28		17–22	14		15–18	11–15
12S tv[3]	Strict clock		29.5–50		28–48	14	21.1	13.9–24	
cytb + *12S*[4]	Linearised tree	(25)	24.3	21.6	19.6	14	16	13.8	10.7
mt genome[5]	Relaxed clock	<38	20						
nucDNA[6]	Relaxed clock	<60	14.4–25	9.7–10.2			5.5–7.8		
Supertree[7]	Fossil ages	32.8							
mt + nucDNA[8]	Relaxed clock	26–35.6	22–29.3	11.6–18	14–24.5	10–15.7	12.5–18.8	13–19.7	12.3
DNA + morphology[9]	Tip and node	30.7	20.6	16.7	16.5	11.5	11.8	12.8	6.5
nucDNA[10]	Relaxed clock	32	20	11–17.3	14–16				
DNA + morphology[11]	Tip and node	22.4–24.9	14.8–20	12.6–14	12.2–13	8.9–11	8.9–9.0	9.0–9.5	4.7

[1] Sarich *et al.* (1982); Baverstock *et al.* (1990).
[2] Painter *et al.* (1995); Krajewski, Blacket *et al.* (1997); Krajewski, Buckley and Westerman (1997); Blacket *et al.* (1999).
[3] Krajewski, Blacket *et al.* (1997); Krajewski, Young *et al.* (1997); Springer *et al.* (1997); Blacket *et al.* (1999).
[4] Krajewski, Wroe and Westerman (2000); a 25 Ma date for the thylacinid-dasyurid split was assumed for calibration.
[5] Nilsson *et al.* (2004).
[6] Drummond *et al.* (2006); Meredith *et al.* (2008).
[7] Bininda-Emonds *et al.* (2007).
[8] Beck (2008); Mitchell *et al.* (2014); Westerman, Krajewski *et al.* (2016); García-Navas *et al.* (2018); Mutton *et al.* (2019).
[9] Kealy and Beck (2017) with 'TipNodeCalib1' priors.
[10] Duchêne *et al.* (2018); Tan *et al.* (2021).
[11] García-Navas *et al.* (2020); Beck *et al.* (2022). Intrafamilial dates from the latter study estimated visually from time tree.

10 Dasyurid macroevolution

So far, we have focused on the 'what' and 'when' of dasyurid cladogenesis in the form of phylogenetic relationships, species boundaries and divergence times. A broader perspective leads to the ultimate questions of 'how' dasyurid evolution followed the course it did to produce the current 75 species with their geographic ranges, habitat affinities and phenotypic adaptations. We address the first two questions in this chapter and the third in Chapter 11. Serious scholarship on dasyurid biogeography post-dates the establishment of continental drift theory and the elaboration of analytical biogeography methods in the 1970s (see Lomolino *et al.* 2017). By this time, dasyurid researchers were aware of the roles that *dispersal*, *vicariance*, *refugia*, *local extinction*, *resource limitations* and *climate change* play in explaining the geography of adaptive radiations. The challenge has been to integrate these processes into a robust theory of dasyurid macroevolution.

CENOZOIC CLIMATE AND GEOGRAPHY OF AUSTRALASIA

The most recent divergence time estimates place the origin of Dasyuromorphia in the Eocene and Dasyuridae in the Oligocene (García-Navas *et al.* 2020; Beck *et al.* 2022). Thus, the history of dasyurids comprises events that took place within approximately the last 56 Ma. In this section, we briefly summarise the physical geography of Australia and New Guinea since the Paleocene; in the next, we relate this history to research on the causes of dasyurid diversification. Detailed accounts of Australian geologic, climatic and vegetation history are given by White (1986), Veevers *et al.* (1991), McGowran *et al.* (2000), Martin (2006), Blewett (2012), Crisp and Cook (2013) and Pillans (2018).

Paleocene (66.0–56.0 Ma) and Eocene (56.0–33.9 Ma)

During the Mesozoic, Australia was part of the southern supercontinent Gondwana and covered by extensive rainforests (White 1986; Veevers *et al.* 1991; Hill 1994; Martin 2006). Until the late Cretaceous, Gondwanan forests were dominated by podocarp and araucarian conifers along with native cypresses like *Casuarina*, but angiosperms were spreading rapidly. By the Paleocene, rifting along the southern margin of Australia began its separation from Antarctica (Fig. 10.1). Beginning in the west, rifting spread slowly eastward until the final isolation, through the South Tasman Rise, was effected in the late Eocene (Veevers *et al.* 1991; Exon *et al.* 2004; Goin *et al.* 2016). During this time, the Australian plate was moving to the north-east from an initial high-latitude position at a rate of ca. 67 mm/year. Overland dispersal of land animals – including the ancestors of Australian marsupials (Eomarsupialia) – from South America to Australia had become difficult well before complete geological isolation.

Global temperatures during the Paleocene-Eocene (66–33.9 Ma) were warm, peaking during the Early Eocene Climate Optimum (EECO, 55–50 Ma). In Australia, wet

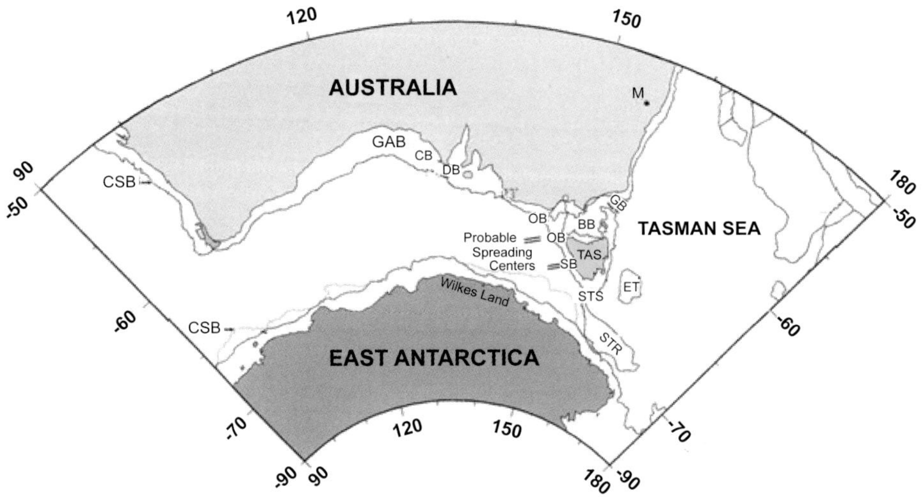

Fig. 10.1. Positions of Australia and East Antarctica at 53 Ma. Abbreviations: BB = Bass Basin; CB = Ceduna Basin; CSB = continental shelf break; DB = Duntroon Basin; ET = East Tasman Rise; GAB = Great Australian Bight; GB = Gippsland Basin; M = Murgon fossil site; OB = Otway Basin; SB = Sorrell Basin; STR = South Tasman Rise; STS = South Tasman Saddle; TAS = Tasmania. Marginal values are latitude and longitude. [Reprinted by permission of Springer Nature from Goin *et al.* (2016).]

forests dominated by conifers and angiosperms covered much of the continent, in which *Nothofagus* (southern beeches) and *Casuarina* (she-oaks) became the dominant elements, indicating a temperate climate with seasonal rainfall (Pillans 2018). Major river systems associated with rainforests drained towards coastal regions and, by the mid-Eocene, separated upland areas with more sclerophyll vegetation (e.g. *Eucalyptus*, *Banksia*) as global temperature declined from the EECO. This sclerophyll flora appears to have developed in Australia in response to infertile soils rather than as an adaptation to dryness (Martin 2006).

During this interval, the eomarsupialian ancestor arrived in Australia from South America where fossil marsupials are known from the early Paleocene (ca. 64 Ma) Tiupampa Fauna of Bolivia (De Muizon 1998; Wroe and Archer 2006; Goin *et al.* 2016). This dispersal was via Antarctica – itself covered by forest at this time – and fossils from Seymour Island (off the Antarctic Peninsula) date to the Eocene. The earliest known marsupial remains in Australia are from the early Eocene (55 Ma) Tingamarra LF in southern Queensland (Godthelp *et al.* 1992). These comprise fragmentary dental, cranial and post-cranial material from a variety of small marsupials, mostly plesiomorphic forms with unknown affinities. Beck (2012) suggested that possible affinities of Tingamarran with South American and African taxa imply a dispersal history more complex than a single ancestor's one-way trek across Antarctica to Australia (see also Eldridge *et al.* 2019). The Tingamarra record is followed by a gap of 30 Ma in which no fossil marsupials are known from Australia.

Oligocene (33.9–23.0 Ma)

The opening of Drake's Passage (between Atlantic and Pacific oceans) and the Tasmanian Gateway (between Indian and Pacific oceans) established the Antarctic Circumpolar Current some 30 Ma, which initiated Antarctic glaciation, falling sea levels and global cooling (Martin 2006; Pillans 2018). One consequence of this was that central Australia became progressively drier, even though the continent continued its northward movement towards the tropics. Pollen data suggest central Australia may have had seasonal dryness as early as 40 Ma, but other parts of the continent still supported extensive wet forests throughout most of the Oligocene. In the mid-Oligocene the northward-moving Australian Plate began to collide with the westward-moving Pacific Plate, leading to the accretion of several terranes along the northern edge of New Guinea and formation of New Guinea's Central Cordillera. These collisions and mountain building affected ocean currents and climate patterns, leading to rain-shadow areas in northern and central Australia. Increasing ice cover in Antarctica also decreased rainfall over Australia, producing seasonal dryness and more open forests. Aridification probably began in near-coastal north-western Australia, where seasonally dry and semi-arid conditions existed by 33 Ma, and spread to the interior of western Australia by mid to late Miocene with the loss of major drainage systems (Mao and Retallack 2019; Hammer *et al.* 2021; Pepper and Keogh 2021). Late Oligocene (25–24 Ma) local faunas associated with the Etadunna Formation of South Australia (Ditjimanka, Ngapakaldi and Ngama) contain the earliest Dasyuromorphia *insertae sedis* fossils (*Apoktesis*, *Ankotarinja*, *Dasylurinja* and *Keeuna*), whereas contemporaneous thylacinids *Badjcinus* and *Nimbacinus dicksoni* are from Riversleigh's Faunal Zone A. Both areas seem to represent habitats with seasonal rainfall and mostly open forest (Archer *et al.* 1995; Travouillon *et al.* 2009; Myers 2017).

Miocene (23.0–5.3 Ma)

Superimposed on a long-term cooling trend, Earth experienced a warm period 17–14 Ma (the Miocene Climate Optimum, or MCO) when temperatures were higher than at any time since the Eocene (Martin 2006; Pillans 2018). Early Miocene rainfall was high in southern Australia, where temperate rainforests were widespread (Fig. 10.2), and there appear to have been large fresh-water lakes in the Eyre Basin. Palaeoecological studies of biotas at Riversleigh in northern Queensland suggest that open forests of the late Oligocene were replaced by wet, temperate, closed forests during the Early-Middle Miocene (Travouillon

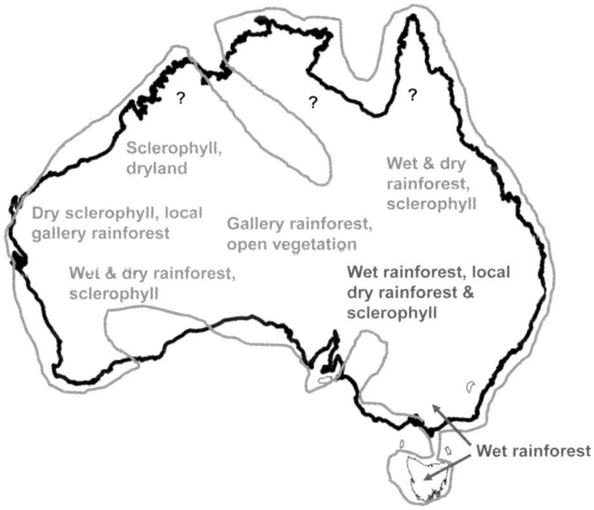

Fig. 10.2. Distribution of Australian vegetation at the Oligocene-Miocene boundary, 23 Ma (from Crisp and Cook 2013). Grey line represents the approximate early Miocene coastline. Text colour shows moisture gradient from drier (orange) to moister (blue); ? = unknown. [Adapted from Crisp and Cook (2013).]

et al. 2009; Woodhead *et al.* 2016). These MCO rainforest communities included the thylacinids *Wabulacinus*, *Ngamalacinus* and *Thylacinus macknessi*, as well as the oldest dasyurid, *Barinya wangala* (16–18 Ma). In contrast, *B. kutjamarpensis*, from the Kutjamarpu LF in South Australia, was part of a drier riparian or sclerophyll forest community (Archer *et al.* 1995; Binfield *et al.* 2017).

After ca. 14 Ma, expansion of the Antarctic ice sheet initiated long-term declines in temperature and sea level, an episode known as the Mid-Miocene Climate Transition. Cooling and drying over Australia were partially offset by its northward movement such that tropical conditions continued in northern parts of the continent, though Riversleigh reverted to more open forests. Eucalypts became increasingly dominant in dry forests and more woodland–savanna habitats appeared. Some grasslands formed in central Australia away from river and lake systems. Rain-shadow effects from the rising New Guinean Cordillera in the north and Great Dividing Range in the east also contributed to drying in north-central Australia. In the south, tropical rainforests retreated towards the coasts, giving way to cool temperate rainforests with different plant communities. Riversleigh dasyuromorphians from the early part of the mid-Miocene (*Maximucinus*, *Muribacinus* and *Joculusium*) represent aseasonal rainforest fauna; *Whollydooleya* and *Nimbacinus dicksoni*, from the later mid-Miocene, lived in drier, more open and possibly seasonal forests. Fossils from the Bullock Creek LF (*Mutpuracinus*, *N. richi*), also in the north, were deposited in a floodplain near a gallery forest (Black *et al.* 2012; Archer, Christmas *et al.* 2016).

More radical changes to the Australian climate occurred in the later Miocene, with the disappearance of many rainforests and a proliferation of eucalypt savannah woodlands plus a few grasslands (Hill 1994; Hill *et al.* 2016), especially around the large lakes of central Australia. This wave of aridification seems to have spread from south to north (Mao and Retallack 2019). By 6 Ma the size of the Antarctic ice sheet increased, rainfall in Australia was further reduced and seasonal climates with summer rain developed in the north. Aided by the rapid uplift in the Central Cordillera of New Guinea (now some 5–8 km high), major contrasts between coastal and inland floras were established. Rainforest and sclerophyll communities existed in a mosaic landscape with araucarian forests still widespread and podocarps common along the eastern seaboard and south-western corner of Australia. *Nothofagus* was probably restricted to highland areas along Australia's east coast and in the mountains of New Guinea. Continued drying and the spread of sclerophyllous vegetation led to an increase in fires across the landscape. Late Miocene dasyuromorphians come from just two localities: Riversleigh's Encore LF (*Mayigriphus*, *Ganbulyani*), and the Northern Territory's Alcoota (*Tyarrpecinus*, *Thylacinus potens*) and Ongeva (*T. megiriani*) LFs. The first locality is thought to represent an open forest (Travouillon *et al.* 2009) and the latter two either subtropical savannas (Murray and Megirian 1992) or sclerophyll forests (Archer *et al.* 1995; Black *et al.* 2012).

Pliocene (5.3–2.6 Ma) and Pleistocene (2.6–0.012 Ma)

The Pliocene saw intensified aridification across much of Australia, featuring the spread of grasslands and restriction of rainforests to the eastern margin, south-western corner, areas of the north-west, and the high elevations in New Guinea. In fact, this epoch is characterised by relatively rapid swings between warm-wet and cool-dry intervals. In pockets of south-western and south-eastern Australia, there was a transient resurgence in *Nothofagus* in rainforests ca. 4.5 Ma, but elsewhere the trend towards *Acacia*, *Casuarina* and *Eucalyptus* forests continued. Low, open communities dominated by sclerophyllous Asteraceae, Chenopodiaceae, Gyrostemonaceae and Poaceae (grasses) occurred in northcentral Australia (Black *et al.* 2012). As cooling intensified throughout the Pliocene, stony gibber deserts formed in central Australia.

Many Pliocene dasyuromorphians come from sites representing mixed forest, woodland and open wetland

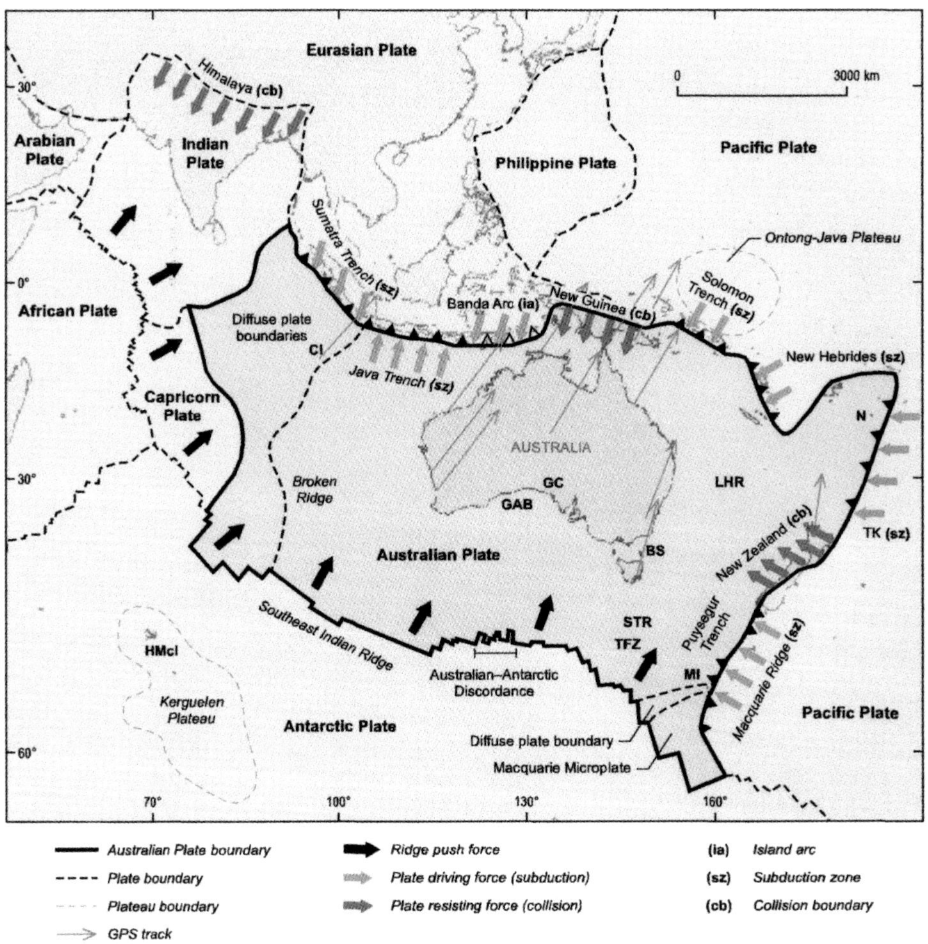

Fig. 10.3. Map of the Australian Plate and its boundaries. Abbreviations: BS = Bass Strait; GAB = Great Australian Bight; GC = Gawler Craton; LHR = Lord Howe Rise; HMcl = Heard and McDonald Islands; MI = Macquarie Island; STR = South Tasman Rise; TFZ = Tasman Fault Zone. [Reprinted by permission of ANU Press from Blewett (2012).]

environments: *Planigale* sp. from Bluff Downs LF, Qld (Archer and Wade 1976); *Thylacinus yorkellus* from the Curramulka LF, SA (Pledge 1992); *Glaucodon* from the Smeaton LF (Turnbull *et al.* 1993); '*Dasycercus*' *worboysi*, *Antechinus* sp. and *Sminthopsis* sp. from the Big Sink LF, NSW (Dawson *et al.* 1999); *T. cynocephalus*, *Archerium* and *Dasyurus dunmalli* from the Chinchilla LF, Qld (Louys and Price 2015); *Dasyuroides achilpatna* from Fisherman's Cliff LF, NSW (Travouillon 2016). A few were part of remnant rainforests, such as *Myoictis* sp. from the Awe LF, PNG (Archer *et al.* 1994) and '*Antechinus*' sp. from the Hamilton LF, VIC (Black *et al.* 2012).

The Australian Pleistocene consisted of extreme aridity and low sea levels during global glacial periods alternating with wetter interglacials; after 1 Ma, these occurred on a ca. 100 ka glacial + 10 ka interglacial cycles. Sand dunes (often stabilised by *Triodia* tussock grasses) and salt lakes developed after one such wet-to-dry shift ca. 1 Ma, with evidence of loess after ca. 0.5 Ma (Byrne *et al.* 2018; Pillans 2018). Thus, the widespread and highly xeric conditions that today characterise Australia's largest biome are of relatively recent origin. Pleistocene climate cycles also coincide with expansions and contractions of forest habitats along the east coast (Byrne *et al.* 2011; Bryant and Krosch 2016) and in south-western Australia (Sniderman *et al.* 2019). Humans arrived in Australia from south-east Asia at least 55 ka, possibly with disastrous consequences for the largest marsupials (see Chapter 12). With the onset of the Holocene, climates, habitats and plate boundaries (Fig. 10.3) approximated their current condition.

Australian biogeographic regions

Climatic conditions of the Holocene and the species distributions they support produced an array of terrestrial communities that biogeographers have characterised with regionalisation schemes at various spatial scales. Archer and Fox (1984) and Ebach (2012) reviewed these efforts, and a current scheme is described by the Australian government's National Reserve System (NRS, www.environment.gov.au/land/nrs). Terrestrial biome characteristics are determined by vegetation, which in

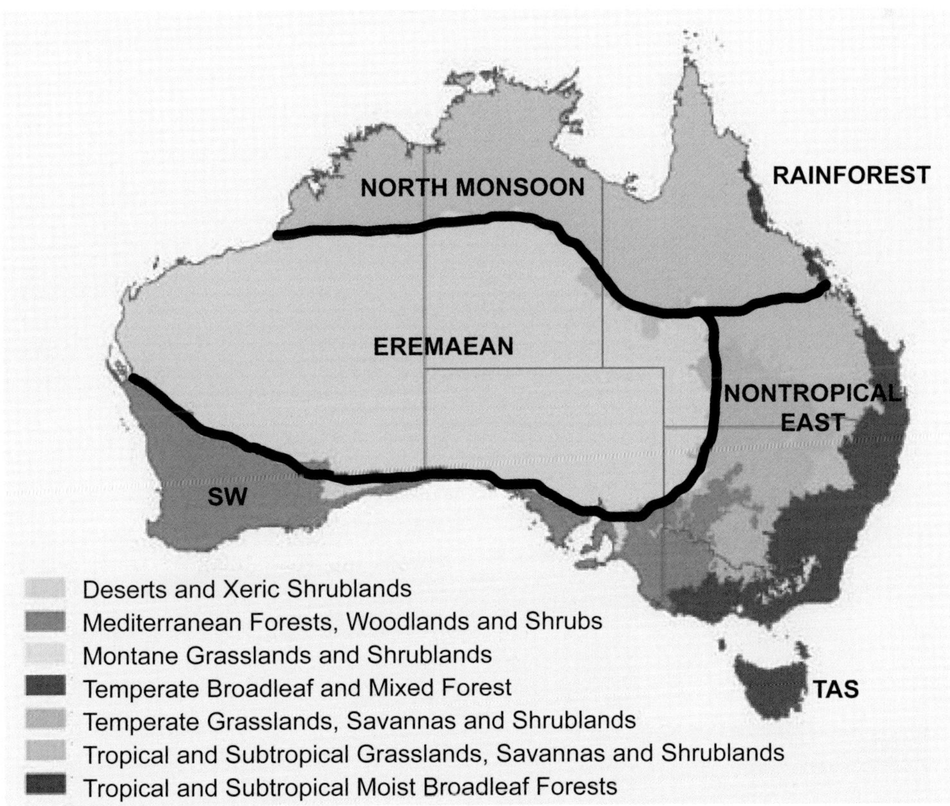

Fig. 10.4. Ecoregions defined by Johnson and Briggs (1975) based on distribution of eucalypts are delimited by bold lines and all-capital labels on a map of ecoregions recognised by the Australian government's National Reserve System. [NRS map reprinted from Australian Department of Sustainability, Environment, Water, Populations and Communities under CC BY-NC 3.0 licence.]

turn is largely determined by annual temperature and precipitation patterns. If stable for suitable periods, biome communities may develop distinct sets of species associations by which they can be recognised. Johnson and Briggs (1975) used the distribution of eucalypt and related species (members of Myrtaceae) to produce a broad-scale regionalisation that has been popular among zoogeographers (Fig. 10.4). In this scheme, central and western Australia are part of the arid Eremaean region, surrounded by the mesic North Monsoon (or Torresian), Non-tropical East (or Bassian) and South-west regions; isolated areas of Rainforest occur along the east coast, and Tasmania is a separate region from the mainland (see also Fig. 2.1).

THE RISE OF DASYURIDS

Explosive evolution and adaptive radiation

A glance at the temporal distribution of dasyurid fossils in Fig. 9.1 or the time trees in Figs 9.13 and 9.14 reveals a striking pattern. There are no definite dasyurid fossils before the early Miocene (ca. 20 Ma) and only three before the Pliocene (ca. 5 Ma), but at least 19 in the Plio-Pleistocene (not counting the many extant species with a Quaternary record). In the time trees, Dasyuridae arises in the Oligocene, splits into two subfamily and four tribal lineages in the early–mid Miocene, but the four tribes diversify into virtually all 75 living species by the mid-Pliocene (i.e. between 2–15 Ma), with most of the latter cladogenesis occurring 5–10 Ma. This burst of speciation created most dasyurid diversity in the late Miocene, followed by much less lineage splitting from the mid-Pliocene to now. Adaptations present in the current tribes appear to have originated within that narrow time window. In *Tempo and Mode in Evolution*, his classic treatise of the evolutionary Modern Synthesis, George Gaylord Simpson referred to such a pattern as 'explosive evolution', caused by the relatively sudden access of an ancestral species to new adaptive zones. It is the 'quantum evolution' end of a continuum, the other end being 'progressive occupation of adaptive zones' with slower rates of branching. Further:

> *The availability of a new adaptive zone does not depend alone on its physical existence and proximity, but also on its being open to other occupants (that is, empty) ... It is a striking fact in the history of life that major changes in the taxonomic groups occupying various ecological positions do not, as a rule, result from direct competition ... On the contrary, the usual sequence is for one dominant group to die out, leaving the zone empty, before the other group becomes abundant and occupies any position in the zone except possibly the marginal fringes. [Simpson 1944: 213]*

Simpson's words are a precis of much subsequent thought on dasyurid macroevolution, begging the questions of what adaptive zones were invaded, which species were there before and what caused the change.

Simpson (1949) described one consequence of explosive speciation as 'adaptive radiation' (Osborn 1902), now a key term in the lexicon of evolutionary biology. Indeed, the radiation of Australian marsupials that produced their iconic convergence on placentals was Simpson's first example of such a process, the course of which he characterised as follows:

> *The times of rapid expansion, high variability, and beginning of adaptive radiation are the 'explosive phases' of evolution ... They are periods when enlarged opportunities are presented to groups able to pursue them ... A quieter period ensues when the basic radiation has been completed, when, so to speak, the opportunities have been parceled out and each has been claimed by some line of populations descended from the expanding group ... Change continues but mainly in two ways: the sharpening and increasing adaptation to the way of life embarked upon, and response to changes made necessary in that way of life as time goes on – for no environment is absolutely static. [Simpson 1951a: 72–73]*

Thus, Simpson's adaptive radiation has two components: a period of explosive diversification (preceded and followed by slower cladogenesis) and new adaptations evolved in response to 'ecological opportunities'. Simpson's (1953) *The Major Features of Evolution* (an update of *Tempo and Mode in Evolution*) developed these ideas in more detail.

Schluter (2000) reformulated adaptive radiation as a hypothesis to be tested, beginning with a more explicit definition: 'Adaptive radiation is the evolution of ecological and phenotypic diversity within a rapidly multiplying lineage. It occurs when a single ancestor diverges into a host of species that uses a variety of environments and that differ in traits used to exploit those environments' (p. 2). Schluter gave four conditions that should be demonstrated to confirm an adaptive radiation: (1) common ancestry of the species involved; (2) rapid diversification of lineages to produce those species; (3) correlation between phenotypic traits (putative adaptations) of those species and the environments in which they occur; and (4) the adaptive utility of those traits. Stroud and Losos (2016) argued that ecological opportunity antecedent to adaptive radiation usually involves invasion of a new habitat that is isolated or otherwise not exploited by current species and includes much unoccupied niche space. Although these characterisations have been amended in more recent literature (e.g. Hernández-Hernández 2019; Gillespie *et al.* 2020), we use them as framework for evaluating the dasyurid radiation.

Regarding dasyurids, the first of Schluter's (2000) conditions (monophyly) has been abundantly demonstrated, regardless of whether we consider their radiation as that of a single clade or four separate lineages (tribes) evolving independently. Evidence for the latter three conditions has been slower to accumulate. Virtually all authors, however, have equated the ecological opportunity (*sensu* Stroud and Losos 2016) of dasyurids with the development of xeric habitats brought on by Neogene climate change.

The arid adaptation hypothesis

Our summary of Australian paleogeography above illustrates that dasyurid diversification took place during the transition from widespread wet forests to drier and more open habitats across north-western and central Australia. Archer (1984a) suggested that this change had three major consequences for the fauna: (1) many wet-forest species went extinct; (2) some wet-forest species survived in refuges (e.g. New Guinea, montane forests of eastern Australia); and (3) some groups evolved arid adaptions that allowed them to survive in drier habitats. Does post-Oligocene drying explain the explosive evolution of dasyurids? Several observations about dasyuromorphians are consistent with this scenario. Most of the Oligo-Miocene thylacinids were rainforest species and all but one went extinct before the Pliocene. The oldest putative dasyurid, *Barinya wangala*, inhabited an Oligocene rainforest at Riversleigh, but its contemporary and congener *B. kutjamarpensis* occurred in drier riparian or sclerophyll forests of South Australia. Most Pliocene fossil dasyurids and many extant dasyurin and sminthopsine lineages occur in xeric habitats, though a few species in these groups and many phascogalins occupy wet forests in New Guinea and Australia. These facts suggest that Archer (1984a) had identified a causal influence on dasyurid diversification, but perhaps only in broad outline. Were early dasyurids restricted to 'the marginal fringes' of Miocene wet forests? If so, was this restriction enforced by rainforest thylacinids or other competitors? Did dasyurids 'explode' because their competitors went extinct, because they were preadapted to drier habitats, or both, or neither? Were the relevant 'arid adaptations' those noted by Archer (1984a) – classic features of desert-dwelling mammals – or traits more suited to the milder conditions of the late Miocene? Answers to these questions were vague prior to recent integrative phylogenetic analyses.

The role of climate change as a driver of biotic turnover in Australian ecosystems was reiterated in later reviews by Archer *et al.* (1995) and McGowran *et al.* (2000), the latter concluding, 'a major late Cainozoic transition in types of vertebrates [including dasyuromorphians] in Australian communities occurred between 6 and 5 million years ago, a time of global cooling and drying' (p. 456). The late Miocene date cited for this transition is consistent through most of the Australian marsupial palaeontology literature. Incorporating early molecular divergence time estimates, Crowther and Blacket (2003: 125) concurred: 'The increasing aridity of the late Miocene was probably a major factor that drove the dasyurid radiation and the diversification of modern dasyurid groups ... these changes probably aided the diversification of dasyurid genera by creating a mosaic of new habitats and niches.' Archer and Hand (2006: 581) echoed the refrain:

> *Now that the Riversleigh fossil dasyurid and dasyurid-like taxa, spanning the period 24–5 m.y.a., have been studied ... it is clear that something profound happened to this group in the late Miocene, between about 10 and 5 million years ago ... The drying out of the continent in the late Miocene probably triggered not only a wave of extinctions of the older types, but stimulated a new evolutionary radiation of the modern lineages.*

As is apparent from the foregoing quotations, the 'arid adaptation hypothesis' for dasyurid diversification began as a rather sweeping generalisation based on rough correlations among continent-wide changes in paleogeography, the spatiotemporal distribution of fossils and molecular divergence dates. The picture became more nuanced after the work of Byrne *et al.* (2008) on the origin of Australia's arid biome. These authors fixed the onset of aridification at 15 Ma but emphasised its patchwork development in space and time, with much regional and local variation of habitats along the rainforest-desert continuum. Their review of molecular divergence dates from many taxa showed that current xeric-adapted lineages arose 2–20 Ma, whereas Pleistocene climate fluctuations and habitat shifts produced mostly genetic subdivision within these species (Fig. 10.5).

> *[S]ome of the major arid-zone radiations, in both plants and animals, originated ... during the Mid- to Late Miocene. It is possible that the environments inhabited by the ancestors of some arid-zone lineages may have pre-adapted them to invasion of arid environments as they developed. But it is also likely that many older arid zone lineages diversified when spatially restricted precursors of the modern arid zone developed and expanded, most likely in central and north-western Australia ... The relatively young nature of many present day arid environments does not necessarily reflect a relatively young biota. [Byrne et al. 2008: 4412]*

Xeric-adapted dasyurid lineages were among those that arose early in this interval, in some cases before 15 Ma. Moreover, the dasyurid 'arid invasion' consisted of several xeric clades independently originating from mesic ancestors, some of them undergoing subsequent diversification within the arid zone. The idiosyncratic history of arid zone colonisation by dasyurids was reiterated by Byrne *et al.* (2018). However, more detailed fossil chronologies and paleohabitat characterisations, particularly of Riversleigh local faunas (Travouillon *et al.* 2009; Arena *et al.* 2016; Woodhead *et al.* 2016), continued to reinforce the temporal association of late Miocene drying with taxon-specific extinction and cladogenesis.

Phylogenetic investigations of arid invasion

As described in Chapters 5 and 9, Mitchell *et al.* (2014) used a molecular data set of mitochondrial genomes and 26 nuclear loci to estimate relationships and divergence times of 193 marsupial species representing 97% of extant genera (see Fig. 5.2). These authors also used Bayesian analysis to estimate ancestral habitats associated with each internal branch on their phylogeny. They assigned each included species to one (or a combination) of three habitat categories: wet-closed (WC), mesic-open (MO), or arid (A). Among several models of habitat evolution investigated, a simple ordered-states model with one rate gave results comparable to more complex schemes. This model can be expressed as follows:

$$WC \overset{q}{\leftrightarrow} MO \overset{q}{\leftrightarrow} A$$

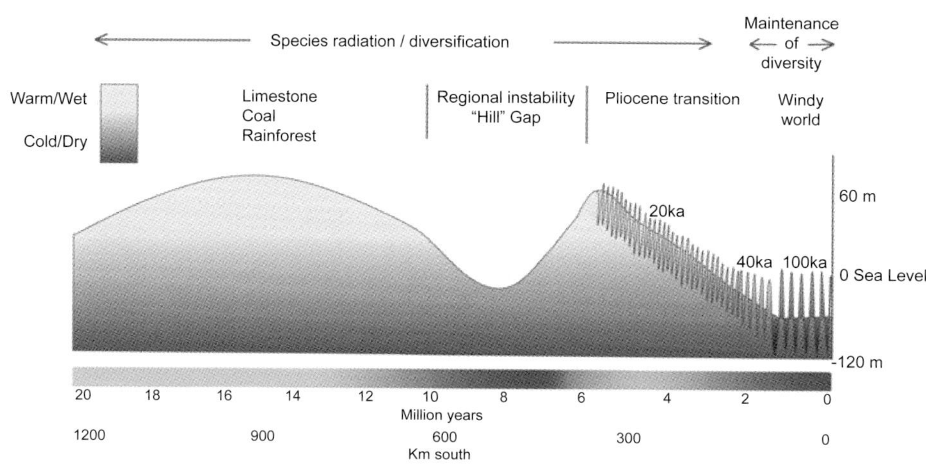

Fig. 10.5. Diversification patterns in Australian arid zone taxa (top) during the past 20 Ma along with indications of continent-wide climate conditions (Byrne *et al.* 2008). Scale at bottom shows the number of kilometres south of its current position that the Australian continent occurred as a function of millions of years before the present. [Reprinted by permission of Springer Nature from Byrne *et al.* (2018).]

where q is the transition rate between habitats over a short time interval. Posterior probabilities of q are obtained with MCMC sampling on estimated trees with branch lengths and fixed terminal states. The method of Pagel *et al.* (2004) approximates the posterior distribution of q over possible trees and branch lengths, thereby accounting for uncertainty in these parameters given the data from which they were estimated. These calculations result in a probability for each ancestor on the phylogeny occupying each of the three habitat types.

Results of the analysis (Fig. 10.6) showed that most early ancestors of Australian marsupials had a high probability of inhabiting wet forests, consistent with their existence in the Eocene-Oligocene when this habitat dominated the continent. The dasyuromorphian and dasyurid ancestors, however, are reconstructed as most likely inhabitants of mesic areas; in fact, arid ancestral habitats are more likely than wet ones. The ancestral dasyurine was also probably in mesic habitats and almost certainly not in arid ones. Dasyurini most likely originated in mesic, not wet, forest, but the opposite is true for Phascogalini. In their study of *Antechinus* evolution, Mutton *et al.* (2019) used a similar method of ancestral state reconstruction to infer a likely wet-closed habitat for the phascogalin ancestor, though their phylogeny differs from Mitchell *et al.* (2014) in having *Antechinus*, rather than *Murexia*, as sister to *Phascogale* (see Figs 6.16 and 9.13). In contrast, Mitchell *et al.* (2014) assigned high probabilities to arid habitats for ancestral sminthopsines and sminthopsins. (These authors treated planigales as a single tip with all three habitats, so no ancestral state could be estimated for Planigalini.) Within tribes, multiple transitions from wet to dry, and dry to wet, environments were inferred.

> *[I]t is generally thought that the modern Australian arid zone biota evolved from ancestral mesic forms and that the ancestral state for these clades should consequently optimize as wet-closed (Byrne et al. 2008). However, departures from this expected*

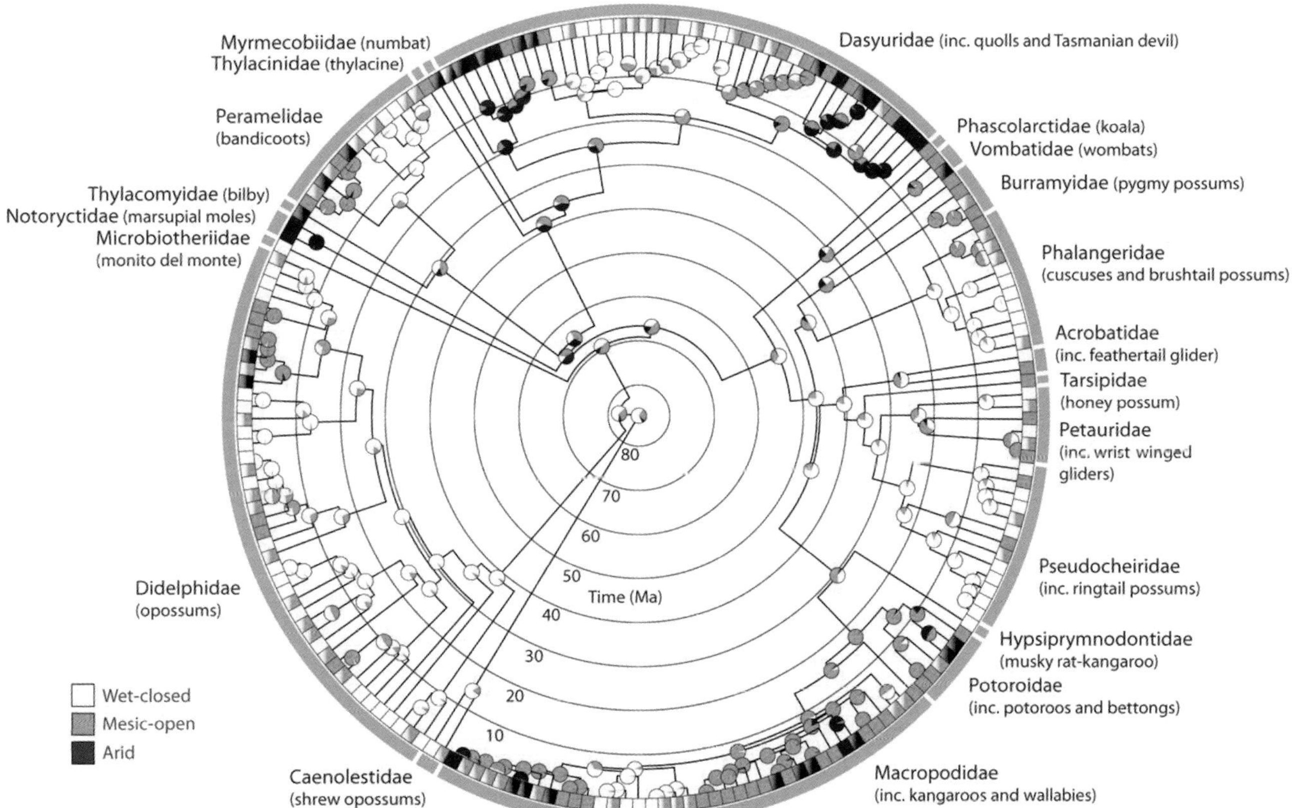

Fig. 10.6. Ancestral habitat reconstruction of marsupials from Mitchell *et al.* (2014). Ancestral states and their probabilities (pie charts) estimated with BayesTraits v1.0 software implementing the method of Pagel *et al.* (2004). [Reprinted by permission of Oxford University Press from Mitchell *et al.* (2014).]

direction of habitat evolution are observed for several Australasian taxa. For example, several clades and individual species appear to have undergone reversions from mesic-open to wet-closed habitats... particularly within Macropodidae and Dasyuridae. Additionally, several groups including Dasyuromorphia ... appear to have made the transition from wet-closed to mesic-open prior to the onset of widespread Miocene aridification. One explanation for this observation is that these older inferred transitions reflect lineages that survived from drier periods in the Oligocene when limited open forest habitats may have been available ... [Mitchell et al. 2014: 2325]

Thus, Mitchell *et al.* (2014) suggest the dasyurid invasion of arid habitats occurred independently within the Miocene tribal radiations from mesic-adapted ancestors. Moreover, and in marked contrast to the original arid adaptation hypothesis, they assign low probability to the Oligocene ancestor of extant dasyurids occupying wet forest areas. Of course, the fossil record documents that some early potential dasyurids (e.g. *Barinya wangala*) lived in rainforests and it is not clear how the inclusion of such information might affect ancestral state reconstructions, assuming the fossil taxa could be accurately located on the phylogeny. Nevertheless, Mitchell *et al.* (2014: 2326) concluded 'that evolution toward mesic-open and arid habitats has been a gradual process rather than a concerted shift ... in association with some abrupt climatic event'.

Westerman, Krajewski *et al.* (2016) amplified this theme, emphasising that their molecular divergence dates placed the origin of dasyurid tribes in the early Miocene, well before the existence of widespread xeric habitats in Australia (see Fig. 9.11). Arid invasions, they argued, occurred multiple times within tribes during the late Miocene.

The timing of dasyurid cladogenesis therefore appears to have been manifold, with initial diversification following mesic habitat differentiation in the Oligocene accompanied by progressive radiation within Thylacinidae and Dasyuridae. Further intraclade divergences occurred in the early and middle Miocene, including the appearance of extant dasyurid genera, which were apparently all in existence by the end of the epoch. Most extant species splits [on the mainland] coincide with the spread of arid-woodland habitats ... from the late Miocene. [Westerman, Krajewski et al. 2016: 697]

These large-alignment analyses of extant species thus cast doubt on whether ancestral dasyurids, or even dasyuromorphians, were rainforest adapted, suggesting instead that they occupied, and initially diversified in, isolated mesic areas before the onset of aridification (and such areas did exist; Martin 2006). In a similar study integrating morphological and molecular data, Kear *et al.* (2016) reached the same conclusion regarding bandicoots: major clades originated before the arid environments now occupied by some of their constituent species.

Tip-and-node dates obtained by Kealy and Beck (2017) moved estimated dates of tribal origins forward to the middle (for dasyurines and sminthopsins) or late Miocene (planigalins), the former coincident with the Middle Miocene Climate Transition and the latter with spread of open habitats (see Fig. 9.14). Although they did not perform ancestral habitat reconstructions, the authors argued that their later divergence estimates reconciled the implications of molecular phylogeny with paleontological explanations of dasyurid cladogenesis:

The fossil record indicates a major turnover in Australian mammals at the end of the middle Miocene ... this turnover was likely connected with the replacement of closed, wet forest by more open, drier forest and woodland, in response to a fall in global temperatures of up to 7°C (the Middle Miocene Climate Transition) ... An increase in diversification in Dasyuridae at this time is congruent with such a turnover event. [Kealy and Beck (2017: 17)]

García-Navas and Westerman (2018) asked whether the evolution of habitat associations in sminthopsin species conformed to predictions of niche divergence or niche conservatism models (see Wiens *et al.* 2010). The former pattern is consistent with descendant species adapting to new environmental conditions relative to their ancestors, whereas the latter suggests that speciation may be driven instead by factors such as altered competitive regimes or vicariance within the ancestral habitat. If, as suggested by Mitchell *et al.* (2014) and Westerman, Krajewski *et al.* (2016), arid invasions took place independently within tribes, were they early or late events in tribal evolution? Did the earliest sminthopsin lineages invade xeric habitats and diversify within them, or did invasion occur later and within specific clades or even individual species? To address this, García-Navas and Westerman (2018) used ecological niche modelling (Elith and Leathwick 2009) based on 11 climatic variables to estimate the habitat requirements of extant sminthopsin species (Fig. 10.7). These data in conjunction with the Westerman, Krajewski *et al.* (2016) time tree allowed estimation of 'predicted niche occupancy' profiles for sminthopsin ancestors, from which patterns of niche overlap among current species and niche disparity over time could be inferred. The upshot of the analyses was that sminthopsins as a whole display phylogenetic niche conservatism (PNC), but the Murina group shows evidence of niche divergence:

> *The evidence of PNC in this speciose group evidences that their diversity cannot be explained by niche shifts. That is, niche differentiation did not boost diversification in the Sminthopsini … [A]lthough in most species nonadaptive processes (such as population isolation and climate fluctuation) drive niche evolution, non-neutral processes and some degree of adaptive divergence may have been involved in the diversification of the Murina species group. [García-Navas and Westerman (2018: 10)]*

These results are consistent with the ancestral habitat reconstructions of Mitchell *et al.* (2014) in suggesting that, for sminthopsins at least, several ancestral lineages made the mesic-to-xeric transition during the mid-Miocene, some of them diversifying *in situ* to form species complexes. Moreover, some xeric-to-mesic and mesic-to-wet reversions also took place in the late Miocene.

In a complementary set of phylogenetic comparative analyses, García-Navas *et al.* (2020) found no evidence that competition structures dasyurid communities at local or regional scales, suggesting instead that the species occurrences are determined by abiotic factors such as aridity (see Chapter 11). For adaptive traits such as body mass and length of the molar tooth row, sympatric dasyurids seem to have experienced convergent evolution, further reinforcing the pattern of niche conservatism documented in the previous study.

In summary, evidence for the hypothesis that climate drying in the mid-to-late Miocene *caused* explosive evolution at the base of dasyurid subfamilies and tribes is equivocal. That dasyurids and other marsupial groups *responded* to aridification is almost certain, but much of this appears to have occurred after tribal lineages originated, during diversification within existing genera.

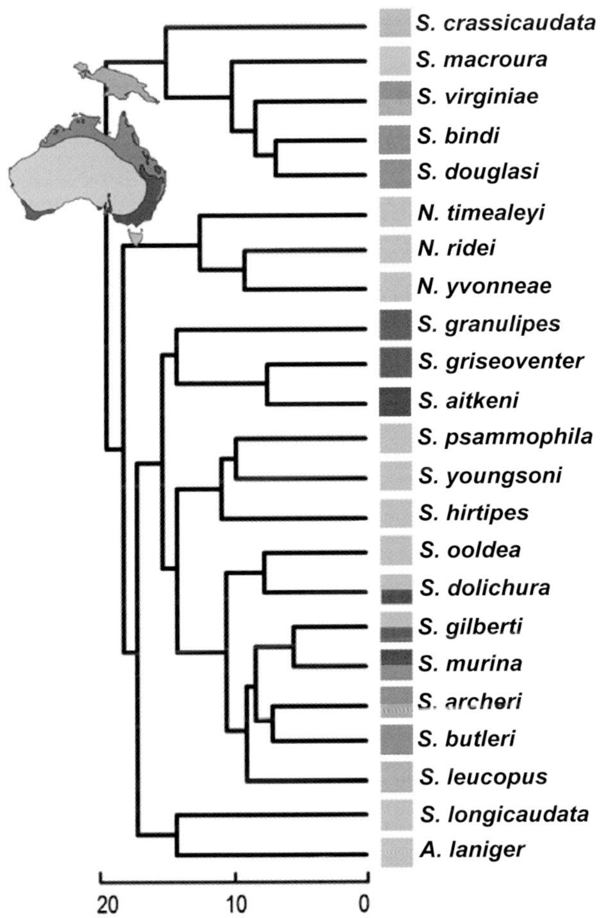

Fig. 10.7. Time tree of Sminthopsini with biogeographic regions of species indicated by colour-coded boxes at tips (García-Navas and Westerman 2018). Orange = Eremaean or arid biome; red = Southwestern mesic biome; blue = Torresian region or monsoonal tropics; purple = Bassian region; green = New Guinea; pink = Tasmania. Note: print book readers refer to the colour version of this figure in the Appendix. [Reproduced with permission of John Wiley and Sons from García-Navas and Westerman (2018), with minor adaptations.]

Competition, faunal turnover and evolutionary innovation

In an early review of paleobiology at the Riversleigh World Heritage site, Archer *et al.* (1991) offered the following remarks on the scarcity of dasyurid fossils in early-to-middle Miocene deposits:

> *Tentatively, Jeanette Muirhead attributes this limited representation to the otherwise extraordinary abundance of similar-sized peroryctid bandicoots in the same faunal assemblages. It is possible that species of the two groups competed for food on the rainforest floor, the numbers of one limiting the numbers of the other. But why ... ? [p. 91]*
>
> *[T]he relatively low diversity of Riversleigh dasyurids may have enabled or been the result of higher bandicoot diversity. In the main, modern bandicoots are omnivorous/carnivorous to herbivorous while similar-sized dasyurids are carnivorous to omnivorous which provides considerable room for competition and, probably, competitive exclusion. [p. 101]*

This idea was articulated more fully by Muirhead and Filan (1995) in their description of *Yarala birchfieldi*, type species of what is now considered a clade (Yaraloidea) distinct from all living bandicoots. Their summary of the hypothesis and its supporting evidence is worth quoting in full:

> Yarala burchfieldi *is smaller than any other known bandicoot and is comparable in size to a small antechinus. Its morphology is also unlike that of any previously described bandicoot. Because its dentition is plesiomorphic compared with that of most bandicoots, and in some regards symplesiomorphic with dasyurids, the diet of* Y. burchfieldi *was probably closer to that of many strictly insectivorous-carnivorous dasyurids than to any of the more typically insectivorous-omnivorous bandicoots. Because of its dentition and size, it is likely that* Y. burchfieldi *competed with dasyurids. This species, together with the many other bandicoots found in the deposit, could have filled insectivorous roles during the Oligo-Miocene that currently are occupied by dasyurids. Supporting the ecological role-change in bandicoots from the Tertiary to the Recent is the apparent change in ratios of bandicoot and dasyurid taxa. The Upper Site Local Fauna as well as other faunas from the Riversleigh deposit show relatively few dasyurid taxa. The sample size of each dasyurid species is also considerably low. In contrast, these faunas contain a high diversity of bandicoots and many of these are represented by large numbers of individuals. In Recent Australian communities, there are far more sympatric dasyurid than bandicoot taxa. Morphological diversity of bandicoots in the Riversleigh faunas indicates that they included both insectivorous-carnivorous as well as more insectivorous-omnivorous types with insectivorous-carnivorous forms being far more commonly occurring. Today bandicoot diversity in Australia is restricted primarily to insectivorous-omnivorous species. It appears that since the Miocene Australian bandicoots may have had the number of potential niches available to them restricted, which consequently channeled bandicoots into a more omnivorous and less carnivorous lifestyle. In contrast, dasyurids have become more diverse, occupying and today filling both insectivorous and carnivorous niches. It is possible that dasyurids have subsequently taken the role that was at least partially exploited by bandicoots. [p. 133]*

Some palaeontologists still endorse this scenario: 'It is possible that as the dasyurids differentiated in the late Miocene they out-competed comparably sized peramelemorphians and in particular the previously diverse yaralids which were extinct by the end of the mid Pliocene' (Black *et al.* 2012: 1008). Gurovich *et al.* (2014: 265) buttressed the case by including the Oligo-Miocene stem bandicoots *Bulungu* and *Galadi* as insectivore and faunivore competitors, respectively, of early dasyurids. Warburton and Travouillon (2017) agreed and Travouillon *et al.* (2017) added the Pliocene bandicoot *Silvicultor* as a late-surviving insectivore-carnivore that was ultimately replaced by dasyurids. These and other authors referred to the Miocene change in abundance of marsupial groups as a 'faunal turnover', though none formally tested it against the predictions of Vrba's (1985) controversial 'turnover-pulse' hypothesis. In contrast to studies supporting a causal role for bandicoot-dasyurid competition in faunal turnover, Kear *et al.* (2016) questioned that hypothesis in light of evidence from fossils and molecular time trees that arid- and forest-adapted bandicoots coexisted from the Eocene-Oligocene until the Quaternary. They noted that if there was no dramatic mid-Miocene faunal turnover within Peramelemorphia, the influence (if any) of competition with dasyurids becomes unclear.

In his description of *Muribacinus*, Wroe (1996: 1041) initiated a parallel narrative about Miocene competition between rainforest thylacinids and dasyurids, predicated on their inferred dietary overlap.

> *Disparate diversity during the Miocene for thylacinids and dasyurids has clearly been reversed in Pliocene to present times. The radiation of dasyurid lineages into niches for small to medium sized carnivores may have been facilitated passively by the widespread extinction of thylacinids following late Miocene climate change … or actively through competitive exclusion.*

Muirhead (1997: 374) focused the 'thylacine competition' hypothesis on large-bodied dasyurines.

> *… prior to the late Miocene, available resources enabled the 'thylacine niche' to be more finely divided. Part of the explanation may be found in the apparent absence from the Riversleigh local faunas of any large dasyurids as specialised for carnivory as the late Cainozoic species of* Glaucodon, Sarcophilus, *and* Dasyurus. *Presence of large carnivorous dasyurids appears inversely correlated with thylacinid diversity. The subsequent rise of these dasyurines may, therefore, have accompanied late Miocene decline in thylacinid diversity.*

In this view, the Miocene small marsupicarnivore niche was contested by comparably sized small thylacinids and large dasyurids. The former prevailed initially but succumbed to extinction, leaving only the hypercarnivorous *Thylacinus* from the Pliocene on, while the latter diversified into domination. Wroe (1997a: 43) articulated the alternative hypotheses with remarkable clarity:

> *Niches filled by small to medium-sized thylacinids during Miocene times have since been occupied by specialized dasyurids. Three possible explanations for this phenomenon are proposed: (1) dasyurids outcompeted thylacinids for existing niche spaces following the derivation of 'key innovations' which conferred an adaptive advantage on Dasyuridae; (2) changing extrinsic factors differentially selected the relatively specialized Thylacinidae for extinction over the generalist, less ecologically constrained, Dasyuridae; or, (3) changing extrinsic factors provided new niche spaces for which dasyurids were preadapted. Each of these hypotheses may stand as an explanation singularly or in some combination with the remaining factor(s).*

Wroe's argument for scenarios 1 and 3 is described below; scenario 2 related the thylacinid-dasyurid turnover to Miocene climate change and loss of wet forest habitat as the extrinsic factors that drove thylacinids to extinction. 'Subsequently, dasyurids may have assumed dominance by default and simply adapted to fill vacated positions' (Wroe 1997a: 44). The distinction here is between (1) changing patterns of active competitive exclusion based on differential adaptation of the two groups versus (2) extinction of one group simply clearing niche space into which the other diversified. Muirhead and Wroe (1998) reiterated the thylacine-dasyurid competition hypothesis in their description of *Badjcinus*, Wroe's (1999b) description of *Barinya* extended it to include bandicoots and it appeared again in Wroe and Musser's (2001) analysis of *Nimbacinus*.

One of the few anatomical synapomorphies that Wroe (1999b) validated for Dasyuridae was an expanded alisphenoid tympanic wing coupled with a periotic hypotympanic sinus (see Chapter 11). Along with development of the petrosal tympanic wing, this has the effect of enlarging and enclosing the middle ear of dasyurids (Fig. 10.8). This in turn improves the ability to detect low frequency sound (Webster and Webster 1975; Mason 2016), which would be adaptive for predator avoidance in the open spaces of semi-arid and arid habitats. Wroe argued that the dasyurid middle ear is an example of a 'key innovation', defined by Heard and Hauser (1995: 152) as 'an evolutionary change in individual trait(s) that is causally linked to an increased diversification rate of the resulting clade (for which it is a synapomorphy)'. Such a trait would be efficacious in either of Wroe's (1999b) adaptative scenarios – allowing dasyurids to prevail in direct competition with thylacinids and bandicoots or as a preadaptation to the empty xeric niches they invaded in the late Miocene. Deciding between these alternatives has so far been impossible given the sparse fossil record and uncertainty about the timing of extinctions and arid invasions (Kealy and Beck 2017).

A classic adaptive radiation?

In most of the literature described above there has been the implicit assumption that the rise of dasyurids constitutes an adaptive radiation in the sense of Simpson (1949). More recently, however, investigators have taken to heart Schluter's (2000) dictum that adaptive radiation is a hypothesis to be tested. But how to do so? Following Felsenstein's (1985b) insight that phylogenetic relationships must be considered when analysing trait correlations among species, a wave of 'phylogenetic comparative methods' (PCMs) have been developed to test hypotheses about character evolution and clade diversification (O'Meara 2012; Garamszegi 2014). Unlike phylogenetic inference methods that implement just a few basic forms of statistical estimation, PCMs constitute a diverse family of statistical techniques tailored to specific questions. Several of these were applied to the dasyurid radiation by Kealy and Beck (2017) and García-Navas *et al.* (2018).

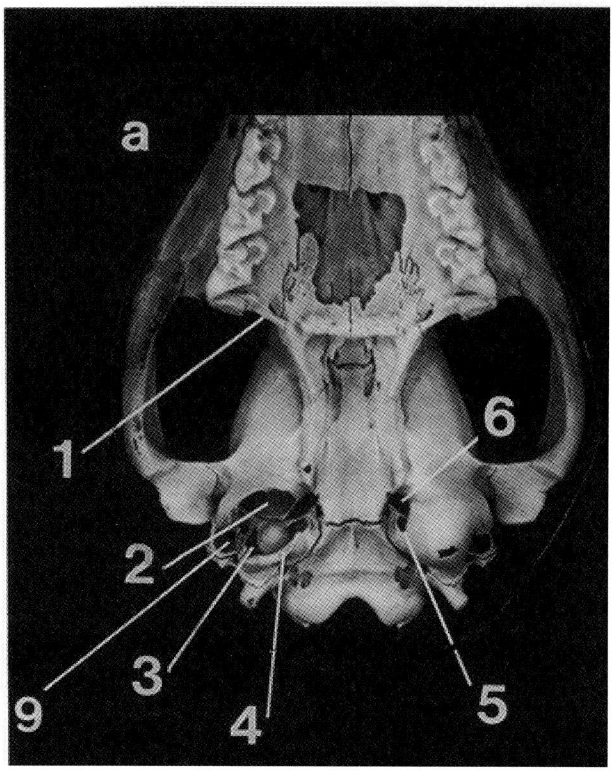

Fig. 10.8. Ventral view of *Dasyurus viverrinus* skull (posterior region) showing the expanded middle ear chamber on the left side of image, with right alisphenoid tympanic wing removed to expose the alisphenoid and periotic tympanic sinuses (from Wroe 1997a). Labelled structures: (1) posterolateral palatal foramen; (2) alisphenoid tympanic sinus; (3) periotic tympanic sinus; (4) periotic tympanic process; (5) tubular foramen; (6) foramen ovale; (9) squamosal epitympanic sinus. [Reprinted by permission of Springer Nature from Wroe (1997a).]

Expected features of a Simpsonian adaptive radiation include a rapid burst of speciation resulting in lineage and phenotypic diversification, followed by a slowdown as newly invaded niche space becomes saturated. Modern analyses have shown that 'early burst' models of phenotypic change fit relatively few actual adaptive radiations (Harmon *et al.* 2010), but much attention has been given to quantifying changes in the rate of lineage accumulation. Nee *et al.* (1992) introduced *lineage through time* (LTT) plots to visualise and estimate rates of diversification within clades from molecular time trees and Harmon *et al.* (2003) pioneered the use of LTTs in testing for adaptive radiation. These graphs show the logarithm of the specific or cumulative number of lineages in existence at time points between the origin of a clade and the present. Because the plots are constructed from trees of extant species, cumulative diversity increases over time, but the rates of increase may change, revealing trends or episodes in macroevolutionary history. The null expectation for an LTT is that lineages will accumulate in either a pure-birth (Yule) or birth–death process at a net rate given by the difference between speciation and extinction rates. If a single net rate applies throughout the history of a clade, lineage accumulation will be steady; if multiple rates apply, the LTT pattern will be irregular. Given an empirically derived LTT, a statistical test or model-selection procedure may be used to determine whether one or several rates best fit the data.

An alternative way to estimate divergence rates derives from Rabosky *et al.*'s (2013) Bayesian Analysis of Macroevolutionary Mixtures (BAMM) model which estimates the posterior distribution of rate 'configurations' on a time tree using the MCMC method. The model accounts for temporal variations in split frequency by fitting discrete rate shifts to individual branches, evaluating their likelihoods and returning a credible set of such configurations, including a point estimate of the configuration with highest posterior probability.

Kealy and Beck (2017) and García-Navas *et al.* (2018) performed LTT and BAMM analyses on dasyurid time trees, the latter using the phylogeny of Westerman, Krajewski *et al.* (2016). Their LTT plots (Fig. 10.9) show a distinct upturn in the rate of lineage accumulation at or just after 15 Ma. Kealy and Beck's (2017) analysis also revealed bursts within dasyurins, phascogalins and sminthopsins at 5–10 Ma, and planigalins at 6–7 Ma.

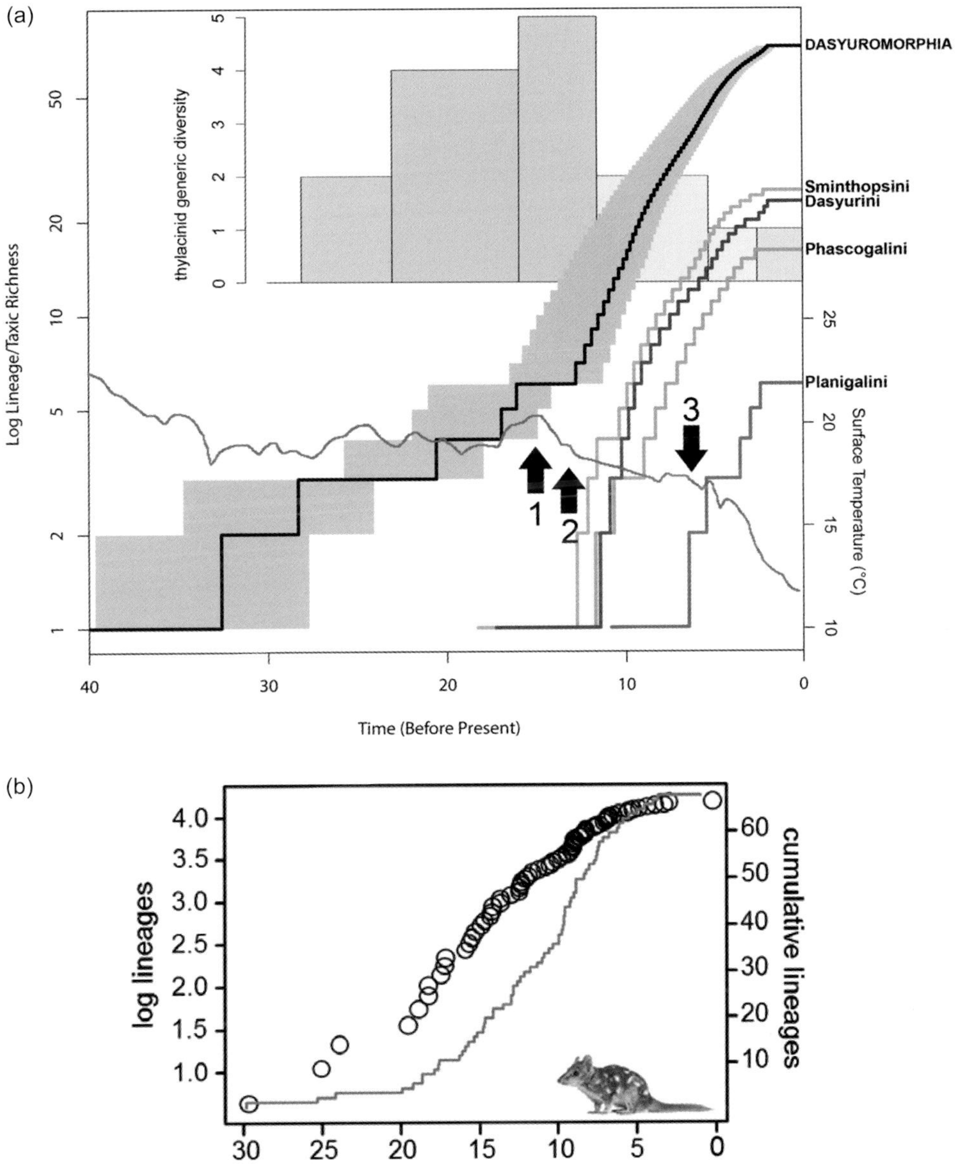

Fig. 10.9. Lineage through time (LTT) plots for dasyurids. (a) From Kealy and Beck (2017), based on their 'TipNodeCalib1' time tree. Separate plots are shown for dasyuromorphians (black curve with grey area representing 95% confidence interval) and each dasyurid tribe. The red curve shows estimated global surface temperatures. Inset histogram shows the number of thylacinid genera from the fossil record in discrete time blocks on the same time axis as the LTT plots. Arrows: 1 = Middle Miocene Climatic Optimum; 2 = Middle Miocene Climatic Transition; 3 = increased frequency of fossil grass pollen in Australia. (b) From García-Navas *et al.* (2018), based on the time tree of Westerman, Krajewski *et al.* (2016). Dots are lineage counts at specific time points and the grey curve shows the cumulative number of species. [(a) Reprinted from Kealy and Beck (2017) under CC BY 4.0 license. (b) Reprinted by permission of Oxford University Press from García-Navas *et al.* (2018).]

These times are consistent with the authors' interpretation of their dated phylogeny, which emphasised, respectively, coincidence with the Mid-Miocene Climate Transition and a Pliocene increase in grass pollen (reflecting the dry, grassy habitats favoured by planigales). García-Navas *et al.* (2018) statistically rejected the hypothesis that the dasyurid LTT pattern could be explained by a constant rate of diversification. Moreover, model comparisons with the Akaike Information Criterion (AIC) suggested that a variable-rate model (logistic diversity-dependent speciation rates) gave a better fit than constant-rate (pure birth and birth-death) models. In both studies, however, BAMM analyses assigned highest probability to a configuration with no rate shifts, possibly due to this method's limited power to detect shifts on trees with fewer than 100 tips (V. García-Navas, personal communication). Kealy and Beck (2017: 14) made a subtle distinction to explain this discordance: 'although the LTT plots show a sharp increase in the diversification of dasyurids … this is not interpreted as a significant change in diversification rate.' García-Navas *et al.* (2018: 786) similarly remarked that 'none of the branches was associated with a speciation-extinction rate slowdown'. Both sets of authors agree that LTT plots indicate an episode of explosive evolution in dasyurid tribes roughly coincident with rapid climate and habitat change in the late Miocene, thus supporting the 'radiation' portion of an adaptive radiation hypothesis for dasyurids.

García-Navas *et al.* (2018) tested two other predictions of Simpson's (1953) 'ecological opportunity' (EO) hypothesis as elaborated by Schluter (2000) and Stroud and Losos (2016). First, they looked for evidence of 'morphological disparity' among dasyurid clades using data on body mass, a trait they argue shows significant adaptive variation along the mesic–xeric habitat gradient. The EO model predicts that body mass is correlated with phylogeny, that rates of body mass evolution show the same acceleration-then-slowdown pattern seen in LTT plots as niche space fills, and (for dasyurids) that body mass is correlated with some aspect of arid habitats. Second, they asked whether dasyurid species show 'phylogenetic clustering' by geographic region, given that regions occupy different positions on the mesic–xeric gradient. Blomberg *et al.*'s (2003) K statistic, which measures phylogenetic signal in a phenotypic trait, was $K = 1.44$ ($P = 0.001$), suggesting that longer branches (older lineages) contribute proportionately more to body size evolution than shorter branches. This is consistent with an 'early burst' model in which most body mass change occurred in old lineages that radiated in mesic or xeric areas. García-Navas *et al.* (2018) applied 10 different tests to the predicted body-mass rate deceleration in dasyurids, four of which they considered to provide strong supportive evidence: (1) the value Harmon *et al.*'s (2003) 'morphological disparity index' was MDI = −0.225, suggesting that body mass varies more among, rather than within, dasyurid clades; (2) the correlation of node heights with phylogenetic independent contrast values of body mass was −0.055 ($P = 0.02$), suggesting that body mass evolution slowed as the number of species increased; (3) AIC comparisons favoured a niche-filling model of body mass evolution over Brownian motion (random evolution) and Ornstein-Uhlenbeck (stabilising selection) models; and (4) the correlation between species body mass and the mean annual temperature of their ranges was negative, suggesting small body size is associated with warm habitats. Finally, García-Navas *et al.* (2018) found significant phylogenetic clustering of dasyurid species in New Guinea and Eremaean regions at opposite ends of the mesic–xeric continuum (see Fig. 9.12). Coupled with Wroe's (1997a) key innovation argument, this evidence suggests that that dasyurid radiation was indeed adaptive.

'So, what makes this radiation special?' asked García-Navas *et al.* (2018: 792). Their answer struck familiar chords:

> *The most important feature is probably the elevated synchronicity of this radiation … Most splits in each of the dasyurid subfamilies appear to coincide with two historical episodes: the spread of more open woodland-grassland habitats during the early to middle Miocene and the waves of dispersal into proto-New Guinea islands … [A] relaxation of interspecific competition (attributable to the progressive decline in thylacinid diversity) in conjunction with the availability of new underexploited areas appears to underlie this rapid and explosive radiation.*

Thus, the early speculation of Archer (1984a) that Australian dasyurids owe their success to invading Neogene xeric habitats has largely been supported by recent phylogenetic comparative analyses. The latter studies, however, suggest that the arid invasion took place later than envisioned by Archer and independently within each tribe. The proposals of Muirhead and Filan (1995) and Wroe (1996) that these invasions were facilitated by the extinction of early bandicoots and thylacines remain plausible but have not been tested rigorously (and are not likely to be until a denser sampling of fossils from all three groups becomes available).

HISTORICAL BIOGEOGRAPHY OF DASYURIDS

Dasyurids in New Guinea

Several dasyurid clades and individual species are endemic to New Guinea: *Murexia* (at least five species); *Myoictis* (four species); *Phascolosorex* + *Neophascogale* (three species); *Dasyurus* (two species); and *Planigale novaeguineae.*

Two species – *Sminthopsis archeri* and *S. virginiae* – occur in both southern New Guinea and northern Australia. Because all these taxa are nested within larger clades having most of their diversity in Australia, there is no possibility that they represent a single radiation or primitive stem lineages to a single Australian clade (see Chapters 4–6). This rejects a phylogenetic interpretation of Archer's (1984c: 636) suggestion: 'Some of the highland New Guinean living forms such as species of *Murexia* and *Neophascogale* resemble the fossil forms from the middle Miocene sites in central Australia … these living genera are in a way relics in a little-changed environment preserved by uplift into cooler wetter altitudes.' If these dasyurids are not early evolving stem lineages isolated in a rainforest refugium, how and when did they get to New Guinea?

Most areas of present-day New Guinea that have been above water for more than 5–6 Ma, such as the south-eastern peninsula, were initially submerged (van Ufford and Cloos 2005). As described in 'Cenozoic climate and geography of Australia', the Australian and Pacific plates initially collided in the Oligocene, leading to the formation of several terranes and later uplift of New Guinea's Central Cordillera (Fig. 10.10). Although terranes such as the East Papua Composite Terrane (EPCT) and those which accreted along the north coast initially constituted archipelagos, over the past 20–30 Ma they merged with each other and with the northern edge of the Australian Craton to form modern New Guinea (Klootwijk *et al.* 2003; Cloos *et al.* 2005; van Ufford and Cloos 2005; Toussaint *et al.* 2014; Rivera *et al.* 2017). At various times during this interval, areas of New Guinea and northern Australia were in proximity or even connected by land as sea levels and emergent coastlines fluctuated. Thus, ancestral dasyurids and other terrestrial taxa had ample opportunities for dispersing into New Guinea, becoming isolated there and perhaps undergoing subsequent speciation. Because each of the six endemic dasyurid groups listed above is sister to an Australian clade, each represents a separate dispersal event. But *when* did these events occur? Were they synchronous or diachronous? Few fossils exist to address these questions, so investigators have turned to molecular divergence dates and time trees.

Ziegler (1977) postulated multiple dispersals of marsupials from Australia to islands of the early New Guinean archipelago, with the descendants of these waifs responding evolutionarily to the fusions and fissions of land masses brought about by crustal movement and changing sea level during the past 30 Ma. He considered all New Guinean endemic dasyurid genera and species to have originated *in situ*; the former 'were very early Australian immigrants to Northern New Guinea, possibly even 'rafting' in before [the pre-Pliocene] merger of the North Australian land mass and portions of Northern New Guinea' (Ziegler 1977: 125). Dow (1977) proposed a paleogeographic model in which a large, proto-New Guinean landmass was continuous with Australia in the late

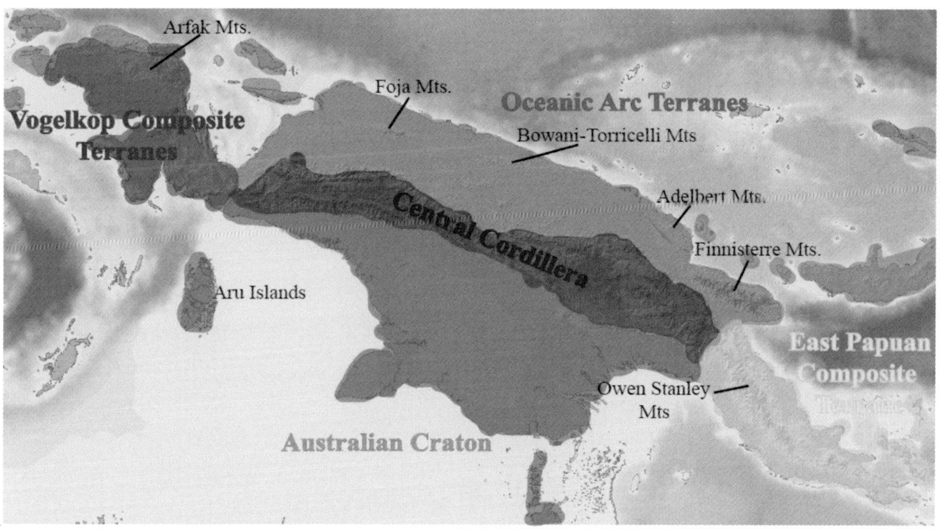

Fig. 10.10. Simplified geological reconstructions of New Guinea showing the four major terranes accreted to the Australian Craton since the early Miocene. Mountain ranges are indicated, with the Owen Stanley Ranges acting as a barrier between lowlands to their north-east and south-east and the Central Cordillera acting as a major barrier between regions to the north and south. [Adapted from Tallowin *et al.* (2018).]

Table 10.1. Divergence date estimates (Ma) for endemic New Guinean dasyurid clades from their Australian sister groups based on DNA sequence studies.

		Relaxed clock			Tip-and-node	
Clade	mtDNA clock	Mitchell *et al.* (2014)	Westerman, Krajewski *et al.* (2016)	García-Navas *et al.* (2018)	Kealy and Beck (2017)	García-Navas *et al.* (2020)
Murexia	13.4–21.1[1]	15	17.4	17	11	7.8
Phascolosoricines	12.0–13.8[2]	12	16.0	16	16	7.6
Myoictis	14.1–14.6[3]	10	13.0	12	9	7.2
D. albopunctatus	6.1[4]	4	7.0	7	3	3.7
D. spartacus	4.0–4.2[2]	2	3.0	3	2	1.4
P. novaeguineae	4.5[4]	–	9.0	8	2	2.2

[1] Springer *et al.* (1994); Krajewski, Wroe and Westerman (2000).
[2] Krajewski, Blacket *et al.* (1997); Krajewski, Wroe, and Westerman (2000).
[3] Krajewski, Wroe, and Westerman (2000); Westerman *et al.* (2006).
[4] Krajewski *et al.* (1993).

Oligocene, then isolated and re-fragmented via subsidence in the early Miocene. Accepting this model, Flannery (1989, 1990) formulated a vicariance hypothesis for the origin of New Guinean mammals in which endemic clades originated in the early Miocene and widespread species (i.e. those also occurring in Australia) arrived during times of low sea level in the Pleistocene. The former included all New Guinean dasyurids except *P. novaeguineae* and the two *Sminthopsis* species, which Flannery considered Pleistocene isolates.

In a single issue of *Science in New Guinea*, three groups of investigators tested Flannery's (1989, 1990) hypothesis with different molecular clocks. These studies, described in Chapter 9, were uniform in rejecting a single Miocene vicariance event. Aplin *et al.* (1993) interpreted their albumin MC′F distances as supporting episodes of isolation in the early Miocene, late Miocene (10–12 Ma) and Pliocene (2.7–4.7 Ma). Kirsch and Springer's (1993) calibrated DNA hybridisation distances implied Miocene isolations spread over 8.1–12.5 Ma. Krajewski *et al.* (1993) used partial *cytb* divergences to estimate New Guinean dasyurid isolations in the range of 4.3–9.3 Ma. Although Springer *et al.*'s (1997) estimate of a 21.1 Ma *Phascogale-Murexia* split from 12S transversions was high, most subsequent estimates from mitochondrial molecular clocks gave divergence times in the middle–late Miocene (Table 10.1). If emergent parts of New Guinea were separated by ocean from northern Australia between the early Miocene and Pleistocene, these dates imply that dasyurid colonists must indeed have arrived via sweepstakes routes as envisioned by Ziegler (1977). The alternatives, that the molecular clocks were wrong or that land connections existed during the Miocene-Pliocene, remained to be evaluated by subsequent studies. Flannery (1995) suggested that some geological evidence might support an emergent dispersal path in the late Miocene.

Estimates of divergence times for endemic New Guinean clades from large-alignment and integrative studies bracket those from mtDNA (Table 10.1), with relaxed clock dates generally older and tip-and-node dates generally younger. With a single exception these estimates agree that endemic groups were isolated during the Miocene and Pliocene. The most recent analyses (Kealy and Beck 2017; Mutton *et al.* 2019; García-Navas *et al.* 2020) date the origin of multispecies lineages *Murexia*, phascolosoricines and *Myoictis* to the later Miocene (12.5–7.8 Ma), while single-species isolates are Pliocene (*D. albopunctatus*, *D. spartacus*,) or early Pleistocene (*P. novaeguineae*).

Such are the currently available answers to questions of how (multiple dispersals) and when (late Miocene–early Pleistocene) dasyurids reached New Guinea. What were the characteristics of these colonists? Ziegler (1977) stressed that most endemic species (*Murexia*, *Phascolosorex*, *Neophascogale*, *Myoictis*, *D. albopunctatus*) are wet-forest inhabitants. Was the same true of their colonist ancestors? The ancestral habitat reconstructions of Mitchell *et al.* (2014) and Mutton *et al.* (2019) shed some light on this (see Figs 10.6 and 9.13). Wet forest habitat had the highest probability for ancestral *Murexia*, as well as the common ancestor of *Murexia* and its putative sisters *Antechinus* or *Antechinus* + *Phascogale* – i.e. ancestral Australian murexias lived in the same kind of habitat as their modern descendants. More recent phylogenies, however, suggest that *Phascogale* is the sister of

Murexia. Given that the ancestral phascogale was probably mesic, the same may have been true for the original *Murexia* under this branching order. The ancestral phascolosoricine was reconstructed as a probable wet-forest dweller, but its common ancestor with *Dasyurus* and *Sarcophilus* was more likely mesic; thus, living in rainforests may be a derived trait for *Phascolosorex* and *Neophascogale*. *Myoictis* species are found in both wet and mesic areas of New Guinea, but their common ancestor with *Parantechinus* was probably mesic, again suggesting secondary adaptation to wet forests. The sister of *D. albopunctatus* among quolls is not resolved, but most studies place it in a clade with *D. viverrinus*, an inhabitant of both wet and mesic areas. *D. spartacus* and *P. novaeguineae* are savanna dwellers, a mesic habitat comparable to that of their sisters (*D. geoffroii* and *P. maculata*, respectively). In sum, our current understanding of phylogeny, divergence times and ancestral areas suggest that most New Guinean rainforest dasyurids evolved from mesic-adapted colonists. They are not relictual lineages isolated in a refugium that preserves the original dasyurid habitat.

The foregoing scenario for *Murexia* is consistent with a radiation facilitated by the creation of new habitats in New Guinea resulting from merging of new terranes (such as the EPCT), mountain building (the Central Cordillera) and later accretions along the northern edge of the New Guinean land mass. Ongoing diversification within *Murexia* may have been associated with colonisation of new habitats on these terranes as they arose. It is probable that one of the first *Murexia* species to differentiate in New Guinea was *M. rothschildi*, a species limited to the EPCT (Woolley *et al.* 2020); subsequent radiations likely took place in a westerly direction as new habitats developed. Presence of late Pleistocene *Murexia*-like fossils ('cf. *M. melanurus*') in the Mt Etna deposits of northern Queensland (Cramb *et al.* 2009) suggest the possibility of a later, over-land, back migration at a time of recent low sea levels.

Areas of endemism and dispersal barriers

When clusters of close phylogenetic relatives from multiple taxa occur within circumscribed regions, those regions are termed 'areas of endemism' (Nelson and Platnick 1981). Such areas reflect the geographic history of the region, evolutionary history of resident taxa and current ecosystem conditions. Because these factors are seldom highly congruent across broad stretches of land, continental areas of endemism have proven difficult to define (Harald and Mooi 1994), although they may coincide with ecoregions (see 'Australian biogeographic regions'). Perhaps the most influential delineation of Australian areas of endemism was that based on the phylogeny and distribution of vertebrates by Cracraft (1991) in which 14 areas subdivide the ecoregions of Johnson and Briggs (1975), some of them separated by long-recognised dispersal barriers (Fig. 10.11). These regions proved roughly consistent with the distributional pattern for many faunal and floral groups studied subsequently, including *Triodia* grasses (Anderson *et al.* 2016), *Tiliqua* lizards (Ansari *et al.* 2019), argiolestid damselflies (Kalkman *et al.* 2018) and the species of eastern tetrapods reviewed by Bryant and Krosch (2016).

A casual survey of the distributions shown in Chapter 2 reveals that many dasyurid species also 'respect' these area boundaries – e.g. *Dasykaluta* in the Pilbara, *Pseudantechinus ningbing* in the Kimberley Plateau, *Antechinus agilis* in the South-eastern Forest, *A. bellus* in Arnhem Land, *A. godmani* in the Atherton Plateau, *A. leo* in the Cape York Peninsula, *Planigale tenuirostris* in the Eastern Desert, and *Sminthopsis granulipes* in the

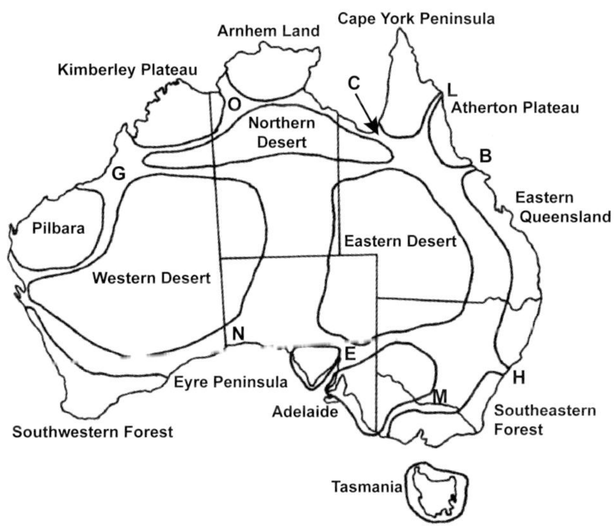

Fig. 10.11. Areas of endemism defined by Cracraft (1991) based on the distribution and phylogeny of vertebrates. Capital letters added to show the approximate locations of named dispersal barriers: B = Burdekin Gap; C = Carpentarian Gap; E = Eyrean Barrier; G= Great Sandy Desert; H = Hunter Valley Gap; L = Laura Basin; M = Murray River; N = Nullarbor Plain; O = Ord Arid Intrusion. [Redrawn with permission of CSIRO Publishing from Cracraft (1991). Permission conveyed through Copyright Clearance Center, Inc.]

Southwestern Forest. Moreover, phylogeographic studies of widespread species whose ranges span multiple areas have found genetic subdivisions among populations in each area (see Chapter 7). Notable among the latter are several species of *Dasyurus* (Firestone *et al.* 2000; Woolley *et al.* 2016; Fig. 10.12), the *Phascogale tapoatafa-pirata* complex (Spencer *et al.* 2001) and *Planigale maculata* (Westerman, Blacket, *et al.* 2016). Indeed, much of the recent revision of *Antechinus* involves recognition of new species in single areas or subdivisions within single areas (Mutton *et al.* 2019). As in the case of New Guinea, however, the timing of evolutionary isolations and dispersals that produced these spatial patterns spans the entire 15 Ma history of the tribal radiations. The takeaway lesson from this body of research seems to be that no single series of biogeographic events (e.g. progressive vicariant subdivision of a once homogenous rainforest Australia) explains modern distributions. Rather, the ranges of dasyurids and other native taxa are the result of long-running interactions between climate change, community assembly, dispersal and adaptation that played out uniquely in each lineage.

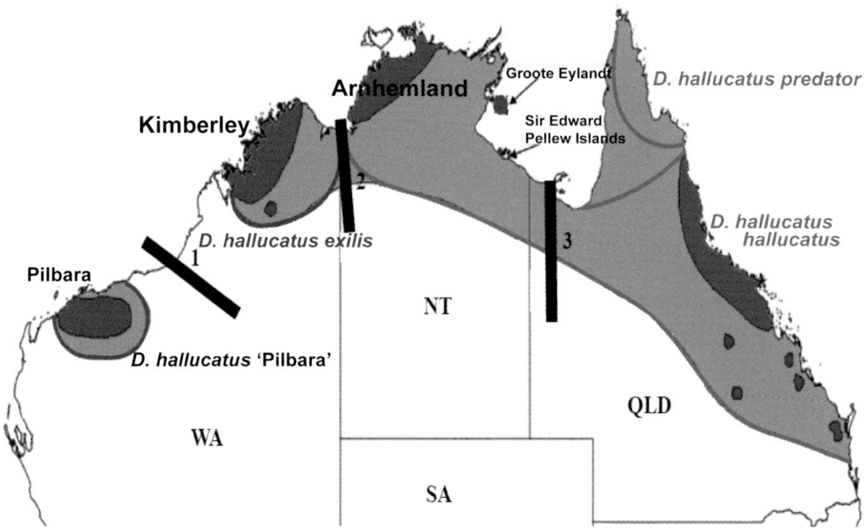

Fig. 10.12. Distributions of previously recognised subspecies of northern quolls (*Dasyurus hallucatus*) together with the locations of known biogeographic breaks for other taxonomic groups that match gaps in the distribution of northern quolls. 1 = Great Sandy Desert; 2 = Ord arid intrusion; 3 = Carpentaria. Dark grey areas indicate present day distributions, lighter grey historical distributions.

11 Evolution of the dasyurid phenotype

Well-supported phylogenetic trees provide frameworks for comparative analyses of biotic history. In Chapter 10 we saw how knowledge of phylogenetic relationships and divergence times has been employed to understand dasyurid macroevolution, leading to hypotheses about adaptive radiation and biogeography. Here we turn our attention to organismal traits. To date, only a few attempts have been made at using phylogeny to understand phenotypic evolution in dasyurids. These involve four types of studies: (1) evaluating potential craniodental synapomorphies for Dasyuridae; (2) tracing the evolution of the unusual penis morphology of some dasyurids; (3) reconciling the diverse array of dasyurid reproductive strategies with phylogenetic history; and (4) placing the conservative pattern of dasyurid chromosome variation in the broader context of marsupial karyotype evolution.

SKULLS AND TEETH

Comparative anatomy is much older than phylogenetic systematics and morphology is the original source of characters for taxonomy. In Chapters 3–8 we describe the use of anatomical traits to develop dasyuromorphian classification and estimate relationships, but here we examine how knowledge of phylogeny (specifically the resolution of a monophyletic Dasyuromorphia comprising thylacinids, myrmecobiids and dasyurids) informed the assessment of potential synapomorphies for Dasyuridae. Such efforts have involved almost exclusively craniodental traits for two reasons: (1) Michael Archer's foundational studies of dasyurid comparative anatomy in the 1970s focused on this division of the skeleton; and (2) the fossil record of dasyuromorphians consists primarily of teeth and skull fragments (see Chapter 8). Before tracing the course of this research, we summarise how uncertainty over marsupial tooth homologies was clarified by comparative ontogenetic studies.

Establishing tooth homologies

Owen's (1840–45) classic treatise *Odontography* established the tooth categories of heterodont mammals as incisors, canines, premolars and molars. Various criteria were proposed to distinguish these types, including jaw position (e.g. upper incisors are rooted in the premaxilla) and shape (e.g. canines are 'conical, pointed and longer than the rest'). Premolars and molars were thought to differ in ontogeny, the former (like incisors and canines) having deciduous ('milk') precursors, but the latter having only a single generation. These generalised criteria, however, gloss over considerable variation among mammal groups. Counts of each tooth type on one side of the upper and lower jaws are conveniently expressed as a dental formula and students of mammalogy are routinely taught that the ancestral dental formula of marsupials is I 5/4, C 1/1, P 3/3, M 4/4 in contrast to the placental formula of I 3/3, C 1/1, P 4/4, M 3/3 (Feldhamer *et al.* 2020).

Flower (1867) was the first to point out the differing counts of premolars and molars in the two clades and he also noted that only the last (most posterior or third) marsupial premolar has a deciduous precursor. This simple observation initiated over a century of confusion about the homology of marsupial cheek teeth that has impacted our understanding of dasyurid dentition.

Owen's (1868) resolution of the discrepancy between marsupial and placental premolar counts was that the homologue of the placental P4 does not develop in marsupials and the homologue of the placental deciduous P4 (dP4) is the marsupial M1. Owen was also aware that some dasyurids (e.g. *Dasyurus*, *Sarcophilus*) have only two adult premolars and he argued that these two were homologous to P2 and P3 of other marsupials (e.g. *Thylacinus*), the P1 having been lost. In contrast, Flower (1869) concluded that it was P3 that had been lost, given the marked reduction of this tooth in several dasyurid species (e.g. *Dasycercus cristicauda*, *Myoictis melas*). Pointing out that some antemolar teeth in some species are not replaced, Flower drew a dual distinction between deciduous versus permanent, and primary versus successional, teeth. Deciduous teeth are replaced; permanent teeth are not. Primary teeth develop first and may or may not be replaced; successional teeth replace primary teeth. Thomas (1887) agreed that *Dasyurus* and *Sarcophilus* had lost their last premolars but disagreed with Owen (1868) about marsupial-placental premolar homologies. Having observed atavistic extra premolars between the first and second positions in specimens of *Dasyurus maculatus* and *Phascolosorex doriae*, Thomas concluded that marsupials had in fact lost the P2 of an ancestral dentition with four premolars. He expressed this homology in his notation for the marsupial cheektooth formula: P1, P3, dP4, M1–4 (dP4 being the milk tooth that is replaced in most species). This notation was adopted by many anatomists.

This was the situation when Archer (1974) published his description of cheektooth development in *Antechinus flavipes*. By then it was known that the ontogeny of primary teeth begins with an infolding of the oral epithelium in both upper and lower jaws to form a dental lamina (Fig. 11.1a). Cell proliferation and differentiation in discrete swellings at the base of this lamina mark the onset of tooth development, which proceeds as the swelling migrates towards the epithelium along the buccal side of the lamina. At the base of each developing tooth, a dermal papilla organises and induces differentiation of dermal odontoblasts to deposit dentine in the core of the future crown, while overlying epidermal ameloblasts secrete an enamel covering (Fig. 11.1b). Morphologically, the tooth goes through so-called bud, cap and bell stages prior to eruption. A replacement tooth, if any, forms later from a successional lamina connected lingually to the overlying enamel epithelium of the primary tooth. In fact, this ontogenetic derivation provides a recognition criterion for a successional tooth (Luckett 1993). Archer (1974) tracked development of canines and cheekteeth in *A. flavipes* through a series of stages he called S (initiation), FE (swelling moved to the buccal side of the dental lamina), C (calcification) and ED (erupted). Archer's primary finding was that these teeth develop in two mesio-distal series, one corresponding to C1-P3 (using modern notation) and the other to dP3-M4 (Fig. 11.2). In *A. flavipes*, dP3 is distinctly molariform.

Archer (1974) drew two conclusions from these observations: (1) there is no evidence for Thomas' (1887) contention that marsupials have a suppressed P2, so the appropriate adult premolar formula is P1–3; and (2) 'there are no true premolars in *Antechinus flavipes* if premolars are defined in the terms of Owen (1840–45) as those post-canine teeth which possess milk predecessors …' (p. 63). Archer (1978) took this a step further, proposing that the single deciduous tooth in marsupials that had been called dP3 or dP4 by previous authors is in fact a molar. Archer (1984b: 590–591) articulated this hypothesis concisely:

> *[A]lthough we do not appear to have the same basis for recognizing molars and premolars in marsupials as we do in placentals, the timing of development and tooth shape appear to enable us to distinguish two series of postcanine teeth: 3 premolariform teeth not unreasonably called premolars; and 5 molariform teeth which can be called molars. Of the molars, M1 is, for some obscure reason lost in the mists of time, deciduous. As P3 erupts, it crowds out and evicts M1.*

Thus, in Archer's (1978, 1984b) view, the complete dental formula of the ancestral marsupial was dI 5/4, I 5/4, dC 1/1, C 1/1, P 3/3, M 5/5. The primary incisors and canines do not erupt and M1 is deciduous. In dasyurids that have lost one premolar, Archer (1976d, 1978) agreed with Flower (1869) that the missing tooth is P3.

Luckett (1993) summarised the results of extensive histological examinations of tooth development in

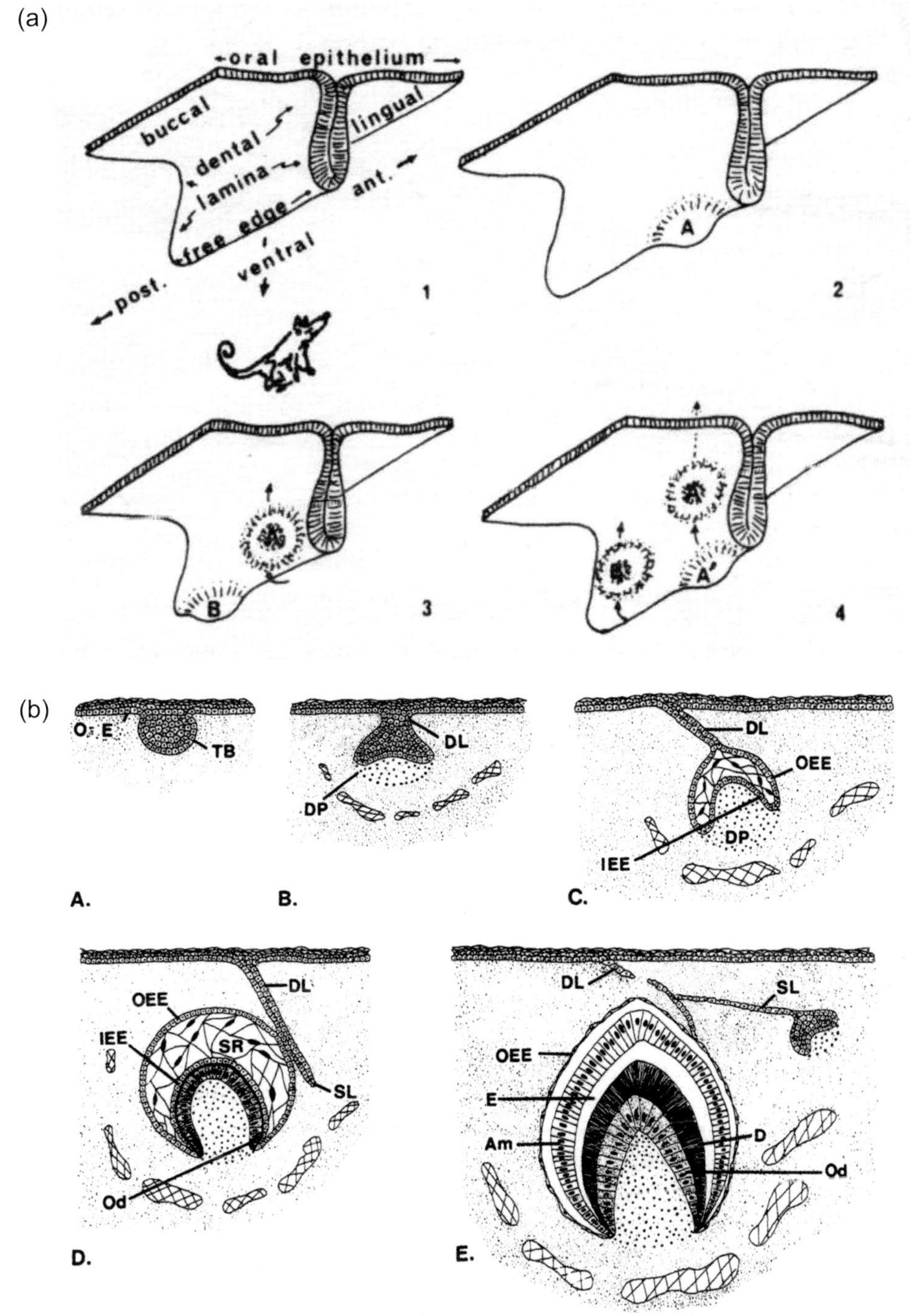

Fig. 11.1. Two models of tooth development. (a) Archer's (1974) illustration of tooth formation at three successive stages (2–4) in the lower jaw. The swelling marked A is a future primary tooth that initially forms at the free edge of the dental lamina but moves upward along the buccal side of the lamina as it develops. A′ represents the successor tooth to A, and B represents the next posterior primary tooth. This depiction of successional tooth formation from the primary dental lamina is not consistent with current understanding. The small animal figure shows orientation. (b) Luckett's (1993) illustration of eutherian tooth development at five successive stages, including bud (A), cap (B) and bell (C–D). The successional tooth develops from a successional lamina (SL) extending from the dental lamina (DL). Abbreviations: AM = ameloblasts; D = dentine; DP = dermal papilla; DL = dental lamina; E = enamel; IEE = inner enamel epithelium; Od = odontoblasts; OE = oral epithelium; OEE = outer enamel epithelium; SL = successional lamina; TB = tooth bud. [(a) Reprinted by permission of Michael Archer from Archer M (1984a). (b) Reprinted by permission of Springer Nature from Luckett (1993).]

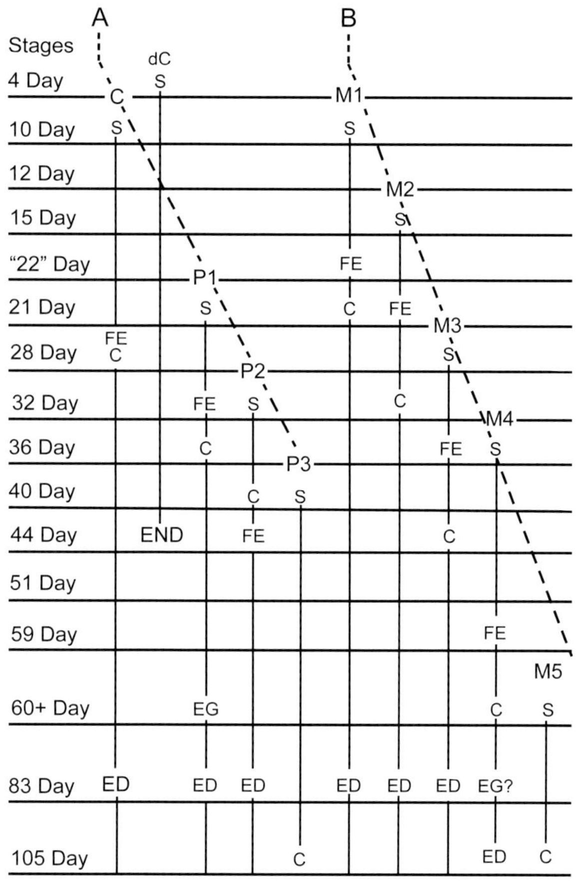

Fig. 11.2. Archer's (1974) timeline of tooth development in *Antechinus flavipes* and hypothesis of marsupial tooth homologies. A and B indicate parallel development of two tooth series, the canine through adult premolar series (A) and the molar series (B). In this diagram, M1 is the tooth previously and currently recognised as dP3 by most anatomists, and M2-M5 correspond to M1-M4. Abbreviations: C = first calcification of tooth crown; ED = tooth erupted; EG = tooth erupting; END = tooth resorbed; FE = free edge of dental lamina visible on lingual side of tooth; S = start of tooth bud development. [Reprinted by permission of Michael Archer from Archer M (1984a).]

species representing 26 placental and four marsupial families (Didelphidae, Dasyuridae, Peramelidae, Phalangeridae). In contrast to Archer (1974), Luckett standardised his comparisons to the bud, cap and bell stages and defined a successional tooth as one that forms from the successional lamina of its deciduous predecessor. His results appeared to clarify a great deal about marsupial tooth homologies, including the following:

- Marsupial incisors and canines develop rudimentary/vestigial primary teeth that never erupt; the adult incisors and canines are successional teeth.
- No successional lamina develop for P1–2 in marsupials; the teeth found at these positions in adults are dP1 and dP2 that are not replaced.
- dP3 has a successional lamina that leads to P3. Archer (1974) had argued that P3 comes from the primary dental lamina anterior to dP3; Luckett (1993) showed that although P3 may be anterior to dP3, the two are connected by a successional lamina (Fig. 11.3). dP3 erupts and is evicted by P3.
- *Dasyurus viverrinus* has a gap between dP1 and dP3 (Fig. 11.3) suggesting that the missing adult premolar is dP2, not P3 (Flower 1869; Archer 1976d) or P1 (Owen 1868). Thus the loss of a premolar in some dasyurids (e.g. *Dasyurus*, *Sarcophilus*) is ontogenetically and evolutionarily unrelated to reduction of the third premolar in others (e.g. *Dasycercus*).
- Marsupial (like placental) molars are primary teeth that have no successors.

These conclusions were confirmed by Luckett and Woolley's (1996) more detailed study of *Sminthopsis virginiae* young examined at 5 day intervals from birth to 60 days, as well as limited comparisons of *Antechinus stuartii*. The authors summarised their results as follows (Fig. 11.4):

> *In the postcanine dentition, only a single tooth position, dP3, was characterized by the later development of a replacing successional tooth (P3), following developmental pathways identical to those in eutherian mammals. In contrast, the anterior dentition is characterized by the formation of rudimentary, nonerupting deciduous incisors and canines, and the accelerated development of normal, erupting successional incisors and canines in both jaws [Luckett and Woolley 1996: 327]*

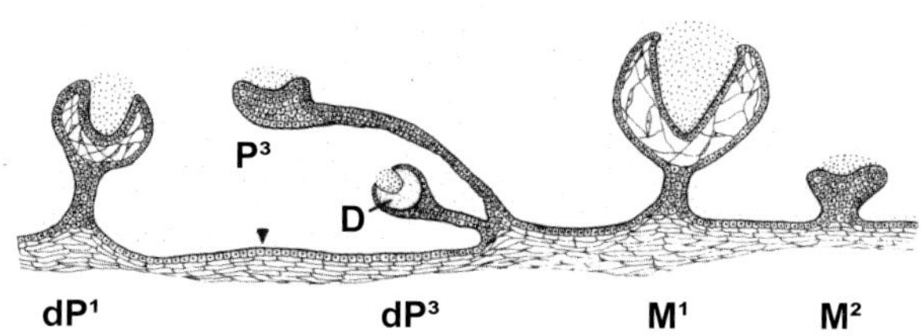

Fig. 11.3. Luckett's (1993) depiction of developing dP^1-M^2 in a sagittal section of *Dasyurus viverrinus*. 'All primary teeth retain their dental lamina connections to the oral epithelium, whereas P^3 arises from a lingual successional lamina in common with the vestigial dP^3. P^3 extends anterior to dP^3 and occupies the position vacated by the lost dP^2 locus (indicated by arrowhead)' (p. 193). D = dentine. [Reprinted by permission of Springer Nature from Luckett (1993).]

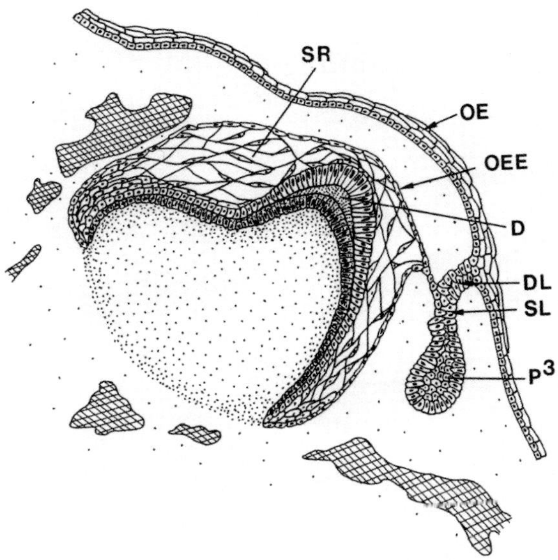

Fig. 11.4. Drawing of a transverse section through dP3 and P3 in a 30 day *Sminthopsis virginiae* pouch young showing the successional lamina (SL) connecting these teeth. Abbreviations as in Fig. 11.1b and SR = stellate reticulum. [Reprinted by permission of Springer Nature from Luckett and Woolley (1996).]

In the dasyuromorphian literature, Luckett's tooth homologies have been adopted by most anatomists commencing with Wroe (1996). Although the adult premolar identities are best expressed as dP1–2, P3, many authors continue to designate them simply as P1–3. More recently, Nasrullah *et al.* (2022) have shown that in the tammar wallaby (*Macropus eugenii*, Macropodidae), P3 develops from the primary dental lamina, suggesting that tooth replacement patterns in marsupials are more diverse than suggested by Luckett (1993).

The search for synapomorphies

As noted in Chapter 3, the foundation of cladistic inference is that members of monophyletic groups share at least one derived trait (synapomorphy) that arose in their common ancestor. Although such a trait may be further derived or even reversed within the clade, the original evolutionary novelty should map to the ancestral branch uniting the descendant species. Making such a determination requires polarising characters – i.e. distinguishing primitive from derived states (plesiomorphies from apomorphies in Hennig's terms). This proved difficult in the absence of a phylogeny relating the ingroup to outgroups and was largely abandoned in favour of unpolarised parsimony analysis for phylogenetic inference, but the value of characterising clades by their synapomorphies has outlived Hennigian argumentation. Among other things, a reliable synapomorphy allows phylogenetic placement of organismal fragments (e.g. fossils) for which few other characters may be available. If an independently derived phylogenetic framework is at hand from molecular or integrative analyses, potential synapomorphies may be evaluated with considerable rigor.

Archer (1984c: 640) provided the first explicit list of putative dasyurid synapomorphies (as distinct from mere apomorphies):

> *As a group, dasyurids have a number of synapomorphic (relative to didelphoids) features including: a large stylar cusp D and small (or absent) stylar cusp C … 4/3 incisors … loss of the foramen ovale and elaboration of the foramen pseudovale for passage out of the braincase of the trigeminal nerve … a well developed squamosal epitympanic sinus … tendency toward suppression of P3 … loss of the posterolateral palatal foramen …*

This list is based on anatomical studies dating to Bensley (1903), but primarily from Archer's own comparative studies of dasyurid basicrania (Archer 1976c) and dentitia (Archer 1976d), as well as his early reviews of fossil dasyuromorphians (Archer 1982a, c). Running through all these papers was the then-conventional assumption that states of craniodental characters in 'didelphoids' are primitive relative to those of Australian marsupials. At the time, 'didelphoids' included all living and fossil American marsupials, and of these the Cretaceous *Alphadon* was considered the most primitive, if not directly ancestral to later groups. Our understanding of marsupial phylogeny has come a long way since the early 1980s: *Alphadon* is now classified as a metatherian stem lineage ('marsupialiform') of uncertain affinity, Cenozoic polydolopidomorphs and sparassodonts are distinct non-marsupial metatherian orders and Didelphimorphia is restricted to didelphids and two extinct families (Eldridge *et al.* 2019). However, to understand how current views on dasyurid synapomorphies developed, we must consider Archer's work in historical context and acknowledge his criterion for character polarity.

Critical evaluation of Archer's (1984c) synapomorphies and their supplementation by new ones appeared in Wroe's (1997a) review and Wroe's (1999b) analysis of *Barinya wangala*, in which polarity decisions were based on outgroup comparisons guided by an updated phylogeny of marsupial families (including the crucial placement of Thylacinidae within Dasyuromorphia), new fossils and functional relationships among characters. Some of Wroe's (1997a) conclusions were challenged or qualified by subsequent paleontological and phylogenetic studies, as we describe below.

Archer's (1984c) dental synapomorphies

Sizes of stylar cusps B, C and D. Stylar cusps form a series of peaks along the buccal margin of the occlusal surface of upper molars and are mesio-distally designated A–E (Fig. 11.5). Acknowledging that cusp homologies among families were not absolutely clear, Archer (1976d: 13) noted: 'In dasyurids the pattern is basically similar to ancestral didelphines except as follows. StD is generally very large in $M^{\underline{1-2}}$... StC in dasyurids is normally absent or miniscule ...' Archer (1982c) noted that this cusp-size asymmetry applied to the Miocene taxa *Keeuna*, *Dasylurinja* and *Ankotarinja*, as well as various Neogene dasyurid fossils, and Archer (1982a) thought that the reduction of stylar cusps in *Thylacinus* made them hard to homologise with those of dasyurids. In his analysis of the thylacinid *Muribacinus*, Wroe (1996) focused on the absolute size of StD on M^3 and took its smallness in *Ankotarinja* and *Keeuna* as plesiomorphic for Dasyuromorphia, the cusp being variously reduced further or enlarged in thylacinids and dasyurids. Given that StD reduction is correlated with carnivory ('carnassialisation')[1] in marsupials, Wroe (1996) concluded that such reduction was an apomorphic reversal in taxa such as *Sarcophilus* and thylacinids. This left open the possibility that ancestral enlargement of StD is a dasyurid synapomorphy, which Wroe (1997a: 40) reformulated in terms of the relative sizes of StD and StB: 'Nearly all dasyurids show a large stylar cusp D which exceeds stylar cusp B in height.' Again noting the reduction of StD in carnivores but its possible utility for insectivores, he assessed the character as follows:

> *[T]he presence of a large stylar cusp D in all insectivorous dasyurids, less specialized thylacinids, and plesiomorphic insectivorous bandicoots ... suggests that the cusp plays some role in cuticle puncturing ... [A] large stylar cusp D ... may be synapomorphic for Peramelemorphia-Dasyuromorphia, or Thylacinidae-Dasyuridae, or may represent an independently derived shared character for each of these clades. [Wroe 1997a: 41]*

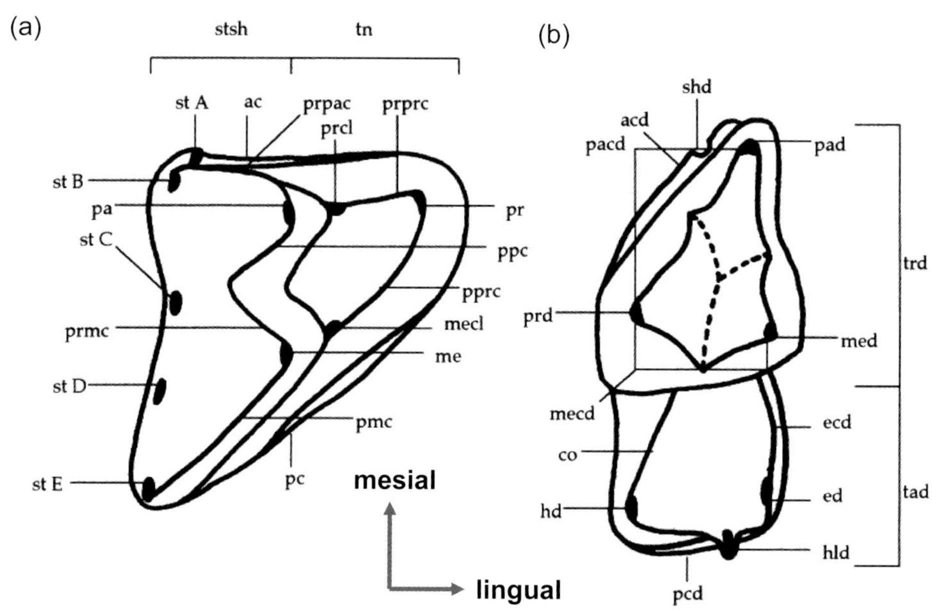

Fig. 11.5. Anatomical structures on the occlusal surfaces of generalised marsupial upper (a) and lower (b) molars. Abbreviations: ac = anterior cingulum; acd= anterior cingulid; co = cristid obliqua; ecd = entocristid; ed = entoconid; hd = hypoconid; hld = hypoconulid; me = metacone; mecd = metacristid; mecl = metaconule; med = metaconid; pa = paracone; pacd = paracristid; pad = paraconid; pc = posterior cingulum; pcd = posterior cingulid; pmc = postmetacrista; ppc = postparacrista; pprc = postprotocrista; pr = protocone; prcl = protoconule; prd = protoconid; prmc = premetacrista; prpac = preparacrista; shd = sulcus for hypoconulid of preceding molar; stA – StE = stylar cusps A–E; stsh = stylar shelf; tad = talonid; trd = trigonid; tn = talon. [Reprinted by permission of John Wiley and Sons from Wroe (1999b).]

Thus, a large StD is at best an equivocal synapomorphy for dasyurids.

Incisor number. All dasyurids have four upper and three lower incisors per quadrant, whereas the didelphid count is 5/4 (Archer 1976d). However, myrmecobiids and thylacinids also have I 4/3. Once it became accepted (after Aplin and Archer 1987) that numbats are not dasyurids and thylacines are dasyuromorphians rather than borhyaenoids, this character could no longer be considered a dasyurid synapomorphy.

Reduction of P_3. Didelphids, microbiotheriids, myrmecobiids, thylacinids and some dasyurids (e.g. *Murexia*, *Antechinomys*) have three lower premolars (Fig. 11.6), often with P_3 smaller than P_2. As noted in 'Establishing tooth homologies', P_3 is markedly reduced in many dasyurids (e.g. *Sminthopsis ooldea*, Archer 1981) and absent in others (e.g. *Sarcophilus*). Tate (1947), Archer (1982a) and Marshall *et al.* (1990) considered the unreduced condition as primitive and Archer (1984c) articulated the derived state as a 'tendency' to reduce P_3. Wroe (1996: 1038) criticised the latter claim:

> *Application of the term synapomorphic implies that diminution of P_3 was a character shown by the common ancestor of all dasyurids. Strong contradictory evidence confronts this proposal. Firstly, it requires the consideration of P_3 reduction as an underlying synapomorphy ... 'close parallelism as a result of inherited factors within a monophyletic group causing incomplete synapomorphy' (Saether 1983, p. 343). The use of underlying synapomorphy in phylogenetic analysis is strongly criticized by a number of authors ...*

If *actual* P_3 reduction was the ancestral dasyurid state, then species showing unreduced premolars must have experienced reversals to the even more primitive state of ancestral dasyuromorphs. Because taxa with an unreduced P_3 are scattered across the dasyurid tree, this reversal must have happened multiple times, including in the Pliocene *Dasyurus dunmalli*, the only quoll known to have a P_3 (Wroe and Mackness 1998). On the other hand, if the ancestral dasyurid had a typical P_3, then reduction or loss of this tooth occurred several times in parallel, perhaps by selection for 'brachycephalisation, permitting retraction of the canines closer to the point of maximum bite force in the tooth row and increasing torsion resistance in the skull' (Wroe 1996: 1038). Interpretation of P_3 reduction is even more complicated if Luckett (1993) is correct in finding that the missing premolar in *Dasyurus* (and probably other two-premolared species) is in fact P_2. In view of these considerations, Wroe (1997a) rejected P_3 reduction as a dasyurid synapomorphy. Nevertheless, a 'tendency toward suppression of P_3' was still cited as a shared derived trait of dasyurids by Archer and Hand (2006).

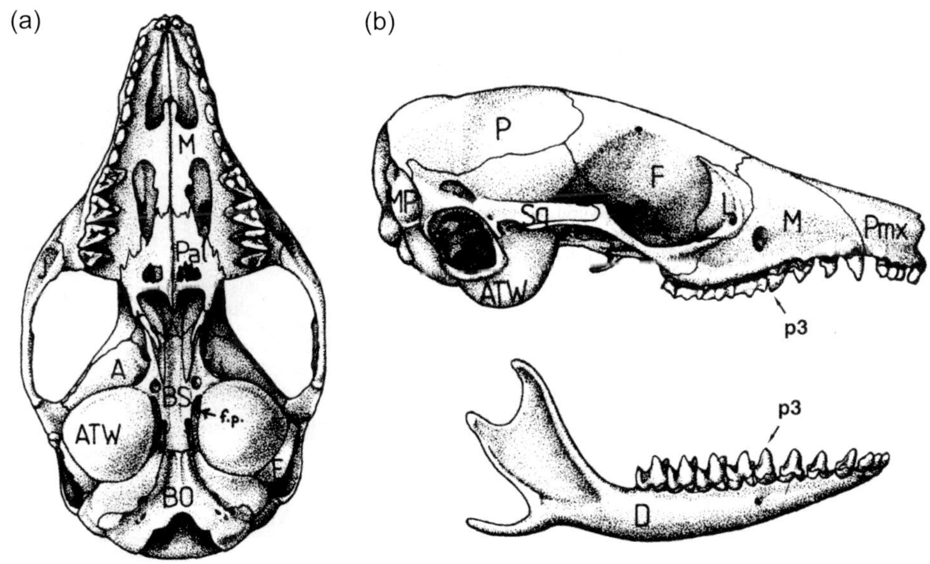

Fig. 11.6. Skull and dentary of *Antechinomys laniger* in (a) ventral and (b) lateral views. Abbreviations: A = alisphenoid; ATW = alisphenoid tympanic wing; BS = basisphenoid; BO = basioccipital; D = dentary; E = ectotympanic; F = frontal; f.p. = foramen pseudovale; J = jugal; L = lachrymal; M = maxilla; MP = mastoid process; N = nasal; P = parietal; p3 = third premolar; Pal = palatine; Pmx = premaxilla; Sq = squamosal. [Reprinted by permission of Michael Archer from Archer and Hand (2006).]

Hypoconulid notch. The anterior cingulids of marsupial lower molars often feature a slight indentation that accommodates the posteriorly protruding hypoconulid of the preceding tooth ('shd' in Fig. 11.5). Archer (1982a) suggested this hypoconulid notch as a dasyurid apomorphy. Wroe (1997a) pointed out, however, that the notch is also prominent in some didelphids and thylacinids but absent in other thylacinids and *Myrmecobius*. It is found in all dasyurids except *Glaucodon* and *Sarcophilus*, though reduced in *Dasyurus maculatus*. Loss of the notch appears correlated with increasing body size and carnassialisation in dasyuromorphs and its degree of development may be influenced by several functional considerations (e.g. lesser role of horizontal shear in mastication). Wroe (1997a: 42) concluded, 'This feature is lost only in dasyurid and thylacinid taxa showing advanced carnassialisation and the termite-eating *Myrmecobius fasciatus*. It is considered likely that a hypoconulid notch was present in the common ancestor of all three families and does not represent a dasyurid synapomorphy.'

The upshot of Wroe's (1996, 1997a) character analyses is that there are no unequivocal dental synapomorphies uniting Dasyuridae. It was for this reason that he recommended reassigning the fossil taxa *Ankotarinja*, *Keeuna*, *Dasylurinja* and *Wakamatha* (represented only or primarily by dental remains) from Dasyuridae to Dasyuromorphia *incertae sedis*. To date, no other dental traits have been forwarded as potential synapomorphies.

Archer's (1984c) basicranial synapomorphies

Foramen pseudovale. Cranial nerve V (the trigeminal) emerges from the ventral hindbrain and ramifies within the skull. One of its branches, the mandibular (V^3), exits the basicranium via the foramen ovale to innervate the lower jaw and masticatory muscles. In most eutherians, the foramen ovale is bounded entirely by the alisphenoid, but in some marsupials it forms as a gap between the alisphenoid and the petrosal portion of the periotic, a condition termed a foramen pseudovale ('f.p.' in Figs 11.6 and 11.7). In still other species, the alisphenoid forms a bony bridge anterior to the foramen ovale, under which V^3 passes after emerging from the braincase; this structure is usually called a secondary foramen ovale. Archer (1982a) noted that dasyurids and the numbat have a foramen pseudovale ('alisphenoid-periotic enclosure of the foramen ovale'), whereas most didelphids and then-known species of *Thylacinus* have a typical foramen ovale. The foramen pseudovale was accepted as a dasyurid synapomorphy by Marshall *et al.* (1990). With the benefit of a resolved Dasyuromorphia and new fossil marsupicarnivores (especially thylacinids), Wroe's (1997a) outgroup analysis located the origin(s) of a foramen pseudovale deeper in marsupial phylogeny:

> *Because an alisphenoid-periotic contribution to the foramen is found in Myrmecobiidae, and basal thylacinids and dasyurids, in outgroups to Recent Theria, as well as in most taxa considered as possible dasyuromorphian outgroups, it is concluded here that alisphenoid-periotic delimitation of the foramen ovale represents a plesiomorphy for Dasyuromorphia. Consequently, the alisphenoid-periotic enclosed foramen ovale of Dasyuridae and Myrmecobiidae is treated as a retained plesiomorphy … [Wroe 1997a: 30]*

Development of a secondary foramen ovale seems related to dietary habit: 'Among dasyuromorphians there is an apparent correlation between the evolution of a secondary foramen ovale, advanced carnassialisation and large size, with the most specialised taxa of both thylacinid and dasyurid clades showing the most advanced development' (Wroe 1997a: 34).

Walls of the tympanic cavity. The mammalian middle ear is housed in the tympanic cavity, the walls of which are formed by parts of the squamosal, sphenoid, periotic and occipital bones of the skull. The contribution of these bones to the cavity walls varies considerably among marsupials and often involves the development of processes ('wings') to create pockets in its roof and floor (epitympanic and hypotympanic sinuses, respectively). Archer (1984b) suggested that a well-developed squamosal epitympanic sinus ('SES' in Fig. 11.7) could be a dasyurid synapomorphy, but this is tempered by his earlier conclusion that the sinus is also found in some didelphids and not all dasyurids: '… enlargement of this sinus in thylacinids and those dasyuroids in which it is well-developed, could have been the result of parallel evolution in response to the development of late Cainozoic aridity in Australia' (Archer 1982a: 462). (As noted in Chapter 10, enlargement and enclosure of the tympanic cavity is a common adaptation to auditory sensitivity in desert-dwelling mammals.) Other than repetition by Archer and Hand (2006), the squamosal epitympanic sinus has not been further discussed as a dasyurid synapomorphy. Archer (1982a, c) posited that an enlarged alisphenoid tympanic wing contribution to the hypotympanic sinus (visible externally as part of the tympanic bulla; 'ATW' in Fig. 11.6 and 'ASTW' in Fig. 11.7) was derived in parallel by numbats and dasyurids (hence, technically, a synapomorphy for the latter). Wroe (1997a, 1999b), however, entertained the possibility that it could be a shared derived feature uniting the two families, consistent with later molecular evidence that they are indeed sisters (see Chapter 8).

Posterolateral palatal foramina. These structures are formed by short processes from the palatine and maxilla that unite to create a bony ring for passage of blood vessels and nerves ('plpf' in Fig. 11.7). They are primitively present in many therians; Archer (1982c) and Marshall *et al.* (1990) cited their reduction or loss as derived in dasyurids. Wroe (1997a: 36) rejected this idea:

> *Posterolateral palatal foramina are incomplete [form a notch at the posterolateral junction of the palatine and maxilla] or lost in all dasyurids excepting* Neophascogale *and some* Dasyurus viverrinus *... Loss of the posterolateral foramina and accessory posterolateral foramina correlates with increased carnassialization in marsupicarnivores ... Given the appearance of posterolateral palatal foramina in the basal dasyurine* Neophascogale*, a high degree of variation at all taxonomic levels, and demonstrable correlation of p.p.f. loss with carnassialization in independent lineages, it is suggested that loss of this character can no longer be applied as a dasyurid synapomorphy.*

Intestinal caecum. As noted by Barbour (1977: 255), 'The large intestine has a caecum in most marsupials, but not in the Dasyuridae, Notoryctidae, *Tarsipes* and *Dromiciops australis*.' Aplin and Archer (1987) considered loss of the caecum a dasyuromorphian synapomorphy, but Marshall *et al.* (1990) interpreted it as a derived trait uniting dasyurids. Wroe (1997a: 42) supported the former position:

> *For extant Australidelphia the intestinal cecum is present in all taxa excepting Notoryctidae, Thylacinidae, Myrmecobiidae, Dasyuridae, and a single diprotodontian taxon,* Tarsipes *... From the distribution of this feature among extant marsupial taxa, it is likely that loss of the intestinal cecum appeared early in dasyuromorphian evolution and probably represents a synapomorphy for the order, rather than just for Dasyuridae.*

Wroe's (1999b) basicranial synapomorphies

Derived traits in *Barinya* and modern dasyurids. (1) The petrous portion of the periotic bone lies at the posterolateral corner of the skull between the sphenoid and occipital bones. In dasyurids, but not numbats or thylacinids, this

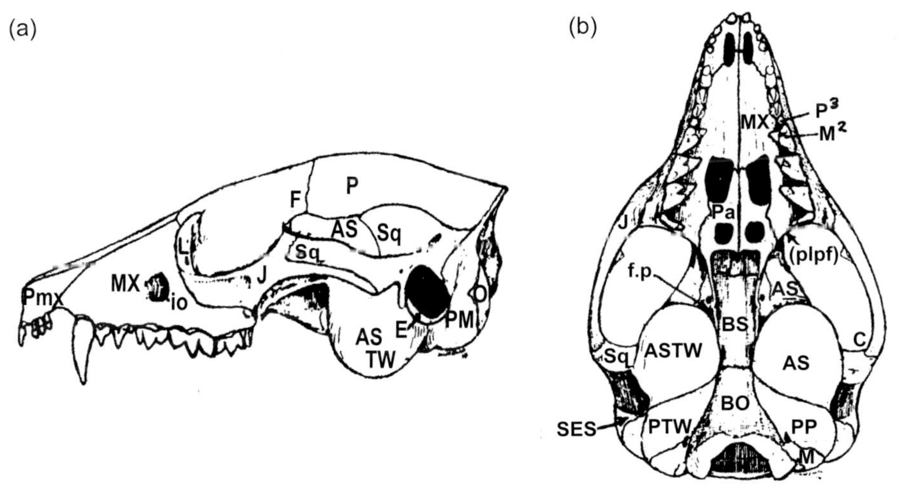

Fig. 11.7. Skull of *Dasycercus cristicauda* in (a) lateral and (b) ventral views. Abbreviations as in Fig. 11.6 except: AS = alisphenoid; ASTW = alisphenoid tympanic wing; io = infraorbital canal; M = mastoid process; M^2 = 'second' molar (this tooth is designated M^1 in publications after Luckett 1993); Mx = maxilla; P^3 = third premolar; plpf = posterolateral palatine foramen (missing in *Dasycercus*); Pmx = premaxilla; PM = mastoid wing of periotic; PP = periotic; PTW = periotic tympanic wing; SES = squamosal epitympanic sinus; Sq = squamosal. [Reprinted by permission of Michael Archer from Archer (1984b).]

bone forms a distinct tympanic wing ('PTW' in Fig. 11.7) that contacts the alisphenoid tympanic wing, floors part of the middle ear cavity and creates a *periotic hypotympanic sinus*. (2) Similarly, the dasyurid paraoccipital bone uniquely forms a *paraoccipital hypotympanic sinus*. (3) Associated with the former trait in dasyurids is development of a distinct *tubular foramen*, bounded by the petrosal and alisphenoid, that carries the Eustachian canal. (4) Dasyurids also show a uniquely derived *jugular canal sulcus*: 'a deeply invasive sulcus [in the internal jugular canal] with the posteroventral lip formed by a mesially directed process of the pars petrosa for passage of the internal jugular' (Wroe 1999b: 501).

Derived traits in crown dasyurids. Wroe (1999b) adduced four traits as synapomorphies for modern dasyurids, to which *Barinya* was the sister group. (1) Cranial nerve VII exits the braincase via a facial nerve canal in the petrosal bone, often leading to a distinct stylomastoid foramen near the mastoid process ('M' in Fig. 11.7). Posterior to the nerve canal is an open sulcus and some marsupials (including numbat and thylacinds) lack a stylomastoid foramen. Uniquely in extant dasyurids, the sulcus becomes enclosed such that the nerve passes through a bony tube and exits via a *fully formed stylomastoid foramen* with a contribution from the periotic. (2) In addition to having a sulcus, the jugal canal in modern dasyurids is fully enclosed by bone dorsally and extends anteriorly more than halfway to the basisphenoid (a *long enclosed jugal canal*). (3) In extant dasyurids, the *paroccipital tympanic process contacts the petrosal* anteriorly to fully enclose the paroccipital hypotympanic sinus. (4) Likewise, in modern dasyurids the *parocciptal tympanic process fuses with the mastoid tympanic process* laterally in the paroccipital hypotympanic sinus.

Evaluation of Wroe's synapomorphies

Subsequent character analyses confirmed that the fossil thylacinids *Badjcinus* (Muirhead and Wroe 1998) and *Nimbacinus* (Wroe and Musser 2001) lack some or all of Wroe's eight dasyurid synapomorphies. Wroe *et al.* (2000) scored 44 dental and 33 cranial characters from 24 taxa to estimate dasyuromorphian relationships. Their numerical parsimony analysis returned 64 minimum-length trees, the strict consensus of which (Fig. 11.8) supported monophyly of Dasyuridae (including *Barinya*) and modern dasyurids. Accelerated transformation ('ACCTRAN') character optimisation identified seven and six state changes, respectively, supporting each node. Each of these transformation sets included the four synapomorphies obtained in Wroe (1999b) from Hennigian analysis, suggesting that they are at least partially robust to quantitative treatment. Reviews by Wroe (2003) and Archer and Hand (2006) cited these traits, though the latter authors also included many of Archer's (1984c) putative synapomorphies for the family. Archer, Christmas *et al.* (2016) refrained from a definite assignment of *Wholleydooleya* to Dasyuridae because the new fossil (a single tooth) lacked basicranial information.

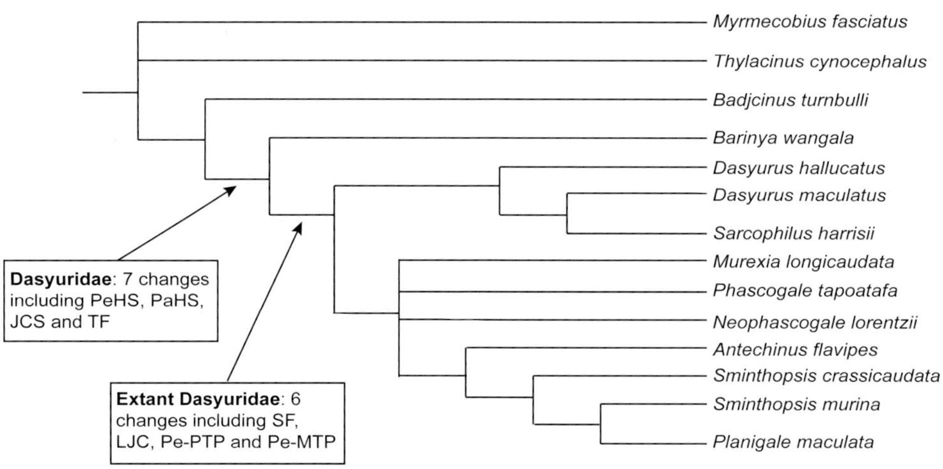

Fig. 11.8. Strict consensus of 64 minimum-length trees from 77 craniodental characters analysed by Wroe *et al.* (2000), showing dasyuromorphian taxa only (see Fig. 8.23a for outgroup branches). Tree obtained with PAUP 3.1.1 using TBR branch swapping. ACCTRAN optimisation locates seven changes to the dasyurid ancestral node and six changes to the modern dasyurid ancestral node, including all eight putative synapomorphies proposed by Wroe (1999b). Abbreviations: PeHS = periotic hypotympanic sinus; PaHS = paroccipital hypotympanic sinus; JCS = jugular canal sulcus; TF = tubular foramen; SF = fully formed stylomastoid foramen; LJC = long enclosed jugal canal; PTP-Pe = parocciptal process contacts petrosal; MTP-Pe = mastoid process contacts petrosal. [Adapted from Wroe *et al.* (2000).]

Several later studies challenged some of Wroe's (1999b) synapomorphies, mostly those uniting *Barinya* with Dasyuridae. In their description of new fossil material from the putative thylacinid *Mutpuracinus archibaldi*, Murray and Megirian (2006) found a close resemblance between this species and *Barinya wangala*, including some character states adduced as dasyurid synapomorphies. For example, they found that *Mutpuracinus* had a periotic hypotympanic sinus, a jugular canal sulcus, a 'rudimentary' tubular foramen and a long enclosed jugal canal. Although Murray and Megirian (2006) considered *Mutpuracinus* an 'undoubted' thylacinid and *Barinya* a dasyurid, their interpretation of character evolution was significantly different from Wroe's (1999b). For example, '... the presence of a distinct tympanic process of the petrosal appears to be a character shared by dasyuromorphians + peramelemorphians, and the feeble expression or absence of it in *T. cynocephalus* and *B. turnbulli* should be therefore characterised as an apomorphic loss, as the 'open' state of the tympanic bulla in *Badjcinus turnbulli* and *Thylacinus cynocephalus* is clearly not homologous to that of didelphids' (Murray and Megirian 2006: 257). Their parsimony analysis of 33 cranial characters from five thylacinids, two dasyurids and the numbat used the bandicoot *Yarala birchfieldi* as an outgroup with all primitive states (because 'it provides a structural link between peramelemorphs and dasyuromorphs', p. 259). Their five most parsimonious trees showed a clade of thylacinids + numbat to which the two dasyurids were successive sister groups (see Fig. 8.24). From this they drew the following radical conclusion: '*Barinya wangala* + extant Dasyuridae represent the plesiomorphic sister group to *Mutpuracinus archibaldi* + all other Thylacinidae. The argument that a member of the Dasyuridae, even as narrowly constrained by Wroe (1997a), could not have given rise to the Thylacinidae is not supported' (p. 260). This seems to be the first suggestion since Bensley (1903) and Tate (1947) that thylacines evolved from dasyurid ancestors.

Although these conflicting results were summarised uncritically by Black *et al.* (2012) and Binfield *et al.* (2017), Murray and Megirian's (2006) phylogeny gained few adherents. The similarity they found between *Mutpuracinus* and *Barinya*, however, was reflected in two subsequent studies. In their phylogenetic analysis of *Malleodectes*, Archer, Hand *et al.* (2016) found that *Mutpuracinus* and *Barinya* fell outside their respective families (see Fig. 8.9). Similarly, in the morphological + molecular tree of Kealy and Beck (2017), *Mutpuracinus* formed a polytomy with dasyurids and *Baryina* + *Myrmecobius* to which an otherwise monophyletic Thylacinidae was sister (Fig. 11.9).

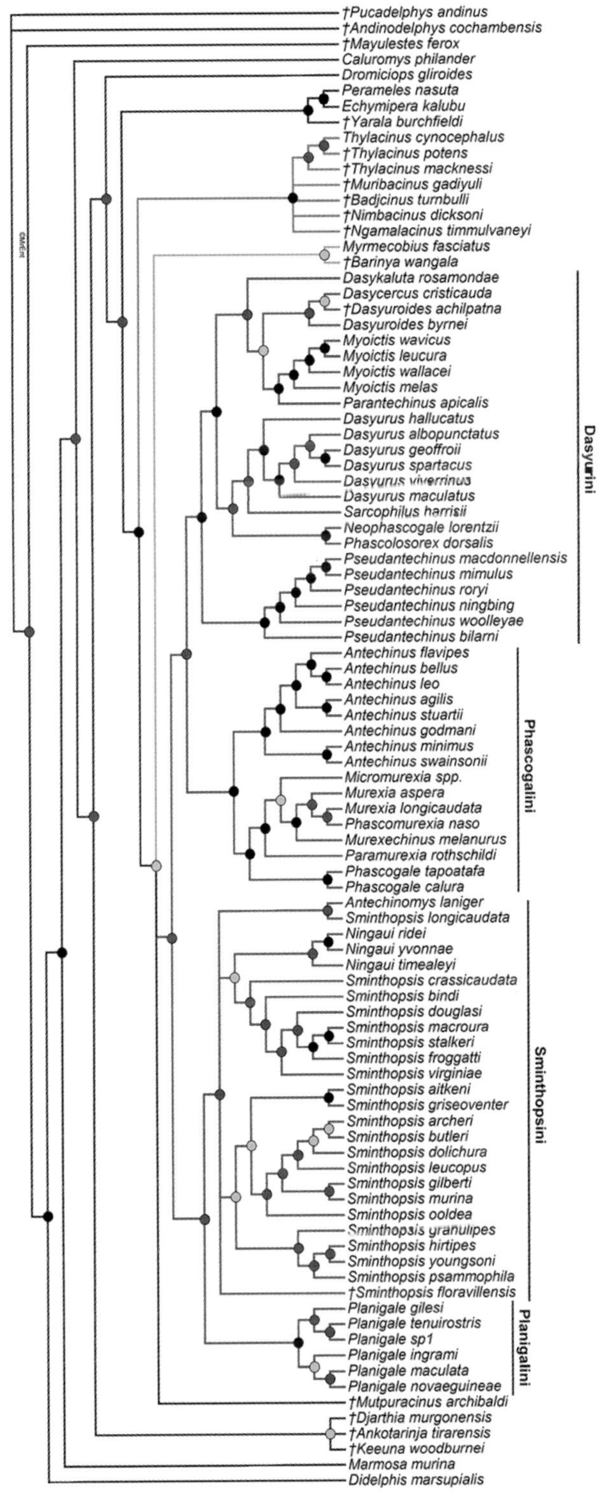

Fig. 11.9. Bayesian estimate of phylogeny from combined morphological (173 characters) and DNA sequence (16.4 kb) data (Kealy and Beck 2017). Dots at nodes indicate posterior probabilities: black = BPP ≥ 0.95, dark grey = 0.74 ≤ BPP ≤ 0.94, light grey = 0.50 ≤ BPP ≤ 0.74. [Reprinted from Kealy and Beck (2017) under license CC BY 4.0.]

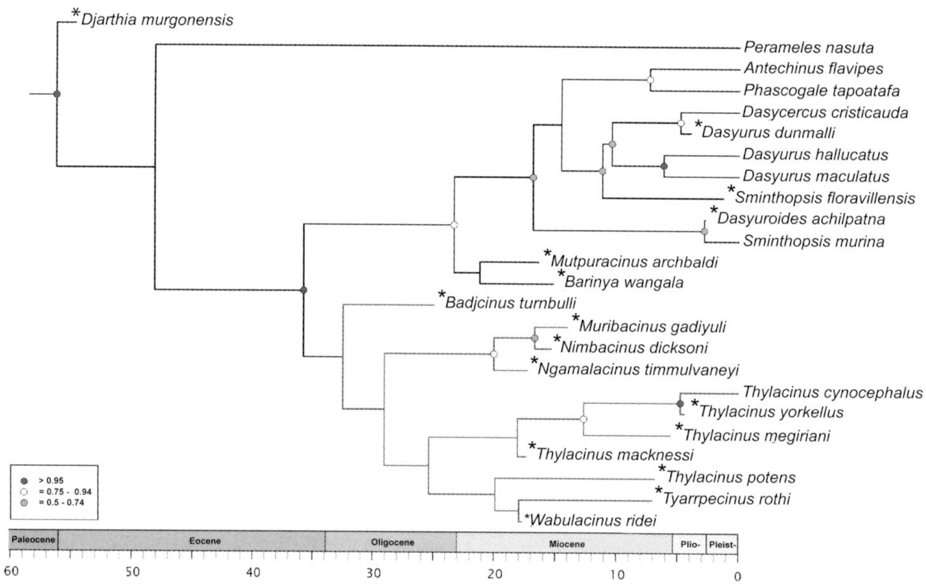

Fig. 11.10. Tip-dated Bayesian phylogeny of dasyuromorphians based on morphological (80 characters) and *12S* DNA sequence (853 bp) data from Rovinsky *et al.* (2019). Asterisks indicate extinct species. [Reprinted with changes from Rovinsky *et al.* (2019) under CC BY 4.0 license.]

Rovinsky *et al.* (2019), also using Bayesian analyses of combined morphological and molecular data, obtained a tree with *Mutpuracinus* and *Barinya* as sisters (but with <0.5 posterior probability; Fig. 11.10). In a methodologically similar study, Beck *et al.* (2022) found that *Mutpuracinus* and *Barinya* grouped with *Myrmecobius* as the sister group to Dasyuridae. These authors listed seven 'unambiguous craniodental synapomorphies' that mapped to the dasyurid ancestor in their parsimony analyses (consistency indices of 0.042–0.625); these include reduction of the interparietal, a stylomastoid foramen formed by the rostral and tympanic processes of the petrosal (*vide* Wroe 1999b), and eruption of P3 after M4. The authors could not consistently score Wroe's (1999b) 'paroccipital tympanic process contacts the petrosal' character. Thus, while the monophyly of modern Dasyuridae has been established by DNA sequence data, the morphological synapomorphies suggested to date do not unequivocally resolve *Barinya* as a member of the family.

ANATOMY OF THE PENIS

Among those intrigued by Tyson's (1698) original dissection of a female opossum (*Didelphis virginiana*; see Chapter 3) was the surgeon-anatomist William Cowper[2] (1666–1709). Cowper obtained the carcass of a male opossum that had been sent alive from Virginia to the Royal Society by William Bird (also the source of Tyson's specimen). Deferring to Tyson on all aspects of general anatomy, Cowper (1704) described the structure of the male reproductive system, prefaced with the following confession: 'The singular contrivances of that Animals Organs renders the Anatomy of them very desirable, I may say entertaining, to those who have tastes for such enquiries' (p. 1576). Cowper noted that the non-erect penis lay within the body cavity between the pubis and the rectum, and on erection was extruded through the preputial opening and cloaca. He noted the four prominent bulbourethral ('Cowper's') glands at the base of the penis (Fig. 11.11a). These glands are adjacent to 'cavernous bodies' or crura that unite to form the erectile corpus cavernosum in the shaft of the penis. Cowper deduced that the penis was withdrawn by a pair of muscles (retractors) that he considered arose from the rectum and inserted onto the penis by a common tendon to the shaft (i.e. onto the tunica albuginea surrounding the corpus cavernosum in the shaft). The retractor muscles are now known to insert onto the vertebral column in the sublumbar region (Nogueira *et al.* 1999). The tip of the penis is forked, or bifid, a condition not seen in any other mammal at the time (but predicted by Tyson based on the female's paired vaginae). Cowper related this to nature's providence: 'For to what end has Nature been at the trouble of making double Emissaries for the *Semen* of the Male *Opossum*, tho she design'd the impregnation of a double *Uterus* of the Female?' (p. 1585). Based on the arrangement of veins and muscles involved in maintaining erection, Cowper also

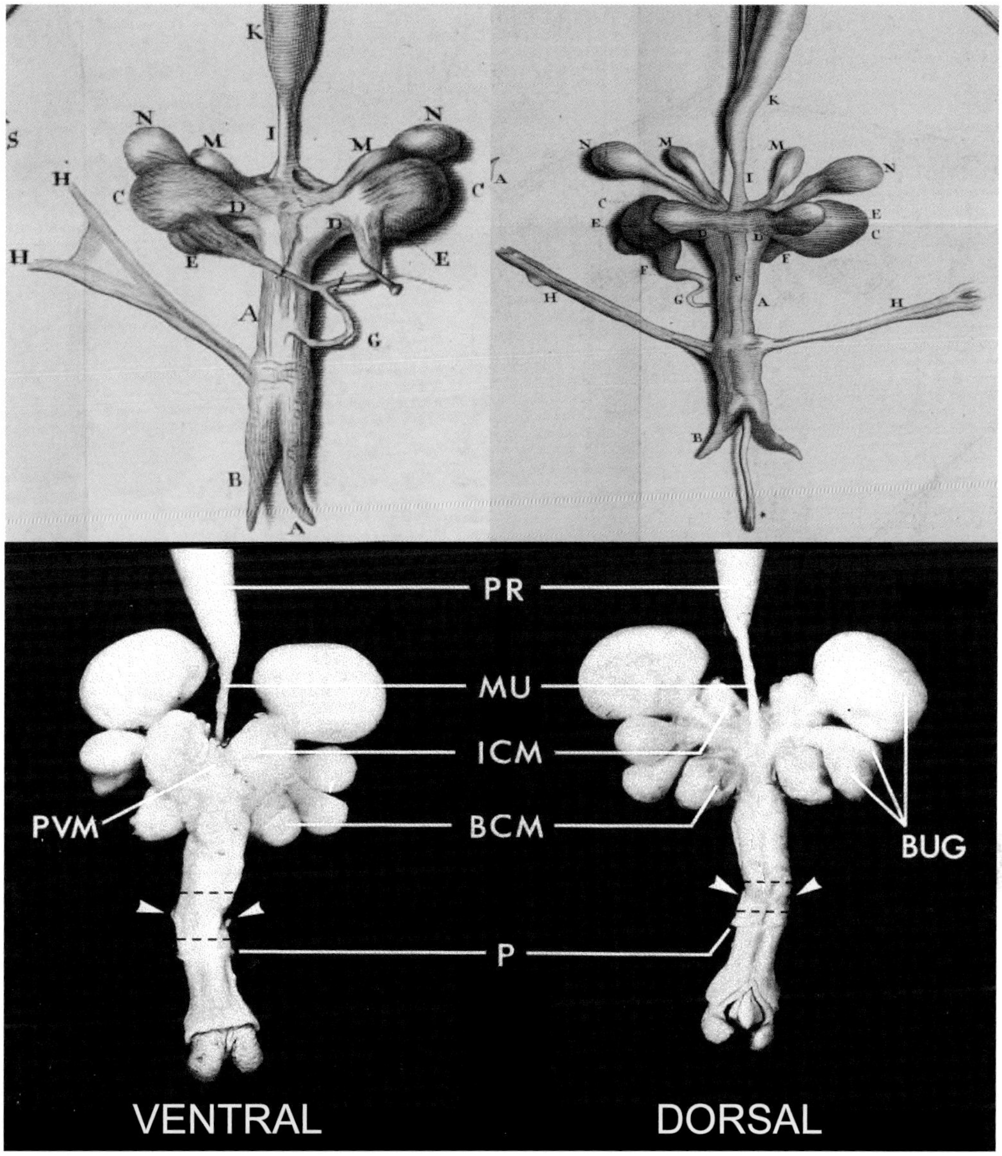

Fig. 11.11. External anatomy of the marsupial penis. Ventral view on left, dorsal view on right. (a) Cowper's (1704) illustration of penis anatomy in *Didelphis virginiana*. Abbreviations: A = body of the penis; B = glans; C = bulbo-cavernosus muscles; D = bulbs (crura) of corpora cavernosa; E = bulbs of corpus spongiosum; f = levator muscles; G = tendon of levator muscles; H = retractor muscles; I = urethra; K = prostate; M, N = bulbourethral glands. (b) Penis of the dasyurid *Antechinus stuartii* (now *A. agilis*). Abbreviations: BCM = bulbo-cavernosus muscle; BUG = bulbourethral glands; ICM = ischio-cavernosus muscle; MU = membranous urethra; P = site of attachment of preputial skin; PR = prostate; PVM = penis vein muscle. [(b) Reprinted by permission of PA Woolley from Woolley and Webb (1977).]

speculated that marsupials engage in lengthy copulation, a prediction that proved correct for many species.

By the time Owen (1839a) reviewed marsupial anatomy, it was known that many species lack a bifid penis and others vary in the degree and form of bifurcation. Owen also described the internal anatomy of the penis, distinguishing the ventrolateral corpus cavernosum (formed by the confluence of the crura) and dorsal corpus spongiosum (originating as paired ventral bulbs at the base surrounding the urethra in the body and glans). The urethra itself may terminate in open 'urethral grooves' at the tips. A modern description of penis anatomy in didelphids was given by Nogueira *et al.* (2004). The dasyurid penis has a basically similar external structure (Fig. 11.11b), though the position of the levator muscles is distinct. In didelphids, these appear to arise variously on the bulbs of the corpus spongiosum, corpus cavernosum and between the two; in dasyurids they arise on the bulbs of the corpus cavernosum and run side by side on the ventral surface of the body of the penis, with the tendons arising mesially.

In the earliest study to include a detailed description of dasyurid penis morphology, Gerhardt (1905) noted a prominent appendage to the urethral penis (Fig. 11.12) in *Dasyurus viverrinus*; Gerhardt (1933) found a similar structure in *Myoictis melas*. Van den Broek (1910) saw the appendage in *Dasyurus viverrinus* and showed that it was erectile and documented its absence in *Antechinus flavipes*. This was the state of knowledge when Woolley and Webb (1977) reported their comparison of penis anatomy in *Antechinus stuartii* and *Parantechinus* (then *Antechinus*) *apicalis*. In short, *A. stuartii* lacks the appendage but *P. apicalis* has it; moreover, the appendage contains branches of the corpora cavernosa (Figs 11.11b and 11.13). From

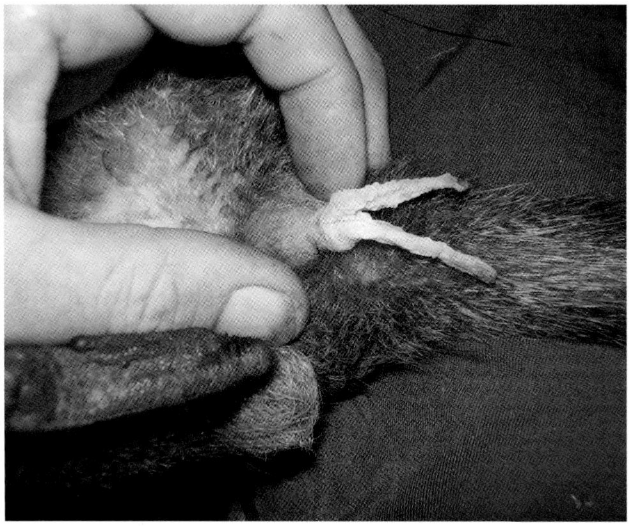

Fig. 11.12. The erect penis and appendage of *Dasyurus spartacus*. [Photo credit: Steven G. Hamilton.]

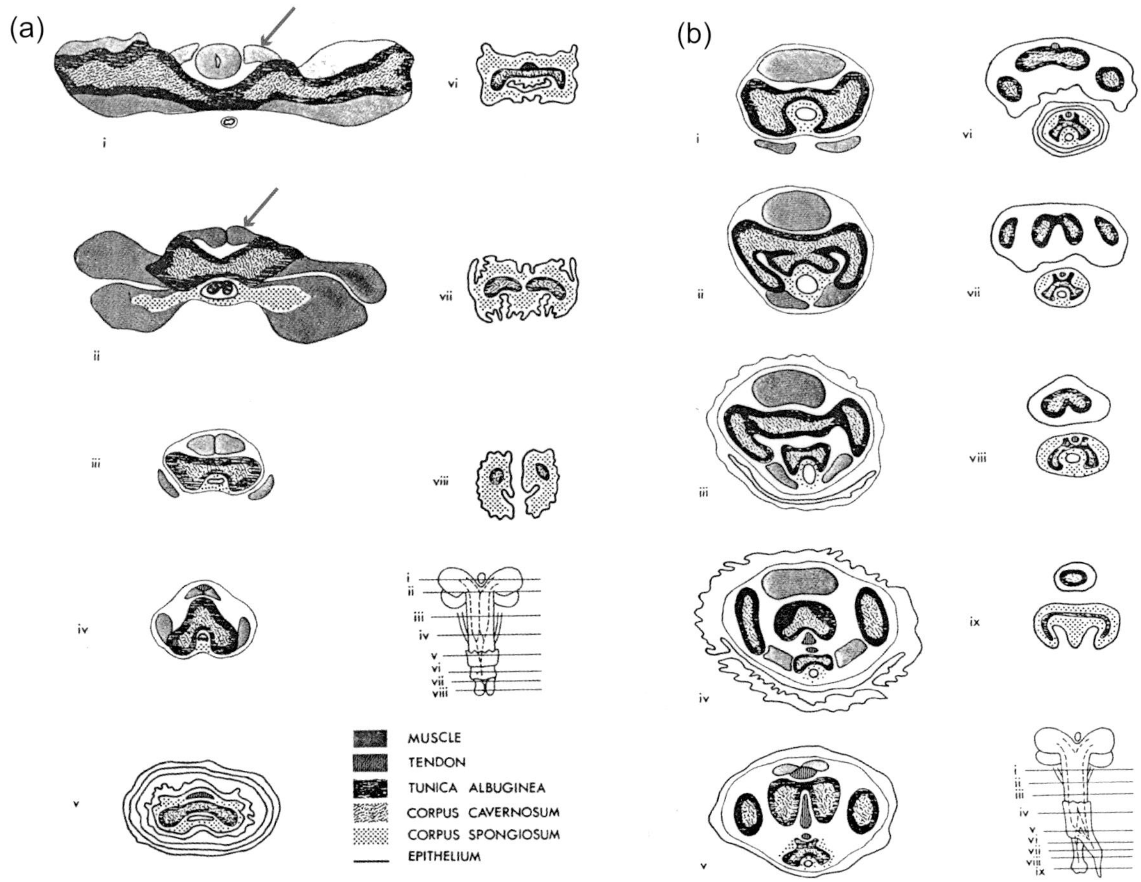

Fig. 11.13. Cross-sections showing internal penis morphology of (a) *Antechinus stuartii* (now *A. agilis*) and (b) *Parantechinus apicalis*. Insets at lower right show ventral views of external anatomy with lines indicating the planes of sections at left and above. Ventral surface at top of each section. The arrow in (a) indicates the penis vein muscle, a structure discovered by Woolley and Webb (1977) and apparently unique to dasyurids. [Reprinted by permission of P.A. Woolley from Woolley and Webb (1977).]

external examinations of specimens and literature reports on 43 dasyurid species, Woolley and Webb (1977) found that the appendage had been documented in only eight species at that time: five *Dasyurus*, *P. apicalis*, *Pseudantechinus macdonnellensis* and *M. melas*. The authors also discovered and named the penis vein muscle (Fig. 11.13a), a structure that appears to occur in all dasyurids but is not known in other marsupials.

Woolley (1982) undertook a survey of penis morphology in Australian species then assigned to *Antechinus*, including those now placed in *Dasykaluta*, *Parantechinus* and *Pseudantechinus*. She found differences between species in the form of the tip, shape of the median lobe, length of the urethral grooves, size of the levator muscles, origins of the tendons of levator muscles and extent of accessory erectile tissue. These traits separated species into two groups (Fig. 11.14a), corresponding to what are now recognised as members of *Antechinus* (Group 1) and 'false antechinuses' (Group 2), a result that weighed heavily in Archer's (1982c, 1984c) decision to isolate members of Group 2 in 'Parantechini' (see Chapter 3). Hierarchical nesting of penis traits revealed subdivisions in both groups (Fig. 11.15) which, except for the association of *Pseudantechinus bilarni* with *Parantechinus apicalis*, correspond to clades identified by integrative phylogenetic analyses (see Chapter 5). Woolley (1987) described the

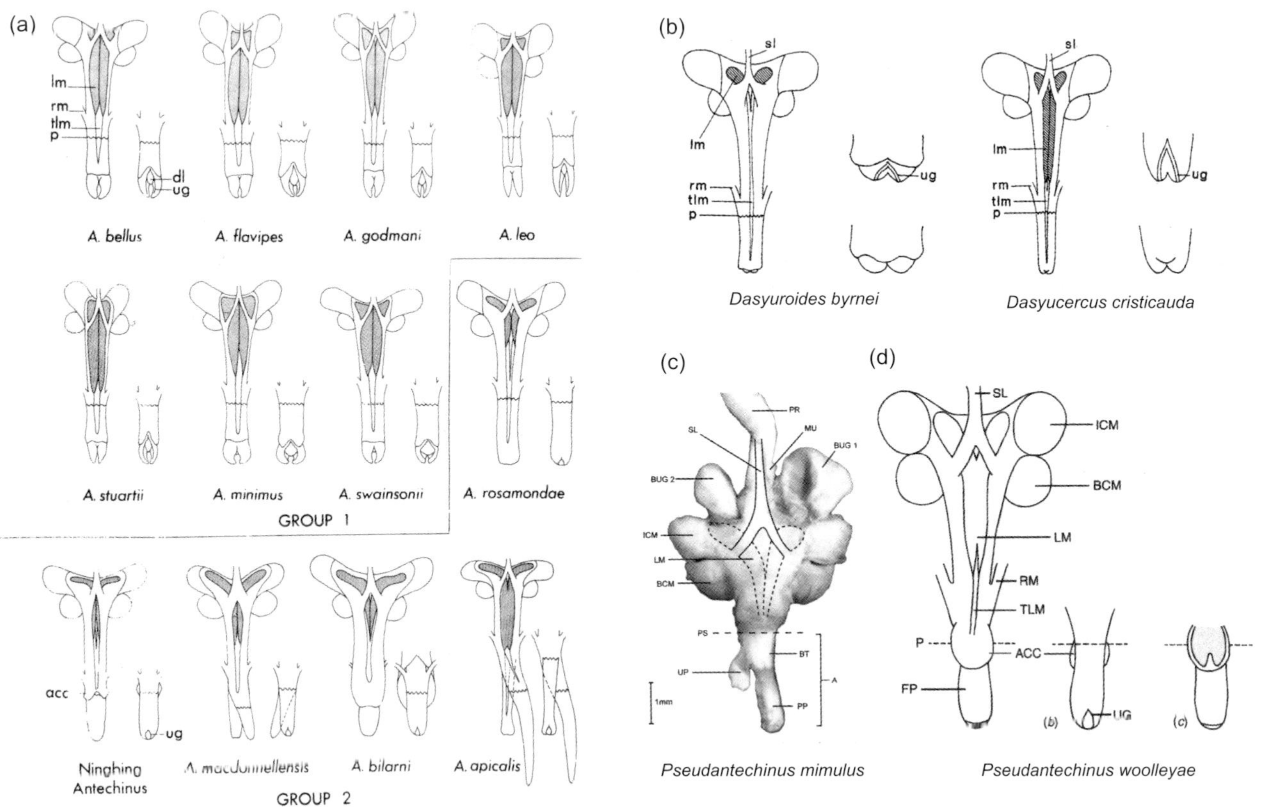

Fig. 11.14. External penis morphology of select dasyurines. (a) Penis of species in *Antechinus*, *Dasykaluta* (*'A.' rosamondae*), *Pseudantechinus* ('Ningbing Antechinus', *'A.' macdonnellensis*, *'A.' bilarni*) and *Parantechinus* (*'A.' apicalis*). Ventral view on left, dorsal view of tip on right. (b) External penis morphology of *Dasyuroides byrnei* and *Dasycercus cristicauda*. Ventral view of entire penis on left, dorsal (above) and ventral (below) views of tip on right. (c) Ventral view of penis of *Pseudantechinus mimulus*. (d) Penis of *Pseudantechinus woolleyae*. Ventral view of entire penis at left; dorsal view of tip at centre; plan view showing cavernous tissue (shaded) at right. Abbreviations: A = appendage; ACC = accessory corpora cavernosa; acc = accessory structure; BCM = bulbo-cavernosus muscle; BT = bifid tissue; dl = dorsal lobe; BUG 1, BUG 2 = bulbourethral glands; FP = free portion of penis; ICM = ischio-cavernosus muscle; LM/lm = levator muscle; MU = membranous urethra; P/p/PS = level of attachment of preputial skin; PP = pocket of preputial skin; PR = prostate; RM/rm = insertion of retractor muscle; SL/sl = suspensory ligament; TLM/tlm = tendon(s) of levator muscles; UG/ug = urethral groove. [(a) Reprinted by permission of the Royal Zoological Society of New South Wales from Woolley (1982). (b) Reprinted by permission of CSIRO Publishing from Woolley (1987). Permission conveyed through Copyright Clearance Center, Inc. (c) Reprinted by permission of CSIRO Publishing from Woolley (2011). (d) Reprinted by permission of CSIRO Publishing from Woolley (2017). Permission conveyed through Copyright Clearance Center, Inc.]

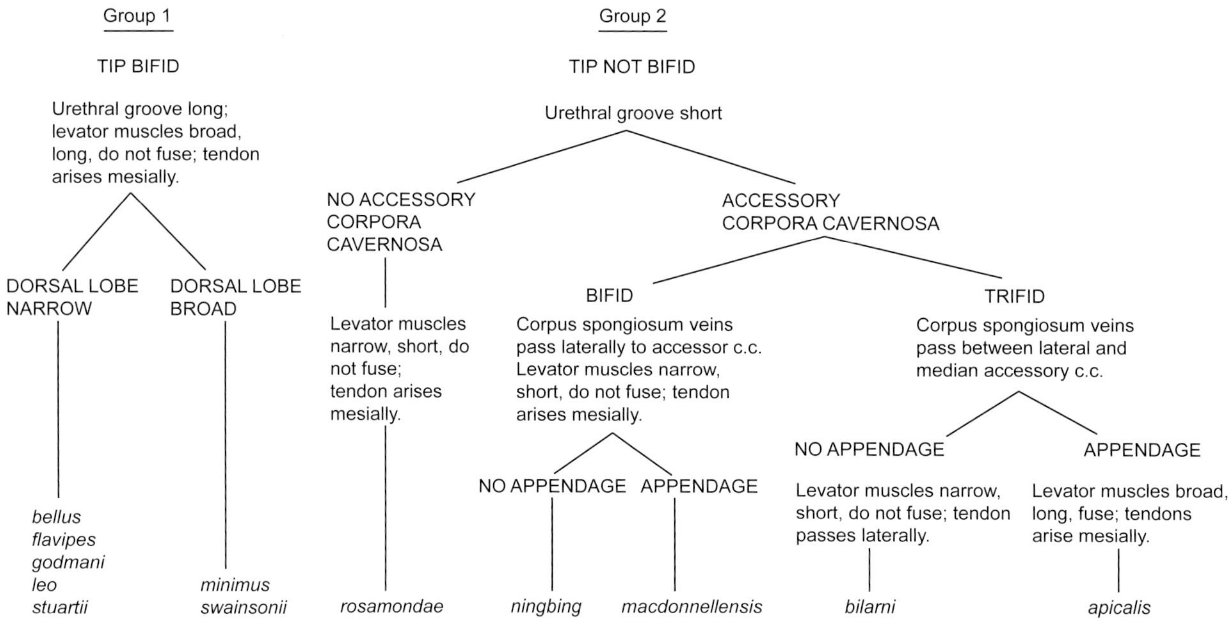

Fig. 11.15. Woolley's (1982) groupings of *'Antechinus'* species based on phallic morphology. [Redrawn with permission of the Royal Zoological Society of New South Wales from Woolley (1982).]

penis structures of *Dasyuroides byrnei* and *Dasycercus cristicauda*, finding that they most closely resembled that of *Dasykaluta* (tip not bifid, no appendage or accessory corpora cavernosa), though all three differed in several respects (Fig. 11.14b). The current consensus is that these genera form a clade with *Myoictis* and *Parantechinus* (see Figs 5.6 and 5.7). Woolley (2011) found that the penis of *Pseudantechinus mimulus* resembles that of *P. macdonnellensis* in that both have accessory erectile tissue and an appendage (Fig. 11.14c). In contrast, that of *P. woolleyae* (Fig. 11.14d) resembles *P. ningbing* and *P. bilarni* in possessing accessory erectile tissue but no appendage (Woolley 2017).

Woolley *et al.* (2007) described the anatomy of the penis for the 19 species of *Sminthopsis*, none of which was found to have an appendage or accessory erectile tissue (Fig. 11.16). However, differences in the structure of the free portion (nature of the tip, presence of a subterminal skin fold, placement of urethra and urethral grooves) distinguished 10 species groups, all but one of which correspond to clades recovered from molecular and integrative analyses (see Chapter 5). The exception is the Murina group (*S. murina*, *S. gilberti*, *S. leucopus*, *S. butleri*, *S. dolichura*, *S. archeri*), species of which collectively display four distinct penis morphologies. Woolley *et al.* (2007) noted that extensive variation in this clade of closely related species suggests an accelerated rate of diversification in penis morphology, the basis of which (adaptive or otherwise) remains obscure. Recall that García-Navas and Westerman (2018) found the Murina group as the only example of adaptive niche divergence among dasyurids (Chapter 10). Krajewski *et al.* (2012) used parsimony and likelihood to estimate ancestral penis morphotypes on a molecular phylogeny of Sminthopsini (see Fig. 6.22). They found that all but the deepest nodes could be reconstructed without ambiguity, suggesting that this character complex tracks phylogeny with relatively little homoplasy. Morphotype 1 in Fig. 11.16 appeared as ancestral for the Macroura group (including *S. crassicaudata*), whereas Morphotype 2 was ancestral for most of the remaining species.

In a phylogenetic study of quolls, Woolley *et al.* (2015) provided descriptions of penis morphology for all *Dasyurus* species, including the first such for *D. albopunctatus* and *D. hallucatus* (Fig. 11.17). A key finding was that all members of *Dasyurus* except *D. hallucatus* possess an accessory appendage. The position of *D. hallucatus* as sister to all other quolls (see Chapter 5) suggests that the appendage is a synapomorphy for the latter clade. Woolley (2019) examined the penis structure of New Guinean *Murexia*, *Phascolosorex*, *Neophascogale* and *Myoictis* species. Members of the first three have no accessory corpora cavernosa

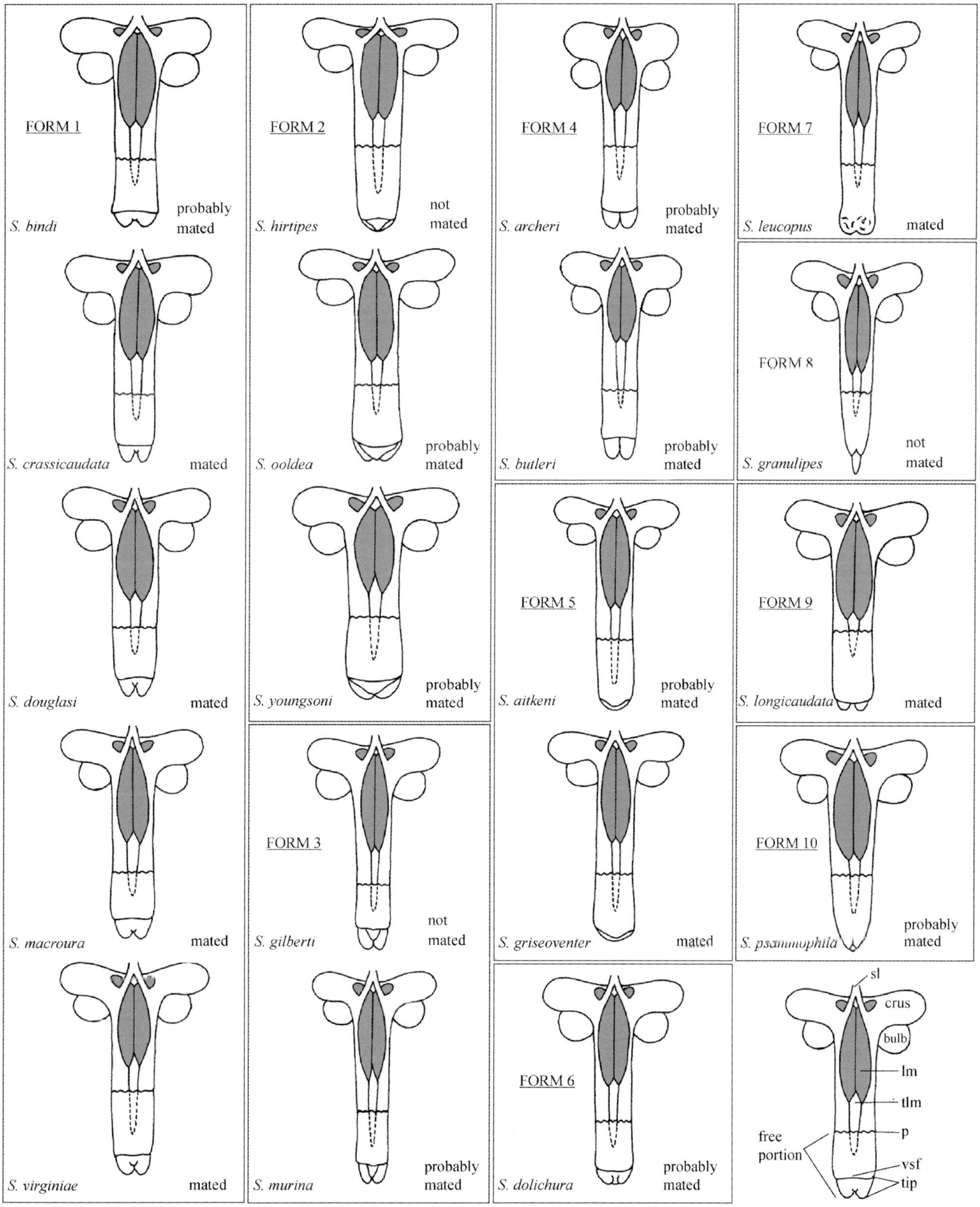

Fig. 11.16. Ventral views of external penis morphology for species of *Sminthopsis* from Woolley *et al.* (2007). Boxes enclose groups of species with similar traits. Abbreviations: lm = levator muscle; p = level of attachment of preputial skin; sl = suspensory ligament; tlm = tendon of levator muscle; vsf = free edge of ventral skin fold. [Reprinted by permission of Oxford University Press from Woolley *et al.* (2007).]

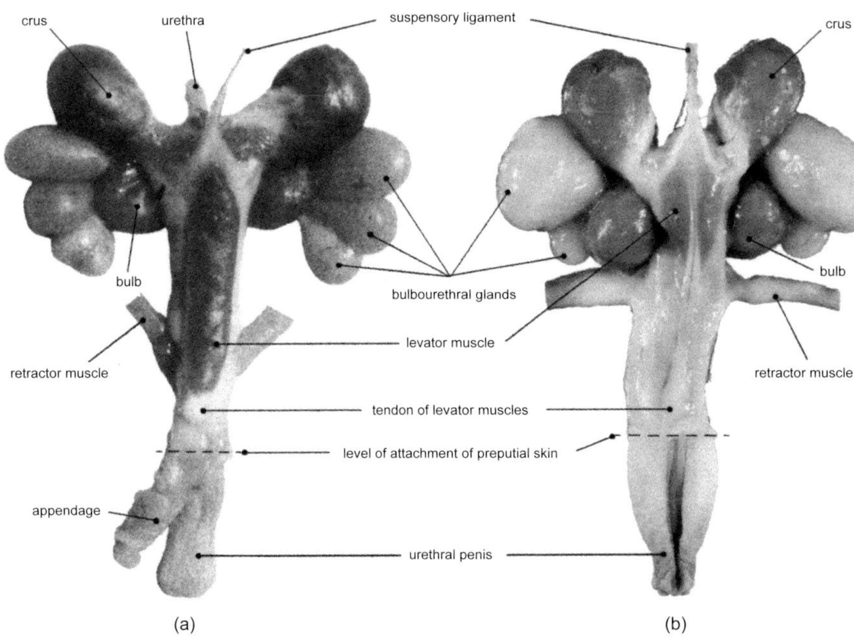

Fig. 11.17. Ventral view of penis anatomy in (a) *Dasyurus albopunctatus* and (b) *D. hallucatus* (right) from Woolley *et al.* (2015). Length of penis: (a) 34 mm and (b) 39 mm. [Reprinted by Permission of Oxford University Press from Woolley *et al.* (2015).]

or appendage (Fig. 11.18), but both structures occur in *Myoictis* (Fig. 11.19) consistent with its close phylogenetic relationship to *Parantechinus*, *Dasyuroides* and *Dasycercus* (see Chapter 5). The penis of *Sarcophilus harrisii* has not been described in detail, but it has a blunt tip and no appendage (Woolley, unpublished observation).

The functional significance of differences in dasyurid, or indeed marsupial, penis morphology has long been unclear. Cowper's (1704) suggestion that a bifid tip is complementary to the paired vaginal canals of females is countered by the observation that, although males of many marsupial species lack a bifid tip, females of all species have lateral vaginae. Nor is there any obvious variation in female reproductive anatomy that correlates with a bifid penis. The appendage is even more mysterious in that no clear description of its disposition during coitus has been recorded. After observing copulation in *Dasyurus maculatus*, Settle (1978) suggested that the appendage is inserted into the female rectum to ensure that the penis entered the urogenital sinus. This provocative hypothesis has not been tested. Whatever the functional consequences of variation in penis structure among dasyurids, it appears that this character system evolves with considerably less homoplasy than craniodental features (Woolley *et al.* 2007; Krajewski *et al.* 2012). As pointed out by Westerman, Krajewski *et al.* (2016), the appendage occurs only in members of Dasyurini (see Fig. 5.4), where it appears to have arisen at least three times (in *Dasyurus*, *Parantechinus* + *Myoictis* and *Pseudantechinus*).

REPRODUCTIVE STRATEGIES

Male semelparity in *Antechinus*

Careful study of dasyurid reproductive biology did not commence until the early 20th century with the work of pioneering embryologist James Peter Hill (1873–1954) at University College London. Hill, who had previously been a university teacher in Sydney, published several early investigations of development in platypus and marsupials. Descriptive studies on the reproductive cycle of *Dasyurus viverrinus* were summarised by Hill and O'Donoghue (1913), documenting (among other things) the unique marsupial features of a delay between the onset of oestrous and ovulation and a post-ovulatory period of uterine

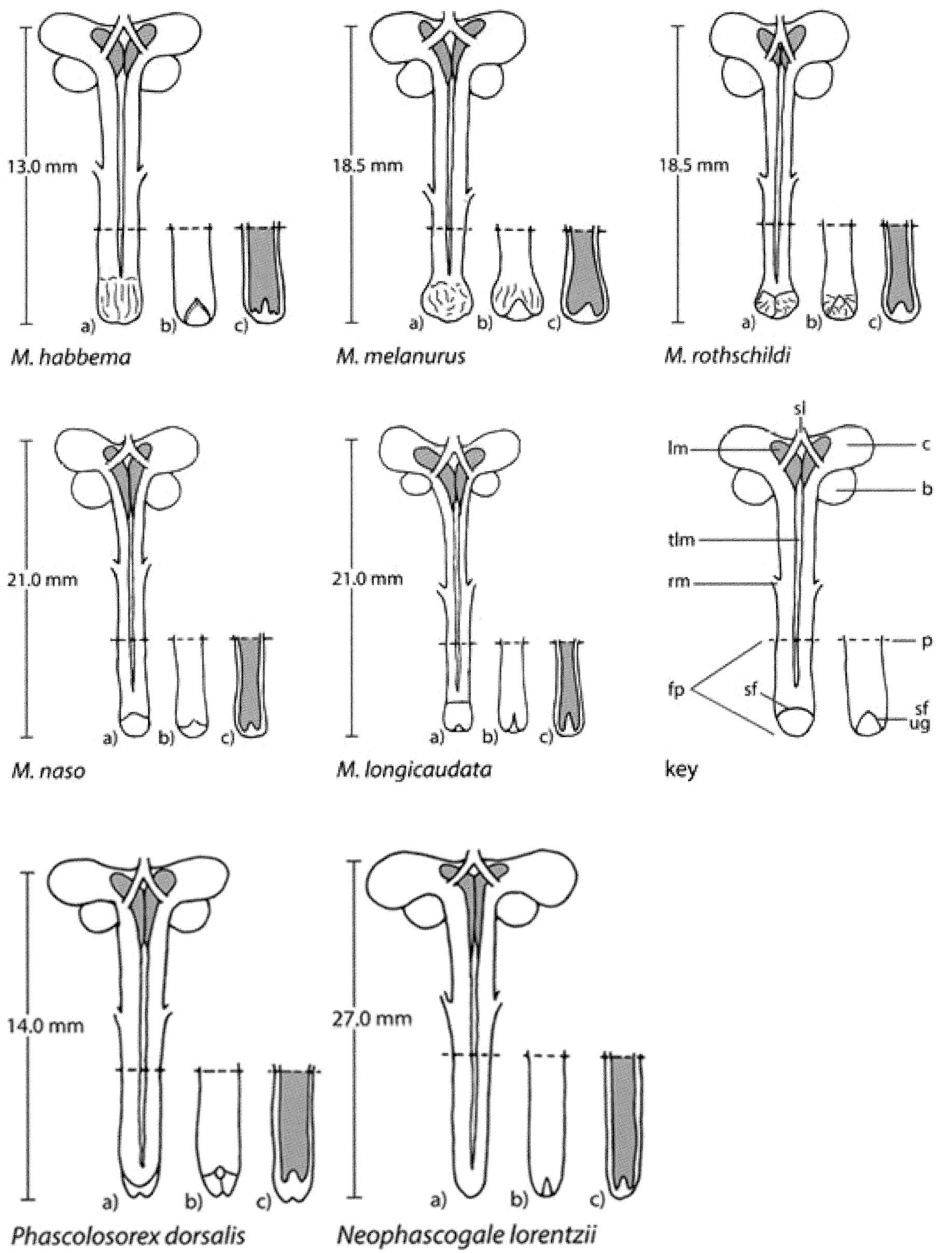

Fig. 11.18. External penis morphology of New Guinean dasyurids lacking an appendage. For each species, a ventral view of the entire penis is shown at left, a dorsal view of the distal portion at centre and a plan diagram of the corpus cavernosum at right. Abbreviations: b = bulb; c = crus; fp = free portion of penis; lm = levator muscle; p = point of attachment of preputial skin; rm = retractor muscle; sl = suspensory ligament; sf = skin fold; sf ug = skin fold over urethral grooves. [Reprinted by Permission of Oxford University Press from Woolley (2019).]

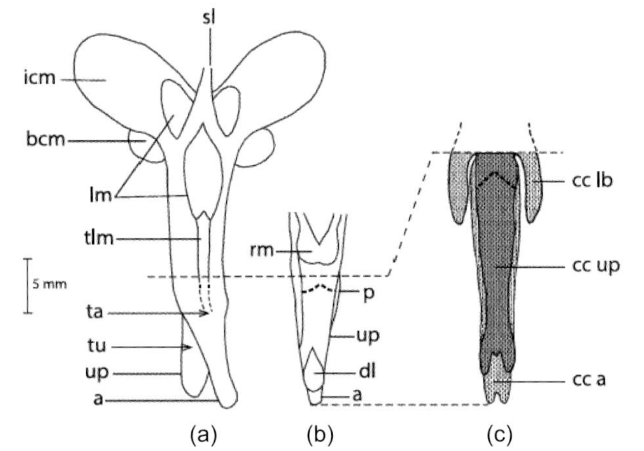

Fig. 11.19. External penis morphology of *Myoictis wallacei*. (a) Ventral view of entire penis. (b) Dorsal view of distal portion. (c) Plan diagram of corpus cavernosum in urethral penis and appendage. Abbreviations as in Fig. 11.14 and as follows: a = appendage; bcm = bulbocavernosus muscle; cc a = ventral branch of corpus cavernosum in appendage; cc lb = lateral branch of corpus cavernosum; cc up = dorsal branch of corpus cavernosum in urethral penis; dl = dorsal lobe; ism = ischio-cavernosus muscle; ta = level of insertion of ventral branch of tendon; tlm = tendon of levator muscle; tu = level of insertion of dorsal tendon; up = urethral penis. [Adapted from Woolley (2019).]

development ('pseudopregnancy') when fertilisation does not occur. Hill and Hill (1955) described development of *D. viverrinus* pouch young. In the interval between and shortly after these two papers, several other studies documented basic reproductive traits of dasyurids (summarised by Woolley 1966). Modern work on dasyurid reproduction began with Marlow's (1961) description of captive breeding in *Antechinus stuartii* (incorrectly identified as *A. flavipes*) from New South Wales. Marlow found that *A. stuartii* breeding is restricted to early August, involved prolonged copulation (>5 hours) and entailed a gestation period averaging 31.5 days (longer than the breeding season).

Woolley (1966) studied reproduction of *Antechinus agilis* (then *stuartii*) and *A. flavipes* collected near Canberra, in both the field and laboratory. Captive *A. agilis* mated in July, August and September. Development of the pouch and weight gain were seen before birth of young (or termination of pseudopregnancy in unmated females) 26–35 days later. Unmated females did not show a recurrence of pouch development or weight gain, suggesting that they were monoestrous. Woolley's observations on males were more surprising. First, although approximately equal numbers of wild males and females were trapped from January to July, only females were caught after August. Following mating in the laboratory, males displayed symptoms of senescence: reduced testis and epididymis weights, cessation of spermatogenesis, reduced size of prostate and Cowper's glands, a flaccid and unretractable penis, lethargy and deterioration of fur. Most captive males died after mating and the few that survived did not return to breeding condition. From these data, Woolley inferred that all males in wild populations died at the end of the breeding season.

Wood's (1970) 3 year trapping study of wild *Antechinus subtropicus* (then *stuartii*) in south-eastern Queensland (Fig. 11.20) confirmed Woolley's (1966) conclusions:

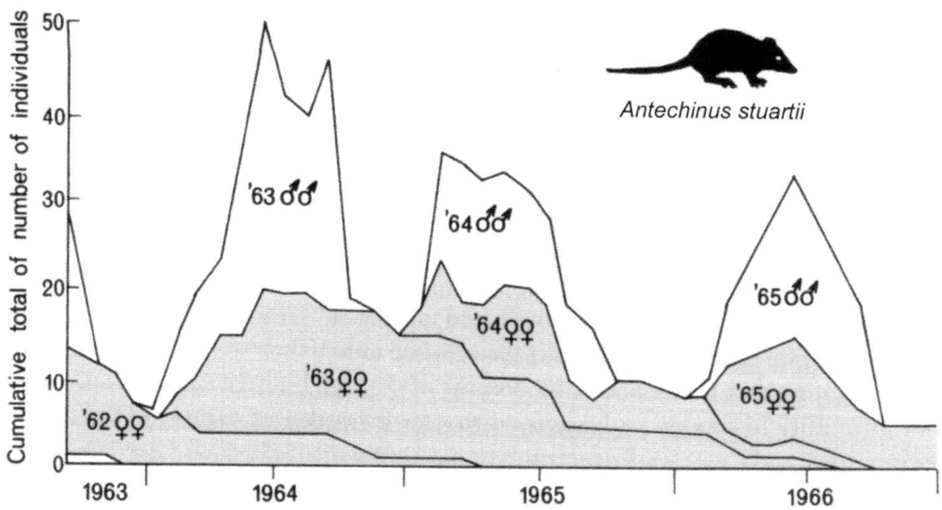

Fig. 11.20. Male die-off in *Antechinus subtropicus* (then *stuartii*) at Mt. Glorious, Qld, during 1964–66 (Wood 1970). The ordinate shows the number of individuals of each sex trapped during each month. Cohorts are identified by year and sex. [Reprinted by permission of CSIRO Publishing from Tyndale-Biscoe (2005). Permission conveyed through Copyright Clearance Center, Inc.]

The population reached a peak each autumn and declined to low numbers each spring. A few females survived to breed a second time but males suffered 100% mortality each breeding season. Death after mating was not inevitable under laboratory conditions. Mating occurred only in the last 2 weeks of September each year in the field. Pregnancy lasted about 1 month, and lactation about 3 months. Juveniles attained sexual maturity at 11 months of age. [Wood 1970: 185]

As later announced by Braithwaite and Lee (1979), these studies documented the first example of mammalian semelparity (a single breeding episode before death).

Lee *et al.* (1977) investigated the proximate causes of male semelparity (or 'die-off' as it came to be known) in *A. subtropicus* from southern Queensland and *A. agilis* from Victoria (both then *A. stuartii*). Because die-off occurred abruptly and synchronously in populations that occupied different climatic settings, the authors argued for a common intrinsic mechanism. Necropsies revealed that dead males suffered from gastrointestinal haemorrhage and pathogen-parasite infections indicative of reduced immune function. Live males during the breeding season showed elevated levels of plasma corticosteroids associated with reduced corticosteroid binding globulin (CBG) (Fig. 11.21a) and a deficiency of IgG and IgA antibodies, the latter a known result of

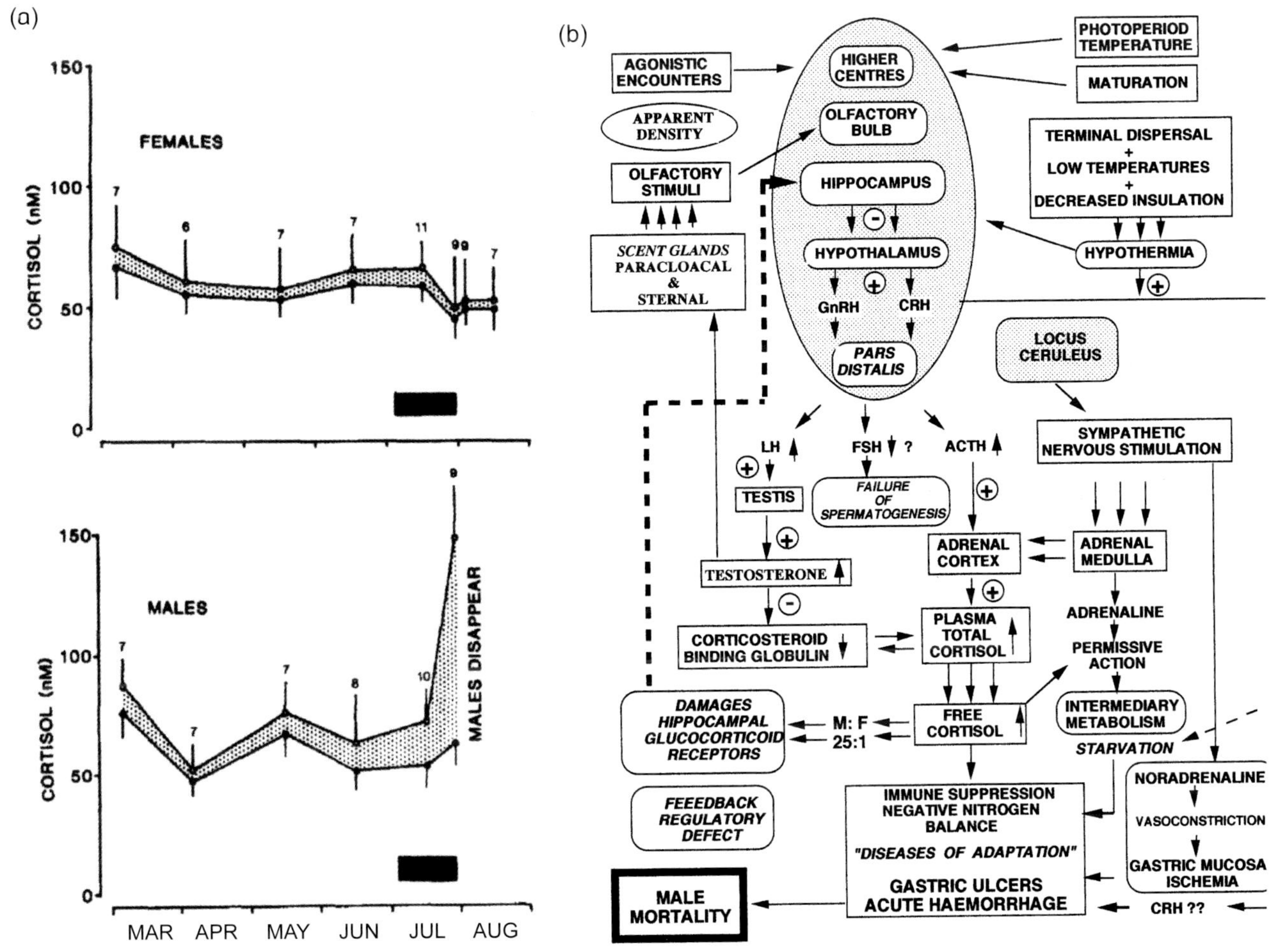

Fig. 11.21. Physiology of male semelparity in *Antechinus* and *Phascogale*. (a) Levels of total (upper lines) and CBG-bound (lower lines) cortisol in the blood of female (top) and male (bottom) *Phascogale calura* during winter (Bradley 1987). The shaded area between the curves represents the amount of free, biologically active hormone. The black bar shows the mating period. (b) Bradley's (2003) model of endocrine changes during the mating season for males. ACTH = adrenocorticotropic hormone; CRH = corticotropin-releasing hormone; FSH = follicle stimulating hormone; GnRH = gonadotropin-releasing hormone; LH = luteinising hormone; arrows inside boxes indicate direction of change; arrows outside boxes indicate causality. [(a) Reprinted by permission of Elsevier from Bradley (1987). (b) Reprinted by permission of CSIRO Publishing from Bradley (2003). Permission conveyed through Copyright Clearance Center, Inc.]

Table 11.1. Classification of dasyurid life history strategies from Lee *et al.* (1982).

Strategy	Oestrous pattern	Male breeding seasons	Duration of breeding	Seasonality of breeding	Age at maturity (months)
I	Monoestrous	Annual	Restricted	Seasonal	11
II	Monoestrous	Perennial	Restricted	Seasonal	11
III	Facultatively polyoestrous	Perennial	Restricted	Seasonal	11
IV	Polyoestrous	Perennial[1]	Extended	Seasonal	6
V	Polyoestrous	Perennial[1]	Extended	Seasonal	8–11
VI	Polyoestrous	Perennial[1]	Extended	Aseasonal	8–11[2]

[1] Incorrectly listed as 'annual' by Lee *et al.* (1982) and several subsequent authors.
[2] Based on information published after 1982.

prolonged exposure to high cortisol. Bradley *et al.* (1980) showed that injection of testosterone or adrenocorticotropic hormone (ACTH) depressed CBG levels and Bradley (1982) demonstrated that large increases in plasma androgens of males during the mating season are facilitated by absence of the steroid hormone binding globulin (SHBG) found in most mammals. This was confirmed by Kerr and Hedges (1983), who also described the cessation of spermatogonial replenishment in the seminiferous tubules of post-mating males first observed by Woolley (1966). Lee *et al.* (1977) speculated that the drop in CBG and concomitant spike in corticosteroids was due to stress during the mating season, a time when males are known to be highly mobile, aggressive and frenetic in attempting copulations. McDonald *et al.* (1986) and Bradley (1990) found that male *A. mimetes* (then *swainsonii*) and *Phascogale calura*, respectively, lack the negative feedback mechanism by which high glucocorticoid levels depress pituitary ACTH secretion in other mammals (Fig. 11.21b). Thus, as males reach maturity near the end of their first year, their testes produce androgens that, not bound by plasma globulins, rise to high levels and stimulate energetic aggressive and mate-seeking behaviour, perhaps at the expense of feeding (Lee and Cockburn 1985). This behaviour causes release of ACTH from the pituitary, triggering production of adrenal corticosteroids, most of which are biologically active as CBG levels decline. This in turn induces the well-known effects of long-term stress, including immune suppression (Letendre *et al.* 2018) and gluconeogenesis. Lacking a negative feedback loop to shut down ACTH production, stressed males succumb to gastric ulcers, infections, starvation, kidney failure and hypothermia after several days (Morton *et al.* 1989; Bradley 2003). Moreover, their testes at this stage become permanently incapable of producing viable sperm. Naylor *et al.* (2008) provided a detailed review of *Antechinus* reproductive physiology.

The discovery of male die-off triggered a surge of interest in dasyurid reproduction. Woolley (1971a) showed that in *Dasyuroides byrnei* females are polyoestrous during a May–October breeding season, gestation lasts 30–35 days, males may survive to breed in a second season and young mature at 8 months. In the closely related *Dasycercus blythi* (then *cristicauda*), females are monoestrous during a 5 week mating period, gestation is 35–44 days, males survive to a second year and young mature at ca. 11 months. Woolley (1971b) established that mainland *Parantechinus* (then *Antechinus*) *apicalis* is a monoestrous, seasonal breeder during March–April, with a gestation period of 44–53 days; males reached maturity at 10–11 months and were found capable of breeding in a second year. Although a field study of *P. apicalis* on Boullanger Island, WA, by Dickman and Braithwaite (1992) showed that no males survived their first breeding season, Woolley's (1991c) analysis of captive individuals from this population suggested that die-off is not obligate as it is in *Antechinus*.

Classification of life history strategies

Publication of the *Carnivorous Marsupials* volumes in 1982 was a watershed for study of dasyurid life histories. Nine chapters addressed descriptive and mechanistic issues raised by the discovery of male die-off and its alternatives. Piecing together the fragmentary literature on dasyurid reproduction available at the time, Lee *et al.* (1982) erected a classification of life history strategies that served as the framework for subsequent investigations on this topic. Their scheme (Table 11.1) recognised six strategies among 30 species based on five traits:

1. **Oestrous pattern.** Species are *monoestrous* (females have one oestrous cycle per breeding season), *polyoestrous* (more than one cycle), or *facultatively polyoestrous* (a second litter may be conceived if the first one fails).
2. **Number of breeding seasons in males.** *Annual* males breed in one season only, dying of intrinsic or extrinsic causes before they can breed again. *Perennial* males may breed in more than one season.
3. **Duration of breeding season.** A *restricted* breeding season is shorter than the species' gestation time; an *extended* season is longer.
4. **Seasonality of breeding.** *Seasonal* species breed only in certain months of the year; *aseasonal* species can breed in all months.
5. **Age at sexual maturity.** Dasyurids appear to fall into three groups, maturing at 6, 8–11, or 11 months of age.

Strategy I is the obligate male die-off discovered in *Antechinus stuartii* and *A. flavipes*, and subsequently documented for other members of the genus (Woolley 1981; Williams and Williams 1982; Wilson and Bourne 1984; Watt 1997; Mutton *et al.* 2017). *Phascogale tapoatafa* (Cuttle 1982) and *P. calura* (Bradley 1987) also display Strategy I, as does *Dasykaluta rosamondae* (Woolley 1991a; Hayes *et al.* 2019).

Strategy II is like Strategy I except male die-off is not inevitable, though in fact few males survive to breed in a second year. This pattern has been found in many dasyurins, including *Parantechinus apicalis* (Woolley 1971b, 1991c; Mills and Bencini 2000), *Pseudantechinus ningbing* (Woolley 1988), *P. macdonnellensis* (Woolley 1991b), *P. bilarni* (Woolley 1995a), *Dasyurus hallucatus* (Dickman and Braithwaite 1992; Oakwood *et al.* 2001) and *Sarcophilus harrisii* (Hughes 1982) (classification of the Tasmanian devil is complicated by its 2 year age at maturity).

Strategy III is like Strategy II except females may enter oestrous a second time in the same breeding season if their first cycle is unsuccessful. It is common in dasyurins such as *Dasyurus viverrinus*, *D. maculatus* (Lee *et al.* 1982), *D. geoffroii* (Soderquist and Serena 1990) and *Dasycercus blythi* (Woolley 1971a) and occurs in a single sminthopsine, *Sminthopsis leucopus* (Woolley and Ahern 1983; Woolley and Gilfillan 1991).

Strategies IV–VI feature polyoestrous females and extended breeding seasons, with males possibly surviving the first season. The young of Strategy IV species mature 6 months after birth and this category is exemplified by many species of *Sminthopsis*, including *S. murina* (Fox and Whitford 1982), *S. crassicaudata* (Morton 1982; Woolley 1984a), *S. macroura* (Woolley 1990a, b), *S. dolichura* (Friend *et al.* 1997), *S. griseoventer* (Crowther *et al.* 1999) and *S. douglasi* (Woolley 2015). Bennett *et al.* (1990), however, noted that age at maturity differs in captive *S. crassicaudata*, with females able to ovulate at about 3 months of age but males not producing sperm until they are 7–8 months old. Lacking data on age at maturity, Woolley (2007) was unable to determine whether *S. archeri* conformed to Strategy IV or V. A single instance of Strategy IV outside of dunnarts has been suggested – the dasyurin *Pseudantechinus mimulus* (Woolley 2011, 2017).

Strategy V is like Strategy IV, but maturity requires 8–11 months. With a single exception (*Dasyuroides byrnei*, Woolley 1971a), it occurs only in sminthopsines: *Antechinomys laniger* (Woolley 1984c); all species of *Ningaui* (Fanning 1982; Kitchener *et al.* 1986); *Sminthopsis longicaudata* (Woolley and Valente 1986) and *S. ooldea* (Aslin 1983); *Planigale gilesi* (Whitford *et al.* 1982), *P. maculata*, *P. ingrami* and *P. tenuirostris* (Denny 1982; Woolley and Elliott 2014).

In Strategy VI breeding takes place year-round and maturity is attained at 8–11 months. It is found mostly in New Guinean dasyurids, including all species of *Murexia*, *Dasyurus albopunctatus*, *Myoictis melas*, *Phascolosorex dorsalis*, *Neophascogale lorentzi* and *Sminthopsis virginiae rufigenis* (Dwyer 1977; Woolley 1994, 2001, 2003). *Planigale maculata sinualis* from the Adelaide River floodplain of the Northern Territory also appears to have aseasonal breeding (Taylor *et al.* 1982). Captive *Sminthopsis v. virginiae* from north-eastern Queensland can breed year-round but, like *S. v. nitela* from the Northern Territory, matures at 6 months (Woolley 1995b).

McAllan (2003) summarised the literature on dasyurid breeding biology and that literature has grown substantially in subsequent years. Nevertheless, the relevant data are extremely difficult to obtain (especially from wild populations) and our knowledge of reproductive traits in many species is piecemeal or lacking. This is especially true for New Guinean dasyurids (e.g. *Dasyurus spartacus*, most *Myoictis*, *Planigale novaeguineae*) and many species of *Sminthopsis* (e.g. *S. aitkeni*, *S. bindi*, *S. butleri*, *S. gilberti*, *S. granulipes*, *S. hirtipes*, *S. psammophila*, *S. youngsoni*). Moreover, while the Lee *et al.* (1982) classification has been a valuable framework for guiding research, it remains unclear whether the six categories are qualitatively distinct or whether gaps between them are bridged by species independently adjusting specific reproductive traits in evolutionary or ecological time. Indeed, following Oakwood *et al.* (2001), most researchers have considered variation in marsupial reproductive strategy as a continuum rather than a set of discrete alternatives.

Evolutionary ecology of reproduction

A reproductive pattern as extreme as Strategy I called for both proximate and ultimate explanations (in the sense of Mayr 1988). The former was provided by the physiological mechanisms described in 'Male semelparity in *Antechinus*' above and the latter developed in parallel. Assuming that male semelparity is an adaptation, how does it maximise individual fitness in the species that evolved it? The same question applies to other strategies. A significant portion of life history theory involves optimising trade-offs between fitness-enhancing traits given an array of organismal and environmental constraints (Stearns 1992). Mature males of Strategy I species allocate nearly all their energy to reproduction and virtually none to survival during the breeding season. Presumably, this optimisation evolved because individuals that possessed it left more offspring than those that did not. Thus, natural selection increased the frequency of those heritable traits that collectively define male die-off at the expense of alternative traits. Why? As for all evolutionary 'why' questions, potential answers take the form of historical explanations (i.e. hypotheses about causes) and several have been suggested for Strategy I and its alternatives.

The basic (and enduring) explanation for male die-off in *Antechinus* was articulated by Lee *et al.* (1977), Braithwaite and Lee (1979), Lee *et al.* (1982), Fox (1982) and Lee and Cockburn (1985) as an application of Williams's (1966) 'costs of reproduction' theory. These authors suggested that Strategy I optimises the trade-off between survival and reproduction in species that must time their breeding so that the energetically demanding period of late lactation coincides with a single, predictable and brief annual peak of food availability, coupled with a low probability of overwinter survival. This simple statement is underpinned by a large volume of natural history information. Female antechinuses usually have as many pouch young (6–12) as they have teats to suckle them. As weaning approaches, the weight of the litter equals or exceeds that of the mother such that suckling the young is energetically demanding and requires readily available food. After weaning, offspring survival during a period of rapid growth also requires food. Antechinuses are insectivores that live in seasonal habitats where insect populations boom in the spring and bust in the winter; other things being equal, it is adaptive to have late lactation and post-weaning growth occur in spring and early summer when insects are abundant. Gestation requires 25–35 days and weaning takes place at 80–140 days (McAllan 2003); lack of variation in these intervals among *Antechinus* species suggests that they are not evolutionarily labile. Thus, constraints on development and food supply create a narrow time window during which fertilisation must occur during winter. Breeding outside that window produces litters when there is insufficient food to rear them and no time for a second litter. Predation pressure on these small marsupials is high and the chance of an individual surviving to breed in a second year is low, placing a high premium on a successful first litter. Antechinuses are promiscuous and males are subject to sperm competition that establishes intense selection pressure for successful fertilisation during the short breeding season (Dickman 1993). While females must survive long enough to provision young until weaning, there is no such constraint on males, who are free to maximise their energy investment in first-year mating. The theory is that they have done so by evolving physiological mechanisms to drive their frenetic mating behaviour at the price of inevitable death when mating is done. Fisher and Blomberg (2011) showed that facultative semelparity in Strategy I females maximises reproductive output over attempts to rear young in a second year. This enhancement is realised primarily by offspring survival in seasonal, predictable habitats, whereas reproductive output (e.g. litter size) responds independently to overall food availability (Collett *et al.* 2018).

Cuttle (1982) extended the Lee *et al.* (1977, 1982) theory to *Phascogale tapoatafa* and noted that the larger body size of this species (ca. 220 g, at least twice the size of an antechinus) made its 1 year lifespan unusual, presumably because larger size improves overwinter survival. Though Cuttle did not say so, this hints at the possibility of a phylogenetic constraint on life history (see 'Phylogenetic analyses of reproduction' below). Later authors extended the theory to explain the adaptive value of other strategies. Relating Strategies II and III to Strategy I, Lee and Cockburn (1985: 103) argued as follows:

> *[T]he survival of some [strategy II] individuals of both sexes to reproduce in a second year provides two possible advantages: an opportunity to spread reproductive effort and so reduce the probability of loss of litters through overcommitment (they generally have small litters), and an insurance against failure at their first attempt at reproduction. Additional insurance against loss of a litter or failure to mate is found in species showing Strategy III, in which some females are polyoestrous, although only producing one litter annually.*

Although Cockburn (1997) warned against formulating a hypothesis for Strategy I evolution based solely on *Antechinus*, the Lee *et al.* (1977) model exemplifies an established generalisation from life-history theory:

> *The evolution of the length of the reproductive lifespan is controlled by the relationships of the mean and variance of adult and juvenile mortality. Increases in the mean and variance of adult mortality should decrease the reproductive lifespan, eventually producing semelparity. Increases in the mean and variance of juvenile mortality can lengthen the lifespan, eventually producing very long-lived organisms with high investments in somatic structures and their repair. [Stearns 1992: 205]*

If living in predictable seasonal habitats is the adaptive context for monoestrous strategies, occupying less predictable, harsh or aseasonal habitats might favour polyoestry. Morton (1978) and Fanning (1982) put this idea forward for *Sminthopsis crassicaudata* and *Ningaui ridei*, respectively. Fox and Whitford (1982), however, documented polyoestry in a *Sminthopsis murina* population living in the same coastal forest as *Antechinus stuartii* and suggested that 'adoption of polyoestry or monoestry would appear more closely related to phylogenetic similarity than predictability in food supply' (p. 47). This insight notwithstanding, the association of polyoestry with seasonal unpredictability continued to impress dasyurid biologists. Morton (1982) pointed out that 17 of 19 arid zone dasyurids are polyoestrous and 'concluded from the available information that selective pressures operating upon these animals are dominated by the considerable variability of the invertebrate populations which constitute their food supply' (p. 117). Indeed, Morton felt that the two arid zone monoestrous species – *Dasykaluta* and *Dasycercus* – were ecological anomalies not fully accounted for by the Lee *et al.* (1977) theory. Lee and Cockburn (1985) had little to add to this explanation for the adaptive value of Strategies IV and V, but echoed Fox's (1982) caution against ecological determinism: '[T]his example should remind us that the evolution of life history strategies is not governed by the seasonal distribution and predictability of the food supply alone' (p. 104). They were equally circumspect about the functional distinction between these strategies:

> *Strategies IV and V were distinguished on the time required to attain sexual maturity, approximately 6 months in the former and 8–11 months in the latter. The more rapid attainment of sexual maturity by species employing Strategy IV was seen to provide an opportunity for reproduction during the season of birth, and therefore a higher reproductive potential. However, there are reasons for suspecting that this advantage is not attained in natural populations ... [Lee and Cockburn 1985: 98]*

Strategy VI (aseasonal breeding) has received the least theoretical attention even though, in the context of dasyurid life histories, it is almost as extreme as Strategy I. Its strong association with the tropical habitats of New Guinea is often taken in support of the notion that year-round food availability permits year-round reproduction, an option largely unavailable to Australian species. However, reproductive data on Strategy VI species are rare and it may not yet be possible to formulate a more specific adaptive explanation.

Relatively few rigorous tests of the Lee *et al.* (1977) theory have been performed, largely due to the logistical challenges of doing so. Dickman (1982) showed that where *A. stuartii* (including what is now *agilis*) is sympatric with *A. mimetes* (then *swainsonii*) in south-eastern Australia, its mating period is shifted away from optimal coincidence with food availability, possibly a result of niche segregation. As noted above, Dickman and Braithwaite (1992) documented intraspecific variability in the extent of extrinsic male die-off in a pair of Strategy II species – geographic variation in *Parantechinus apicalis* and temporal (among years) variation in *Dasyurus hallucatus*. Both studies showed that reproductive cycles can be fine-tuned by complex selection pressures associated with local ecology. Holleley *et al.* (2006) tested two potential selective forces – intermale competition for access to females and sperm competition after copulation – favouring Strategy I in *Antechinus stuartii.* Using scrotum size as a proxy for male access and microsatellite markers to identify paternity, they found support for both factors. This is consistent with the enhanced mating success of larger *A. stuartii* males documented by Fisher and Cockburn (2005). Fisher *et al.* (2013) found additional evidence for the importance of sperm competition in male die-off species and argued that this form of sexual selection is the primary driver of male reproductive strategy evolution in dasyurids. Hayes *et al.* (2019) demonstrated multiple paternity and positive testis allometry in *Dasykaluta*, further supporting the importance of sperm competition in Strategy I species.

Phylogenetic analyses of reproduction

In describing how social dominance may increase the variance of reproductive success among male *Antechinus stuartii*, Braithwaite (1979: 517) remarked, 'The life history of this species appears to be geared to predictable seasonal food resources, but it may not be optimal throughout the present range and the species appears to be evolutionarily trapped.' This was the earliest expression of concern that, as later stated by Lee and Cockburn (1985) and quoted above, seasonal food availability might not explain everything about the evolution of Strategy I. Later authors suggested other relevant ecological factors, as described in the previous section, but what else could influence the distribution of strategies among species? Cuttle (1982) indirectly, and Fox and Whitford (1982) directly, called attention to possible phylogenetic constraints, but Friend *et al.* (1997) put it bluntly: 'The observed sympatry [of polyoestrous members of the *S. murina* species group] with highly seasonal monoestrous dasyurids of the genus *Antechinus* is postulated to occur through spatial and temporal selection of different microhabitats, but also *suggests that phylogenetic factors may be at least as important as predictability of climate and food resources in explaining the evolution of different reproductive strategies*' (italics added). In other words, a species may display a particular fitness-enhancing trait because (1) it independently evolved that trait as an adaptation to its specific environment; (2) it inherited that trait from an ancestor that it shares with other species; or (3) both ancestry and species-specific selection have influenced the trait. Phylogenetic effects are implicated when most members of a clade share the same adaptation, suggesting that the adaptation arose in their ancestor. This is the insight behind Felsenstein's (1985b) dictum that species and their traits are not statistically independent data points for correlational analysis. Boonstra (2005) credited phylogenetic and ecological constraints as interacting causes of semelparity in marsupials. The implications of phylogeny for understanding the evolution of dasyurid reproductive strategies were explored in two molecular studies.

Krajewski *et al.* (1996) used partial *cytb* sequences to estimate relationships among 10 species within Phascogalini and obtained a tree (Fig. 11.22) on which *Antechinus* and *Murexia* were weakly supported (73% bootstrap) as sisters apart from *Phascogale*. As discussed in Chapter 4, this was one of several early studies indicating that New Guinean '*Antechinus*' and *Murexia* (*sensu lato*) were a single clade within Phascogalini. Krajewski *et al.* (1996) drew attention to the occurrence of the two most extreme reproductive strategies in closely related sister genera. Given that *Phascogale* and *Antechinus* share Strategy I but did not appear as sisters, simple parsimony would suggest that Strategy I was ancestral for phascogalins, with Strategy VI evolving from Strategy I in *Murexia*. However, citing the then-current hypothesis that tribal ancestors occupied Miocene rainforests in Australia (see Chapter 10), the authors favoured a scenario in which Strategy VI was ancestral and Strategy I arose

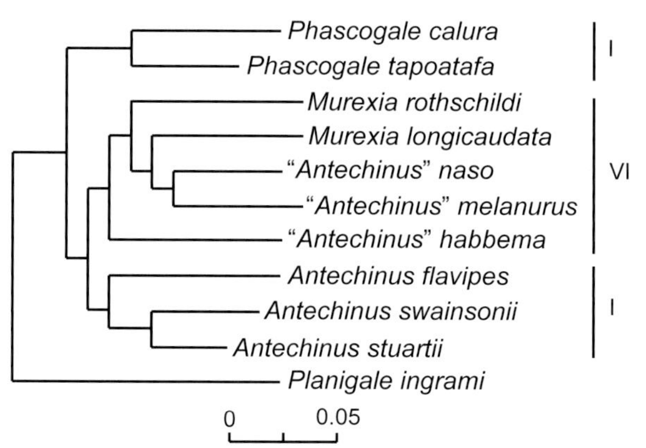

Fig. 11.22. Phylogenetic analyses of phascogalin reproductive strategies by Krajewski *et al.* (1996). This best-fit distance tree is derived from Kimura two-parameter distances estimated with an expected transition:transversion ratio of 3.6, obtained with FITCH algorithm of PHYLIP 3.5. Species labelled *"Antechinus"* are now recognised as *Murexia. Planigale ingrami* is the outgroup. Branch length scale is substitutions/site. Roman numerals at right indicate reproductive strategy. Bootstrap support for the *Antechinus-Murexia* clade (node 6) is 73%. [Redrawn by permission of Springer Nature from Krajewski *et al.* (1996).]

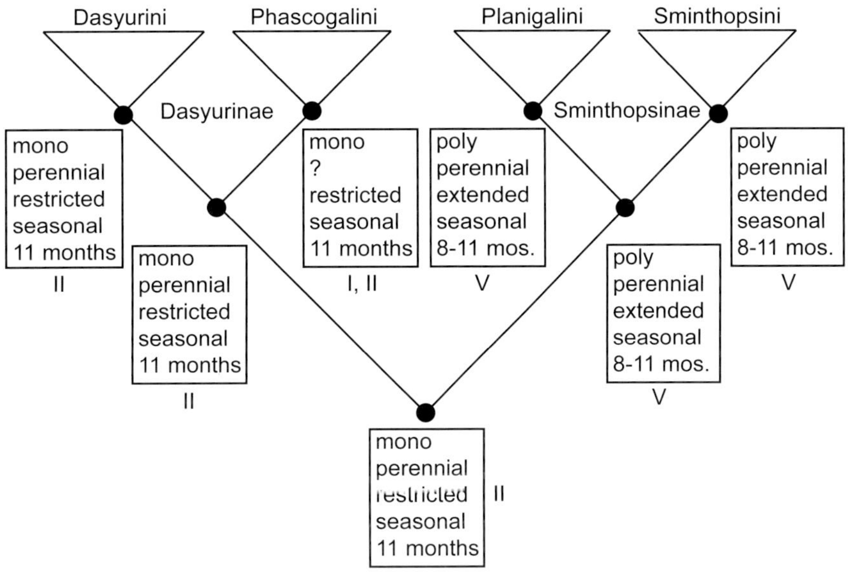

Fig. 11.23. Summary of parsimony reconstruction of ancestral states and strategies on a molecular phylogeny by Krajewski, Woolley and Westerman (2000). Reproductive traits (in boxes) were treated as independently evolving characters on the phylogeny of Krajewski, Wroe and Westerman (2000). Traits (in order from top to bottom within boxes): oestrous pattern = monoestrous, polyoestrous; male breeding season = annual or perennial; duration of breeding season = restricted or extended; seasonality of breeding = seasonal or aseasonal; age at maturity = 6, 8–11 or 11 months. [Redrawn by Permission of Oxford University Press from Krajewski, Woolley and Westerman (2000).]

independently in *Phascogale* and *Antechinus*. The most recent integrative phylogenies of dasyurids (Kealy and Beck 2017; Mutton *et al.* 2019; García-Navas *et al.* 2020) disagree about intergeneric relationships within Phascogalini (see Figs 5.6, 5.7 and 6.16), but the argument for parallel evolution of Strategy I still applies. It is also plausible that both Strategies I and VI were derived from a Strategy II–III phascogalin ancestor.

Krajewski, Woolley and Westerman (2000) used the dasyurid phylogeny of Krajewski, Wroe and Westerman (2000) based on complete *cytb*, *12S* and *P1* DNA sequences (see Fig. 4.14) as framework on which to map reproductive traits using parsimony reconstruction of ancestral states. Data were tabulated for all species that could be placed into one of the six Lee *et al.* (1982) categories and reconstructions performed by treating each of the five component traits as independent. Data from *Thylacinus cynocephalus* and *Myrmecobius fasciatus* were used to help fix the ingroup root state. The results (Fig. 11.23) were remarkably clear. States mapped to the ancestor of living dasyurids formed Strategy II, which persisted on the dasyurine side of the subfamily split to the dasyurin and perhaps phascogalin ancestors (Strategies I and II were equally parsimonious at the latter node). Strategy V arose on the sminthopsine side and was retained in both planigalin and sminthopsin ancestors. In this scenario, Strategies III, IV, VI and perhaps I appear as late evolving (intratribal) modifications of older ancestral patterns. States assigned to the phascogalin ancestor were inconsistent with Krajewski *et al.*'s (1996) suggestion that this species employed Strategy VI; Braithwaite's (1979) contention that Strategy I is irreversible favours the Strategy II alternative for this ancestor.

Krajewski, Woolley and Westerman (2000) suggested that the basal split between Strategies II and V may reflect divergence of dasyurine and sminthopsine ancestors into forest and non-forest habitats, respectively. Although they also mapped biogeographic regions of individual species onto the phylogeny, this revealed no clear association between habitat and reproduction after the basal split. Indeed, the amount of within-tribe parallelism and convergence in both traits is considerable, consistent with optimisation of species' reproductive patterns to recent ecological conditions superimposed on their phylogenetic legacy.

CHROMOSOMES

Chromosomes and karyotypes

The word 'chromosome' derives from the coloured bodies ([Gr.] *chromo* + *soma*) in a cell nucleus visible under the microscope after chemical staining. These structures carry the organism's genetic information – the Boveri-Sutton chromosomal theory of inheritance (Sutton 1903; Boveri 1904). As with all other eukaryotes, most of the genetic material (DNA) of marsupials occurs in the cell nucleus complexed with histone proteins, around which the DNA is wrapped, to form nucleosomes. Non-histone proteins act as a scaffold to fold the long DNA molecule (1–20 cm in eukaryotes), condensing it by several orders of magnitude into chromosomes (ca. 1–10 μm) that can be seen during cell division (Fig. 11.24). The folding process is highly specific for each chromosome and allows individual genes to be accessed for transcription during interphase. Not all chromosomal DNA is coding, however, and not all genes are expressed in every cell. Centromeres (primary constrictions) are involved in chromosome movement during mitosis and meiosis, and telomeres form the ends of chromosomes; both are examples of functional non-coding DNA. The entire set of chromosomes in a nucleus is known as the karyotype, which John and Lewis (1968) referred to as an organism's 'endophenotype'.

Early characterisations of karyotypes consisted of counts and morphological descriptions of chromosomes. Animal species carry a characteristic number of chromosomes in the nuclei of their diploid ('2n') somatic cells. In therian mammals these include a pair of sex

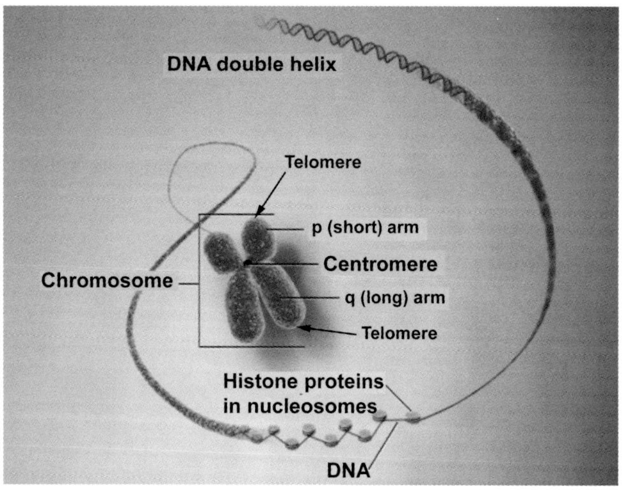

Fig. 11.24. How DNA is packed into a eukaryote chromosome. The illustration shows general features of a eukaryotic chromosome including long arm (q), short arm (p) and centromere (specialised sequences which help move chromosomes during cell divisions). The ends of chromosomes (telomeres) are made up of specialised repeat sequences $(TTAGGG)_n$. [Courtesy U.S. National Library of Medicine.]

chromosomes (XX in females, XY in males) and a variable number of autosomes. Autosomes are numbered in decreasing order of size as they appear when condensed during metaphase (1 = largest, 2 = next largest, etc.). In contrast to the diploid nuclei in somatic cells, the nuclei of gametes (eggs and sperm) carry a half or haploid ('n') complement of chromosomes. Gametic fusion at fertilisation restores the diploid number of an embryo. In eutherian mammals, diploid numbers range from 6 (Indian muntjak, *Muntiacus muntjak*) to 104 (red viscacha rat, *Tympanoctomys barrerae*), but the range is much smaller in marsupials (see below). The gross morphology of a chromosome is described by reference to centromere location. Chromosomes with the centromere near the middle are metacentric, those with the centromere closer to one end than the other (creating short and long arms) are submetacentric or acrocentric, and those with the centromere at the end are telocentric.

Since the mid-1900s, staining has been used to study chromosome homology and evolution (Sessions 1996). Techniques include simple staining with orcein (Fig. 11.25a), as well as G-, C- and N-banding (Fig. 11.25b–d). In G-banding, condensed metaphase chromosomes are digested with trypsin to remove proteins and treated with Giemsa stain (methylene blue, eosin and azure B dyes). The stain binds preferentially to regions rich in adenine and thymine, which tend to have low gene densities ('heterochromatin'), such that they stain darkly; gene-rich regions ('euchromatin') with more guanine (G) and cytosine (C) are lighter. Eldridge and Metcalfe (2006) present a collection of G-banded marsupial karyotypes. C- and N-banding use a similar approach, but differentially stain centromeric and other 'constitutive' heterochromatin (mostly repetitive DNA) or nucleolar organiser region (NOR, or secondary constriction) sequences, respectively. NORs are the sites of ribosomal RNA genes. More recently, fluorescent *in situ* hybridisation (FISH) techniques have been developed to study chromosome organisation. For interspecies chromosome painting ('Zoo-FISH'), DNA from individual flow-sorted chromosomes or their parts are labelled with fluorescent dyes and hybridised back to chromosomes of the same or different species to ascertain variation in chromatin arrangement (Fig. 11.26a). Probe sequences can reveal the chromosomal location of a gene (Fig. 11.26b, c), telomeric DNA or centromeres (Fig. 11.27). Each of these techniques has played a significant role in studying the evolution of marsupial chromosomes.

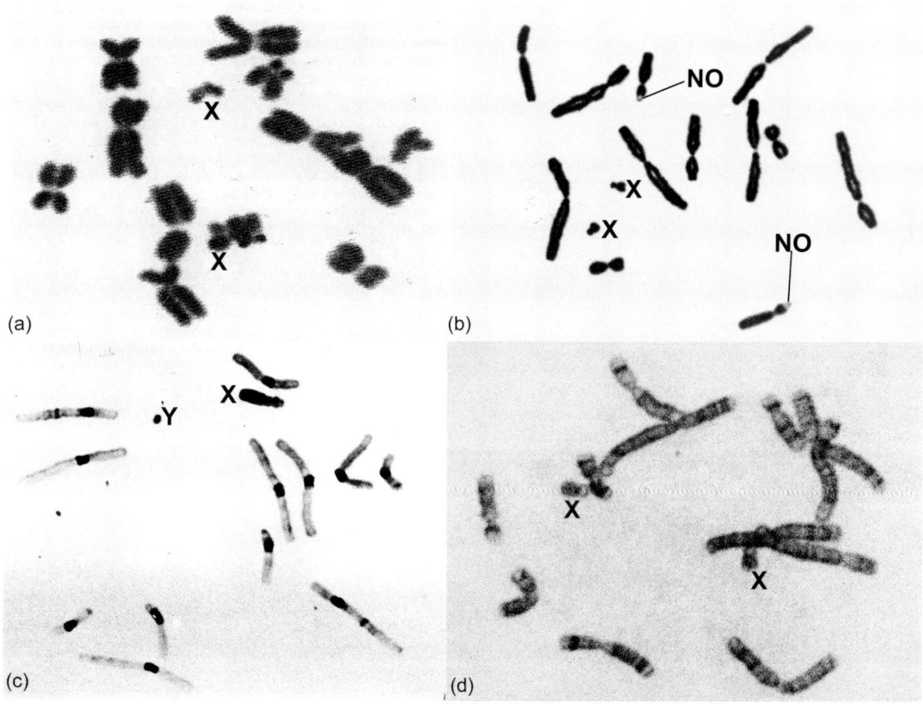

Fig. 11.25. Dasyurid chromosomes stained at metaphase. (a) *Sminthopsis archeri* female stained with orcein showing X chromosomes. (b) *Murexia melanurus* female N-banded karyotype showing nucleolar organising (NO) regions on short arm of chromosome 5. (c) *Parantechinus apicalis* male C-bands. (d) *Dasyurus spartacus* female G-bands. [Photo credit: Mike Westerman.]

Marsupial karyotypes

Chromosome number, size and shape

Marsupial chromosomes were some of the first to be studied, over a century ago, when Benda (1906) illustrated those from a species of *Perameles*.[3] Marsupials have relatively large chromosomes and low diploid numbers, the latter ranging from 10 in swamp wallabies (*Wallabia bicolor*) to 32 in rufous rat-kangaroos (*Aepyprymnus rufescens*), although most species have either 2n = 22 or 2n = 14 (Sharman 1973; Hayman and Martin 1974; Reig *et al.* 1977; Rofe and Hayman 1985). The combination of low number and large size made marsupial chromosomes, particularly those of dasyurids, easy to study. To date over 50 dasyurid species have been karyotyped and show virtually identical 2n = 14 complements (Hayman *et al.* 1987; Hayman 1990; Westerman and Woolley 1990, 1993; M. Westerman, unpublished observations). Since the same 2n = 14 karyotype is found in *Myrmecobius* (Sharman 1961) and may also characterise *Thylacinus* (P.A. Woolley, unpublished), it appears to have been remarkably stable over the evolutionary history of dasyuromorphians (Fig. 11.28).

Dasyurid karyotypes comprise six pairs of metacentric or submetacentric autosomes of varying size, along with two smaller X chromosomes in females or an X and even smaller (punctiform) Y in males (Figs 11.26c and 11.27). The six autosomes are readily identifiable by their size, shape and G-band pattern (Fig. 11.28). All dasyurid species have an NOR visible as a secondary constriction at the distal end of the short arm of chromosome 5 (Fig. 11.25b). As in all marsupials, dasyurid X and Y chromosomes do not form a synaptonemal complex during meiosis and therefore do not undergo recombination (Sharp 1982). Bennett *et al.* (1990) found striking sex differences in recombination rates for two 3-locus autosomal linkage groups in colony-bred *Sminthopsis crassicaudata*, with interlocus recombination frequencies of 0–12% in females but 16–51% in males. Interestingly, chiasmata occur only at the ends of female autosomes but all along the arms in males.

Although arm length measurements and G-band patterns suggest karyotypic stability in dasyurids, several studies reported variable C-banding patterns and DNA content among species (Timms *et al.* 1982; Young *et al.*

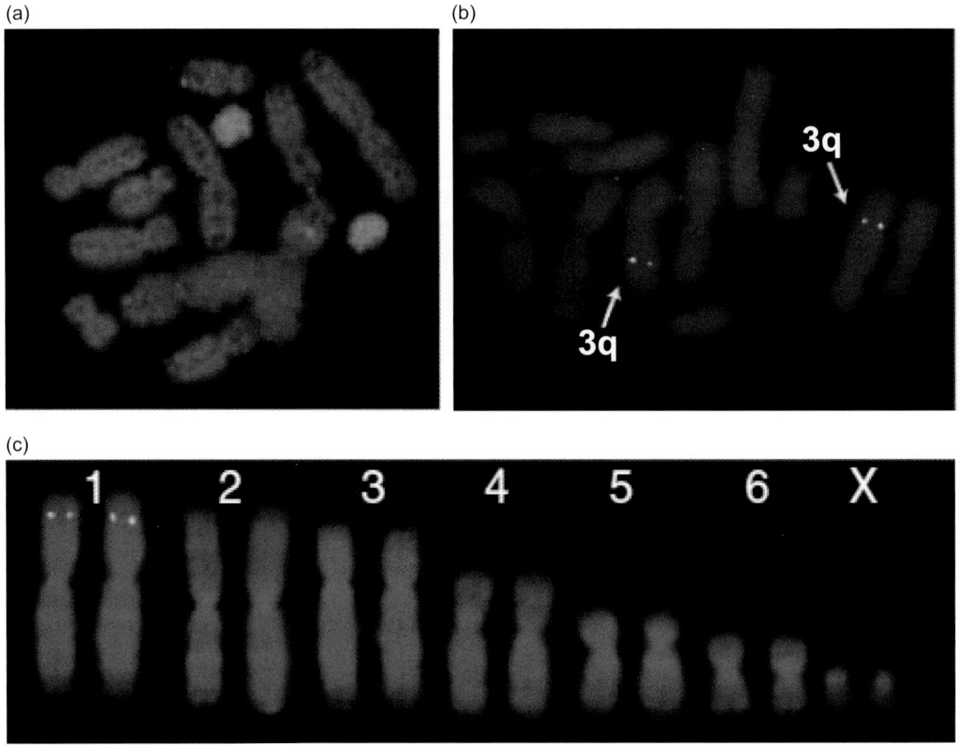

Fig. 11.26. Cross-species chromosome painting in dasyurid marsupials. (a) *Sminthopsis crassicaudata* female chromosomes painted with probes of *Trichosurus vulpecula* chromosome 1 (red) and X (green). (b) Localisation of beta globin gene on long arm of chromosome 3 (3q) in *Sminthopsis crassicaudata*. (c) Localisation of a marker gene to chromosome 1 of *Sarcophilus harrisii*. [(a) Reprinted by permission of Springer Nature from Rens *et al.* (1999). (b) Reprinted by permission of Karger Publishers from De Leo *et al.* (2005). (c) Reprinted by permission of Surrey Beatty & Sons from Potter and Deakin (2018). Permission conveyed through Copyright Clearance Center, Inc.]

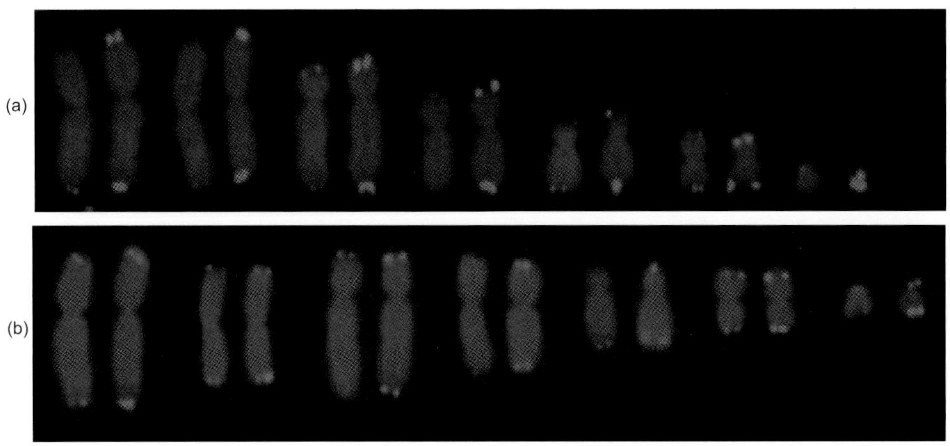

Fig. 11.27. Differential size of telomeres on maternally and paternally derived homologues of (a) *Sarcophilus harrisii* and (b) *Dasyurus hallucatus*. The maternal homologue is on the left in each pair. [(a) Reprinted from Bender *et al.* (2012) under CC BY license. (b) Reprinted from Ingles and Deakin (2016) under CC BY 4.0 license.]

1982; Westerman and Woolley 1990, 1993). C-banding reveals 'constitutive heterochromatin', usually consisting of highly repetitive DNA (hrDNA) sequences, often near the centromere. With few exceptions, such as *Dasykaluta rosamondae*, only small centromeric C-bands were observed on dasyurid autosomes, though telomeric bands were sometimes present on arms 3q and 5p. The small size of the C-bands is consistent with small amounts of

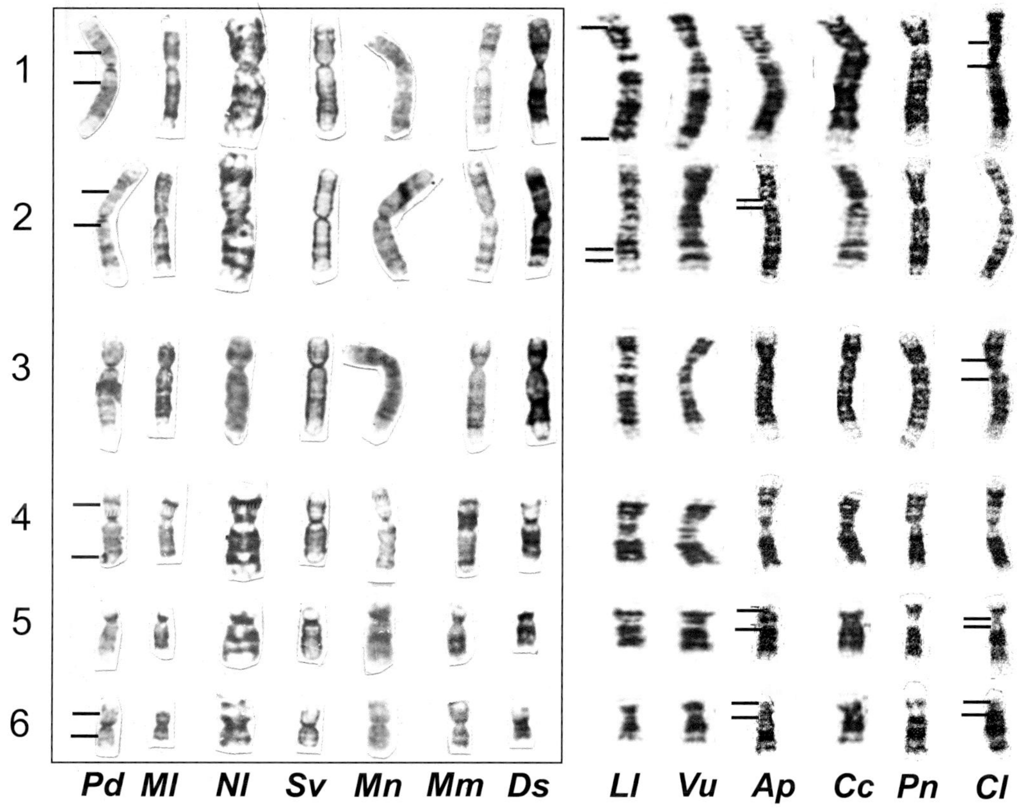

Fig. 11.28. G-banded autosomes of representative dasyurid species and other marsupials with 2n = 14 karyotypes. Box shows dasyurid autosomes 1–6 for *Pd* = *Phascolosorex dorsalis; Ml* = *Murexia longicaudata; Nl* = *Neophascogale lorentzii; Sv* = *Sminthopsis virginiae; Mn* = *Murexia naso; Mm*= *Murexia melanurus; Ds* = *Dasyurus spartacus.* Other marsupials: *Ll* = *Lasiorhinus latifrons; Vu* = *Vombatus ursinus; Ap* = *Acrobates pygmeus; Cc* = *Cercartetus concinnus; Pn* = *Perameles nasuta; Cl* = *Caluromys lanatus.* Lines show inversions in particular autosomes. [Photo credit: Mike Westerman.]

hrDNA in dasyurid genomes. In *Parantechinus apicalis* and *Ningaui timealeyi*, C-bands are more extensive on the enlarged X chromosomes. Although Hayman and Martin (1974) reported only small differences (ca. 1.2%) in DNA content ('C-value') for seven dasyurid species, much greater differences were noted by Westerman and Woolley (1990, 1993), with *Dasyuroides byrnei* having >30% more DNA than *Dasykaluta rosamondae*. This large difference in C-values has no apparent effect on G-band patterns, however, suggesting that the extra DNA is scattered throughout the chromosomes rather than forming major blocks.

Sex chromosomes and sex determination

In mammalian XX/XY sex determination the Y chromosome carries a testis-determining gene (*SRY*), expression of which causes differentiation of the gonadal ridge into testes, which in turn release hormones to stimulate later-developing characters such as a scrotum and descended testicles. An *SRY* gene has been identified on marsupial Y chromosomes (Foster *et al.* 1992), but studies of intersex individuals, including dasyurids, showed that sex differentiation in marsupials differs from that in eutherians and does not depend solely on androgen production by the developing testis (Shaw *et al.* 1990). Woolley *et al.* (2003) showed that in *Sminthopsis douglasi* presence of a Y chromosome is insufficient for complete sex determination. An XXY individual had abdominal ovary-like gonads with a gonadal cord and external hemiscrotum on the right side, but a hemipouch on the left (Fig. 11.29). In contrast, XO marsupials lack testes, pouch and teats, but often have a scrotum. Thus, although the Y chromosome determines testis development, other sexual characteristics are controlled independently by X-linked 'pouch-mammary-scrotum (PMS) switch' genes that act via a dosage or imprinting mechanism. PMS switch genes were initially proposed by Sharman *et al.* (1990b) from a study of intersex macropodids, in which one copy of the X

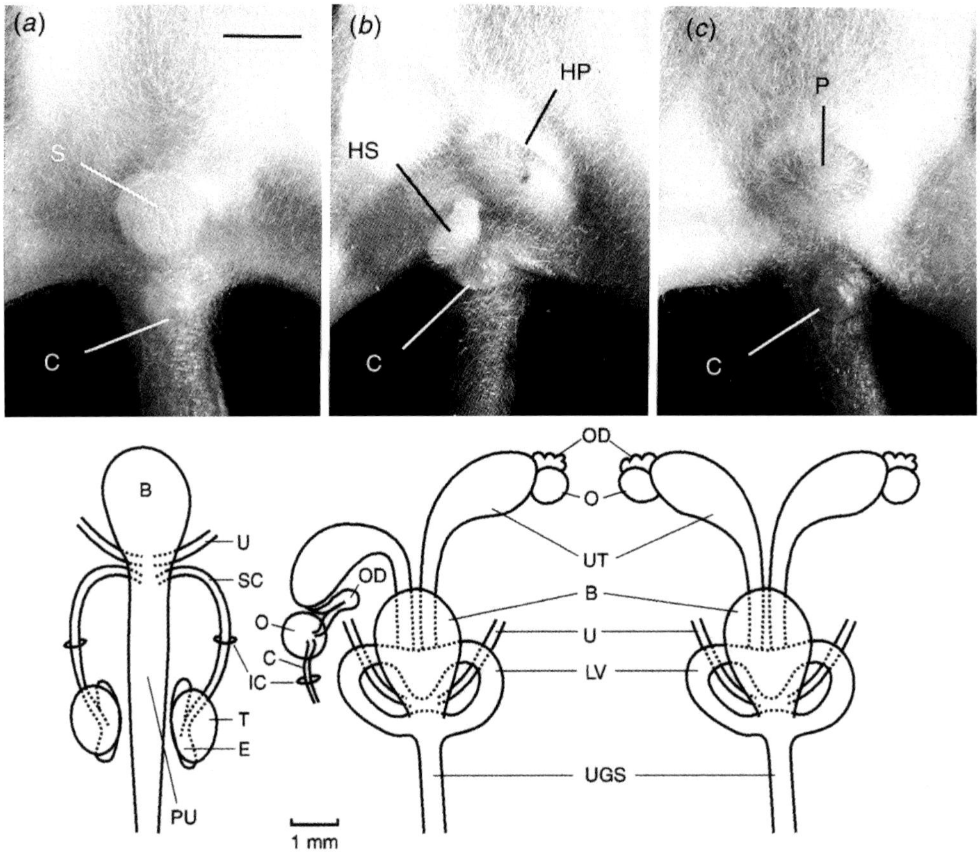

Fig. 11.29. External (above) and internal (below) reproductive structures of (a) male, (b) intersexual (XXY) and (c) female 50-day-old pouch young of *Sminthopsis douglasi*. Abbreviations: B = urinary bladder; C (top) = cloacal aperture; C (bottom) = gonadal cord; E = epididymis; HP = hemipouch with partial ridge; HS = hemiscrotum; IC = inguinal canal; LV = lateral vagina; O = ovary; OD = oviduct; P = pouch area defined by a circular ridge of skin; PU = prostatic urethra; S = scrotum; SC = spermatic cord; T= testis; U = ureter; UGS = urogenital sinus; UT = uterus. Nipples are too small to be visible in the photographs. Scale bar = 2 mm. [Reprinted by permission of CSIRO Publishing from Woolley *et al.* (2003). Permission conveyed through Copyright Clearance Center, Inc.]

induces scrotum development, but two copies result in a pouch (dosage effect). In contrast, presence of a paternally derived X produces a pouch (imprinting).

Marsupials, like eutherians, inactivate one X chromosome in female somatic cells as part of a mechanism that equalises the expression of X-linked genes between males (which have one X) and females (which have two). In eutherians, inactivation is random with respect to the maternal or paternal origin of the X and tightly controlled by some combination of methylation and late replication (Riggs 1990). In marsupials, the paternal X is preferentially inactivated, often incompletely (Cooper *et al.* 1977, 1990, 1993). Marsupial X-inactivation involves deacetylation but not methylation (Graves and Westerman 2003) and seems to be under the control of the *RSX* ('RNA on Silent X') locus, which plays a role similar to *XIST* ('X inactive transcript') in eutherians (Grant *et al.* 2012). In peramelids and *Dromiciops*, the paternal X may be physically eliminated from the nucleus in some female tissues (Hayman and Martin 1974; Hayman 1977; Gallardo and Patterson 1987; Johnston *et al.* 2002) and the Y eliminated from the same tissues in males (Hayman and Martin 1974). There is no evidence for such elimination in dasyurids. X chromosomes from different dasyurid species vary in size and shape but are always the smallest members of the complement, making up less than 3% of the female genome (Hayman and Martin 1974).

As noted above, the marsupial Y is very small, containing only ca. 10–12 Mb of DNA in *Sminthopsis macroura* (Toder *et al.* 2000), and is often hard to identify in chromosome spreads. Although comprised mostly of non-coding repetitive DNA, the dasyurid Y carries several genes in addition to *SRY* (Sharman *et al.* 1990b; Cooper *et al.* 1993), including *RBMY* (RNA binding motif protein of Y), *UBE1Y* (ubiquitin-activating enzyme E1 on Y) and *ATRY* (alpha-thalassemia mental retardation syndrome on Y), all of which have homologs on the X (Delbridge *et al.* 1999). Other Y genes do not have such homologs and are expressed exclusively in the testis. Although the number of genes on the dasyurid Y is unknown, 16 have been identified in didelphids (*Didelphis* and *Monodelphis*), of which 10 are also found in the tammar wallaby (*Notomacropus eugenii*) (Bellot *et al.* 2014). All are involved in testis determination.

Telomere length dimorphism in dasyurids

Vertebrate telomeres comprise thousands of copies of short sequences (5′-TTAGGG-3′) which are eroded and replenished during DNA replication, providing a protective buffer for genes upstream on the chromosome arms (Hartwell *et al.* 2011). The paternal homologue of each dasyurid chromosome pair has substantially longer telomeres (i.e. more repeats) than the maternal homologue (Fig. 11.27); no such dimorphism occurs in peramelemorphs, diprotodonts, or didelphimorphs (Bender *et al.* 2012; Ingles and Deakin 2016; Deakin 2019). First discovered in the Tasmanian devil (*Sarcophilus harrisii*), telomere dimorphism has now been found in *Dasyurus hallucatus*, *D. maculatus*, *Phascogale calura*, *Sminthopsis macroura*, *S. crassicaudata* and *S. douglasi*, suggesting that it arose early in dasyurid evolution. Marin-Gual *et al.* (2022) demonstrated that elongation is initiated in the late zygotene stage of meiotic prophase I and reaches a maximum in late pachytene, producing long telomeres in all spermatids and sperm. Its functional significance, if any, is unknown.

Chromosomal positioning and translocation

Multiple studies have shown that, in many organisms, chromosomes are not randomly arranged within the cell nucleus (Cremer and Cremer 2010), but occupy specific territories during interphase, anchored to the nuclear matrix by non-histone proteins. These territories are dynamic during the cell cycle: chromosome regions containing particular genes change position as they are activated or inactivated (Chuang *et al.* 2006; Deakin 2019). Chromosome territories in the sperm heads of *Sminthopsis crassicaudata* were demonstrated by Greaves *et al.* (2001) using Zoo-FISH techniques (Fig. 11.30). One consequence of fixed positioning within the nucleus is that certain pairs of non-homologous chromosomes always lie closer together than others, predisposing them to participate in translocations (Westerman 1968). The mechanism of such translocation involves cellular repair systems that act on damaged DNA. The size of the repair enzyme complex is such that if translocation is to result from the repair process, the damage sites in the two non-homologous chromosomes must be in proximity so that the complex can 'mis-repair' them. Preferential involvement of specific chromosomes in translocation has been observed in macropodids (see 'Insights from chromosome painting' below).

Towards an ancestral dasyurid karyotype

Insights from chromosome morphology and banding

The stability of dasyurids in chromosome number, size, shape and relative arm length is also seen in G- and R-band ('reverse G-band') patterns (Rofe 1979; Young *et al.* 1982; Rofe and Hayman 1985; Westerman and Woolley 1990,

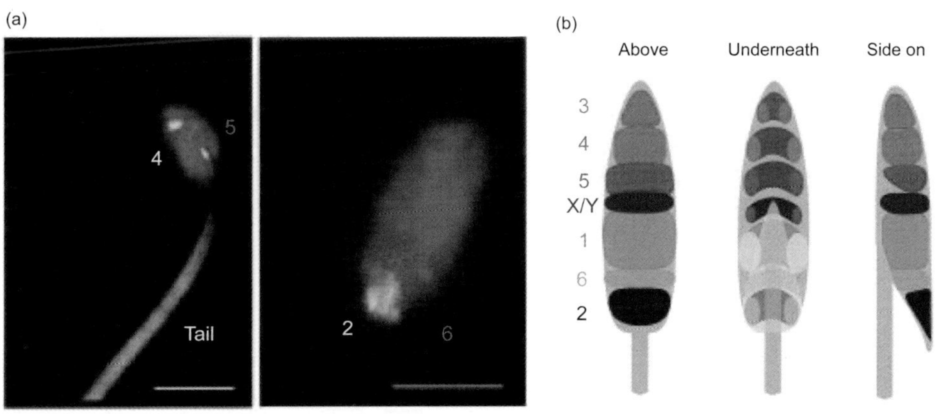

Fig. 11.30. Non-random positioning of chromosomes in *Sminthopsis crassicaudata* sperm heads (Greaves *et al.* 2001). (a) Two-colour chromosome painting shows positions of chromosomes 2, 4, 5 and 6. (b) Diagram showing arrangement of all chromosomes. [Reprinted by permission of Springer Nature from Greaves *et al.* (2001).]

1993; Fig. 11.28). Although the amount and location of C-band heterochromatin may differ considerably among species (Young *et al.* 1982), this chromosome component seems to have little functional significance. Only four structural variations have been revealed by G- and R-banding: (1) *Antechinomys laniger* has a unique paracentric (i.e. not including the centromere) inversion or small insertion on the long arm of chromosome 6 (Rofe and Hayman 1985); (2) *Ningaui yvonneae* and *N. ridei* have pericentric (i.e. including the centromere) inversions on chromosome 6 (Baverstock *et al.* 1983); (3) *Sarcophilus harrisii* is polymorphic for a small pericentric inversion in chromosome 5 (but this has only been detected in cells of devil facial tumour disease; Pyecroft *et al.* 2007); and (4) *Pseudantechinus macdonnellensis* shows variation in the arm lengths of chromosome 2 and in amounts of centric heterochromatin (Young *et al.* 1982). Conservation of chromosome structure presumably reflects stability in gene location and perhaps evolutionary constraints on karyotypic change. Karyotype stability in dasyuromorphians contrasts with greater variability in other marsupial groups such as Phalangeriformes and Macropodiformes, in which extensive chromosomal reshuffling has taken place (Hayman 1990; Sharman *et al.* 1990a; Eldridge and Close 1997; Westerman *et al.* 2010).

Hayman and Martin (1974), Hayman (1977) and Rofe and Hayman (1985) argued that 2n = 14 represents the ancestral dasyurid – and probably ancestral marsupial – chromosome complement. The G-banding study of Rofe and Hayman (1985) suggested that the 2n = 14 karyotypes of dasyurids, bandicoots, vombatids and some South American species are relatable via a small number of changes such as pericentric inversions (Fig. 11.31).[4] Westerman *et al.* (2010) supported this view by mapping chromosome traits onto a well-resolved molecular phylogeny (Fig. 11.32); Suarez-Villota *et al.* (2016) did the same for South American marsupials. If the ancestral australidelphian karyotype was like that of *Dromiciops*, wombats, bandicoots and pygmy possums, only four pericentric inversions (on chromosomes 1, 2, 4 and 6) are required to produce the dasyurid G-band structure in Fig. 11.31. Later evolution resulted in occasional reversions to the ancestral form of some chromosomes, such as the submetacentric chromosome 6 of *Ningaui*.

Insights from chromosome painting

Marsupial chromosome painting (Ferguson-Smith 1997; Rens *et al.* 2001, 2003) has supported and extended Hayman's (1990) evolutionary scenario based on G-banding.

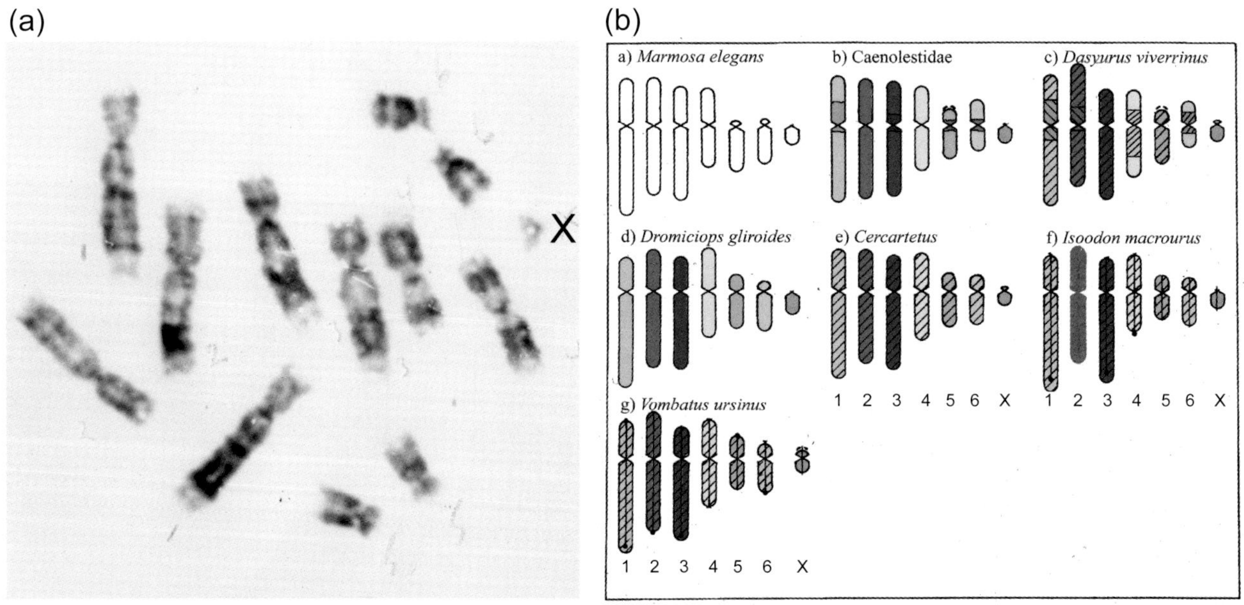

Fig. 11.31. Relationships between superficially similar 2n = 14 karyotypes in various marsupial groups. (a) G-banded karyotype of male *Dromiciops gliroides* (minute Y chromosome not visible). (b) Probable relationships of karyotypes of members of major marsupial families. Cross-hatched chromosomes have similar banding patterns to other cross-hatched chromosomes of the same number. [(a) Photo credit: Mike Westerman. (b) Redrawn with permission of the Zoological Society of New South Wales from Sharman (1982).]

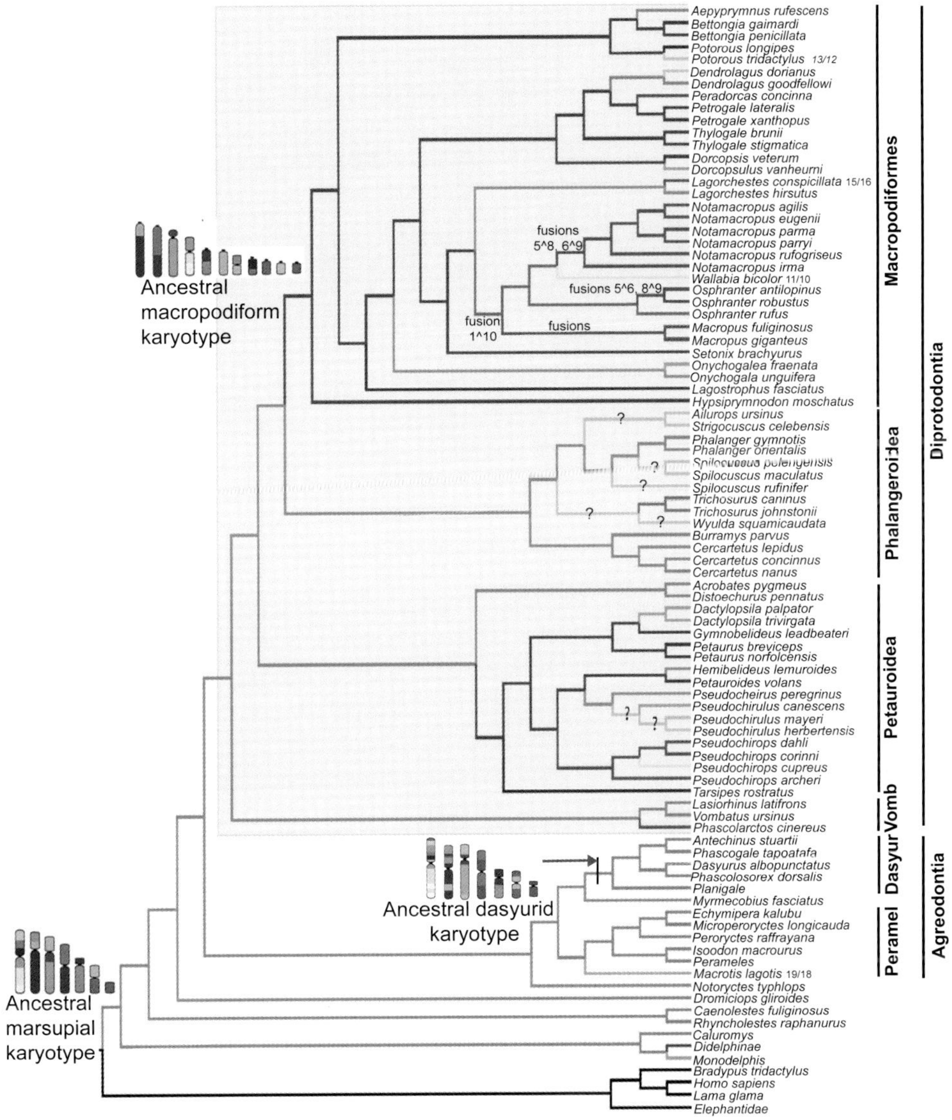

Fig. 11.32. Evolution of chromosome number in marsupials as inferred by Westerman *et al.* (2010). Diploid numbers for species of marsupials are plotted on a molecular phylogeny. Branch colour coding: yellow 2n = 10–11; dark green 2n = 14; red 2n = 16; pink 2n = 18; purple 2n = 22; blue 2n = 24; light green 2n = 32. Chromosome fusions (^) inferred from G-banding studies. Ancestral arrangements of conserved chromosome blocks (C1–20) are shown for ancestral marsupial, macropodiform and dasyurid branches. [Reprinted by Permission of Oxford University Press from Westerman *et al.* (2010).]

Painting effectively resolved small karyotypic differences between taxa that were not easily detected with G-banding and thereby confirmed the conservation of chromosome structure across Marsupialia. Indeed, all karyotypic variation in marsupials can now be accounted for by rearrangements of only 20 conserved chromatin blocks: C1–18 on autosomes, C19 on the X and C20 on the Y (Rens *et al.* 1999, 2001, 2003). The probable arrangements of these blocks in relation to G-band patterns of a dasyurid (*Sminthopsis crassicaudata*) and a didelphid (*Monodelphis domestica*) are shown in Fig. 11.33.

Graphodatsky *et al.* (2011) used patterns of association among syntenic blocks revealed by cross-species painting to reconstruct a putative ancestral eutherian karyotype (AEK). In similar fashion, analysis of G-bands and the 20 syntenic blocks of marsupials allows inference of a probable ancestral marsupial karyotype (AMK), from which the ancestral dasyurid complement can be derived (Figs 11.34

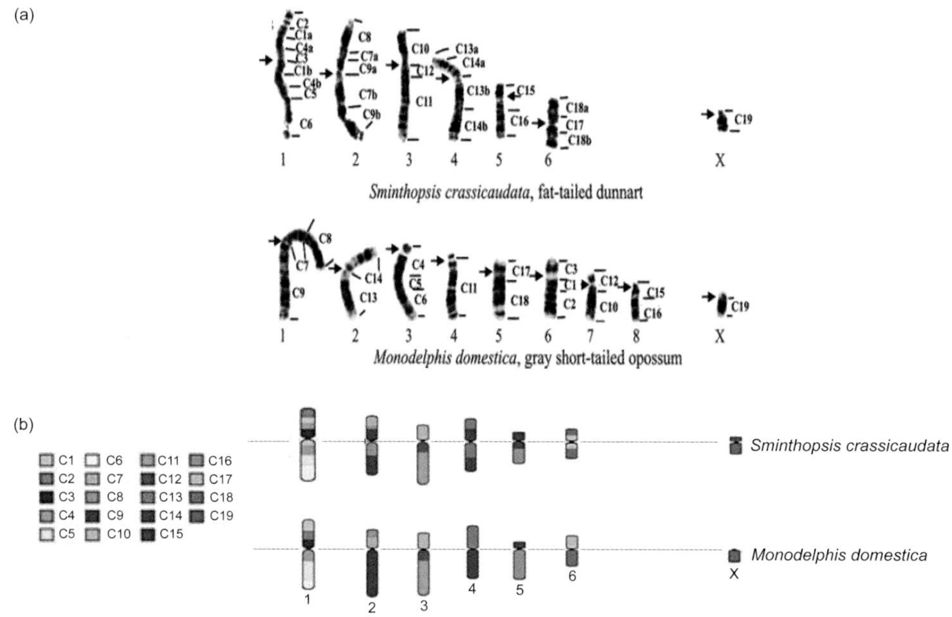

Fig. 11.33. Conserved blocks of genomic material in marsupial chromosomes. (a) Conserved blocks C1–19 in the G-banded chromosomes of *Sminthopsis crassicaudata* (2n = 14) and *Monodelphis domestica* (2n = 18). Arrows indicate centromere location. Block C20 is on the Y chromosome and not shown here. (b) Stylised colouring to show the arrangement of blocks in (a). [(a) Reprinted by permission of S. Karger AG from Rens *et al.* (2003). (b) Reprinted by permission of Springer Nature from Deakin and Kruger-Andrzejewska (2016).]

and 11.35). Consistent with the G-band results described above, changes from the AMK in chromosomes 1 (two pericentric inversions and a probable centromere shift), 2, 4 and 6 (one pericentric inversion each) yield the dasyurid karyotype. Studies of numbat chromosomes should reveal whether these changes predate the dasyurid ancestor.

Putative ancestral karyotypes have now been derived for all marsupial groups (Fig. 11.35). Although some groups retain the AMK or something like it, phalangerids, petauroids and macropodiforms show many karyotypic changes, including those giving rise to the 2n = 32 karyotype of *Aepyprymnus rufescens*. The ancestral karyotype has been retained in some species from all three families of Macropodoidea. Other species show major changes in chromosome number and shape that often involve the same six syntenic blocks of the ancestral macropodoid karyotype: A1 (= AMK blocks C7, C8), A5 (= C15, C16), A6 (= C10), A8 (= C2, C3), A9 (= C18) and A10 (= C17) (Fig. 11.34). This suggests that these chromosomes are more susceptible to exchanges, possibly due to their proximity in the interphase nucleus (see 'Chromosomal positioning and translocation' above) or to properties of their DNA sequences. Derivation of the ancestral didelphimorphian karyotype from the AMK involved inversions in autosomes 1, 3, 5 and 6 to give shapes like those in caluromyids (Fig. 11.35). These initial inversions were followed by other, often radical, changes (including centric fissions, fusions and interchanges) to produce the varied karyotypes of modern opossums.

Despite differences in chromosome number and shape between Australian and South American marsupials, conserved features of the ancestral marsupial karyotype are apparent in both groups. G-and R-band patterns in *Sminthopsis crassicaudata* and *Monodelphis domestica* show that *Sminthopsis* chromosomes 1 and 3 have extensive homology to *Monodelphis* chromosomes 3 + 6 and 4 + 7, respectively (Fig. 11.33).

Chromosomal rearrangements and speciation

The idea that chromosomal rearrangements play a major role in speciation runs through much of the early literature of evolutionary genetics (Mayr 1970; King 1993; Potter *et al.* 2017), its popularity perhaps peaking with White's (1973, 1978) classic reviews. Chromosomal differences between parents may cause suppression of recombination in heterozygous offspring due to faulty pairing of homologues at meiosis. Such hybrid infertility reduces gene flow between karyotypically different parental populations and allows the accumulation of additional genetic incompatibilities, ultimately leading to

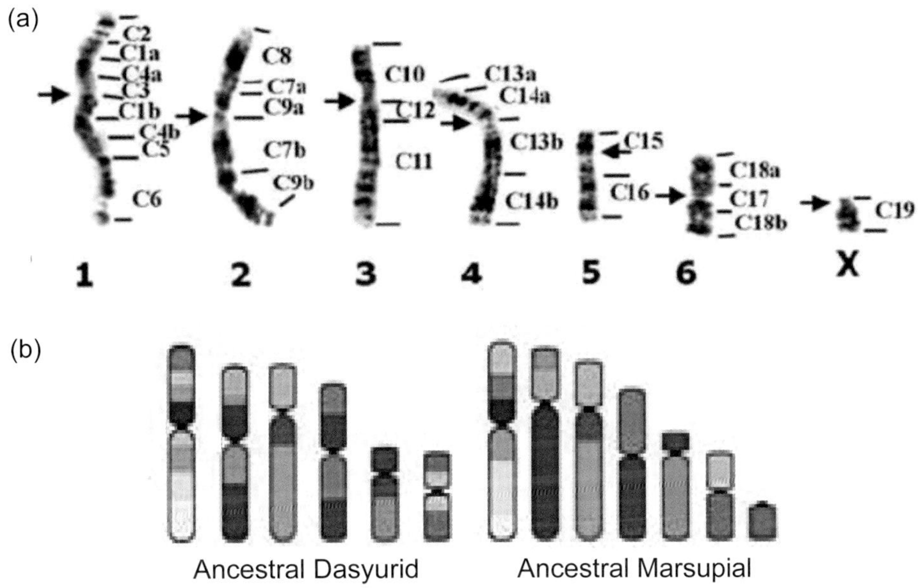

Fig. 11.34. Dasyurid karyotype showing (a) G-band pattern in terms of a conserved chromatin blocks C1-C19 and (b) how these may have been arranged in the common ancestor of dasyurids and all marsupials. Arrows show centromere location. [(a) Reprinted by permission of S. Karger AG from Rens *et al.* (2003). (b) Reprinted from Deakin *et al.* (2012) under CC BY license.]

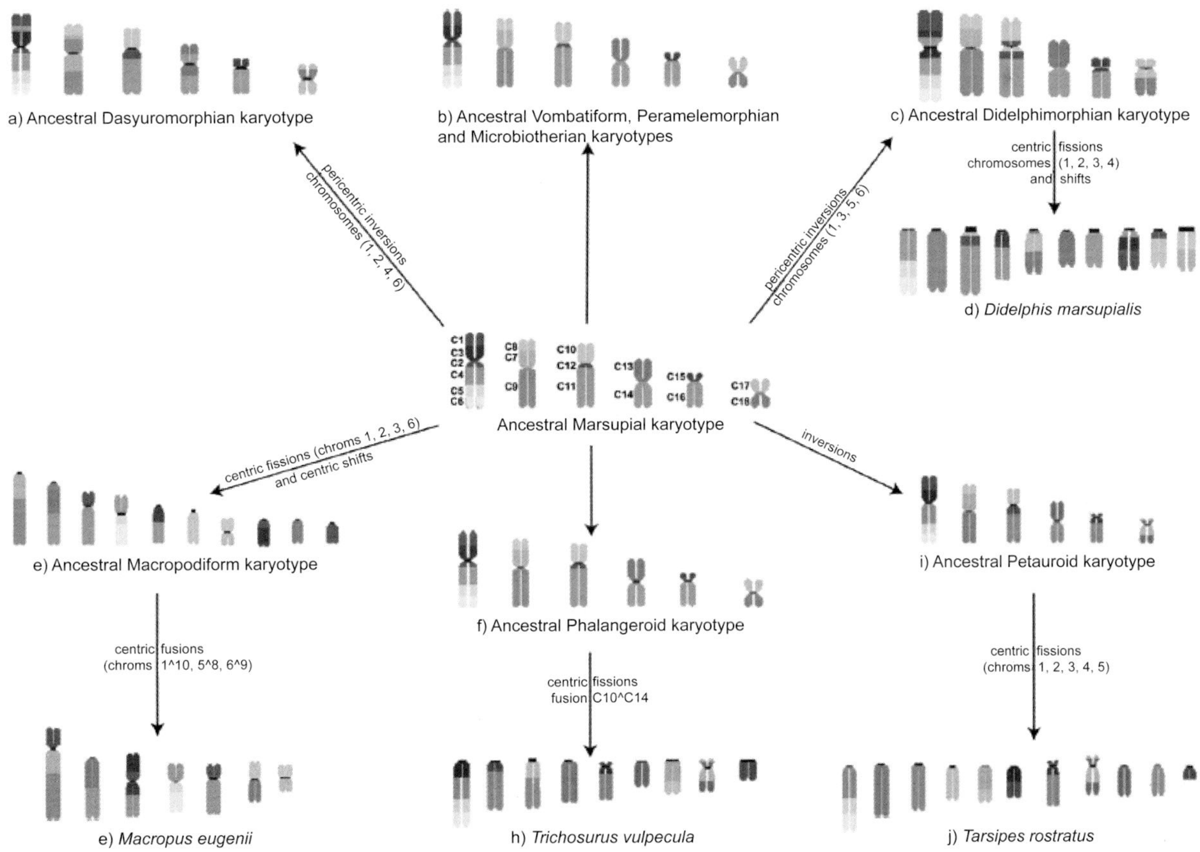

Fig. 11.35. Ancestral marsupial karyotype (centre), conserved chromatin blocks C1–18 on autosomes, and the possible evolutionary transformations leading to modern karyotypes (arrows). [Reprinted by Permission of Oxford University Press from Westerman *et al.* (2010).]

reproductive isolation and speciation. In recent years, this scenario has been abandoned as the major mechanism of speciation, although chromosome differences may contribute to genetic isolation (Deakin *et al.* 2019). Currently, the question is whether karyotypic differences among closely related species are a cause or a consequence of speciation.

In dasyurids, the 2n = 14 karyotype seen in virtually all species differs from the ancestral karyotype by pericentric inversions in autosomes 1, 2, 4 and 6. However, these chromosomal differences are not associated with taxonomic differences at the species or genus level, casting doubt on their significance as an isolating mechanism (White 1978) or facilitators of adaptive evolution (Patton and Sherwood 1983). Groups such as Microbiotheridae, Acrobatidae, Burramyidae and Peramelidae also appear to have karyotypes little changed from the 2n = 14 form of their common ancestor over 60 Ma. In marked contrast, differences in number and shape of chromosomes among closely related species are common in Didelphidae, Petauridae, Phalangeridae, Pseudocheiridae and Macropodiformes as noted above (Fig. 11.35). Karyotypic

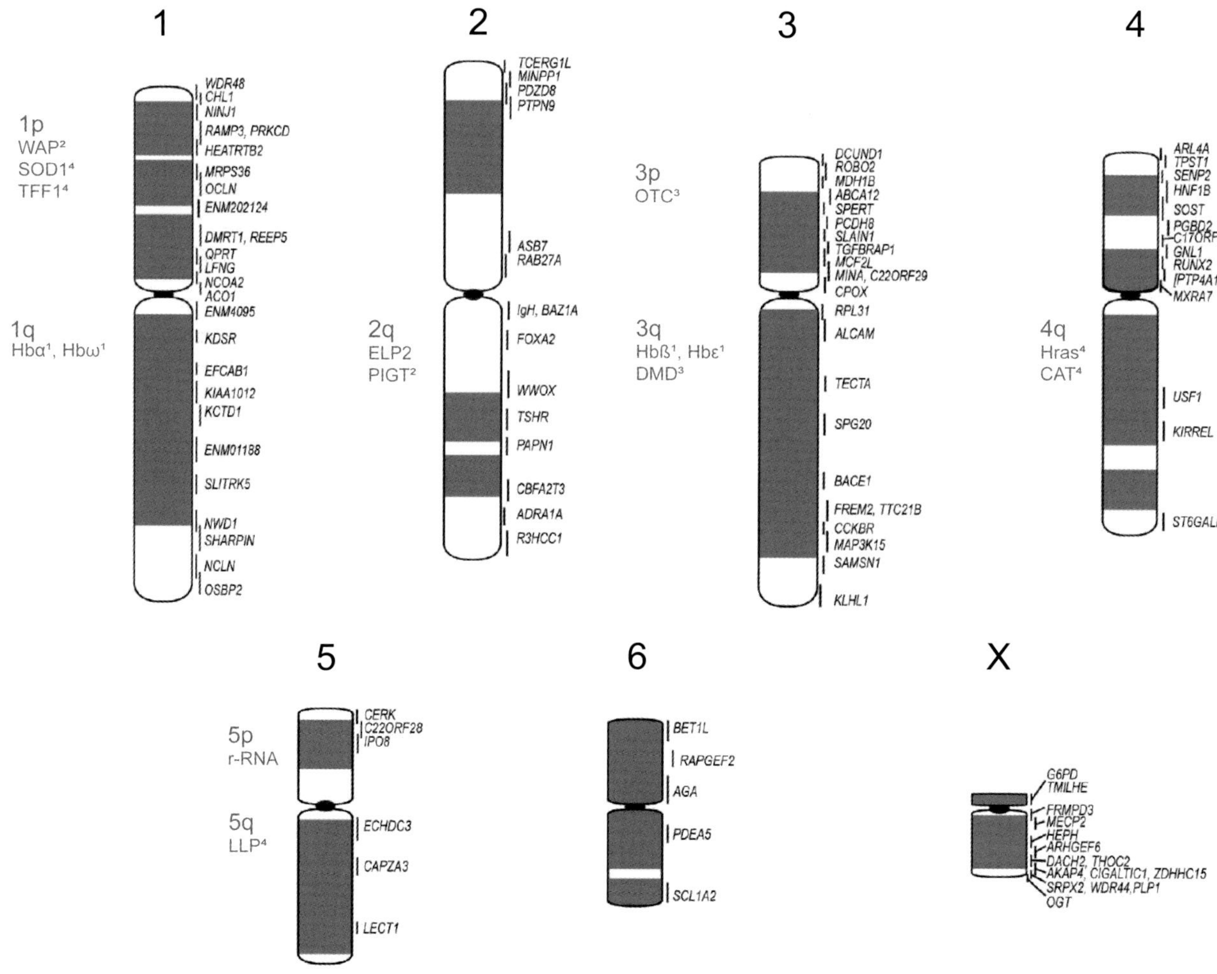

Fig. 11.36. Gene maps of *Sarcophilus harrisii* chromosomes. Fluorescent banding patterns are shown in grey, and genes contained within the same BAC clone are indicated by brackets. Other genes mapped to autosomal arms shown in red. Abbreviations: p = short arm; q = long arm; CAT = catalase; DMD = Duchene muscular dystrophy; ELP = early lactation protein; Hbα, Hbβ, Hbε, Hbω = haemoglobins α, β, ε, ω; Hras = Hras proto-oncogene; LLP = late lactation protein; OTC = ornithine transcarbamylase; PIGT = phosphatidylinositol glycan anchor biosynthesis class T; SOD = superoxide dismutase; TFF1 = trefoil factor 1; WAP = whey acidic protein. Nuclear genes most often used in dasyurid phylogenetics have been mapped only to individual chromosomes: P1 (chromosome 1); ApoB, bfib7, IRBP (chromosome 2); BRCA1 (chromosome 4); vWF (chromosome 5); RAG1 (chromosome 6). Sources: [1] DeLeo *et al.* (2005); [2] DeLeo *et al.* (2006); [3] Samollow & Graves (1998); [4] Westerman *et al.* (1991). [Redrawn from Deakin *et al.* (2012) under CC BY license.]

changes associated with speciation have been particularly important in macropodids (Sharman 1974, 1989; Sharman *et al.* 1990a; Eldridge and Close 1997; Waugh O'Neil *et al.* 1999; Potter *et al.* 2012). Such variation is also observed among families of Carnivora (O'Brien *et al.* 1988; Ferguson-Smith and Trifonov 2007), but its causes are unknown.

Dasyurid gene maps

While morphology and banding patterns suggest karyotypic stability in dasyurids, only a few studies have explored chromosomal organisation at the molecular level. Hayman and Martin (1974) reported that all seven dasyurid species they examined had similar DNA amounts, deviating only by ca. 1.2% on either side of the mean. This estimate is in marked contrast with the 35% range reported by Westerman and Woolley (1993) for 11 dasyurid species.

G-banding has been a generally reliable indicator of chromosome homology in that chromatin regions with the same banding patterns usually carry the same genes. Linkage groups on single chromosomes were classically defined by breeding experiments (Bennett *et al.* 1990), but such research has rarely been conducted on marsupials because few large-scale, long-term breeding colonies have been successful. However, somatic cell genetic techniques have been used in conjunction with gene probes to locate particular sequences on chromosomes in hybrid cells (Ruddle 1981; Graves *et al.* 1990). Recently, bacterial artificial chromosomes (BACs) have been constructed containing genes from opossum or wallaby conserved blocks that can be used as targets for chromosome painting. Deakin *et al.* (2012) used 100 BAC clones distributed across all six autosomes and the X of the Tasmanian devil (*Sarcophilus harrisii*) to construct a detailed map of 90 autosomal and 15 X-linked genes (Fig. 11.36). Mapping *Sarcophilus* BAC clones onto the chromosomes of other dasyurid species should resolve whether the genomic stability implied by G-band studies extends to the DNA sequence level.

ENDNOTES

1. Carnassials are cheek teeth (premolars and molars) specialised for carnivory, in extreme form resulting in high, blade-like shearing crests. Lang *et al.* (2022: 92–93) provide a more anatomically precise characterisation: carnassials are teeth that show 'enlarged distal upper (metacrista) and mesial lower (paracristid) cutting blades and the reduction of subordinate cutting structures (prevallum/postvallid shearing) as well as the reduction of the crushing protocone-talonid functional complex'. The evolution of such teeth, usually from a more generalised ancestral form, is called 'carnassialisation'.
2. Cowper, an established British anatomist, is infamous for republishing, without acknowledgement, illustrations from the Dutch physician Govert Bidloo's (1685) *Anatomia Humani Corporis* in his 1698 volume *Anatomy of the Humane Bodies*. The incident has been held to exemplify the widespread practice of plagiarism in anatomy publications of the 17th and 18th centuries (Cole 1944).
3. Benda erred in reporting only 12 chromosomes; we now know that peramelids have 2n = 14.
4. Noting that then-known karyotypes of dasyurids differed slightly from the presumed ancestral G-band pattern, Sharman (1982) predicted that the ancestral pattern might be found in 'primitive' New Guinean dasyurids. Westerman and Woolley (1990, 1993) showed that New Guinean species have the same karyotype as most Australian forms, again refuting the notion that the former are primitive in any general sense.

12 Dasyurids past, present and future

As shown in the preceding chapters, dasyurid evolution has been studied from many different perspectives. While it is easy to get lost in the morass of methodology, conclusions, criticisms and caveats, these details reveal the process of scientific inference and cannot be neglected or oversimplified. Here we synthesise what is known about dasyurid evolution in the spirit of O'Hara's (1988) 'evolutionary chronicle' (phylogenetic sequence of events) and 'evolutionary narrative' (historical explanations of those events). We certainly know more of the dasyurid chronicle than the narrative, in part because a consensus on phylogeny and divergence times has only recently been achieved – i.e. we are just coming to understand exactly what must be explained. Our recital has two parts, the first a retelling of conclusions reached in previous chapters about how dasyurids evolved, and the second an assessment of the present and future of dasyurids considering human impacts.

OLIGOCENE TO PLEISTOCENE: EVOLUTION WITHOUT HUMANS

Dasyurid ancestors

According to the most recent divergence date estimates, the common ancestor of dasyurids existed during the Oligocene ca. 27 Ma. There is no fossil evidence of this species, but several Oligocene dasyuromorphians are known from teeth and skull fragments. Among these are *Ankotarinja*, *Keeuna*, *Dasylurinja* and *Apoktesis* from Late Oligocene local faunas in South Australia. These animals were relatively small-bodied, antechinus–mulgara size (<200 g), with sharp-cusped teeth indicating an insectivorous diet (Fig. 12.1). Amador and Giannini (2021) used parsimony to reconstruct ancestral marsupial diets on the phylogeny of Mitchell *et al.* (2014), confirming the insectivory of both dasyuromorphian and dasyurid ancestors (Fig. 12.2). The small thylacinid *Badjcinus*, a carnivore from Riversleigh Faunal Zone A, also lived in the Late Oligocene. Indeed, thylacinids represent the transition from insectivory to carnivory among australidelphians (Amador and Giannini 2021). Oligocene dasyuromorphians occupied open forest habitats with seasonal rainfall. Unfortunately, no pre-Pleistocene fossils exist for myrmecobiids and the highly specialised phenotype of modern numbats tells us little about the dasyurid ancestor. However, it is plausible that this ancestor resembled the *incertae sedis* dasyuromorphians and occupied a similar habitat.

Some phenotypic features of the original dasyurid can be inferred from analyses of extant forms. It would have had several traits that arose in earlier dasyuromorphians: insectivory; a dental formula of I4/3, C1/1, P3/3, M4/4; a large stylar cusp D on the upper molars for puncturing insect cuticle; a hypoconulid notch in the lower molars; a foramen pseudovale; a squamosal epitympanic sinus; an alisphenoid tympanic wing; and no intestinal caecum.

Fig. 12.1. Artist's reconstruction of the Oligocene dasyuromorphian *Ankotarinja tirarensis*. The ancestral dasyurid may have had a similar body form and diet. [Illustration by Anne Musser.]

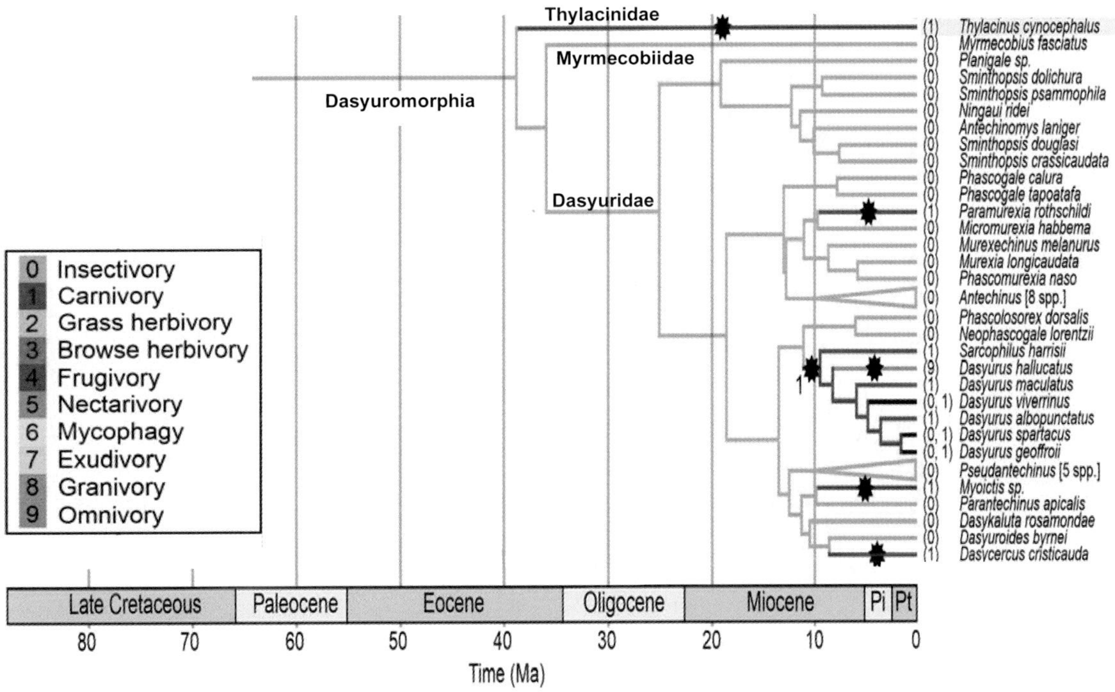

Fig. 12.2. Parsimony reconstruction of ancestral diets from those of extant species by Amador and Giannini (2021) on the phylogeny of Mitchell *et al.* (2014). Stars indicate state transitions. [Reproduced with permission of John Wiley and Sons from Amador and Giannini (2021), with minor adaptations.]

Four newly derived basicranial characters likely evolved in this, or perhaps an earlier, ancestor: a periotic tympanic wing, paraoccipital tympanic sinus, tubular foramen and jugular canal sulcus. Its penis would have lacked an appendage and accessory erectile tissue. Living in a habitat with seasonal rainfall, it likely employed reproductive Strategy II (a restricted breeding season with monoestrous females and males able to survive and breed in a second year). Pericentric inversions on chromosomes 1, 2, 4 and 6 would have arisen in this species, distinguishing its 2n = 14 G-band karyotype from the ancestral australidelphian complement.

Dasyurids appear in the fossil record of the late Early Miocene with *Barinya wangala* from the rainforest at

Riversleigh and *B. kutjamarpensis* from drier forests of eastern South Australia, though there is some question whether these species are indeed dasyurids. Regardless, they were carnivores about the size of a small northern quoll (300–500 g), with crania resembling the dasyurid ancestor just described. After separation of *Barinya*, the ancestor of crown group Dasyuridae evolved several new basicranial traits: a fully formed stylomastoid foramen through which cranial nerve VII exits the skull; a long, enclosed jugal canal; a paraoccipital tympanic process that contacted the periotic bone anteriorly and fused with the mastoid tympanic process laterally; and a reduced interparietal. Some of these traits contribute to the enclosed middle ear of dasyurids, a feature which improved their ability to perceive low-frequency sounds and avoid predators in open environments. Although we know nothing else about this ancestor, the growing diversity of thylacinids known from early Miocene rainforests (*Thylacinus macknessi*, *Ngamalacinus* and *Wabulacinus*) accompanied by only a single putative dasyurid (*B. wangala*) hint that crown dasyurids may have favoured drier areas.

The great Miocene radiations

Although the date of divergence between *Barinya* and crown dasyurids is unknown, recent molecular clock estimates place the split between crown dasyurids and myrmecobiids at ca. 25 Ma in the Oligocene. The deepest split among living dasyurids – separation of Dasyurinae and Sminthopsinae – took place ca. 20 Ma, in the early Miocene coincident with the spread of seasonal habitats and open forests across Australia. The only phenotypic trait that has been correlated with this split is the origin of reproductive Strategy V in sminthopsines – i.e. a shift to polyoestry and an extended breeding season. This appears to be adaptive in habitats where annual rainfall is relatively unpredictable, though it is not clear how prevalent such habitats were in the Early Miocene. Moreover, departures from Strategy V in sminthopsines appear mostly on terminal or subterminal branches of the phylogeny, so we do not know *when* the transition from Strategy II to V took place in the ancestral lineage; it may have been well after the split from dasyurines.

Sminthopsins diverged from planigalins, and dasyurins from phascogalins, 12–16 Ma. No middle Miocoene fossils record these lineages, but thylacinid diversity peaked at this time, represented by *Maximucinus*, *Muribacinus*, *Mutpuracinus* (if it was a thylacinid) and *Nimbacinus*. The latter three were quoll-sized, but *Maximacinus* resembled a small modern thylacine. Other contemporary dasyuromorphians were *Wholleydoolea* (a devil-sized hypercarnivore), *Joculusium* (a small insectivore) and *Malleodectes* (a snail eater weighing ca. 900 g). *Maximucinus*, *Muribacinus* and *Joculusium* lived in Riversleigh rainforests; the other taxa were from seasonal, open-forest habitats. Absence of dasyurid remains from mid-Miocene deposits suggests that tribal ancestors may have occupied drier areas where fossilisation is rare, that they were minor elements of the fauna, or both. These ancestors appear to have retained the penis morphologies, reproductive strategies and karyotypes of their subfamily ancestors. No anatomical synapomorphies for the tribes have been adduced.

Except for planigalins, diversification within tribes began 9–12 Ma, just after the Miocene Climate Optimum. They were well under way by the late Miocene and largely complete by the onset of the Pliocene (5 Ma). The delayed radiation of planigales in the late Miocene coincided with the spread of their open grassy habitats. *Ganbulyani* is known from open forests at Riversleigh some 10–12 Ma. *Mayigriphus*, a planigale-like dasyuromorphian, occurs in the same deposits. Generic diversity of Late Miocene thylacinids fell to two: *Tyarrpecinus* and *Thylacinus* (*T. potens* and *T. megiriani*) from open forest or savanna habitats in the Northern Territory. Although only *Ganbulyani* records the presence of dasyurids in the late Miocene, both paleontological and molecular analyses suggest that this was the time when thylacinids and archaic bandicoots experienced a reversal of fortune with dasyurids. As closed wet-forest habitats shrank towards the continental margins, the thylacinids and yaralids they supported declined. Dasyurids came to dominate the new open forests and savanna woodlands of central Australia; some may have filled the small carnivore niche vacated by thylacines, and others diversified lower in the food chain as micro-predators. Their success in the more open spaces of the late Miocene was facilitated by their derived middle ear.

Fossils document the presence of 10 dasyurid genera in the Pliocene (eight of them still extant) and most modern genera (plus *Myrmecobius*) in the Pleistocene. Thylacinid diversity dwindled to just two species of *Thylacinus* in the Pliocene, with only *T. cynocephalus* surviving to the Recent. Indeed, after the Miocene, dasyurid cladogenesis was mostly phylogeographic structuring within fragmented habitats, with evidence for recent or ongoing allopatric speciation in only a few genera (*Antechinus*, *Pseudantechinus*, *Sminthopsis* and *Planigale*).

The products of adaptive radiation

The Late Miocene diversification of dasyurids was an instance of Simpson's (1951a) 'explosive evolution', preceded and followed by lower rates of speciation. It was also an *adaptive* radiation, generating new phenotypic diversity to exploit ecological opportunities. The adaptations produced are evident in the morphology, physiology and natural history of extant species. Among these, the most studied features are habitat affinities, arid adaptations, penis morphology and reproductive strategies.

Dasyurids invade rainforests and arid zones

Ancestral dasyurines retained the mesic habitat preference of their predecessors and transmitted it to their dasyurin descendants. The phascogalin ancestor, however, found its way to mid-Miocene coastal forests of Australia and diversified there to produce *Antechinus* species in wetter areas and *Phascogale* species in drier ones. At about the same time, the *Murexia* ancestor arrived in eastern New Guinea and began to radiate in its wet forests. Meanwhile, the deepest split within Dasyurini separated quolls, devil and phascolosoricines from other lineages. Within the former, *Dasyurus* and *Sarcophilus* ancestors maintained an affinity for mesic, open forest habitats; the progenitor of *Phascolosorex* and *Neophascogale*, however, reached New Guinea at approximately the same time as *Murexia* and likewise adapted to life in the closed forest. Near the end of the Miocene *Dasyurus albopunctatus* separated from its sister, colonised New Guinea and evolved a rainforest affinity. The other deep clade of dasyurins transitioned from mesic to arid areas of Australia (see below), but one of its lineages – *Myoictis* – entered New Guinea contemporaneously with *Murexia* and phascolosoricine ancestors and evolved rainforest descendants. Importantly, these New Guinean dasyurins invaded *new* habitats, not those previously dominated by thylacinids and archaic bandicoots; they did so independently and from *mesic adapted* predecessors whose heritage extended back to the original dasyurid. In no meaningful sense are they primitive relicts of bygone dasyurid ancestors in Oligo-Miocene rainforests, though some (e.g. *Murexia*) re-evolved dental traits found in early dasyuromorphians.

Arid environments in the Late Miocene were not as widespread as they are today, but early dasyurins nevertheless exploited them. The ancestor of the subclade containing *Myoictis* produced lineages leading to arid-associated species of *Dasycercus*, *Dasyuroides* and *Dasykaluta* from western and central Australia. This group includes the single mesic-dwelling species *Parantechinus apicalis* from south-western heathlands. *Pseudantechinus* is sister to the *Myoictis* subclade and their ancestor appears to have been associated with arid environments, a trait that still characterises most of its modern species. Even those *Pseudantechinus* (*P. bilarni*, *P. ningbing*) occurring in mesic macrohabitats prefer relatively dry, rocky areas. The most recent speciation events within *Pseudantechinus* (e.g. *P. ningbing-macdonnellensis-mimulus*) are Pliocene, suggesting cladogenesis associated with the rapidly expanding arid zone.

In contrast to dasyurines, the sminthopsine, sminthopsin and perhaps planigalin ancestors were likely arid-land dwellers, such that modern species inhabiting dry areas are 'primitive' at the subfamily level in this respect. Although some early sminthopsin lineages (e.g. *Antechinomys* + *S. longicaudata*, *Ningaui*) retain strong dry-country preferences, many *Sminthopsis* groups that arose in the latest Miocene include members that straddle the mesic–xeric boundary, though the mesic habitats in which they occur tend towards dry. Indeed, García-Navas and Westerman (2018) showed that *Sminthopsis* evolution entailed phylogenetic niche conservation: diversification within the clade had more to do with range fragmentation and community processes than with adaptation to new habitats. A few species however, such as *S. leucopus*, reinvaded wet forests. The *Planigale* radiation produced mesic–xeric species with a preference for cracking clay soils and grassy regions, though the widespread *P. maculata* has broader habitat affinities (and may comprise more than one species; see Chapters 6 and 7). For both sminthopsin tribes, the most recent speciation events were probably Pliocene, including colonisation of New Guinea by *S. archeri*, *S. virginiae* and *P. novaeguineae*.

Habitats and phenotypes

A large literature exists on mammalian adaptations to arid conditions (e.g. Schmidt-Nielsen 1964; Degen 1997), many of which are exemplified by desert-dwelling dasyurids. Archer (1984b) highlighted four selection pressures on desert vertebrates: (1) extreme cold during the night; (2) extreme heat during the day; (3) scarcity of water; and (4) unpredictability of good conditions. Heat loss at night and dehydration during the day are exacerbated by the high surface/volume ratios of small-bodied marsupials (Geiser 2003). Although the dorsoventrally flattened body of planigales enhances heat loss by conduction, the deep soil cracks frequented by these animals are buffered from daily temperature extremes. Moreover, maintaining the high metabolic rate of an endotherm demands a large and steady supply of food, especially in smaller species. Some desert dwellers, such as *Sminthopsis crassicaudata* and *Dasycercus cristicauda*, store fat at the base of their incrassated tails (Wallis 1982). *S. crassicaudata* undergoes daily torpor during winter, reducing its night-time metabolic rate and allowing body temperature to drop 15–20°C below its daytime high (Warneke *et al.* 2008). These animals bask, ectotherm-like, in the morning sun to rewarm and return to shelter in cool crevices late in the day. Torpor has been documented in xeric-dwelling species of *Planigale, Ningaui, Sminthopsis, Antechinomys, Dasycercus* and *Dasyuroides*, but also in some mesic *Antechinus, Phascogale* and *Dasyurus* (Geiser 2003). Cold periods are spent in nests or burrows, often huddling in groups of several individuals to share body warmth. During the summer, nocturnal activity patterns reduce exposure to daytime heat. Large pinnae and long tails dissipate heat rapidly in some desert-living dasyurids (e.g. *Antechinomys, S. longicaudata*).

If resources are unpredictable, a capacity to exploit them when they are available is adaptive and the ultimate fitness benefit is producing offspring in the face of resource uncertainty. It has been argued that polyoestrous reproductive strategies IV and V represent adaptations to such unpredictability (see Chapter 11), implying that a preponderance of arid-adapted species should display these strategies. Krajewski, Woolley and Westerman (2000) found no strong association between habitat and reproductive strategy across all dasyurids, though Strategies IV and V predominate in Sminthopsinae (many of which are arid-adapted), with Strategy V also occurring in the arid-dwelling *Dasyuroides*. However, Strategy I (monoestrous females and die-off in males) – supposedly an adaptation for highly predictable seasonal resources as experienced by *Antechinus* – also occurs in *Dasykaluta* from the decidedly arid Pilbara. The distribution of Strategies II–V among dasyurids more strongly reflects phylogeny (II–III prevailing among dasyurines, IV–V among sminthopsines) than habitat association. Strategy I occurs mostly, but not exclusively, in species occupying forests with seasonal resource pulses; however, all these species (of *Antechinus, Phascogale, Dasykaluta*) are dasyurines. Strategy VI (aseasonal breeding) is more strongly associated with habitat, occurring mostly (but also not exclusively) in species from rainforests; however, all but one of these species are also dasyurines. While II↔III and IV↔V transitions are common *within* subfamilies, only four occurred *between* subfamilies: the arid-adapted dasyurin *Dasyuroides* with Strategy V; the mesic-adapted sminthopsins *Sminthopsis griseoventer* and *S. leucopus* with Strategies II and III, respectively; and *Planigale maculata sinualis* from northern floodplains with Strategy VI. These departures from phylogenetic inertia occur exclusively at tips of the dasyurid tree. Clearly, the causal connection between reproductive strategy, environment and phylogeny remains to be fully elucidated.

Clades and characters

In addition to a radiation of arid-adapted genera, four independent invasions of New Guinea, and a parallel derivation of Strategy I, evolution in Dasyurini produced the hypercarnivorous clade of *Dasyurus, Glaucodon* and *Sarcophilus*. Some craniodental correlates of hypercarnivory were noted in Chapter 11, but perhaps the most obvious feature of these taxa is large body size. With adult weights of ca. 400–7000 g for quolls and 5000–9000 g for the devil, these animals outweigh most other dasyurids by several hundred grams. Dasyurins also provide the context in which accessory erectile tissue of the penis arose, often accompanied by an appendage. This trait complex appears to have originated independently in the ancestor of quolls other than *Dasyurus hallucatus*, in the ancestor of *Parantechinus* and *Myoictis*, and in *Pseudantechinus*. In the first two instances, the two penile traits strictly co-occur; in the last, accessory erectile tissue appeared in the *Pseudantechinus* ancestor, but the appendage only in the *P. macdonnellensis-mimulus* clade. Because the function of this trait complex is unknown, explaining its origin is difficult. It is not obviously correlated with habitat, geography, or any organismal feature. Without implying anything about inherent synapomorphies, one might say that a 'tendency to evolve accessory erectile tissue and a penile appendage' is characteristic of dasyurins.

Penile morphology also tracks phylogeny in *Sminthopsis* (see Fig. 6.22), though estimates of ancestral states suggest only two unambiguous internal nodes: penis Form 1 in the ancestor of the Macroura group + *S. crassicaudata;* and Form 2 in the ancestor of most other *Sminthopsis* (including the Murina group). Form 1 occurs in all species of the expanded Macroura lineage, but the Murina-related lineage features eight distinct penile forms among 13 species, seven of them found only in single species or sister-pairs. Both lineages began diversifying ca. 7 Ma (see Chapter 9), so their histories of penis evolution are in stark contrast. Again, however, the functional significance of this trait complex has yet to be demonstrated and drivers of its variation are unknown. Finally, whereas large body sizes arose in dasyurins, small ones are derived in sminthopsines. With adult weights of <10 g, *Planigale ingrami* and *Ningaui timealeyi* are the smallest marsupials. Such tiny bodies, with their high surface/volume ratios, create significant heat- and water-loss challenges for endotherms, making it all the more remarkable that they occur in arid habitats. Even with the lower metabolic rates of marsupials relative to eutherians, the reputation of planigales and ningauis as 'ferocious little predators' is born of energetic necessity.

WHITHER DASYURIDS?

Man meets marsupial in the Pleistocene

Humans arrived in Sahul (Australia + New Guinea) from south-east Asia in the late Pleistocene, largely by overland dispersal at a time of low sea level (Groves 2006). The timing of this dispersal has been the subject of much study, with early archaeological dates of 45–60 ka. Unfortunately, such ages are near the limit of accuracy for radiocarbon dating and, coupled with taphonomic and stratigraphic uncertainties, required validation by other methods. Clarkson *et al.* (2017) used optically stimulated thermoluminescence (OSL) to date quartz grains from horizons bearing human artefacts at the Madjedbebe rock shelter in Arnhem Land. OSL dates agreed with those from radiocarbon analysis of charcoal in suggesting that human occupation began at 65 ka. This is currently the oldest well-dated site of human occupation in Australia, but its validity has been disputed. Other archaeological sites track the dispersal of people across Sahul. New Guinea is thought to have been colonised ca. 50 ka (O'Connell and Allan 2003), with the oldest archaeological sites from the Ivane Valley of the Papua Peninsula dated as 43–49 ka (Summerhayes *et al.* 2010). Hamm *et al.* (2016) reported OSL dates from charcoal and avian eggshells showing that the Warratyi rock shelter in South Australia harboured humans ca. 49 ka. Colonisation of Tasmania is thought to coincide with formation of a land bridge to the mainland ca. 43 ka (Johnson *et al.* 2016). A stochastic model based on human ecology, archaeology, landscape and climate data (Bradshaw, Norman *et al.* 2021) found equal support for human arrival dates in Sahul of 50 and 75 ka, with dispersal complete (except to Tasmania) in less than 5000 years. Crabtree *et al.* (2021) coupled human bioenergetics with topographic data to simulate geographic pathways of human dispersal across Sahul, finding good agreement with the locations of archaeological sites (Fig. 12.3). Thus, it appears that humans took possession of the continent sometime between 75 and 43 ka. What they did along the way has been controversial.

The first human colonists encountered a Sahul biota with not only those species familiar in historical times, but also a spectacular array of large-bodied amniotes known as the 'Pleistocene megafauna'. Similarly big species occurred at this time on other continents and islands. The Australian megafauna (Wroe 1999a; Clode 2009) included the carnivorous 'Megalania' lizard *Varanus priscus* (>1900 kg), the dromornithid 'thunderbird' *Genyornis* (ca. 220 kg) and the giant long-beaked echidna *Zaglossus hacketti* (ca. 30 kg), as well as a diverse group of marsupials such as the herbivore *Diprotodon* (>2500 kg) and the carnivore *Thylacoleo* (100–160 kg). What qualifies as 'mega' has not been consistently defined, with various authors setting the bar between 10 and 1000 kg. Applying the lowest cut-off, the only mega-dasyuromorphians were *Thylacinus cynocephalus* (10–30 kg) and *Sarcophilus laniarius* (ca. 10 kg). Regardless of the definition, however, most of these giant species were victims of what has been called the 'Pleistocene megafaunal extinction'. The question is: what caused it?

The Pleistocene (2.58–0.012 Ma) was the most recent episode of widespread continental glaciation in Earth history, with multiple cycles of ice advance during cold intervals and retreat during warm ones in the northern hemisphere. Although southern continents did not experience such massive glaciation, they did suffer dramatic temperature and precipitation shifts as Earth alternated between cold, dry glacial periods and warm, wet interglacials. Not surprisingly, such stressful climate change was (and is) considered a likely cause of megafaunal extinction. However, American paleoanthropologist Paul S. Martin (1928–2010) pointed out that extinctions were concentrated in the late Pleistocene even though the megafauna had survived more than 2 Ma of intense climate oscillations prior to their demise (Martin 1966). What was different during the last 100 ka? Martin suggested that the spread of humans around the globe during this period, coupled with our species' penchant for

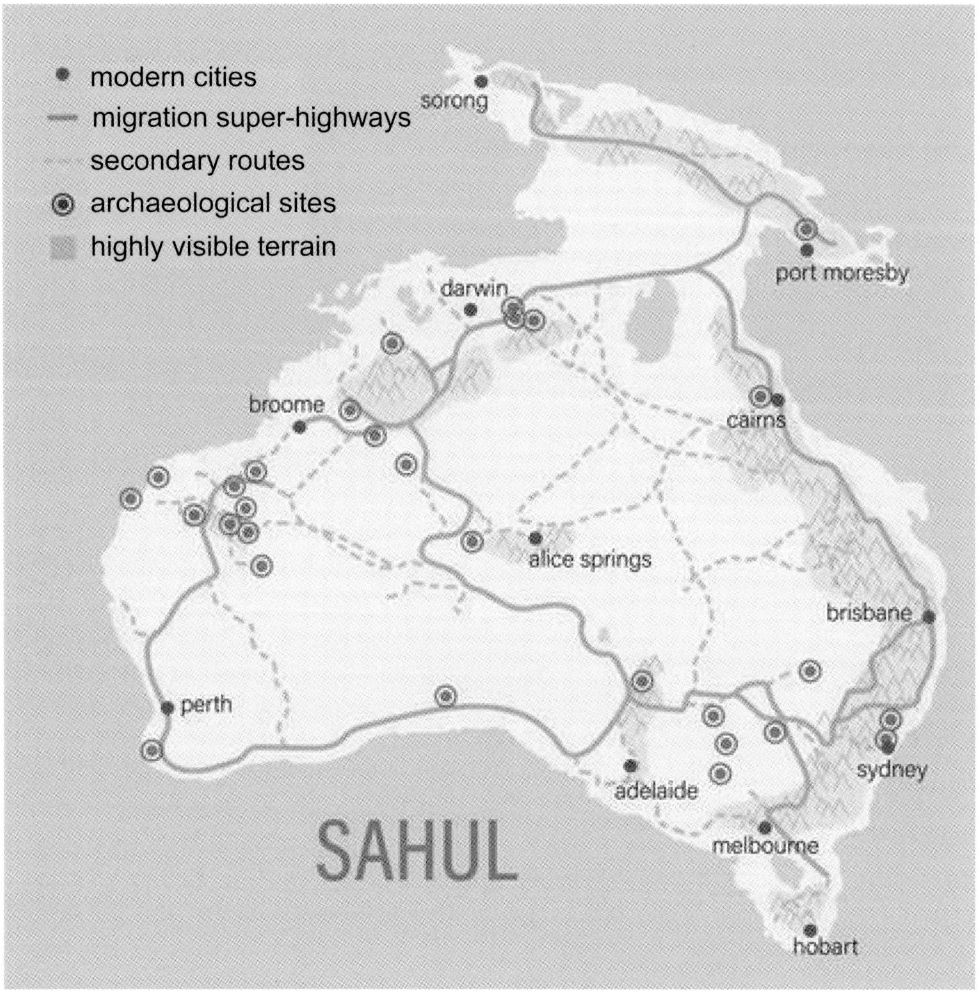

Fig. 12.3. Simulated pathways of human dispersal during the past 65 ka, based on human bioenergetics and landscape topography (Crabtree *et al.* 2021). [Illustration by Megan Hotchkiss Davidson. Used with permission of Sandia National Laboratories.]

environmental alteration, spelled doom for the megafauna. The most direct effect may have been hunting; large amniotes, particularly the less dangerous herbivores, are high-value food sources, and those outside Africa in the Pleistocene would have been particularly vulnerable to the introduction of human predators. Martin's 'Pleistocene overkill' hypothesis was that human hunting drove megaherbivores extinct, producing a trophic cascade that resulted in the loss of megacarnivores. Alternatively, newly arrived humans may have altered the environment in ways that made it unable to support a megafauna. Other possible causes have been proposed, but debate for the past 50 years has focused on climate versus human hypotheses. These factors are not mutually exclusive, but until recently few studies have considered 'double whammy' scenarios. The debate has not been fully resolved for any continent; some recent reviews (e.g. Bartlett *et al.* 2016) favour humans as the dominant driver, but other studies make a strong case for climate (e.g. Cooper *et al.* 2015). What happened in Australia is unclear.

In some parts of the world there is abundant direct evidence for a human 'blitzkrieg' on megafauna. Flannery (1994) recounted how hunting pressure led to the loss of New Zealand's moas (10–230 kg) by 1440 CE, little more than a century after arrival of the first Polynesians. Evidence for human hunting comes from multiple kill sites with many moa bones and butchering artefacts. The primary natural predator of moas, the giant Haast eagle (*Hieraaetus moorei*, 30 kg), quickly followed them into extinction. In Australia, however, no such kill sites have been found. According to Brook and Johnson (2006) and

Johnson (2006), this may be expected in that even modest hunting pressure would suffice to exterminate large-bodied species with low reproductive rates ('imperceptible overkill'). One of the few carefully documented examples of human-megafauna interaction is from Warratyi Cave (49 ka) in South Australia, which includes both human artefacts and *Diprotodon* bones. Hamm *et al.* (2016: 281) reported, 'The age of these fossils, together with the absence of carnivore tooth marks and the position of the shelter on a steep escarpment unsuitable for climbing by individuals of *D. optatum* … indicate co-occurrence of these taxa with humans who were probably involved in the accumulation of their remains'. Johnson *et al.* (2016) cited charred *Genyornis* and emu eggshells 54–47 ka as evidence of human hunting pressure. Given that such data are rare, however, research on Pleistocene extinctions in Australia has emphasised correlating the times of human arrival, anthropogenic environmental alteration and climate change with the extinctions themselves.

As noted above, estimates of human arrival and dispersal in Sahul are in the range of 75 to 43 ka. One of the most significant environmental effects that early humans had in Australia involved the use of fire. As described by Flannery (1994), Australia's impoverished soil selects for ecosystems that perform rapid nutrient cycling. Prior to human arrival, these ecosystems included mesic forests, woodlands and savannas (see Chapter 10), as well as a megafauna that quickly returned nutrients to the soil as dung and carcasses. Aboriginal people, however, used local, low-intensity burning to facilitate agriculture and hunting, eventually leading to the practice of 'fire-stick farming' later observed by Europeans. In this process, fire performs the nutrient-cycling role of megaherbivores but, in the long term, selects for fire-resistant sclerophyll vegetation. Miller *et al.* (2005) used ^{13}C stable isotope analysis to argue that fire-induced vegetational change markedly altered the nature and amount of plant food for emus and wombats, which they considered strong evidence for ecosystem collapse after human arrival. Increased burning may also be reflected in late Pleistocene charcoal records, but its direct effects on the megafauna have long been unclear (Calaby 1984). Bird *et al.* (2013) concluded that altered fire regimes in tropical Australia could have affected megafaunal survival but questioned whether such an effect could be stronger than rapid climate change. In contrast, Rule *et al.* (2012) argued that the loss of mega-herbivores allowed the build up of dry vegetation, which altered the fire regime by fuelling more extensive burning.

As long ago as 1959, anthropologist N.B. Tindale reasoned 'that in process of developing a state of equilibrium with this [Pleistocene] Australia flora and fauna the [human] newcomers must very quickly have eliminated the more vulnerable animals' (Tindale 1959: 42). Reviewing subsequent research on human impacts, Archer (1984a: 154) emphasised three points:

> *1) there is no evidence that Aborigines directly hunted any species to extinction; 2) there is good evidence that extensive use of fire by Aborigines regionally altered vegetation patterns in Australia and New Guinea, and consequently it is probable that this stressed or even led to the extinction of some of the smaller ecologically restricted vertebrate species; 3) the late Pleistocene species of the megafauna that became extinct (and not all appear to have done so) are as reasonably considered to be the victims of an unfortunate intersection of climatic and physiological factors as they are the victims of Aboriginal fire-induced habitat alteration.*

Archer's parenthetic comment refers to species (e.g. *Sarcophilus laniarius* and some kangaroos) that may have responded to late Pleistocene environmental change by evolving smaller body size. His favoured scenario for extinction was one in which aridification reduced the amount of plant food for megaherbivores, thereby eliminating these species and their predators. Flannery (1994), however, agreed with Tindale; arguing from analogy with human behaviour in other historical settings, he posited that a hunting blitzkrieg shortly after human arrival exterminated the megafauna and that firestick farming subsequently developed as a survival strategy in nutrient-starved habitats. Indeed, Flannery's (1994) 'future eater' model of Australian historical ecology is based on the prevalence of infertile soils. One piece of evidence claimed in support of this model was the apparent paucity of large mammalian predators in Australia's past. Although this was later refuted by the diversity of large carnivores discovered in Riversleigh deposits (Wroe and Myers 1998), nutrient-deficient soil has been widely acknowledged as a major determinant of Australian ecosystem structure.

A major source of uncertainty in early scenarios of human and climatic effects on the megafauna was imprecise estimates of extinction times. In the early 2000s, researchers began to refine these estimates and compare them with the arrival date of humans. Roberts *et al.* (2001) used optical and $^{230}Th/^{234}U$ dating to obtain burial ages of megafaunal (>45 kg) remains at 28 sites around Australia. These centred on 46.4 ka (95% confidence interval = 51.2–39.8 ka), well before the extreme aridity of the last glacial maximum (ca. 22 ka). Ruling out aridity as the primary cause of extinction, these authors left open the possible influence of overkill, human habitat disruption and other climate effects. Wroe *et al.* (2013) used a liberal definition of megafauna (>10 kg) and found that the distribution of last fossil occurrence dates of 88 species revealed four staggered pulses of extinction beginning >400 ka, all but the last ending before human arrival (Fig. 12.4). In addition, Wroe *et al.* (2013: 8779) interpreted Antarctic ice core data as indicating 'a distinct change in glacial-interglacial cycles after ~450 ka … In short, the temperature swings of the last few glacial cycles were without precedent and the last glacial cycle exhibits unusual variability'. Thus, they argued, climate change caused the megafaunal extinction.

Saltré *et al.* (2016) restricted their analysis of Pleistocene extinctions to 659 fossils of 16 genera with body weights >40 kg from sites that had been dated with high precision. They used statistical hindcasting to estimate extinction times and regression analysis to relate those times to climate changes during the past 120 ka. The mean extinction peak occurred at 42.1 ka (range 61–35 ka), shortly after the arrival of humans (which they took to be 55.6 ka). Extinctions were not significantly correlated with climate change; indeed, conditions 59–24 ka appeared less extreme than those of 120–80 ka. This conclusion, they argued, is robust to an earlier date (62 ka) for human arrival. Johnson *et al.* (2016) further pointed out that evidence from cave deposits, eggshells and fungal spores associated with herbivore dung showed no decline in species abundance prior to ca. 50 ka. Van der Kaars *et al.* (2017) extracted data for seven proxies of environmental conditions during the past 150 ka in south-western Australia from an offshore sediment core and related them to megafaunal abundance (Fig. 12.5):

> *We establish that substantial changes in vegetation and fire regime occurred ~70,000 years ago under a climate much drier than today. We record high levels of the dung fungus* Sporormiella, *a proxy for herbivore biomass, from 150,000 to 45,000 years ago, then a marked decline indicating megafaunal population collapse, from 45,000 to 43,100 years ago, placing the extinctions within 4,000 years of human dispersal across Australia [assuming arrival at 47 ka]. These findings rule out climate change, and implicate humans, as the primary extinction cause. [Van der Kaars et al. (2017: 1)]*

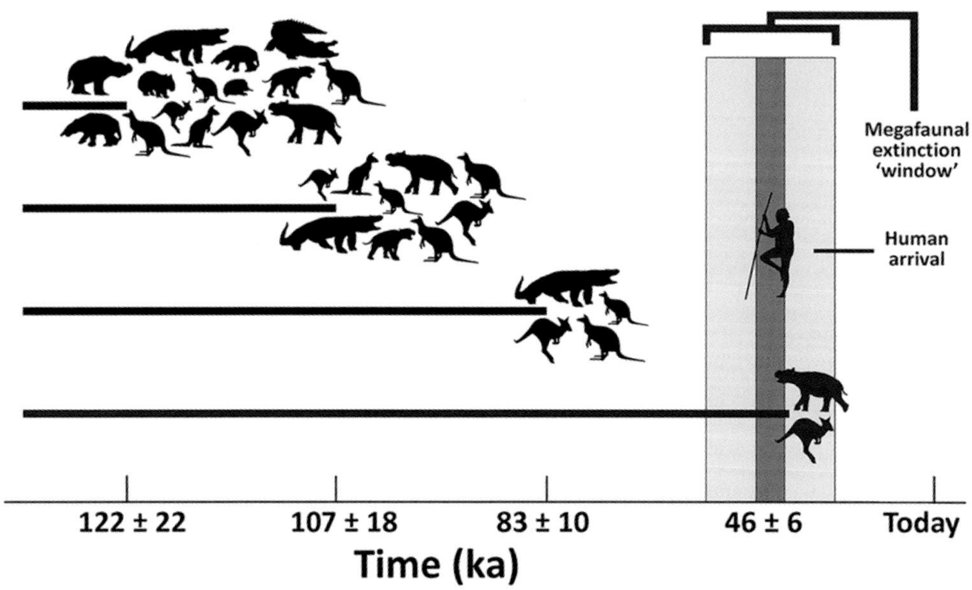

Fig. 12.4. Timeline of Pleistocene megafaunal extinctions estimated by Wroe *et al.* (2013) for 88 Australian species with body weights >10 kg. Dates are estimates of last occurrences of fossils for each species. The 'Megafaunal extinction "window"' is the age range postulated by previous authors to support a human-driven extinction event. [Reprinted with permission of the National Academy of Sciences from Wroe *et al.* (2013).]

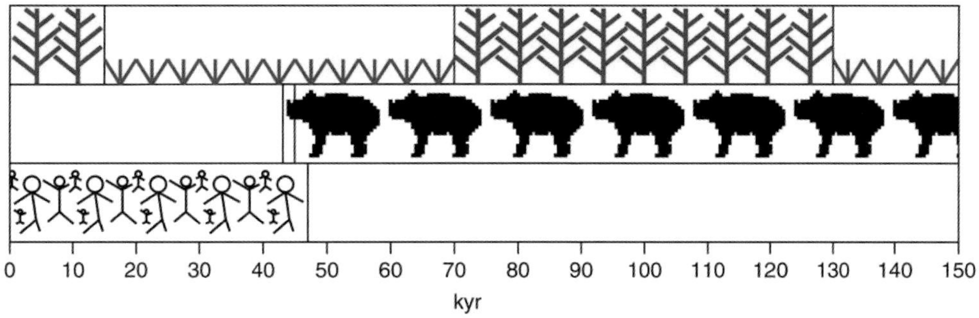

Fig. 12.5. Timeline of environmental change, megafaunal abundance and human presence in the late Pleistocene of south-western Australia inferred from offshore sediment core data by van der Kaars *et al.* (2017). Top panel shows relative dominance of trees versus herbs; middle and lower panels show the coincidence of megafaunal disappearance with human arrival. [Reprinted from van der Kaars *et al.* (2017) under license CC BY 4.0.]

Subsequent studies of regional and local deposits have found evidence both for (e.g. Hocknull *et al.* 2020) and against (e.g. David *et al.* 2021) climate-induced megafaunal extinction in Australia, but the case for human agency remains circumstantial. In perhaps the most comprehensive theoretical analysis to date, Bradshaw, Johnson *et al.* (2021) used stochastic age-structured demographic models for 13 extinct megafaunal (>45 kg) and eight extant species to evaluate the probable contribution to extinction of five potential causes (Fig. 12.6). Using the predicted extinction sequence of species as a measure of causal influence, no single factor emerged as dominant across all taxa, though generation length was consistently associated with extinction risk. The authors concluded 'that the actual mechanisms leading to the observed extinction chronology were unlikely related to variation in demographic susceptibility *per se* but were possibly driven instead by finer-scale variation in climate change and/or human prey choice and relative hunting success' (Bradshaw, Johnson *et al.* 2021: 1). This perspective, also found in other recent studies of megafaunal extinctions around the world, transcends the human–climate dichotomy in favour of more complex causal scenarios that have yet to be fully developed.

Whatever environmental stressors impacted the Australian fauna 40–50 ka, diminutive dasyurid carnivores seem to have come through it relatively unscathed. Indeed, the maintenance of habitat mosaics and prevention of high intensity burns by firestick farming seems to have provided considerable benefit to dasyurids and other marsupials in more xeric areas. Discontinuation of this practice after European arrival has been associated with species decline (see 'Modern atrocities' below).

Dingos and dasyuromorphians in the Holocene

The only known dasyurid extinctions in the Pleistocene were several forms of *Antechinus*, possibly *Sminthopsis floravillensis* and *Sarcophilus laniarius*. The latter may have evolved into the extant *S. harrisii* via dwarfing (Marshall and Corruccini 1978) but, regardless, fossil and subfossil materials document that *S. harrisii* occurred widely on the Australian mainland until the late Holocene (Westaway *et al.* 2019). Its disappearance there coincided with that of *Thylacinus cynocephalus*, also known from the Holocene of New Guinea. That these two extirpations were synchronous was confirmed by White *et al.* (2018b) using statistical analyses of fossil radiocarbon ages, which placed their date at ca. 3.2 ka (Fig. 12.7). This supports an earlier analysis by Brown (2006) in refuting widely cited claims that devils persisted on the mainland until ca. 450 years ago. That both species vanished there simultaneously, but survived in Tasmania, begs the question of a common cause for their fates.

Archer (1984a) noted that thylacine extirpation roughly coincides with the first dingo (*Canis lupus dingo*) fossils in Australia, which are ca. 3.3 ka (Balme *et al.* 2018).[1] Dingos originated in south-east Asia and were likely carried to Australia by seafaring humans. Archer suggested that competition between these carnivores may have caused the thylacine's decline. Flannery (1994: 276) extended the hypothesis to include devils:

> *Prior to the introduction of the dingo some 3500 years ago, Australia had experienced a period of relative ecological stability. It had suffered no extinctions since the great wave that carried off the megafauna over 30 000 years earlier. Three species fell victim to the dingo on the*

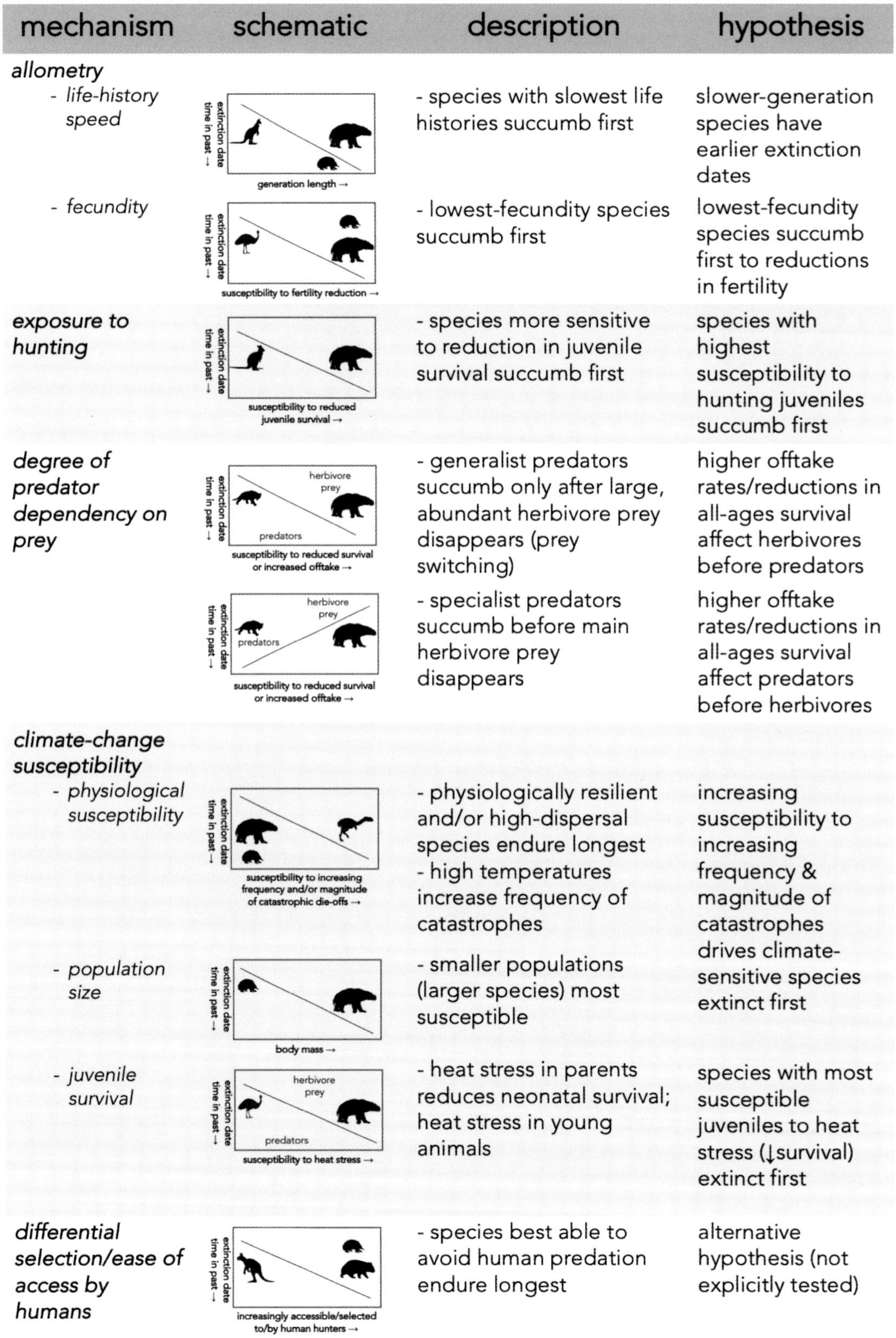

Fig. 12.6. Possible causes of megafaunal extinction evaluated by the stochastic modelling analysis of Bradshaw, Johnson *et al.* (2021a). [Reprinted from Bradshaw, Johnson *et al.* (2021a) under CC BY license.]

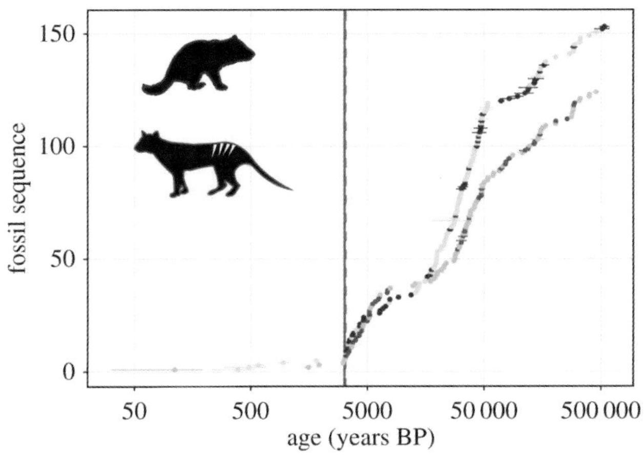

Fig. 12.7. Ages of mainland fossils from *Thylacinus cynocephalus* (red) and *Sarcophilus harrisii* (blue) obtained by White *et al.* (2018b). Fossils are arrayed in order from youngest to oldest on the vertical axis; the horizontal axis is logarithmic. Heavy dots represent dates validated by quality filtering criteria; light dots are published dates that were not validated. Vertical lines show the estimated dates of mainland extinction. [Reprinted by permission of the Royal Society (Great Britain) from White *et al.* (2018b). Permission conveyed through Copyright Clearance Center, Inc.]

> *Australian mainland, but all survived in Tasmania, where the dingo was never present. The Tasmanian tiger* (Thylacinus cynocephalus) *and Tasmanian devil* (Sarcophilus harrisii) *were probably affected through direct competition for resources and became extinct on the mainland soon after the arrival of the dingo. Their extinction may indicate that dingos had greatly reduced the prey available to both carnivorous marsupials and humans.*

Flannery further suggested that use of dogs for hunting by humans may have contributed to the thylacine's demise in New Guinea. Wroe (1999a) accepted the dingo competition scenario, but Paddle (2000) found it less plausible than anthropogenically driven declines of thylacines and devils on the mainland. Corbett (1995, 2006) attempted to explain the competitive superiority of dingos:

> *The demise of the thylacine ... on the Australian mainland about 3000 years ago... and the Tasmanian devil ... has been attributed to competition with dingoes according to the 'superior adaptability' hypothesis ... This hypothesis hinges on the superior social organization of dingoes during critical periods when food supplies were scarce, widely dispersed, or clumped, which usually occurs during drought or after extensive wildfire. Only dingoes form large integrated packs and cooperate to catch large prey, and to defend carcasses, water and other crucial resources. On the other hand, thylacines hunted alone or in pairs and devils were essentially solitary, so that neither would successfully compete against the weight of dingo numbers during those critical periods. [Corbett (2006): 748]*

Biomechanical analyses of dingo and thylacine skulls by Wroe *et al.* (2007) confirmed the potential for feeding niche overlap of the species as solitary predators, though social hunting by dingos gave them access to larger prey than were available to thylacines. That competition would be most intense during times of food scarcity allows the possibility of a synergistic climate effect. Until recently, dingos were widely cited as the primary driver of mainland thylacine decline, though their impact on devils seemed equivocal (Johnson and Wroe 2003).

Prowse *et al.* (2014) constructed a dynamic model of predator–prey interactions coupled with human population growth and climate change in the late Holocene. Their results suggested that 'human intensification' (i.e. increased population size and enhanced technology) and climate change were more important than dingo competition in driving the decline of mainland thylacines and devils. Indeed, Brown (2006) had suggested that climate variability brought on by the El Niño-Southern Oscillation (ENSO) after 6 ka impacted mainland devil populations by reducing carrion availability; because Tasmania is less affected by ENSO, it became a devil refuge. Ancient mtDNA (Brüniche-Olsen *et al.* 2018) documents a devil population bottleneck in Tasmania roughly coincident with the onset of ENSO (see Fig. 7.8). In fact, the phylogeographic structure of both mainland and Tasmanian devils resembles that of thylacines (White *et al.* 2018a) and is consistent with human, dingo and climate effects on mainland populations. Thus, as for megafaunal extinctions, current explanations of thylacine and devil declines emphasise multifactorial models.

Modern atrocities

Threats to biodiversity

Archer (1984a: 155) introduced his summary of the consequences of European 'invasion' for Australian vertebrates as follows: 'In the space of less than 300 years of Australian occupation, Europeans have accomplished more vertebrate exterminations than Aborigines appear to a have accomplished in over 100 000 years.' Updating Archer's prehistoric timeframe and giving due weight to the possibility of negative human impacts on the Pleistocene megafauna, his point remains valid – especially considering the growing, but largely ineffectual, conservation awareness of the last 60 years. Nor is it restricted to Australia in that virtually every region of the Earth has experienced drastic declines in biodiversity attributable to human activity during the last three centuries. However, the modern extinction rate of Australia's endemic land mammals – more than 10% of 273 species in 200 years (Woinarski *et al.* 2015, 2019) – is the highest in the world. Archer (1984d) referred to the biodiversity-destroying actions of European colonists as 'atrocities' (acts of enormous wickedness). Here we summarise those acts and their consequences for dasyuromorphians.

Modern causes of species decline worldwide are now elements of common knowledge ubiquitously repeated in elementary textbooks: habitat destruction and degradation, overexploitation, invasive species and disease, pollution and climate change. The last factor acts both independently and in synergy with the others, and its results are not yet fully understood (but unlikely to be trivial). All of them have impacted, and continue to impact, Australian marsupials, but the first three have been especially deadly. Woinarski *et al.* (2015: 4531) summarised the situation as follows: 'In contrast to general patterns of extinction on other continents where the main cause is habitat loss, hunting and impacts of development by humans, particularly in areas of high and increasing human population pressures, the loss of Australian land mammals is most likely due primarily to predation by introduced species, particularly the feral cat, *Felis catus*, and European red fox, *Vulpes*, and changed fire regimes.' These stressors have been present for variable periods. Feral cats in Australia date to ca. 1820 (Abbott 2002); foxes were introduced ca. 1871 (Fairfax 2019); modern fire regimes replaced low-intensity, small scale, human-lit burns with high-intensity, large scale wildfires as Aboriginal land management was restricted after European colonisation (Mariani *et al.* 2022). The effect of just these three factors on dasyurids has been devastating: cats and foxes are estimated to kill tens of millions of Australian dasyurids per year (Murphy *et al.* 2019; Stobo-Wilson *et al.* 2022); the 2019–20 wildfires alone burned 38–95% of the ranges of *Sminthopsis aitkeni*, *S. leucopus*, *Antechinus stuartii* and *A. mimetes* (Legge *et al.* 2021).

In addition to extrinsic factors, several intrinsic ones have been suggested as predisposing species to extinction. Among these are restricted distributions, narrow ecological niches, low reproductive rates, high trophic position and body size. Burbidge and McKenzie (1989) pointed out that medium-sized (35–5500 g) terrestrial herbivore-omnivores from arid or semi-arid habitats in Australia were over-represented among recently extinct and threatened species. They called this the 'critical weight range' and the elevated threat to such species was widely cited in subsequent reviews (e.g. Kennedy 1992; Maxwell *et al.* 1996; Burbidge 2008; Woinarski *et al.* 2014; Murphy *et al.* 2019; Woolley *et al.* 2019); it may represent the optimal prey size for foxes and cats. Johnson and Isaac (2009), however, found that the critical weight range was not strongly associated with population declines of arboreal and wet-zone marsupials. Fisher *et al.* (2014) also rejected it for tropical species, but reanalysis of the data by Murphy and Davies (2014) supported the original correlation. Wilson *et al.* (2003) noted that occupancy of only one or two habitat types was predictive of endangerment for small (<500 g) dasyurids.

In describing anthropogenic effects on dasyuromorphians, we focus on the current conservation status of extant and recently extinct species. These are taken from the International Union for the Conservation of Nature (IUCN) Red List (IUCN 2021): Extinct (EX), Extinct in the Wild (EW), Critically Endangered (CR), Endangered (EN), Vulnerable (VU), Near Threatened (NT), Least Concern (LC) and Data Deficient (DD). The implications of placing a species in most of these categories are self-evident, but the intermediate levels CR, EN and VU (collectively termed 'threatened') merit more careful description because they contain species tending towards extinction and most urgently in need of conservation. Table 12.1 shows the five main criteria for classification in the threatened categories. Threat levels are based on indicators of declining population size and geographic dispersion, as well as estimated extinction probabilities. Threatened and Near Threatened dasyuromorphians are listed in Tables 12.2 and 12.3; fractions in each category are shown in Fig. 12.8.

Extinction of the Tasmanian thylacine

The story of thylacine genocide at the hands of European colonists has been told in detail by Guiler (1985) and Paddle (2000). Unlike the indirect effects of human activities on

Table 12.1. IUCN (2012) criteria for classification of species in threatened categories.
Notation 'xY, zG' means x years or z generations, whichever is longer.

Criterion*	Category		
	Critically Endangered	**Endangered**	**Vulnerable**
A. Population reduction in last 10Y or 3G, or next 10Y or 3G	≥90, ≥80%	≥70%, ≥50%	≥50%, ≥30%
B. Extent of occurrence, or area of occupancy (km^2); evidence of fragmentation, decline, or fluctuation	<100, <10	<5000, <500	<20 000, <2000
C. Population size of mature individuals, rate of decline; evidence of fragmentation, isolation, or fluctuation	<250, 25%/(3Y, 1G)	<2500, 20%/(5Y, 2G)	<10 000, 10%/(10Y, 3G)
D. Population restriction in number of mature individuals [area of occupancy (km^2), or number of locations]	<50	<250	<1000 [<20, <5]
E. Extinction probability	>50% in 10Y, 3G	>20% in 20Y, 5G	>10% in 100Y

* Each criterion includes multiple contingencies in addition to those listed here. Application of the higher percentages for criterion A in each category requires that the cause of population reduction be clearly reversible, understood and ceased.

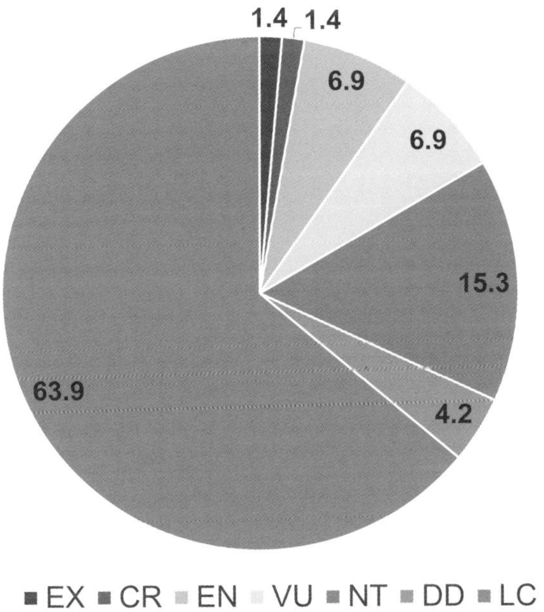

Fig. 12.8. Percentages of the 72 dasyuromorphian species recognised in the IUCN Red List classified in each conservation status category. EX = Extinct; CR = Critically Endangered; EN = Endangered; VU = Vulnerable; NT = Near Threatened; DD = Data Deficient; LC = Least Concern.

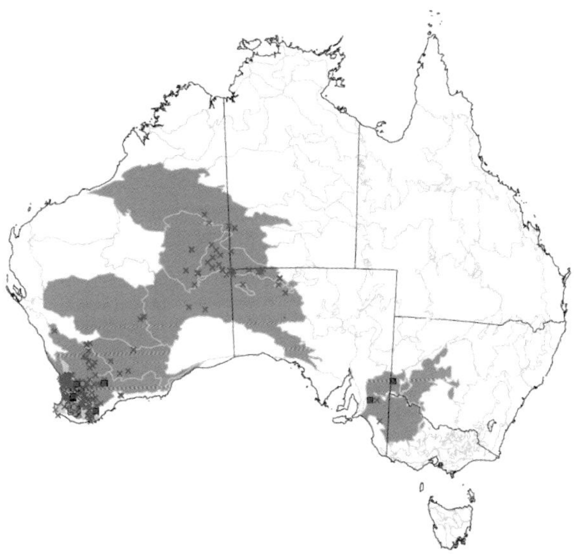

Fig. 12.9. Range contraction of the numbat (*Myrmecobius fasciatus*). Grey area show distribution at time of European settlement (1788), red crosses are localities of records between 1788 and 1992, green dots are current localities and blue squares are successful translocations. [Reprinted by permission of CSIRO Publishing from Woinarski *et al.* (2014). Permission conveyed through Copyright Clearance Center, Inc.]

current species, the Tasmanian thylacine was deliberately exterminated because of its falsely perceived threat to domestic sheep and an accompanying campaign of misinformation. As summarised by Paddle (2000: 212):

> *The list of causal factors involved in the extinction of the thylacine includes, (1) the destruction of the original Tasmanian environment and its replacement with an agricultural community; (2) the concomitant destruction of the thylacines' native prey species through environmental alterations, or for food or the fur trade; (3) human predation through the direct killing of thylacines for protection of agricultural industry or for 'sport'; and (4) non-human predation involving competition from introduced carnivores and the effects of introduced disease micro-organisms. It becomes of relatively minor*

Table 12.2. Extinct and threatened dasyuromorphian species and principal stressors.

	Species	Status; trend[1]	Threats[1]
Thylacinidae	*Thylacinus cynocephalus*	Extinct	Agriculture (sheep replaced native prey); hunting and trapping
Myrmecobiidae	*Myrmecobius fasciatus*	Endangered; decreasing	Invasive species (fox, cat); intense fire destroys logs; predation by native raptors
Dasyuridae **Dasyurinae** **Dasyurini**	*Dasyurus hallucatus*	Endangered; decreasing	Residential and commercial development; agriculture; fire; invasive species (cane toads, cats, dogs) and disease (from cats); vehicles and pesticide poisoning
	Dasyurus viverrinus	Endangered; decreasing	Roads and railroads; invasive species (cats, foxes) and disease; climate change; vehicles
	Sarcophilus harrisii	Endangered; decreasing	Hunting and trapping; invasive species (dogs, recently foxes) and disease (devil tumour facial disease); road kills, human persecution, low genetic diversity
	Parantechinus apicalis	Endangered; decreasing	Fire (vulnerable to frequent, intense fires); invasive species (fox, cat, mice) and disease (*Phytophthora cinnamomi* plant pathogen degrades habitat)
	Dasyuroides byrnei	Vulnerable; decreasing	Livestock grazing; habitat loss
Phascogalini	*Phascogale pirata*[2]	Vulnerable; unknown	Tree plantations; livestock grazing; invasive species and disease
	Antechinus bellus[2]	Vulnerable; decreasing	Habitat; fire (frequent and intensive); invasive species (cats)
Sminthopsinae **Sminthopsini**	*Sminthopsis aitkeni*	Critically Endangered; decreasing	Fire; fungal destruction of heathland; invasive species (cats); agriculture
	Sminthopsis butleri	Vulnerable; decreasing	Wood plantations (on Melville Is.), habitat destruction; fire; invasive species (cats, dogs)
	Sminthopsis psammophila	Vulnerable; decreasing	Invasive species (cats, foxes); fire

[1] From IUCN Red List of Threatened Species (IUCN 2021), Van Dyck and Strahan (2008) and other sources.
[2] Note the danger of reproductive Strategy I combined with small population size: reproductive failure in any one year results in local extinction (i.e. no males in the next year).

> *academic interest to ask what additional factors, unrelated to human activities, need to be included in the list.*

Of these, outright trapping had the greatest impact. Beginning with the institution of a bounty on thylacines by the Van Dieman's Land Company in 1836 and by the Tasmanian government in 1840, thousands of thylacines were killed before the bounty was lifted in 1914 (Smith 1982). The depleted thylacine population (and other dasyurids) may have suffered a distemper-like disease outbreak ca. 1905–10 (Mooney and Rounsevell 2008) and the last known thylacine died in the Hobart Zoo in 1933.

Decline of the numbat

> *When Europeans first arrived on this continent, Numbats extended from their present stronghold [in south-western Australia], north into central Western Australia, east through central Australia and even into the eastern states. Their decline has been nothing short of catastrophic. And because the reasons for this decline are not well understood, the same factors may well be threatening the last surviving Numbats. [Archer (1984c: 642)]*

Some of that decline (Fig. 12.9) appears to have been due to high-intensity fires reducing the availability of hollow logs that numbats use for shelter (Archer 1984d) and to predation by foxes (Kennedy 1992). The numbat's range contracted from east to west, the species disappearing from the Murray–Darling area by 1850 and accelerating with the westward spread of foxes after 1900 (Friend and Thomas 2003; Friend 2008). Fox control (by baiting and fencing) has been a key component in the maintenance of numbat populations in south-western Australia, although agricultural development in the region has left only isolated patches of habitat with suitable ground cover

Table 12.3. Near Threatened dasyurids and principal stressors.

	Species	Population trend*	Threats*
Dasyurinae Dasyurini	*Dasyurus maculatus*	Decreasing	Hunting and trapping; logging; invasive species and disease (early 1900s); fire; non-target poisoning and trapping; competition/predation by foxes, cats, dogs; cane toads; persecution near poultry
	Dasyurus albopunctatus	Decreasing	Residential and commercial development; slash and burn agriculture; logging; hunting and trapping; invasive species (dogs, possibly cats)
	Dasyurus geoffroii	Stable	Invasive species (fox, cat); vehicles; non-target trapping
	Dasyurus spartacus	Unknown	Hunting (including non-target with dogs) and trapping; fire; invasive species (feral dogs, cats, *Mimosa*)
	Pseudantechinus mimulus	Decreasing	Habitat degradation by livestock, mining and quarrying; fire; invasive species (fox, cat, cane toad)
	Dasycercus cristicauda	Stable	Habitat degradation from livestock and feral herbivores; fire; invasive species (predation/competition by foxes, cats)
Phascogalini	*Phascogale calura*	Stable	Forest clearing and fragmentation; livestock farming; fire; invasive species (cats)
	Phascogale tapoatafa	Decreasing	Forest clearing and fragmentation; fire; invasive species (cats, foxes)
	Murexia rothschildi	Unknown	Forest clearing; mining; invasive species (dogs)
Sminthopsinae Sminthopsini	*Sminthopsis bindi*	Decreasing	Fire; invasive species (cats, possibly cane toads)
	Sminthopsis douglasi	Decreasing	Invasive species (cats, foxes, livestock, prickly acacia); mining

*From IUCN Red List of Threatened Species (IUCN 2021), Van Dyck and Strahan (2008) and other sources.

and termite availability. An aggressive management program includes captive breeding and reintroduction of numbats into protected areas of the south-west (George and Brown 1992), as well as sanctuaries in South Australia and New South Wales. The early success of these efforts prompted Friend and Thomas (2003) to express optimism that the numbat's status could be upgraded from Endangered to Vulnerable, but the total number of mature individuals remains fewer than 1000 with evidence of decline (Woinarski and Burbidge 2016).

Endangered dasyurids

One dunnart (*S. aitkeni*) is Critically Endangered and four dasyurins (*Dasyurus hallucatus, D. viverrinus, Sarcophilus harrisii, Parantechinus apicalis*) are Endangered (Table 12.2). *S. aitkeni* is a small population of <500 individuals restricted to Kangaroo Island, SA. It is under perennial threat of extinction due to loss of native vegetation and wildfires; the fires of 2019–20 destroyed over 95% of its habitat. Jones *et al.* (2003: 422) summarised threats to quolls and the devil as follows:

> *While the causes of decline are multiple and complex, the overriding factor is probably habitat loss, degradation and fragmentation, and the associated factors: loss of protective cover from predators, reduced food availability, and increased contact with humans. Introduced predators [cats, foxes, dingos] ... play a not well understood role. Other important mortality factors include persecution, non-target poisoning and road mortality.*

That northern quolls can be burned out of an area by intense fire was demonstrated in a survey of the Abydos-Woodstock Reserve in Western Australia by How *et al.* (1991). Genetic studies have shown marked phylogeographic subdivision among the now isolated populations of this species in the Pilbara, Kimberley, Northern Territory and Queensland (Firestone 2003; How *et al.* 2009; Hohnen *et al.* 2016; Moore *et al.* 2021). Poisoning of northern quolls by invasive cane toads (*Rhinella marina*) in the eastern half of their range has been a significant source of decline (Oakwood 2008; Woinarski *et al.* 2015). These toads were deliberately released in Queensland in 1935 to control beetle pests on sugarcane plantations and have since colonised much of northern Australia, appearing as far west as the Kimberley region in 2009. They have negatively impacted many other native species, though some (including *D. hallucatus*) appear to have heritable toad-avoidance behaviour segregating within populations (Kelly and Phillips 2018).

Eastern quolls declined on the mainland (Ride 1970; Archer 1984d) and were extirpated there in recent times, surviving only in Tasmania (Kennedy 1992; Jones *et al.* 2003). In addition to other impacts, Dickman and Woodford-Ganf (2007) noted that mainland *D. viverrinus* may have been affected by eating rabbits poisoned with strychnine. New cause for concern stems from the possible arrival of foxes in Tasmania (Jones 2008).

Tasmanian devils were relatively abundant before the epidemic of devil facial tumour disease (DFTD) first detected in 1996 (McCallum *et al.* 2007). DFTD is caused by transmission of cancerous cells between individuals, most often a result of the facial biting characteristic of devils. These cells become established as tissue grafts in infected individuals and are not subject to immune rejection. They develop into cancerous lesions that are nearly always fatal. The DFTD epidemic has caused a ca. 80% reduction of the total Tasmanian devil population (Woods *et al.* 2018). Although transmission rates of the original clone (DFT1) have declined such that it is no longer 'spreading' (i.e. it is becoming endemic), a second clone (DFT2) was discovered in south-eastern Tasmania on which few data are available (Patton *et al.* 2020). Prospects for devil recovery are complicated by their low but geographically structured genetic diversity (Hendricks *et al.* 2017).

Prior to 1967, no dibbler specimen had been collected since 1884 and the species was presumed extinct. Morcombe (1967) and Ride (1970) described its rediscovery in the coastal sandplain near Cheyne Beach, WA. The dibbler's range once extended into Queensland and South Australia (Archer 1984d), but wild populations are now limited to Fitzgerald National Park on the mainland, as well as Boullanger and Whitlock islands. Range contraction on the mainland may pre-date European arrival (Woolley 2008), but subsequent decline seems to have been driven by over-burning, fragmentation and pathogen-induced degradation of heathland habitats, along with cat and fox predation. Island populations may be impacted by invasive house mice (*Mus musculus*) (Kennedy 1992; Wilson *et al.* 2003). Captive breeding and translocations have now reintroduced dibblers to several mainland and island localities in Western Australia, but the total population remains below 1000 individuals (Burbidge and Woinarski 2016).

Vulnerable and Near Threatened species

Dasyuroides byrnei, *Phascogale pirata*, *Antechinus bellus*, *Sminthopsis butleri* and *S. psammophila* are currently classified as Vulnerable (Table 12.2). In addition to ubiquitous predation by foxes and cats, all these have experienced habitat loss due to fire and agriculture (livestock grazing, crop farming, timber plantations). Grazing livestock remove sheltering vegetation and destroy burrows on which small dasyurids depend; fire and timber harvesting cause loss of tree hollows, substrates and understorey crucial for arboreal phascogales, which may also ingest cane toads (Wilson *et al.* 2003; Rhind *et al.* 2008). Dickman and Woodford Ganf (2007) noted that as the Great Artesian Basin Sustainability Initiative, begun in 1999, pumps water into the kowari's naturally arid range to service livestock, numbers of cats and foxes there are likely to increase.

An additional 11 dasyurid species are in the Near Threatened category (Table 12.3), including all remaining *Dasyurus* and *Phascogale* species, two arid zone dasyurins (*Pseudantechinus mimulus*, *Dasycercus cristicauda*), two dunnarts (*Sminthopsis bindi*, *S. douglasi*) and a single New Guinean endemic (*Murexia rothschildi*). A glance down the 'Threats' column reveals the usual suspects and Wilson *et al.* (2003) showed that the ecological characteristics of (small-bodied) Near Threatened species resemble those in the 'Threatened' categories more than those of 'Least Concern'. Many of the former (*D. geoffroii*, *D. maculatus*, *P. calura*, *P. tapoatafa*, *D. cristicauda*) underwent major range contractions after European settlement (Fig. 12.10a) and others (*D. spartacus*, *P. mimulus*, *S. butleri*) have very restricted distributions (Archer 1984d; Kennedy 1992). Loss of hummock grasslands to fire is implicated in the decline of western quolls and crest-tailed mulgaras, though fox predation appears to have driven the former to its last refuge in south-western Australia (Morris *et al.* 2003). Uzqueda *et al.* (2020) documented loss of suitable habitat as responsible for the decrease of spotted-tail quoll numbers in Queensland to probably <300 individuals (Fig. 12.10b). Fusco *et al.* (2017) used fossils to trace the decline of dasyurids (including

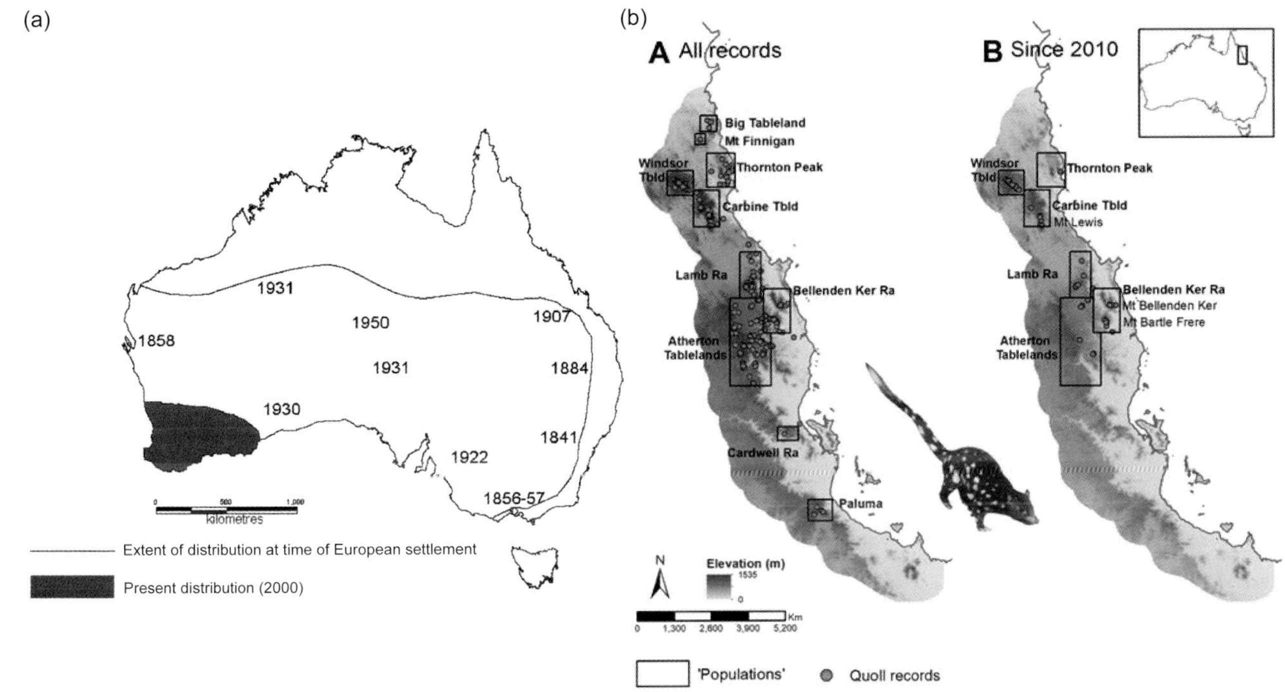

Fig. 12.10. Range contraction in quolls. (a) *Dasyurus geoffroii* (Morris *et al.* 2003). Dates refer to most recent specimen records at each locality. (b) *Dasyurus maculatus* in Queensland (Uzqueda *et al.* 2020). [(a) Reprinted by permission of CSIRO Publishing from Morris *et al.* (2003). Permission conveyed through Copyright Clearance Center, Inc. (b) Reprinted by permission of Oxford University Press from Uzqueda *et al.* (2020).]

Dasycercus) and other mammals of the Murray–Darling Basin during the past two centuries. Overland stock routes, pastoralism, vegetation clearing and invasive species reduced the diversity of native mammals in this region by 50–60%.

Consequences of conservation action and inaction

A series of conservation action plans for marsupials (Kennedy 1992; Maxwell *et al.* 1996; Burbidge 2008; Woinarski *et al.* 2014) consistently highlighted general themes relating to dasyurid protection, most of them noted above. One is the importance of islands as habitat refuges for species such as *Dasyurus viverrinus*, *D. maculatus*, *Sarcophilus harrisii* and *Antechinus minimus* (Tasmania), *D. hallucatus* (Groote Eylandt, Astell and Pobasso Islands), *Parantechinus apicalis* (Boullanger and Whitlock Islands), *Pseudantechinus mimulus* (Sir Edward Pellew Group), *Phascogale pirata* (Melville Island), *Sminthopsis aitkeni* (Kangaroo Island) and *S. butleri* (Tiwi Islands) (Griffiths *et al.* 2017; Legge *et al.* 2018). Small islands are often free of introduced predators and have relatively little habitat destruction; maintaining them as such is a critical conservation strategy. Thus, developments such as the possible arrival of foxes in Tasmania, house mice on Boullanger Island and the expansion of forestry plantations on Melville Island are serious concerns. Northern quolls on Groote Eylandt exploit savanna woodlands in addition to the rocky escarpments characteristic of mainland populations (Thomas *et al.* 2021).

Despite the plethora of threats facing dasyurids, there have been a few conservation success stories. Fox control, genetically informed captive breeding, translocation and reintroduction programs in Western Australia may have deferred the extinction of numbats and western quolls (West *et al.* 2020). Indeed, *D. geoffroii* is the only threatened dasyurid to show improvement in status between 1992 and 2012 (Woinarski *et al.* 2014). Similarly, new populations of eastern quolls (*D. viverrinus*) have been established in the ACT (Wilson *et al.* 2020), dibblers (*Parantechinus apicalis*) on Jurien Island and in the mainland Peniup Nature Preserve and Tasmanian devils on Maria Island (with ongoing discussion of reintroducing them to the mainland; Hunter *et al.* 2015). Expanding such programs to other species has begun (Old and Stannard 2020). Monitoring, captive breeding, translocations and the existence of a captive 'insurance population', coupled with declining transmission rates of DFT1, have moderated predictions about the imminent extinction of Tasmanian devils (Woods *et al.* 2018). Conservation efforts have not been uniformly successful, however. Northern quolls born on Astell Island in Arnhem Land lacked predator-avoidance behaviour when reintroduced to the mainland and succumbed to dingo predation. An attempt at retraining them to recognise the scents of dingos and domestic dogs failed (Jolly *et al.* 2020), suggesting that captive-reared and refugial quolls may be at a selective disadvantage in their former mainland habitats. Translocated dibblers at Peniup Creek (Beringa) Nature Preserve experienced a >10-fold reduction in effective population size, with consequent potential inbreeding effects, relative to their source populations in Fitzgerald River National Park (Thavornkanlapachai *et al.* 2021). Woinarski *et al.* (2015: 4538) noted that 'no broadscale and enduring mechanism with acceptably small nontarget effects has yet been developed for the control of feral cats'.

Individual triumphs are dwarfed by the slow pace at which conservation policies and practices have been implemented by governments at all levels. Although, as noted above, the status of *D. geoffroii* improved due to rescue efforts between 1992 and 2012, that of six other dasyurids and the numbat declined over the same interval (Woinarski *et al.* 2014). Although action plans offered specific, evidence-based recommendations for conservation initiatives – including research, habitat preservation, predator control and climate change mitigation (Woinarski *et al.* 2015) – few of these have been put into practice. One of the most prominent legislative attempts to institute conservation oversight was Australia's *Environmental Protection and Biodiversity Conservation Act* (EPBC Act) of 1999. However, Ward *et al.* (2019) offered a stinging indictment of the EPBC Act's effectiveness, documenting that less than 7% of the 7.7 million ha of native habitat destroyed between 2000 and 2017 was assessed under the Act, even though 84% of 1390 threatened species in these areas suffered significant habitat losses. In a similar vein, Wintle *et al.* (2019: 1) compared Australian government budget data with estimated conservation costs to conclude 'that annual spending on targeted threatened species recovery is around US$92m (AU$122m) which is around one tenth of that spent by the U.S. endangered species recovery program and about 15% of what is needed to avoid extinctions and recover threatened species'. In Papua New Guinea, where biodiversity protection is constitutionally mandated, >50 protected areas and additional wildlife management areas have been established, but the value of these has been compromised by ineffective administration (Seri 1992; Leverington *et al.* 2017). Little published information is available on conservation efforts in Indonesian New Guinea; an early prospectus by Ramono and Nash (1992) described the design of a reserve system that was predicted to encounter opposition from mining, agriculture, forestry and wildlife trade interests. Gaveau *et al.* (2021) developed a simulation based on data from 2001 to 2019 that revealed the prospect of substantial forest loss in this region by 2036 due to road development, timber extraction and industrial plantation.

The prospect of extinction

It is a truism that all species go extinct, but the balance between extinction and speciation rates sets the level of biodiversity that fuels ecosystem functioning around the world. That recent, anthropogenic species losses amount to a Holocene mass extinction is widely accepted (Ripple *et al.* 2017; but see Vaidyanathan 2021) and cogent reasons for mitigating it have been put forward. Broadly, the latter include maintenance of Earth as habitable for future human generations and the intrinsic value of biological species. In opposition is the allure of short-term economic prosperity derived from exploitation of natural resources, intensified by the subsistence demands of a human population (at the time of writing 7.9 billion) that may exceed Earth's carrying capacity. In the roughly six decades since the rise of environmentalism, the wildlife conservation agenda has not fared well. Dasyuromorphians are no exception, as the previous sections demonstrate. But what of the future? Certainly, the immediate future will include global climate change driven by greenhouse warming.

> *For the Australian land mammal fauna, some climate change impacts are probably already being realized, but impacts will be magnified substantially in the next few decades, with modeling predicting severe to catastrophic losses due to climate change in the diverse and highly distinctive mammal faunas of the already limited area of high-altitude tropical rainforests and alpine environments, for the koala, for the mammal fauna of the extensive tropical savannas of northern Australia, for mammals of coastal areas, and island species. The mechanisms for this impact include*

marked habitat loss or change, likely increased fire intensity and frequency, increased incidence of severe drought and days of extreme heat, and changes in foliage nutrient composition. [Woinarski et al. (2015): 4536]

Geyle *et al.* (2018) analysed multiple expert assessment schemes of threatened birds and mammals native to Australia to rank species by extinction risk over the next 20 years. Among the 20 most imperilled mammals were six dasyurids (likelihood of extinction in parentheses): *Phascogale tapoatafa kimberleyensis* (0.28), *P. pirata* (0.23), *Sminthopsis aitkeni* (0.22), *Antechinus argentus* (0.20), *A. arktos* (0.17), *A. vandycki* (0.14). The short time frame of this analysis means that the relatively low extinction risks are no cause for optimism, though they are in line with IUCN classifications. That dasyurids account for 30% of the most imperilled Australian mammals (only rodents have as a large a fraction) is even more sobering.

Throughout this book we have used phylogeny as the framework in which to understand past dasyurid evolution. In closing, we offer a phylogenetic perspective on the future, in terms of both extinction and diversification. Following the exposition of Vane-Wright *et al.* (1991) that taxonomy imposes an 'agony of choice' on assigning conservation priorities, several authors (Crozier 1992; Faith 1992; Weitzman 1992) proposed indices of 'phylogenetic diversity' (PD). These measures differed in detail, but all attempted to quantify the fraction of phylogenetic history that is lost when a species goes extinct. Conservationists have embraced this perspective, up to a point, and PD is now a recognised component of biodiversity. Application of the concept to conservation planning has involved Faith's (1992) approach, a conceptually simplified version of which equates the PD contribution of a species or set of species with the fraction of total branch lengths represented by those species and their ancestors on a phylogeny. Of course, there are many ways to define terminals, many levels of phylogenetic inclusiveness and many branch-length scales – and choice of these significantly affects the analytical outcome (Krajewski 1994). Putting these issues aside, however, we illustrate how extinction may impact dasyuromorphians by calculating PDs on the time-scaled branches of the phylogeny in García-Navas *et al.* (2020) (see Fig. 5.6).

Figure 12.11 shows the proportionate losses of dasyuromorphian phylogenetic diversity given four cumulative

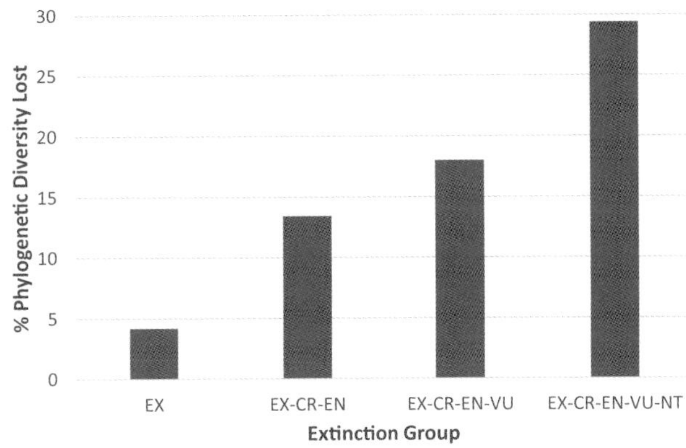

Fig. 12.11. Cumulative loss of dasyuromorphian phylogenetic diversity due to extinction of species in Extinct (EX), Threatened (CR, EN, VU) and Near Threatened (NT) categories based on the time tree of García-Navas *et al.* (2020) (see Fig. 5.6). See Tables 12.2 and 12.3 for species in each category.

levels of extinction corresponding to loss of species in the EX, EN, VU and NT categories. These extinctions eliminate ca. 5–30% of dasyuromorphian phylogenetic history, with loss of *Thylacinus* and *Myrmecobius* (species occupying long branches) responsible for the greatest single decrements. The symmetric phylogeny of Dasyuridae ensures that no one dasyurid species contributes disproportionately to PD, but group extinctions take a significant toll. Chief among these are loss of the *Dasyurus-Sarcophilus* and *Phascogale* clades; indeed, Dasyurinae is much more decimated by catastrophic extinction than is Sminthopsinae. Faith (1992) emphasised that loss of species entails loss of their characteristics ('feature diversity'). Extinction of all threatened dasyurids has the notable effect of removing all large, carnivorous species and the unique numbat, leaving a family composed entirely of 'faunivorous' (generally insectivorous) species weighing less than 255 g.

In the bleakest of futures, then, Dasyuromorphia in the 22nd century becomes a clade of small- to medium-sized dasyurids. The surviving species are still distributed among all major habitat types and species of all tribes remain in both Australia and New Guinea. However, the only dasyurids left in Tasmania are antechinuses and the white-footed dunnart (*Sminthopsis leucopus*); only *S. virginiae* and *Planigale maculata* remain in Arnhem Land. The historical contingency of macroevolution means that we cannot predict which, if any, of these post-apocalyptic survivors will be the ancestors of future dasyurids. If each has the same chance of generating descendants (a gratuitous assumption), the odds favour widespread and abundant sminthopsines as the family's hope for a brighter tomorrow. Such speculations have little value except to underscore the importance of avoiding the scenario from which they arise. We have lost the thylacine, but we still have the numbat, dibbler, quolls, phascogales and the rest. Does humanity have the collective will to save them?

> *Sadly, there is nothing unique about the history of the destruction of the thylacine ... predictable and disturbing parallels in the treatment of indigenous carnivores may be found almost at will in the invasive history of human colonization over the past millennium. European scientists and their colonial and post-colonial counterparts have proven pathetically ineffective in countering such destruction. It is questionable whether, granted a knowledge of the extinction of the thylacine, Australian zoologists and comparative psychologists are any better equipped to deal with conservative political beliefs and aggressive economic rationalism today ... When powerful economic pressure groups are arraigned against scientists with their knowledge claims it is difficult to see how scientists will ever win the battle. [Paddle (2000: 237)]*

Speaking as scientists, we hope Paddle was wrong. Our goal for this volume is to put the rich history of research on dasyurids, and the knowledge to which it has given rise, into the hands of current and future students of dasyurid biology. May your efforts yield a world that retains *all* these magnificent species and achieves an even deeper understanding of them.

ENDNOTE

1 Confusion on this point was introduced by Zhang *et al.* (2020), who analysed genome sequences of dingos and other dogs from south-east Asia. They used a Bayesian method to estimate the coalescent split of dingo haplotypes from their sister-clade (Indonesian village dogs) at 8.3 ka ago but reported this as the date of dingos' arrival in Australia. Fossils document the presence of dingos in Thailand and Vietnam ca. 5.5 ka (Corbett 2006) but the oldest dingo fossils in Australia are ca. 3.0–3.5 ka from Madura Cave (Balme *et al.* 2018).

Appendix: colour images

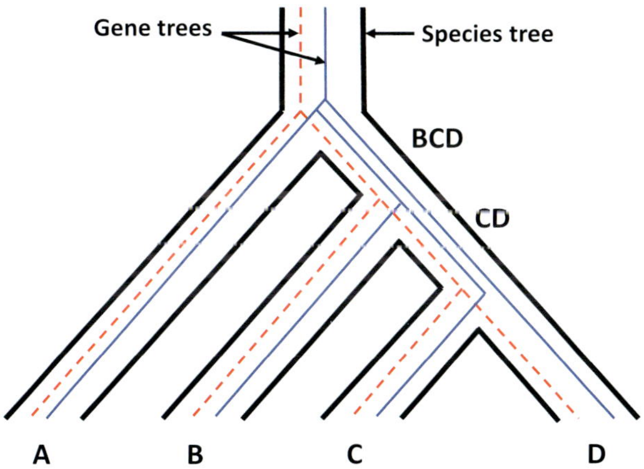

Fig. 1.9. Incomplete lineages sorting causes conflict between gene trees (light lines) and the species tree (heavy lines). The gene tree for the red locus (light dashed line) matches the species tree, but that for the blue locus (light solid line) does not. Note that lineage sorting at the blue locus in the ABCD ancestor is incomplete and two blue alleles are transmitted to ancestral species BCD and CD.

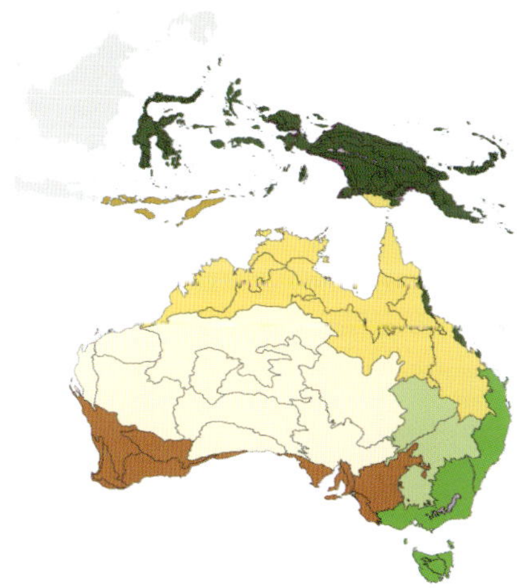

Fig. 2.1. Ecoregions of Australia and New Guinea, colour-coded by biome. Dark green = tropical and subtropical moist broadleaf forests; light brown = tropical and subtropical dry broadleaf forests; yellow = tropical and subtropical grasslands, savannas and shrublands; green = temperate broadleaf and mixed forests; light green = temperate grasslands, savannas and shrublands; light purple = montane grasslands and shrublands; brown= Mediterranean forests, woodlands and scrub; beige = deserts and xeric shrublands; magenta = mangroves. [Image by Terpsichores – own work, licensed under CC BY-SA 3.0 via Wikimedia Commons.]

Fig. 2.2. *Dasyurus*. (a) Current distribution of each of the six species. (b) *D. viverrinus* colour phases. (c) *D. albopunctatus*. (d) Sleeping *D. maculatus* with erect penis and penis appendage visible (see Chapter 11 for discussion). [Photo credits: (b) T.P. Fletcher; (c) P.A. Woolley and D. Walsh; (d) J. Seebeck.]

Fig. 2.3. *Sarcophilus*. (a) *S. harrisii*, the Tasmanian devil, is the only extant member of this genus and is restricted to Tasmania (inset map). (b) Woodland area where devils have been sighted. [Photo credits: (a) C. Krajewski; (b) Amy Edwards.]

Fig. 2.4. *Neophascogale*. (a) *N. lorentzii* (inset, underside of hind foot showing the long claws). (b) Distribution (inset map) and the village garden near Porgera, Enga Province where the animal was found. [Photo credits: (a) P.A. Woolley and D. Walsh**;** (b) P.A. Woolley.]

Fig. 2.5. *Phascolosorex*. (a) *P. dorsalis*. (b) North summit of Mt Kaindi, Morobe Province where this species is close to eastern limit of distribution. (c) Distribution of each of the two species. (d) Site of collection of an individual on Mt Kaindi arrowed. [Photo credits: (a) P.A. Woolley and D. Walsh; (b, d) P.A. Woolley.]

Fig. 2.6. *Pseudantechinus.* (a) Distribution of each of the five species. (b) *P. macdonnellensis*. This species lives in both rock outcrops and termite mounds in different parts of its range such as (c) the Sedimentaries (foreground) and at the base of Uluru (on horizon) in central Australia and (d) termite mounds in the Tanami desert. (e) *P. bilarni*. (f) Rocky habitat of *P. bilarni* in Arnhem Land. [Photo credits: (b, e) P.A. Woolley and D. Walsh; (c, d, f) P.A. Woolley.]

Fig. 2.7. *Myoictis*. (a) *M. wavicus*. (b) *M. wallacei*. (c) Mt Missim (2962 m), as seen looking across the Wau valley from Mt Kaindi. *M. wavicus* has been found on the southern slopes at 1450 m. (d) Kobroor, one of the low-lying Aru Islands where *M. wallacei* occurs. (e) Distribution of each of the four species. (f) Site of collection of *M. wallacei*. [Photo credits: (a, b) P.A. Woolley and D. Walsh; (c, d, f) P.A. Woolley.]

Fig. 2.8. *Parantechinus*. (a) *P. apicalis*. (b) Distribution (inset map) and habitat (low *Banksia* shrubland in foreground) at site of rediscovery (Morcombe 1967) of the species at Cheyne Beach, near Albany, WA. [Photo credits: (a) P.A. Woolley and D. Walsh; (b) P.A. Woolley.]

Fig. 2.9. *Dasykaluta*. (a) *D. rosamondae*. (b) Habitat and distribution (inset map). [Photo credits: (a) P.A. Woolley and D. Walsh; (b) P.A. Woolley.]

Fig. 2.10. *Dasyuroides*. (a) *D. byrnei* and inset map of distribution. (b) Burrow at base of shrub in sandy patch on gibber plain. White pegs mark holes associated with the burrow system. [Photo credits: (a, b) P.A. Woolley.]

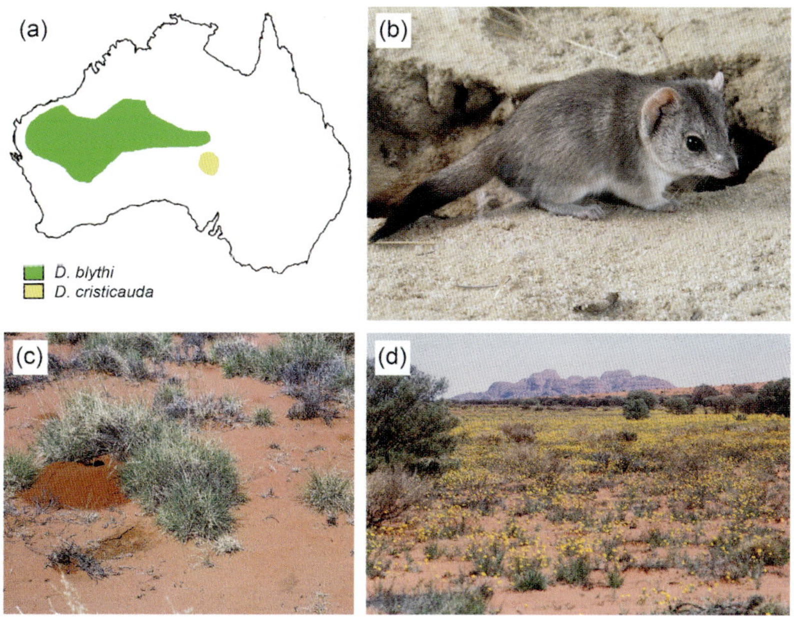

Fig. 2.11. *Dasycercus*. (a) Distribution of each of the two species. (b) *D. blythi*. (c) Entrance to burrow of a *D. blythi* female with newborn young, found on an interdune swale (d) near Uluru (Ayers Rock) in central Australia. Kata Tjuta (the Olgas) on horizon. [Photo credits: (b) P.A. Woolley and D. Walsh; (c, d) P.A. Woolley.]

Fig. 2.12. *Phascogale*. (a) *P. calura*. (b) Distribution of each of the three species. (c) *P. tapoatafa*. [Photo credits: (a, c) Rohan Clarke.]

Fig. 2.13. *Antechinus*. (a) *A. leo*. (b) Habitat of *A. leo*, far north Queensland. (c) *A. mimetes*. (d) Habitat of *A. mimetes*, Victoria Valley, Southern Grampians, Vic. (e) Distribution of each of the 15 species of *Antechinus*. [Photo credits: (a) P.A. Woolley and D. Walsh; (b) P.A. Woolley; (c, d) Rohan Clarke.]

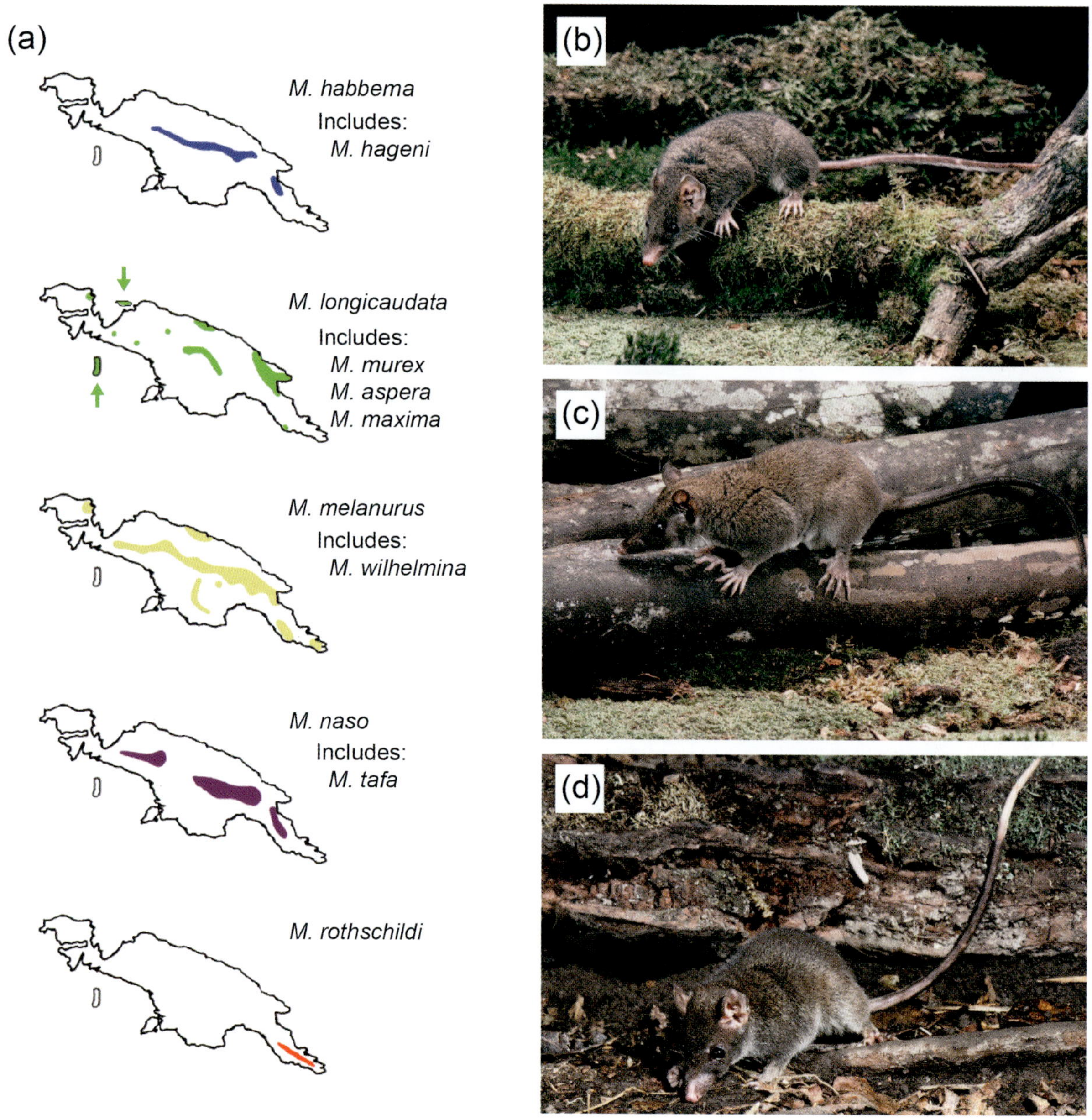

Fig. 2.14. *Murexia.* (a) Distribution of each of the five species of *Murexia* (including additional named forms) based on Woolley *et al.* (2020). (b) *M. habbema*. (c) *M. longicaudata* (*murex* form). (d) *M. naso* (*tafa* form). [Photo credits: (b, c, d) P.A. Woolley and D. Walsh.]

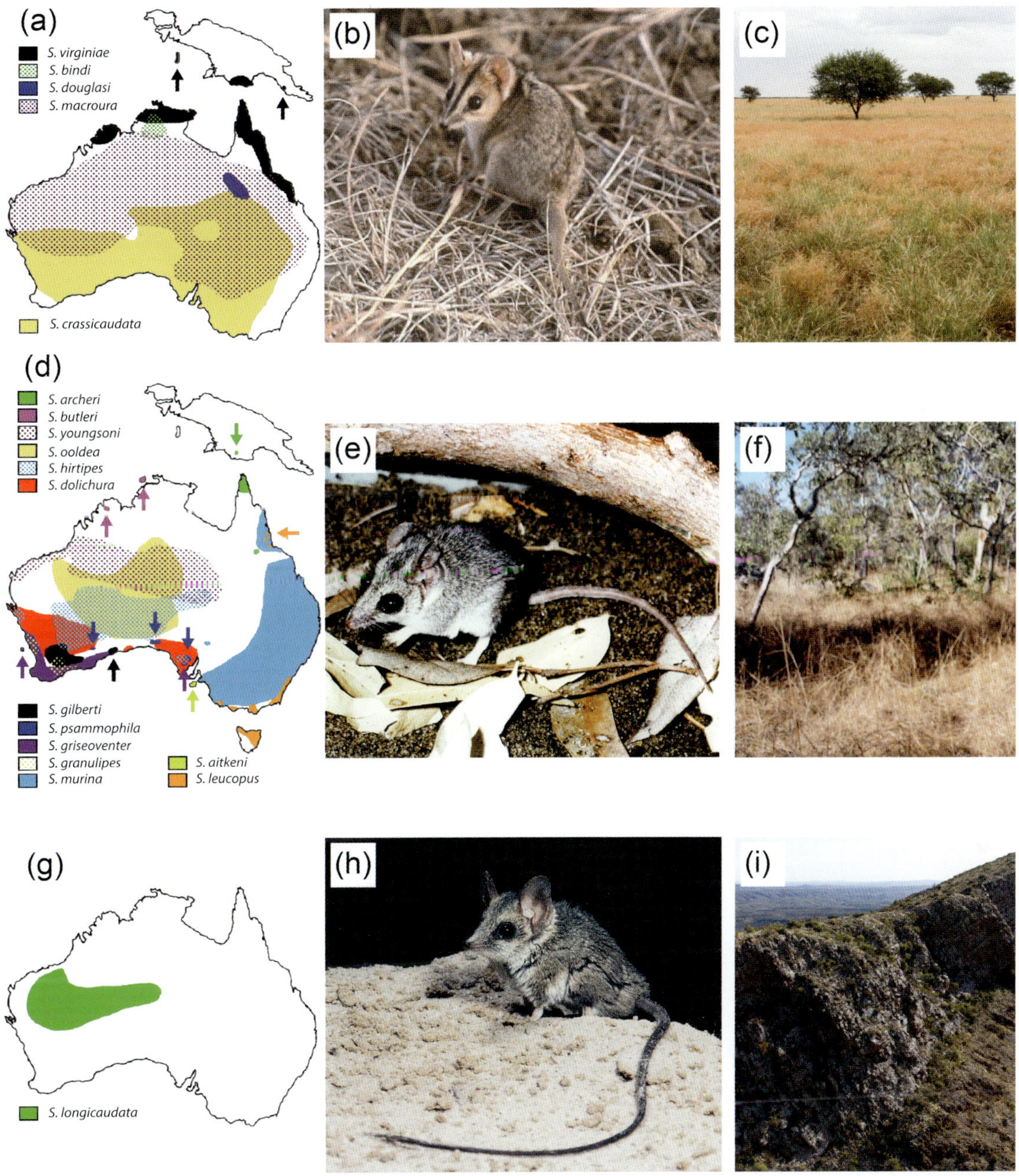

Fig. 2.15. *Sminthopsis*. (a) Distribution of each of the five species of the '*S. macroura* group'. (b) *S. douglasi*. (c) Habitat of *S. douglasi*, Mitchell grass downs with *Acacia nilotica* trees introduced for stock food. (d) Distribution of each of the 13 species of the '*S. murina* group'. (e) *S. butleri*, type specimen photographed live. (f) Habitat of *S. butleri* in the locality of collection of the type specimen at Kalumburu, WA. (g) Distribution of *S. longicaudata*. (h) *S. longicaudata*. (i) Habitat of *S. longicaudata* in the Macdonnell Ranges, central Australia. (See Chapter 6 for description of *Sminthopsis* species groups.) [Photo credits: (b, c, f) P.A. Woolley; (e) J.A.W. Kirsch; (h) P.A. Woolley and D. Walsh; (i) Rohan Clarke.]

Fig. 2.16. *Antechinomys.* (a) *A. laniger*. Pair found mating at 0200 hours beside vehicle track in central Australia. (b) Distribution. (c) Habitat, far west of range. [Photo credits: (a) Rohan Clarke; (c) P.A. Woolley.]

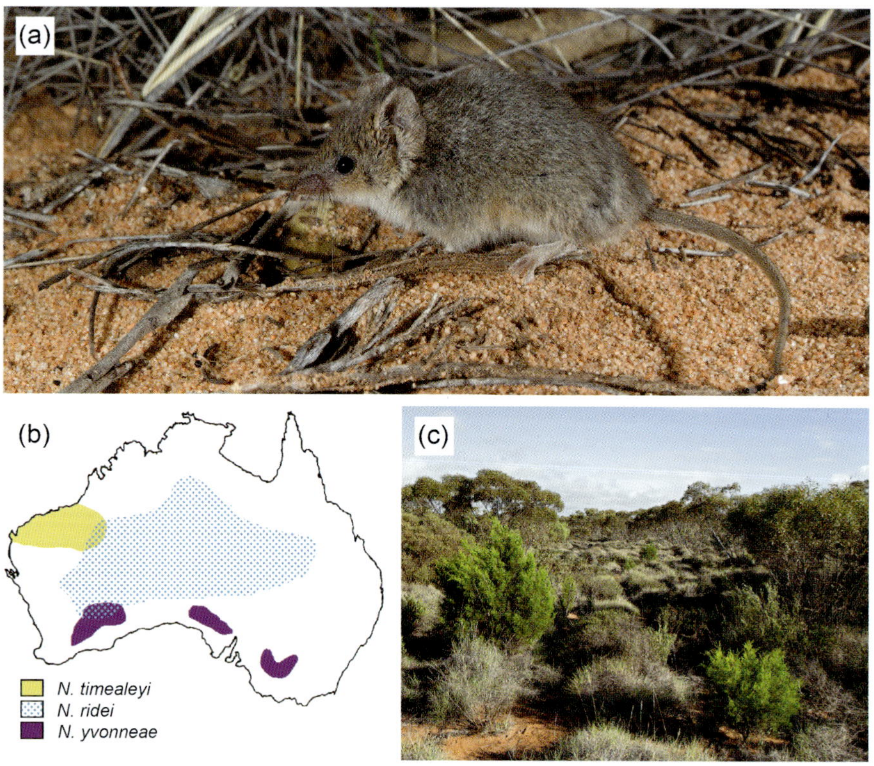

Fig. 2.17. *Ningaui.* (a) *Ningaui yvonneae*. (b) Distribution of each of the three species. (c) Mallee habitat of *N. yvonneae* on the Eyre Peninsula, SA. [Photo credits: (a, c) Rohan Clarke.]

Fig. 2.18. *Planigale*. (a) Distribution of each of the five species. (b) Scree slope where *P. ingrami*, the long-tailed planigale shown in (c), was collected in Arnhem Land, NT, and (d) *P. ingrami* from the cracking clay soils of the Mitchell Grass Downs in Queensland (see Fig. 2.15c). [Photo credits: (b, d) P.A. Woolley; (c) P.A. Woolley and D. Walsh.]

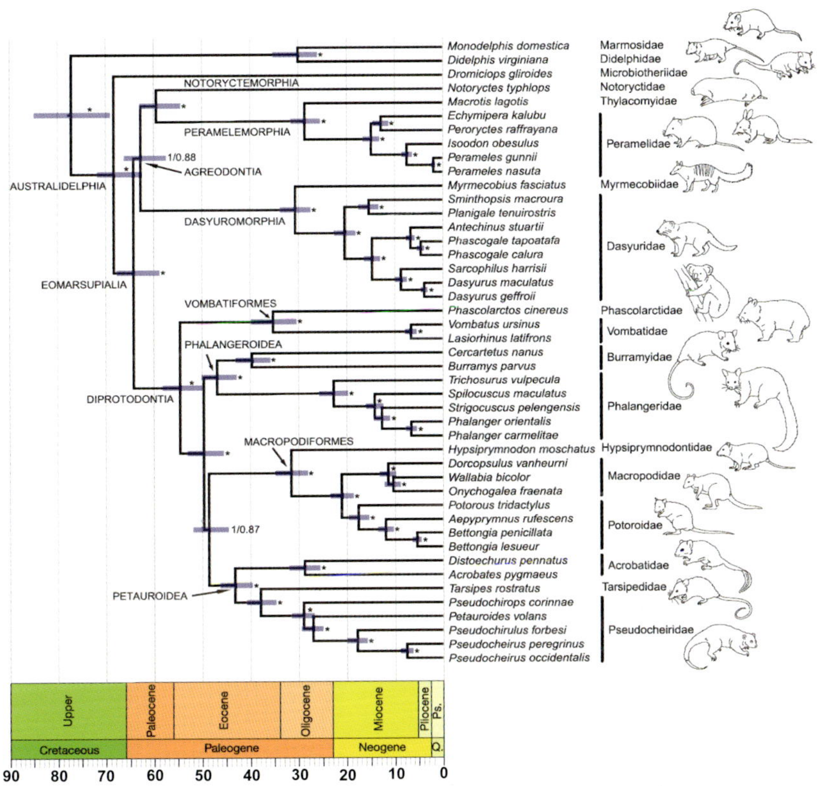

Fig. 5.5. Phylogeny and divergence times of marsupial families from Duchêne *et al.* (2018). Species tree estimated with ASTRAL-II from gene trees for 1550 loci. Divergence time estimates used the fossil calibration dates in Mitchell *et al.* (2014) and the Bayesian MCMCTREE program with a GTR + Γ substitution model and an uncorrelated lognormal relaxed-clock model. Blue bars at nodes represent 95% credibility intervals on divergence dates. [Reprinted by permission of Oxford University Press from Duchêne *et al.* (2018).]

Fig. 5.6. Dasyuromorphian phylogeny from García-Navas *et al.* (2020). Tree is based on partitioned (by locus) Bayesian analysis of the mitochondrial and nuclear DNA sequence alignment of Westerman, Krajewski *et al.* (2016) plus *Ω-globin*, *RUNX2* and *col2A1* loci (17 968 bp). Nodal values are posterior probabilities. [Redrawn with permission of John Wiley and Sons from García-Navas *et al.* (2020).]

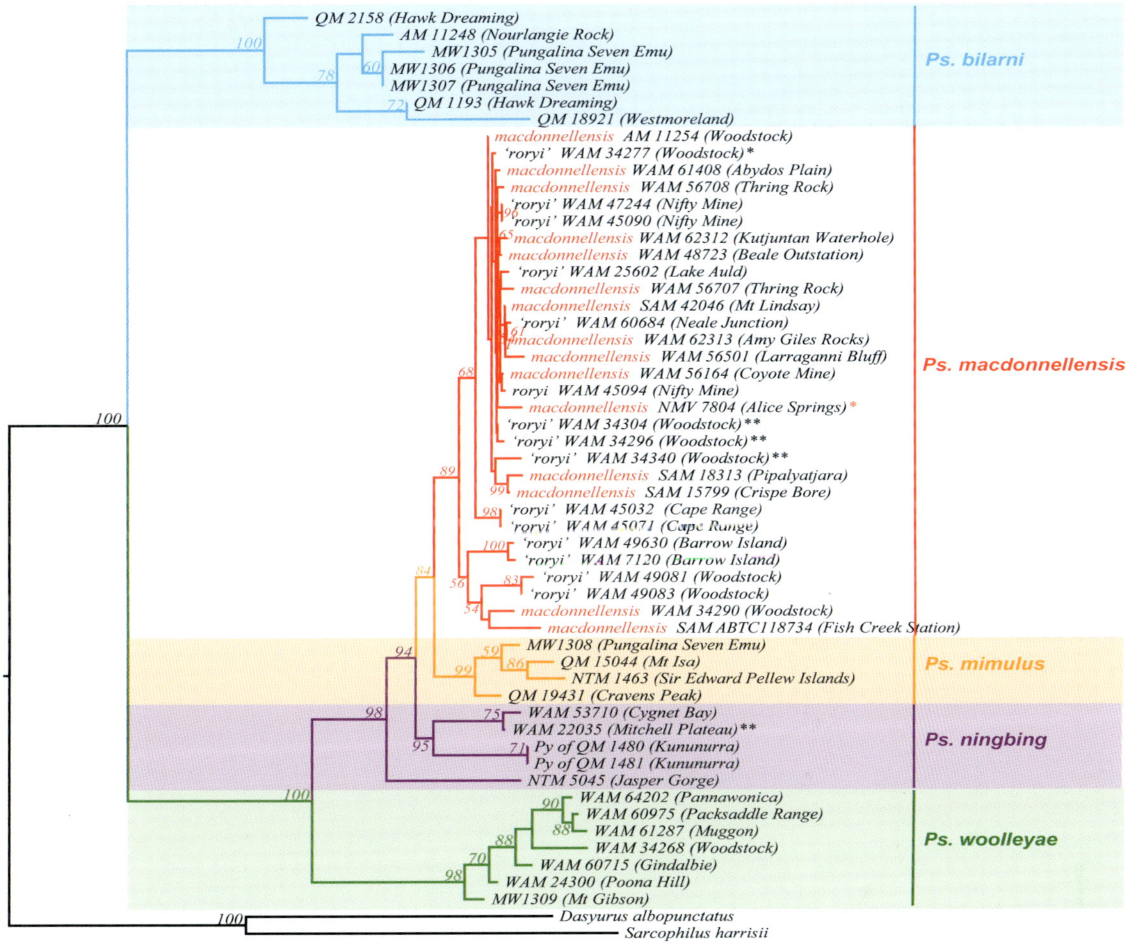

Fig. 6.11. Phylogeny of *Pseudantechinus* from Umbrello *et al.* (2017). Tree estimated by Bayesian analysis of concatenated *cytb*, *12S*, *16S*, *IRBP*, *bfib7* and *e-globin* DNA sequences; each locus was treated as a partition with distinct parameter values under the GTR + Γ + I model. Tip labels are specimen identification numbers with localities in parentheses. Asterisk (*) indicates sequence from a holotype specimen; double asterisks (**) indicate paratypes. Branch lengths are scaled to substitutions/site; nodal values are posterior probabilities × 100. [Reprinted by permission of CSIRO Publishing from Umbrello *et al.* (2017). Permission conveyed through Copyright Clearance Center, Inc.]

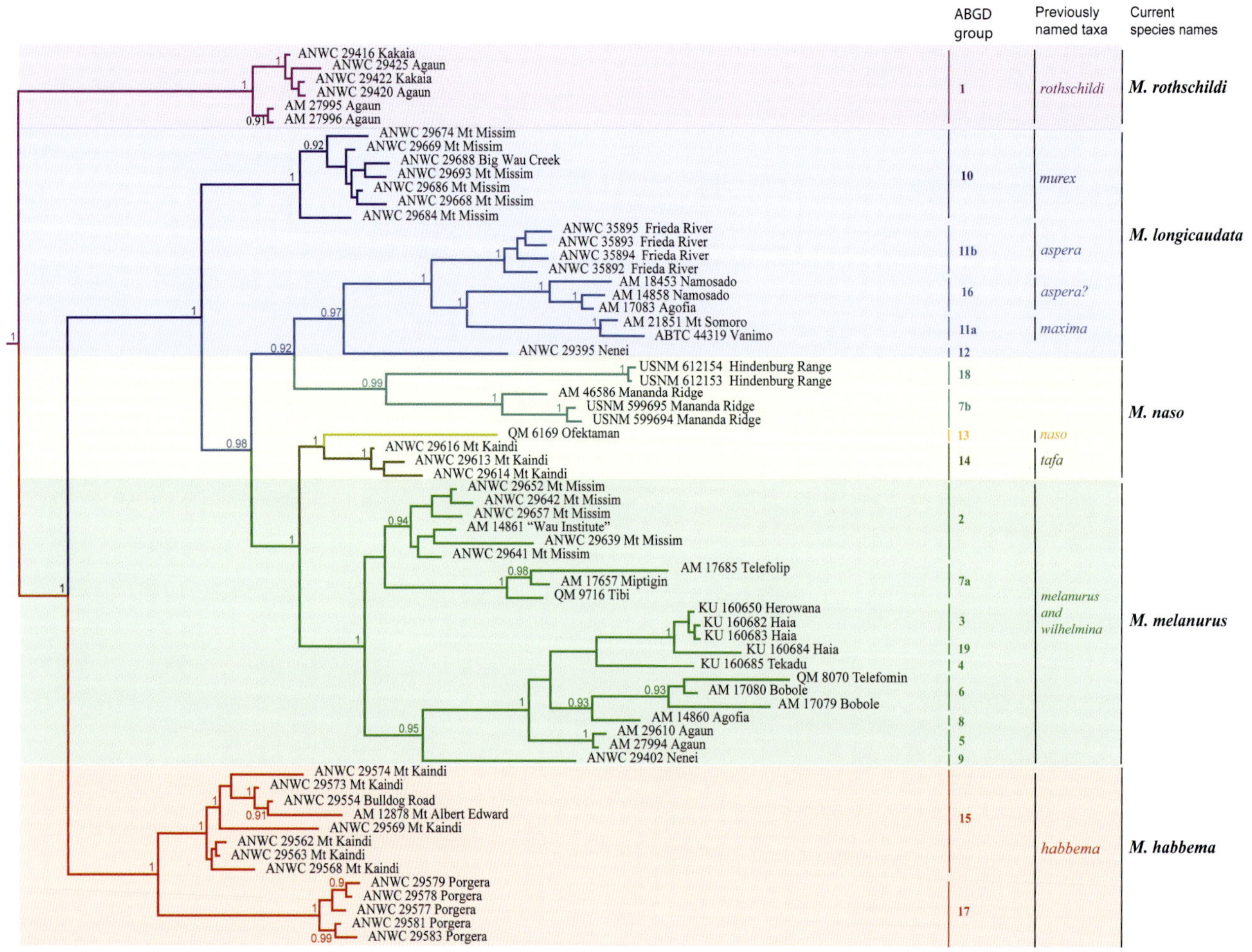

Fig. 6.18. Phylogeographic tree for *Murexia* specimens (Woolley *et al.* 2020). Tree estimated with Bayesian analysis of concatenated, partitioned *cytb, 12S, 16S, IRBP, bfib7 and e-globin* DNA sequences (ca. 12 kb). Tip labels are specimen numbers followed by collecting localities; nodal values are posterior probabilities; branch length scale is substitutions per site. 'ABGD group' shows haplotype clusters identified as potential species by the ABGD method. 'Previously named taxa' identifies tips that correspond to forms described in earlier taxonomic literature. 'Current species names' shows the assignments of specimens to the five species currently recognised. [Reprinted by permission of CSIRO Publishing from Woolley *et al.* (2020). Permission conveyed through Copyright Clearance Center, Inc.]

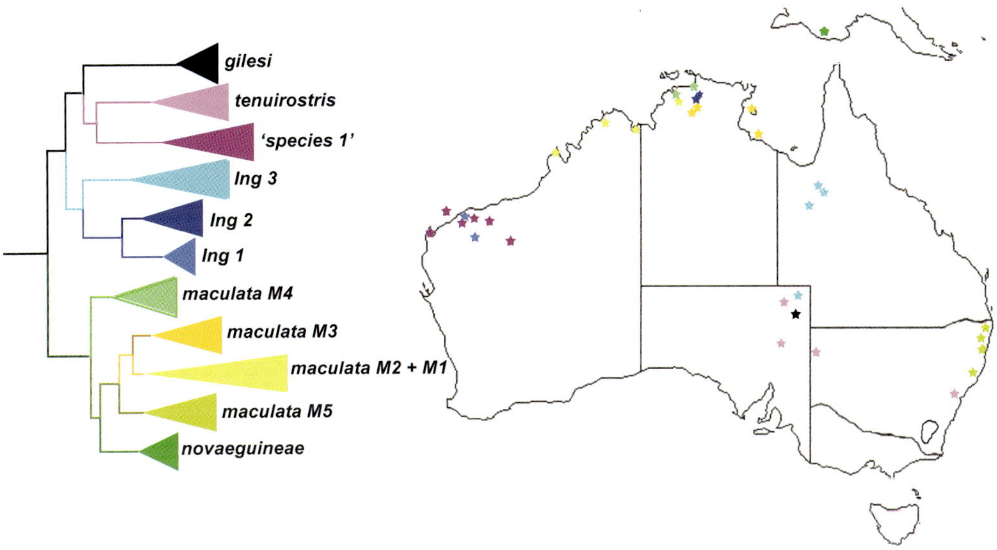

Fig. 6.29. Phylogeny of *Planigale* from Westerman, Blacket *et al.* (2016). The tree summarises Bayesian analyses of partitioned *cytb, 12S, 16S, CR, P1, IRBP, bfib7* and *e-globin* DNA sequences from multiple specimens representing each species. All clades on this tree have posterior probabilities ≥0.98 except that linking 'species 1' to *P. tenuirostris* (0.94). Each terminal triangle represents a resolved haplogroup. Stars on map correspond to collecting localities of specimens associated with each haplogroup. 'Ing' = *Planigale ingrami*. [Redrawn with permission of CSIRO Publishing from Westerman, Blacket *et al.* (2016). Permission conveyed through Copyright Clearance Center, Inc.]

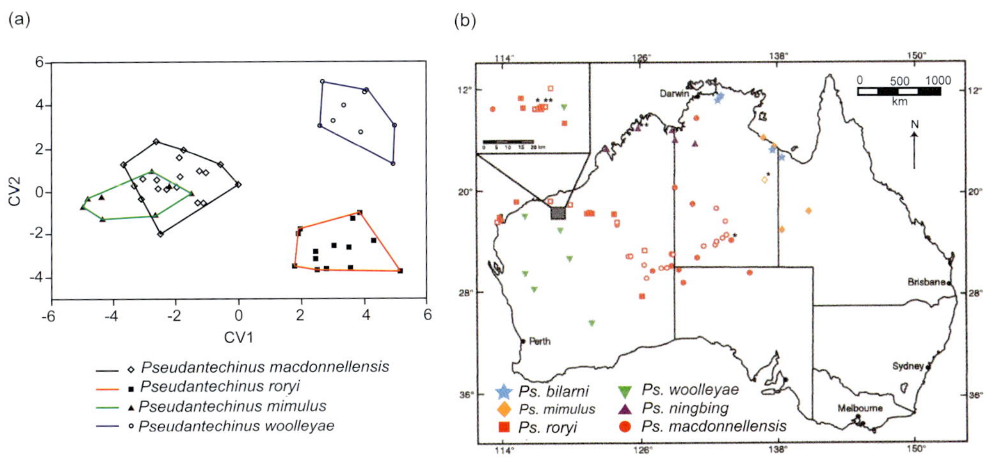

Fig. 7.10. (a) Discriminant analysis of *Pseudantechinus* exemplars based on five cranial characters statistically selected for best group separation from 19 total measurements (Cooper *et al.* 2000a). (b) *Pseudantechinus* collecting localities from Umbrello *et al.* (2017) that outline the distributions of five currently recognised species and *P. 'roryi'*. Open symbols are localities from Cooper *et al.* (2000a). [(a) Redrawn with permission of the Western Australian Museum from Cooper *et al.* (2000a). (b) Reproduced with permission of CSIRO Publishing from Umbrello *et al.* (2017), with minor adaptations. Permission conveyed through Copyright Clearance Center, Inc.]

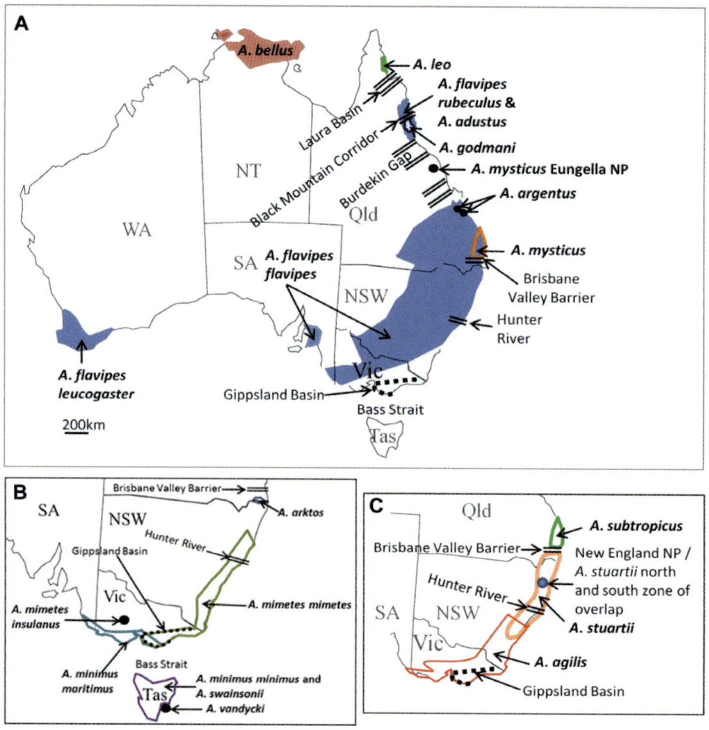

Fig. 7.12. Geographic distributions of currently recognised *Antechinus* species (Mutton *et al.* 2019). [Reprinted by permission of Oxford University Press from Mutton *et al.* (2019).]

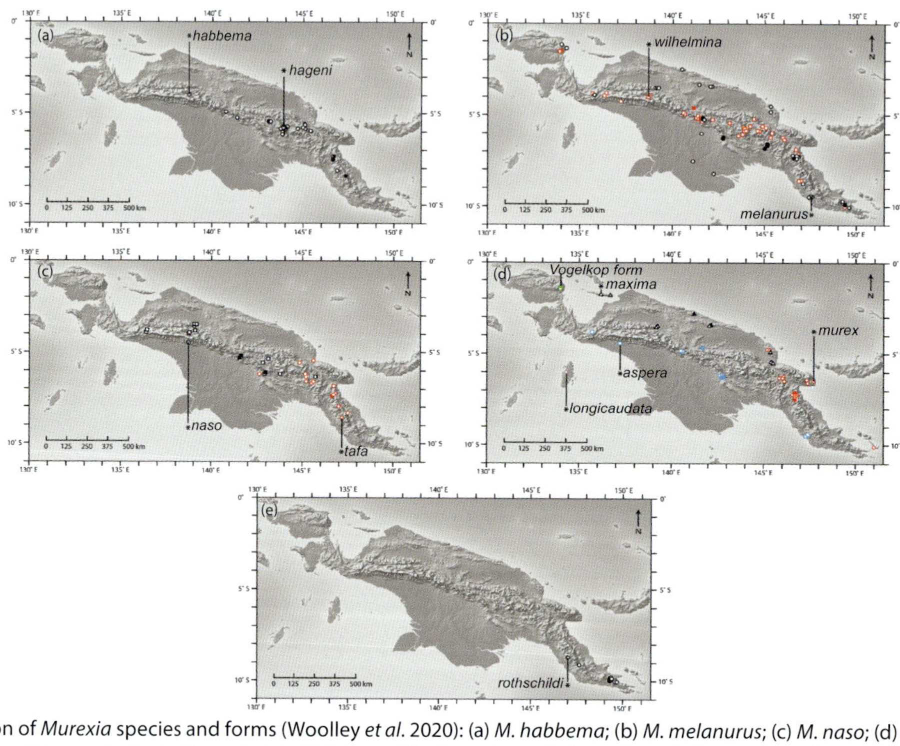

Fig. 7.13. Distribution of *Murexia* species and forms (Woolley *et al.* 2020): (a) *M. habbema*; (b) *M. melanurus*; (c) *M. naso*; (d) *M. longicaudata*; (e) *M. rothschildi*. Type localities are indicated by asterisks (*) followed by form name. Coloured symbols correspond to named forms. Solid symbols show localities of specimens included in molecular analyses. [Reprinted by permission of CSIRO Publishing from Woolley *et al.* (2020). Permission conveyed through Copyright Clearance Center, Inc.]

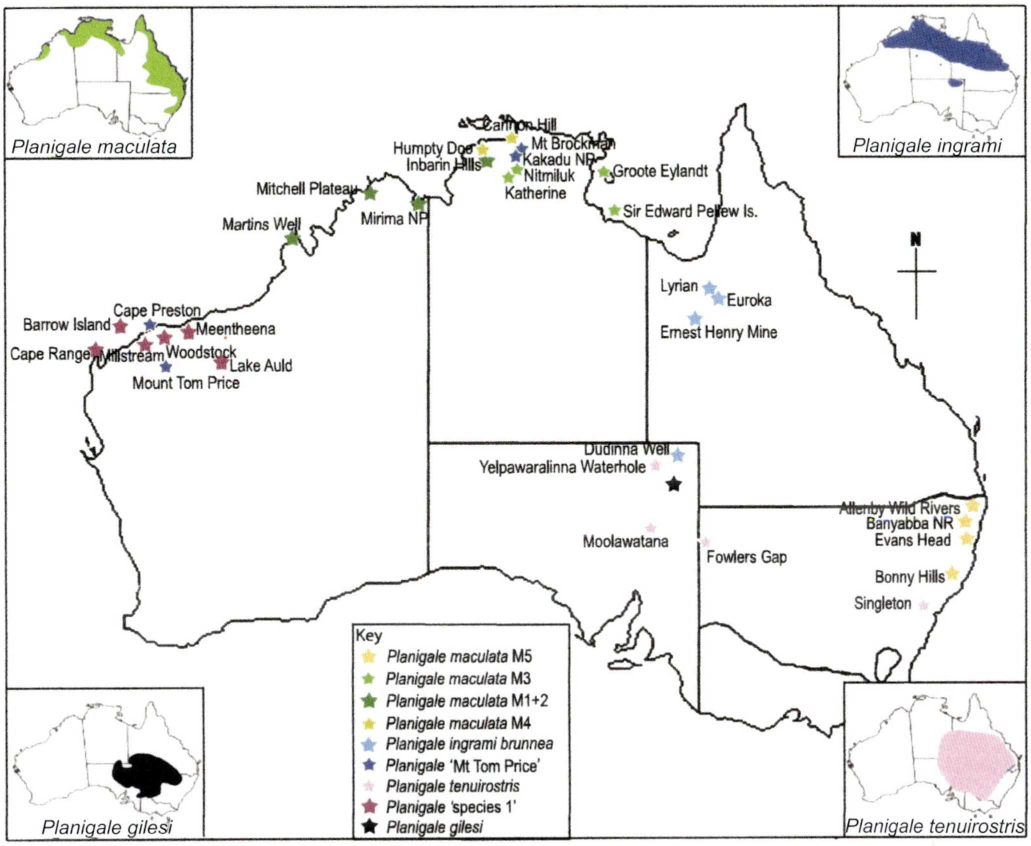

Fig. 7.20. Collecting localities that map the geographic distribution of *Planigale* clades shown in Fig. 7.19 from Westerman, Blacket *et al.* (2016). Inset maps show distributions of currently recognised species. [Reprinted by permission of CSIRO Publishing from Westerman, Blacket *et al.* (2016). Permission conveyed through Copyright Clearance Center, Inc.]

(a)

Fig. 8.1. (a) The numbat *Myrmecobius fasciatus*.

Fig. 8.2. The Tasmanian thylacine, *Thylacinus cynocephalus*. [Colourised version of image dated 1933 by an unknown photographer, from Wikimedia Commons.]

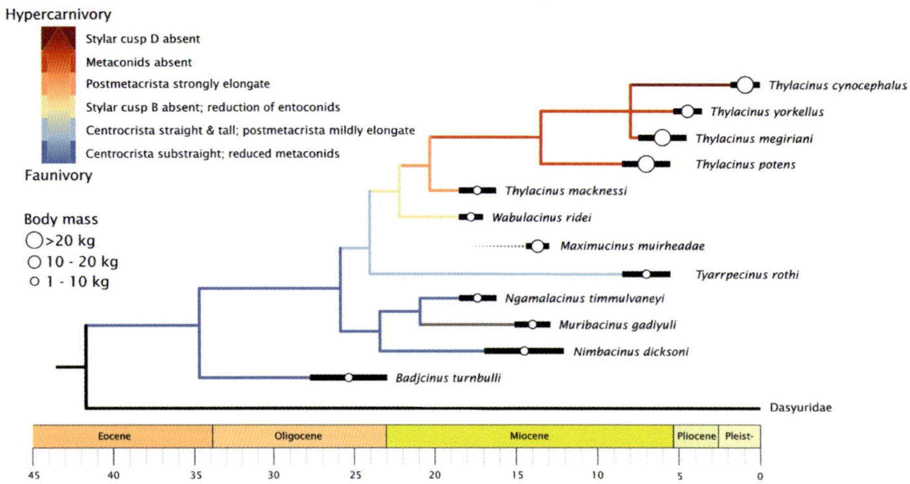

Fig. 8.7. Time tree of thylacinids estimated with Bayesian analysis of 80 morphological characters and *12S* DNA sequences (for *T. cynocephalus* and selected extant outgroups, not shown). Tip dating employed a gamma-rates clock model and fossilised birth-death process model. [Reprinted from Rovinsky *et al.* (2019), licensed under CC BY 4.0.]

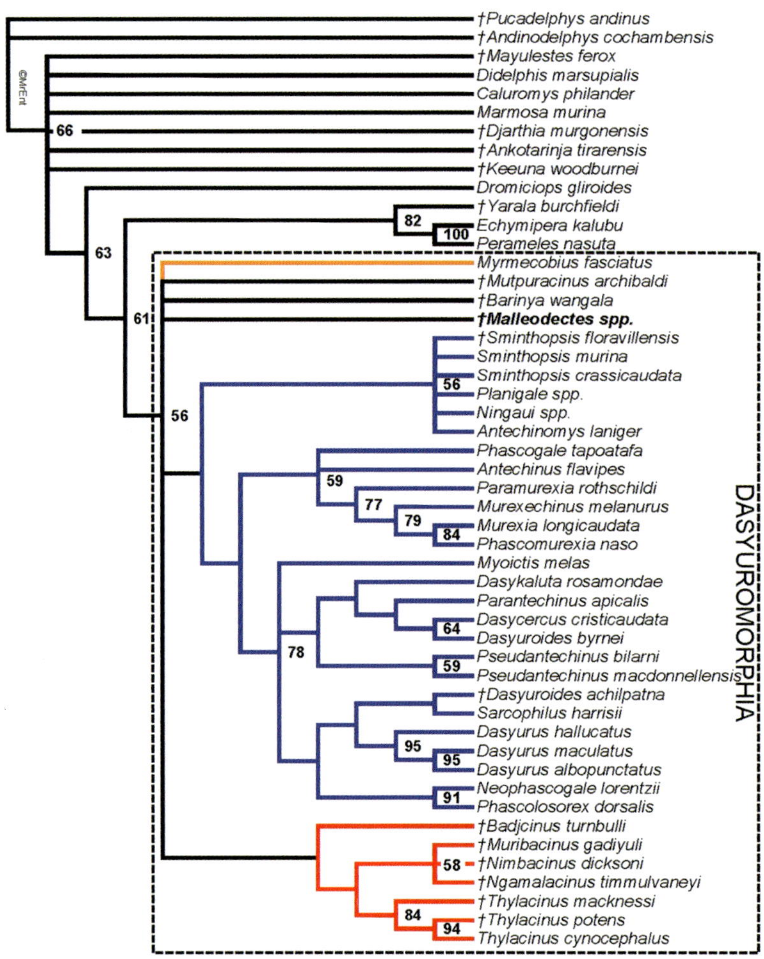

Fig. 8.9. Phylogeny of *Malleodectes*. Tree is a strict consensus of 18 619 minimum-length trees for 62 craniodental characters, with relationships among extant taxa constrained to those recovered from molecular analyses by Mitchell *et al.* (2014), Westerman, Krajewski *et al.* (2016) and Jansa *et al.* (2014). Coloured branches indicate recognised families Myrmecobiidae (orange), Dasyuridae (blue) and Thylacinidae (red). Nodal values are bootstrap percentages ≥50%. [Reprinted from Archer, Hand *et al.* (2016) under CC BY 4.0 International License.]

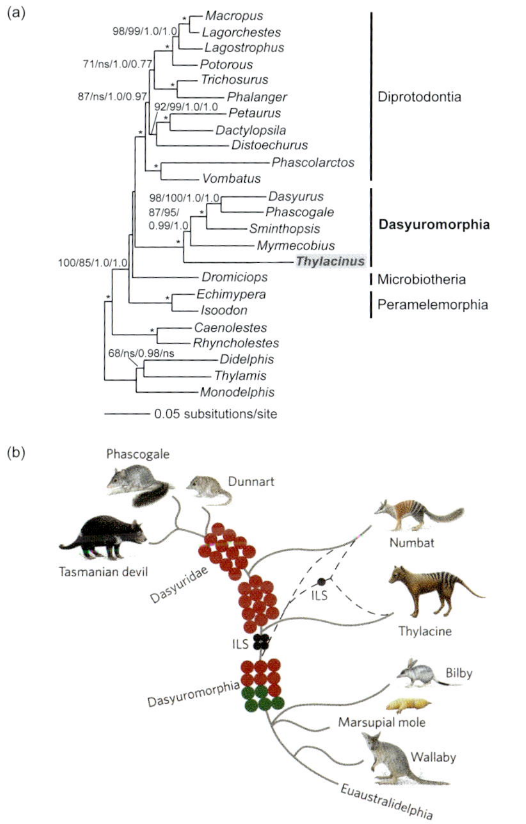

Fig. 8.26. Genomic resolution of dasyuromorphian family relationships. (a) Maximum likelihood tree based on complete mitochondrial genomes from Miller *et al.* (2009). Analysis based on GTR +I + Γ model with 'reduced bias coding' (i.e. omitting third positions of protein-coding genes, considering transversions only in RNA genes and transitions only at first positions of protein-coding genes). Nodal values are (left to right): bootstrap percentage with reduced bias coding, bootstrap percentage with raw sequences, posterior probability with reduced bias coding, and posterior probability with raw sequences. Asterisks (*) indicate branches with bootstrap = 100 and posterior probability = 1.0. (b) Retroposon phylogeny from Feigin *et al.* (2018). Circles indicate shared insertions of retroposons WSINE1/1a (red) and WALLSI1A (green). ILS indicates inferred instances of incomplete lineage sorting suggesting relationships discordant with most insertions. [(a) Redrawn from Miller *et al.* (2009) under CC BY 4.0 license. (b) Reprinted from Feigin *et al.* (2018) under CC BY 4.0 license.]

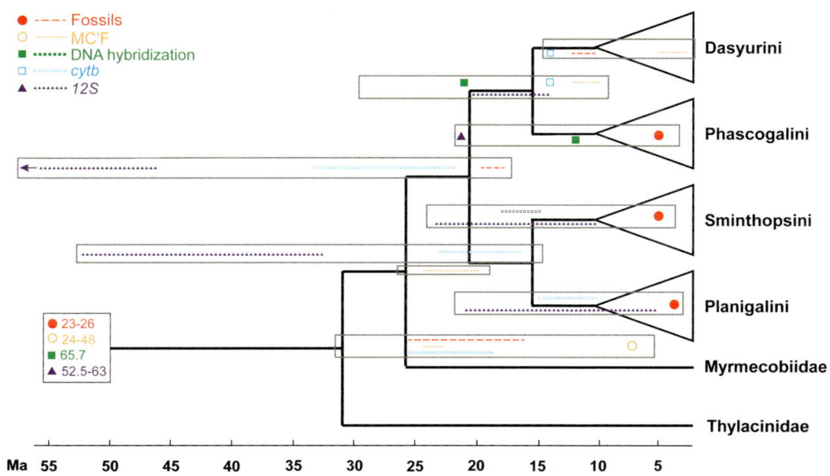

Fig. 9.2. Estimates of divergence times made prior to 2000. Each box encloses a phylogenetic split and its estimated dates (dots and lines). Fossil dates are earliest clade occurrences. Dates at the root are estimates for the origin of Dasyuromorphia (including stem lineages). Apparent dates of branch points on the tree itself have no significance. See text for description of sources.

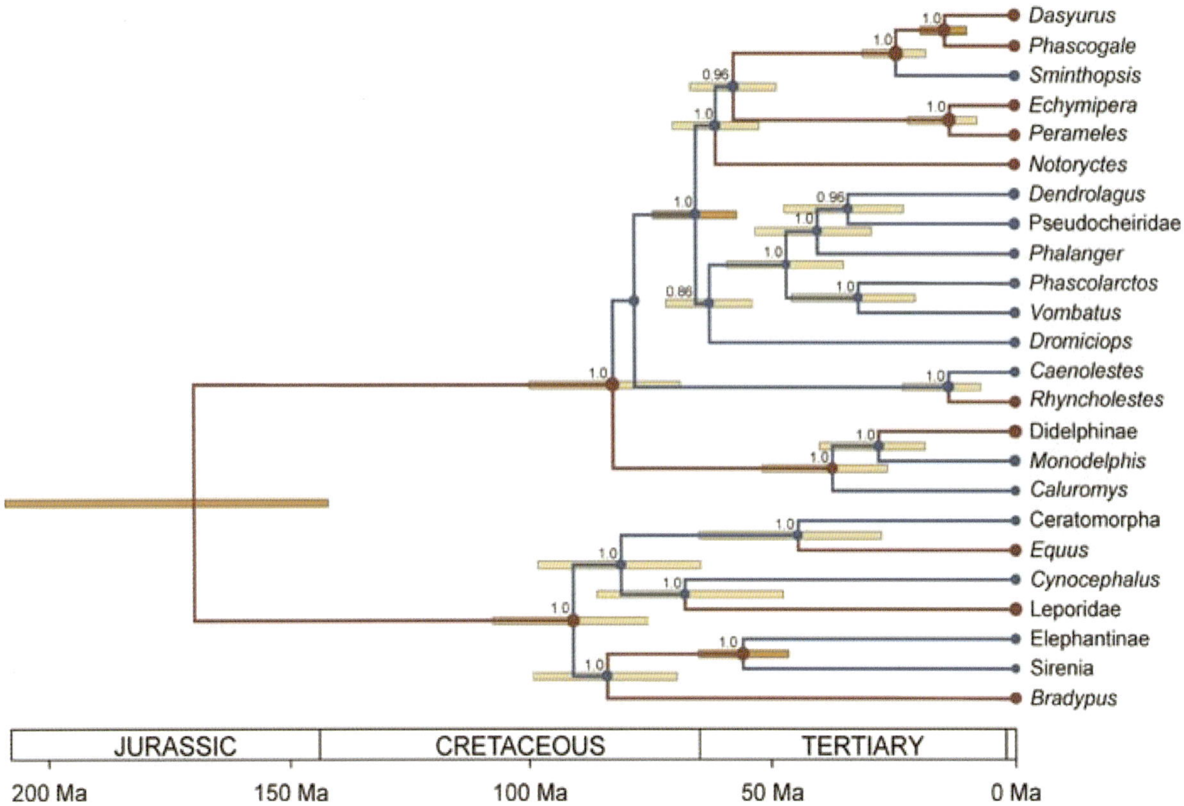

Fig. 9.8. Marsupial time tree based on DNA sequences of *APOB*, *RAG1*, *IRBP*, *vWF* and *BRCA1* (Drummond *et al.* 2006). Topology and dates estimated with BEAST 1.4 assuming an unpartitioned GTR + Γ model and uncorrelated log-normal rate distributions among branches. Nodal values are posterior probabilities >0.5. Yellow bars are 95% highest probability densities for divergence dates. Orange bars show nodes with normal calibration priors. Red branches have above-average, and blue branches below-average, rates. [Reprinted from Drummond *et al.* (2006) under CC BY license.]

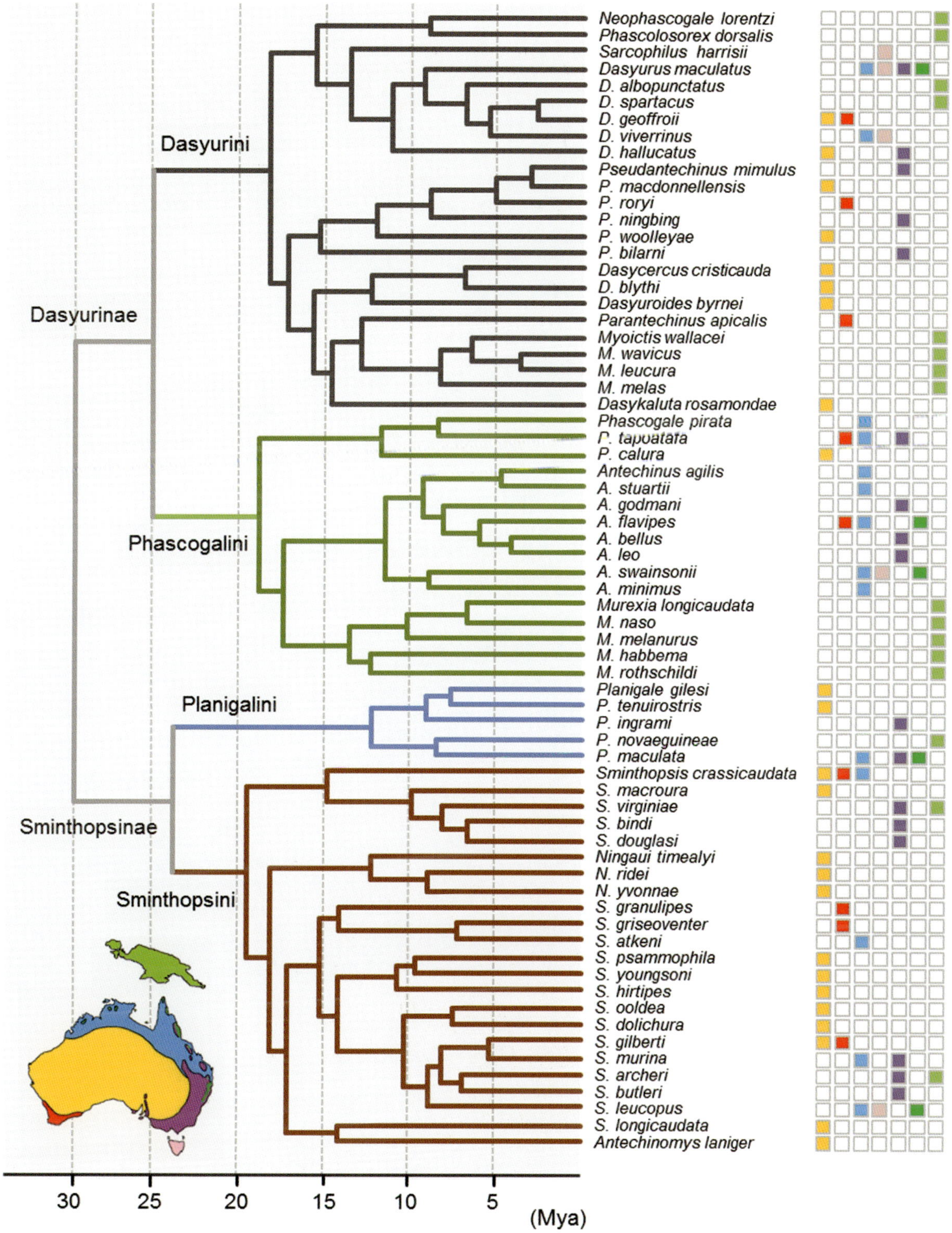

Fig. 9.12. Dasyurid time tree with mapping of species by geographic region (García-Navas *et al.* 2018). Branches are coloured by tribe. Biogeographic regions inhabited by species are shown in the grid at right: orange = Eremean; red = south-west; purple = Bassian; pink = Tasmania; blue = Torresian; dark green = Irian; light green = New Guinea. [Reprinted by permission of Oxford University Press from García-Navas *et al.* (2018).]

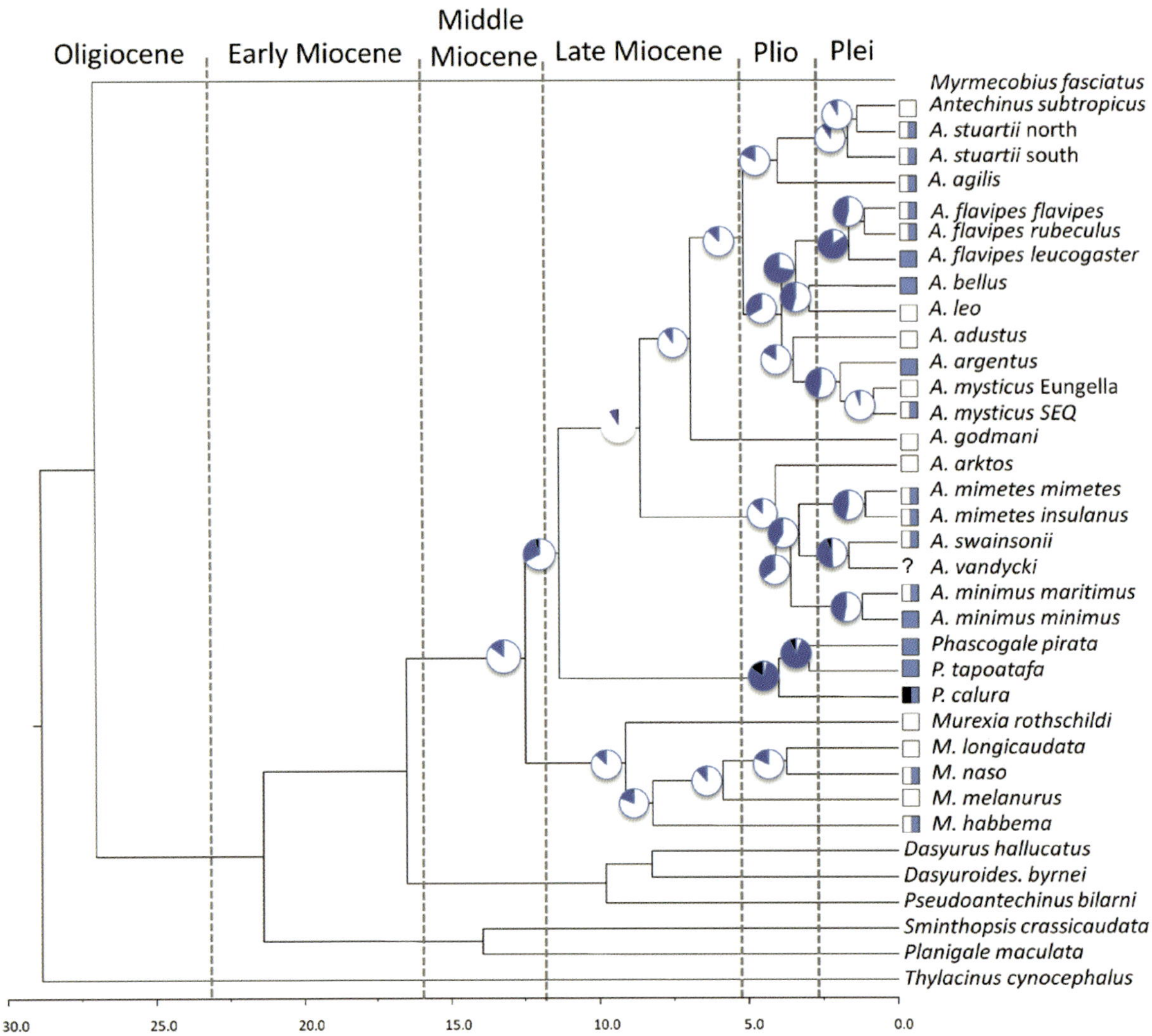

Fig. 9.13. Time tree of Mutton *et al.* (2019). Results based on analysis of two mitochondrial (*cytb, 16S*) and four nuclear (*IRBP, ApoB, vWF, BRCA1*) gene segments with a UCLN rates model in BEAST, using fossil age constraints from Westerman, Krajewski *et al.* (2016). Squares at tips are habitat codes for current species, circles on internodes are pie charts of ancestral habitat probabilities estimated with BayesTraits. White = wet-closed habitat; blue = wet-open habitat; black = arid habitat. [Reprinted by permission of Oxford University Press from Mutton *et al.* (2019).]

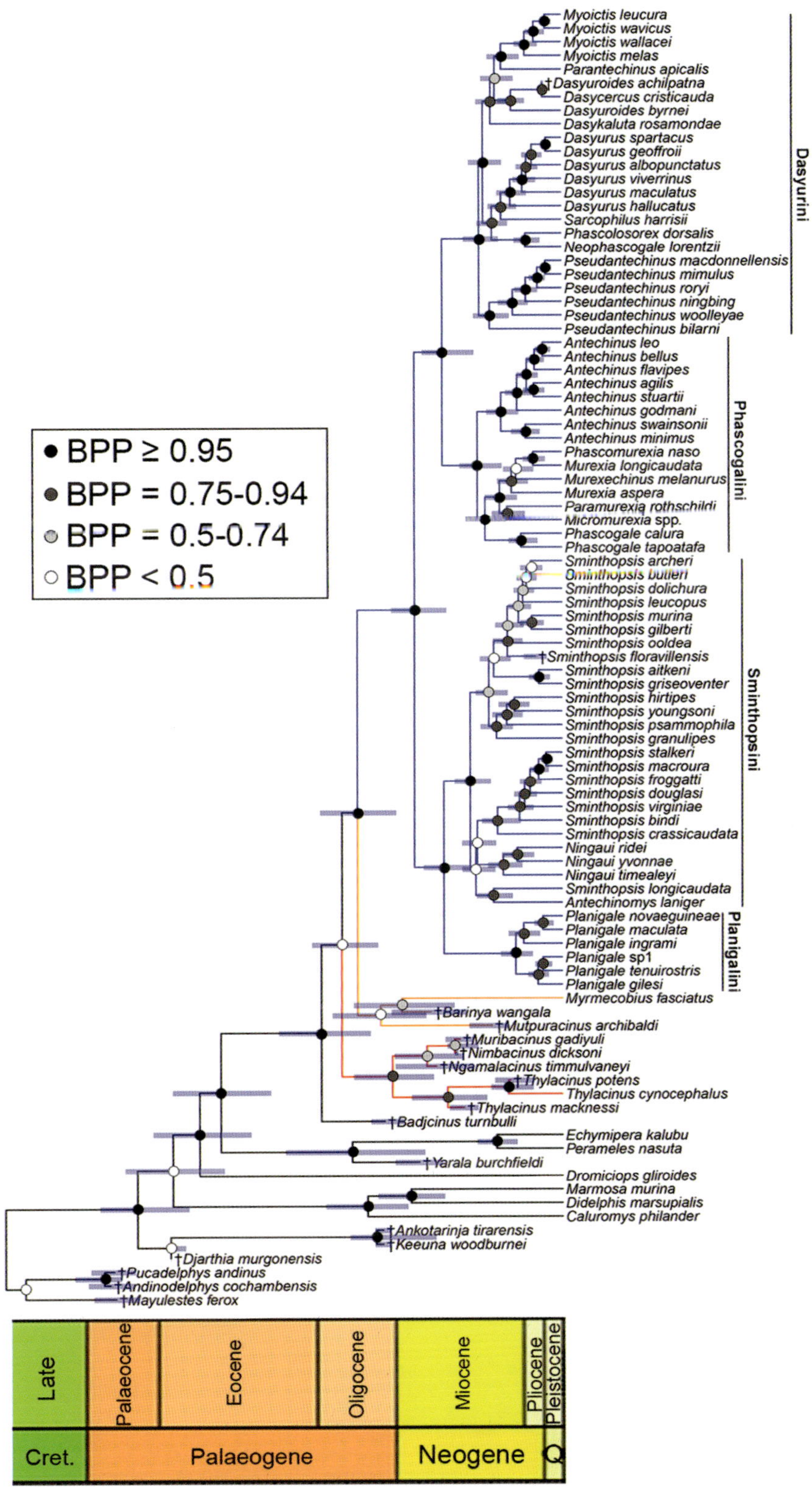

Fig. 9.14. Dasyuromorphian time tree from Kealy and Beck (2017). Tree and divergence times obtained with Bayesian inference (MrBayes analysis of 16.4 kb of DNA sequence data and 173 morphological characters) using tip and node dating with IGR clock model and 'TipNodeCalib1' calibration priors (see text for explanation). Bars at nodes are 95% HPDs. Branch colour coding: orange = Myrmecobiidae; red = Thylacinidae; blue = Dasyuridae. [Reprinted from Kealy & Beck (2017) under CC BY 4.0 license.]

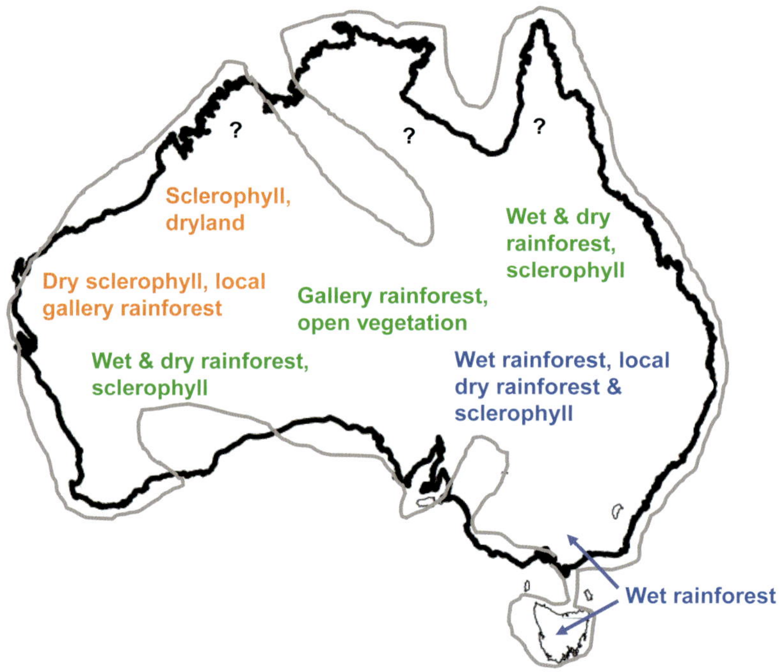

Fig. 10.2. Distribution of Australian vegetation at the Oligocene-Miocene boundary, 23 Ma (from Crisp and Cook 2013). Grey line represents the approximate early Miocene coastline. Text colour shows moisture gradient from drier (orange) to moister (blue); ? = unknown. [Adapted from Crisp and Cook (2013).]

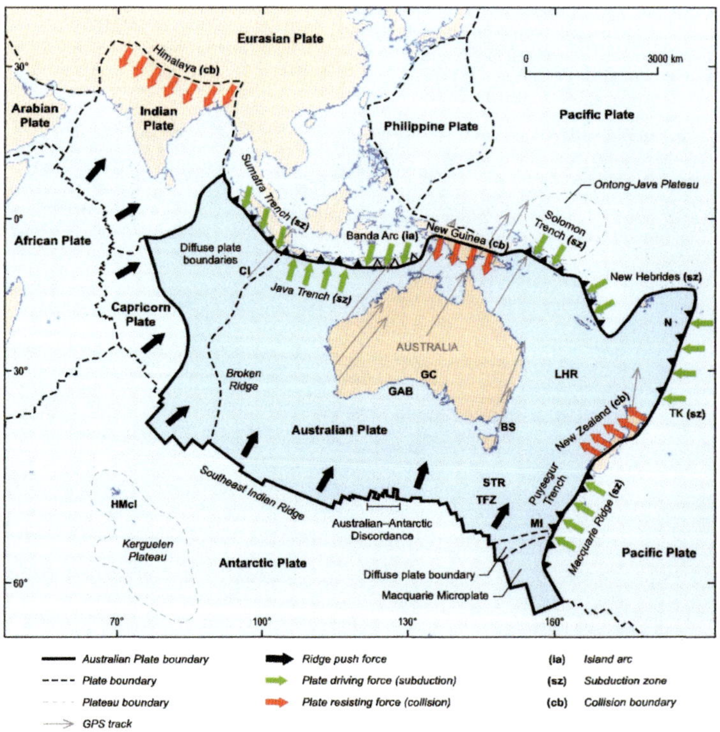

Fig. 10.3. Map of the Australian Plate and its boundaries. Abbreviations: BS = Bass Strait; GAB = Great Australian Bight; GC = Gawler Craton; LHR = Lord Howe Rise; HMcI = Heard and McDonald Islands; MI = Macquarie Island; STR = South Tasman Rise; TFZ = Tasman Fault Zone. [Reprinted by permission of ANU Press from Blewett (2012).]

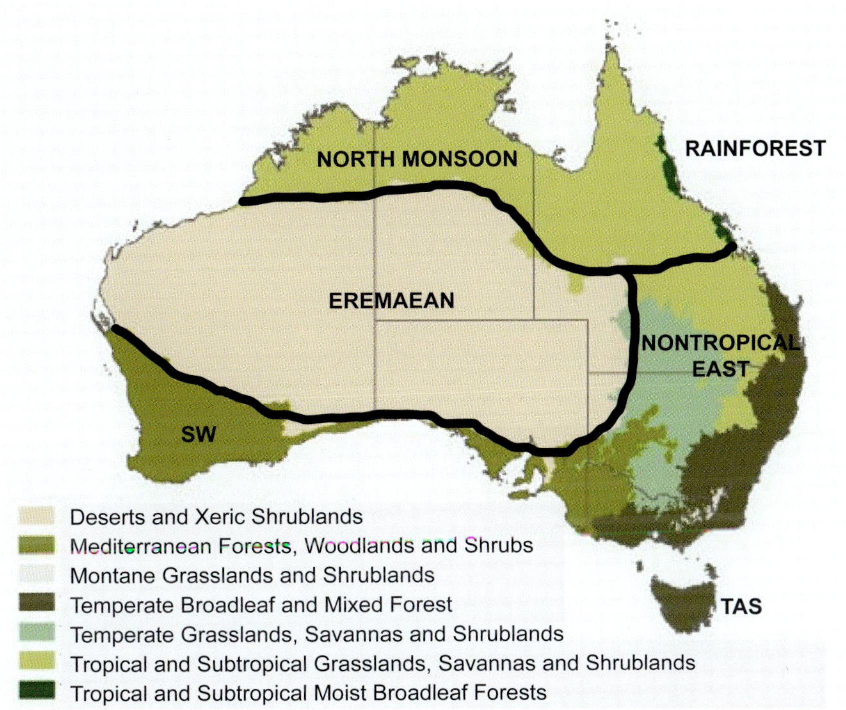

Fig. 10.4. Ecoregions defined by Johnson and Briggs (1975) based on distribution of eucalypts are delimited by bold lines and all-capital labels on a map of ecoregions recognised by the Australian government's National Reserve System. [NRS map reprinted from Australian Department of Sustainability, Environment, Water, Populations and Communities under CC BY-NC 3.0 licence.]

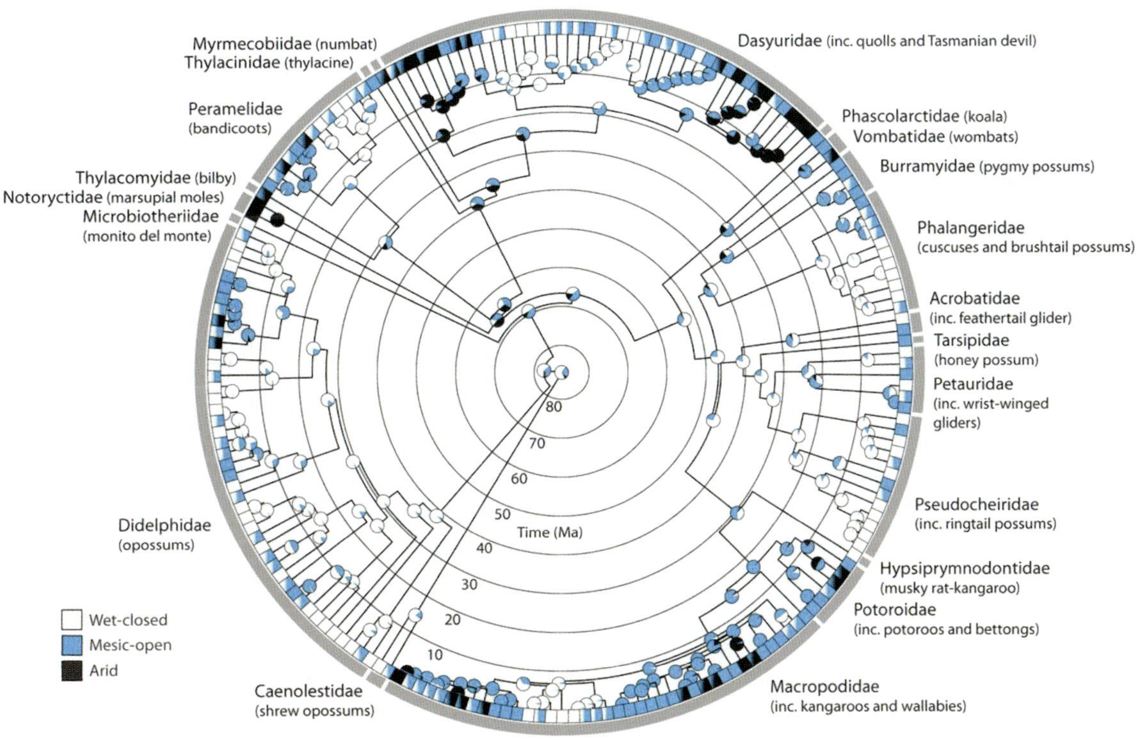

Fig. 10.6. Ancestral habitat reconstruction of marsupials from Mitchell *et al.* (2014). Ancestral states and their probabilities (pie charts) estimated with BayesTraits v1.0 software implementing the method of Pagel *et al.* (2004). [Reprinted by permission of Oxford University Press from Mitchell *et al.* (2014).]

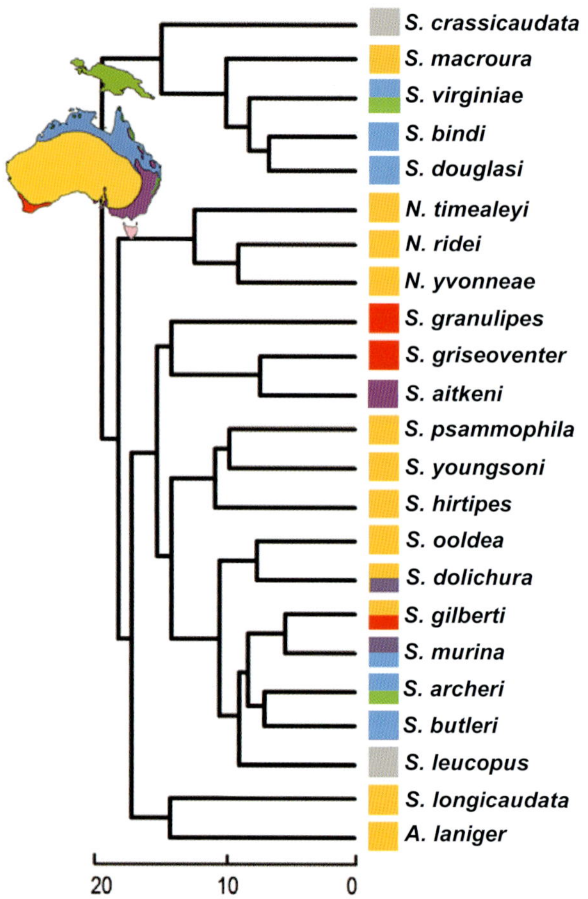

Fig. 10.7. Time tree of Sminthopsini with biogeographic regions of species indicated by colour-coded boxes at tips (García-Navas and Westerman 2018). Orange = Eremaean or arid biome; red = South-western mesic biome; blue = Torresian region or monsoonal tropics; purple = Bassian region; green = New Guinea; pink = Tasmania. [Reproduced with permission of John Wiley and Sons from García-Navas and Westerman (2018), with minor adaptations.]

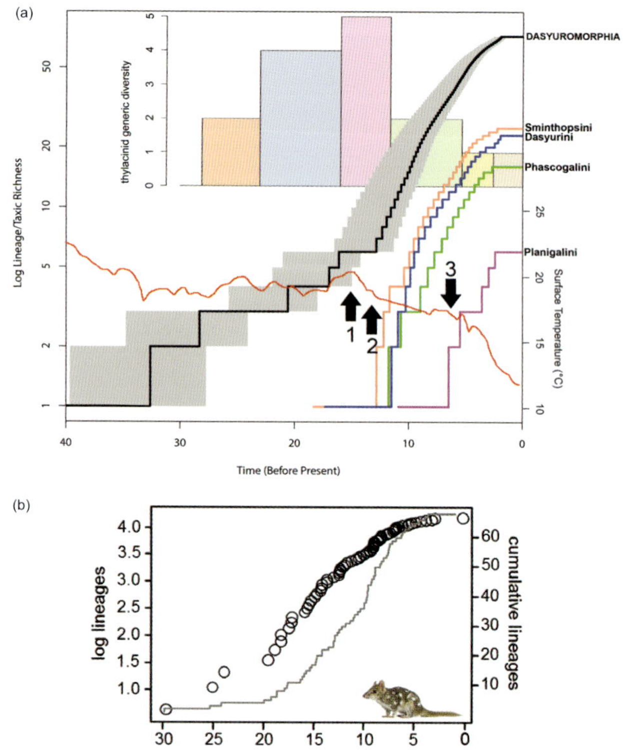

Fig. 10.9. Lineage through time (LTT) plots for dasyurids. (a) From Kealy and Beck (2017), based on their 'TipNodeCalib1' time tree. Separate plots are shown for dasyuromorphians (black curve with grey area representing 95% confidence interval) and each dasyurid tribe. The red curve shows estimated global surface temperatures. Inset histogram shows the number of thylacinid genera from the fossil record in discrete time blocks on the same time axis as the LTT plots. Arrows: 1 = Middle Miocene Climatic Optimum; 2 = Middle Miocene Climatic Transition; 3 = increased frequency of fossil grass pollen in Australia. (b) From García-Navas *et al.* (2018), based on the time tree of Westerman, Krajewski *et al.* (2016). Dots are lineage counts at specific time points and the grey curve shows the cumulative number of species. [(a) Reprinted from Kealy and Beck (2017) under CC BY 4.0 license. (b) Reprinted by permission of Oxford University Press from García-Navas *et al.* (2018).]

Fig. 10.10. Simplified geological reconstructions of New Guinea showing the four major terranes accreted to the Australian Craton since the early Miocene. Mountain ranges are indicated, with the Owen Stanley Ranges acting as a barrier between lowlands to their north-east and south-east and the Central Cordillera acting as a major barrier between regions to the north and south. [Adapted from Tallowin *et al.* (2018).]

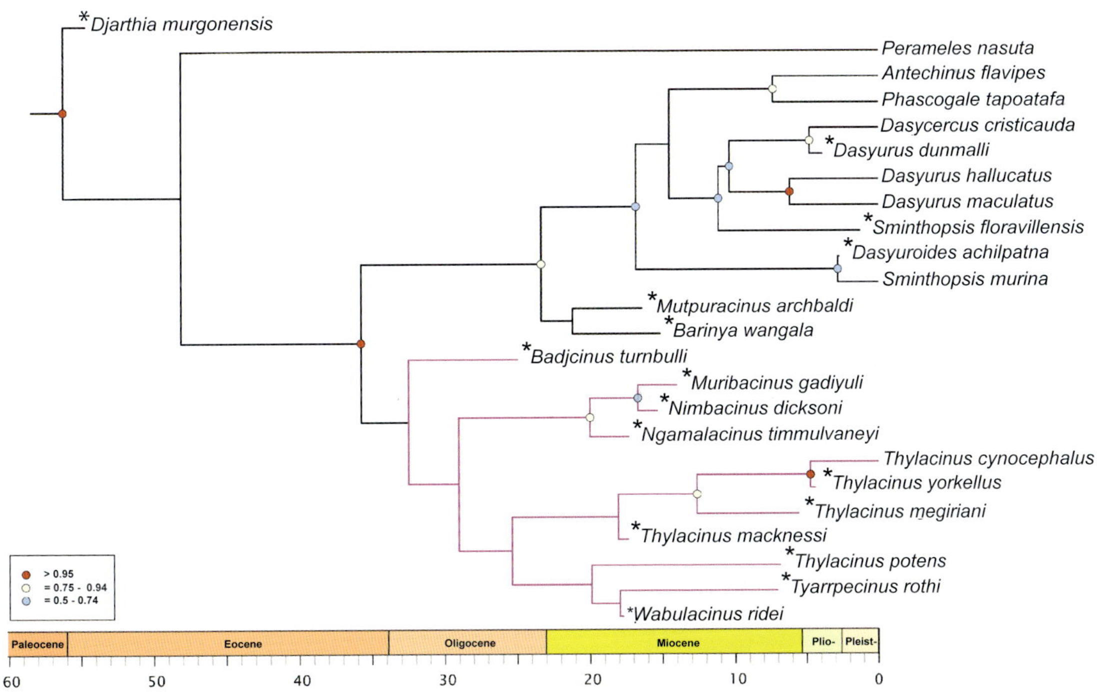

Fig. 11.10. Tip-dated Bayesian phylogeny of dasyuromorphians based on morphological (80 characters) and *12S* DNA sequence (853 bp) data from Rovinsky *et al.* (2019). Asterisks indicate extinct species. [Reprinted with changes from Rovinsky *et al.* (2019) under CC BY 4.0 license.]

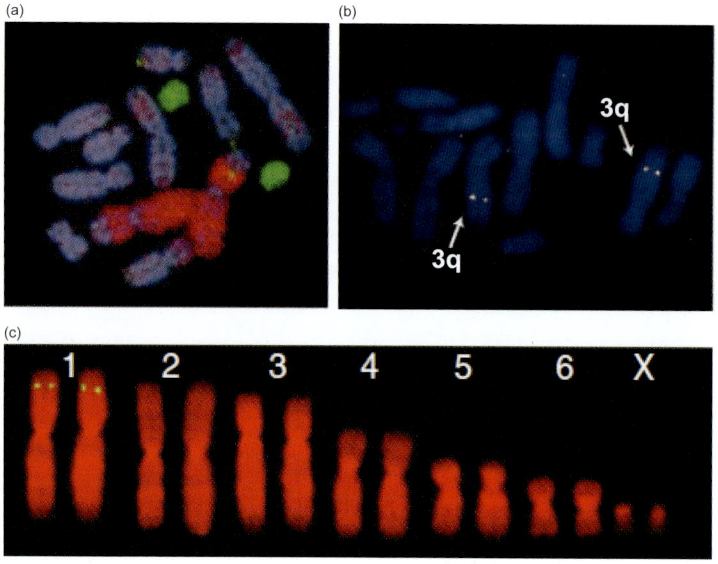

Fig. 11.26. Cross-species chromosome painting in dasyurid marsupials. (a) *Sminthopsis crassicaudata* female chromosomes painted with probes of *Trichosurus vulpecula* chromosome 1 (red) and X (green). (b) Localisation of beta globin gene on long arm of chromosome 3 (3q) in *Sminthopsis crassicaudata.* (c) Localisation of a marker gene to chromosome 1 of *Sarcophilus harrisii.* [(a) Reprinted by permission of Springer Nature from Rens *et al.* (1999). (b) Reprinted by permission of Karger Publishers from De Leo *et al.* (2005). (c) Reprinted by permission of Surrey Beatty & Sons from Potter and Deakin (2018). Permission conveyed through Copyright Clearance Center, Inc.]

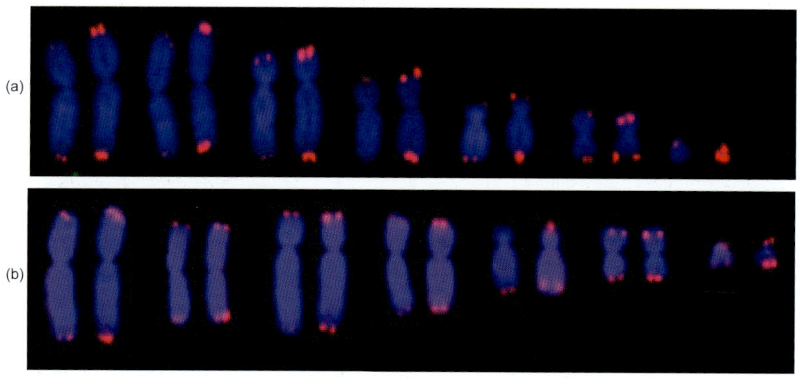

Fig. 11.27. Differential size of telomeres on maternally and paternally derived homologues of (a) *Sarcophilus harrisii* and (b) *Dasyurus hallucatus*. The maternal homologue is on the left in each pair. [(a) Reprinted from Bender *et al.* (2012) under CC BY license. (b) Reprinted from Ingles and Deakin (2016) under CC BY 4.0 license.]

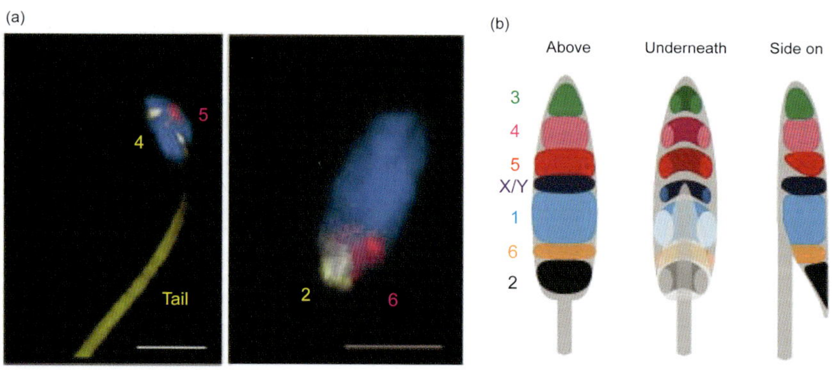

Fig. 11.30. Non-random positioning of chromosomes in *Sminthopsis crassicaudata* sperm heads (Greaves *et al.* 2001). (a) Two-colour chromosome painting shows positions of chromosomes 2, 4, 5 and 6. (b) Diagram showing arrangement of all chromosomes. [Reprinted by permission of Springer Nature from Greaves *et al.* (2001).]

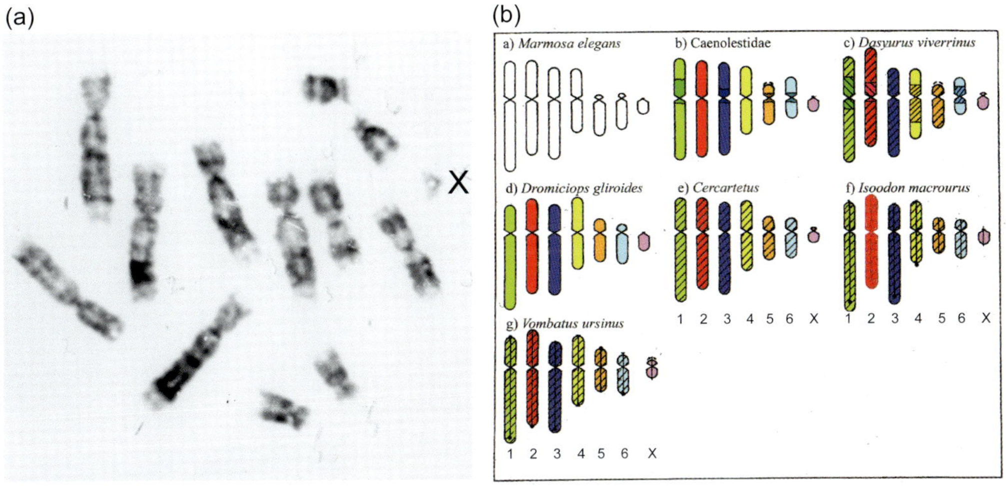

Fig. 11.31. Relationships between superficially similar 2n = 14 karyotypes in various marsupial groups. (a) G-banded karyotype of male *Dromiciops gliroides* (minute Y chromosome not visible). (b) Probable relationships of karyotypes of members of major marsupial families. Cross-hatched chromosomes have similar banding patterns to other cross-hatched chromosomes of the same number. [(a) Photo credit: Mike Westerman. (b) Redrawn with permission of the Zoological Society of New South Wales from Sharman (1982).]

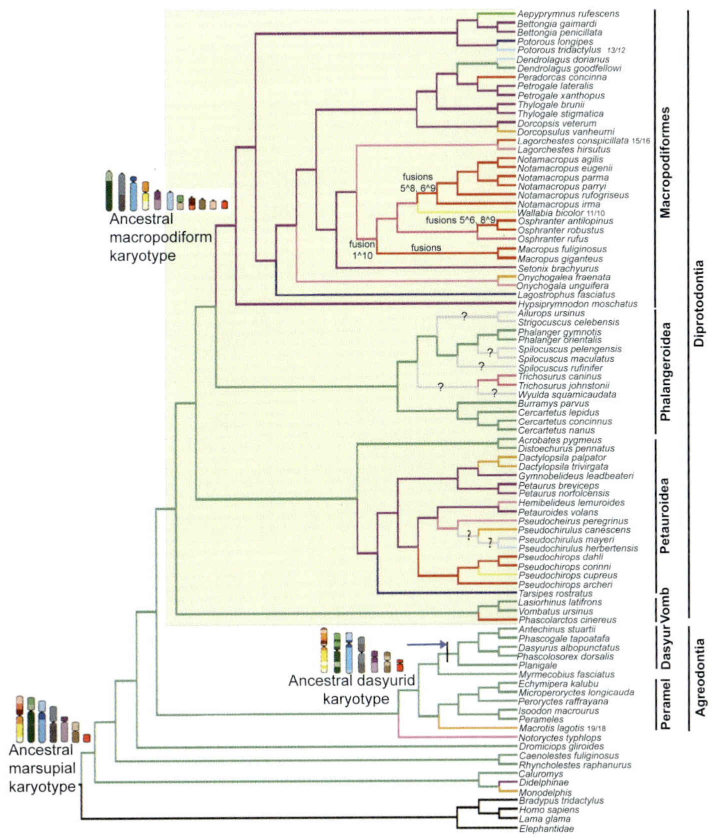

Fig. 11.32. Evolution of chromosome number in marsupials as inferred by Westerman *et al.* (2010). Diploid numbers for species of marsupials are plotted on a molecular phylogeny. Branch colour coding: yellow 2n = 10–11; dark green 2n = 14; red 2n = 16; pink 2n = 18; purple 2n = 22; blue 2n = 24; light green 2n = 32. Chromosome fusions (^) inferred from G-banding studies. Ancestral arrangements of conserved chromosome blocks (C1–20) are shown for ancestral marsupial, macropodiform and dasyurid branches. [Reprinted by Permission of Oxford University Press from Westerman *et al.* (2010).]

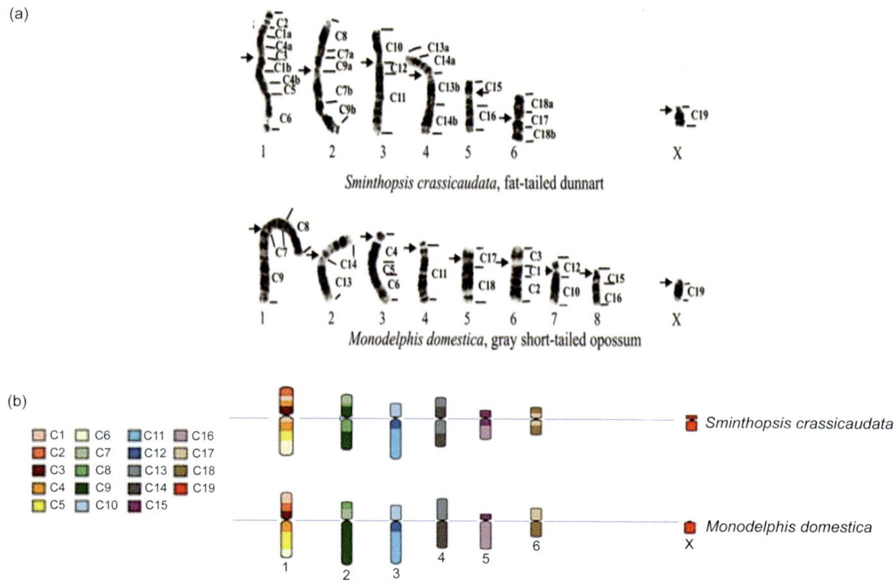

Fig. 11.33. Conserved blocks of genomic material in marsupial chromosomes. (a) Conserved blocks C1–19 in the G-banded chromosomes of *Sminthopsis crassicaudata* (2n = 14) and *Monodelphis domestica* (2n = 18). Arrows indicate centromere location. Block C20 is on the Y chromosome and not shown here. (b) Stylised colouring to show the arrangement of blocks in (a). [(a) Reprinted by permission of S. Karger AG from Rens *et al.* (2003). (b) Reprinted by permission of Springer Nature from Deakin and Kruger-Andrzejewska (2016).]

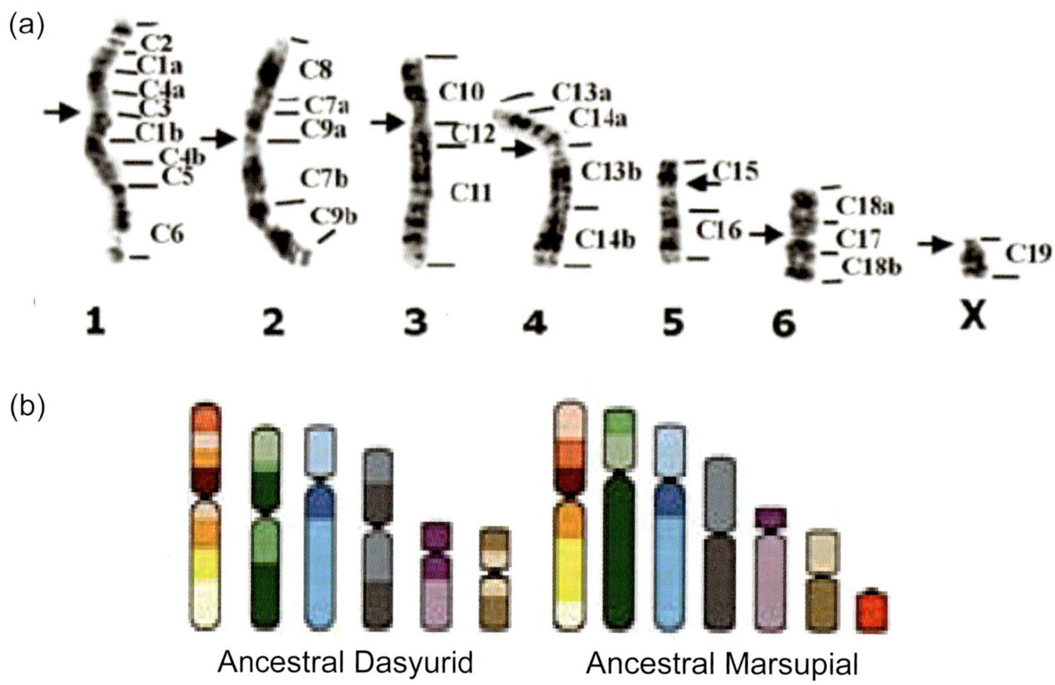

Fig. 11.34. Dasyurid karyotype showing (a) G-band pattern in terms of a conserved chromatin blocks C1-C19 and (b) how these may have been arranged in the common ancestor of dasyurids and all marsupials. Arrows show centromere location. [(a) Reprinted by permission of S. Karger AG from Rens *et al.* (2003). (b) Reprinted from Deakin *et al.* (2012) under CC BY license.]

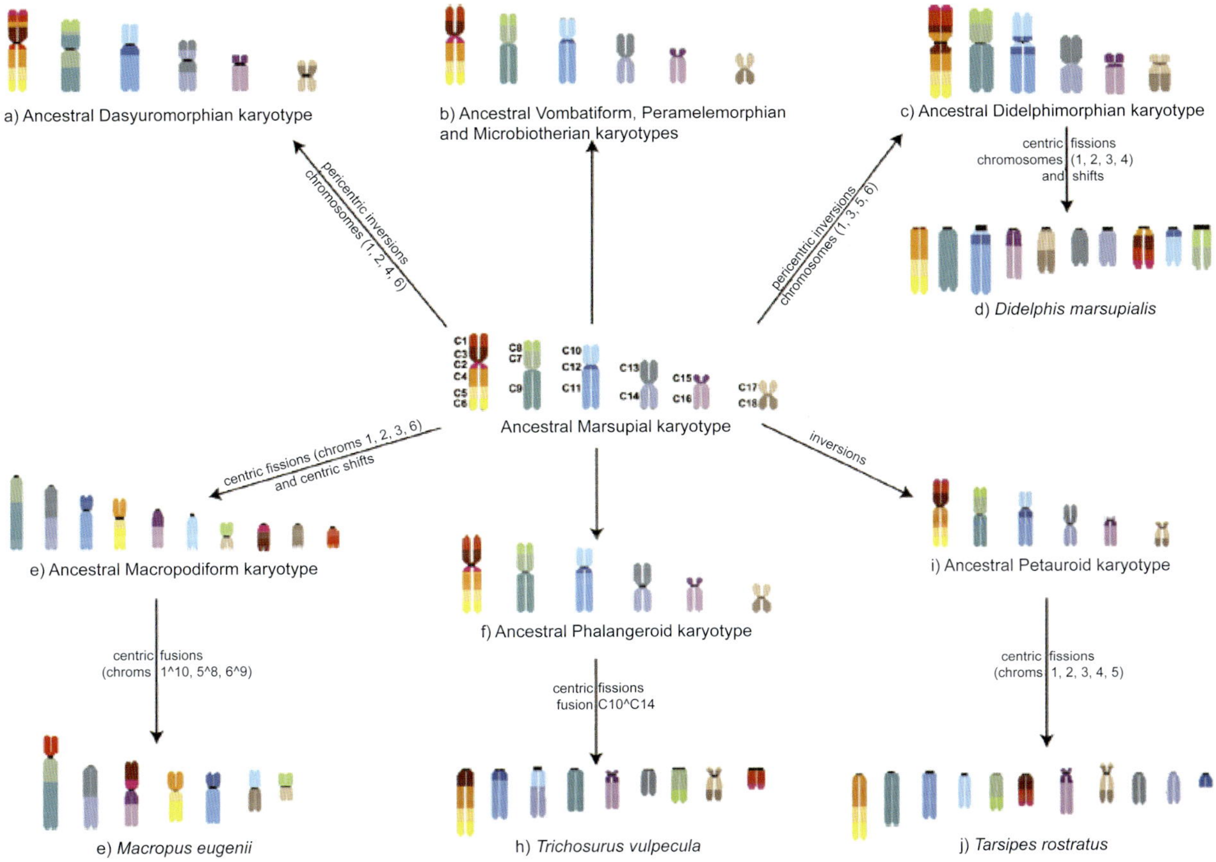

Fig. 11.35. Ancestral marsupial karyotype (centre), conserved chromatin blocks C1–18 on autosomes, and the possible evolutionary transformations leading to modern karyotypes (arrows). [Reprinted by Permission of Oxford University Press from Westerman *et al.* (2010).]

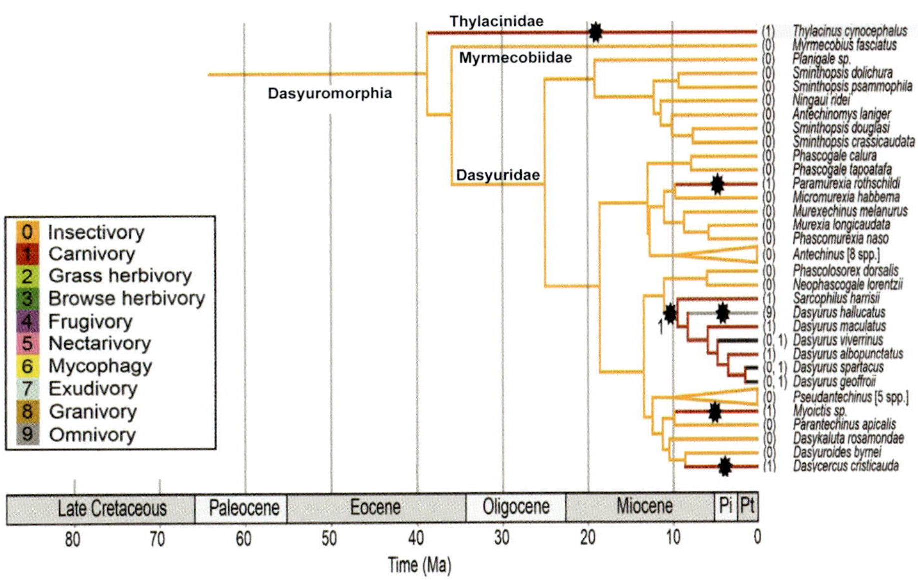

Fig. 12.2. Parsimony reconstruction of ancestral diets from those of extant species by Amador and Giannini (2021) on the phylogeny of Mitchell *et al.* (2014). Stars indicate state transitions. [Reproduced with permission of John Wiley and Sons from Amador and Giannini (2021), with minor adaptations.]

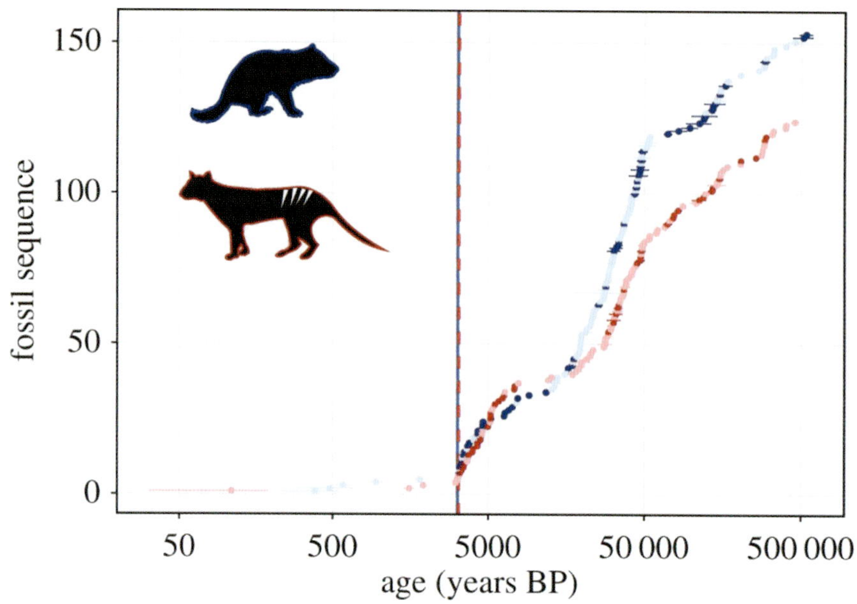

Fig. 12.7. Ages of mainland fossils from *Thylacinus cynocephalus* (red) and *Sarcophilus harrisii* (blue) obtained by White *et al.* (2018b). Fossils are arrayed in order from youngest to oldest on the vertical axis; the horizontal axis is logarithmic. Heavy dots represent dates validated by quality filtering criteria; light dots are published dates that were not validated. Vertical lines show the estimated dates of mainland extinction. [Reprinted by permission of the Royal Society (Great Britain) from White *et al.* (2018b). Permission conveyed through Copyright Clearance Center, Inc.]

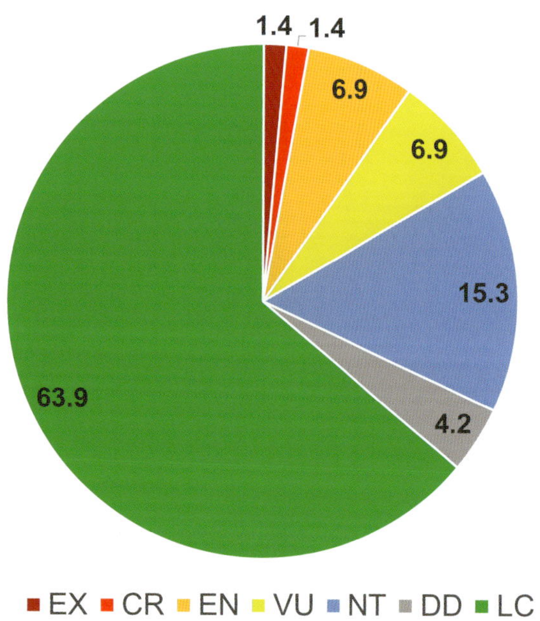

Fig. 12.8. Percentages of the 72 dasyuromorphian species recognised in the IUCN Red List classified in each conservation status category. EX = Extinct; CR = Critically Endangered; EN = Endangered; VU = Vulnerable; NT = Near Threatened; DD = Data Deficient; LC = Least Concern.

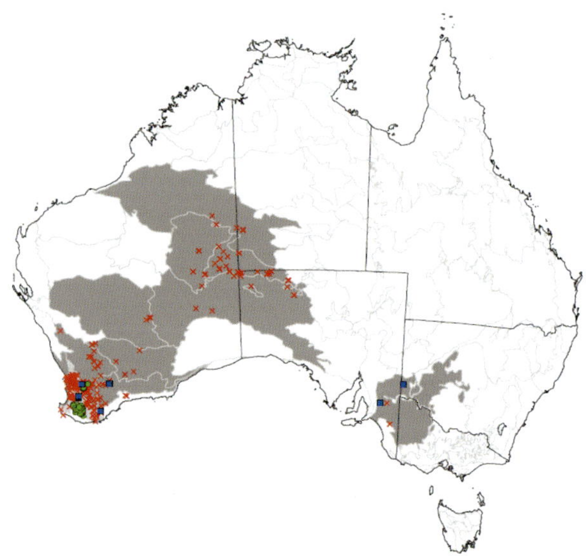

Fig. 12.9. Range contraction of the numbat (*Myrmecobius fasciatus*). Grey area show distribution at time of European settlement (1788), red crosses are localities of records between 1788 and 1992, green dots are current localities and blue squares are successful translocations. [Reprinted by permission of CSIRO Publishing from Woinarski *et al.* (2014). Permission conveyed through Copyright Clearance Center, Inc.]

References

Abbott I (2002) Origin and spread of the cat, *Felis catus*, on mainland Australia, with a discussion of the magnitude of its early impact on native fauna. *Wildlife Research* **29**, 51–74. doi:10.1071/WR01011

Adams M, Cooper N, Armstrong J (2000) 'Revision of *Dasycercus* systematics'. Unpublished report to the South Australian Department for Environment and Heritage, Adelaide.

Aitken PF (1972) *Planigale gilesi* (Marsupialia, Dasyuridae): a new species from the interior of south eastern Australia. *Records of the South Australia Museum* **16**, 1–14, <https://www.biodiversitylibrary.org/page/40861770>.

Amadon D (1949) The seventy-five percent rule for subspecies. *Condor* **51**, 250–258. doi:10.2307/1364805

Amador LI, Giannini NP (2021) Evolution of diet in extant marsupials: emergent patterns from a broad phylogenetic perspective. *Mammal Review* **51**, 178–192. doi:10.1111/mam.12223

Amrine-Madsen H, Scally M, Westerman M, Stanhope MJ, Krajewski C, *et al.* (2003) Nuclear gene sequences provide evidence for the monophyly of Australidelphia marsupials. *Molecular Phylogenetics and Evolution* **28**, 186–196. doi:10.1016/s1055-7903(03)00122-2

Anderson BM, Barrett MD, Krauss SL, Thiele K (2016) Untangling a complex of arid zone grasses (*Triodia*) reveals patterns congruent with co-occurring animals. *Molecular Phylogenetics and Evolution* **101**, 142–162. doi:10.1016/j.ympev.2016.05.014

Ansari MH, Cooper SJB, Schwarz MP, Ebrahimi M, Dolman G, *et al.* (2019) Plio-Pleistocene diversification and biogeographic barriers reflected in the phylogeography of a widespread and common lizard species. *Molecular Phylogenetics and Evolution* **133**, 107–119. doi:10.1016/j.ympev.2018.12.014

Anthony HE (1954) Obituary notice: George H.H. Tate (1894–1953). *Journal of Mammalogy* **35**, 281–282. doi:10.2307/1376064

Aplin K, Archer M (1987) Recent advances in marsupial systematics with a new syncretic classification. In *Possums and Opossum: Studies in Evolution*. (Ed. M Archer) pp. xv–lxxii. Surrey Beatty & Sons and the Royal Society of New South Wales, Sydney.

Aplin K, Baverstock PR, Donnellan SC (1993) Albumin immunological evidence for the time and mode of origin of the New Guinean terrestrial mammal fauna. *Science in New Guinea* **19**, 131–145.

Aplin KP, Rhind SG, Ten Have J, Chesser RT (2015) Taxonomic revision of *Phascogale tapoatafa* (Meyer, 1793) (Dasyuridae; Marsupialia), including descriptions of two new subspecies and confirmation of *P. pirata* Thomas, 1904 as a 'Top End' endemic. *Zootaxa* **4055**, 1–73. doi:10.11646/zootaxa.4055.1.1

Archer M (1974) The development of the cheek teeth in *Antechinus flavipes* (Marsupialia, Dasyuridae). *Journal of the Royal Society of Western Australia* **57**, 54–63, <https://www.biodiversitylibrary.org/page/5078721>.

Archer M (1975a) The development of premolar and molar crowns of *Antechinus flavipes* (Marsupialia, Dasyuridae). *Journal of the Royal Society of Western Australia* **57**, 118–125, <https://www.biodiversitylibrary.org/item/206149#page/132/mode/1u>.

Archer M (1975b) *Ningaui*, a new genus of tiny dasyurids (Marsupialia) and two new species, *N. timealeyi* and *N. ridei*, from arid Western Australia. *Memoirs of the Queensland Museum* **17**, 237–249, <https://www.biodiversitylibrary.org/page/4877110>.

Archer M (1976a) Miocene marsupicarnivores (Marsupialia) from central South Australia, *Ankotarinja tirarensis* gen. et sp. nov., *Keeuna woodburnei* gen et sp. nov., and their significance in terms of early marsupial radiations. *Transactions of the Royal Society of South Australia* **100**, 58–73, <https://www.biodiversitylibrary.org/page/41078384>.

Archer M (1976b) Revision of the marsupial genus *Planigale* Troughton (Dasyuridae). *Memoirs of the Queensland Museum* **17**, 341–365, <https://www.biodiversitylibrary.org/page/48804334>.

Archer M (1976c) The basicranial region of marsupicarnivores (Marsupialia), interrelationships of carnivorous marsupials, and affinities of the insectivorous marsupial peramelids. *Zoological Journal of the Linnean Society* **59**, 217–322. doi:10.1111/j.1096-3642.1976.tb01016.x

Archer M (1976d) The dasyurid dentition and its relationships to that didelphids, thylacinids, borhyaenids (Marsupicarnivora) and peramelids (Peramelina: Marsupialia). *Australian Journal of Zoology Supplemental Series* **24** (39), 1–34. doi:10.1071/AJZS039

Archer M (1977) Revision of the dasyurid marsupial genus *Antechinomys* Krefft. *Memoirs of the Queensland Museum* **18**, 17–29, <https://www.biodiversitylibrary.org/page/48741764>.

Archer M (1978) The nature of the molar-premolar boundary in marsupials and reinterpretation of the homology of marsupial cheekteeth. *Memoirs of the Queensland Museum* **18**, 157–164, <https://www.biodiversitylibrary.org/page/54219486>.

Archer M (1979) Two new species of *Sminthopsis* Thomas (Dasyuridae: Marsupialia) from northern Australia, *S. butleri* and *S. douglasi*. *Australian Zoologist* **20**, 327–345, <https://www.biodiversitylibrary.org/page/39036838>.

Archer M (1981) Results of the Archbold Expeditions No. 104. Systematic revision of the marsupial dasyurid genus *Sminthopsis* Thomas. *Bulletin of the American Museum of

Natural History **168**, 61–224, <http://hdl.handle.net/2246/1050>.

Archer M (1982a) A review of Miocene thylacinids (Thylacinidae, Marsupialia), the phylogenetic position of the Thylacinidae and the problem of apriorisms in character analysis. In *Carnivorous Marsupials.* (Ed. M Archer) pp. 445–476. Royal Zoological Society of New South Wales, Mosman, NSW.

Archer M (1982b) Genesis: and in the beginning there was an incredible carnivorous mother. In *Carnivorous Marsupials.* (Ed. M Archer) pp. vii–x. Royal Zoological Society of New South Wales, Mosman, NSW.

Archer M (1982c) Review of the dasyurid (Marsupialia) fossil record, integration of data bearing on phylogenetic interpretation and suprageneric classification. In *Carnivorous Marsupials.* (Ed. M Archer) pp. 397–443. Royal Zoological Society of New South Wales, Mosman, NSW.

Archer M (1984a) Evolution of arid Australia and its consequences for vertebrates. In *Vertebrate Zoogeography and Evolution in Australasia.* (Eds M Archer and G Clayton) pp. 97–108. Hesperian Press, Carlyle, WA.

Archer M (1984b) Origins and early radiations of marsupials. In *Vertebrate Zoogeography and Evolution in Australasia.* (Eds M Archer and G Clayton) pp. 585–625. Hesperian Press, Carlisle, WA.

Archer M (1984c) The Australian marsupial radiation. In *Vertebrate Zoogeography and Evolution in Australasia.* (Eds M Archer and G Clayton) pp. 633–808. Hesperian Press, Carlisle, WA.

Archer M (1984d) The status of Australian dasyurids, thylacinids, and myrmecobiids. In *Vertebrate Zoogeography and Evolution in Australasia.* (Eds M Archer and G Clayton) pp. 1015–1021. Hesperian Press, Carlisle, WA.

Archer M, Christmas O, Hand SJ, Black KH, Creaser P, *et al.* (2016) Earliest known record of a hypercarnivorous dasyurid (Marsupialia), from newly discovered carbonates beyond the Riversleigh World Heritage Area, north Queensland. *Memoirs of Museum Victoria* **74**, 137–150. doi:10.24199/j.mmv.2016.74.13

Archer M, Fox B (1984) Background to vertebrate zoogeography in Australia. In *Vertebrate Zoogeography and Evolution in Australasia.* (Eds M Archer and G Clayton) pp. 1–15. Hesperian Press, Carlyle, WA.

Archer M, Godthelp HJ, Hand SJ (1994) 6 patterns in the history of Australia's mammals and inferences about paleohabitats. In *History of the Australian Vegetation.* (Ed. RS Hill) pp. 80–103. University of Adelaide Press, Adelaide. doi:10.20851/australian-vegetation.

Archer M, Hand SJ (2006) The Australian marsupial radiation. In *Evolution and Biogeography of Australasian Vertebrates.* (Eds JR Merrick, M Archer, GM Hickey, and MSY Lee) pp. 575–636. Auscipub, Oatlands, NSW.

Archer M, Hand SJ, Godthelp HJ (1991) *Rivesleigh: the story of animals in ancient rainforests of Australia.* Reed Books, Balgowlah.

Archer M, Hand SJ, Godthelp HJ (1995) Tertiary environmental and biotic change in Australia. In *Palaeoclimate and Evolution, with Emphasis on Human Origins.* (Eds E Vrba, GH Denton, TC Partridge, and LH Burckle) pp. 77–90. Yale University Press, New Haven, CT.

Archer M, Hand SJ, Black KH, Beck RDM, Arena DA, *et al.* (2016) A new family of bizarre durophagus carnivorous marsupials from Miocene deposits in the Riversleigh World Heritage Area, northwestern Queensland. *Scientific Reports* **6**, 26911. doi:10.1038/srep26911

Archer M, Kirsch JAW (1977) The case for Thylacomyidae and Myrmecobiidae, Gill 1872, or why are marsupial families so extended? *Proceedings of the Linnean Society of New South Wales* **102**, 18–25, <https://www.biodiversitylibrary.org/page/34874076>.

Archer M, Rich TH (1979) *Wakamatha tasselli* gen. et sp. nov., a fossil dasyurid (Marsupialia) from South Australia convergent on modern *Sminthopsis. Memoirs of the Queensland Museum* **19**, 309–317, <https://www.biodiversitylibrary.org/page/52405090>.

Archer M, Wade M (1976) Results of the Ray E. Lemley Expeditions, Part 1. The Allingham Formation and a new Pliocene vertebrate fauna from northern Queensland. *Memoirs of the Queensland Museum* **17**, 379–397, <https://www.biodiversitylibrary.org/page/48804394>.

Archibald JD (2012) Darwin's two competing phylogenetic trees: marsupials as ancestors or sister taxa? *Archives of Natural History* **39**, 217–233. doi:10.3366/anh.2012.0091

Archie JW (1989) A randomization test for phylogenetic information in systematic data. *Systematic Zoology* **38**, 219–252. doi:10.2307/2992285

Arena DA, Archer M, Goldthelp H, Hand SJ, Hocknull S (2011) Hammer-toothed 'marsupial skinks' from Australian Cenozoic. *Proceedings of the Royal Society B* **278**, 3529–3533. doi:10.1098/rspb.2011.0486

Arena DA, Travouillon KJ, Beck RMD, Black KH, Gillespie AK, Myers TJ, Archer M, Hand SJ (2016) Mammalian lineages and the biostratigraphy and biochronology of Cenozoic faunas from the Riversleigh World Heritage Area, Australia. *Lethaia* **49**, 43–60.

Armstrong LA, Krajewski C, Westerman M (1998) Phylogeny of the marsupial genus *Antechinus* based on cytochrome-*b*, 12S-rRNA, and protamine P1 genes. *Journal of Mammalogy* **79**, 1379–1389. doi:10.2307/1383028

Asher RJ, Novacek MJ, Geisler JH (2003) Relationships of endemic African mammals and their fossil relatives based on morphological and molecular evidence. *Journal of Mammalian Evolution* **10**, 131–194. doi:10.1023/A:1025504124129

Asher RJ, Geisler JH, Sànchez-Villagra MR (2008) Morphology, paleontology, and placental mammal phylogeny. *Systematic Biology* **57**, 311–317. doi:10.1080/10635150802033022

Aslin HJ (1983) Reproduction in *Sminthopsis ooldea* (Marsupialia: Dasyuridae). *Australian Mammalogy* **6**, 93–95.

Attard MRG, Parr WCH, Wilson LAB, Archer M, Hand SJ, *et al.* (2014) Virtual reconstruction and prey size preference in the mid-Cenozoic thylacinid, Nimbacinus dicksoni (Thylacinidae, Marsupialia). *PLoS ONE* **9**, e93088. doi:10.1371/journal.pone.0093088

Averianov AO (2015) Taxonomic revision of tribosphenic mammals from the Lower Cretaceous Antlers Formation of Texas and Oklahoma, USA. *Proceedings of the Zoological Institute RAS* **319**, 141–181, <https://www.zin.ru/journals/trudyzin/doc/vol_319_2/TZ_319_2_Averianov.pdf>.

Avise JC (1994) *Molecular Markers, Natural History, and Evolution*. Chapman & Hall, New York.

Avise JC, Arnold J, Ball RM, Bermingham E, Lamb T, *et al.* (1987) Intraspecific phylogeography: the mitochondrial DNA bridge between population genetics and systematics. *Annual Review of Ecology and Systematics* **18**, 489–522. doi:10.1146/annurev.es.18.110187.002421

Avise JC, Ball RM (1990) Principles of genealogical concordance in species concepts and biological taxonomy. In *Oxford Surveys in Evolutionary Biology 7*. (Eds D Futuyma and J Antonovics) pp. 45–67. Oxford University Press, Oxford, UK.

Avise JC, Robinson TJ (2008) Hemiplasy: a new term in the lexicon of phylogenetics. *Systematic Biology* **57**, 503–507. doi:10.1080/10635150802164587

Baker AM (2012) What's the difference? Describing a new dasyurid species. *Wildlife Australia Magazine* **49**, 12–15.

Baker AM, Mutton TY, Van Dyck S (2012) A new dasyurid marsupial from eastern Queensland, Australia: the Buff-footed Antechinus, *Antechinus mysticus* sp. nov. (Marsupialia: Dasyuridae). *Zootaxa* **3515**, 1–37. doi:10.11646/zootaxa.3515.1.1

Baker AM, Mutton TY, Hines HB (2013) A new dasyurid marsupial from Kroombit Tops, south-east Queensland, Australia: the silver-headed antechinus, *Antechinus argentus* sp. nov. (Marsupialia: Dasyuridae). *Zootaxa* **3746**, 201–239. doi:10.11646/zootaxa.3746.2.1

Baker AM, Mutton TY, Hines HB, Van Dyck S (2014) The Black-tailed Antechinus, *Antechinus arktos* sp. nov.: a new species of carnivorous marsupial from montane regions of the Tweed Volcano caldera, eastern Australia. *Zootaxa* **3765**, 101–133. doi:10.11646/zootaxa.3765.2.1

Baker AM, Mutton TY, Mason ED, Gray EL (2015) A taxonomic assessment of the Australian Dusky Antechinus complex: a new species, the Tasman Peninsula Dusky Antechinus (*Antechinus vandycki* sp. nov.) and an elevation to species of the Mainland Dusky Antechinus (*Antechinus swainsonii mimetes* (Thomas)). *Memoirs of the Queensland Museum – Nature* **59**, 75–126. doi:10.17082/j.2204-1478.59.2015.2014-10

Baker AM, Van Dyck S (2013) Taxonomy and redescription of the Fawn Antechinus, *Antechinus bellus* (Thomas) (Marsupialia: Dasyuridae). *Zootaxa* **3613**, 201–228. doi:10.11646/zootaxa.3613.3.1

Baker AM, Van Dyck S (2015) Taxonomy and redescription of the Swamp Antechinus, *Antechinus minimus* (E. Geoffroy) (Marsupialia: Dasyuridae). *Memoirs of the Queensland Museum – Nature* **59**, 127–170. doi:10.17082/j.2204-1478.59.2015.2014-11

Baker RJ, Bradley RD (2006) Speciation in mammals and the Genetic Species Concept. *Journal of Mammalogy* **87**, 643–662. doi:10.1644/06-MAMM-F-038R2.1

Balme J, O'Connor S, Fallon S (2018) New dates on dingo bones from Madura cave provide oldest firm evidence for arrival of the species in Australia. *Scientific Reports* **8**, 9933. doi:10.1038/s41598-018-28324-x

Barbour RA (1977) Anatomy of marsupials. In *The Biology of Marsupials*. (Eds B Stonehouse and D Gilmore) pp. 237–272. Macmillan Press, London.

Bartholomai A (1971) *Dasyurus dunmalli*, a new species of fossil marsupial (Dasyuridae) in the upper Cainozoic deposits of Queensland. *Memoirs of the Queensland Museum* **16**, 19–26, <https://www.biodiversitylibrary.org/page/48735058>.

Bartlett LJ, Williams DR, Prescott GW, Balmford A, Green RE, *et al.* (2016) Robustness despite uncertainty: regional climate data reveal the dominant role of humans in explaining global extinctions of the Late Quaternary megafauna. *Ecography* **39**, 152–161. doi:10.1111/ecog.01566

Baum DA, Smith SD (2013) *Tree Thinking: An Introduction to Phylogenetic Biology*. Roberts and Company, Greenwood Village, CO.

Baverstock PR, Archer M, Adams M, Richardson BJ (1982) Genetic relationships among 32 species of Australian dasyurid marsupials. In *Carnivorous Marsupials*. (Ed. M Archer) pp. 641–650. Royal Zoological Society of New South Wales, Mosman, NSW.

Baverstock PR, Adams M, Archer M, McKenzie NL, How R (1983) An electrophoretic and chromosomal study of the dasyurid marsupial genus *Ningaui*. *Australian Journal of Zoology* **31**, 381–392. doi:10.1071/ZO9830381

Baverstock PR, Adams M, Archer M (1984) Electrophoretic resolution of species boundaries in the *Sminthopsis murina* complex (Dasyuridae). *Australian Journal of Zoology* **32**, 823–832. doi:10.1071/ZO9840823

Baverstock PR, Krieg M, Birrell J (1989) Evolutionary relationships of Australian marsupials as assessed by albumin immunology. *Australian Journal of Zoology* **37**, 273–287. doi:10.1071/ZO9890273

Baverstock PR, Krieg M, Birrell J (1990) Evolutionary relationships of Australian marsupials as assessed by albumin immunology. In *Mammals from Pouches and Eggs: Genetics, Breeding and Evolution of Marsupials and Monotremes*. (Eds JA Marshall Graves, RM Hope, and DW Cooper) pp. 131–145. CSIRO, Australia. doi:10.1071/ZO9890273

Beck RMD (2008) A dated phylogeny of marsupials using a molecular supermatrix and multiple fossil constraints. *Journal of Mammalogy* **89**, 175–189. doi:10.1644/06-MAMM-A-437.1

Beck RMD (2012) An 'ameridelphian' marsupial from the early Eocene of Australia supports a complex model of Southern Hemisphere marsupial biogeography. *Naturwissenshaften* **99**, 715–729. doi:10.1007/s00114-012-0953-x

Beck RMD, Godthelp H, Weisbecker V, Archer M, Hand SJ (2008) Australia's oldest marsupial fossils and their biogeographic implications. *PLoS ONE* **3**, e1858. doi:10.1371/journal.pone.0001858

Beck RMD, Voss RS, Jansa SA (2022) Craniodental morphology and phylogeny of marsupials. *Bulletin of the American Museum of Natural History* **147**, 1–350, <http://hdl.handle.net/2246/7298>.

Bei M (2009) Molecular genetics of tooth development. *Current Opinion in Genetics & Development* **19**, 504–510. doi:10.1016/j.gde.2009.09.002

Bellott DW, Hughes JF, Skaletsky H, Brown LG, Pyntikova T, *et al.* (2014) Mammalian Y chromosomes retain widely expressed dosage-sensitive regulators. *Nature* **508**, 494–499. doi:10.1038/nature13206

Benda C (1906) Die Spermiogenese de Marsupialier. *Denkschriften der Medicinisch-Naturwissenschaftlichen Geselschaft zu Jena* **6**, 439–458, <https://www.biodiversitylibrary.org/page/39295707>.

Bender HS, Murchison EP, Pickett HA, Deakin JE, Strong MA, *et al.* (2012) Telomere length dimorphism in the Tasmanian Devil and related marsupials suggests parental control of telomere length. *PLoS ONE* **7**, e416105. doi:10.1371/journal.pone.0046195

Bennett JH, Breed WG, Hayman DL, Hope RM (1990) Reproductive and genetical studies with a laboratory colony of the dasyurid marsupial *Sminthopsis crassicaudata*. In *Mammals from Pouches and Eggs: Genetics, Breeding and Evolution of Marsupials and Monotremes.* (Eds JA Marshall Graves, RM Hope, and DW Cooper) pp. 65–80. CSIRO, Australia. doi:10.1071/ZO989020

Bensasson D, Zhang D-X, Hartl DL, Hewitt GM (2001) Mitochondrial pseudogenes: evolution's misplaced witnesses. *Trends in Ecology & Evolution* **16**, 314–321. doi:10.1016/s0169-5347(01)02151-6

Bensley BA (1901) On the question of an arboreal ancestry of the Marsupialia, and the interrelationships of the mammalian subclasses. *American Naturalist* **35**, 117–138. doi:10.1086/277873

Bensley BA (1903) On the evolution of the Australian Marsupialia: with remarks on the relationships of marsupials. *Transactions of the Linnean Society of London (Second Series)* **9**, 83–217. doi:10.1111/j.1096-3642.1903.tb00447.x

Benton MJ, Donoghue PCJ (2007) Paleontological evidence to date the tree of life. *Molecular Biology and Evolution* **24**, 26–53. doi:10.1093/molbev/msl150

Berkovitz B, Shellis P (2018) *The Teeth of Mammalian Vertebrates.* Elsevier, Amsterdam.

Bickford D, Lohman DJ, Sodhi NS, Ng PK, Meier R, *et al.* (2007) Cryptic species as a window on diversity and conservation. *Trends in Ecology & Evolution* **22**, 148–155. doi:10.1016/j.tree.2006.11.004

Bidloo G (1685) *Anatomia humani corporis.* Sumptibus viduae Joannis à Someren, haeredum Joannis à Dyk, Henrici & viduae Theodori Boom, Amsterdam.

Binfield P, Archer M, Hand SJ, Black KH, Myers TJ, *et al.* (2017) A new Miocene carnivorous marsupial, *Barinya kutjamarpensis* (Dasyuromorphia), from central Australia. *Alcheringa* **41**, 46–53. doi:10.1080/03115518.2016.1180029

Bininda-Emonds ORP, Gittleman JL, Steel MA (2002) The (super)tree of life: procedures, problems, and prospects. *Annual Review of Ecology and Systematics* **33**, 265–289. doi:10.1146/annurev.ecolsys.33.010802.150511

Bininda-Emonds ORP, Cardillo M, Jones KE, MacPhee RDE, Beck RMD, *et al.* (2007) The delayed rise of present-day mammals. *Nature* **446**, 507–512. doi:10.1038/nature05634

Bird MI, Hutley LB, Lawes MJ, Lloyd J, Luly JG, *et al.* (2013) Humans, megafauna and environmental change in tropical Australia. *Journal of Quaternary Science* **28**, 439–452. doi:10.1002/jqs.2639

Black KH, Archer M, Hand SJ, Godthelp H (2012) The rise of Australian marsupials: a synopsis of biostratigraphic, phylogenetic, palaeoecologic, and palaeobiogeographic understanding. In *Earth and Life: Global Biodiversity, Extinction Intervals and Biogeographic Perturbations Through Time.* (Ed. JA Talent) pp. 983–1078. International Year of Planet Earth Sciences, Springer Verlag, Berlin. doi:10.1007/978-90-481-3428-1_35

Blacket M, Krajewski C, Labrinidis A, Cambron B, Cooper S, *et al.* (1999) Systematic relationships within the dasyurid marsupial tribe Sminthopsini – a multigene approach. *Molecular Phylogenetics and Evolution* **12**, 140–155. doi:10.1023/A:1011322031747

Blacket MJ, Adams M, Krajewski C, Westerman M (2000) Genetic variation within the dasyurid marsupial genus *Planigale. Australian Journal of Zoology* **48**, 443–456. doi:10.1071/ZO000020

Blacket MJ, Adams M, Cooper SJB, Krajewski C, Westerman M (2001) Systematics and evolution of the dasyurid marsupial genus *Sminthopsis*: I. Species Group Macroura. *Journal of Mammalian Evolution* **8**, 149–168. doi:10.1023/A:1011322031747

Blacket MJ, Cooper SJB, Krajewski C, Westerman M (2006) Systematics and evolution of the dasyurid marsupial genus *Sminthopsis*: II. The Murina species group. *Journal of Mammalian Evolution* **13**, 125–138. doi:10.1007/s10914-006-9008-4

Blackwelder RE (1967) *Taxonomy: Text and Reference Book.* John Wiley & Sons, New York.

Blewett R (2012) *Shaping a Nation: A Geology of Australia.* Geoscience Australia and ANU E-Press, Acton, ACT. doi:10.22459/SN.08.2012

Blomberg SP, Garland T Jr, Ives AR (2003) Testing for phylogenetic signal in comparative data: behavioral traits are more labile. *Evolution* **57**, 717–745. doi:10.1111/j.0014-3820.2003.tb00285.x

Blumenbach JF (1791) *Handbuch der Naturegeschichte.* Bei Heinrich Dieterich, Göttingen, <https://www.biodiversitylibrary.org/page/18185533>.

Bonaparte CL (1837) New systematic arrangement of vertebrated animals. *Transactions of the Linnean Society* **18**, 247–304. doi:10.1111/j.1095-8339.1838.tb00177.x

Bonaparte CL (1838) Synopsis vertebratorum systematis. *Novi Annali delle Scienze Naturali, Bologna* **2** 105–133.

Boonstra R (2005) Equipped for life: the adaptive role of the stress axis in male mammals. *Journal of Mammalogy* **86**, 236–247. doi:10.1644/BHE-001.1

Boveri TH (1904) *Ergebnisse über die Konstitution der chromatischen Substanz des Zelkerns.* Fisher, Jena, <https://www.biodiversitylibrary.org/page/21714986>.

Boyden A (1942) Systematic serology: a critical appreciation. *Physiological Zoölogy* **15**, 109–145. doi:10.1086/physzool.15.2.30151814

Bradley AJ (1982) Steroid binding proteins in the plasma of dasyurid marsupials. In *Carnivorous Marsupials.* (Ed. M

Archer) pp. 651–657. Royal Zoological Society of New South Wales, Sydney.

Bradley AJ (1987) Stress and mortality in the Red-Tailed Phascogale, *Phascogale calura* (Marsupialia: Dasyuridae). *General and Comparative Endocrinology* **67**, 85–100. doi:10.1016/0016-6480(87)90208-5

Bradley AJ (1990) Failure of glucocorticoid feedback during breeding in the male Red-tailed Phascogale *Phascogale calura* (Marsupialia: Dasyuridae). *Journal of Steroid Biochemistry and Molecular Biology* **37**, 155–163. doi:10.1016/0960-0760(90)90384-w

Bradley AJ (2003) Stress hormones and mortality in small carnivorous marsupials. In *Predators with Pouches: The Biology of Carnivorous Marsupials.* (Eds M Jones, C Dickman, and M Archer) pp. 254–267. CSIRO Publishing, Melbourne.

Bradley AJ, McDonald IR, Lee AK (1980) Stress and mortality in a small marsupial (*Antechinus stuartii* Macleay). *General and Comparative Endocrinology* **40**, 188–200. doi:10.1016/0016-6480(80)90122-7

Bradshaw CJA, Johnson CN, Llewelyn J, Weisbecker V, Strona G, *et al.* (2021) Relative demographic susceptibility does not explain the extinction chronology of Sahul's megafauna. *eLife* **10**, e63870. doi:10.7554/eLife.63870

Bradshaw CJA, Norman K, Ulm S, Williams AN, Clarkson C, *et al.* (2021) Stochastic modeling supports rapid peopling of late Pleistocene Sahul. *Nature Communications* **12**, 2440. doi:10.1038/s41467-021-21551-3

Braithwaite RW (1979) Social dominance and habitat utilization in *Antechinus stuartii* (Marsupialia). *Australian Journal of Zoology* **27**, 517–528. doi:10.1071/ZO9790517

Braithwaite RW, Lee AK (1979) A mammalian example of semelparity. *American Naturalist* **113**, 151–155. doi:10.1086/283372

Brisson MJ (1762) *Regnum animale in classes IX distributum sive Synopsis methodica.* Haak, Paris, <https://www.biodiversitylibrary.org/page/28278568>.

Britten RJ (1986) Rates of DNA sequence evolution differ between taxonomic groups. *Science* **231**, 1393–1398. doi:10.1126/science.3082006

Britten RJ, Kohne DE (1966) Nucleotide sequence repetition in DNA. *Carnegie Institution of Washington Yearbook* **65**, 78–106.

Brocklehurst N, Benevento GL (2020) Dental characters used in phylogenetic analyses of mammals show higher rates of evolution, but not reduced independence. *PeerJ* **8**, e8744. doi:10.7717/peerj.8744

Bromham L (2011) The genome as a life-history character: why rate of molecular evolution varies between mammal species. *Philosophical Transactions of the Royal Society B: Biological Sciences* **366**, 2503–2513. doi:10.1098/rstb.2011.0014

Bromham L, Rambaut A, Harvey PH (1996) Determinants of rate variation in mammalian DNA sequence evolution. *Journal of Molecular Evolution* **43**, 610–621. doi:10.1007/BF02202109

Brook BW, Johnson CN (2006) Selective hunting of juveniles as a cause of the imperceptible overkill of the Australian Pleistocene megafauna. *Alcheringa* **30**, 39–48. doi:10.1080/03115510609506854

Brown OJF (2006) Tasmanian devil (*Sarcophilus harrisii*) extinction on the Australian mainland in the mid-Holocene: multicausality and ENSO intensification. *Alcheringa* **30**, 49–57. doi:10.1080/03115510609506855

Brüniche-Olsen A, Jones ME, Burridge CP, Murchison EP, Holland BR, *et al.* (2018) Ancient DNA tracks the mainland extinction and island survival of the Tasmanian devil. *Journal of Biogeography* **45**, 963–976. doi:10.1111/jbi.13214

Bryant LM, Krosch MN (2016) Lines in the land: a review of evidence for eastern Australia's major biogeographic barriers to closed forest taxa. *Biological Journal of the Linnean Society* **119**, 238–264. doi:10.1111/bij.12821

Buffon GL (1749–1804) *Histoire naturelle, générale et particulière.* 44 volumes. Imprimerie Royale, Paris, <https://www.biodiversitylibrary.org/page/41188467>.

Burbidge AA (2008) Conservation of Australian mammals. In *The Mammals of Australia.* 3rd edn. (Eds S Van Dyck and R Strahan) pp. 26–27. Reed New Holland, Sydney.

Burbidge AA, McKenzie NL (1989) Patterns in the modern decline of Western Australia's vertebrate fauna: causes and conservation implications. *Biological Conservation* **50**, 143–198. doi:10.1016/0006-3207(89)90009-8

Burbidge AA, Woinarski J (2016) *Parantechinus apicalis. IUCN Red List of Threatened Species* **201,** e.T16138A21944584. doi:10.2305/IUCN.UK.2016-2.RLTS.T16138A21944584.en

Burgin CJ, Colella JP, Kahn PL, Upham NS (2018) How many species of mammals are there? *Journal of Mammalogy* **99**, 1–11. doi:10.1093/jmammal/gyx147

Byrne M, Steane DA, Joseph L, Yeates DK, Jordan GJ, *et al.* (2011) Decline of a biome: fragmentation, extinction, and invasion of the Australian mesic zone biota. *Journal of Biogeography* **38**, 1635–1656. doi:10.1111/j.1365-2699.2011.02535.x

Byrne M, Joseph L, Yeates DK, Roberts JD, Edwards D (2018) Evolutionary history. In *On the Ecology of Australia's Arid Zone.* (Ed. H Lambers) pp. 45–75. Springer, Switzerland.

Byrne M, Yeates DK, Joseph L, Kearney M, Bowler J, *et al.* (2008) Birth of a biome: insights into the assembly and maintenance of the Australian arid zone. *Molecular Ecology* **17**, 4398–4417. doi:10.1111/j.1365-294X.2008.03899.x

Cabrera A (1927) Datos para el conocimiento de los dasyuroideos fósiles Argentinos. *Revista del Museo de La Plata* **30**, 271–315, <https://publicaciones.fcnym.unlp.edu.ar/rmlp/article/view/1409>.

Calaby J (1984) Changing the face of a continent. In *Vertebrate Zoogeography and Evolution in Australasia.* (Eds M Archer and G Clayton) pp. 1011–1014. Hesperian Press, Carlisle, WA.

Callender-Crowe LM, Sansom RS (2022) Osteological characters of birds and reptiles are more congruent with molecular phylogenies than soft characters are. *Zoological Journal of the Linnean Society* **194**, 1–13. doi:10.1093/zoolinnean/zlaa136

Campbell CR (1976) Tertiary Dasyuridae and Peramelidae (Marsupialia) from the Tirari Desert, South Australia. PhD Dissertation, University of California, Berkeley.

Cardillo M, Bininda-Emonds ORP, Boakes E, Purvis A (2004) A species-level phylogenetic supertree of marsupials. *Journal of Zoology (London)* **264**, 11–31. doi:10.1017/S0952836904005539

Cascini M, Mitchell KJ, Cooper A, Phillips MJ (2019) Reconstructing the evolution of giant kangaroos: comparing the utility of DNA, morphology, and total evidence. *Systematic Biology* **68**, 520–537. doi:10.1093/sysbio/syy080

Chaplin K, Sumner J, Hipsley CA, Melville J (2020) An integrative approach using phylogenomics and high-resolution X-ray computed tomography for species delimitation in cryptic taxa. *Systematic Biology* **69**, 294–307. doi:10.1093/sysbio/syz048

Chuang C, Carpenter AD, Fuchsova B, Johnson R, de Lanerolle P, *et al.* (2006) Long-range directional movement of an interphase chromosome site. *Current Biology* **16**, 825–831. doi:10.1016/j.cub.2006.03.059

Churchill TJ, Archer M, Hand SJ, Myers T, Gillespie A, Beck RMD (2023) A new diminutive durophagous Miocene dasyuromorphian (Marsupialia, Malleodectidae) from the Riversleigh World Heritage Area, northern Australia. *Journal of Vertebrate Paleontology*. doi:10.1080/02724634.2023.2170804

Clarkson C, Jacobs Z, Marwick B, Fullagar R, Wallis L, *et al.* (2017) Human occupation of northern Australia by 65,000 years ago. *Nature* **547**, 306–310. doi:10.1038/nature22968

Clode D (2009) *Prehistoric Giants: The Megafauna of Australia*. Museum of Victoria, Melbourne.

Cloos M, Sapiie B, van Ufford AQ, Weiland RJ, Warren PO, *et al.* (2005) Collisional delamination in New Guinea: The geotectonics of subducting slab breakoff. *Geological Society of America Special Paper* **400**, 1–51, <https://u1lib.org/book/1251319/37da62>.

Cockburn A (1997) Living slow and dying young: senescence in marsupials. In *Marsupial Biology: Recent Research, New Perspectives*. (Eds N Saunders and L Hinds) pp. 163–171. University of New South Wales Press, Sydney.

Cole FJ (1944) *A History of Comparative Anatomy from Aristotle to the Eighteenth Century*. Macmillan, London.

Collett RA, Baker AM, Fisher DO (2018) Prey productivity and predictability drive different axes of life-history variation in carnivorous marsupials. *Proceedings of the Royal Society B* **285**, 20181291. doi:10.1098/rspb.2018.1291

Cooper A, Turney C, Hughen KA, Brook BW, McDonald HG, *et al.* (2015) Abrupt warming drove Late Pleistocene Holarctic megafaunal turnover. *Science* **349**, 602–606. doi:10.1126/science.aac4315

Cooper CE (2011) *Myrmecobius fasciatus* (Dasyuromorphia: Myrmecobiidae). *Mammalian Species* **43**, 129–140. doi:10.1644/881.1

Cooper DW, Edwards C, James E, Sharman GB, VandeBerg JL, *et al.* (1977) Studies on metatherian sex chromosomes VI. A third state of an X-linked gene: partial activity for the paternally derived Pgk-A allele in cultured fibroblasts of *Macropus giganteus* and *M. parryi*. *Australian Journal of Biological Science* **30**, 431–443. doi:10.1071/BI9770431

Cooper DW, Johnston PG, VandeBerg JL, Robinson ES (1990) X-chromosome inactivation in marsupials. In *Mammals from Pouches and Eggs: Genetics, Breeding and Evolution of Marsupials and Monotremes*. (Eds JA Marshall Graves, RM Hope, and DW Cooper) pp. 269–275. CSIRO, Australia. doi:10.1071/ZO989041

Cooper DW, Johnson PG, Watson JM, Graves JAM (1993) X-inactivation on marsupials and monotremes. *Seminars in Developmental Biology* **4**, 117–128. doi:10.1006/sedb.1993.1014

Cooper DW, Woolley PA (1983) Confirmation of a new species of small dasyurid marsupial by electrophoretic analysis of enzymes and proteins. *Australian Journal of Zoology* **31**, 743–751. doi:10.1071/ZO9830743

Cooper NK, Aplin KP, Adams M (2000a) A new species of false antechinus (Marsupialia: Dasyuridae) from the Pilbara region, Western Australia. *Records of the Western Australian Museum* **20**, 115–136, <https://museum.wa.gov.au/sites/default/files/A%20NEW%20SPECIES%20OF%20FALSE%20ANTECHINUS%20(MARSUPIALIA%20DASYUROMORPHIA%20DASYURIDAE)%20FROM%20THE%20PILBARA%20REGION,%20WESTERN%20AUSTRALIA.pdf>.

Cooper SJB, Adams M, Labrinidis A (2000b) Phylogeography of the Australian dunnart *Sminthopsis crassicaudata* (Marsupialia: Dasyuridae). *Australian Journal of Zoology* **48**, 461–473. doi:10.1071/ZO00014

Cope ED (1888) The multituberculata monotremes. *American Naturalist* **22**, 259, <https://www.journals.uchicago.edu/doi:/pdf/10.1086/274676>.

Corbett L (1995) *The Dingo in Australia and Asia*. University of New South Wales Press, Sydney.

Corbett L (2006) The Australian dingo. In *Evolution and Biogeography of Australasian Vertebrates*. (Eds JR Merrick, M Archer, GM Hickey and MSY Lee) pp. 745–753. Auscipub, Oatlands, NSW.

Cowper W (1704) Account of the anatomy of those parts of a male opossum that differ from the female. *Philosophical Transactions of the Royal Society* **24**, 1576–1590, <https://www.jstor.org/stable/102927>.

Crabb P (1982) Pleistocene dasyurids from southwestern New South Wales. In *Carnivorous Marsupials*. (Ed. M Archer) pp. 511–516. Royal Zoological Society of New South Wales, Sydney.

Crabtree SA, White DA, Bradshaw CJA, Saltré F, Williams AN, *et al.* (2021) Landscape rules predict optimal superhighways for the first peopling of Sahul. *Nature Human Behavior*. doi:10.1038/s41562-021-01106-8

Cracraft J (1983) Species concepts and speciation analysis. *Current Ornithology* **1**, 159–187. doi:10.1007/978-1-4615-6781-3_6

Cracraft J (1991) Patterns of diversification within continental biotas: hierarchical congruence among the areas of endemism of Australian vertebrates. *Australian Systematic Botany* **4**, 211–227. doi:10.1071/SB9910211

Cramb J, Hocknull S (2010) Two new species of *Antechinus* Macleay (Dasyuridae: Marsupialia) from mid-Pleistocene cave deposits in east central Queensland. *Australian Mammalogy* **32**, 127–144. doi:10.1071/AM09025

Cramb J, Hocknull S, Webb GE (2009) High diversity Pleistocene rainforest dasyurid assemblages with implications for the radiation of the Dasyuridae. *Austral Ecology* **34**, 663–669. doi:10.1111/j.1442-9993.2009.01972.x

Crisp MD, Cook LG (2013) How was the Australian flora assembled over the last 65 million years? A molecular phylogenetic perspective. *Annual Review of Ecology, Evolution, and Systematics* **44**, 303–324. doi:10.1146/annurev-ecolsys-110512-135910

Cremer T, Cremer M (2010) Chromosome territories. *Cold Spring Harbor Perspectives in Biology* **2**, a003889. doi:10.1101/cshperspect.a003889

Crowther MS (2002a) Distributions of species of the *Antechinus stuartii-A. flavipes* complex as predicted by bioclimatic modelling. *Australian Journal of Zoology* **50**, 77–91. doi:10.1071/zo01031

Crowther MS (2002b) Morphological variation within *Antechinus agilis* and *Antechinus stuartii* (Marsupialia: Dasyuridae). *Australian Journal of Zoology* **50**, 339–356. doi:10.1071/ZO01030

Crowther MS, Blacket MJ (2003) Biogeography and speciation in Dasyuridae: Why are there so many kinds of dasyurids? In *Predators with Pouches: The Biology of Carnivorous Marsupials*. (Eds M Jones, C Dickman, and M Archer) pp. 124–130. CSIRO Publishing, Melbourne.

Crowther MS, Dickman CR, Lynam AJ (1999) *Sminthopsis griseoventer boullangerensis* (Marsupialia: Dasyuridae), a new subspecies in the *S. murina* complex from Boullanger Island, Western Australia. *Australian Journal of Zoology* **47**, 215–243. doi:10.1071/ZO98062

Crowther MS, Spencer PBS, Alpers D, Dickman CR (2002) Taxonomic status of the mardo, *Antechinus flavipes leucogaster* (Marsupialia: Dasyuridae): a morphological, molecular, reproductive, and bioclimatic approach. *Australian Journal of Zoology* **50**, 627–647. doi:10.1071/ZO02030

Crowther MS, Sumner J, Dickman CR (2003) Speciation in *Antechinus stuartii* and *A. subtropicus* (Marsupialia: Dasyuridae) in eastern Australia: molecular and morphological evidence. *Australian Journal of Zoology* **51**, 443–462. doi:10.1071/ZO02009

Crozier RH (1992) Genetic diversity and the agony of choice. *Biological Conservation* **61**, 11–15. doi:10.1016/0006-3207(92)91202-4

Cuttle P (1982) Life history strategy of the dasyurid marsupial *Phascogale tapoatafa*. In *Carnivorous Marsupials*. (Ed. M Archer) pp. 13–22. Royal Zoological Society of New South Wales, Sydney.

Cuvier G (1804) Mémoire sur le squelette Presque entier d'un petit quadrupède du genre de Sarigues, trouvé dans le pierre à plâtre des environs de Paris. *Annales du Muséum Nationale 'Histoire Naturelle* **5,** 277–292.

Cuvier G (1817) *Le Règne Animal, distribué d'aprés son organization*. Chez Deterville, Paris, <https://www.biodiversitylibrary.org/page/1844622>.

Darwin C (1847) A Natural History of the Mammalia. By G.R. Waterhouse, Esq., of the British Museum. Illustrated with engravings on wood and coloured plates. London, H. Baillière. [A Review.] *Annals and Magazine of Natural History* **19**, 53–56, <http://darwin-online.org.uk/converted/published/1847_waterhouse_F1675.html>.

Darwin C (1859) *On the Origin of Species*. John Murray, London, <http://darwin-online.org.uk/content/frameset?itemID=F373&viewtype=text&pageseq=1>.

Darwin C (1872) *The Origin of Species*. 6th edn. John Murray, London, <http://darwin-online.org.uk/content/frameset?itemID=F391&viewtype=text&pageseq=1>.

Darwin F (1899) *The Life and Letters of Charles Darwin*. D. Appleton and Company, New York, <http://darwin-online.org.uk/content/frameset?viewtype=text&itemID=F1452.2&pageseq=1>.

Davalos LM, Cirranello AL, Geisler JH, Simmons NB (2012) Understanding phylogenetic incongruence: lessons from phyllostomid bats. *Biological Reviews of the Cambridge Philosophical Society* **87**, 991–1024. doi:10.1111/j.1469-185X.2012.00240.x

Davalos LM, Velazco PM, Warsi OM, Smits PD, Simmons NB (2014) Integrating incomplete fossils by isolating conflicting signal in saturated and nonindependent morphological characters. *Systematic Biology* **63**, 582–600. doi:10.1093/sysbio/syu022

David B, Arnold LJ, Delannoy J-J, Fresløv J, Urwin C, *et al.* (2021) Late survival of megafauna refuted for Cloggs Cave, SE Australia: implications for the Australian Late Pleistocene megafauna extinction debate. *Quaternary Science Reviews* **253**, 106781. doi:10.1016/j.quascirev.2020.106781

Dawson L (1982) Taxonomic status of fossil devils (*Sarcophilus*, Dasyuridae, Marsupialia) from Late Quaternary eastern Australian localities. In *Carnivorous Marsupials*. (Ed. M Archer) pp. 517–525. Royal Zoological Society of New South Wales, Sydney.

Dawson L, Muirhead J, Wroe S (1999) The Big Sink Local Fauna: a lower Pliocene mammalian fauna from the Wellington Caves Complex, Wellington, New South Wales. *Records of the Western Australian Museum, Supplement* **57**, 265–290, <https://museum.wa.gov.au/research/records-supplements/records/big-sink-local-fauna-lower-pliocene-mammalian-fauna-wellington->.

Dayhoff MO, Schwartz RM, Orcutt BC (1978) A model of evolutionary change in proteins. In *Atlas of Protein Sequence and Structure Volume 5, Supplement 3*. (Ed. MO Dayhoff) pp. 345–352. National Biomedical Research Foundation, Maryland.

Dayrat B (2003) The roots of phylogeny: how did Haeckel build his trees? *Systematic Biology* **52**, 515–527. doi:10.1080/10635150390218277

Deakin JE (2019) Marsupial Chromosome Evolution. In *eLS*. John Wiley & Sons, Chichester, UK. doi:10.1002/9780470015902.a002828

Deakin JE, Bender HS, Pearse AM, Rens W, O'Brien PCM, *et al.* (2012) Genomic restructuring in the Tasmanian devil facial tumour: chromosome painting and gene mapping provide clues to evolution of a transmissible tumour. *PLoS Genetics* **8**, e1002483. doi:10.1371/journal.pgen.1002483

Deakin JE, Kruger-Andrzejewska M (2016) Marsupials as models for understanding the role of chromosome rearrangements in evolution and disease. *Chromosoma* **12**, 633–644. doi:10.1007/s00412-016-0603-8

Deakin JE, Potter S, O'Neill R, Ruiz-Herrera A, Cio MB, *et al.* (2019) Chromosomics: bridging the gap between genomes and chromosomes. *Genes* **10**, 627. doi:10.3390/genes10080627

De Blaineville HMD (1816) Prodrome d'une nouvelle distribution systematique du regne animal. *Bulletin des Sciences* **8**, 113–124, <https://www.biodiversitylibrary.org/page/4439803>.

Degen AA (1997) *Ecophysiology of Small Desert Mammals.* Springer, Berlin.

Delbridge ML, Lingenfelter PA, Disteche CM, Marshall Graves JA (1999) The candidate spermatogenesis gene *RBMY* has a homologue on the human X chromosome. *Nature Genetics* **22**, 223–224. doi:10.1038/10279

De Leo AA, Wheeler D, Lefevre C, Cheng J-F, Hope R, *et al.* (2005) Sequencing and mapping hemoglobin gene clusters in the Australian model dasyurid marsupial *Sminthopsis macroura*. *Cytogenetic and Genome Research* **108**, 333–341. doi:10.1159/000081528

De Leo AA, Lefevre C, Pharo E, Cheng JF, Frappell P, *et al.* (2006) Characterisation of two whey protein genes in the Australian dasyurid marsupial, the stripe-faced dunnart (*Sminthopsis macroura*). *Cytogenetics and Genome Research* **115**, 62–69. doi:10.1159/000094802

De Muizon C (1998) *Mayulestes ferox*, a borhyaenoid (Metatheria, Mammalia) from the early Paleocene of Bolivia. Phylogenetic and paleobiologic implications. *Geodiversitas* **20**, 19–142, <https://sciencepress.mnhn.fr/sites/default/files/articles/pdf/g1998n1a2.pdf>.

Denny MJS (1982) Review of *Planigale* (Dasyuridae, Marsupialia) ecology. In *Carnivorous Marsupials.* (Ed. M Archer) pp. 131–138. Royal Zoological Society of New South Wales, Sydney.

De Queiroz K (1998) The general lineage concept of species, species criteria, and the process of speciation: a conceptual unification and terminological recommendations. In *Endless Forms: Species and Speciation.* (Eds DJ Howard and SJ Berlocher) pp. 57–75. Oxford University Press, Oxford, <https://repository.si.edu/bitstream/handle/10088/4652/VZ_1998deQueirozEndlessForms.pdf>.

De Queiroz K (2007) Species concepts and species delimitation. *Systematic Biology* **56**, 879–886. doi:10.1080/10635150701701083

De Queiroz K, Gauthier J (1994) Toward a phylogenetic system of biological nomenclature. *Trends in Ecology & Evolution* **9**, 27–31. doi:10.1016/0169-5347(94)90231-3

Dickman C, Woodford Ganf R (2007) *A Fragile Balance: The Extraordinary Story of Australian Marsupials.* University of Chicago Press, Chicago.

Dickman CR (1982) Some ecological aspects of seasonal breeding in *Antechinus* (Dasyuridae, Marsupialia). In *Carnivorous Marsupials.* (Ed. M Archer) pp. 139–150. Royal Zoological Society of New South Wales, Sydney.

Dickman CR (1986) Niche compression: two tests of an hypothesis using narrowly sympatric predator species. *Australian Journal of Ecology* **11**, 121–134. doi:10.1111/j.1442-9993.1986.tb01383.x

Dickman CR (1993) Evolution of semelparity in male dasyurid marsupials: a critique, and an hypothesis of sperm competition. In *The Biology and Management of Australasian Carnivorous Marsupials.* (Eds M Roberts, J Carnio, G Crawshaw, and M Hutchins) pp. 25–38. Metropolitan Toronto Zoo, Ontario, and the American Association of Zoological Parks and Aquariums, Washington, DC.

Dickman CR, Braithwaite RW (1992) Postmating mortality of males in the dasyurid marsupials, *Dasyurus* and *Parantechinus*. *Journal of Mammalogy* **73**, 143–147. doi:10.2307/1381875

Dickman CR, Parnaby HE, Crowther MS, King DH (1998) *Antechinus agilis* (Marsupialia: Dasyuridae), a new species from the *A. stuartii* complex in south-eastern Australia. *Australian Journal of Zoology* **46**, 1–26. doi:10.1071/ZO97036

Donoghue MJ, Kadereit JW (1992) Walter Zimmerman and the growth of phylogenetic theory. *Systematic Biology* **41**, 74–85. doi:10.1093/sysbio/41.1.74

Doolittle RF, Blombäck B (1964) Amino acid sequence investigations of fibrinopeptides from various mammals: evolutionary implications. *Nature* **202**, 147–152. doi:10.1038/202147a0

Dow DB (1977) *A geological synthesis of Papua New Guinea.* Bulletin, Bureau of Mineral Resources, Geology, and Geophysics Number 201. Australian Government Publishing Service, Canberra.

Drew LW (2011) Are we losing the science of taxonomy? As need grows, numbers and training are failing to keep up. *BioScience* **61**, 942–946. doi:10.1525/bio.2011.61.12.4

Drummond AJ, Ho SYW, Phillips MJ, Rambaut A (2006) Relaxed phylogenetics and dating with confidence. *PLoS Biology* **4(5)**, e88. doi:10.1371/journal.pbio.0040088

Duchêne DA, Bragg JG, Duchêne S, Neaves LE, Potter S, *et al.* (2018) Analysis of phylogenomic tree space resolves relationships among marsupial families. *Systematic Biology* **67**, 400–412. doi:10.1093/sysbio/syx076

Dwyer PD (1977) Notes on *Antechinus* and *Cercartetus* (Marsupialia) in the New Guinea highlands. *Proceedings of the Royal Society of Queensland* **88**, 69–73.

Eastman CR (1915) Early portrayals of the opossum. *American Naturalist* **49**, 585–594. doi:10.1086/279504

Ebach M (2012) A history of biogeographic regionalization in Australia. *Zootaxa* **3392**, 1–34. doi:10.11646/zootaxa.3392.1.1

Edman P, Högfeldt E, Sillén LG, Kinell P-O (1950) Method for determination of the amino acid sequence in peptides. *Acta Chemica Scandinavica* **4**, 283–293. doi:10.3891/acta.chem.scand.04-0283

Edwards SV, Liu L, Pearl DK (2007) High-resolution species trees without concatenation. *Proceedings of the National Academy of Sciences USA* **104**, 5936–5941. doi:10.1073/pnas.0607004104

Eldridge MDB, Beck RMD, Croft DA, Travouillon KJ, Fox BJ (2019) An emerging consensus in the evolution, phylogeny, and systematics of marsupials and their fossil relatives (Metatheria). *Journal of Mammalogy* **100**, 802–837. doi:10.1093/jmammal/gyz018

Eldridge MDB, Close RL (1997) Chromosomes and evolution in rock-wallabies, *Petrogale* (Marsupialia: Macropodidae).

Australian Mammalogy **19**, 123–136. doi:10.13140/2.1.4353.0886

Eldridge MDB, Deakin JE, MacDonald AJ, Byrne M, Fitzgerald A, *et al.* (2020) The Oz Mammal Genomics initiative: developing genomic resources for mammal conservation at a continental scale. *Australian Zoologist* **40**, 505–509. doi:10.7882/AZ.2020.003

Eldridge MDB, Metcalfe CJ (2006) Marsupialia. In *Atlas of Mammalian Chromosomes.* (Eds SJ O'Brien, J Menninger and W Nash) pp. 9–21. Wiley, New York.

Elith J, Leathwick JR (2009) Species distribution models: ecological explanation and prediction across space and time. *Annual Review of Ecology, Evolution, and Systematics* **40**, 677–697. doi:10.1146/annurev.ecolsys.110308.120159

Exon NF, Kennett JP, Malone MJ (2004) 1 Leg 189 synthesis: Cretaceous – Holocene history of the Tasmanian Gateway. In *Proceeding of the Ocean Drilling Program, Scientific Results, Volume 189.* (Eds NF Exon, JP Kennett and MJ Malone) pp. 1–37, <http://www-odp.tamu.edu/publications/189_SR/synth/synth.htm>.

Faircloth BC, McCormack JE, Crawford NG, Harvey MG, Brumfield RT, Glenn TC (2012) Ultraconserved elements anchor thousands of genetic markers spanning multiple evolutionary timescales. *Systematic Biology* **61**, 717–726. doi:10.1093/sysbio/sys004

Fairfax RJ (2019) Dispersal of the introduced red fox (*Vulpes vulpes*) across Australia. *Biological Invasions* **21**, 1259–1268. doi:10.1007/s10530-108-1897-7

Faith DP (1992) Conservation evaluation and phylogenetic diversity. *Biological Conservation* **61**, 1–10. doi:10.1016/0006-3207(92)91201-3

Fanning FD (1982) Reproduction, growth and development in *Ningaui* sp. (Dasyuridae, Marsupialia) from the Northern Territory. In *Carnivorous Marsupials.* (Ed. M Archer) pp. 23–37. Royal Zoological Society of New South Wales, Sydney.

Farris JS (1983) The logical basis of phylogenetic analysis. In *Advances in Cladistics, Volume 2.* (Eds N Platnick and V Funk) pp. 7–36. Columbia University Press, New York.

Feigin CY, Newton AH, Doronina L, Schmitz J, Hipsley CA, *et al.* (2018) Genome of the Tasmanian tiger provides insights into the evolution and demography of an extinct marsupial carnivore. *Nature Ecology & Evolution* **2**, 182–192. doi:10.1038/s41559-017-0417-y

Feigin CY, Frankenburg S, Pask A (2022) A chromosome-scale hybrid genome assembly of the extinct Tasmanian tiger (*Thylacinus cynocephalus*). *Genome Biology and Evolution* **14**, evac048. doi:10.1093/gbe/evac048

Feldhamer GA, Merritt JF, Krajewski C, Rachlow JL, Stewart KM (2020) *Mammalogy: Adaptation, Diversity, Ecology.* 5th edn. Johns Hopkins University Press, Baltimore.

Felsenstein J (1978) The number of phylogenetic trees. *Systematic Biology* **27**, 27–33. doi:10.2307/2412810

Felsenstein J (1981) Evolutionary trees from DNA sequences: a maximum likelihood approach. *Journal of Molecular Evolution* **17**, 368–376. doi:10.1007/BF01734359

Felsenstein J (1983) Methods for inferring phylogenies: a statistical view. In *Numerical Taxonomy.* (Ed. J Felsenstein) pp. 315–334. Springer-Verlag, Heidelberg.

Felsenstein J (1984) Distance methods for inferring phylogenies: a justification. *Evolution* **38**, 16–24. doi:10.2307/2408542

Felsenstein J (1985a) Confidence limits on phylogenies: an approach using the bootstrap. *Evolution* **39**, 783–791. doi:10.1111/j.1558-5646.1985.tb00420.x

Felsenstein J (1985b) Phylogenies and the comparative method. *American Naturalist* **125**, 1–15. doi:10.1086/284325

Felsenstein, J (1993) *PHYLIP (Phylogenetic Inference Package).* Department of Genetics, University of Washington, Seattle. <https://evolution.genetics.washington.edu/phylip.html>

Felsenstein J (2004) *Inferring Phylogenies.* Sinauer Associates, Sunderland, MA.

Feng S, Bai M, Rivas-González S, Li C, Liu S, *et al.* (2022) Incomplete lineage sorting and phenotypic evolution in marsupials. *Cell* **185**, 1646–1660. doi:10.1016/j.cell.2022.03.034

Ferguson-Smith MA (1997) Genetic analysis by chromosome sorting and painting: phylogenetic and diagnostic applications. *European Journal of Human Genetics* **5**, 253–265. doi:10.1159/000484775

Ferguson-Smith MA, Trifonov V (2007) Mammalian karyotype evolution. *Nature Reviews Genetics* **8**, 950–962. doi:10.1038/nrg2199

Firestone KB (2000) Phylogenetic relationships among quolls revisited: the mtDNA control region as a useful tool. *Journal of Mammalian Evolution* **7**, 1–22. doi:10.1023/A1009425815903

Firestone KB (2003) The application of genetic research to conservation management in carnivorous marsupials with special emphasis on dasyurids. In *Predators with Pouches: The Biology of Carnivorous Marsupials.* (Eds M Jones, C Dickman and M Archer) pp. 475–486. CSIRO Publishing, Melbourne.

Firestone KB, Elphinstone MS, Sherwin WB, Houlden BA (1999) Phylogeographical structure of tiger quolls, *Dasyurus maculatus* (Dasyuridae: Marsupialia), an endangered carnivorous marsupial. *Molecular Ecology* **8**, 1613–1625. doi:10.1046/j.1365-294x.1999.00745.x

Firestone KB, Houlden BA, Sherwin WB, Geffen E (2000) Variability and differentiation of microsatellites in the genus *Dasyurus* and conservation implications for the large Australian carnivorous marsupials. *Conservation Genetics* **1**, 115–133. doi:10.1023/A:1026578821339

Fisher DO, Blomberg SP (2011) Costs of reproduction and terminal investment by females in a semelparous marsupial. *PLoS ONE* **6**, e15226. doi:10.1371/journal.pone.0015226

Fisher DO, Cockburn A (2005) The large-male advantage in brown antechinuses: female choice, male dominance, and delayed male death. *Behavioral Ecology* **17**, 164–171. doi:10.1093/beheco/arj012

Fisher DO, Dickman CR, Jones ME, Blomberg SP (2013) Sperm competition drives the evolution of suicidal reproduction in mammals. *Proceedings of the National Academy of Sciences USA* **110**, 17910–17914. doi:10.1073/pnas.1310691110

Fisher DO, Johnson CN, Lawes MJ, Fritz SA, McCallum H, *et al.* (2014) The current decline of tropical marsupials in

Australia: is history repeating? *Global Ecology and Biogeography* **23**, 181–190. doi:10.1111/geb.12088

Fitch WM (1971) Toward defining the course of evolution: minimum change for a specified tree topology. *Systematic Zoology* **20**, 406–416. doi:10.1093/sysbio/20.4.406

Fitch WM (1976) Molecular evolutionary clocks. In *Molecular Evolution*. (Ed. FJ Ayala) pp. 160–178. Sinauer Associates, Sunderland, MA.

Fitch WM, Margoliash E (1967) Construction of phylogenetic trees. *Science* **155**, 279–284. doi:10.1126/science.155.3760.279

Fitzhugh K (2006) The 'requirement of total evidence' and its role in phylogenetic systematics. *Biology and Philosophy* **21**, 309–351. doi:10.1007/s10539-005-7325-2

Flannery TF (1989) Origins of the Australo-Pacific land mammal fauna. *Australian Zoological Reviews* **1**, 15–24.

Flannery T (1990) *Mammals of New Guinea*. Robert Brown & Associates, Carina, Queensland.

Flannery T (1994) *The Future Eaters*. Reed Books, Chatswood, NSW.

Flannery T (1995) *Mammals of New Guinea, Revised and Updated Edition*. Reed Books, Chatswood, NSW.

Flower WH (1867) On the development and succession of teeth in the Marsupialia. *Philosophical Transactions of the Royal Society* **157**, 631–641, <https://www.jstor.org/stable/108984>.

Flower WH (1869) Remarks on the homologies and notation of the teeth of the Mammalia. *Journal of Anatomy and Physiology* **3**, 262–278. PMCID: PMC1318682.

Foote M, Hunter JP, Janis CM, Sepkoski JJ (1999) Evolutionary and preservational constraints on origins of biologic groups: divergence times of eutherian mammals. *Science* **283**, 1310–1314. doi:10.1126/science.283.5406.1310

Foster JW, Brennan FE, Hampikian GK, Goodfellow PN, Sinclair AH, *et al.* (1992) Evolution of sex determination and the Y chromosome: SRY-related sequences in marsupials. *Nature* **359**, 531–533. doi:10.1038/359531a0

Fox BJ (1982) A review of dasyurid ecology and speculation on the role of limiting similarity in community organization. In *Carnivorous Marsupials*. (Ed. M Archer) pp. 97–116. Royal Zoological Society of New South Wales, Sydney.

Fox BJ, Whitford D (1982) Polyoestry in a predictable coastal environment: reproduction, growth and development in *Sminthopsis murina* (Dasyuridae, Marsupialia). In *Carnivorous Marsupials*. (Ed. M Archer) pp. 39–48. Royal Zoological Society of New South Wales, Sydney.

Frankham GJ, Thompson S, Ingleby S, Soderquist T, Eldridge MDB (2017) Does the 'extinct' eastern quoll (*Dasyurus viverrinus*) persist in Barrington Tops, N.S.W.? *Australian Mammalogy* **39**, 243–247. doi:10.1071/AM18024

Friend GR, Johnson BW, Mitchell DS, Smith GT (1997) Breeding, population dynamics and habitat relationships of *Sminthopsis dolichura* (Marsupialia: Dasyuridae) in semi-arid shrublands of Western Australia. *Wildlife Research* **24**, 245–262. doi:10.1071/WR96070

Friend JA (1989) Myrmecobiidae. In *Fauna of Australia, Volume 1B*. (Eds DW Walton and BJ Richardson) Chapter 22, pp. 1–18. Australian Government Publishing Service, Canberra, <https://www.awe.gov.au/sites/default/files/env/pages/a117ced5-9a94-4586-afdb-1f333618e1e3/files/22-ind.pdf>.

Friend JA (2008) Family Myrmecobiidae. In *The Mammals of Australia*. 3rd edn. (Eds S Van Dyck and R Strahan) pp. 162–165. Reed New Holland, Sydney.

Friend JA, Thomas ND (2003) Conservation of the numbat (*Myrmecobius fasciatus*). In *Predators with Pouches: The Biology of Carnivorous Marsupials*. (Eds M Jones, C Dickman and M Archer) pp. 452–463. CSIRO Publishing, Melbourne.

Fusco DA, McDowell MC, Medlin G, Prideaux GJ (2017) Fossils reveal late Holocene diversity and post-European decline of the terrestrial mammals of the Murray-Darling Depression. *Wildlife Research* **44**, 60–71. doi:10.1071/WR16134

Gallardo MH, Patterson B (1987) An additional 14-chromosome karyotype and sex chromosome mosaicism in South American marsupials. In *Studies in Neotropical Mammalogy: Essays in Honour of Philip Herschkowitz*. (Eds BD Patterson and RM Timm) pp. 111–115. Fieldiana Zoology New Series Number 39, <https://www.biodiversitylibrary.org/page/4465235>.

Gallus S, Janke A, Kumar V, Nilsson MA (2015) Disentangling the relationship of the Australian marsupial orders using retrotransposon and evolutionary network analysis. *Genome Biology and Evolution* **7**, 985–992. doi:10.1093/gbe/evv052

Galtier N (2007) A model of horizontal gene transfer and the bacterial phylogeny problem. *Systematic Biology* **56**, 633–642. doi:10.1080/10635150701546231

Garamszegi LZ (Ed.) (2014) *Modern Phylogenetic Comparative Methods and their Application in Evolutionary Biology*. Springer-Verlag, Berlin Heidelberg.

García-Navas V, Kear BP, Westerman M (2020) The geography of speciation in dasyurid marsupials. *Journal of Biogeography* **47**, 2042–2053. doi:10.1111/jbi.13852

García-Navas V, Rodríguez-Rey M, Westerman M (2018) Bursts of morphological and lineage diversification in dasyurids, a 'classic' adaptive radiation. *Biological Journal of the Linnean Society* **123**, 782–795. doi:10.1093/biolinnean/bly064

García-Navas V, Westerman M (2018) Niche conservatism and phylogenetic clustering in a tribe of arid-adapted marsupial mice, the Sminthopsini. *Journal of Evolutionary Biology* **31**, 1204–1215. doi:10.1111/jeb.13297

Garnett ST, Christidis L (2017) Taxonomy anarchy hampers conservation. *Nature* **546**, 25–27. doi:10.1038/546025a

Gatesy J, O'Grady P, Baker RH (1999) Corroboration among data sets in simultaneous analysis: hidden support for phylogenetic relationships among higher level artiodactyl taxa. *Cladistics* **15**, 271–313. doi:10.1111/j.1096-0031.1999.tb00268.x

Gatesy J, Springer MS (2014) Phylogenetic analysis at deep timescales: unreliable gene trees, bypassed hidden support, and the coalescence/concatalescence conundrum. *Molecular Phylogenetics and Evolution* **80**, 231–266. doi:10.1016/j.ympev.2014.08.013

Gaveau DLA, Santos L, Locatelli B, Salim MA, Husnayaen H, *et al.* (2021) Forest loss in Indonesian New Guinea (2001–2019): trends, drivers, and outlook. *Biological Conservation* **261**, 109225. doi:10.1016/j.biocon.2021.109225

Geiser F (2003) Thermal biology and energetics of carnivorous marsupials. In *Predators with Pouches: The Biology of Carnivorous Marsupials.* (Eds M Jones, C Dickman, and M Archer) pp. 238–253. CSIRO Publishing, Melbourne.

Geoffroy Sainte-Hilaire E (1796) Dissertation sure les animaux à bourse (*Didelphis* L.). Imprimerie du Magasin encyclopédique, (1796), Paris.

Geoffroy Sainte-Hilaire E (1803) Note sur les expèces du genre dasyure. *Bulletin des Sciences de la Société Philomatique Paris* **3**, 258–259.

George GG, Brown PR (1992) Conservation management of small populations. In *Australasian Marsupials and Monotremes: An Action Plan for their Conservation.* (Ed. M Kennedy) pp. 38–42. IUCN, Gland, Switzerland, <https://portals.iucn.org/library/node/6133>.

Gerhardt U (1905) Morphologische und biolgische Studien über die Kopulationsorgane der Säugetiere. *Jenaische Zeitschrift für Naturwissenschaft* **39**, 43–118, <https://www.biodiversitylibrary.org/page/11955672>.

Gerhardt U (1933) Kloake and Begattungsorgane. (b) Marsupialier, Beuteltiere. In *Handbuch der vergleichenden Anatomie der Wirbeltiere.* (Eds L Bolk, E Göppert, E Kallius, and W Lubosch) pp. 318–321. Bd vi. Urban and Schwarzenberg, Berlin.

Gerdtz WR, Archbold NW (2003) *Glaucodon ballaratensis* (Marsupialia, Dasyuridae), a late Pliocene 'devil' from Batesford, Victoria. *Proceedings of the Royal Society of Victoria* **115**, 35–44.

Geyle HM, Woinarski JC, Baker GB, Dickman CR, Dutson G, *et al.* (2018) Quantifying extinction risk and forecasting the number of impending Australian bird and mammal extinctions. *Pacific Conservation Biology* **24**, 157–167. doi:10.1071/PC18006

Ghiselin MT (1966) On psychologism in the logic of taxonomic controversies. *Systematic Zoology* **15**, 207–215. doi:10.2307/2411392

Gill T (1872) Arrangement of the families of mammals. *Smithsonian Miscellaneous Collections* **230**, 1–98, <https://www.biodiversitylibrary.org/page/14512573>.

Gill TN (1870) On the relations of the orders of mammals. In *Proceedings of the American Association for the Advancement of Science, 19th meeting.* (Ed. J Lovering) pp. 267–270. Lovering, Cambridge, MA.

Gillespie RG, Bennett GM, De Meester L, Feder JL, Fleischer RC, *et al.* (2020) Comparing adaptive radiations across time, space, and taxa. *Journal of Heredity* **111**, 1–20. doi:10.1093/jhered/esz064

Godthelp H, Archer M, Cifelli R, Hand SJ, Gilkeson CF (1992) Earliest known Australian Tertiary mammal fauna. *Nature* **356**, 514–516. doi:10.1038/356514a0

Godthelp H, Wroe S, Archer M (1999) A new marsupial from the Early Eocene Tingamarra Local Fauna of Murgon, southeastern Queensland: a prototypical Australian marsupial? *Journal of Mammalian Evolution* **6**, 289–313. doi:10.1023/A:1020517808869

Goin FJ, Woodburne MO, Zimicz AN, Martin GM, Chornogubsky L (2016) *A Brief History of South American Metatherians: Evolutionary Contexts and Intercontinental Dispersals.* Springer Earth System Sciences, Springer, Dordrecht, <https://u1lib.org/book/2619119/93e8d1>.

Goldfuss GA (1820) *Handbuch der Zoologie, Volume 2.* JL Schrag, Nurnberg, <https://catalog.hathitrust.org/Record/001997627>.

Goloboff PA, Carpenter JM, Arias JS, Esquivel DRM (2008) Weighting against homoplasy improves phylogenetic analysis of morphological data sets. *Cladistics* **14**, 1–16. doi:10.1111/j.1096-0031.2008.00209.x

Goloboff PA, Pittman M, Pol D, Xu X (2019) Morphological data sets fit a common mechanism much more poorly than DNA sequences and call into question the Mkv model. *Systematic Biology* **68**, 494–504. doi:10.1093/sysbio/syy077

Goswami A. 2006. Cranial modularity shifts during mammalian evolution. *American Naturalist* **168**, 270–280. doi:10.1086/505758

Gould J (1863) *The Mammals of Australia, Volume 1.* Taylor and Francis, London, <https://www.biodiversitylibrary.org/page/49738000>.

Gould SJ (1985) A clock of evolution. *Natural History* **94**(4), 12–25.

Grant J, Mahadevaiah SK, Sangrthi MN, Royo H, Duckworth J, *et al.* (2012) Rsx is a metatherian RNA with Xist-like properties in X-chromosome inactivation. *Nature* **487**, 254–258. doi:10.1038/nature11171

Graphodatsky AS, Trifonov VA, Stanyon R (2011) The genome diversity and evolution of mammals. *Molecular Cytogenetics* **4**, 22. doi:10.1186/1755-8166-4-22

Graur D (2016) *Molecular and Genome Evolution.* Sinauer Associates, Sunderland, MA.

Graves JAM, Sinclair AH, Spencer JA (1990) Marsupial gene mapping and the evolution of mammalian sex chromosome form and function. In *Mammals from Pouches and Eggs: Genetics, Breeding and Evolution of Marsupials and Monotremes.* (Eds JA Marshall Graves, RM Hope, and DW Cooper) pp. 223–241. CSIRO, Australia. doi:10.1071/ZO989036

Graves JAM, Westerman M (2003) Marsupial genetics and genomics. *Trends in Genetics* **18**, 517–521. doi:10.1016/s0168-9525(02)02772-5

Gray JE (1858) List of species of Mammalia sent from the Aru Islands by Mr. A.R. Wallace to the British Museum. *Proceedings of the Zoological Society of London* **26**, 106–113, <https://www.biodiversitylibrary.org/page/32271594>.

Grealy A, McDowell M, Retallick C, Bunce M, Peacock D (2019) Novel mitochondrial haplotype of spotted-tail quoll (*Dasyurus maculatus*) present on Kangaroo Island (South Australia) prior to extirpation. *The Holocene* **30**, 136–144. doi:10.1177/0959683619875805

Greaves IK, Svartman M, Wakefield M, Taggart D, De Leo A, *et al.* (2001) Chromosomal painting detects non-random chromosome arrangement in dasyurid marsupial sperm. *Chromosome Research* **9**, 251–259. doi:10.1023/A:1016656722134

Gregory ME (2008) *Evolutionism in Eighteenth-Century French Thought*. Peter Lang, New York.

Gregory WK (1910) The orders of mammals. *Bulletin of the American Museum of Natural History* **27**, 1–104, <https://www.biodiversitylibrary.org/page/15495336>.

Gregory WK (1920) Studies in comparative myology and osteology, No. IV. A review of the evolution of the lachrymal bone. *Bulletin of the American Museum of Natural History* **42**, 95–263, <https://digitallibrary.amnh.org/handle/2246/1061>.

Griffiths AD, Rankmore B, Brennan K, Woinarski JCZ (2017) Demographic evaluation of translocating the threatened northern quoll to two Australian islands. *Wildlife Research* **44**, 238–247. doi:10.1071/WR16165

Grossek G, Woolley PA, Menzies JI (2010) The diet of some New Guinean dasyurid marsupials. *Australian Mammalogy* **32**, 145–155. doi:10.1071/AM10002

Groves CP (2005) Order Dasyuromorphia. In *Mammal Species of the World*. (Eds DE Wilson and DM Reeder) pp. 23–37. Johns Hopkins University Press, Baltimore.

Groves CP (2006) Humans among primates. In *Evolution and Biogeography of Australasian Vertebrates*. (Eds JR Merrick, M Archer, GM Hickey and MSY Lee) pp. 779–831. Auscipub, Oatlands, NSW.

Guiler ER (1985) *Thylacine: The Tragedy of the Tasmanian Tiger*. Oxford University Press, Melbourne.

Gurovich Y, Travouillon KJ, Beck RMD, Muirhead J, Archer M (2014) Biogeographical implications of a new mouse-sized fossil bandicoot (Marsupialia: Peramelemorphia) occupying a dasyurid-like ecological niche across Australia. *Journal of Systematic Palaeontology* **12**, 265–290. doi:10.1080/14772019.2013.776646

Haeckel E (1866) *Generelle Morphologie der Organismen*. Reimer, Berlin, <https://www.biodiversitylibrary.org/page/3138555>.

Haeckel E (1874) *Anthropogenie*. Engelmann, Leipzig, <https://www.biodiversitylibrary.org/page/59675763>.

Hamm G, Mitchell P, Arnold LJ, Prideaux GJ, Questiaux D, *et al.* (2016) Cultural innovation and megafauna interaction in the early settlement of arid Australia. *Nature* **539**, 280–283. doi:10.1126/science.283.5399.205

Hammer TA, Renton M, Mucina L, Thiele K A (2021) Arid Australia as a source of plant diversity: the origin and climatic evolution of *Ptilotis* (Amaranthaceae). *Australian Systematic Botany* **34**, 570–586. doi:10.1071/SB21012

Haouchar D, Haile J, McDowell MC, Murray DC, White NE, *et al.* (2014) Thorough assessment of DNA preservation from fossil bone and sediments excavated from a late Pleistocene-Holocene cave deposit on Kangaroo Island, South Australia. *Quaternary Science Reviews* **84**, 56–64, <https://researchrepository.murdoch.edu.au/id/eprint/20309/1/DNA-preservation-from-fossil-bone-and-sediments.pdf>.

Harald AS, Mooi RD (1994) Areas of endemism: definition and recognition criteria. *Systematic Zoology* **43**, 261–266. doi:10.1093/sysbio/43.2.261

Harjunmaa E, Seidel K, Häkkinen T, Renvoisé E, Corfe IJ, *et al.* (2014) Replaying evolutionary transitions from the dental fossil record. *Nature* **512**, 44–49. doi:10.1038/nature13613

Harmon LJ, Shulte II JA, Larson A, Losos JB (2003) Tempo and mode of evolutionary radiation in iguanian lizards. *Science* **301**, 961–964. doi:10.1126/science.1084786

Harmon LJ, Losos JB, Davies J, Gillespie RG, Gittleman JL, *et al.* (2010) Early bursts of body size and shape evolution are rare in comparative data. *Evolution* **64**, 2385–2396. doi:10.1111/j.1558.2010.01025.x

Harris GP (1808) Description of two new species of *Didelphis* from Van Dieman's Land. *Transactions of the Linnean Society of London* **9**, 174–178. doi:10.1111/j.1096-3642.1818.tb00336.x

Hartman CG (1920) Studies in the development of the opossum: V. The phenomenon of parturition and the method of transfer to the pouch. *Anatomical Record* **19**, 251–261. doi:10.1002/ar.1090190502

Hartman CG (1952) *Possums*. University of Texas Press, Austin.

Hartwell LH, Hood L, Goldberg ML, Reynolds AE, Silver LM (2011) *Genetics: From Genes to Genomes*. McGraw-Hill, New York.

Hasegawa M, Kishino H, Yano T (1985) Dating of the human-ape splitting by a molecular clock of mitochondrial DNA. *Journal of Molecular Evolution* **21**, 160–174. doi:10.1007/BF02101694

Hayes GLT, Simmons LW, Dugand RJ, Mills HR, Roberts JD, *et al.* (2019) Male semelparity and multiple paternity confirmed in an arid-zone dasyurid. *Journal of Zoology* **308**, 266–273. doi:10.1111/jzo.12672

Hayman DL (1977) Chromosome number – constancy and variation. In *The Biology of Marsupials*. (Eds B Stonehouse and D Gilmore) pp. 27–48. Macmillan Press, London, <https://link.springer.com/content/pdf/bfm%3A978-1-349-02721-7%2F1.pdf>.

Hayman DL (1990) Marsupial cytogenetics. *Australian Journal of Zoology* **37**, 331–339. doi:10.1071/ZO9890331

Hayman DL, Martin PG (1974) Mammalia I: Monotremata and Marsupialia. In *Animal Cytogenetics, Volume 4, Chordata 4*. (Ed. B. John) pp. 1–110. Gebruder Borntraeger, Berlin.

Hayman DL, Rofe RH, Sharp PJ (1987) Chromosome evolution in marsupials. In *Chromosomes Today, Volume 9*. (Eds A Stahl, JM Luciani and AM Vagner-Capodano) pp. 91–102. Allen and Unwin, London.

Hazkani-Covo E (2022) A burst of numt insertion in the Dasyuridae family during marsupial evolution. *Frontiers in Ecology and Evolution* **10**. doi:10.3389/fevo.2022.844443

Heard SB, Hauser DI (1995) Key evolutionary innovations and their ecological mechanisms. *Historical Biology* **10**, 151–173. doi:10.1080/10292389509380518

Heath TA, Huelsenbeck JP, Stadler T (2014) The fossilized birth-death process for calibration of divergence time estimates. *Proceedings of the National Academy of Sciences USA* **111**, E2957–E2966. doi:10.1073/pnas.1319091111

Hecht MK (1976) Phylogenetic inference and methodology as applied to the vertebrate record. In *Evolutionary Biology, Volume 9*. (Eds T Dohzhansky, M Hecht and W Steere) pp. 335–363. Appleton-Century-Crofts, New York.

Hecht MK, Edwards JL (1976) The determination of parallel or monophyletic relationships: the proteid salamanders – a test case. *American Naturalist* **110**, 653–677. doi:10.1086/283096

Hendricks S, Epstein B, Schönfeld B, Wiench C, Hamede R, *et al.* (2017) Conservation implications of low genetic diversity and population structure in Tasmanian devils (*Sarcophilus harrisii*). *Conservation Genetics* **18**, 977–982. doi:10.1007/s10592-017-0939-5

Hennig W (1950) *Grundzüge einer Theorie der phylogenetischen Systematik*. Deutscher Zentralverlag, Berlin.

Hennig W (1966) *Phylogenetic Systematics*. University of Illinois Press, Urbana-Champaign.

Hernández-Hernández T (2019) Evolutionary rates and adaptive radiations. *Biology and Philosophy* **34**, 41. doi:10.1007/s10539-019-9694-y

Hill JE (1990) A memoir and bibliography of Michael Rogers Oldfield Thomas, F.R.S. *Bulletin of the British Museum of Natural History (Historical Series)* **18**, 25–113, <https://www.biodiversitylibrary.org/page/4126598>.

Hill JP, Hill WCO (1955) The growth stages of the pouch young of the Native Cat (*Dasyurus viverrinus*) together with observations on the anatomy of the new born young. *Transactions of the Zoological Society of London* **28**, 349–453. doi:10.1111/j.1096-3642.1955.tb00003.x

Hill JP, O'Donoghue CH (1913) The reproductive cycle in the marsupial *Dasyurus viverrinus*. *Quarterly Journal of Microscopical Science* **59**, 133–174, <https://biodiversitylibrary.org/page/49662660>.

Hill RS (Ed) (1994) *History of the Australian Vegetation: Cretaceous to Recent*. Cambridge University Press, Cambridge, <https://www.adelaide.edu.au/press/system/files/media/documents/2019-04/uap-australian-vegetation-ebook.pdf>.

Hill RS, Beer YK, Hill KE, Maciunas E, Tarran MA, *et al.* (2016) Evolution of the eucalypts–an interpretation from the macrofossil record. *Australian Journal of Botany* **16**, 600–608. doi:10.1071/BT16117

Hillis DM, Moritz C (1990) An overview of applications of molecular systematics. In *Molecular Systematics*. (Eds DM Hillis and C Moritz) pp. 502–515. Sinauer Associates, Sunderland, MA.

Hocknull SA, Lewis R, Arnold LJ, Pietsch T, Joannes-Boyau R, *et al.* (2020) Extinction of eastern Sahul megafauna coincides with sustained environmental deterioration. *Nature Communications* **11**, 2550. doi:10.1038/s41467-020-15785-w

Hohnen R, Tuft KD, Legge S, Hillyer M, Spencer PBS, *et al.* (2016) Rainfall and topography predict gene flow among populations of the declining northern quoll (*Dasyurus hallucatus*). *Conservation Genetics* **17**, 1213–1228. doi:10.1007/s10592-016-0856-z

Holleley CE, Dickman CR, Crowther MS, Oldroyd BP (2006) Size breeds success: multiple paternity, multivariate selection and male semelparity in a small marsupial, *Antechinus stuartii*. *Molecular Ecology* **15**, 3439–3448. doi:10.1111/j.1365-294X.2006.03001.x

Hope RM, Bennett JH, Chesson CM (1986) Genetic variation in natural and laboratory populations of the marsupial *Sminthopsis crassicaudata*. *Biochemical Genetics* **24**, 597–613. doi:10.1007/BF00504337

How RA, Dell J, Cooper NK (1991) Ecological survey of Abydos-Woodstock Reserve, Western Australia: vertebrate fauna. *Records of the Western Australian Museum, Supplement* **37**, 78–125, <https://museum.wa.gov.au/research/records-supplements/records/ecological-survey-abydos-woodstock-reserve-pilbara-region-wes-3>.

How RA, Cooper NK, Girardi L, Bow BG (2002) The mardo: an examination of geographic variation in morphology and reproductive potential in *Antechinus flavipes* in southwestern Australia. *Records of the Western Australian Museum* **20**, 441–447, <https://museum.wa.gov.au/research/records-supplements/records/mardo-examination-geographic-variation-morphology-and-reproduct>.

How RA, Spencer PBS, Schmitt LH (2009) Island populations have high conservation value for northern Australia's top marsupial predator ahead of a threatening process. *Journal of Zoology (London)* **278**, 206–217. doi:10.1111/j.1469-7998.2009.00569.x

Hubby JL, Throckmorton LH (1965) Protein differences in *Drosophila*. II. Comparative species genetics and evolutionary problems. *Genetics* **52**, 203–215. doi:10.1093/genetics/52.1.203

Huelsenbeck JP, Ronquist F (2001) MRBAYES: Bayesian inference of phylogenetic trees. *Bioinformatics* **17**, 754–755. doi:10.1093/bioinformatics/17.8.754

Hughes RL (1982) Reproduction in the Tasmanian Devil *Sarcophilus harrisii* (Dasyuridae, Marsupialia). In *Carnivorous Marsupials*. (Ed. M Archer) pp. 49–63. Royal Zoological Society of New South Wales, Sydney.

Hull DL (1976) Are species really individuals? *Systematic Zoology* **25**, 174–191. doi:10.2307/2412744

Hull DL (1978) A matter of individuality. *Philosophy of Science* **45**, 335–360. doi:10.1086/288811

Hull DL (1988) *Science as a Process*. University of Chicago Press, Chicago.

Hunter DO, Britz T, Jones M, Letnic M (2015) Reintroduction of Tasmanian devils to mainland Australia can restore top-down control in ecosystems where dingoes have been extirpated. *Biological Conservation* **191**, 428–435. doi:10.1016/j.biocon.2015.07.030

Huxley J (1942) *Evolution: The Modern Synthesis*. Allen & Unwin, London.

Huxley TH (1872) *A Manual of the Anatomy of Vertebrated Animals*. D. Appleton and Company, New York, <https://www.biodiversitylibrary.org/page/35088574>.

Huxley TH (1880) On the application of the laws of evolution to the arrangement of the Vertebrata, and more particularly of the Mammalia. *Proceedings of the Zoological Society of London* **1880**, 649–662, <https://www.biodiversitylibrary.org/page/28522869>.

ICZN (1999) *International Code of Zoological Nomenclature*. 4th edn. The International Trust for Zoological Nomenclature, London, <https://www.iczn.org/the-code/the-code-online/>.

Illiger JKW (1811) *Prodromus Systematis Mammalium et Avium*. Berolini, Berlin, <https://www.biodiversitylibrary.org/page/48453690>.

Ingles ED, Deakin JE (2016) Telomeres, species differences and unusual telomeres in vertebrates: presenting challenges and opportunities. *AIMS Genetics* **3**, 1–24. doi:10.3934/genet.2016.1.1

Innis MA, Gelfand DH (1990) Optimization of PCRs. In *PCR Protocols: A Guide to Methods and Applications*. (Eds MA Innis, DH Gelfand, JJ Sininsky and TJ White) pp. 3–12. Academic Press, San Diego.

Iredale T, Troughton ELG (1934) A check-list of the mammals recorded from Australia. *Australian Museum Memoir* **6**, 1–122. doi:10.3853/j.0067-1967.6.1934.516

Irwin DM, Kocher TD, Wilson AC (1991) Evolution of the cytochrome *b* gene in mammals. *Journal of Molecular Evolution* **32**, 128–144. doi:10.1007/BF02515385

Isaac NJB, Mallet J, Mace GM (2004) Taxonomic inflation: its influence on macroecology and conservation. *Trends in Ecology & Evolution* **19**, 464–469. doi:10.1016/j.tree.2004.06.004

IUCN (2012) *IUCN Red List Categories and Criteria*, version 3.1. 2nd edn. Gland and Cambridge, Cambridge, <https://portals.iucn.org/library/node/10315>.

IUCN (2021) The IUCN Red List of Threatened Species, Version 2021-1, <https://www.iucnredlist.org>.

Jansa SA, Barker FK, Voss RS (2014) The early diversification of didelphid marsupials: a window into South America's 'Splendid Isolation'. *Evolution* **68**, 684–695. doi:10.1111/evo.12290

Jenner RA (2004) Accepting partnership by submission? Morphological phylogenetics in a molecular millennium. *Systematic Biology* **53**, 333–342. doi:10.1080/10635150490423962

Jentink FA (1911) New and interesting mammals of the Dutch New-Guinea-Expedition to the Snow Mountains. *Notes from the Leyden Museum* **33**, 233–238, <https://www.biodiversitylibrary.org/page/13206533>.

Jernvall J, Thesleff I (2012) Tooth shape formation and tooth renewal: evolving with the same signals. *Development* **139**, 3487–3497. doi:10.1242/dev.085084

Jiang X, Edwards SV, Liu L (2020) The multispecies coalescent model outperforms concatenation across diverse phylogenomic data sets. *Systematic Biology* **69**, 795–812. doi:10.1093/sysbio/syaa008

John B, Lewis K (1968) *The Chromosome Complement*. Protoplasmalogia (Handbuch der Protoplasmaforschung), Volume 6, Springer Verlag, Vienna, <https://www.scribd.com/document/368898695/Protoplasmatologia-6-a-B-John-K-R-Lewis-Auth-The-Chromosome-Complement-Springer-Verlag-Wien-1968>.

Johnson LAS, Briggs BG (1975) On the Proteaceae – evolution and classification of a southern family. *Botanical Journal of the Linnaean Society* **70**, 83–182. doi:10.1111/j.1095-8339.1975.tb01644.x

Johnson CN (2006) *Australia's Mammal Extinctions: a 50,000 Year History*. Cambridge University Press, Cambridge.

Johnson CN, Alroy J, Beetong NJ, Bird MI, Brook BW, *et al.* (2016) What caused extinction of the Pleistocene megafauna of Sahul? *Proceedings of the Royal Society B* **283**, 2015399. doi:10.1098/rspb/2015.2399

Johnson CN, Isaac JL (2009) Body mass and extinction risk in Australian marsupials: the 'critical weight range' revisited. *Austral Ecology* **34**, 35–40. doi:10.1111/j.1442-9993.2008.01878.x

Johnson CN, Wroe S (2003) Causes of extinction of vertebrates during the Holocene of mainland Australia: arrival of the dingo, or human impact? *The Holocene* **13**, 941–948. doi:10.1191/0959683603hl682fa

Johnson DH (1954) A new marsupial of the genus *Antechinus* from Northern Australia. *Proceedings of the Biological Society of Washington* **67**, 77–79, <https://upload.wikimedia.org/wikipedia/commons/9/9a/Proceedings_of_the_Biological_Society_of_Washington_%28IA_proceedingsofbio67biol%29.pdf>.

Johnston PG, Watson CM, Adams M, Paull DJ (2002) Sex chromosome elimination, X chromosome inactivation and reactivation in the southern brown bandicoot *Isoodon obesulus* (Marsupialia: Peramelidae). *Cytogenetics and Genome Research* **99**, 119–124. doi:10.1159/000071583

Jolly CJ, Webb JK, Gillespie GR, Phillips BL (2020) Training fails to elicit behavioral change in a marsupial suffering evolutionary loss of antipredator behaviors. *Journal of Mammalogy* **101**, 1108–1116. doi:10.1093/jmammal/gyaa060

Jones M (2008) Eastern Quoll. In *The Mammals of Australia*. 3rd edn. (Eds S Van Dyck and R Strahan) pp. 62–64. Reed New Holland, Sydney.

Jones ME, Oakwood M, Belcher CA, Morris K, Murray AJ, *et al.* (2003) Carnivore concerns: problems, issues and solutions for conserving Australasia's marsupial carnivores. In *Predators with Pouches: The Biology of Carnivorous Marsupials*. (Eds M Jones, C Dickman, and M Archer) pp. 422–434. CSIRO Publishing, Melbourne.

Jones ME, Paetkau D, Geffen E, Moritz C (2004) Genetic diversity and population structure of Tasmanian devils, the largest marsupial carnivore. *Molecular Ecology* **13**, 2197–2209. doi:10.1111/j.1365-294X.2004.02239.x

Jukes TH, Cantor CR (1969) Evolution of protein molecules. In *Mammalian Protein Metabolism*. (Ed. HN Munro) pp. 21–132. Academic Press, New York.

Kalkman VJ, Dijkstra K-DB, Dow RA, Stokvis FR, van Tol J (2018) Out of Australia: the Argiolestidae reveal the Melanesian Arc System and East Papua Composite Terrane as possible ancient dispersal routes to the Indo-Australian archipelago (Odonata: Argiolestidae). *International Journal of Odonatology* **21**, 1–14. doi:10.1080/13887890.2017.1402825

Kangas AT, Evans AR, Thesleff I, Jernvaal J (2004) Nonindependence of dental characters. *Nature* **432**, 211–214. doi:10.1038/nature02927

Kealy S, Beck R (2017) Total evidence phylogeny and evolutionary timescale for Australian faunivorous marsupials (Dasyuromorphia). *BMC Evolutionary Biology* **17**, 240. doi:10.1186/s12862-017-1090-0

Kear BP, Aplin KP, Westerman M (2016) Bandicoot fossils and DNA elucidate antiquity amongst xeric-adapted Australasian marsupials. *Scientific Reports* **6**, 37537. doi:10.1038/srep37537

Kelley NP, Motani R (2015) Trophic convergence drives morphological convergence in marine tetrapods. *Biology Letters* **11**, 2014709. doi:10.1098/rsbl.2014.0709

Kelly E, Phillips BL (2018) Targeted gene flow and rapid adaptation in an endangered marsupial. *Conservation Biology* **33**, 112–121. doi:10.1111/cobi.13149

Kemper CM, Cooper SJB, Medlin GC, Adams M, Stemmer D, *et al.* (2011) Cryptic grey-bellied dunnart (*Sminthopsis griseoventer*) discovered in South Australia: genetic, morphological, and subfossil analyses show the value of collecting voucher material. *Australian Journal of Zoology* **59**, 127–144. doi:10.1071/ZO11037

Kennedy M (Ed) (1992) *Australasian Marsupials and Monotremes: An Action Plan for their Conservation.* IUCN, Gland, Switzerland, <https://portals.iucn.org/library/node/6133>.

Kerr JB, Hedges MP (1983) Spontaneous spermatogenic failure in the marsupial mouse *Antechinus stuartii* Macleay (Dasyuridae: Marsupialia). *Australian Journal of Zoology* **31**, 445–466. doi:10.1071/ZO9830445

Kimura M (1968) Evolutionary rate at the molecular level. *Nature* **217**, 624–626. doi:10.1038/217624a0

Kimura M (1980) A simple method for estimating evolutionary rate of base substitutions through comparative studies of nucleotide sequences. *Journal of Molecular Evolution* **16**, 111–120. doi:10.1007/BF01731581

Kimura M (1983) *The Neutral Theory of Molecular Evolution.* Cambridge University Press, Cambridge.

King CE, Jukes TH (1969) Non-Darwinian evolution. *Science* **164**, 788–798. doi:10.1126/science.164.3881.788

King M (1993) *Species Evolution: The Role of Chromosome Changes.* Cambridge University Press, Cambridge, UK.

Kingman JFC (1982) On the genealogy of large populations. *Journal of Applied Probability* **19**, 27–43. doi:10.2307/3213548

Kirsch JAW (1977) The comparative serology of Marsupialia, and a classification of marsupials. *Australian Journal of Zoology Supplemental Series* **52**, 1–152. doi:10.1071/AJZS052

Kirsch JAW (1982) The builder and the bricks: notes toward a philosophy of characters. In *Carnivorous Marsupials.* (Ed. M Archer) pp. 587–594. Royal Zoological Society of New South Wales, Mosman, Australia.

Kirsch JAW, Archer M (1982) Polythetic cladistics, or, when parsimony's not enough: the relationships of carnivorous marsupials. In *Carnivorous Marsupials.* (Ed. M Archer) pp. 595–619. Royal Zoological Society of New South Wales, Mosman, Australia.

Kirsch JAW, Krajewski C, Springer M, Archer M (1990) DNA-DNA hybridization studies of marsupials II. Relationships among dasyurids (Marsupialia: Dasyuridae). *Australian Journal of Zoology* **38**, 673–696. doi:10.1071/ZO9900673

Kirsch JAW, Lapointe F-J, Springer MS (1997) DNA-hybridisation studies of marsupials and their implications for metatherian classification. *Australian Journal of Zoology* **45**, 211–280. doi:10.1071/ZO96030

Kirsch JAW, Springer MS (1993) Timing of the molecular evolution of New Guinean marsupials. *Science in New Guinea* **19**, 147–156.

Kishino H, Hasegawa M (1989) Evaluation of the maximum likelihood estimate of the evolutionary tree topologies from DNA sequence data, and the branching order in Hominoidea. *Journal of Molecular Evolution* **29**, 170–179. doi:10.1007/BF02100115

Kishino H, Thorne JL, Bruno WJ (2001) Performance of a divergence time estimation method under a probabilistic model of rate evolution. *Molecular Biology and Evolution* **18**, 352–361. doi:10.1093/oxfordjournals.molbev.a003811

Kitchener DJ (1988) A new species of false antechinus (Marsupialia: Dasyuridae) from the Kimberley, Western Australia. *Records of the Western Australian Museum* **14**, 61–71, <https://museum.wa.gov.au/sites/default/files/A%20NEW%20SPECIES%20OF%20FALSE%20ANTECHINUS%20%28MARSUPIALIA%20DASYURIDAE%29%20FROM%20THE%20KIMBERLEY%2C%20WESTERN%20AUSTRALIA.pdf>.

Kitchener DJ (1991) *Pseudantechinus mimulus* (Thomas, 1906) (Marsupialia: Dasyuridae): rediscovery and redescription. *Records of the Western Australian Museum* **15**, 191–202, <https://museum.wa.gov.au/sites/default/files/PSEUDANTECHINUS%20MIMULUS%20(THOMAS,%201906)%20(MARSUPIALIA,%20DASYURIDAE)%20REDISCOVERY%20AND%20REDESCRIPTION.pdf>.

Kitchener DJ, Caputi N (1988) A new species of false antechinus (Marsupialia: Dasyuridae) from Western Australia, with remarks on the generic classification within Parantechini. *Records of the Western Australian Museum* **14**, 35–59, <https://museum.wa.gov.au/sites/default/files/A%20NEW%20SPECIES%20OF%20FALSE%20ANTECHINUS%20(MARSUPIALIA%20DASYURIDAE)%20FROM%20WESTERN%20AUSTRALIA,%20WITH%20REMARKS%20ON%20THE%20GENERIC%20CLASSIFICATI.pdf>.

Kitchener DJ, Stoddart J, Henry J (1983) A taxonomic appraisal of the genus *Ningaui* Archer (Marsupialia: Dasyuridae), including description of a new species. *Australian Journal of Zoology* **31**, 361–379. doi:10.1071/ZO9830361

Kitchener DJ, Stoddart J, Henry J (1984) A taxonomic revision of the *Sminthopsis murina* complex (Marsupialia: Dasyuridae) in Australia, including descriptions of four new species. *Records of the Western Australian Museum* **11**, 201–248, <https://museum.wa.gov.au/sites/default/files/A%20TAXONOMIC%20REVISION%20OF%20THE%20SMINTHOPSIS%20MURINA%20COMPLEX%20(MARSUPIALIA,%20DASYURIDAE)%20IN%20AUSTRALIA,%20INCLUDING%20DESCRIPTIONS%20OF%20FO.pdf>.

Kitchener DJ, Cooper N, Bradley A (1986) Reproduction in male *Ningaui* (Marsupialia: Dasyuridae). *Australian Wildlife Research* **13**, 13–25. doi:10.1071/WR9860013

Klootwijk C, Giddings C, Pigram C, Loxton C, Davies H, *et al.* (2003) North Sepik region of Papua New Guinea: palaeomagnetic constraints on arc accretion and deformation. *Tectonophysics* **362**, 273–301. doi:10.1016/S0040-1951(02)00641-8

Kluge AJ (1989) A concern for evidence and a phylogenetic hypothesis of relationships among *Epicrates* (Boidae, Serpentes). *Systematic Zoology* **38**, 7–25. doi:10.1093/sysbio/38.1.7

Kocher TD, Thomas WK, Meyer A, Edwards SV, Pääbo S, *et al.* (1989) Dynamics of mitochondrial DNA evolution in animals: amplification and sequencing with conserved primers. *Proceedings of the National Academy of Sciences USA* **86**, 6196–6200. doi:10.1073/pnas.86.16.6196

Krajewski C (1994) Phylogenetic measures of biodiversity: a comparison and critique. *Biological Conservation* **69**, 33–39. doi:10.1016/0006-3207(94)90326-3

Krajewski C, Anderson FE, Woolley PA, Westerman M (2012) Molecular evidence for a deep clade of dunnarts (Marsupialia: Dasyuridae: *Sminthopsis*). *Journal of Mammalian Evolution* **19**, 265–276. doi:10.1007/s10914-012-9204-3

Krajewski C, Blacket M, Buckley L, Westerman M (1997) A multigene assessment of phylogenetic relationships within the dasyurid marsupial subfamily Sminthopsinae. *Molecular Phylogenetics and Evolution* **8**, 236–248. doi:10.1006/mpev.1997.0421

Krajewski C, Blacket MJ, Westerman M (2000) DNA sequence analysis of familial relationships among dasyuromorphian marsupials. *Journal of Mammalian Evolution* **7**, 95–108. doi:10.1023/A:1009454515671

Krajewski C, Buckley L, Westerman M (1997) DNA phylogeny of the marsupial wolf resolved. *Proceedings of the Zoological Society of London B* **264**, 911–917. doi:10.1098/rspb.1997.0126

Krajewski C, Buckley L, Woolley PA, Westerman M (1996) Phylogenetic analysis of cytochrome *b* sequences in the dasyurid marsupial subfamily Phascogalinae: systematics and the evolution of reproductive strategies. *Journal of Mammalian Evolution* **3**, 81–91. doi:10.1007/BF01454254

Krajewski C, Dickerman AW (1990) Bootstrap analysis of phylogenetic trees derived from DNA hybridization distances. *Systematic Zoology* **39**, 383–390. doi:10.2307/2992358

Krajewski C, Driskell AC, Baverstock PR, Braun MJ (1992) Phylogenetic relationships of the thylacine (Mammalia: Thylacinidae) among dasyuroid marsupials: evidence from cytochrome *b* DNA sequences. *Proceedings of the Royal Society of London B* **250**, 19–27. doi:10.1098/rspb.1992.0125

Krajewski C, Fain MG, Buckley L, King DG (1999) Dynamically heterogenous partitions and phylogenetic inference: an evaluation of analytical strategies with cytochrome *b* and ND6 gene sequences in cranes. *Molecular Phylogenetics and Evolution* **13**, 302–313. doi:10.1006/mpev.1999.0646

Krajewski C, King DG (1996) Molecular divergence and phylogeny: rates and patterns of cytochrome *b* evolution in cranes. *Molecular Biology and Evolution* **13**, 21–30. doi:10.1006/mpev.1999.0646

Krajewski C, Moyer GR, Sipiorski JT, Fain MG, Westerman M (2004) Molecular systematics of the enigmatic 'phascolosoricine' marsupials of New Guinea. *Australian Journal of Zoology* **52**, 389–415. doi:10.1071/ZO04020

Krajewski C, Painter J, Buckley L, Westerman M (1994) Phylogenetic structure of the marsupial family Dasyuridae based on cytochrome-b DNA sequences. *Journal of Mammalian Evolution* **2**, 25–35. doi:10.1007/BF01464348

Krajewski C, Painter J, Driskell AC, Buckley L, Westerman M (1993) Molecular systematics of New Guinean dasyurids. *Science in New Guinea* **19**, 157–168, <https://www.researchgate.net/publication/281705575_Molecular_systematics_of_New_Guinean_dasyurids_Marsupialia_Dasyuridae#fullTextFileContent>.

Krajewski C, Torunsky R, Sipiorski JT, Westerman M (2007) Phylogenetic relationships of the dasyurid marsupial genus *Murexia*. *Journal of Mammalogy* **88**, 696–705. doi:10.1644/06-MAMM-A-310R.19R1.1

Krajewski C, Westerman M (2003) Molecular systematics of Dasyuromorphia. In *Predators with Pouches: The Biology of Carnivorous Marsupials*. (Eds M Jones, C Dickman, and M Archer) pp. 3–20. CSIRO Publishing, Melbourne.

Krajewski C, Woolley PA, Westerman M (2000) The evolution of reproductive strategies in dasyurid marsupials: implications of molecular phylogeny. *Biological Journal of the Linnean Society* **71**, 417–435. doi:10.1111/j.1095-8312.2000.tb01267.x

Krajewski C, Wroe S, Westerman M (2000) Molecular evidence for the pattern and timing of cladogenesis in dasyurid marsupials. *Zoological Journal of the Linnean Society* **130**, 375–404. doi:10.1006/zjls.1999.0222

Krajewski C, Young J, Buckley L, Woolley PA, Westerman M (1997) Reconstructing the evolutionary radiation of dasyurine marsupials with cytochrome *b*, 12S rRNA, and protamine P1 gene trees. *Journal of Mammalian Evolution* **4**, 217–236. doi:10.1023/A:1027349725642

Krause WJ, Krause WA (2006) *The Opossum: Its Amazing Story*. Department of Pathology and Anatomical Sciences, University of Missouri, Columbia, <https://mospace.umsystem.edu/xmlui/bitstream/handle/10355/15130/OpossumItsAmazingStory.pdf?sequence=1&isAllowed=y-8169">.

Krefft G (1867) On the classification of the small Dasyuridae of Australia, with descriptions of two new genera and one new species. *Proceedings of the Zoological Society of London* **1866**, 431–435, <https://www.biodiversitylibrary.org/page/28627927>.

Krefft G (1872) Natural History. The native cat family, or Dasyuridae (continued). *The Sydney Mail and New South Wales Advertiser (new series)* **XIV** (645), 598. [9 November], <http://news.google.com/newspapers?id=yPRhAAAAIBAJ&sjid=P5MDAAAAIBAJ&pg=7262%2C7746776>.

Kubato LS, Degnan JH (2007) Inconsistency of phylogenetic estimates from concatenated data under coalescence. *Systematic Biology* **56**, 17–24. doi:10.1080/10635150601146041

Kumar S (2005) Molecular clocks: four decades of evolution. *Nature Reviews Genetics* **6**, 654–662. doi:10.1038/nrg1659

Kumar S, Tamura K, Nei M (1993) *MEGA: Molecular Evolutionary Genetics Analysis*. Pennsylvania State University, University Park, PA

Labrinidis A, Cooper SJB, Adams M, Baczocha N (1998) Systematic affinities of island and mainland populations of the dunnart *Sminthopsis griseoventer* in Western Australia: data from allozymes and mitochondrial DNA. *Pacific Conservation Biology* **4**, 289–295. doi:10.1071/PC980289

Ladeveze S, De Muizon C (2010) Evidence of early evolution of Australidelphia (Metatheria, Mammalia) in South America: phylogenetic relationships of the metatherians from the Late Paleocene of Itaboraí (Brazil) based on teeth and petrosal bones. *Zoological Journal of the Linnean Society* **159**, 746–784. doi:10.1111/j.1096-3642.2009.00577.x

Lamarck JB (1809) *Zoological Philosophy* (H. Elliot, translator). Hafner Publishing, New York, <http://www.blc.arizona.edu/courses/schaffer/449/Lamarck/Lamarck%20Zoological%20Philosophy.pdf>.

Lanave C, Preparata G, Saccone C, Serio G (1984) A new method for calculating evolutionary substitution rates. *Journal of Molecular Evolution* **20**, 86–93. doi:10.1007/BF02101990

Lang AJ, Engler T, Martin T (2022) Dental topographic and three-dimensional geometric morphometric analysis of carnassialization in different clades of carnivorous mammals (Dasyuromorphia, Carnivora, Hyaenodonta). *Journal of Morphology* **283**, 91–108. doi:10.1002/jmor.21429

Lanyon SM (1985) Detecting internal inconsistencies in distance data. *Systematic Zoology* **34**, 397–403. doi:10.1093/sysbio/34.4.397

Larget B, Simon DL (1999) Markov chain Monte Carlo algorithms for the Bayesian analysis of phylogenetic trees. *Molecular Biology and Evolution* **16**, 750–759. doi:10.1093/oxfordjournals.molbev.a026160

Lavery TH, Collett R, Fisher DO, Hoskin CJ, Rowland J (2022) White-footed dunnarts (*Sminthopsis leucopus*) in Queensland's wet tropics, with the description of a new subspecies. *Australian Mammalogy* doi:10.1071/AM22002

Lee AK, Bradley AJ, Braithwaite RW (1977) Corticosteroid levels and male mortality in *Antechinus stuartii*. In *The Biology of Marsupials*. (Eds B Stonehouse and D Gilmore) pp. 209–220. MacMillan, London.

Lee AK, Cockburn A (1985) *Evolutionary Ecology of Marsupials*. Cambridge University Press, Cambridge.

Lee AK, Woolley P, Braithwaite RW (1982) Life history strategies of dasyurid marsupials. In *Carnivorous Marsupials*. (Ed. M Archer) pp. 1–11. Royal Zoological Society of New South Wales, Sydney.

Lee MSY, Palci A (2015) Morphological phylogenetics in the genomic age. *Current Biology* **25**, R922–R929. doi:10.1016/j.cub.2015.07.009

Legge SM, Woinarski JCZ, Burbidge AA, Palmer A, Ringma J, *et al.* (2018) Havens for threatened Australian mammals: the contributions of fenced areas and offshore islands to protecting mammal species that are susceptible to introduced predators. *Wildlife Research* **45**, 627–644. doi:10.1071/WR17172

Legge S, Woinarski J, Scheele B, Garnett ST, Lintermans M, *et al.* (2021) Rapid assessment of the biodiversity impacts of a megafire to guide urgent management intervention and recovery. *Diversity and Distributions* [online early]. doi:10.1111/ddi.13428

Letendre C, Sawyer E, Young LJ, Old JM (2018) Immunosenescence in a captive semelparous marsupial, the red-tailed phascogale (*Phascogale calura*). *BMC Zoology* **3**, 1–14. doi:10.1186/s40850-018-0036-3

Leverington F, Peterson A, Peterson G, Jano W, Sabi J, Wheatley A (2017) 'Assessment of management effectiveness for Papua New Guinea's protected areas 2017'. Secretariat of the Pacific Regional Environment Programme, Apia, Samoa, <https://rris.biopama.org/sites/default/files/2019-09/Leverington2017%20SPREP%20Protected%20area%20management%20effectiveness%20assessment%20Papua%20New%20Guinea.pdf>.

Lewis PO (2001) A likelihood approach to estimating phylogeny from discrete morphological character data. *Systematic Biology* **56**, 913–925. doi:10.1080/106351501753462876

Lidicker WZ Jr, Marlow BJ (1970) A review of the dasyurid marsupial genus *Antechinomys* Krefft. *Mammalia* **34**, 212–227. doi:10.1515/mamm.1970.34.2.212

Linnaeus C (1758) Systema Naturae per Regna Tria Naturae Secundum Classes, Ordines, Genera, Species, Cum Characteribus, Differentia, Synonimis, Locis. Tomus I. Editio decima, reformata. Holmiæ, <https://www.biodiversitylibrary.org/page/726886>.

Loewenthal G, Rapoport D, Avram O, Moshe O, Wygoda E, *et al.* (2021) A probabilistic model for indel evolution: differentiating insertions from deletions. *Molecular Biology and Evolution* **38**, 5769–5781. doi:10.1093/molbev/msab266

Lomolino MV, Riddle BR, Whittaker RJ (2017) *Biogeography: Biological Diversity Across Space and Time*. Sinauer Associates, Sunderland, MA.

Long J, Archer M, Flannery T, Hand S (2002) *Prehistoric Mammals of Australia and New Guinea*. Johns Hopkins University Press, Baltimore.

Louys J, Price GJ (2015) The Chinchilla Local Fauna: an exceptionally rich and well-preserved Pliocene vertebrate assemblage from fluvatile deposits of south-eastern Queensland, Australia. *Acta Paleontologica Polonica* **60**, 551–572. doi:10.4202/app.00042.2013

Lovette I (2004) Mitochondrial dating and mixed support for the '2% rule' in birds. *The Auk* **121**, 1–6. doi:10.1093/auk/121.1.1

Lowenstein JM, Sarich VM, Richardson BJ (1981) Albumin systematics of the extinct mammoth and Tasmanian wolf. *Nature* **291**, 409–411. doi:10.1038/291409a0

Luckett WP (1993) An ontogenetic assessment of dental homologies in therian mammals. In *Mammal Phylogeny: Mesozoic Differentiation, Multituberculates, Monotremes, Early Therians, and Marsupials*. (Eds FS Szalay, MJ Novacek, and MC McKenna) pp. 182–204. Springer-Verlag, New York.

Luckett WP, Woolley PA (1996) Ontogeny and homology of the dentition in dasyurid marsupials: development in *Sminthopsis virginiae*. *Journal of Mammalian Evolution* **3**, 327–364. doi:10.1007/BF02077449

Lydekker R (1899) The dental formula of the marsupial and placental carnivora. *Proceedings of the Royal Society of London* **1899**, 922–928, <https://www.biodiversitylibrary.org/page/31572920>.

Machado DJ, Castroviejo-Fisher S, Grant T (2021) Evidence of absence treated as absence of evidence: the effects of variation in the number and distribution of gaps treated as missing data on the results of standard maximum likelihood analysis. *Molecular Phylogenetics and Evolution* **154**, 106966. doi:10.1016/j.ympev.2020.106966

Macleay WS (1841) Notice of a new genus of Mammalia discovered by J. Stuart, Esq., in New South Wales. *Annals and*

Magazine of Natural History **8**, 241–243, <https://www.biodiversitylibrary.org/page/2338118>.

Mackness BS, Whitehead P, McNamara G (2000) A new potassium–argon date in relation to the Pliocene Bluff Downs Local Fauna, northern Australia. *Australian Journal of Earth Sciences* **47**, 807–811. doi:10.1046/j.1440-0952.2000.00812.x

Mackness BS, Wroe S, Wilkinson C, Wilkinson C (2002) Confirmation of *Thylacinus* from the Pliocene Chinchilla Local Fauna. *Australian Mammalogy* **24**, 237–242. doi:10.1071/AM02237

Maddison WP (1993) Missing data versus missing characters in phylogenetic analysis. *Systematic Biology* **42**, 576–581. doi:10.2307/2992490

Maddison WP (1997) Gene trees in species trees. *Systematic Biology* **46**, 523–536. doi:10.1093/sysbio/46.3.523

Maddison WP, Knowles LL (2006) Inferring phylogeny despite incomplete lineage sorting. *Systematic Biology* **55**, 21–30. doi:10.1080/10635150500354928

Mahoney JA, Ride WDL (1984) The identity of Captain Cook's quoll *Mustela quoll* Zimmerman 1783. *Australian Mammalogy* **7**, 57–62.

Mahoney JA, Ride WDL (1988) Dasyuridae. In *Zoological Catalogue of Australia, Volume 5, Mammalia.* (Ed. DW Walton) pp. 14–33. Australian Government Publishing Service, Canberra.

Mao X, Retallack GJ (2019) Late Miocene drying of Central Australia. *Palaeogeography, Palaeoclimatology, Palaeoecology* **514**, 292–304. doi:10.1016/j.palaeo.2018.10.008

Mao Y, Catacchio CR, Hillier LW, Porubsky D, Li R, *et al.* (2021) A high-quality bonobo genome refines the analysis of hominid evolution. *Nature* **594**, 77–81. doi:10.1038/s41586-021-03519-x

Margoliash E (1963) Primary structure and evolution of cytochrome c. *Proceedings of the National Academy of Sciences USA* **50**, 672–679. doi:10.1073/pnas.50.4.672

Mariani M, Connor SE, Theuerkauf M, Herbert A, Kuneš P, *et al.* (2022) Disruption of cultural burning promotes shrub encroachment and unprecedented wildfires. *Frontiers in Ecology and the Environment* [Online early]. doi:10.1002/fee.2395

Marin-Gual L, González-Rodelas L, Pujol G, Vara C, Martin-Ruiz M, *et al.* (2022) Strategies for meiotic chromosome dynamics and telomeric elongation in marsupials. *PLoS Genetics* **18**, e1010040. doi:10.1371/journal.pgen.1010040

Markova EA, Sibiryakov PA, Kartavtseva IV, Lapin AS, Morozkina AV, *et al.* (2019) What can invasive species tell us about evolution? A study of dental variation in disjunctive populations of *Microtus rossiaemeridionalis* (Arvicolinae, Rodentia). *Journal of Mammalian Evolution* **26**, 267–282. doi:10.1007/s10914-017-9401-1

Marlow BJ (1961) Reproductive behaviour of the marsupial mouse *Antechinus flavipes* (Waterhouse) (Marsupialia) and the development of the pouch young. *Australian Journal of Zoology* **9**, 203–218. doi:10.1071/ZO9610203

Marshall LG (1973) Fossil vertebrate faunas from the Lake Victoria region, S.W. New South Wales, Australia. *Memoirs of the National Museum of Victoria* **34**, 151–181. doi:10.24199/j.mmv.1973.34.03

Marshall LG (1977a) Cladistic analysis of borhyaenoid, dasyuroid, didelphoid, and thylacinid (Marsupialia: Mammalia) affinity. *Systematic Zoology* **26**, 410–425. doi:10.1093/sysbio/26.4.410

Marshall LG (1977b) *Evolution of the Borhyaenidae, extinct South American predaceous marsupials.* University of California Press, Berkeley.

Marshall LG (1987) Systematics of Itaborian (middle Paleocene) age 'opossum-like' marsupials from the limestone Quarry at Sao Jose de Itaborai, Brazil. In *Possums and Opossums: Studies in Evolution.* (Ed. M Archer) pp. 91–160. Royal Zoological Society of New South Wales, Sydney.

Marshall LG, Case JA, Woodburne MO (1990) Phylogenetic relationships of the families of marsupials. In *Current Mammalogy.* (Ed. HH Genoways) pp. 433–505. Plenum Press, New York.

Marshall LG, Corruccini RS (1978) Variability, evolutionary rates, and allometry in dwarfing lineages. *Paleobiology* **4**, 101–119. doi:10.1017/S0094837300005790

Martin HA (2006) Cenozoic climatic change and the development of arid vegetation in Australia. *Journal of Arid Environments* **66**, 533–563. doi:10.1016/j.jaridenv.2006.01.009

Martin PS (1966) Africa and Pleistocene overkill. *Nature* **212**, 339–342. doi:10.1038/212339a0

Mason MJ (2016) Structure and function of the mammalian middle ear. I: large middle ears in small desert mammals. *Journal of Anatomy* **228**, 284–299. doi:10.1111/joa.12313

Matschie P (1916) Die Verbreitung der Beuteltier auf Neuguinea mit einigen Bemerkungen über ihre Einteilung in Untergattungen. *Mitteilungen Zoologischen Museum, Berlin* **8**, 258–309, <https://archive.org/details/mitteilungenausd8191517zool/page/256/mode/2up>.

Matthew WD (1915) Climate and evolution. *Annals of the New York Academy of Sciences* **24**, 171–318. doi:10.1111/j.1749-6632.1914.tb55346.x

Maxam AM, Gilbert W (1977) A new method for sequencing DNA. *Proceedings of the National Academy of Sciences USA* **74**, 560–564. doi:10.1073/pnas.74.2.560

Maxson LR, Maxson RD (1990) Proteins II: immunological techniques. In *Molecular Systematics.* (Eds DM Hillis and C Moritz) pp. 127–155. Sinauer Associates, Sunderland, MA.

Maxson RD, Maxson LR (1986) Micro-complement fixation: a quantitative estimator of protein evolution. *Molecular Biology and Evolution* **3**, 375–388. doi:10.1093/oxfordjournals.molbev.a040405

Maxson LR, Sarich VM, Wilson AC (1975) Continental drift and the use of albumin as an evolutionary clock. *Nature* **255**, 397–399. doi:10.1038/255397a0

Maxwell S, Burbidge AA, Morris K (Eds) (1996) *The 1996 Action Plan for Australian Marsupials and Monotremes.* Wildlife Australia, Canberra.

May RM 1990. Taxonomy as destiny. *Nature* **347**, 129–130. doi:10.1038/347129a0

May-Collado LJ, Kirkpatrick CW, Agnarsson I (2015) Mammals from 'down under': a multi-gene species-level phylogeny of marsupial mammals (Mammalia, Metatheria). *PeerJ* **3**, e805. doi:10.7717/peerj.805

Mayden RL (1997) A hierarchy of species concepts: the denouement in the saga of the species problem. In *Species: The Units of Diversity*. (Eds MF Claridge, HA Dawah, and MR Wilson) pp. 381–423. Chapman and Hall, London.

Mayr E (1942) *Systematics and the Origin of Species*. Columbia University Press, New York.

Mayr E (1959) Drawing and the evolutionary theory in biology. In *Evolution and Anthropology: A Centennial Appraisal*. (Ed. BJ Meggers) pp. 409–412. Anthropological Society of Washington, Washington, DC.

Mayr E (1970) *Populations, Species, and Evolution*. Harvard University Press, Cambridge, MA.

Mayr E (1975) *Evolution and the Diversity of Life*. Harvard University Press, Cambridge, MA.

Mayr E (1982a) Of what use are subspecies? *The Auk* **99**, 593–595. doi:10.1093/auk/99.3.593a

Mayr E (1982b) *The Growth of Biological Thought: Diversity, Evolution, and Inheritance*. Belknap Press, Cambridge, MA.

Mayr E (1988) *Toward a New Philosophy of Biology: Observations of an Evolutionist*. Harvard University Press, Cambridge, MA.

Mayr E, Ashlock PD (1991) *Principles of Systematic Zoology*. McGraw-Hill, New York.

Mayr E, Linsley EG, Usinger RL (1953) *Methods and Principles of Systematic Zoology*. McGraw-Hill, New York.

McAllan B (2003) Timing of reproduction in carnivorous marsupials. In *Predators with Pouches: The Biology of Carnivorous Marsupials*. (Eds M Jones, C Dickman, M Archer) pp. 147–168. CSIRO Publishing, Melbourne.

McCallum H, Tompkins DM, Jones M, Lachish S, Marvanek S, *et al.* (2007) Distribution and impacts of Tasmanian Devil Facial Tumour Disease. *EcoHealth* **4**, 318–325. doi:10.1007/s10393-007-0118-0

McDonald IR, Lee AK, Than KA, Martin RW (1986) Failure of glucocorticoid feedback in males of a population of small marsupials (*Antechinus swainsonii*) during the period of mating. *Journal of Endocrinology* **108**, 63–68. doi:10.1677/joe.0.1080063

McGowran B, Archer M, Bock P, Darragh TA, Godthelp H, *et al.* (2000) Australian palaeogeography: the Palaeogene and Neogene record. *Memoir of the Association of Australian Paleontologists* **23**, 405–470.

McKenzie NL, Archer M (1982) *Sminthopsis youngsoni* (Marsupialia: Dasyuridae) the lesser hairy-footed dunnart, a new species from arid Australia. *Australian Mammalogy* **5**, 267–279.

McLean AL, Cooper SJB, Lancaster ML, Gaikhorst G, Lambert C, *et al.* (2019) Small marsupial, big dispersal? Broad and fine-scale genetic structure of an endangered marsupial from the Australian arid zone. *Australian Journal of Zoology* **66**, 214–227. doi:10.1071/ZO18054

Megirian D, Prideaux G, Murray PF, Smith N (2010) An Australian land mammal age biochronology scheme. *Paleobiology* **36**, 658–671. doi:10.1666/09047.1

Menzies JL (1991) *A Handbook of New Guinea Marsupials and Monotremes*. Kristen Press, Madang.

Menzies J (2011) *A Handbook of New Guinea's Marsupials and Monotremes*. University of Papua New Guinea Press, Port Moresby.

Meredith RW, Westerman M, Case JA, Springer MS (2008) A phylogeny and timescale for marsupial evolution based on sequences for five nuclear genes. *Journal of Mammalian Evolution* **15**, 1–36. doi:10.1007/s10914-007-9062-6

Meredith RW, Janečka JE, Gatesy J, Ryder OA, Fisher CA, *et al.* (2011) Impacts of the Cretaceous Terrestrial Revolution and KPg extinction on mammalian diversification. *Science* **334**, 521–524. doi:10.1126/science.1211028

Miller GH, Fogel ML, Magee JW, Gagan ML, Clarke SJ, *et al.* (2005) Ecosystem collapse in Pleistocene Australia and a human role in megafaunal extinction. *Science* **309**, 287–290. doi:10.1126/science.1111288

Miller W, Drautz DI, Janecka JE, Lesk AM, Ratan A, *et al.* (2009) The mitochondrial genome sequence of the Tasmanian tiger (*Thylacinus cynocephalus*). *Genome Research* **19**, 213–220. doi:10.1101/gr.082628.108

Mills HR, Bencini R (2000) New evidence for facultative male die-off in island populations of dibblers, *Parantechinus apicalis*. *Australian Journal of Zoology* **48**, 501–510. doi:10.1071/ZO00025

Mills HR, Moro D, Spencer PBS (2004) Conservation significance of island versus mainland populations: a case study of dibblers (*Parantechinus apicalis*) in Western Australia. *Animal Conservation* **7**, 387–395. doi:10.1017/S1367943004001568

Mirarab S, Warnow T (2015) ASTRAL-II: coalescent-based species tree estimation with many hundreds of taxa and thousands of genes. *Bioinformatics* **31**, 44–52. doi:10.1093/bioinformatics/btv234

Mitchell KJ, Pratt RC, Watson LN, Gibb GC, Llamas B, *et al.* (2014) Molecular phylogeny, biogeography, and habitat preference evolution of marsupials. *Molecular Biology and Evolution* **31**, 2322–2330. doi:10.1093/molbev/msu176

Mooney N, Rounsevell DE (2008) Thylacine. In *Mammals of Australia*. 3rd edn. (Eds S Van Dyck and R Strahan) pp. 167–168. Reed New Holland, Sydney.

Moore HA, Dunlop JA, Jolly CJ, Kelly E, Woinarski JCZ, *et al.* (2021) A brief history of the northern quoll (*Dasyurus hallucatus*): a systematic review. *Australian Mammalogy* [online early]. doi:10.1071/AM21002

Moore JA (1993) *Science as a Way of Knowing: The Foundations of Modern Biology*. Harvard University Press, Cambridge, Massachusetts.

Moore WS (1995) Inferring phylogenies from mtDNA variation: mitochondrial-gene trees versus nuclear-gene trees. *Evolution* **49**, 718–726. doi:10.1111/j.1558-5646.1995.tb02308.x

Morcombe MK (1967) The rediscovery after 83 years of the Dibbler *Antechinus apicalis* (Marsupialia, Dasyuridae). *Western Australian Naturalist* **10**, 103–111, <https://www.biodiversitylibrary.org/page/59343557>.

Moritz C (1994) Defining 'evolutionarily significant units' for conservation. *Trends in Ecology & Evolution* **9**, 373–375. doi:10.1016/0169-5347(94)90057-4

Moritz C, Dowling TE, Brown WM (1987) Evolution of animal mitochondrial DNA: relevance for population biology and systematics. *Annual Review of Ecology and Systematics* **18**, 269–292. doi:10.1146/annurev.es.18.110187.001413

Morris K, Johnson B, Orell P, Gaikhorst G, Wayne A, *et al.* (2003) Recovery of the threatened chuditch (*Dasyurus geoffroii*): a case study. In *Predators with Pouches: The Biology of Carnivorous Marsupials*. (Eds M Jones, C Dickman, and M Archer) pp. 435–451. CSIRO Publishing, Melbourne.

Morrison III, WR, Lohr JL, Duchen P, Wilches R, Trujillo D, *et al.* (2009) The impact of taxonomic change on conservation: Does it kill, can it save, or is it just irrelevant. *Biological Conservation* **142**, 3201–3206. doi:10.1016/j.biocon.2009.07.019

Morton SR (1978) An ecological study of *Sminthopsis crassicaudata* (Marsupialia: Dasyuridae) III. Reproduction and life-history. *Australian Wildlife Research* **5**, 183–211. doi:10.1071/WR9780183

Morton SR (1982) Dasyurid marsupial of the Australian arid zone: an ecological review. In *Carnivorous Marsupials*. (Ed. M Archer) pp. 117–130. Royal Zoological Society of New South Wales, Sydney.

Morton SR, Alexander G (1982) Geographic variation in the external morphology of *Sminthopsis crassicaudata* (Dasyuridae, Marsupialia). In *Carnivorous Marsupials*. (Ed. M Archer) pp. 695–698. Royal Zoological Society of New South Wales, Sydney.

Morton SR, Armstrong MD, Braithewaite RW (1987) The breeding season of *Sminthopsis virginiae* (Marsupialia: Dasyuridae) in the Northern Territory. *Australian Mammalogy* **10**, 41–42. doi:10.1071/AM87009

Morton SR, Dickman CR, Fletcher TP (1989) Dasyuridae. In *Fauna of Australia*. (Eds DW Walton and BJ Richardson) pp. 560–582. Australian Government Publishing Service, Canberra, <https://www.awe.gov.au/sites/default/files/env/pages/a117ced5-9a94-4586-afdb-1f333618e1e3/files/21-ind.pdf>.

Mounce RC, Sansom R, Wills MA (2016) Sampling diverse characters improves phylogenies: craniodental and postcranial characters of vertebrates often imply different trees. *Evolution* **70**, 666–686. doi:10.1111/evo.12884

Muirhead J (1992) A specialized thylacinid, *Thylacinus macknessi* (Marsupialia: Thylacinidae) from Miocene deposits of Riversleigh, northwestern Queensland. *Australian Mammalogy* **15**, 67–76. doi:10.1071/AM92009

Muirhead J (1997) Two new thylacines (Marsupialia: Thylacinidae) from early Miocene sediments of Riversleigh, northwestern Queensland and a revision of the family Thylacinidae. *Memoirs of the Queensland Museum* **41**, 367–377, <https://www.biodiversitylibrary.org/page/40090851>.

Muirhead J, Archer M (1990) *Nimbacinus dicksoni*, a plesiomorphic thylacine (Marsupialia, Thylacinidae) from Tertiary deposits of Queensland and the Northern Territory. *Memoirs of the Queensland Museum* **28**, 203–221, <https://www.biodiversitylibrary.org/page/40652748>.

Muirhead J, Filan SL (1995) *Yarala birchfieldi*, a plesiomorphic bandicoot (Marsupialia, Peramelemorphia) from Oligo-Miocene deposits of Riversleigh, Northwestern Queensland. *Journal of Paleontology* **69**, 127–134, <https://www.jstor.org/stable/1306286>.

Muirhead J, Wroe S (1998) A new genus and species, *Badjcinus turnbulli* (Thylacinidae: Marsupialia) from the Late Oligocene of Riversleigh, Northern Australia, and an investigation of thylacinid phylogeny. *Vertebrate Paleontology* **18**, 612–626, <https://www.jstor.org/stable/4523932>.

Mulvaney DJ (1990) Spencer, Sir Walter Baldwin (1860–1929). Australian Dictionary of Biography. National Center of Biography, Australian National University, <http://adb.anu.au/biography/spencer-sir-walter-baldwin-8606/text15031>.

Murphy BP, Davies HF (2014) There is a critical weight range for Australia's declining tropical mammals. *Global Ecology and Biogeography* **23**, 1058–1061, <https://onlinelibrary.wiley.com/doi:/pdf/10.1111/geb.12173>.

Murphy BP, Woolley LA, Geyle HM, Legge SM, Palmer R, *et al.* (2019) Introduced cats (*Felis catus*) eating a continental mammal fauna: the number of individuals killed in Australia. *Biological Conservation* **237**, 28–40. doi:10.1016/j.biocon.2019.06.013

Murray PF (1997) *Thylacinus megiriani*, a new species of thylacine (Marsupialia: Thylacinidae) from the Ongeva Local Fauna of central Australia. *Records of the South Australia Museum* **30**, 43–61, <https://www.biodiversitylibrary.org/page/40728207>.

Murray P, Megirian D (1992) Continuity and contrast in middle and late Miocene vertebrate communities from the Northern Territory. *The Beagle, Records Northern Territory Museum of Arts and Sciences* **9**, 195–218, <https://www.biodiversitylibrary.org/page/55752481>.

Murray PF, Megirian D (2000) Two new genera and three new species of Thylacinidae (Marsupialia) from the Miocene of the Northern Territory, Australia. *The Beagle, Records of the Museums and Art Galleries of the Northern Territory* **16**, 145–162. doi:10.5962/p.254543

Murray PF, Megirian D (2006) Cranial morphology of the Miocene thylacinid *Mutpuracinus archibaldi* (Thylacinidae, Marsupialia) and relationships within the Dasyuromorphia. *Alcheringa* **30**(S1), 229–276. doi:10.1080/03115510609506865

Mustonen T, Pispa J, Mikkola ML, Pummila M, Kangas AT, *et al.* (2003) Stimulation of ectodermal organ development by ectodysplasin-A1. *Developmental Biology* **259**, 123–136. doi:10.1016/s0012-1606(03)00157-x

Mutton TY, Gray EL, Fuller SJ, Baker AM (2017) Life history, breeding biology and movement in a new species of carnivorous marsupial, the buff-footed antechinus (*Antechinus mysticus*) and a sympatric congener, the subtropical antechinus (*Antechinus subtropicus*). *Mammal Research* **62**, 373–385. doi:10.1007/s13364-017-0325-1

Mutton TY, Phillips MJ, Fuller SJ, Bryant LM, Baker AM (2019) Systematics, biogeography and ancestral state of the Australian marsupial genus *Antechinus* (Dasyuromorphia: Dasyuridae). *Zoological Journal of the Linnean Society* **186**, 553–568. doi:10.1093/zoolinnean/zly062

Myers T (2017) Taphonomy and palaeoecology of the White Hunter Local Fauna. *PeerJ Reprints* **5**, e3314v1.

Nasrullah Q, Renfree M, Evans AR (2022) From embryo to adult: the complete development and unusual replacement of the dentition of the tammar wallaby (*Macropus eugenii*). *Journal of Mammalian Evolution* [Online early] doi:10.1007/s10914-021-09597-y

Naylor R, Richardson SJ, McAllan BM (2008) Boom and bust: a review of the physiology of the marsupial genus *Antechinus*. *Journal of Comparative Physiology, B. Biochemical, Systematic, and Environmental Physiology* **178**, 545–562. doi:10.1007/s00360-007-0250-8

Nee S, Mooers AO, Harvey PH (1992) Tempo and mode of evolution revealed from molecular phylogenies. *Proceedings of the National Academy of Sciences USA* **89**, 8322–8326. doi:10.1073/pnas.89.17.8322

Nei M (1972) Genetic distance between populations. *American Naturalist* **106**, 283–292. doi:10.1086/282771

Nei M, Kumar S (2000) *Molecular Evolution and Genetics*. Oxford University Press, Oxford.

Nelson GM, Platnick NI (1981) *Systematics and Biogeography: Cladistics and Vicariance*. Columbia University Press, New York.

Nilsson MA, Arnason U, Spencer PBS, Janke A (2004) Marsupial relationships and a timeline for marsupial radiation in Gondwana. *Gene* **340**, 189–196. doi:10.1016/j.gene.2004.07.040

Nilsson MA, Churakov G, Sommer M, Van Tran N, Zemann A, *et al.* (2010) Tracking marsupial evolution using archaic genomic retroposon insertions. *PLoS Biology* **8**, e10000436. doi:10.1371/journal.pbio.1000436

Nogueira JC, Martinelli PM, Costa SF, Carvalho GA, Câmara BGO (1999) The penis morphology of *Didelphis, Lutreolina, Metachirus,* and *Caluromys* (Marsupialia, Didelphidae). *Mammalia* **63,** 79–92. doi:10.1515/mamm.1999.63.1.79

Nogueira JC, Castro ACS, Câmara EVC, Câmara BGO (2004) Morphology of the male genital system of *Chironectes minimus* and comparison to other didelphid marsupials. *Journal of Mammalogy* **85**, 834–841. doi:10.1644/207

Norell MA (1992) Taxic origin and temporal diversity: the effect of phylogeny. In *Extinction and Phylogeny*. (Eds MJ Novacek and QD Wheeler) pp. 88–118. Columbia University Press, New York.

Nunn GB, Stanley SE (1998) Body size effects and rate of cytochrome *b* evolution in tube-nosed seabirds. *Molecular Biology and Evolution* **15**, 1360–1371. doi:10.1093/oxfordjournals.molbev.a025864

Nute M, Chou J, Molloy EK, Warnow T (2018) The performance of coalescent-based species tree estimation methods under models of missing data. *BMC Genomics* **19**, 286. doi:10.1186/s12864-018-4619-8

Nuttall GHF (1904) *Blood Immunity and Blood Relationship: A Demonstration of Certain Blood-Relationships Amongst Animals by Means of the Precipitin Test for Blood*. Cambridge University Press, Cambridge, <https://www.biodiversitylibrary.org/page/8509137>.

Oakwood M (2008) Northern Quoll. In *The Mammals of Australia*. 3rd edn. (Eds S Van Dyck and R Strahan) pp. 57–59. Reed New Holland, Sydney.

Oakwood M, Bradley AJ, Cockburn A (2001) Semelparity in a large marsupial. *Proceedings of the Royal Society of London B* **268**, 407–411. doi:10.1098/rspb.2000.1369

O'Brien SJ, Seuanez H, Womack JE (1988) Mammalian genome organization: an evolutionary overview. *Annual Review of Genetics* **22**, 323–351. doi:10.1146/annurev.ge.22.120188.001543

O'Connell JF, Allen J (2003) Dating the colonization of Sahul (Pleistocene Australia-New Guinea): a review of recent research. *Journal of Archaeological Science* **31**, 835–853. doi:10.1016/j.jas.2003.11.005

O'Hara RJ (1988) Homage to Clio, or, toward an historical philosophy for evolutionary biology. *Systematic Zoology* **37**, 142–155. doi:10.2307/2992272

O'Hara RJ (1993) Systematic generalization, historical fate, and the species problem. *Systematic Biology* **42**, 231–246. doi:10.2307/2992462

O'Hara RJ (1997) Population and tree thinking in systematics. *Zoologica Scripta* **26**, 323–329. doi:10.1111/j.1463-6409.1997.tb00422.x

Old JM, Stannard HJ (2020) Conservation of quolls (*Dasyurus* spp.) in captivity – a review. *Australian Mammalogy* [online early]. doi:10.1071/AM20033

O'Meara BC (2012) Evolutionary inferences from phylogenies: a review of methods. *Annual Review of Ecology, Evolution, and Systematics* **43**, 267–285. doi:10.1146/annurev-ecolsys-110411-160331

O'Reilly JE, Donoghue PCJ (2016) Tips and nodes are complementary not competing approaches to the calibration of molecular clocks. *Biology Letters* **12**, 20150795. doi:10.1098/rsbl.2015.0975

O'Reilly JE, dos Reis M, Donoghue PCJ (2015) Dating tips for divergence-time estimation. *Trends in Genetics* **31**, 637–650. doi:10.106/j.tig.2015.08.001

Osborn HF (1902) The law of adaptive radiation. *American Naturalist* **36**, 92–96, <https://www.journals.uchicago.edu/doi/pdf/10.1086/278137>.

Osborn HF (1910) *The Age of Mammals in Europe, Asia, and North America*. MacMillan, New York, <https://www.biodiversitylibrary.org/page/27395772>.

Osgood WH (1921) A monographic study of the American marsupial Caenolestes. Field Museum of Natural History Publication 207, XIV, No. 1, <https://www.biodiversitylibrary.org/page/2662999>.

Owen R (1834) On the generation of the marsupial animals, with a description of the impregnated uterus of the kangaroo. *Philosophical Transactions of the Royal Society* **124**, 333–364. doi:10.1098/rstl.1834.0019

Owen R (1838) Fossil remains from Wellington Valley, Australia. Marsupialia. Appendix to Three Expeditions into the Interior of Eastern Australia, with Descriptions of the Recently Explored Region of Australia Felix, and of the Present Colony of New South Wales (T.L. Mitchell). *T. and W. Boone,* London 2, 359–369, <https://www.gutenberg.org/files/13033/13033-h/13033-h.htm>.

Owen R (1839a) Marsupialia. In *Cyclopedia of Anatomy and Physiology, Volume 3*. (Ed. RB Todd) pp. 257–329. Sherwood, Gilbert and Piper, London, <https://www.biodiversitylibrary.org/page/848262>.

Owen R (1839b) Outlines of a classification of the Marsupialia. *Proceedings of the Zoological Society of London* **7**, 5–19, <https://www.biodiversitylibrary.org/page/30571244>.

Owen R (1840–1845) *Odontography; or, a Treatise on the Comparative Anatomy of Teeth; Their Physiological Relations, Mode of Development, and Microscopic Structure, in the Vertebrate Animals.* Hippolyte Baillère, London, <https://www.biodiversitylibrary.org/page/16276768>.

Owen R (1868) *On the Anatomy of Vertebrates, Volume III Mammals.* Longmans, Green, and Co., London, <https://www.biodiversitylibrary.org/page/839872>.

Owen R (1871) *Monograph of the Fossil Mammalia of the Mesozoic Formation.* Palaeontographical Society of London, London, <https://www.biodiversitylibrary.org/page/14836431>.

Owen R (1877) *Researches on the Fossil Remains of the Extinct Mammals of Australia; With a Notice on the Extinct Marsupials of England.* J. Erxleben, London, <https://www.biodiversitylibrary.org/page/43607564>.

Paddle R (2000) *The Last Tasmanian Tiger: The History and Extermination of the Thylacine.* Cambridge University Press, Cambridge.

Padial JM, de la Riva I (2006) Taxonomic inflation and the stability of species lists: the perils of ostrich's behavior. *Systematic Biology* **55**, 859–867. doi:10.1080/10635150600081588

Padial JM, Miralles A, de la Riva I, Vences M (2010) The integrative future of taxonomy. *Frontiers in Zoology* **7**, 16. doi:10.1186/1742-9994-7-16

Pagel M, Meade A, Barker D (2004) Bayesian estimation of ancestral character states on phylogenies. *Systematic Biology* **53**, 673–684. doi:10.1080/10635150490522232

Painter J, Krajewski C, Westerman M (1995) Molecular phylogeny of the marsupial genus *Planigale* (Dasyuridae). *Journal of Mammalogy* **76**, 406–413. doi:10.2307/1382351

Parnaby H, Ingleby S, Divljan A (2015) Taxonomic status of *Podabrus albocaudatus* Krefft, 1872 and declaration of *Sminthopsis granulipes* Troughton 1932 (Marsupialia: Dasyuridae) as a protected name for the White-tailed Dunnart from Western Australia. *Zootaxa* **3904**(2), 283–292. doi:10.11646/zootaxa.3904.2.7

Parrish SS (1997) The female opossum and the nature of the New World. *William and Mary Quarterly* **54**(3), 475–514. doi:10.2307/2953837

Pask AJ, Behringer RR, Renfree MB (2008) Resurrection of DNA function *in vivo* from an extinct genome. *PLoS ONE* **3**, e2240. doi:10.1371/journal.pone.0002240

Patten MA (2015) Subspecies and the philosophy of science. *The Auk* **132**, 481–485. doi:10.1642/AUK-15-1.1

Patton AH, Lawrance MF, Margres MJ, Kozakiewicz CP, Hamede R, *et al.* (2020) A transmissible cancer shifts from emergence to endemism in Tasmania devils. *Science* **370**, eabb9772. doi:10.1126/science.abb9772

Patton JL, Sherwood SW (1983) Chromosome evolution and speciation in rodents. *Annual Review of Ecology and Systematics* **14**, 139–158. doi:10.1146/annurev.es.14.110183.001035

Pennant T (1781) *History of Quadrupeds, Volume I.* B & J White, London, <https://www.biodiversitylibrary.org/page/49329134>.

Pepper M, Keogh JS (2021) Life in the 'dead heart' of Australia: the geohistory of the Australian deserts and its impact on genetic diversity of arid zone lizards. *Journal of Biogeography* **48**, 716–746. doi:10.111.jbi.14063.

Peters W, Doria G (1876) Diagnosi di tre nuove specie de Mammiferi della Nuova Guine e di Salawatti. *Annali del Museo Civico di Storia Naturale di Genova* **8**, 335–336, <https://www.biodiversitylibrary.org/page/7916095>.

Phillips MJ, Westerman M, Cascini M (2021) The value of updating GenBank accessions for supermatrix phylogeny: the case of the New Guinean marsupial carnivore genus *Myoictis*. *Molecular Phylogenetics and Evolution* **166**, 107328. doi:10.1016/j.ympev.2021.107328.

Pillans BJ (2018) Seeing red: some aspects of the geological and climatic history of the Australian arid zone. In *On the Ecology of Australia's Arid Zone*. (Ed. H Lambers) pp. 5–43. Springer, Switzerland, <https://vdoc.pub/documents/on-the-ecology-of-australias-arid-zone-1alkvlrmlt78>.

Pledge NS (1992) The Curramulka Local Fauna: a new late Tertiary fossil assemblage from the Yorke Peninsula, South Australia. *The Beagle, Records of the Northern Territory Museum of Arts and Sciences* **9**, 115–142.

Pocock RI (1926) The external characters of *Thylacinus, Sarcophilus*, and some related marsupials. *Proceedings of the Zoological Society of London* **96**, 1037–1084. doi:10.1111/j.1096-3642.1926.tb02233.x

Pons J, Barraclough TG, Gomes-Zurita J, Cardoso A, Duran DP, *et al.* (2006) Sequence-based delimitation for the DNA taxonomy of undescribed insects. *Systematic Biology* **55**, 595–609. doi:10.1080/10635150600852011

Popper KR (1959) *The Logic of Scientific Discovery.* Hutchinson, London.

Potter S, Bragg JG, Blom MPK, Deakin JE, Kirkpatrick M, *et al.* (2017) Chromosomal speciation in the genomic era: disentangling phylogenetic evolution of rock-wallabies. *Frontiers in Genetics* **8(10)**. doi:10.3389/fgene.2017.00010

Potter, S, Cooper SJB, Metcalfe CJ, Taggart DA, Eldridge MDB (2012) Phylogenetic relationships of rock-wallabies, *Petrogale* (Marsupialia: Macropodidae) and their biogeographic history within Australia. *Molecular Phylogenetics and Evolution* **62**, 640–652. doi:10.1016/j.ympev.2011.11.005

Potter S, Deakin JE (2018) Cytogenetics: an important inclusion in the conservation genetics toolbox. *Pacific Conservation Biology* **24**, 280–288. doi:10.1071/PC18016

Poulton EB (1889) The true teeth and horny plates of *Ornithorhynchus*. *Quarterly Journal of Microscopical Science* **29**, 9–48, <https://www.biodiversitylibrary.org/page/49015717>.

Prager EM, Wilson AC (1976) Congruency of phylogenies derived from different proteins. A molecular analysis of the phylogeny of cracid birds. *Journal of Molecular Evolution* **9**, 45–57. doi:10.1007/BF01796122

Prevosti FJ, Forasiepi M (2018) *Evolution of South American Mammalian Predators During the Cenozoic: Paleobiogeographic and Paleoenvironmental Contingencies.* Springer, Cham, Switzerland.

Prowse TAA, Johnson CN, Bradshaw CJA, Brook BW (2014) An ecological regime shift resulting from disrupted predator-prey interactions in Holocene Australia. *Ecology* **95**, 693–702. doi:10.1890/13-0746.1

Puillandre N, Lambert A, Brouillet S, Achaz G (2012) ABGD, Automatic Barcode Gap Discovery for primary species delimitation. *Molecular Ecology* **21**, 1864–1877. doi:10.1111/j.1365-294X.2011.05239.x

Pyecroft SB, Pearse AM, Loh R, Swift K, Belov K, *et al.* (2007) Towards a case definition for devil facial tumour disease: what is it? *EcoHealth* **4**, 346–351. doi:10.1007/S10393-007-0126-0

Pyron RA (2011) Divergence time estimation using fossils as terminal taxa and the origins of Lissamphibia. *Systematic Biology* **60**, 466–481. doi:10.1093/sysbio/syr047

Pyron RA (2015) Post-molecular systematics and the future of phylogenetics. *Trends in Ecology & Evolution* **30**, 384–389. doi:10.1046/j.tree.2015.04.016

Rabosky DL, Santini F, Eastman J, Smith SA, Sidlauskas B, *et al.* (2013) Rates of speciation and morphological evolution are correlated across the largest vertebrate radiation. *Nature Communications* **4**, 1958, doi:10.1038/ncomms2958

Rambaut A (2000) Estimating the rate of molecular evolution: incorporating non-contemporaneous sequences into maximum likelihood phylogenies. *Bioinformatics* **16**, 395–399. doi:10.1093/bioinformatics/16.4.395

Ramono WS, Nash SV (1992) Conservation of marsupials and monotremes in Indonesia. In *Australasian Marsupials and Monotremes: An Action Plan for their Conservation.* (Ed. M Kennedy) pp. 5–11. IUCN, Gland, Switzerland, <https://portals.iucn.org/library/node/6133>.

Ramsay EP (1888) Notes on the fauna of the Bellenden-Ker ranges. *Proceedings of the Linnean Society of New South Wales* **2**, 1295–1299, <https://www.biodiversitylibrary.org/page/36220177>.

Randi E 1996. A mitochondrial cytochrome *b* phylogeny of *Alectoris* partirdges. *Molecular Phylogenetics and Evolution* **6**, 214–227. doi:10.1006/mpev.1996.0072

Rannala B, Yang Z (1996) Probability distribution of molecular evolutionary trees: a new method of phylogenetic inference. *Journal of Molecular Evolution* **43**, 304–311. doi:10.1007/BF02338839

Rannala B, Yang Z (2007) Inferring speciation times under an episodic clock. *Systematic Biology* **56**, 453–466. doi:10.1080/10635150701420643

Raposo do Amaral F, Neves LG, Resende MFR Jr, Mobili F, Miyaki CY, *et al.* (2015) Ultraconserved elements sequencing as a low-cost source of complete mitochondrial genomes and microsatellite markers in non-model amniotes. *PLoS ONE* **10**, e0138446. doi:10.1371/journal.pone.0138446

Read DG, Fox BJ, Whitford D (1983) Notes on breeding in *Sminthopsis* (Marsupialia: Dasyuridae). *Australian Mammalogy* **6**, 89–92. doi:10.1071/AM83013

Reig OA, Gardner AL, Bianchi NO, Patton JL (1977) The chromosomes of the Didelphidae (Marsupialia) and their evolutionary significance. *Biological Journal of the Linnean Society* **9**, 191–216. doi:10.1111/j.1095-8312.1977.tb00265.x

Reig OA, Kirsch JAW, Marshall L (1987) Systematic relationships of the living and neocenozoic American opossum-like marsupials (Suborder Didelphimorphia), with comments on the classification of these and of the Cretaceous and Paleogene New World and European metatherians. In *Possums and Opossums: Studies in Evolution, Volume I.* (Ed. M Archer) pp. 1–89. Surrey Beatty & Sons and The Royal Society of New South Wales, Sydney.

Rens W, O'Brien PCM, Yang F, Graves JAM, Ferguson-Smith MA (1999) Karyotype relationships between four distantly related marsupials revealed by reciprocal chromosome painting. *Chromosome Research* **7**, 461–474. doi:10.1023/A:1009249813617

Rens W, O'Brien PCM, Yang F, Solansky N, Perelman P, *et al.* (2001) Karyotype relationships between distantly related marsupials from South America and Australia. *Chromosome Research* **9**, 301–308. doi:10.1023/A:1016646629889

Rens W, O'Brien PCM, Fairclough H, Harman L, Graves JAM, *et al.* (2003) Reversal and convergence in marsupial chromosome evolution. *Cytogenetics and Genome Research* **102**, 282–290. doi:10.1159/000075764

Retief JD, Krajewski C, Westerman M, Dixon GH (1995a) The evolution of protamine P1 genes in dasyurid marsupials. *Journal of Molecular Evolution* **41**, 549–555. doi:10.1007/BF00175812

Retief JD, Krajewski C, Westerman M, Winkfein RJ, Dixon GH (1995b) Molecular phylogeny and evolution of marsupial protamine P1 genes. *Proceedings of the Royal Society of London B* **259**, 7–14. doi:10.1098/rspb.1995.0002

Retief JD, Rees JS, Westerman M, Dixon GH (1995c) Convergent evolution of cysteine residues in sperm proteins of one genus of marsupials, the *Planigales*. *Molecular Biology and Evolution* **12**, 708–712. doi:10.1093/oxfordjournals.molbev.a040247

Reydon TAC, Kunz W (2021) Classification below the species level: when are infraspecific groups biologically meaningful? *Biological Journal of the Linnean Society* **134**, 246–260. doi:10.1093/biolinnean/blab067

Rhind SG, Bradley JS, Cooper NK (2001) Morphometric variation and taxonomic status of brush-tailed phascogales, *Phascogale tapoatafa* (Meyer, 1793) (Marsupialia: Dasyuridae). *Australian Journal of Zoology* **49**, 345–368. doi:10.1071/ZO01002

Rhind S, Woinarski J, Aplin KP (2008) Northern Brush-tailed Phascogale. In *The Mammals of Australia.* 3rd edn. (Eds S Van Dyck and R Strahan) pp. 103–104. Reed New Holland, Sydney.

Ride WDL (1964a) *Antechinus rosamondae*, a new species of dasyurid marsupial from the Pilbara District of Western Australia; with remarks on the classification of Antechinus. *Western Australian Naturalist* **9**, 58–65, <https://www.biodiversitylibrary.org/page/59343245>.

Ride WDL (1964b) A review of Australian fossil marsupials. *Journal of the Proceedings of the Royal Society of Western*

Australia **47**, 97–131, <https://www.biodiversitylibrary.org/page/49414264>.

Ride WDL (1965) Locomotion in the Australian Marsupial *Antechinomys*. *Nature* **205**, 199.

Ride WDL (1970) *A Guide to the Native Mammals of Australia*. Oxford University Press, Melbourne.

Ridley M (1986) *Evolution and Classification: The Reformation of Cladism*. Longman, Harlow, UK.

Riggs AD (1990) Marsupials and mechanisms of X chromosome inactivation. In *Mammals from Pouches and Eggs: Genetics, Breeding and Evolution of Marsupials and Monotremes*. (Eds JA Marshall Graves, RM Hope, and DW Cooper) pp. 277–299. CSIRO, Australia. doi:10.1071/ZO989041

Ripple WJ, Wolf C, Newsome TM, Galetti M, Alamgir M, *et al.*, 15,364 scientist signatories from 184 countries (2017) World scientists' warning to humanity: a second notice. *BioScience* **67**, 1026–1028. doi:10.1093/biosci/bix125

Rivera JA, Kraus F, Allison A, Butler MA (2017) Molecular phylogenetics and dating of the problematic New Guinea microhylid frogs (Amphibia: Anura) reveals elevated speciation rates and need for taxonomic reclassification. *Molecular Phylogenetics and Evolution* **112**, 1–11. doi:10.1016/j.ympev.2017.04.008

Roberts RG, Flannery TF, Ayliffe LK, Yoshida H, Olley JM, *et al.* (2001) New ages for the last Australian megafauna: continent-wide extinction about 46,000 years ago. *Science* **292**, 1888–1892. doi:10.1126/science.1060264

Rofe RH (1979) G-banding and chromosomal evolution in Australian marsupials. PhD dissertation, University of Adelaide, Australia, <https://digital.library.adelaide.edu.au/dspace/handle/2440/21182>.

Rofe R, Hayman DA (1985) G-banding evidence for a conserved complement in the Marsupialia. *Cytogenetics and Cell Genetics* **39**, 40–50. doi:10.1159/000132101

Rokas A, Holland PW (2000) Rare genomic changes as a tool for phylogenetics. *Trends in Ecology & Evolution* **15**, 454–459. doi:10.1016/s0169-5347(00)01967-4

Rooney AP, Zhang J, Nei M (2000) An unusual form of purifying selection in a sperm protein. *Molecular Biology and Evolution* **17**, 278–283. doi:10.1093/oxfordjournals.molbev.a026307

Ronquist F, Klopfstein S, Vilhelmsen L, Schulmeister S, Murray DL, *et al.* (2012) A total-evidence approach to dating with fossils, applied to the early radiation of the Hymenoptera. *Systematic Biology* **61**, 973–999. doi:10.1093/sysbio/sys058

Rose RK, Pemberton DA, Mooney NJ, Jones ME (2017) *Sarcophilus harrisii* (Dasyuromorphia: Dasyuridae). *Mammalian Species* **49**, 1–17. doi:10.1093/mspecies/sex001

Rovinsky DS, Evans AR, Adams JW (2019) The pre-Pleistocene fossil thylacinids (Dasyuromorphia: Thylacinidae) and the evolutionary context of the modern thylacine. *PeerJ* **7**, e7457. doi:10.7717/peerj.7457

Ruddle FH (1981) A new era in mammalian gene mapping: somatic cell genetics and recombinant DNA methodologies. *Nature* **294**, 115–120. doi:10.1038/294115a0

Rule S, Brook BW, Haberle SG, Turney CSM, Kershaw AP, *et al.* (2012) The aftermath of megafaunal extinction: ecosystem transformation in Pleistocene Australia. *Science* **335**, 1483–1486. doi:10.1126/science.1214261

Saether OA (1983) The canalized evolutionary potential: inconsistencies in phylogenetic reasoning. *Systematic Zoology* **32**, 343–359. doi:10.2307/2413162

Saiki RK, Scharf S, Faloona F, Mullis KB, Horn GT, *et al.* (1985) Enzymatic amplification of β-globin genomic sequences and restriction-site analysis for diagnostic sickle cell anemia. *Science* **230**, 1350–1354. doi:10.1126/science.2999980

Saitou N, Nei M (1987) The neighbor-joining method: a new method for reconstructing phylogenetic trees. *Molecular Biology and Evolution* **4**, 406–425. doi:10.1093/oxfordjournals.molbev.a040454

Saltré F, Rodriguez-Rey M, Brook BW, Johnson CN, Turney CS, *et al.* (2016) Climate change not to blame for late Quaternary megafauna extinctions in Australia. *Nature Communications* **7**, 10511. doi:10.1038/ncomms10511

Samollow PB, Graves JAM (1998) Gene maps of marsupials. *ILAR Journal* **39**, 203–224. doi:10.1093/ilar.9.2-3.203

Sánchez-Villagra MR, Williams BA (1998) Levels of homoplasy in the evolution of the mammalian skeleton. *Journal of Mammalian Evolution* **5**, 113–126. doi:10.1023/A:1020549505177

Sanderson MJ (1997) A nonparametric approach to estimating divergence times in the absence of rate constancy. *Molecular Biology and Evolution* **14**, 1218–1232. doi:10.1093/oxfordjournals.molbev.a025731

Sanderson MJ (2002) Estimating absolute rates of evolution and divergence times: a penalized likelihood approach. *Molecular Biology and Evolution* **19**, 101–109. doi:10.1093/oxfordjournals.molbev.a003974

Sanger F, Nicklen S, Coulson AR (1977) DNA sequencing with chain-terminating inhibitors. *Proceedings of the National Academy of Sciences USA* **74**, 5463–5467. doi:10.1073/pnas.74.12.5463

Sansom RS, Wills MA, Williams T (2017) Dental data perform relatively poorly in reconstructing mammal phylogenies: morphological partitions evaluated with molecular benchmarks. *Systematic Biology* **66**, 813–822. doi:10.1093/sysbio/syw116

Sarich VM (1985) Rodent macromolecular systematics. In *Evolutionary Relationships Among Rodents: A Multidisciplinary Analysis*. (Eds WP Luckett and J-L Hartenberger) pp. 423–452. NATO Science Series A: Life Sciences, Volume 92. Springer Nature, Basingstoke, UK.

Sarich VM, Lowenstein JM, Richardson BJ (1982) Phylogenetic relationships of the thylacine (*Thylacinus cynocephalus*, Marsupialia) as reflected in comparative serology. In *Carnivorous Marsupials*. (Ed. M Archer) pp. 707–709. Royal Zoological Society of New South Wales, Mosman, NSW.

Sarich VM, Wilson AC (1966) Quantitative immunochemistry and the evolution of primate albumins: micro-complement fixation. *Science* **154**, 1563–1566. doi:10.1126/science.154.3756.1563

Sarich VM, Wilson AC (1967) Immunological time scale for hominid evolution. *Science* **158**, 1200–1203. doi:10.1126/science.158.3805.1200

Schlegel H (1866) Observations zoologiques. III. *Mammifères. Nederlandsch Tydschrift Voor de Dierkunde* **iii**, 350–358, <https://www.biodiversitylibrary.org/page/49109851>.

Schlegel H (1880) On *Dasyurus albopunctatus. Notes from the Leyden Museum* **2**, 51–53, <https://repository.naturalis.nl/pub/509241>.

Schluter D (2000) *The Ecology of Adaptive Radiation.* Oxford University Press, Oxford.

Schmidt-Nielsen K (1964) *Desert Animals.* Oxford University Press, Oxford.

Schuster SC (2008) Next-generation sequencing transforms today's biology. *Nature Methods* **5**, 16–18. doi:10.1038/nmeth1156

Scotland RW, Olmstead RG, Bennett JR (2003) Phylogeny reconstruction: the role of morphology. *Systematic Biology* **52**, 539–548. doi:10.1080/10635150390223613

Scott WB (1913) *History of Land Mammals of the Western Hemisphere.* MacMillan, New York, <https://www.biodiversitylibrary.org/page/14839931>.

Seri L (1992) Conservation of marsupials and monotremes in Papua New Guinea. In *Australasian Marsupials and Monotremes: An Action Plan for their Conservation.* (Ed. M Kennedy) pp. 12–25. IUCN, Gland, Switzerland, <https://portals.iucn.org/library/node/6133>.

Sessions SK (1996) Chromosomes: molecular cytogenetics. In *Molecular Systematics.* 2nd edn. (Eds DM Hillis, C Moritz, and BK Mable) pp. 121–168. Sinauer Associates, Sunderland, MA.

Settle GA (1978) The quiddity of tiger quolls. *Australian Natural History* **19**, 165–169.

Sharman GB (1961) The mitotic chromosomes of marsupials and their bearing on taxonomy and phylogeny. *Australian Journal of Zoology* **9**, 38–60. doi:10.1071/ZO9610038

Sharman GB (1973) The chromosomes of non-eutherian mammals. In *Cytotaxonomy and Vertebrate Evolution.* (Eds AB Chiarelli and E Capanna) pp. 485–530. Academic Press, New York.

Sharman GB (1974) Marsupial taxonomy and phylogeny. *Australian Mammalogy* **1**, 136–154.

Sharman GB (1982) Karyotypic similarities between *Dromiciops australis* (Microbiotheriidae, Marsupialia) and some Australian marsupials. In *Carnivorous Marsupials.* (Ed. M Archer) pp. 711–714. Royal Zoological Society of New South Wales, Sydney.

Sharman GB (1989) A chromosome phylogeny of kangaroos. In *Kangaroos, Wallabies and Rat-Kangaroos.* (Eds G Grigg, P Jarman, and I Hume) pp. v–xi. Surrey Beatty and Sons Pty Ltd. Chipping Norton, NSW.

Sharman GB, Close RL, Maynes GM (1990a) Chromosomal evolution, phylogeny and speciation of rock wallabies (*Petrogale*: Macropodidae). *Australian Journal of Zoology* **37**, 351–363. doi:10.1071/ZO9890351

Sharman GB, Hughes RL, Cooper DW (1990b) The chromosomal basis of sex-differentiation in marsupials. *Australian Journal of Zoology* **37**, 451–466. doi:10.1071/ZO9890451

Sharp PG (1982) Sex chromosome pairing during male meiosis in marsupials. *Chromosoma* **86**, 27–47. doi:10.1007/BF00330728

Shaw G, Renfree MB, Short RV (1990) Primary genetic control of sexual differentiation in marsupials. In *Mammals from Pouches and Eggs: Genetics, Breeding and Evolution of Marsupials and Monotremes.* (Eds JA Marshall Graves, RM Hope, and DW Cooper) pp. 301–308. CSIRO, Australia. doi:10.1071/ZO989044

Sibley CG, Ahlquist JE (1983) Phylogeny and classification of birds based on the data of DNA-DNA hybridization. In *Current Ornithology, Volume 1.* (Ed. RF Johnston) pp. 245–292. Plenum Publishing, New York.

Sibley CG, Ahlquist JE (1990) *Phylogeny and Classification of Birds: A Study in Molecular Evolution.* Yale University Press, New Haven, CT.

Simpson GG (1941) The affinities of Borhyaenidae. *American Museum Novitates* **1118**, 1–6, <https://digitallibrary.amnh.org/handle/2246/2252>.

Simpson GG (1944) *Tempo and Mode in Evolution.* Columbia University Press, New York.

Simpson GG (1945) The principles of classification and a classification of mammals. *Bulletin of the American Museum of Natural History* **85**, 1–350, <https://digitallibrary.amnh.org/handle/2246/1104>.

Simpson GG (1948) The beginning of the age of mammals in South America; Part 1. Introduction, Systematics: Marsupialia, Edentata, Condylarthra, Litopterna, and Notoprogonia. *Bulletin of the American Museum of Natural History* **91**, 1–232, <https://digitallibrary.amnh.org/handle/2246/1632>.

Simpson GG (1949) *The Meaning of Evolution: A Study of the History of Life and Its Significance for Man.* Yale University Press, New Haven, CT.

Simpson GG (1951a) *The Meaning of Evolution: A Special Revised and Abridged Edition.* Mentor Books, New York.

Simpson GG (1951b) The species concept. *Evolution* **5**, 285–298. doi:10.1111/j.1558-5646.1951.tb02788.x

Simpson GG (1953) *The Major Features of Evolution.* Simon and Schuster, New York.

Simpson GG (1961) *Principles of Animal Taxonomy.* Columbia University Press, New York.

Sinclair WJ (1905) The marsupial fauna of the Santa Cruz beds. *Proceedings of the American Philosophical Society* **44**, 73–81, <https://www.biodiversitylibrary.org/page/26854456>.

Slowinski JB, Crother BI (1998) Is the PTP Test useful? *Cladistics* **14**, 297–302. doi:10.1006/clad.1998.0065

Smith M (1982) Review of the thylacine (Marsupialia, Thylacinidae). In *Carnivorous Marsupials.* (Ed. M Archer) pp. 237–253. Royal Zoological Society of New South Wales, Sydney.

Smith AMA (1983) The subspecific biochemical taxonomy of *Antechinus minimus, A. swainsonii,* and *Sminthopsis leucopus* (Marsupialia: Dasyuridae). *Australian Journal of Zoology* **32**, 753–762. doi:10.1071/ZO9830753

Smith ND, Turner AH (2005) Morphology's role in phylogeny reconstruction: perspectives from paleontology. *Systematic Biology* **54**, 166–173. doi:10.1080/10635150590906000

Sneath PHA, Sokal RR (1973) *Numerical Taxonomy.* W.H. Freeman, San Francisco.

Sniderman JMK, Hellstrom J, Woodhead JD, Drysdale RN, Bajo P, *et al.* (2019) Vegetation and climate change in southwestern Australia during the last glacial maximum. *Geophysical Research Letters* **46**, 1709–1720. doi:10.1029/2018GL080832

Sober E (1988) *Reconstructing the Past.* MIT Press, Cambridge, MA.

Soderquist TR, Serena M (1990) Occurrence and outcome of polyoestry in wild western quolls, *Dasyurus geoffroii* (Marsupialia: Dasyuridae). *Australian Mammalogy* **13**, 205–208. doi:10.1071/AM90021

Spencer PBS, Rhind SG, Eldridge MDB (2001) Phylogeographic structure within *Phascogale* (Marsupialia: Dasyuridae) based on partial cytochrome *b* sequence. *Australian Journal of Zoology* **49**, 369–377. doi:10.1071/ZO00080

Spencer WB (1895) Preliminary notice of two new species of marsupials from Central Australia. *Proceedings of the Royal Society of Victoria* [new series] **7**, 222–224, <https://www.biodiversitylibrary.org/page/8659827>.

Spencer WB (1896) Mammalia. In *Report on the Work of the Horn Scientific Expedition to Central Australia, Part 2, Zoology.* (Ed. WB Spencer) pp. 1–52. Melville, Mullen, & Slade, Melbourne, <https://books.google.com/books?printsec=frontcover&vid=LCCN16017552#v=onepage&q&f=false>.

Spencer WB (1909) Description of a new species of *Sminthopsis. Proceedings of the Royal Society of Victoria* [new series] **21**, 449–451, <https://www.biodiversitylibrary.org/page/8499322>.

Springer MS (1995) Molecular clocks and the incompleteness of the fossil record. *Journal of Molecular Evolution* **41**, 531–538. doi:10.1007/BF00175810

Springer MS, Burk-Herrick A, Meredith R, Eizirik E, Teeling E, *et al.* (2007) The adequacy of morphology for reconstructing the early history of placental mammals. *Systematic Biology* **56**, 673–684. doi:10.1080/10635150701491149

Springer MS, DeBry RW, Douady C, Amrine HM, Madsen O, *et al.* (2001) Mitochondrial versus nuclear gene sequences in deep-level mammalian phylogeny reconstruction. *Molecular Biology and Evolution* **18**, 132–143. doi:10.1093/oxfordjournals.molbev.a003787

Springer MS, Douzery E (1996) Secondary structure and patterns of evolution among mammalian mitochondrial 12S rRNA molecules. *Journal of Molecular Evolution* **43**, 357–373. doi:10.1007/BF02339010

Springer MS, Kirsch JAW (1989) Rates of single-copy DNA evolution in phalangeriform marsupials. *Molecular Biology and Evolution* **6**, 331–341. doi:10.1007/BF01464363

Springer MS, Kirsch JAW, Case JA (1997) The chronicle of marsupial evolution. In *Molecular Evolution and Adaptive Radiation.* (Eds TJ Givnish and KJ Sytsma) pp. 129–161. Cambridge University Press, Cambridge.

Springer M, Krajewski C (1989a) Additive distances, rate variation, and the perfect fit theorem. *Systematic Zoology* **38**, 371–375. doi:10.2307/2992402

Springer M, Krajewski C (1989b) DNA hybridization in animal taxonomy: a critique from first principles. *Quarterly Review of Biology* **64**, 291–318. doi:10.1086/416360

Springer MS, Meredith RW, Eizirik E, Teeling E, Murphy WJ (2008) Morphology and placental mammal phylogeny. *Systematic Biology* **57**, 499–503. doi:10.1080/10635150802164504

Springer MS, Westerman M, Kirsch JAW (1994) Relationships among orders and families of marsupials based on 12S ribosomal DNA sequences and the timing of the marsupial radiation. *Journal of Mammalian Evolution* **2**, 85–115. doi:10.1007/BF01464363

Stadler T, Yang Z (2013) Dating phylogenies with sequentially sampled tips. *Systematic Biology* **62**, 674–688. doi:10.1093/sysbio/syt030

Start AN, Moro D, Adams M, Bencini R (2006) Dunnarts from Boullanger Island: new evidence and reassessment of a taxonomic issue with resource implications. *Australian Mammalogy* **28**, 51–58. doi:10.1071/AM06006

Stearns SC (1992) *The Evolution of Life History Strategies.* Oxford University Press, Oxford.

Stein G (1933) Weitere Mitteil-ungen zur Systematik papuanischer Säuger. *Zeitschrift für Säugertierkunde* **8**, 87–95, <https://www.biodiversitylibrary.org/page/45342790>.

Stirton RA (1957) Tertiary marsupials from Victoria, Australia. *Memoirs of the National Museum of Victoria* **21**, 129–133, <https://museumsvictoria.com.au/media/5640/jmmv19572109.pdf>.

Stobo-Wilson AM, Murphy BP, Crawford HM, Dawson SJ, Dickman CR, *et al.* (2021) Sharing meals: predation on Australian mammals by the introduced European red fox compounds and complements predation by feral cats. *Biological Conservation* **261**, 109284. doi:10.1016/j.biocon.2021.109284

Storr G (1958) Are marsupials 'second class' mammals? *The Western Australian Naturalist* **6**, 179–183, <https://www.biodiversitylibrary.org/page/59019558>.

Storr GCC (1780) *Prodromus Methodi Mammalium.* Tubingae, Germany, <https://books.google.com/books/about/Prodromus_methodi_mammalium.html?id=IU5YmgEACAAJ>.

Stroud JT, Losos JB (2016) Ecological opportunity and adaptive radiation. *Annual Review of Ecology and Systematics* **47**, 507–532. doi:10.1146/annurev-ecolsys-121415-032254

Suarez-Villota EK, Haro RE, Vargas RA, Gallardo MH (2016) The ancestral chromosomes of *Dromiciops gliroides* (Microbiotheriidae) and its bearing on the karyotypic evolution of American marsupials. *Molecular Cytogenetics* **9**, 59. doi:10.1186/s13039-016-0270-8

Suchard MA, Lemey P, Baele G, Ayers DL, Drummond AJ, *et al.* (2018) Bayesian phylogenetic and phylodynamic data integration using BEAST 1.10. *Virus Evolution* **4**, vey016. doi:10.1093/ve/vey016

Summerhayes GR, Leavesly M, Fairbairn A, Mandui H, Field J, *et al.* (2010) Human adaptation and plant use in highland New Guinea 49,000 to 44,000 years ago. *Science* **330**, 78–81. doi:10.1126/science.1193130

Sumner J, Dickman CR (1998) Distribution and identity of species in the *Antechinus stuartii-A. flavipes* group (Marsupialia: Dasyuridae) in south-eastern Australia. *Australian Journal of Zoology* **46**, 27–41. doi:10.1071/ZO94055

Sutton WS (1903) The chromosomes in heredity. *Biological Bulletin* **4**, 231–251, <http://www.esp.org/foundations/genetics/classical/holdings/s/wss-03.pdf>.

Swofford DL (1999) *PAUP*: Phylogenetic Analysis Using Parsimony (and Other Methods). Version 4.* Sinauer Associates, Sunderland, MA.

Swofford DL, Olsen GJ (1990) Phylogeny reconstruction. In *Molecular Systematics.* (Eds DM Hillis and C Moritz) pp. 411–501. Sinauer Associates, Sunderland, MA.

Swofford DL, Olsen GJ, Waddell PJ, Hillis DM. 1996. Phylogenetic inference. In *Molecular Systematics.* 2nd edn. (Eds DM Hillis, C Moritz, and BK Mable) pp. 407–514. Sinauer Associates, Sunderland, MA.

Szalay FS (1982) A new appraisal of marsupial phylogeny and classification. In *Carnivorous Marsupials.* (Ed. M Archer) pp. 621–640. Royal Zoological Society of New South Wales, Mosman, NSW.

Takezaki N, Rhzetsky A, Nei M (1995) Phylogenetic test of the molecular clock and linearized trees. *Molecular Biology and Evolution* **12**, 823–833. doi:10.1093/oxfordjournals.molbev.a04025

Tallowin OJS, Tamar K, Meiri A, Allison A, Kraus F, *et al.* (2018) Early insularity and subsequent mountain uplift were complementary drivers of diversification in a Melanesian lizard radiation (Gekkonidae: *Cyrtodactylus*). *Molecular Phylogenetics and Evolution* **125**, 29–39. doi:10.1016/j.ympev.2018.03.020

Tamura K (1992) Estimation of the number of nucleotide substitutions when there are strong transition-transversion and GC-content biases. *Molecular Biology and Evolution* **9**, 678–687. doi:10.1093/oxfordjournals.molbev.a040752

Tan R, Han K, Geng Y, Yang C, Shi C, *et al.* (2021) A chromosome-level genome of *Antechinus flavipes* provides a reference for an Australian marsupial genus with male death after mating. *Molecular Ecology Resources* **22**, 740–754. doi:10.1111/1755-0998.13501

Taplin LW (1980) Some observations on the reproductive biology of *Sminthopsis virginiae* (Tarragon) (Marsupialia: Dasyuridae). *Australian Zoologist* **20**, 407–418, <https://www.biodiversitylibrary.org/page/39036924>.

Tarragon L (1847) Description du male adulte du *Colobe guerza* et du *Phascogale virginiae. Revue de Zoologique* **1847**, 177–178, <https://www.biodiversitylibrary.org/page/2271683>.

Tate GHH (1947) Results of the Archbold Expeditions. No. 56. On the anatomy and classification of Dasyuridae (Marsupialia). *Bulletin of the American Museum of Natural History* **88**, 99–155, <https://digitallibrary.amnh.org/handle/2246/1836>.

Tate GHH (1952) Results of the Archbold expedition, no. 66. Mammals of Cape York Peninsula, with notes on the occurrence of rainforest in Queensland. *Bulletin of the American Museum of Natural History* **98**, 563–616, <https://digitallibrary.amnh.org/handle/2246/1834>.

Tate GHH, Archbold R (1937) Results of the Archbold Expeditions. 16. Some marsupials of New Guinea and Celebes. *Bulletin of the American Museum of Natural History* **73**, 331–476, <https://digitallibrary.amnh.org/handle/2246/1835>.

Taylor JM, Calaby JH, Redhead TD (1982) Breeding in wild populations of the marsupial-mouse *Planigale maculata sinualis* (Dasyuridae, Marsupialia). In *Carnivorous Marsupials.* (Ed. M Archer) pp. 83–87. Royal Zoological Society of New South Wales, Mosman, Australia.

Temminck CJ (1824) Sur les mammifères du genre dasyure. In *Monographies de Mammalogie.* (Ed. CJ Temminck) pp. 55–72. G. Dufour et E. D'Ocagne, Paris, <https://www.biodiversitylibrary.org/page/52681164>.

Templeton AR (1983) Phylogenetic inference from restriction endonuclease cleavage maps with particular reference to humans and apes. *Evolution* **37**, 221–244. doi:10.1111/j.1558-5646.1983.tb05533.x

Thavornkanlapachai R, Mills HR, Ottewell K, Friend JA, Kennington WJ (2021) Temporal variation in the genetic composition of an endangered marsupial reflects reintroduction history. *Diversity* **13**, 257. doi:10.3390/d13060257

Thesleff I (2003) Epithelial-mesenchymal signaling regulating tooth morphogenesis. *Journal of Cell Science* **116**, 1647–1648. doi:10.1242/jcs.00410

Thomas H, Cameron SF, Campbell HA, Micheli-Campbell MA, Kirke EC, *et al.* (2021) Rocky escarpment versus savanna woodlands: comparing diet and body condition as indicators of habitat quality for the endangered northern quoll (*Dasyurus hallucatus*). *Wildlife Research* **48**, 434–443. doi:10.1071/WR20032

Thomas O (1886a) Diagnosis of a new species of Phascogale. *Annali del Museo Civico di Storia Naturale de Genova* [Series 2] **24**, 208, <https://www.biodiversitylibrary.org/page/29872112>.

Thomas O (1886b) On the specimens of *Phascogale* in the Museo Civico, Genoa, with notes on the allied species of the genus. *Annali del Museo Civico di Storia Naturale de Genova* [Series 2] **24**, 502–511, <https://www.biodiversitylibrary.org/page/29872406>.

Thomas O (1887) The homologies and succession of the teeth in the Dasyuridae, with an attempt to trace the history of evolution of mammalian teeth in general. *Philosophical Transactions of the Royal Society* **178**, 443–462. doi:10.1098/rstb.1887.0015

Thomas O (1888) *Catalogue of the Marsupialia and Monotremata in the collection of the British Museum (Natural History).* Taylor and Francis, London, <https://www.biodiversitylibrary.org/page/37986443>.

Thomas O (1898) Descriptions of three new mammals from the East Indian Archipelago and Australia. *Novitates Zoologicae* **5**, 1–4, <https://www.biodiversitylibrary.org/page/3219599>.

Thomas O (1899) Description of a new species of Phascogale from British New Guinea, obtained by Dr. L. Loria. *Annali del Museo Civico di Storia Naturale de Genova* [Series 2] **40**, 191–192, <https://www.biodiversitylibrary.org/page/7991474>.

Thomas O (1902) Two new Australian small mammals. *Annals and Magazine of Natural History* **10**, 491–492, <https://www.biodiversitylibrary.org/page/19339300>.

Thomas O (1904) On a collection of mammals made by Mr. J.T. Tunney in Arnhem Land, Northern Territory of South Australia. *Novitates Zoologicae* **11**, 222–229, <https://www.biodiversitylibrary.org/page/3858949>.

Thomas O (1906a) On mammals collected in South-west Australia for Mr. W.E. Balston. *Proceedings of the Zoological Society of London* **1906**, 468–478, <https://www.biodiversitylibrary.org/page/31208179>.

Thomas O (1906b) A collection of mammals made by Mr. W. Stalker in the Northern Territory of South Australia, and presented to the National Museum by Sir William Ingram, Bart., and the Hon. *John Forrest. Abstracts of the Proceedings of the Zoological Society of London* **32**, 6.

Thomas O (1923) The Godman Exploration Fund: list of mammals from North Queensland collected by Mr. T.V. Sherrin. *Annals and Magazine of Natural History* [Series 9] **11**, 170–178, <https://www.biodiversitylibrary.org/page/34355412>.

Thomas O (1926) The local races of *Dasyurus hallucatus. Annals and Magazine of Natural History* **9**, 543–544. doi:10.1080/00222932608633550

Thomas RH, Schaffner W, Wilson AC, Pääbo S (1989) DNA phylogeny of the extinct marsupial wolf. *Nature* **340**, 465–467. doi:10.1038/340465a0

Thorne JL, Kishino H (2002) Divergence time and evolutionary rate estimation with multilocus data. *Systematic Biology* **51**, 689–702. doi:10.1080/10635150290102456

Thorne JL, Kishino H, Painter IS (1998) Estimating the rate of evolution of the rate of evolution. *Molecular Biology and Evolution* **15**, 1647–1657. doi:10.1093/oxfordjournals.molbev.a025892

Timms M, Westerman M, Düring B (1982) Studies on the DNA of some species of *Antechinus* (Dasyuridae, Marsupialia). In *Carnivorous Marsupials*. (Ed. M Archer) pp. 715–721. Royal Zoological Society of New South Wales, Sydney.

Tindale NB (1959) Ecology of primitive aboriginal man in Australia. In *Biogeography and Ecology in Australia*. (Eds A Keast, RL Crocke and CS Christian) pp. 36–51. Uitgeverij Dr. W. Junk, The Hague.

Toder R, Wakefield MJ, Graves JAM (2000) The minimal mammalian Y chromosome – the marsupial Y as a model system. *Cytogenetics and Cell Genetics* **91**, 285–292. doi:10.1159/000056858

Toussaint EFA, Hall R, Monaghan MT, Sagata K, Ibalim S, *et al.* (2014) The towering orogeny of New Guinea as a trigger for arthropod megadiversity. *Nature Communications* **5**, 4001. doi:10.1038/ncomms5001

Travouillon KJ (2016) Oldest fossil remains of the enigmatic pig-footed bandicoot show rapid herbivorous evolution. *Royal Society Open Science* **3**, 160089. doi:10.1098/rsos.160089

Travouillon KJ, Beck RMD, Hand SJ, Archer M (2013) The oldest fossil record of bandicoots (Marsupialia: Peramelemorphia) from the late Oligocene of Australia. *Paleontologica Electronica* **16**(2), 13a. doi:10.26879/363

Travouillon KJ, Legendre S, Archer M, Hand SJ (2009) Palaeoecological analyses of Riversleigh's Oligo-Miocene sites: Implications for Oligo-Miocene climate change in Australia. *Palaeogeography, Palaeoclimatology, Palaeoecology* **276**, 24–37. doi:10.1016/j.palaeo.2009.02.025

Travouillon KJ, Louys J, Price GJ, Archer M, Hand SJ, *et al.* (2017) A review of the Pliocene bandicoots of Australia, and descriptions of a new genus and species. *Journal of Vertebrate Paleontology* **37**, e1360894. doi:10.1080/02724634.2017.1360894

Troughton E (1947) *Furred Animals of Australia*. Charles Scribner's Sons, New York.

Troughton ELG (1928) A new genus, species, and subspecies of marsupial mice (Family Dasyuridae). *Records of the Australian Museum* **16**, 281–288, <https://journals.australian.museum/troughton-1928-rec-aust-mus-166-281288/>.

Troughton ELG (1932) A new species of fat-tailed marsupial mouse, and the status of *Antechinus froggatti* Ramsay. *Records of the Australian Museum* **18**, 349–353, <https://journals.australian.museum/troughton-1932-rec-aust-mus-186-349354/>.

Troughton ELG (1959) The marsupial fauna: its origin and radiation. In *Biogeography and Ecology in Australia*. (Eds A Keast, RL Crocker, and CS Christian) pp. 69–88. Junk, The Hague.

Turnbull WD, Lundelius EL, Tedford RH (1993) Fossil vertebrate locality at Smeaton, Victoria. *Memoirs of the Association of Australasian Palaeontologists* **15**, 429–440, <https://www.researchgate.net/publication/292807586_Fossil_vertebrate_locality_at_Smeaton_Victoria>.

Turnbull WD, Lundelius EL Jr., Archer M (2003) Dasyurids, perameloids, phalangeroids, and vombatoids from the early Pliocene Hamilton Fauna, Victoria. *Bulletin of the American Museum of Natural History* **279**, 513–540. doi:10.1206/0003-0090(2003)279<0513:C>2.0CO;2

Tyndale-Biscoe H (2005) *Life of Marsupials*. CSIRO Publishing, Collingwood.

Tyson E (1698) Carigueya, seu Marsupiale americanum OR The anatomy of an opossum, dissected at Gresham College. *Philosophical Transactions of the Royal Society* **20**, 105–164, <https://quod.lib.umich.edu/e/eebo2/A64095.0001.001?view=toc>.

Umbrello LS (2018) Evolution and diversification of dasyurid marsupials of the arid zone. PhD dissertation, University of Western Australia, Australia. doi:10.26182/5ca6cffb21fc6

Umbrello LS, Potter LC, Westerman M, Woolley PA (2022) First record of *Pseudantechinus macdonnellensis* in the Kimberley region of Western Australia. *Australian Mammalogy* **44**, 404–406.

Umbrello LS, Woolley PA, Westerman M (2017) Species relationships in the dasyurid marsupial genus *Pseudantechinus* (Marsupialia: Dasyuridae): a re-examination of the taxonomic status of *Pseudantechinus roryi. Australian Journal of Zoology* **65**, 240–247. doi:10.1071/ZO17059

Unmack PJ, Adams M, Hammer MP, Johnson JB, Gruber B, *et al.* (2021) Plotting for change: an analytical framework to aid decisions on which lineages are candidate species in phylogenomic species discovery. *Biological Journal of the Linnean Society* **135**, 117–137. doi:10.1093/biolinnean/blab095

Uzqueda A, Burnett S, Bertola LV, Hoskin CJ (2020) Quantifying range decline and remaining populations of the large marsupial carnivore of Australia's tropical rainforest. *Journal of Mammalogy* **101**, 1021–1034. doi:10.1093/jmammal/gyaa077

Vaidyanathan G (2021) The world's species are playing musical chairs: how will it end? *Nature* **596**, 22–25. doi:10.1038/d41586-021-02088-3

Valente A (1984) Reproductive biology of *Antechinomys laniger* (*spenceri* form) (Marsupialia: Dasyuridae). PhD Dissertation, La Trobe University, Melbourne.

Valente A, Woolley PA (1982) The Dibbler, *Parantechinus apicalis*: failure to locate populations in four regions in the south of Western Australia. *Australian Mammalogy* **5**, 241–245. doi:10.1071/AM82025

Van den Broek AJP (1910) Untersuchungen über den Bau der männlichen Geschlechtsorgane der Beuteltiere. *Gegenbaurs morphologisches Jahrbuch* **41**, 347–436, <https://www.biodiversitylibrary.org/page/13025705>.

Van der Kaars S, Miller GH, Turney CSM, Cook EJ, Nürnberg D, *et al.* (2017) Humans rather than climate the primary cause of Pleistocene megafaunal extinction in Australia. *Nature Communications* **8**, 14142. doi:10.1038/ncomms14142

Van Duzer C (2013) *Sic sunt dracones*: the geography and cartography of monsters. In *The Ashgate Research Companion to Monsters and the Monstrous.* (Eds AS Mittman and PJ Dendle) pp. 387–436. Routledge Taylor & Francis Group, London. doi:10.4324/9781315241197.ch1

Van Dyck S (1980) The Cinnamon Antechinus, *Antechinus leo* (Marsupialia: Dasyuridae) a new species from the vine-forests of Cape York Peninsula. *Australian Mammalogy* **3**, 5–17.

Van Dyck S (1982a) *Antechinus puteus* (Marsupialia: Dasyuridae), a new fossil species from the Texas Caves, southeastern Queensland. *Australian Mammalogy* **5**, 59–68.

Van Dyck S (1982b) The relationships of *Antechinus stuartii* and *A. flavipes* (Dasyuridae, Marsupialia) with special reference to Queensland. In *Carnivorous Marsupials.* (Ed. M Archer) pp. 723–766. Royal Zoological Society of New South Wales, Sydney.

Van Dyck S (1985) *Sminthopsis leucopus* (Marsupialia: Dasyuridae) in north Queensland rainforest. *Australian Mammalogy* **8**, 53–60. doi:10.1071/AM85003

Van Dyck S (1986) *Sminthopsis archeri* (Marsupialia: Dasyuridae), a new species from the savannah of Papua New Guinea and Cape York Peninsula, Australia. *Australian Mammalogy* **9**, 111–124.

Van Dyck S (1987) The bronze quoll, *Dasyurus Spartacus* (Marsupialia: Dasyuridae), a new species from the savannahs of Papua New Guinea. *Australian Mammalogy* **11**, 145–146.

Van Dyck S (2002) Morphology-based revision of *Murexia* and *Antechinus* (Marsupialia: Dasyuridae). *Memoirs of the Queensland Museum* **48**, 239–330, <https://www.biodiversitylibrary.org/page/40460830>.

Van Dyck S, Crowther MS (2000) Reassessment of northern representatives of the *Antechinus stuartii* complex (Marsupialia: Dasyuridae): *A. subtropicus* sp. nov. and *A. adustus* elevated. *Memoirs of the Queensland Museum* **45**, 611–635, <https://www.biodiversitylibrary.org/page/40462416>.

Van Dyck S, Strahan R (Eds) (2008) *Mammals of Australia.* 3rd edn. Reed New Holland, Sydney.

Van Dyck S, Woinarski JCZ, Press AJ (1994) The Kakadu Dunnart, *Sminthopsis bindi* (Marsupialia: Dasyuridae), a new species from the stony woodlands of the Northern Territory. *Memoirs of the Queensland Museum* **37**, 311–323, <https://www.biodiversitylibrary.org/page/40127182>.

Vane-Wright RI, Humphries CJ, Williams PH (1991) What to protect? – Systematics and the agony of choice. *Biological Conservation* **55**, 235–254. doi:10.1016/0006-3207(91)90030-D

Van Ufford AQ, Cloos M (2005) Cenozoic tectonics of New Guinea. *American Association of Petroleum Geologists Bulletin* **89**, 119–140. doi:10.1306/08300403073

Veevers JJ, Powell C McA, Roots SR (1991) Review of seafloor spreading around Australia. I. synthesis of the patterns of spreading. *Australian Journal of Earth Sciences* **38**, 373–389. doi:10.1080/08120099108727979

Viacava P, Baker AM, Blomberg SP, Phillips MJ, Weisbecker V (2021) Using 3D geometric morphometrics to aid taxonomic and ecological understanding of a recent speciation event within a small Australian marsupial (genus *Antechinus*). *Zoological Journal of the Linnean Society* [online early] zlab048. doi:10.1093/zoolinnean/zlab048

Vicq d'Azyr F (1792) *Système anatomique des Quadrupèdes.* Panckoucke, Paris.

Vrba ES (1985) Environment and evolution: alternative causes of the temporal distribution of evolutionary events. *South African Journal of Science* **81**, 229–236.

Wagner PJ (2000) Exhaustion of morphologic character states among fossil taxa. *Evolution* **54**, 365–386. doi:10.1111/j.0014-3820.2000.tb00040.x

Waite ER (1904) Description of a new *Phascogale* from North Western Australia. *Records of the Australian Museum* **5**, 122–124,<https://journals.australian.museum/waite-1904-rec-aust-mus-52-122124/>.

Waithman J (1979) A report on a collection of mammals from southwest Papua, 1972-1973. *Australian Zoologist* **20**, 313–326, <https://www.biodiversitylibrary.org/page/39036824>.

Wakefield NA, Warneke RM (1963) Some revision in *Antechinus* (Marsupialia) – 1. *Victorian Naturalist* **80**, 194–219, <https://www.biodiversitylibrary.org/page/40918718>.

Wakefield NA, Warneke RM (1967) Some revision in *Antechinus* (Marsupialia) – 2. *Victorian Naturalist* **84**, 69–99, <https://www.biodiversitylibrary.org/page/46800936>.

Wakeley J (2006) *An Introduction to Coalescent Theory.* Roberts and Company, Greenwood Village, CO.

Wallace AR (1876) *The Geographical Distribution of Animals.* Macmillan and Company, London, <http://darwin-online.org.uk/converted/Ancillary/1876_GeographicalDistribution_S718.1/1876_GeographicalDistribution_S718.1.html>.

Wallis RL (1982) Adaptation to low environmental temperatures in the carnivorous marsupials. In *Carnivorous Marsupials.* (Ed. M Archer) pp. 285–291. Royal Zoological Society of New South Wales, Sydney.

Wanninger A (2015) Morphology is dead – long live morphology! Integrating MorphoEvoDevo into molecular EvoDevo and phylogenomics. *Frontiers in Ecology and Evolution* **3**, 54. doi:10.3389/fevo.2015.00054.

Warburton NM, Travouillon KJ (2017) The biology and paleontology of the Peramelemorphia: a review of current knowledge and future research directions. *Australian Journal of Zoology* **64**, 151–181. doi:10.1071/ZO16003

Ward MS, Simmonds JS, Reside AE, Watson JEM, Rhodes JR, *et al.* (2019) Lots of loss with little scrutiny: the attrition of habitat critical for threatened species in Australia. *Conservation Science and Practice* **1**, e117. doi:10.1111/csp2.117

Warneke L, Turner JM, Geiser F (2008) Torpor and basking in a small arid zone marsupial. *Naturwissenshaften* **95**, 73–78. doi:10.1007/s00114-007-0293-4

Waterhouse GR (1836) Description of a new genus of mammiferous animals from New Holland, which will probably be found to belong to the marsupial type. *Proceedings of the Zoological Society of London* **4**, 69–70, <https://www.biodiversitylibrary.org/page/12860053>.

Waterhouse GR (1838) Characters of some new species of the genera *Mus* and *Phascogale*. *Proceedings of the Zoological Society of London* **1837**, 75–77, <https://www.biodiversitylibrary.org/page/12860217>.

Waterhouse GR (1841) Marsupialia or pouched animals. In *The Naturalist's Library, Mammalia, Vol. XXIV.* (Ed. W Jardine) pp. 45–309. W.H. Lizars, London, <https://www.biodiversitylibrary.org/page/18064573>.

Waterhouse GR (1846) *The Natural History of the Mammalia. Volume I: Marsupiata, or Pouched Animals.* Wilson and Oglivy, London, <https://www.biodiversitylibrary.org/page/58844221>.

Watt A (1997) Population ecology and reproductive seasonality in three species of *Antechinus* (Marsupialia: Dasyuridae) in the wet tropics of Queensland. *Wildlife Research* **24**, 531–547. doi:10.1071/WR96033

Waugh O'Neill RJ, Eldridge MDB, Toder R, Fergusson-Smith MA, O'Brien PC, *et al.* (1999) Chromosome evolution in kangaroos (Marsupialia: Macropodidae): Cross species chromosome painting between the tammar wallaby and rock wallaby spp. with the $2n = 22$ ancestral macropodid karyotype. *Genome* **42**, 525–530. doi:10.1139/g98-159

Webb S (2013) *Corridors to Extinction and the Australian Megafauna.* Elsevier, London.

Webster DB, Webster M (1975) Auditory systems of Heteromyidae: functional morphology and evolution of the middle ear. *Journal of Morphology* **146**, 343–376. doi:10.1002/jmor.1051460304

Weitzman ML (1992) On diversity. *Quarterly Journal of Economics* **107**, 363–405. doi:10.2307/2118476

Werdelin L (1987) Some observations on *Sarcophilus lanarius* and the evolution of *Sarcophilus*. *Records of the Queen Victoria Museum* **90**, 1–27, <https://www.biodiversitylibrary.org/page/57425436>.

West RS, Tilley L, Moseby KE (2020) A trial reintroduction of the western quoll to a fenced conservation reserve: implications of returning native predators. *Australian Mammalogy* **42**, 257–265. doi:10.1071/AM19041

Westaway MC, Price G, Miscamble T, McDonald J, Cramb J, *et al.* (2019) A palaeontological perspective on the proposal to reintroduce Tasmanian devils to mainland Australia to suppress invasive predators. *Biological Conservation* **232**, 187–193. doi:10.1016/j.biocon.2019.02.002

Westerman M (1968) The effect of X-irradiation on male meiosis in *Schistocerca gregaria* (Forskal). II. The induction of chromosome mutations. *Chromosoma (Berlin)* **24**, 17–36. doi:10.1007/BF00329604

Westerman M, Blacket MJ, Hintz A, Armstrong K, Woolley PA, *et al.* (2016) A plethora of planigales: genetic variability and cryptic species in a genus of dasyurid marsupials from northern Australia. *Australian Journal of Zoology* **64**, 303–311. doi:10.1071/ZO16052

Westerman M, Janczewski DN, O'Brien SJ (1990) DNA-DNA hybridization studies and marsupial phylogeny. *Australian Journal of Zoology* **37**, 315–323. doi:10.1071/ZO9890315

Westerman M, Krajewski C, Kear BP, Meehan L, Meredith RW, *et al.* (2016) Phylogenetic relationships of dasyuromorphian marsupials revisited. *Zoological Journal of the Linnean Society* **176**, 686–701. doi:10.1111/zoj.12323

Westerman M, Meredith RW, Springer MS (2010) Cytogenetics meets phylogenetics: A review of karyotype evolution in diprotodontian marsupials. *Journal of Heredity* **101**, 690–702. doi:10.1093/jhered/esq076

Westerman M, Spencer J, Collet C (1991) Chromosomal localization of the gene for late lactation protein (LLP) in the tammar wallaby (*Macropus eugenii*). *Cytogenetic and Genome Research* **56**, 182–184. doi:10.1159/000133083

Westerman M, Umbrello L, Woolley PA (2023) On the composition of *Antechinomys* (Marsupialia: Dasyuridae): how many species? *Australian Journal of Zoology* **70**, 95–103. doi:10.1071/ZO2204

Westerman M, Woolley PA (1990) Cytogenetics of some New Guinean dasyurids and genome evolution in the Dasyuridae (Marsupialia). *Australian Journal of Zoology* **37**, 521–531. doi:10.1071/ZO9890521

Westerman M, Woolley PA (1993) Chromosomes and the evolution of dasyurid marsupials. *Science in New Guinea* **19**, 123–130.

Westerman M, Young J, Donnellan S, Woolley PA, Krajewski C (2006) Molecular relationships of the New Guinean three-striped dasyures (*Myoictis*, Marsupialia: Dasyuridae). *Journal of Mammalian Evolution* **13**, 211–222. doi:10.1007/s10914-006-9023-5

Westerman M, Young J, Krajewski C (2008) Molecular relationships of *Pseudantechinus*, *Parantechinus*, and *Dasykaluta* (Marsupialia: Dasyuridae). *Australian Mammalogy* **29**, 201–212. doi:10.1071/AM07024

White LC, Mitchell KJ, Austin JJ (2018a) Ancient mitochondrial genomes reveal the demographic history and phylogeography of the extinct, enigmatic thylacine. *Journal of Biogeography* **45**, 1–13. doi:10.1111/jbi.13101

White LC, Saltré F, Bradshaw CJA, Austin JJ (2018b) High quality fossil dates support a synchronous, Late Holocene extinction of devils and thylacines in mainland Australia. *Biology Letters* **14**, 20170642. doi:10.1098/rsbl.2017.0642

White ME (1986) *The Greening of Gondwana.* Reed Books, Frenchs Forest.

White MJD (1973) *Animal Cytology and Evolution.* 3rd edn. Cambridge University Press, Cambridge, <https://wellcomecollection.org/works/meqhhay5/items>.

White MJD (1978) *Chromosomal Speciation.* W.H. Freeman, San Francisco.

Whitford D, Fanning FD, White AW (1982) Some information on reproduction, growth and development in *Planigale gilesi* (Dasyuridae, Marsupialia). In *Carnivorous Marsupials.* (Ed. M Archer) pp. 77–81. Royal Zoological Society of New South Wales, Sydney.

Wiens JJ (2001) Character analysis in morphological phylogenetics: problems and solutions. *Systematic Biology* **50**, 689–699. doi:10.1080/106351501753328811

Wiens JJ (2004) The role of morphological data in phylogeny reconstruction. *Systematic Biology* **53**, 653–661. doi:10.1080/10635150490472959

Wiens JJ, Ackerly DD, Allen AP, Anacker BL, Buckley LB, *et al.* (2010) Niche conservatism as an emerging principle in ecology and conservation biology. *Ecology Letters* **13**, 1310–1324. doi:10.1111/j.1461-0248.2010.01515.x

Willi Y, Kristensen TN, Sgrò CM, Weeks AR, Ørsted M, *et al.* (2022) Conservation genetics as a management tool: the five best-supported paradigms to assist the management of threatened species. *Proceedings of the National Academy of Sciences USA* **119**, e2105076119. doi:10.1073/pnas.2105076119

Williams GC (1966) Natural selection, the costs of reproduction, and a refinement of Lack's principle. *American Naturalist* **100**, 687–690. doi:10.1086/282461

Williams R, Williams A (1982) The life cycle of *Antechinus swainsonii* (Dasyuridae, Marsupialia). In *Carnivorous Marsupials.* (Ed. M Archer) pp. 89–95. Royal Zoological Society of New South Wales, Sydney.

Wilson AC, Carlson SS, White TJ (1977) Biochemical evolution. *Annual Review of Biochemistry* **46**, 473–639. doi:10.1146/annurev.bi.46.070177.003041

Wilson AC, Cann RL, Carr SM, George M Jr, Gyllensten R, *et al.* (1985) Mitochondrial DNA and two perspectives on evolutionary genetics. *Biological Journal of the Linnean Society* **26**, 375–400. doi:10.1111/j.1095-8312.1985.tb02048.x

Wilson BA, Bourne AR (1984) Reproduction in the male dasyurid *Antechinus minimus maritimus* (Marsupialia: Dasyuridae). *Australian Journal of Zoology* **32**, 311–318. doi:10.1071/ZO9840311

Wilson BA, Dickman CR, Fletcher TP (2003) Dasyurid dilemmas: problems and solutions for conserving Australia's small carnivorous marsupials. In *Predators with Pouches: The Biology of Carnivorous Marsupials.* (Eds M Jones, C Dickman, and M Archer) pp. 407–421. CSIRO Publishing, Melbourne.

Wilson BA, Evans MJ, Batson WG, Banks SC, Gordon IJ, *et al.* *et al.* (2020) Adapting reintroduction tactics in successive trials increases the likelihood of establishment for an endangered carnivore in a fenced sanctuary. *PLoS ONE* **15**, e0234455. doi:10.1371/journal.pone.0234455

Wilson DE, Reeder DM (Eds) (2005) *Mammal Species of the World: A Taxonomic and Geographic Reference.* 3rd edn. Johns Hopkins University Press, Baltimore.

Winge H (1941) *The Interrelationships of the Mammalian Genera. Volume 1: Monotremata, Marsupialia, Insectivora, Chiroptera, Edentata.* C.A. Reitzels Forlag, Copenhagen.

Wintle BA, Cadenhead NCR, Morgain RA, Legge SM, Bekessy SA, *et al.* (2019) Spending to save: what will it cost to halt Australia's extinction crisis? *Conservation Letters* **12**, e12682. doi:10.1111/conl.12682

Woinarski J, Burbidge AA. 2016. *Myrmecobius fasciatus.* The IUCN Red List of Threatened Species 2016: e.T14222A21949380. doi:10.2305/IUCN.UK.2016-2.RLTS.T14222A21949380.en

Woinarski JCZ, Burbidge AA, Harrison PL (2014) *The Action Plan for Australian Mammals 2012.* CSIRO Publishing, Melbourne.

Woinarski JCZ, Burbidge AA, Harrison PL (2015) Ongoing unraveling of a continental fauna: decline and extinction of Australian mammals since European settlement. *Proceedings of the National Academy of Sciences USA* **112**, 4531–4540. doi:10.1073/pnas.1417301112

Woinarski JCZ, Braby MF, Burbidge AA, Coates D, Garnett ST, *et al.* (2019) Reading the black book: the number, timing, distribution and causes of listed extinctions in Australia. *Biological Conservation* **239**, 108261. doi:10.1016/j.biocon.2019.108261

Wood DH (1970) An ecological study of *Antechinus stuartii* (Marsupialia) in a south-east Queensland rainforest. *Australian Journal of Zoology* **18**, 185–207. doi:10.1071/ZO9700185

Wood HE (1924) The position of the 'sparassodonts': with notes on the relationships and history of the Marsupialia. *Bulletin of the American Museum of Natural History* **51**, 77–101, <http://hdl.handle.net/2246/1316>.

Woodburne MO (1967) The Alcoota Fauna, central Australia: an integrated paleontological study. *Bureau of Mineral Resources Australian Bulletin* **87**, 1–187, <http://pid.geoscience.gov.au/dataset/ga/166>.

Woodburne MO, Macfadden BJ, Case JA, Springer MS, Pledge NS, *et al.* (1994) Land mammal biostratigraphy and magnetostratigraphy of the Etadunna Formation (Late Oligocene) of South Australia. *Journal of Vertebrate Paleontology* **13**, 483–515. doi:10.1080/02724634.1994.10011527

Woodhead J, Hand SJ, Archer M, Graham I, Sniderman K, *et al.* (2016) Developing a radiometrically-dated chronological sequence for Neogene biotic change in Australia, from the Riversleigh World Heritage Area of Queensland. *Gondwana Research* **29**, 153–167. doi:10.1016/j.gr.2014.10.004

Woods GM, Fox S, Flies AS, Tovar CD, Jones M, *et al.* (2018) Two decades of the impact of Tasmanian Devil Facial Tumor Disease. *Integrative and Comparative Biology* **58**, 1043–1054. doi:10.1093/icb/icy118

Woolley L-A, Geyle HM, Murphy BP, Legge SM, Palmer R, *et al.* (2019) Introduced cats (*Felis catus*) eating a continental fauna: inventory and traits of Australian mammal species killed. *Mammal Review* **49**, 354–368. doi:10.1111/mam.12167

Woolley P (1966) Reproduction in *Antechinus* spp. and other dasyurid marsupials. *Symposia of the Zoological Society of London* **15**, 281–294.

Woolley P (1971a) Maintenance and breeding of laboratory colonies of *Dasyuroides byrnei* and *Dasycercus cristicauda*. *International Zoo Yearbook* **11**, 351–354. doi:10.1111/j.1748-1090.1971.tb01932.x

Woolley P (1971b) Observations on the reproductive biology of the dibbler, *Antechinus apicalis* (Marsupialia: Dasyuridae). *Journal of the Royal Society of Western Australia* **54**, 99–102, <https://www.biodiversitylibrary.org/page/49025350>.

Woolley P (1974) The pouch of *Planigale subtilissima* and other dasyurid marsupials. *Journal of the Royal Society of Western Australia* **57**, 11–15, <https://www.biodiversitylibrary.org/page/50787168>.

Woolley PA (1981) *Antechinus bellus*, another dasyurid marsupial with post-mating mortality of males. *Journal of Mammalogy* **62**, 381–382. doi:10.2307/1380717

Woolley PA (1982) Phallic morphology of the Australian species of *Antechinus* (Dasyuridae, Marsupialia): a new taxonomic tool? In *Carnivorous Marsupials*. (Ed. M Archer) pp. 767–781. Royal Zoological Society of New South Wales, Sydney.

Woolley PA (1984a) Observations on a captive outdoor breeding colony of a small dasyurid marsupial, *Sminthopsis crassicaudata*. *Australian Wildlife Research* **11**, 249–254. doi:10.1071/WR9840249

Woolley PA (1984b) Phallic morphology of the New Guinean species of Antechinus. *Bulletin of the Australian Mammal Society* **8**, 182.

Woolley PA (1984c) Reproduction in *Antechinomys laniger* ('spenceri' form) (Marsupialia: Dasyuridae): field and laboratory investigations. *Australian Wildlife Research* **11**, 481–489. doi:10.1071/WR9840481

Woolley PA (1987) Phallic morphology of *Dasyuroides byrnei* and *Dasycercus cristicauda* (Marsupialia: Dasyuridae). *Australian Journal of Zoology* **35**, 535–540. doi:10.1071/ZO9870535

Woolley PA (1988) Reproduction in the Ningbing antechinus (Marsupialia: Dasyuridae): field and laboratory observations. *Australian Wildlife Research* **15**, 149–156. doi:10.1071/WR9880149

Woolley PA (1989) Nest location by spool-and-line tracking of dasyurid marsupials in New Guinea. *Journal of Zoology London* **218**, 689–700. doi:10.1111/j.1469-7998.1989.tb05010.x

Woolley PA (1990a) Reproduction in *Sminthopsis macroura* (Marsupialia: Dasyuridae) I. The female. *Australian Journal of Zoology* **38**, 187–205. doi:10.1071/ZO9900187

Woolley PA (1990b) Reproduction in *Sminthopsis macroura* (Marsupialia: Dasyuridae) II. The male. *Australian Journal of Zoology* **38**, 207–217. doi:10.1071/ZO9900207

Woolley PA (1991a) Reproduction in *Dasykaluta rosamondae* (Marsupialia: Dasyuridae): field and laboratory observations. *Australian Journal of Zoology* **39**, 549–568. doi:10.1071/ZO9910549

Woolley PA (1991b) Reproduction in *Pseudantechinus macdonnellensis* (Marsupialia: Dasyuridae): field and laboratory observations. *Wildlife Research* **18**, 13–25. doi:10.1071/WR9910013

Woolley PA (1991c) Reproductive pattern of captive Boullanger Island Dibblers, *Parantechinus apicalis* (Marsupialia: Dasyuridae). *Wildlife Research* **18**, 157–163. doi:10.1071/WR9910157

Woolley PA (1994) The dasyurid marsupials of New Guinea: use of museum specimens to assess seasonality of breeding. *Science in New Guinea* **20**, 49–55.

Woolley PA (1995a) Observations on reproduction in captive *Parantechinus bilarni* (Marsupialia: Dasyuridae). *Australian Mammalogy* **18**, 83–85. doi:10.1071/AM95083

Woolley PA (1995b) Red-cheeked Dunnart. In *The Mammals of Australia*. (Ed. R Strahan) pp. 196–197. Reed Books, Chatswood, NSW.

Woolley PA (2001) Observations on the reproductive biology of *Myoictis wallacei, Neophascogale lorentzi, Dasyurus albopunctatus* and *Dasyurus spartacus*, dasyurid marsupials endemic to New Guinea. *Australian Mammalogy* **23**, 63–66. doi:10.1071/AM01063

Woolley PA (2003) Reproductive biology of some dasyurid marsupials of New Guinea. In *Predators with Pouches: The Biology of Carnivorous Marsupials*. (Eds M Jones, C Dickman, and M Archer) pp. 169–182. CSIRO Publishing, Melbourne.

Woolley PA (2005a) Revision of the three-striped dasyures, genus *Myoictis* (Marsupialia: Dasyuridae), of New Guinea, with description of a new species. *Records of the Australian Museum* **57**, 321–340. doi:10.3853/j.0067-1975.57.2005.1450

Woolley PA (2005b) The species of Dasycercus Peters, 1875 (Marsupialia: Dasyuridae). *Memoirs of Museum Victoria* **62**, 213–221. doi:10.24199/j.mmv.2005.62.8

Woolley PA (2006) Studies on the crest-tailed mulgara *Dasycercus cristicauda* and the brush-tailed mulgara *Dasycercus blythi* (Marsupialia: Dasyuridae). *Australian Mammalogy* **28**, 117–120. doi:10.1071/AM06018

Woolley PA (2007) Observations on the oestrous cycle and breeding season of the chestnut dunnart, *Sminthopsis archeri*, a little known dasyurid marsupial. *Australian Mammalogy* **29**, 89–96. doi:10.1071/AM07011

Woolley PA (2008) Dibbler. In *The Mammals of Australia*. 3rd edn. (Eds S Van Dyck and R Strahan) pp. 65–66. Reed New Holland, Sydney.

Woolley PA (2011) *Pseudantechinus mimulus*: a little known dasyurid marsupial. *Australian Mammalogy* **33**, 57–67. doi:10.1071/AM10043

Woolley PA (2015) The Julia Creek dunnart *Sminthopsis douglasi* (Marsupialia: Dasyuridae): breeding of a threatened species in captivity and in wild populations. *Australian Journal of Zoology* **63**, 411–423. doi:10.1071/ZO15056

Woolley PA (2017) Aspects of reproduction, and morphology of the penis, of *Pseudantechinus woolleyae* (Marsupialia: Dasyuridae). *Australian Journal of Zoology* **65**, 357–361. doi:10.1071/ZO17086

Woolley PA (2019) Male members of some endemic New Guinean dasyurid marsupials: they help resolve relationships. *Journal of Mammalogy* **100**, 142–149. doi:10.1093/jmammal/gyy176

Woolley PA (2020) The type specimen of *Murexia longicaudata* (Dasyuromorphia: Dasyuridae): an enigma. *Australian Mammalogy* **42**, 359–360. doi:10.1071/AM19054.

Woolley PA, Ahern LD (1983) Observations on the ecology and reproduction of *Sminthopsis leucopus* (Marsupialia: Dasyuridae). *Proceedings of the Royal Society of Victoria* **95**, 169–180, <https://www.biodiversitylibrary.org/page/57468043>.

Woolley PA, Elliott C (2014) Breeding in wild populations of a small dasyurid marsupial, *Planigale ingrami*, in northwestern Queensland using a novel method for collection of specimens. *Australian Mammalogy* **36**, 81–89. doi:10.1071/AM13027

Woolley PA, Gilfillan SL (1991) Confirmation of polyoestry in captive white-footed dunnarts, *Sminthopsis leucopus* (Marsupialia: Dasyuridae). *Australian Mammalogy* **14**, 137–138. doi:10.1071/AM91016

Woolley PA, Guedelha N, Graves JAM (2003) Anatomy and chromosomes of two intersexual dasyurid marsupials. *Reproduction Fertility and Development* **15**, 293–301. doi:10.1071/RD03045

Woolley PA, Krajewski C, Westerman M (2015) Phylogenetic relationships within *Dasyurus* (Dasyuromorphia: Dasyuridae): quoll systematics based on molecular evidence and male characteristics. *Journal of Mammalogy* **96**, 37–46. doi:10.1093/jmammal/gyu028

Woolley PA, Krajewski C, Westerman M (2020) The endemic New Guinean genus *Murexia* (Dasyuromorphia, Dasyuridae). How many species? An analysis based on morphological, distributional and molecular data. *Australian Journal of Zoology* **67**, 134–144. doi:10.1071/ZO20013

Woolley PA, Leary T, Seri L, Flannery T, Wright D, *et al.* (2016) *Phascolosorex dorsalis*. The IUCN Red List of Threatened Species e.16894A21943541, <https://www.iucnredlist.org/species/16894/21943541>.

Woolley PA, Raftopoulos SA, Coleman GJ and Armstrong SM (1991) A comparative study of the circadian activity patterns of two New Guinean dasyurid marsupials, *Phascolosorex dorsalis* and *Antechinus habbema*. *Australian Journal of Zoology* **39**, 661–671. doi:10.1071/ZO9910661

Woolley PA, Valente A (1986) Reproduction in *Sminthopsis longicaudata* (Marsupialia: Dasyuridae): laboratory observations. *Australian Wildlife Research* **13**, 7–12. doi:10.1071/WR9860007

Woolley PA, Valente A (1992) Hair structure of the dasyurid marsupials of New Guinea. *Science in New Guinea* **18**, 29–39.

Woolley P, Webb SJ (1977) The penis of dasyurid marsupials. In *The Biology of Marsupials*. (Eds B Stonehouse and D Gilmore) pp. 307–323. Macmillan, London.

Woolley PA, Westerman M, Krajewski C (2007) Interspecific affinities within the genus *Sminthopsis* (Dasyuromorphia: Dasyuridae) based on morphology of the penis: congruence with other anatomical and molecular data. *Journal of Mammalogy* **88**, 1381–1392. doi:10.1644/06-MAMM-A-443R.1

Wright AM, Hillis DM (2014) Bayesian analysis using a simple likelihood model outperforms parsimony for estimation of phylogeny from discrete morphological data. *PLoS ONE* **9**, e109210. doi:10.1371/journal.pone.0109210

Wroe S (1996) *Muribacinus gadiyuli*, (Thylacinidae: Marsupialia), a very plesiomorphic thylacinid from the Miocene of Riversleigh, northwestern Queensland, and the problem of paraphyly for the Dasyuridae (Marsupialia). *Journal of Paleontology* **70**, 1032–1044. doi:10.1017/S0022336000038737

Wroe S (1997a) A reexamination of proposed morphology-based synapomorphies for the families of Dasyuromorphia (Marsupialia). I. Dasyuridae. *Journal of Mammalian Evolution* **4**, 19–52. doi:10.1023/A:1027379610196

Wroe S (1997b) *Mayigriphus orbus* gen et sp. nov., a Miocene dasyuromorphian from Riversleigh, northwestern Queensland. *Memoirs of the Queensland Museum* **41**, 439–448, <https://www.biodiversitylibrary.org/page/40090921>.

Wroe S (1998) A new 'bone-cracking' dasyurid (Marsupialia), from the Miocene of Riversleigh, northwestern Queensland. *Alcheringa* **22**, 277–284. doi:10.1080/03115519808619205

Wroe S (1999a) Killer kangaroos and other murderous marsupials. *Scientific American* **280**, 68–74. doi:10.1038/scientificamerican0599-68

Wroe S (1999b) The geologically oldest dasyurid, from the Miocene of Riversleigh, northwestern Queensland. *Paleontology* **42**, 501–527. doi:10.1111/1475-4983.00082

Wroe S (2001a) A new genus and species of dasyuromorphian from the Miocene of Riversleigh, northern Australia. *Memoirs of the Australian Association of Palaeontologists* **25**, 53–59.

Wroe S (2001b) *Maximucinus muirheadeae*, gen. et sp. nov. (Thylacinidae: Marsupialia), from the Miocene of Riversleigh, north-western Queensland, with estimates of body weights for fossil thylacinids. *Australian Journal of Zoology* **49**, 603–614. doi:0.1071/ZO01044

Wroe S (2003) Australian marsupial carnivores: recent advances in palaeontology. In *Predators with Pouches: The Biology of Carnivorous Marsupials*. (Eds M Jones, C Dickman, and M Archer) pp. 102–123. CSIRO Publishing, Melbourne.

Wroe S, Archer M (2006) Origins and early radiations of marsupials. In *Evolution and Biogeography of Australasian Vertebrates*. (Eds JR Merrick, M Archer, GM Hickey and MSY Lee) pp. 551–574. Auscipub, Oatlands, NSW.

Wroe S, Clausen P, McHenry C, Moreno K, Cunningham E (2007) Computer simulation of feeding behaviour in the thylacine and dingo as a novel test for convergence and niche overlap. *Proceedings of the Royal Society B* **274**, 2819–2828. doi:10.1098/rspb.2007.0906

Wroe S, Ebach M, Ahyong S, de Muizon C, Muirhead J (2000) Cladistic analysis of dasyuromorphian (Marsupialia) phylogeny using cranial and dental characters. *Journal of Mammalogy* **81**, 1008–1024. doi:10.1644/1545-1542(2000)081<1008:CAODMP>2.0.CO;2

Wroe S, Field JH, Archer M, Grayson DK, Price GJ, *et al.* (2013) Climate change frames debate over extinction of megafauna in Sahul (Pleistocene Australia-New Guinea). *Proceedings of the National Academy of Sciences USA* **110**, 8777–8781. doi:10.1073/pnas.1302698110

Wroe S, Mackness BS (1998) Revision of the Pliocene dasyurid, *Dasyurus dunmalli* (Dasyuridae: Marsupialia). *Memoirs of the Queensland Museum* **42**, 605–612, <https://www.biodiversitylibrary.org/page/52465120>.

Wroe S, Mackness BS (2000) A new genus and species of dasyurid from the Pliocene Chinchilla Local Fauna of south-eastern Queensland. *Alcheringa* **24**, 319–326. doi:10.1080/03115510008619535

Wroe S, Musser A (2001) The skull of *Nimbacinus dicksoni* (Thylacinidae: Marsupialia). *Australian Journal of Zoology* **49**, 487–514. doi:10.1071/ZO00032

Wroe S, Myers T (1998) Fallacy & future-eating. *Australasian Science* **19**, 27–29.

Wu CI, Li W-H (1985) Evidence for higher rates of nucleotide substitution in rodents than in man. *Proceedings of the National Academy of Sciences USA* **82**, 1741–1745. doi:10.1073/pnas.82.6.1741

Yang Z (2014) *Molecular Evolution: A Statistical Approach.* Oxford University Press, Oxford.

Yang Z, Rannala B (2006) Bayesian estimation of sequence divergence times under a molecular clock using multiple calibrations with soft bounds. *Molecular Biology and Evolution* **23**, 212–226. doi:10.1093/molbev/msj024

Yates AM (2014) New craniodental remains of *Thylacinus potens* (Dasyuromorphia: Thylacinidae), a carnivorous marsupial from the Late Miocene Alcoota Local Fauna of central Australia. *PeerJ* **2**, e547. doi:10.7717/peerj.547

Yates AM (2015) *Thylacinus* (Marsupialia: Thylacinidae) from the Mio-Pliocene boundary and the diversity of Late Neogene thylacinids in Australia. *PeerJ* **3**, e931. doi:10.7717/peerj.931

Young GJ, Graves JAM, Barbieri I, Woolley PA, Cooper DW, *et al.* (1982) The chromosomes of dasyurids (Marsupialia). In *Carnivorous Marsupials.* (Ed. M Archer) pp. 783–795. Royal Zoological Society of New South Wales, Sydney.

Zachos FE (2015) Taxonomic inflation, the phylogenetic species concept and lineages in the Tree of Life – a cautionary comment on species splitting. *Journal of Zoological Systematics and Evolutionary Research* **53**, 180–184. doi:10.1111/jzs.12088

Zemann A, Churakov G, Donnellan S, Grützner F, Zhao F, *et al.* (2013) Ancestry of the Australian termivorous numbat. *Molecular Biology and Evolution* **30**, 1041–1045. doi:10.1093/molbev/mst032

Zhang C, Stadler T, Klopfstein S, Heath TA, Ronquist F (2016) Total evidence dating under the fossilized birth-death process. *Systematic Biology* **65**, 228–249. doi:10.1093/sysbio/syv080

Zhang S, Wang G-D, Ma P, Zhang L, Yin T-T, *et al.* (2020) Genomic regions under selection in the feralization of the dingoes. *Nature Communications* **11**, 671. doi:10.1038/s41467-020-14515-6

Ziegler AC (1977) Evolution of New Guinea's marsupial fauna in response to a forested environment. In *The Biology of Marsupials.* (Eds B Stonehouse and D Gilmore) pp. 117–138. Macmillan, London.

Ziegler A (1982) An ecological check list of New Guinea recent mammals. In *Biogeography and Ecology of New Guinea.* (Ed. GL Gressitt) pp. 863–894. Junk, The Hague.

Zimmerman EAW (1783) *Gestichte des Menschen, und der allgemein verbretiteten vierfüssigen Thiere, mit einer hiezu gehörigen Zoologischen Wetcharte, Volume 3.* Weygandschen Buchhandlung, Leipzeig.

Zou Z, Zhang J (2016) Morphological and molecular convergences in mammalian phylogenetics. *Nature Communications* **7**, 12758. doi:10.1038/ncomms12758

Zuckerkandl E, Pauling L (1962) Molecular disease, evolution, and genetic heterogeneity. In *Horizons in Biochemistry.* (Eds M Kasha and B Pullman) pp. 189–225. Academic Press, New York.

Zuckerkandl E, Pauling L (1965) Evolutionary divergence and convergence in proteins. In *Evolving Genes and Proteins.* (Eds V Bryson and HJ Vogel) pp. 97–166. Academic Press, New York, <https://authors.library.caltech.edu/5456/1/hrst.mit.edu/hrs/evolution/public/papers/zuckerkandlpauling1965/zuckerkandlpauling1965.pdf>.

Index